The Ultimate Guide to
U.S. Army
Survival
Skills, Tactics, and Techniques

Edited by
Jay McCullough

Skyhorse Publishing

Skyhorse Publishing books may be purchased in bulk at special discounts for sales promotion, corporate gifts, fund-raising, or educational purposes. Special editions can also be created to specifications. For details, contact the Special Sales Department, Skyhorse Publishing, 307 West 36th Street, 11th Floor, New York, NY 10018 or info@skyhorsepublishing.com.

Skyhorse® and Skyhorse Publishing® are registered trademarks of Skyhorse Publishing, Inc.®, a Delaware corporation.

www.skyhorsepublishing.com

10 9 8 7 6 5 4 3

Library of Congress Cataloging-in-Publication Data

The ultimate guide to U.S. Army survival skills, tactics, and techniques / edited by Jay McCullough.
   p. cm.
 ISBN-13: 978-1-60239-050-8 (alk. paper)
 ISBN-10: 1-60239-050-9 (alk. paper)
 1. Unified operations (Military science) 2. Operational art (Military science) I. McCullough, Jay.

U260.U48 2007
355.5′4—dc22

                   2007016415

Special edition ISBN: 978-1-61608-327-4

Printed in Canada

CONTENTS

Part V
FIRECRAFT, TOOLS, CAMOUFLAGE, TRACKING, MOVEMENT, AND COMBAT SKILLS

Part VI
ENVIRONMENT-SPECIFIC SURVIVAL

INTRODUCTION

The U.S. has the best-equipped and well-trained Army in the world. Barring unforeseen political considerations or a catastrophic act of nature, it can deploy nearly anywhere and accomplish almost any reasonable mission. This is due in large part to the Army's experience as an institution in a variety of wartime environments. Its hard-earned wisdom about how to cope with almost every imaginable scenario, on a soldier-by-soldier basis, distinguishes it as a service of excellence whose individuals are highly adaptable. They are well prepared, they accomplish the mission, and when the circumstances are truly unfavorable to life itself, they are survivors.

The keys to this preparation are contained in the Army's many sensible, well-written, voluminous, and scattered publications. They address nearly every aspect of running, provisioning, or being in the Army, but they are especially useful for their tips on how to stay alive under any circumstances. The task of culling every bit of useful information about survival from every U.S. Army publication would take months however, so I've done it here for you in *The Ultimate Guide to U.S. Army Survival Skills, Tactics, and Techniques*.

Where subjects are duplicated throughout the literature, I've created a single clearinghouse for that information. For instance, almost every Army manual remotely connected to survival seems to include the same basic instructions about how to make a poncho lean-to, so you'll find a single discussion about that, and related information, in the Shelters section.

In other instances, a subject may be discussed in-depth in a more generalized or comprehensive manner, say for example venomous snakes as a subcategory of dangerous animals. But the subject of snakes also merits inclusion in other categories, especially within the contexts of those categories; jungle, desert, and medical manuals add valuable information not otherwise contained in a herpetologist's catalog of snake habitats, habits, and geographical ranges.

I've tried to make the selections useful to a general reader who may find him- or herself in a survival situation, whether they are alone or in a small group, probably unarmed. Some sections are invaluable; nearly every aspect of first aid will be useful to someone at some time. When in doubt about whether particular passages provide pertinent information, I've included them in the hope that they may serve as a useful reference, comfort the afflicted, or perhaps even save a life. As an example, you will find a caution in the first aid section that warns you not to apply a tourniquet to someone's neck.

Will you ever dig a defensive position with a sloping floor and a grenade trench? Probably not. But everyone who has done so probably never thought about it until they were up to their shoulders in dirt, wondering how much further they had to dig. I should hope that you never find yourself in that circumstance, or one like it. If you do, the best advice is contained in the first three chapters, particularly in regard to your state of mind. Whatever your condition, keep a positive outlook, keep your sense of humor, keep your humanity and sense of decency. Realize the conditions for what they are, be flexible, adapt, and never say die.

Jay McCullough
January 2007
New Haven, Connecticut

PART I
General Survival Skills

CHAPTER 1

Psychology of Survival

INTRODUCTION

This manual is based entirely on the keyword SURVIVAL. The letters in this word can help guide you in your actions in any survival situation. Whenever faced with a survival situation, remember the word SURVIVAL.

SURVIVAL ACTIONS

The following paragraphs expand on the meaning of each letter of the word survival. Study and remember what each letter signifies because you may some day have to make it work for you.

S - Size Up the Situation. If you are in a combat situation, find a place where you can conceal yourself from the enemy. Remember, security takes priority. Use your senses of hearing, smell, and sight to get a feel for the battlefield. What is the enemy doing? Advancing? Holding in place? Retreating? You will have to consider what is developing on the battlefield when you make your survival plan.

Size Up Your Surroundings. Determine the pattern of the area. Get a feel for what is going on around you. Every environment, whether forest, jungle, or desert, has a rhythm or pattern. This rhythm or pattern includes animal and bird noises and movements and insect sounds. It may also include enemy traffic and civilian movements.

Size Up Your Physical Condition. The pressure of the battle you were in or the trauma of being in a survival situation may have caused you to overlook wounds you received. Check your wounds and give yourself first aid. Take care to prevent further bodily harm. For instance, in any climate, drink plenty of water to prevent dehydration. If you are in a cold or wet climate, put on additional clothing to prevent hypothermia.

Size Up Your Equipment. Perhaps in the heat of battle, you lost or damaged some of your equipment. Check to see what equipment you have and what condition it is in.

Now that you have sized up your situation, surroundings, physical condition, and equipment, you are ready to make your survival plan. In doing so, keep in mind your basic physical needs—water, food, and shelter.

U - Use All Your Senses, Undue Haste Makes Waste. You may make a wrong move when you react quickly without thinking or planning. That move may result in your capture or death. Don't move just for the sake of taking action. Consider all aspects of your situation (size up your situation) before you make a decision and a move. If you act in haste, you may forget or lose some of your equipment. In your haste you may also become disoriented so that you don't know which way to go. Plan your moves. Be ready to move out quickly without endangering yourself if the enemy is near you. Use all your senses to evaluate the situation. Note sounds and smells. Be sensitive to temperature changes. Be observant.

R - Remember Where You Are. Spot your location on your map and relate it to the surrounding terrain. This is a basic principle that you must always follow. If there are other persons with you, make sure they also know their location. Always know who in your group, vehicle, or aircraft has a map and compass. If that person is killed, you will have to get the map and compass from him. Pay close attention to where you are and to where you are going. Do not rely on others in the group to keep track of the route. Constantly orient yourself. Always try to determine, as a minimum, how your location relates to—

- The location of enemy units and controlled areas.
- The location of friendly units and controlled areas.
- The location of local water sources (especially important in the desert).
- Areas that will provide good cover and concealment.

This information will allow you to make intelligent decisions when you are in a survival and evasion situation.

V - Vanquish Fear and Panic. The greatest enemies in a combat survival and evasion situation are fear and panic. If uncontrolled, they can destroy your ability to make an intelligent decision. They may cause you to react to your feelings and imagination rather than to your situation. They can drain your energy and thereby cause other negative emotions. Previous survival and evasion training and self-confidence will enable you to vanquish fear and panic.

I - Improvise. In the United States, we have items available for all our needs. Many of these items are cheap to replace when damaged. Our easy come, easy go, easy-to-replace culture makes it unnecessary for us to improvise. This inexperience in improvisation can be an enemy in a survival situation. Learn to improvise. Take a tool designed for a specific purpose and see how many other uses you can make of it.

Learn to use natural objects around you for different needs. An example is using a rock for a hammer. No matter how complete a survival kit you have with you, it will run out or wear out after a while. Your imagination must take over when your kit wears out.

V - Value Living. All of us were born kicking and fighting to live, but we have become used to the soft life. We have become creatures of comfort. We dislike inconveniences and discomforts. What happens when we are faced with a survival situation with its stresses, inconveniences, and discomforts? This is when the will to live—placing a high value on living—is vital. The experience and knowledge you have gained through life and your Army training will have a bearing on your will to live. Stubbornness, a refusal to give in to problems and obstacles that face you, will give you the mental and physical strength to endure.

A - Act Like the Natives. The natives and animals of a region have adapted to their environment. To get a feel of the area, watch how the people go about their daily routine. When and what do they eat? When, where, and how do they get their food? When and where do they go for water? What time do they usually go to bed and get up? These actions are important to you when you are trying to avoid capture.

Animal life in the area can also give you clues on how to survive. Animals also require food, water, and shelter. By watching them, you can find sources of water and food.

WARNING
Animals cannot serve as an absolute guide to what you can eat and drink. Many animals eat plants that are toxic to humans.

Keep in mind that the reaction of animals can reveal your presence to the enemy.

If in a friendly area, one way you can gain rapport with the natives is to show interest in their tools and how they get food and water. By studying the people, you learn to respect them, you often make valuable friends, and, most important, you learn how to adapt to their environment and increase your chances of survival.

L - Live by Your Wits, But for Now, Learn Basic Skills. Without training in basic skills for surviving and evading on the battlefield, your chances of living through a combat survival and evasion situation are slight.

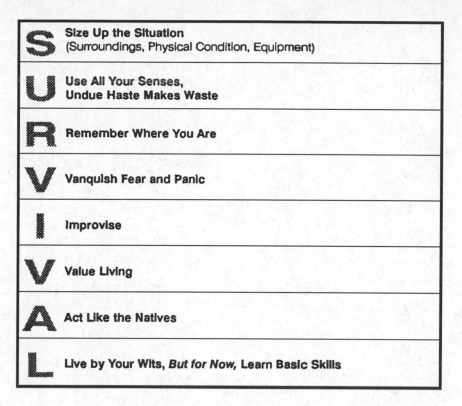

S	**Size Up the Situation** (Surroundings, Physical Condition, Equipment)
U	**Use All Your Senses,** **Undue Haste Makes Waste**
R	**Remember Where You Are**
V	**Vanquish Fear and Panic**
I	**Improvise**
V	**Value Living**
A	**Act Like the Natives**
L	**Live by Your Wits, *But for Now*, Learn Basic Skills**

Learn these basic skills now—not when you are headed for or are in the battle. How you decide to equip yourself before deployment will impact on whether or not you survive. You need to know about the environment to which you are going, and you must practice basic skills geared to that environment. For instance, if you are going to a desert, you need to know how to get water in the desert.

Practice basic survival skills during all training programs and exercises. Survival training reduces fear of the unknown and gives you self-confidence. It teaches you to *live by your wits*.

PATTERN FOR SURVIVAL

Develop a survival pattern that lets you beat the enemies of survival. This survival pattern must include food, water, shelter, fire, first aid, and signals placed in order of importance. For example, in a cold environment, you would need a fire to get warm; a shelter to protect you from the cold, wind, and rain or snow; traps or snares to get food; a means to signal friendly aircraft; and first aid to maintain health. If injured, first aid has top priority no matter what climate you are in.

Change your survival pattern to meet your immediate physical needs as the environment changes.

As you read the rest of this manual, keep in mind the keyword SURVIVAL and the need for a survival pattern.

CHAPTER 2

Psychology of Survival

It takes much more than the knowledge and skills to build shelters, get food, make fires, and travel without the aid of standard navigational devices to live successfully through a survival situation. Some people with little or no survival training have managed to survive life-threatening circumstances. Some people with survival training have not used their skills and died. A key ingredient in any survival situation is the mental attitude of the individual(s) involved. Having survival skills is important; having the will to survive is essential. Without a desire to survive, acquired skills serve little purpose and invaluable knowledge goes to waste.

There is a psychology to survival. The soldier in a survival environment faces many stresses that ultimately impact on his mind. These stresses can produce thoughts and emotions that, if poorly understood, can transform a confident, well-trained soldier into an indecisive, ineffective individual with questionable ability to survive. Thus, every soldier must be aware of and be able to recognize those stresses commonly associated with survival. Additionally, it is imperative that soldiers be aware of their reactions to the wide variety of stresses associated with survival. This chapter will identify and explain the nature of stress, the stresses of survival, and those internal reactions soldiers will naturally experience when faced with the stresses of a real-world survival situation. The knowledge you, the soldier, gain from this chapter and other chapters in this manual, will prepare you to come through the toughest times alive.

A LOOK AT STRESS

Before we can understand our psychological reactions in a survival setting, it is helpful to first know a little bit about stress.

Stress is not a disease that you cure and eliminate. Instead, it is a condition we all experience. Stress can be described as our reaction to pressure. It is the name given to the experience we have as we physically, mentally, emotionally, and spiritually respond to life's tensions.

Need for Stress. We need stress because it has many positive benefits. Stress provides us with challenges; it gives us chances to learn about our values and strengths. Stress can show our ability to handle pressure without breaking; it tests our adaptability and flexibility; it can stimulate us to do our best. Because we usually do not consider unimportant events stressful, stress can also be an excellent indicator of the significance we attach to an event—in other words, it highlights what is important to us.

We need to have some stress in our lives, but too much of anything can be bad. The goal is to have stress, but not an excess of it. Too much stress can take its toll on people and organizations. Too much stress leads to distress. Distress causes an uncomfortable tension that we try to escape and, preferably, avoid. Listed below are a few of the common signs of distress you may find in your fellow soldiers or yourself when faced with too much stress:

- Difficulty making decisions
- Angry outbursts
- Forgetfulness
- Low energy level
- Constant worrying
- Propensity for mistakes

- Thoughts about death or suicide
- Trouble getting along with others
- Withdrawing from others
- Hiding from responsibilities
- Carelessness

As you can see, stress can be constructive or destructive. It can encourage or discourage, move us along or stop us dead in our tracks, and make life meaningful or seemingly meaningless. Stress can inspire you to operate successfully and perform at your maximum efficiency in a survival situation. It can also cause you to panic and forget all your training. Key to your survival is your ability to manage the inevitable stresses you will encounter. The survivor is the soldier who works with his stresses instead of letting his stresses work on him.

Survival Stressors. Any event can lead to stress and, as everyone has experienced, events don't always come one at a time. Often, stressful events occur simultaneously. These events are not stress, but they produce it and are called "stressors." Stressors are the obvious cause while stress is the response. Once the body recognizes the presence of a stressor, it then begins to act to protect itself.

In response to a stressor, the body prepares either to "fight or flee." This preparation involves an internal SOS sent throughout the body. As the body responds to this SOS, several actions take place. The body releases stored fuels (sugar and fats) to provide quick energy; breathing rate increases to supply more oxygen to the blood; muscle tension increases to prepare for action; blood clotting mechanisms are activated to reduce bleeding from cuts; senses become more acute (hearing becomes more sensitive, eyes become big, smell becomes sharper) so that you are more aware of your surrounding and heart rate and blood pressure rise to provide more blood to the muscles. This protective posture lets a person cope with potential dangers; however, a person cannot maintain such a level of alertness indefinitely.

Stressors are not courteous; one stressor does not leave because another one arrives. Stressors add up. The cumulative effect of minor stressors can be a major distress if they all happen too close together. As the body's resistance to stress wears down and the sources of stress continue (or increase), eventually a state of exhaustion arrives. At this point, the ability to resist stress or use it in a positive way gives out and signs of distress appear. Anticipating stressors and developing strategies to cope with them are two ingredients in the effective management of stress. It is therefore essential that the soldier in a survival setting be aware of the types of stressors he will encounter. Let's take a look at a few of these.

Injury, Illness, or Death. Injury, illness, and death are real possibilities a survivor has to face. Perhaps nothing is more stressful than being alone in an unfamiliar environment where you could die from hostile action, an accident, or from eating something lethal. Illness and injury can also add to stress by limiting your ability to maneuver, get food and drink, find shelter, and defend yourself. Even if illness and injury don't lead to death, they add to stress through the pain and discomfort they generate. It is only by controlling the stress associated with the vulnerability to injury, illness, and death that a soldier can have the courage to take the risks associated with survival tasks.

Uncertainty and Lack of Control. Some people have trouble operating in settings where everything is not clear-cut. The only guarantee in a survival situation is that nothing is guaranteed. It can be extremely stressful operating on limited information in a setting where you have limited control of your surroundings. This uncertainty and lack of control also add to the stress of being ill, injured, or killed.

Environment. Even under the most ideal circumstances, nature is quite formidable. In survival, a soldier will have to contend with the stressors of weather, terrain, and the variety of creatures inhabiting an area. Heat, cold, rain, winds, mountains, swamps, deserts, insects, dangerous reptiles, and other animals are just a few of the challenges awaiting the soldier working to survive. Depending on how a soldier handles the stress of his environment, his surroundings can be either a source of food and protection or can be a cause of extreme discomfort leading to injury, illness, or death.

Hunger and Thirst. Without food and water a person will weaken and eventually die. Thus, getting and preserving food and water takes on increasing importance as the length of time in a survival setting increases. For a soldier used to having his provisions issued, foraging can be a big source of stress.

Fatigue. Forcing yourself to continue surviving is not easy as you grow more tired. It is possible to become so fatigued that the act of just staying awake is stressful in itself.

Isolation. There are some advantages to facing adversity with others. As soldiers we learn individual skills, but we train to function as part of a team. Although we, as soldiers, complain about higher headquarters, we become used to the information and guidance it provides, especially during times of confusion. Being in contact with others also provides a greater sense of security and a feeling someone is available to help if problems occur. A significant stressor in survival situations is that often a person or team has to rely solely on its own resources.

The survival stressors mentioned in this section are by no means the only ones you may face. Remember, what is stressful to one person may not be stressful to another. Your experiences, training, personal outlook on life, physical and mental conditioning, and level of self-confidence contribute to what you will find stressful in a survival environment. The object is not to avoid stress, but rather to manage the stressors of survival and make them work for you.

We now have a general knowledge of stress and the stressors common to survival; the next step is to examine our reactions to the stressors we may face.

NATURAL REACTIONS

Man has been able to survive many shifts in his environment throughout the centuries. His ability to adapt physically and mentally to a changing world kept him alive while other species around him gradually died off. The same survival mechanisms that kept our forefathers alive can help keep us alive as well! However, these survival mechanisms that can help us can also work against us if we don't understand and anticipate their presence.

It is not surprising that the average person will have some psychological reactions in a survival situation. We will now examine some of the major internal reactions you and anyone with you might experience with the survival stressors addressed in the earlier paragraphs. Let's begin.

Fear. Fear is our emotional response to dangerous circumstances that we believe have the potential to cause death, injury, or illness. This harm is not just limited to physical damage; the threat to one's emotional and mental well-being can generate fear as well. For the soldier trying to survive, fear can have a positive function if it encourages him to be cautious in situations where recklessness could result in injury. Unfortunately, fear can also immobilize a person. It can cause him to become so frightened that he fails to perform activities essential for survival. Most soldiers will have some degree of fear when placed in unfamiliar surroundings under adverse conditions. There is no shame in this! Each soldier must train himself not to be overcome by his fears. Ideally, through realistic training, we can acquire the knowledge and skills needed to increase our confidence and thereby manage our fears.

Anxiety. Associated with fear is anxiety. Because it is natural for us to be afraid, it is also natural for us to experience anxiety. Anxiety can be an uneasy, apprehensive feeling we get when faced with dangerous situations (physical, mental, and emotional). When used in a healthy way, anxiety urges us to act to end, or at least master, the dangers that threaten our existence. If we were never anxious, there would be little motivation to make changes in our lives. The soldier in a survival setting reduces his anxiety by performing those tasks that will ensure his coming through the ordeal alive. As he reduces his anxiety, the soldier is also bringing under control the source of that anxiety—his fears. In this form, anxiety is good; however, anxiety can also have a devastating impact. Anxiety can overwhelm a soldier to the point where he becomes easily confused and has difficulty thinking. Once this happens, it becomes more and more difficult for him to make good judgments and sound decisions. To survive, the soldier must learn techniques to calm his anxieties and keep them in the range where they help, not hurt.

Anger and Frustration. Frustration arises when a person is continually thwarted in his attempts to reach a goal. The goal of survival is to stay alive until you can reach help or until help can reach you. To achieve this goal, the soldier must complete some tasks with minimal resources. It is inevitable, in trying to do these tasks, that something will go wrong; that something will happen beyond the soldier's control; and that with one's life at stake, every mistake is magnified in terms of its importance. Thus, sooner or later, soldiers will have to cope with frustration when a few of their plans run into trouble. One outgrowth of this frustration is anger. There are many events in a survival situation that can frustrate or anger a soldier. Getting lost, damaged or forgotten equipment, the weather, inhospitable terrain, enemy patrols, and physical limitations are just a few sources of frustration and anger. Frustration and anger encourage impulsive reactions, irrational behavior, poorly thought-out decisions, and, in some instances, an "I quit" attitude (people sometimes avoid doing something they can't master). If the soldier can harness and properly channel the emotional intensity associated with anger and frustration, he can productively act as he answers the challenges of survival. If the soldier does not properly focus his angry feelings, he can waste much energy in activities that do little to further either his chances of survival or the chances of those around him.

Depression. It would be a rare person indeed who would not get sad, at least momentarily, when faced with the privations of survival. As this sadness deepens, we label the feeling "depression." Depression is closely linked with frustration and anger. The frustrated person becomes more and more angry as he fails to reach his goals. If the anger does not help the person to succeed, then the frustration level goes even higher. A destructive cycle between anger and frustration continues until the person becomes worn down—physically, emotionally, and mentally. When a person reaches this point, he starts to give up, and his focus shifts from "What can I do" to "There is nothing I can do." Depression is an expression of this hopeless, helpless feeling. There is nothing wrong with being sad as you temporarily think about your loved ones and remember what life is like back in "civilization" or "the world." Such thoughts, in fact, can give you the desire to try harder and live one more day. On the other hand, if you allow yourself to sink into a depressed state, then it can sap all your energy and, more important, your will to survive. It is imperative that each soldier resist succumbing to depression.

Loneliness and Boredom. Man is a social animal. This means we, as human beings, enjoy the company of others. Very few people want to be alone all the time! As you are aware, there is a distinct chance of isolation in a survival setting. This is not bad. Loneliness and boredom can bring to the surface qualities you thought only others had. The extent of your imagination and creativity may surprise you. When required to do so, you may discover some hidden talents and abilities. Most of all, you may tap into a reservoir of inner strength and fortitude you never knew you had. Conversely, loneliness and boredom can be another source of depression. As a soldier surviving alone, or with others, you must find ways to keep your mind productively occupied. Additionally, you must develop a degree of self-sufficiency. You must have faith in your capability to "go it alone."

Guilt. The circumstances leading to your being in a survival setting are sometimes dramatic and tragic. It may be the result of an accident or military mission where there was a loss of life. Perhaps you were the only, or one of a few, survivors. While naturally relieved to be alive, you simultaneously may be mourning the deaths of others who were less fortunate. It is not uncommon for survivors to feel guilty about being spared from death while others were not. This feeling, when used in a positive way, has encouraged people to try harder to survive with the belief they were allowed to live for some greater purpose in life. Sometimes, survivors tried to stay alive so that they could carry on the work of those killed. Whatever reason you give yourself, do not let guilt feelings prevent you from living. The living who abandon their chance to survive accomplish nothing. Such an act would be the greatest tragedy.

PREPARING YOURSELF

Your mission as a soldier in a survival situation is to stay alive. As you can see, you are going to experience an assortment of thoughts and emotions. These can work for you, or they can work to your downfall. Fear, anxiety, anger, frustration, guilt, depression, and loneliness are all possible reactions to the many stresses common to survival. These reactions, when controlled in a healthy way, help to increase a soldier's likelihood of surviving. They prompt the soldier to pay more attention in training, to fight back when scared, to take actions that ensure sustenance and security, to keep faith with his fellow soldiers, and to strive against large odds. When the survivor cannot control these reactions in a healthy way, they can bring him to a standstill. Instead of rallying his internal resources, the soldier listens to his internal fears. This soldier experiences psychological defeat long before he physically succumbs. Remember, survival is natural to everyone; being unexpectedly thrust into the life and death struggle of survival is not. Don't be afraid of your "natural reactions to this unnatural situation." Prepare yourself to rule over these reactions so they serve your ultimate interest—staying alive with the honor and dignity associated with being an American soldier.

It involves preparation to ensure that your reactions in a survival setting are productive, not destructive. The challenge of survival has produced countless examples of heroism, courage, and self-sacrifice. These are the qualities it can bring out in you if you have prepared yourself. Below are a few to help prepare yourself psychologically for survival. Through studying this manual and attending survival training you can develop the survival attitude.

Know Yourself. Through training, family, and friends take the time to discover who you are on the inside. Strengthen your stronger qualities and develop the areas that you know are necessary to survive.

Anticipate Fears. Don't pretend that you will have no fears. Begin thinking about what would frighten you the most if forced to survive alone. Train in those areas of concern to you. The goal is not to eliminate the fear, but to build confidence in your ability to function despite your fears.

Be Realistic. Don't be afraid to make an honest appraisal of situations. See circumstances as they are, not as you want them to be. Keep your hopes and expectations within the estimate of the situation. When you go into a survival setting with unrealistic expectations, you may be laying the groundwork for bitter disappointment. Follow the adage, "Hope for the best, prepare for the worst." It is much easier to adjust to pleasant surprises about one's unexpected good fortunes than to be upset by one's unexpected harsh circumstances.

Adopt a Positive Attitude. Learn to see the potential good in everything. Looking for the good not only boosts morale, it also is excellent for exercising your imagination and creativity.

Remind Yourself What Is at Stake. Remember, failure to prepare yourself psychologically to cope with survival leads to reactions such as depression, carelessness, inattention, loss of confidence, poor decision-making, and giving up before the body gives in. At stake is your life and the lives of others who are depending on you to do your share.

Train. Through military training and life experiences, begin today to prepare yourself to cope with the rigors of survival. Demonstrating your skills in training will give you the confidence to call upon them should the need arise. Remember, the more realistic the training, the less overwhelming an actual survival setting will be.

Learn Stress Management Techniques. People under stress have a potential to panic if they are not well-trained and not prepared psychologically to face whatever the circumstances maybe. While we often cannot control the survival circumstances in which we find ourselves, it is within our ability to control our response to those circumstances. Learning stress management techniques can enhance significantly your capability to remain calm and focused as you work to keep yourself and others alive. A few good techniques to develop include relaxation skills, time management skills, assertiveness skills, and cognitive restructuring skills (the ability to control how you view a situation).

Remember, "the will to survive" can also be considered to be "the refusal to give up."

CHAPTER 3

Survival Planning and Survival Kits

Survival planning is nothing more than realizing something could happen that would put you in a survival situation and, with that in mind, taking steps to increase your chances of survival. Thus, survival planning means preparation.

Preparation means having survival items and knowing how to use them. People who live in snow regions prepare their vehicles for poor road conditions. They put snow tires on their vehicles, add extra weight in the back for traction, and they carry a shovel, salt, and a blanket. Another example of preparation is finding the emergency exits on an aircraft when you board it for a flight. Preparation could also mean knowing your intended route of travel and familiarizing yourself with the area. Finally, emergency planning is essential.

IMPORTANCE OF PLANNING

Detailed prior planning is essential in potential survival situations. Including survival considerations in mission planning will enhance your chances of survival if an emergency occurs. For example, if your job requires that you work in a small, enclosed area that limits what you can carry on your person, plan where you can put your rucksack or your load-bearing equipment. Put it where it will not prevent you from getting out of the area quickly, yet where it is readily accessible.

One important aspect of prior planning is preventive medicine. Ensuring that you have no dental problems and that your immunizations are current will help you avoid potential dental or health problems. A dental problem in a survival situation will reduce your ability to cope with other problems that you face. Failure to keep your shots current may mean your body is not immune to diseases that are prevalent in the area.

Preparing and carrying a survival kit is as important as the considerations mentioned above. All Army aircraft normally have survival kits on board for the type area(s) over which they will fly. There are kits for over-water survival, for hot climate survival, and an aviator survival vest (see Tables 3-1 to 3-6 for a description of these survival kits and their contents). If you are not an aviator, you will probably not have access to the survival vests or survival kits. However, if you know what these kits contain, it will help you to plan and to prepare your own survival kit.

Even the smallest survival kit, if properly prepared, is invaluable when faced with a survival problem. Before making your survival kit, however, consider your unit's mission, the operational environment, and the equipment and vehicles assigned to your unit.

SURVIVAL KITS

The environment is the key to the types of items you will need in your survival kit. How much equipment you put in your kit depends on how you will carry the kit. A kit carried on your body will have to be smaller than one carried in a vehicle. Always layer your survival kit, keeping the most important items on your body. For example, your map and compass should always be on your body. Carry less important items on your load-bearing equipment. Place bulky items in the rucksack.

In preparing your survival kit, select items you can use for more than one purpose. If you have two items that will serve the same function, pick the one you can use for another function. Do not duplicate items, as this increases your kit's size and weight.

Table 3-1: Cold Climate Kit.

• Food packets	• Survival fishing kit
• Snare wire	• Plastic spoon
• Smoke, illumination signals	• Survival Manual (AFM 64-5)
• Waterproof match box	• Poncho
• Saw/knife blade	• Insect headnet
• Wood matches	• Ejector snap
• First aid kit	• Attaching strap
• MC-1 magnetic compass	• Kit, outer case
• Pocket knife	• Kit, inner case
• Saw-knife-shovel handle	• Shovel
• Frying pan	• Water bag
• Illuminating candles	• Packing list
• Compressed trioxane fuel	• Sleeping bag
• Signaling mirror	

Your survival kit need not be elaborate. You need only functional items that will meet your needs and a case to hold the items. For the case, you might want to use a Band-Aid box, a first aid case, an ammunition pouch, or another suitable case. This case should be—

- Water repellent or waterproof
- Easy to carry or attach to your body
- Suitable to accept varisized components
- Durable

Table 3-2: Hot Climate Kit.

• Canned drinking water	• Snare wire
• Waterproof matchbox	• Frying pan
• Plastic whistle	• Wood matches
• Smoke, illumination signals	• Insect headnet
• Pocket knife	• Reversible sun hat
• Signaling mirror	• Tool kit
• Plastic water bag	• Kit, packing list
• First aid kit	• Tarpaulin
• Sunburn-preventive cream	• Survival manual (AFM 64-5)
• Plastic spoon	• Kit, inner case
• Food packets	• Kit, outer case
• Compressed trioxane fuel	• Attaching strap
• Fishing tackle kit	• Ejector snap
• MC-1 magnetic compass	

Table 3-3: Overwater Kit.

• Kit, packing list	• Plastic spoon
• Raft boat paddle	• Pocket knife
• Survival manual (AFM 64-5)	• Food packets
• Insect headnet	• Fluorescent sea marker
• Reversible sun hat	• Frying pan
• Water storage bag	• Seawater desalter kit
• MC-1 magnetic compass	• Compressed trioxane fuel
• Boat bailer	• Smoke, illumination signals
• Sponge	• Signaling mirror
• Sunburn-preventive cream	• Fishing tackle kit
• Wood matches	• Waterproof match box
• First aid kit	• Raft repair kit

In your survival kit, you should have—

- First aid items
- Water purification tablets or drops
- Fire starting equipment
- Signaling items
- Food procurement items
- Shelter items

Some examples of these items are–

- Lighter, metal match, waterproof matches
- Snare wire
- Signaling mirror
- Wrist compass
- Fish and snare line
- Fishhooks
- Candle
- Small hand lens
- Oxytetracycline tablets (diarrhea or infection)
- Water purification tablets
- Solar blanket
- Surgical blades
- Butterfly sutures
- Condoms for water storage
- Chap Stick
- Needle and thread
- Knife

Include a weapon only if the situation so dictates. Read about and practice the survival techniques in this manual. Consider your unit's mission and the environment in which your unit will operate. Then prepare your survival kit.

Table 3-4: Individual survival kit with general and medical packets.

NSN	DESCRIPTION	QTY/UI
1680-00-205-0474	SURVIVAL KIT, INDIVIDUAL SURVIVAL VEST (OV-1), large, SC 1680-97-CL-A07	
1680-00-187-5716	SURVIVAL KIT, INDIVIDUAL SURVIVAL VEST (OV-1), small, SC 1680-97-CL-A07	
	Consisting of the following components:	
7340-00-098-4327	KNIFE, HUNTING: 5 in. lg blade, leather handle, w/sheath	1 ea
5110-00-526-8740	KNIFE, POCKET: one 3-1/16 in. lg cutting blade, & one 1-25/32 in. lg hook blade, w/safety lock & clevis	1 ea
4220-00-850-8655	LIFE PRESERVER, UNDERARM: gas or orally inflated, w/gas cyl, adult size, 10 in. h, orange color, shoulder & chest type harness w/quick release buckle & clip	1 ea
6230-00-938-1778	LIGHT, MARKER, DISTRESS: plastic body, rd, 1 in. w, accom 1 flashtube; one 5.4 v dry battery required	1 ea
6350-00-105-1252	MIRROR, EMERGENCY SIGNALING: glass, circular clear window in center or mirror for sighting, 3 in. lg, 2 in. w, 1/8 in. thk, w/o case, w/lanyard	1 ea
1370-00-490-7362	SIGNAL KIT, PERSONNEL DISTRESS: w/7 rocket cartridges & launcher	1 ea
6546-00-478-6504	SURVIVAL KIT, INDIVIDUAL	1 ea
4240-00-152-1578	GENERAL PACKET, INDIVIDUAL SURVIVAL KIT: w/ mandatory pack bag; 1 pkg ea of coffee & fruit flavored candy; 3 pkg chewing gum; 1 water storage container; 2 flash guards, w/infrared & blue filters; 1 mosquito headnet & pr mittens; 1 instruction card; 1 emergency signaling mirror; 1 fire starter & tinder; 5 safety	1 ea

(continued)

Table 3-4: *(Continued)*

NSN	DESCRIPTION	QTY/UI
	pins; 1 small straight-type surgical razor; 1 rescue/signal/medical instruction panel; 1 tweezer, & 1 wrist compass, strap & lanyard	
6545-00-231-9421	MEDICAL PACKET, INDIVIDUAL SURVIVAL KIT: w/carrying bag; 1 tube insect repellent & sun screen ointment; 1 medical instruction card; 1 waterproof receptacle, 1 bar soap & following items:	1 ea
6510-00-926-8881	ADHESIVE TAPE, SURGICAL: white rubber coating, 1/2 in. w, 360 in. lg, porous woven	1 ea
6505-00-118-1948	ASPIRIN TABLETS, USP: 0.324 gm, individually sealed in roll strip container	10 ea
6510-00-913-7909	BANDAGE, ADHESIVE: flesh, plastic coated, 3/4 in. w, 3 in. lg	1 ea
6510-00-913-7906	BANDAGE, GAUZE, ELASTIC: white, sterile, 2 in. w, 180 in. lg	1 ea
6505-00-118-1914	DIPHENOXYLATE HYDROCHLORIDE AND ATROPINE SULFATE TABLETS, USP: 0.025 mg atropine sulfate & 2.500 mg diphenoxylate hydrochloride active ingredients, individually sealed, roll strip container	10 ea
6505-00-183-9419	SULFACETAMIDE SODIUM OPHTHALMIC OINTMENT, USP: 10 percent	3.5 gm
6850-00-985-7166	WATER PURIFICATION TABLET, IODINE: 8 mg	50 ea
	VEST, SURVIVAL: nylon duck	1 ea
8415-00-201-9098	large size	1 ea
8415-00-201-9097	small size	1 ea
8465-00-254-8803	WHISTLE, BALL: plastic, olive drab w/lanyard	1 ea

The Army has several basic survival kits, primarily for issue to aviators. There are kits for cold climates, hot climates, and overwater. There is also an individual survival kit with general packet and medical packet. The cold climate, hot climate, and overwater kits are in canvas carrying bags. These kits are normally stowed in the helicopter's cargo/passenger area.

An aviator's survival vest, worn by helicopter crews, also contains survival items.

U.S. Army aviators flying fixed-wing aircraft equipped with ejection seats use a survival vest. The individual survival kits are stowed in the seat pan. Like all other kits, the rigid seat survival kit (RSSK) used depends on the environment.

Table 3-5: Aviator's Survival Vest.

NSN	DESCRIPTION
8465-00-177-4819	Survival Vest
6515-00-383-0565	Tourniquet
5820-00-782-5308	AN/PRC-90 Survival Radio
1305-00-301-1692	.38 caliber tracer ammunition
1305-00-322-6391	.38 caliber ball ammunition
1005-00-835-9773	Revolver, .38 caliber
9920-00-999-6753	Lighter, butane
6350-00-105-1252	Mirror, signaling
6545-00-782-6412	Survival kit, individual tropical
1370-00-490-7362	Signal kit, foliage penetrating
6230-00-938-1778	Light, distress marker, SDU-5/E
8465-00-634-4499	Bag, storage, drinking water
5110-00-162-2205	Knife, pocket
4240-00-300-2138	Net, gill, fishing
6605-00-151-5337	Compass, magnetic, lensatic

Table 3-6: Rigid Seat Survival Kits.

NSN	DESCRIPTION
1680-00-148-9233	Survival kit, cold climate (RSSK OV-1)
1680-00-148-9234	Survival kit, hot climate (RSSK OV-1)
1680-00-965-4702	Survival kit, overwater (RSSK OV-1)

PART II
Survival Medicine

Introduction

Foremost among the many problems that can compromise a survivor's ability to return to safety are medical problems resulting from parachute descent and landing, extreme climates, ground combat, evasion, and illnesses contracted in captivity.

Many evaders and survivors have reported difficulty in treating injuries and illness due to the lack of training and medical supplies. For some, this led to capture or surrender.

Survivors have related feeling of apathy and helplessness because they could not treat themselves in this environment. The ability to treat themselves increased their morale and cohesion and aided in their survival and eventual return to friendly forces.

One man with a fair amount of basic medical knowledge can make a difference in the lives of many. Without qualified medical personnel available, it is you who must know what to do to stay alive.

Part II meets the emergency medical training needs of individual soldiers. Because medical personnel will not always be readily available, the non medical soldiers will have to rely heavily on their own skills and knowledge of life-sustaining methods to survive on the integrated battlefield. This manual also addresses first aid measures for other life-threatening situations. It outlines both self-treatment (self-aid) and aid to other soldiers (buddy aid). More importantly, this manual emphasizes prompt and effective action in sustaining life and preventing or minimizing further suffering. First aid is the emergency care given to the sick, injured, or wounded before being treated by medical personnel. The Army Dictionary defines first aid as "urgent and immediate life saving and other measures which can be performed for casualties by non medical personnel when medical personnel are not immediately available." Non medical soldiers have received basic first aid training and should remain skilled in the correct procedures for giving first aid. A combat lifesaver is a non medical soldier who has been trained to provide emergency care. This includes administering intravenous infusions to casualties as his combat mission permits. Normally, each squad, team, or crew will have one member who is a combat lifesaver. This manual is directed to all soldiers. The procedures discussed apply to all types of casualties and the measures described are for use by both male and female soldiers.

Part II has been designed to provide a ready reference for the individual soldier on first aid. Only the information necessary to support and sustain proficiency in first aid has been boxed and the task number has been listed.

Commercial products (trade names or trademarks) mentioned in this publication are to provide descriptive information and for illustrative purposes only. Their use does not imply endorsement by the Department of Defense.

REQUIREMENTS FOR MAINTENANCE OF HEALTH: OVERVIEW

To survive, you need water and food. You must also have and apply high personal hygiene standards.

Water. Your body loses water through normal body processes (sweating, urinating, and defecating). During average daily exertion when the atmospheric temperature is 20 degrees Celsius (C) (68 degrees Fahrenheit), the average adult loses and therefore requires 2 to 3 liters of water daily. Other factors, such as heat exposure, cold exposure, intense activity, high altitude, burns, or illness, can cause your body to lose more water. You must replace this water.

Dehydration results from inadequate replacement of lost body fluids. It decreases your efficiency and, if injured, increases your susceptibility to severe shock. Consider the following results of body fluid loss:

- A 5 percent loss of body fluids results in thirst, irritability, nausea, and weakness.
- A 10 percent loss results in dizziness, headache, inability to walk, and a tingling sensation in the limbs.
- A 15 percent loss results in dim vision, painful urination, swollen tongue, deafness, and a numb feeling in the skin.
- A loss greater than 15 percent of body fluids may result in death.

The most common signs and symptoms of dehydration are—

- Dark urine with a very strong odor.
- Low urine output.
- Dark, sunken eyes.
- Fatigue.
- Emotional instability.
- Loss of skin elasticity.
- Delayed capillary refill in fingernail beds.
- Trench line down center of tongue.
- Thirst. Last on the list because you are already 2 percent dehydrated by the time you crave fluids.

You replace the water as you lose it. Trying to make up a deficit is difficult in a survival situation, and thirst is not a sign of how much water you need.

Most people cannot comfortably drink more than 1 liter of water at a time. So, even when not thirsty, drink small amounts of water at regular intervals each hour to prevent dehydration.

If you are under physical and mental stress or subject to severe conditions, increase your water intake. Drink enough liquids to maintain a urine output of at least 0.5 liter every 24 hours.

In any situation where food intake is low, drink 6 to 8 liters of water per day. In an extreme climate, especially an arid one, the average person can lose 2.5 to 3.5 liters of water per hour. In this type of climate, you should drink 14 to 30 liters of water per day.

With the loss of water there is also a loss of electrolytes (body salts). The average diet can usually keep up with these losses but in an extreme situation or illness, additional sources need to be provided. A mixture of 0.25 teaspoon of salt to 1 liter of water will provide a concentration that the body tissues can readily absorb.

Of all the physical problems encountered in a survival situation, the loss of water is the most preventable. The following are basic guidelines for the prevention of dehydration:

- *Always drink water when eating.* Water is used and consumed as a part of the digestion process and can lead to dehydration.
- *Acclimatize.* The body performs more efficiently in extreme conditions when acclimatized.
- *Conserve sweat not water.* Limit sweat-producing activities but drink water.
- *Ration water.* Until you find a suitable source, ration your water sensibly. A daily intake of 500 cubic centimeter (0.5 liter) of a sugar-water mixture (2 teaspoons per liter) will suffice to prevent severe dehydration for at least a week, provided you keep water losses to a minimum by limiting activity and heat gain or loss.

You can estimate fluid loss by several means. A standard field dressing holds about 0.25 liter (one-fourth canteen) of blood. A soaked T-shirt holds 0.5 to 0.75 liter.

You can also use the pulse and breathing rate to estimate fluid loss. Use the following as a guide:

- With a 0.75 liter loss the wrist pulse rate will be under 100 beats per minute and the breathing rate 12 to 20 breaths per minute.
- With a 0.75 to 1.5 liter loss the pulse rate will be 100 to 120 beats per minute and 20 to 30 breaths per minute.
- With a 1.5 to 2 liter loss the pulse rate will be 120 to 140 beats per minute and 30 to 40 breaths per minute. Vital signs above these rates require more advanced care.

Food. Although you can live several weeks without food, you need an adequate amount to stay healthy. Without food your mental and physical capabilities will deteriorate rapidly, and you will become weak. Food replenishes the substances that your body burns and provides energy. It provides vitamins, minerals, salts, and other elements essential to good health. Possibly more important, it helps morale.

The two basic sources of food are plants and animals (including fish). In varying degrees both provide the calories, carbohydrates, fats, and proteins needed for normal daily body functions.

Calories are a measure of heat and potential energy. The average person needs 2,000 calories per day to function at a minimum level. An adequate amount of carbohydrates, fats, and proteins without an adequate caloric intake will lead to starvation and cannibalism of the body's own tissue for energy.

Plant Foods. These foods provide carbohydrates—the main source of energy. Many plants provide enough protein to keep the body at normal efficiency. Although plants may not provide a balanced diet, they will sustain you even in the arctic, where meat's heat-producing qualities are normally essential. Many plant foods such as nuts and seeds will give you enough protein and oils for normal efficiency. Roots, green vegetables, and plant food containing natural sugar will provide calories and carbohydrates that give the body natural energy.

The food value of plants becomes more and more important if you are eluding the enemy or if you are in an area where wildlife is scarce. For instance—

- You can dry plants by wind, air, sun, or fire. This retards spoilage so that you can store or carry the plant food with you to use when needed.
- You can obtain plants more easily and more quietly than meat. This is extremely important when the enemy is near.

Animal Foods. Meat is more nourishing than plant food. In fact, it may even be more readily available in some places. However, to get meat, you need to know the habits of, and how to capture, the various wildlife.

To satisfy your immediate food needs, first seek the more abundant and more easily obtained wildlife, such as insects, crustaceans, mollusks, fish, and reptiles. These can satisfy your immediate hunger while you are preparing traps and snares for larger game.

Personal Hygiene. In any situation, cleanliness is an important factor in preventing infection and disease. It becomes even more important in a survival situation. Poor hygiene can reduce your chances of survival.

A daily shower with hot water and soap is ideal, but you can stay clean without this luxury. Use a cloth and soapy water to wash yourself. Pay special attention to the feet, armpits, crotch, hands, and hair as these are prime areas for infestation and infection. If water is scarce, take an "air" bath. Remove as much of your clothing as practical and expose your body to the sun and air for at least 1 hour. Be careful not to sunburn.

If you don't have soap, use ashes or sand, or make soap from animal fat and wood ashes, if your situation allows. To make soap—

- Extract grease from animal fat by cutting the fat into small pieces and cooking them in a pot.
- Add enough water to the pot to keep the fat from sticking as it cooks.

- Cook the fat slowly, stirring frequently.
- After the fat is rendered, pour the grease into a container to harden.
- Place ashes in a container with a spout near the bottom.
- Pour water over the ashes and collect the liquid that drips out of the spout in a separate container. This liquid is the potash or lye.
- Another way to get the lye is to pour the slurry (the mixture of ashes and water) through a straining cloth.
- In a cooking pot, mix two parts grease to one part potash.
- Place this mixture over a fire and boil it until it thickens.

After the mixture—the soap—cools, you can use it in the semiliquid state directly from the pot. You can also pour it into a pan, allow it to harden, and cut it into bars for later use.

Keep Your Hands Clean. Germs on your hands can infect food and wounds. Wash your hands after handling any material that is likely to carry germs, after visiting the latrine, after caring for the sick, and before handling any food, food utensils, or drinking water. Keep your fingernails closely trimmed and clean, and keep your fingers out of your mouth.

Keep Your Hair Clean. Your hair can become a haven for bacteria or fleas, lice, and other parasites. Keeping your hair clean, combed, and trimmed helps you avoid this danger.

Keep Your Clothing Clean. Keep your clothing and bedding as clean as possible to reduce the chance of skin infection as well as to decrease the danger of parasitic infestation. Clean your outer clothing whenever it becomes soiled. Wear clean underclothing and socks each day. If water is scarce, "air" clean your clothing by shaking, airing, and sunning it for 2 hours. If you are using a sleeping bag, turn it inside out after each use, fluff it, and air it.

Keep Your Teeth Clean. Thoroughly clean your mouth and teeth with a toothbrush at least once each day. If you don't have a toothbrush, make a chewing stick. Find a twig about 20 centimeters long and 1 centimeter wide. Chew one end of the stick to separate the fibers. Now brush your teeth thoroughly. Another way is to wrap a clean strip of cloth around your fingers and rub your teeth with it to wipe away food particles. You can also brush your teeth with small amounts of sand, baking soda, salt, or soap. Then rinse your mouth with water, salt water, or willow bark tea. Also, flossing your teeth with string or fiber helps oral hygiene.

If you have cavities, you can make temporary fillings by placing candle wax, tobacco, aspirin, hot pepper, toothpaste or powder, or portions of a ginger root into the cavity. Make sure you clean the cavity by rinsing or picking the particles out of the cavity before placing a filling in the cavity.

Take Care of Your Feet. To prevent serious foot problems, break in your shoes before wearing them on any mission. Wash and massage your feet daily. Trim your toenails straight across. Wear an insole and the proper size of dry socks. Powder and check your feet daily for blisters.

If you get a small blister, do not open it. An intact blister is safe from infection. Apply a padding material around the blister to relieve pressure and reduce friction. If the blister bursts, treat it as an open wound. Clean and dress it daily and pad around it. Leave large blisters intact. To avoid having the blister burst or tear under pressure and cause a painful and open sore, do the following:

- Obtain a sewing-type needle and a clean or sterilized thread.
- Run the needle and thread through the blister after cleaning the blister.
- Detach the needle and leave both ends of the thread hanging out of the blister. The thread will absorb the liquid inside. This reduces the size of the hole and ensures that the hole does not close up.
- Pad around the blister.

Get Sufficient Rest. You need a certain amount of rest to keep going. Plan for regular rest periods of at least 10 minutes per hour during your daily activities. Learn to make yourself comfortable under less than ideal conditions. A change from mental to physical activity or vice versa can be refreshing when time or situation does not permit total relaxation.

Keep Camp Site Clean. Do not soil the ground in the camp site area with urine or feces. Use latrines, if available. When latrines are not available, dig "cat holes" and cover the waste. Collect drinking water upstream from the camp site. Purify all water.

MEDICAL EMERGENCIES

Medical problems and emergencies you may be faced with include breathing problems, severe bleeding, and shock.

Breathing Problems. Any one of the following can cause airway obstruction, resulting in stopped breathing

- Foreign matter in mouth of throat that obstructs the opening to the trachea.
- Face or neck injuries.
- Inflammation and swelling of mouth and throat caused by inhaling smoke, flames, and irritating vapors or by an allergic reaction.
- "Kink" in the throat (caused by the neck bent forward so that the chin rests upon the chest) may block the passage of air.
- Tongue blocks passage of air to the lungs upon unconsciousness.
- When an individual is unconscious, the muscles of the lower jaw and tongue relax as the neck drops forward, causing the lower jaw to sag and the tongue to drop back and block the passage of air.

Severe Bleeding. Severe bleeding from any major blood vessel in the body is extremely dangerous. The loss of 1 liter of blood will produce moderate symptoms of shock. The loss of 2 liters will produce a severe state of shock that places the body in extreme danger. The loss of 3 liters is usually fatal.

Shock. Shock (acute stress reaction) is not a disease in itself. It is a clinical condition characterized by symptoms that arise when cardiac output is insufficient to fill the arteries with blood under enough pressure to provide an adequate blood supply to the organs and tissues.

LIFESAVING STEPS

Control panic, both your own and the victim's. Reassure him and try to keep him quiet.

Perform a rapid physical exam. Look for the cause of the injury and follow the ABCs of first aid, starting with the airway and breathing, but be discerning. A person may die from arterial bleeding more quickly than from an airway obstruction in some cases.

Open Airway and Maintain. You can open an airway and maintain it by using the following steps.

Step 1. Check if the victim has a partial or complete airway obstruction. If he can cough or speak, allow him to clear the obstruction naturally. Stand by, reassure the victim, and be ready to clear his airway and perform mouth-to-mouth resuscitation should he become unconscious. If his airway is completely obstructed, administer abdominal thrusts until the obstruction is cleared.

Step 2. Using a finger, quickly sweep the victim's mouth clear of any foreign objects, broken teeth, dentures, sand.

Step 3. Using the jaw thrust method, grasp the angles of the victim's lower jaw and lift with both hands, one on each side, moving the jaw forward. For stability, rest your elbows on the surface on which the victim is lying. If his lips are closed, gently open the lower lip with your thumb (Figure I-1).

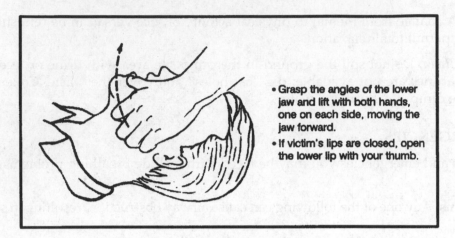

Figure I-1: Jaw thrust method.

Step 4. With the victim's airway open, pinch his nose closed with your thumb and forefinger and blow two complete breaths into his lungs. Allow the lungs to deflate after the second inflation and perform the following:

- Look for his chest to rise and fall.
- Listen for escaping air during exhalation.
- Feel for flow of air on your cheek.

Step 5. If the forced breaths do not stimulate spontaneous breathing, maintain the victim's breathing by performing mouth-to-mouth resuscitation.

Step 6. There is danger of the victim vomiting during mouth-to-mouth resuscitation. Check the victim's mouth periodically for vomit and clear as needed.

Note: Cardiopulmonary resuscitation (CPR) may be necessary after cleaning the airway, but only after major bleeding is under control.

Control Bleeding. In a survival situation, you must control serious bleeding immediately because replacement fluids normally are not available and the victim can die within a matter of minutes. External bleeding falls into the following classifications (according to its source):

- *Arterial.* Blood vessels called arteries carry blood away from the heart and through the body. A cut artery issues bright red blood from the wound in distinct spurts or pulses that correspond to the rhythm of the heartbeat. Because the blood in the arteries is under high pressure, an individual can lose a large volume of blood in a short period when damage to an artery of significant size occurs. Therefore, arterial bleeding is the most serious type of bleeding. If not controlled promptly, it can be fatal.
- *Venous.* Venous blood is blood that is returning to the heart through blood vessels called veins. A steady flow of dark red, maroon, or bluish blood characterizes bleeding from a vein. You can usually control venous bleeding more easily than arterial bleeding.
- *Capillary.* The capillaries are the extremely small vessels that connect the arteries with the veins. Capillary bleeding most commonly occurs in minor cuts and scrapes. This type of bleeding is not difficult to control.

You can control external bleeding by direct pressure, indirect (pressure points) pressure, elevation, digital ligation, or tourniquet.

Direct Pressure. The most effective way to control external bleeding is by applying pressure directly over the wound. This pressure must not only be firm enough to stop the bleeding, but it must also be maintained long enough to "seal off" the damaged surface.

If bleeding continues after having applied direct pressure for 30 minutes, apply a pressure dressing. This dressing consists of a thick dressing of gauze or other suitable material applied directly over the wound and held in place with a tightly wrapped bandage (Figure I-2). It should be tighter than an ordinary compression bandage but not so tight that it impairs circulation to the rest of the limb. Once you apply the dressing, do not remove it, even when the dressing becomes blood soaked.

Leave the pressure dressing in place for 1 or 2 days, after which you can remove and replace it with a smaller dressing.

In the long-term survival environment, make fresh, daily dressing changes and inspect for signs of infection.

Elevation. Raising an injured extremity as high as possible above the heart's level slows blood loss by aiding the return of blood to the heart and lowering the blood pressure at the wound. However, elevation alone will not control bleeding entirely; you must also apply direct pressure over the wound. When treating a snakebite, however, keep the extremity lower than the heart.

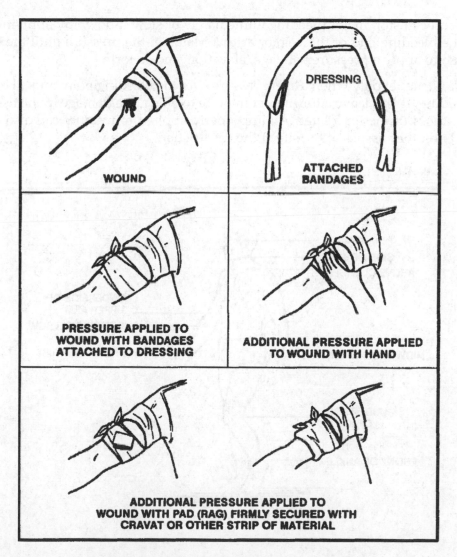

Figure I-2: Application of a pressure dressing.

Pressure Points. A pressure point is a location where the main artery to the wound lies near the surface of the skin or where the artery passes directly over a bony prominence (Figure I-3). You can use digital pressure on a pressure point to slow arterial bleeding until the application of a pressure dressing. Pressure point control is not as effective for controlling bleeding as direct pressure exerted on the wound. It is rare when a single major compressible artery supplies a damaged vessel.

If you cannot remember the exact location of the pressure points, follow this rule: Apply pressure at the end of the joint just above the injured area. On hands, feet, and head, this will be the wrist, ankle, and neck respectively.

> **WARNING**
> Use caution when applying pressure to the neck. Too much pressure for too long may cause unconsciousness or death. Never place a tourniquet around the neck.

Maintain pressure points by placing a round stick in the joint, bending the joint over the stick, and then keeping it tightly bent by lashing. By using this method to maintain pressure, it frees your hands to work in other areas.

Digital Ligation. You can stop major bleeding immediately or slow it down by applying pressure with a finger or two on the bleeding end of the vein or artery. Maintain the pressure until the bleeding stops or slows down enough to apply a pressure bandage, elevation, and so forth.

Tourniquet. Use a tourniquet only when direct pressure over the bleeding point and all other methods did not control the bleeding. If you leave a tourniquet in place too long, the damage to the tissues can progress to gangrene, with a loss of the limb later. An improperly applied tourniquet can also cause permanent damage to nerves and other tissues at the site of the constriction.

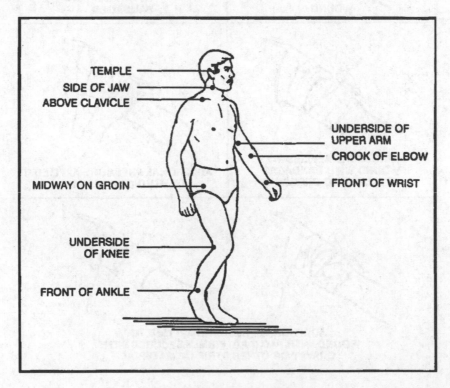

Figure I-3: Pressure points.

If you must use a tourniquet, place it around the extremity, between the wound and the heart, 5 to 10 centimeters above the wound site (Figure I-4). Never place it directly over the wound or a fracture. Use a stick as a handle to tighten the tourniquet and tighten it only enough to stop blood flow. When you have tightened the tourniquet, bind the free end of the stick to the limb to prevent unwinding.

After you secure the tourniquet, clean and bandage the wound. A lone survivor does not remove or release an applied tourniquet. In a buddy system, however, the buddy can release the tourniquet pressure every 10 to 15 minutes for 1 or 2 minutes to let blood flow to the rest of the extremity to prevent limb loss.

Prevent and Treat Shock. Anticipate shock in all injured personnel. Treat all injured persons as follows, regardless of what symptoms appear (Figure I-5):

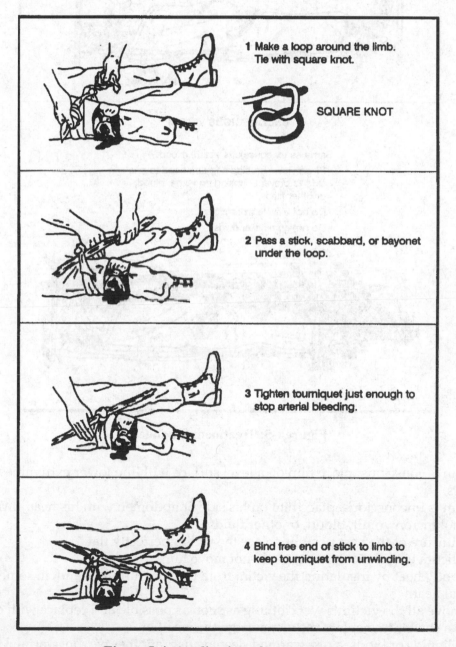

Figure I-4: Application of a tourniquet.

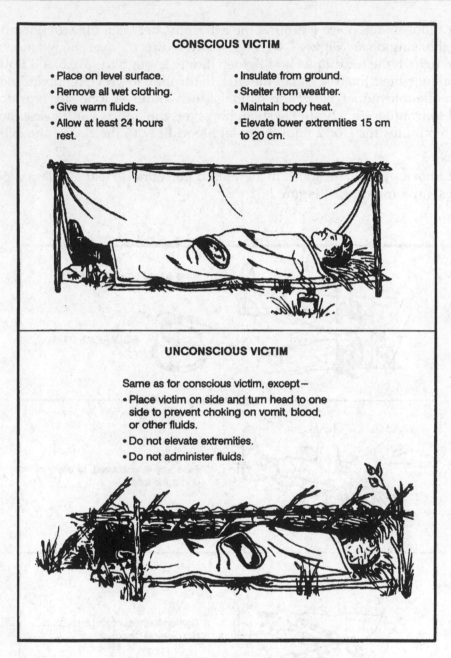

CONSCIOUS VICTIM

- Place on level surface.
- Remove all wet clothing.
- Give warm fluids.
- Allow at least 24 hours rest.

- Insulate from ground.
- Shelter from weather.
- Maintain body heat.
- Elevate lower extremities 15 cm to 20 cm.

UNCONSCIOUS VICTIM

Same as for conscious victim, except—
- Place victim on side and turn head to one side to prevent choking on vomit, blood, or other fluids.
- Do not elevate extremities.
- Do not administer fluids.

Figure I-5: Treatment for shock.

- If the victim is conscious, place him on a level surface with the lower extremities elevated 15 to 20 centimeters.
- If the victim is unconscious, place him on his side or abdomen with his head turned to one side to prevent choking on vomit, blood, or other fluids.
- If you are unsure of the best position, place the victim perfectly flat.
- Once the victim is in a shock position, do not move him.
- Maintain body heat by insulating the victim from the surroundings and, in some instances, applying external heat.
- If wet, remove all the victim's wet clothing as soon as possible and replace with dry clothing.
- Improvise a shelter to insulate the victim from the weather.
- Use warm liquids or foods, a pre-warmed sleeping bag, another person, warmed water in canteens, hot rocks wrapped in clothing, or fires on either side of the victim to provide external warmth.

- If the victim is conscious, slowly administer small doses of a warm salt or sugar solution, if available.
- If the victim is unconscious or has abdominal wounds, do not give fluids by mouth.
- Have the victim rest for at least 24 hours.
- If you are a lone survivor, lie in a depression in the ground, behind a tree, or any other place out of the weather, with your head lower than your feet.
- If you are with a buddy, reassess your patient constantly.

BONE AND JOINT INJURY

You could face bone and joint injuries that include fractures, dislocations, and sprains.

Fractures. There are basically two types of fractures: open and closed. With an open (or compound) fracture, the bone protrudes through the skin and complicates the actual fracture with an open wound. After setting the fracture, treat the wound as any other open wound.

The closed fracture has no open wounds. Follow the guidelines for immobilization, and set and splint the fracture.

The signs and symptoms of a fracture are pain, tenderness, discoloration, swelling deformity, loss of function, and grating (a sound or feeling that occurs when broken bone ends rub together).

The dangers with a fracture are the severing or the compression of a nerve or blood vessel at the site of fracture. For this reason minimum manipulation should be done, and only very cautiously. If you notice the area below the break becoming numb, swollen, cool to the touch, or turning pale, and the victim shows signs of shock, a major vessel may have been severed. You must control this internal bleeding. Rest the victim for shock, and replace lost fluids.

Often you must maintain traction during the splinting and healing process.

You can effectively pull smaller bones such as the arm or lower leg by hand. You can create traction by wedging a hand or foot in the V-notch of a tree and pushing against the tree with the other extremity. You can then splint the break.

Very strong muscles hold a broken thighbone (femur) in place making it difficult to maintain traction during healing. You can make an improvised traction splint using natural material (Figure I-6) as follows:

- Get two forked branches or saplings at least 5 centimeters in diameter. Measure one from the patient's armpit to 20 to 30 centimeters past his unbroken leg. Measure the other from the groin to 20 to 30 centimeters past the unbroken leg. Ensure that both extend an equal distance beyond the end of the leg.
- Pad the two splints. Notch the ends without forks and lash a 20- to 30-centimeter cross member made from a 5-centimeter diameter branch between them.
- Using available material (vines, cloth, rawhide), tie the splint around the upper portion of the body and down the length of the broken leg. Follow the splinting guidelines.
- With available material, fashion a wrap that will extend around the ankle, with the two free ends tied to the cross member.
- Place a 10- by 2.5-centimeter stick in the middle of the free ends of the ankle wrap between the cross member and the foot. Using the stick, twist the material to make the traction easier.
- Continue twisting until the broken leg is as long or slightly longer than the unbroken leg.
- Lash the stick to maintain traction.

Note: Over time you may lose traction because the material weakened. Check the traction periodically. If you must change or repair the splint, maintain the traction manually for a short time.

Dislocations. Dislocations are the separations of bone joints causing the bones to go out of proper alignment. These misalignments can be extremely painful and can cause an impairment of nerve or circulatory function below the area affected. You must place these joints back into alignment as quickly as possible.

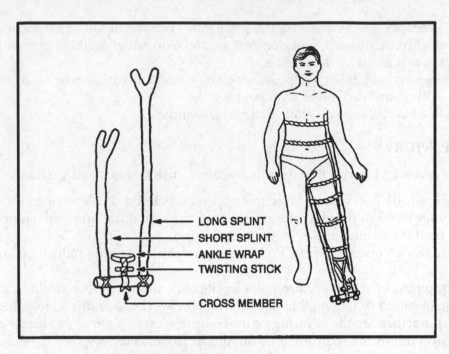

Figure I-6: Improvised traction splint.

Signs and symptoms of dislocations are joint pain, tenderness, swelling, discoloration, limited range of motion, and deformity of the joint. You treat dislocations by reduction, immobilization, and rehabilitation.

Reduction or "setting" is placing the bones back into their proper alignment. You can use several methods, but manual traction or the use of weights to pull the bones are the safest and easiest. Once performed, reduction decreases the victim's pain and allows for normal function and circulation. Without an X ray, you can judge proper alignment by the look and feel of the joint and by comparing it to the joint on the opposite side.

Immobilization is nothing more than splinting the dislocation after reduction. You can use any field-expedient material for a splint or you can splint an extremity to the body. The basic guidelines for splinting are—

- Splint above and below the fracture site.
- Pad splints to reduce discomfort.
- Check circulation below the fracture after making each tie on the splint.

To rehabilitate the dislocation, remove the splints after 7 to 14 days. Gradually use the injured joint until fully healed.

Sprains. The accidental overstretching of a tendon or ligament causes sprains. The signs and symptoms are pain, swelling, tenderness, and discoloration(black and blue).

When treating sprains, think RICE—

R—Rest injured area.
I—Ice for 24 hours, then heat after that.
C—Compression-wrapping and/or splinting to help stabilize. If possible, leave the boot on a sprained ankle unless circulation is compromised.
E—Elevation of the affected area.

BITES AND STINGS

Insects and related pests are hazards in a survival situation. They not only cause irritations, but they are often carriers of diseases that cause severe allergic reactions in some individuals. In many parts of the world you will be exposed to serious, even fatal, diseases not encountered in the United States.

Ticks can carry and transmit diseases, such as Rocky Mountain spotted fever common in many parts of the United States. Ticks also transmit the Lyme disease.

Mosquitoes may carry malaria, dengue, and many other diseases.

Flies can spread disease from contact with infectious sources. They are causes of sleeping sickness, typhoid, cholera, and dysentery.

Fleas can transmit plague.

Lice can transmit typhus and relapsing fever.

The best way to avoid the complications of insect bites and stings is to keep immunizations (including booster shots) up-to-date, avoid insect-infested areas, use netting and insect repellent, and wear all clothing properly.

If you get bitten or stung, do not scratch the bite or sting, it might become infected. Inspect your body at least once a day to ensure there are no insects attached to you. If you find ticks attached to your body, cover them with a substance, such as Vaseline, heavy oil, or tree sap, that will cut off their air supply. Without air, the tick releases its hold, and you can remove it. Take care to remove the whole tick. Use tweezers if you have them. Grasp the tick where the mouth parts are attached to the skin. Do not squeeze the tick's body. Wash your hands after touching the tick. Clean the tick wound daily until healed.

Treatment. It is impossible to list the treatment of all the different types of bites and stings. Treat bites and stings as follows:

- If antibiotics are available for your use, become familiar with them before deployment and use them.
- Predeployment immunizations can prevent most of the common diseases carried by mosquitoes and some carried by flies.
- The common fly-borne diseases are usually treatable with penicillins or erythromycin.
- Most tick-, flea-, louse-, and mite-borne diseases are treatable with tetracycline.
- Most antibiotics come in 250 milligram (mg) or 500 mg tablets. If you cannot remember the exact dose rate to treat a disease, 2 tablets, 4 times a day for 10 to 14 days will usually kill any bacteria.

Bee and Wasp Stings. If stung by a bee, immediately remove the stinger and venom sac, if attached, by scraping with a fingernail or a knife blade. Do not squeeze or grasp the stinger or venom sac, as squeezing will force more venom into the wound. Wash the sting site thoroughly with soap and water to lessen the chance of a secondary infection.

If you know or suspect that you are allergic to insect stings, always carry an insect sting kit with you. Relieve the itching and discomfort caused by insect bites by applying—

- Cold compresses.
- A cooling paste of mud and ashes.
- Sap from dandelions.
- Coconut meat.
- Crushed cloves of garlic.
- Onion.

Spider Bites and Scorpion Stings. The black widow spider is identified by a red hourglass on its abdomen. Only the female bites, and it has a neurotoxic venom. The initial pain is not severe, but severe local pain rapidly develops. The pain gradually spreads over the entire body and settles in the abdomen and legs. Abdominal cramps and progressive nausea, vomiting, and a rash may occur. Weakness, tremors, sweating, and salivation may occur. Anaphylactic reactions can occur. Symptoms begin to regress after several hours and are usually gone in a few days. Treat for shock. Be ready to perform CPR. Clean and dress the bite area to reduce the risk of infection. An antivenom is available.

The funnel web spider is a large brown or gray spider found in Australia. The symptoms and the treatment for its bite are as for the black widow spider.

The brown house spider or brown recluse spider is a small, light brown spider identified by a dark brown violin on its back. There is no pain, or so little pain, that usually a victim is not aware of the bite. Within a few hours a painful red area with a mottled cyanotic center appears. Necrosis does not occur in all bites, but usually in 3 to 4 days, a star-shaped, firm area of deep purple discoloration appears at the bite site. The area turns dark and mummified in a week or two. The margins separate and the scab falls off, leaving an open ulcer. Secondary infection and regional swollen lymph glands usually become visible at this stage. The outstanding characteristic of the brown recluse bite is an ulcer that does not heal but persists for weeks or months. In addition to the ulcer, there is often a systemic reaction that is serious and may lead to death.

Reactions (fever, chills, joint pain, vomiting, and a generalized rash) occur chiefly in children or debilitated persons.

Tarantulas are large, hairy spiders found mainly in the tropics. Most do not inject venom, but some South American species do. They have large fangs. If bitten, pain and bleeding are certain, and infection is likely. Treat a tarantula bite as for any open wound, and try to prevent infection. If symptoms of poisoning appear, treat as for the bite of the black widow spider.

Scorpions are all poisonous to a greater or lesser degree. There are two different reactions, depending on the species:

- Severe local reaction only, with pain and swelling around the area of the sting. Possible prickly sensation around the mouth and a thick-feeling tongue.
- Severe systemic reaction, with little or no visible local reaction. Local pain may be present. Systemic reaction includes respiratory difficulties, thick-feeling tongue, body spasms, drooling, gastric distention, double vision, blindness, involuntary rapid movement of the eyeballs, involuntary urination and defecation, and heart failure. Death is rare, occurring mainly in children and adults with high blood pressure or illnesses.

Treat scorpion stings as you would a black widow bite.

Snakebites. The chance of a snakebite in a survival situation is rather small, if you are familiar with the various types of snakes and their habitats. However, it could happen and you should know how to treat a snakebite. Deaths from snakebites are rare. More than one-half of the snakebite victims have little or no poisoning, and only about one-quarter develop serious systemic poisoning. However, the chance of a snakebite in a survival situation can affect morale, and failure to take preventive measures or failure to treat a snakebite properly can result in needless tragedy.

The primary concern in the treatment of snakebite is to limit the amount of eventual tissue destruction around the bite area.

A bite wound, regardless of the type of animal that inflicted it, can become infected from bacteria in the animal's mouth. With nonpoisonous as well as poisonous snakebites, this local infection is responsible for a large part of the residual damage that results.

Snake venoms not only contain poisons that attack the victim's central nervous system (neurotoxins) and blood circulation (hemotoxins), but also digestive enzymes (cytotoxins) to aid in digesting their prey.

These poisons can cause a very large area of tissue death, leaving a large open wound. This condition could lead to the need for eventual amputation if not treated.

Shock and panic in a person bitten by a snake can also affect the person's recovery. Excitement, hysteria, and panic can speed up the circulation, causing the body to absorb the toxin quickly. Signs of shock occur within the first 30 minutes after the bite.

Before you start treating a snakebite, determine whether the snake was poisonous or nonpoisonous. Bites from a nonpoisonous snake will show rows of teeth. Bites from a poisonous snake may have rows of teeth showing, but will have one or more distinctive puncture marks caused by fang penetration. Symptoms of a poisonous bite may be spontaneous bleeding from the nose and anus, blood in the urine, pain at the site of the bite, and swelling at the site of the bite within a few minutes or up to 2 hours later.

Breathing difficulty, paralysis, weakness, twitching, and numbness are also signs of neurotoxic venoms. These signs usually appear 1.5 to 2 hours after the bite.

If you determine that a poisonous snake bit an individual, take the following steps:

- Reassure the victim and keep him still.
- Set up for shock and force fluids or give an intravenous (IV).
- Remove watches, rings, bracelets, or other constricting items.
- Clean the bite area.
- Maintain an airway (especially if bitten near the face or neck) and be prepared to administer mouth-to-mouth resuscitation or CPR.
- Use a constricting band between the wound and the heart.
- Immobilize the site.
- Remove the poison as soon as possible by using a mechanical suction device or by squeezing.

Do not—

- Give the victim alcoholic beverages or tobacco products.
- Give morphine or other central nervous system (CNS) depressors.
- Make any deep cuts at the bite site. Cutting opens capillaries that in turn open a direct route into the blood stream for venom and infection.

Note: If medical treatment is over one hour away, make an incision(no longer than 6 millimeters and no deeper than 3 millimeters) over each puncture, cutting just deep enough to enlarge the fang opening, but only through the first or second layer of skin. Place a suction cup over the bite so that you have a good vacuum seal. Suction the bite site 3 to 4 times. Use mouth suction only as a last resort and only if you do not have open sores in your mouth. Spit the envenomed blood out and rinse your mouth with water. This method will draw out 25 to 30 percent of the venom.

- Put your hands on your face or rub your eyes, as venom may be on your hands. Venom may cause blindness.
- Break open the large blisters that form around the bite site.

After caring for the victim as described above, take the following actions to minimize local effects:

- If infection appears, keep the wound open and clean.
- Use heat after 24 to 48 hours to help prevent the spread of local infection. Heat also helps to draw out an infection.
- Keep the wound covered with a dry, sterile dressing.
- Have the victim drink large amounts of fluids until the infection is gone.

WOUNDS

An interruption of the skin's integrity characterizes wounds. These wounds could be open wounds, skin diseases, frostbite, trench foot, and burns.

Open Wounds. Open wounds are serious in a survival situation, not only because of tissue damage and blood loss, but also because they may become infected. Bacteria on the object that made the wound, on the individual's skin and clothing, or on other foreign material or dirt that touches the wound may cause infection.

By taking proper care of the wound you can reduce further contamination and promote healing. Clean the wound as soon as possible after it occurs by—

- Removing or cutting clothing away from the wound.
- Always looking for an exit wound if a sharp object, gun shot, or projectile caused a wound.
- Thoroughly cleaning the skin around the wound.
- Rinsing (not scrubbing) the wound with large amounts of water under pressure. You can use fresh urine if water is not available.

The "open treatment" method is the safest way to manage wounds in survival situations. Do not try to close any wound by suturing or similar procedures. Leave the wound open to allow the drainage of any pus resulting from infection. As long as the wound can drain, it generally will not become life-threatening, regardless of how unpleasant it looks or smells.

Cover the wound with a clean dressing. Place a bandage on the dressing to hold it in place. Change the dressing daily to check for infection.

If a wound is gaping, you can bring the edges together with adhesive tape cut in the form of a "butterfly" or "dumbbell" (Figure I-7).

In a survival situation, some degree of wound infection is almost inevitable. Pain, swelling, and redness around the wound, increased temperature, and pus in the wound or on the dressing indicate infection is present.

To treat an infected wound—

- Place a warm, moist compress directly on the infected wound.
- Change the compress when it cools, keeping a warm compress on the wound for a total of 30 minutes. Apply the compresses three or four times daily.

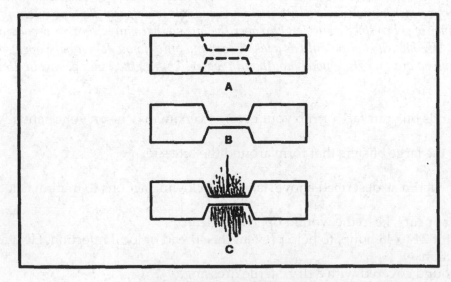

Figure I-7: Butterfly closure.

- Drain the wound. Open and gently probe the infected wound with a sterile instrument.
- Dress and bandage the wound.
- Drink a lot of water.

Continue this treatment daily until all signs of infection have disappeared.

If you do not have antibiotics and the wound has become severely infected, does not heal, and ordinary debridement is impossible, consider maggot therapy, despite its hazards:

- Expose the wound to flies for one day and then cover it.
- Check daily for maggots.
- Once maggots develop, keep wound covered but check daily.
- Remove all maggots when they have cleaned out all dead tissue and before they start on healthy tissue. Increased pain and bright red blood in the wound indicate that the maggots have reached healthy tissue.
- Flush the wound repeatedly with sterile water or fresh urine to remove the maggots.
- Check the wound every four hours for several days to ensure all maggots have been removed.
- Bandage the wound and treat it as any other wound. It should heal normally.

Skin Diseases and Ailments. Although boils, fungal infections, and rashes rarely develop into a serious health problem, they cause discomfort and you should treat them.

Boils. Apply warm compresses to bring the boil to a head. Then open the boil using a sterile knife, wire, needle, or similar item. Thoroughly clean out the pus using soap and water. Cover the boil site, checking it periodically to ensure no further infection develops.

Fungal Infections. Keep the skin clean and dry, and expose the infected area to as much sunlight as possible. Do not scratch the affected area. During the Southeast Asian conflict, soldiers used antifungal powders, lye soap, chlorine bleach, alcohol, vinegar, concentrated salt water, and iodine to treat fungal infections with varying degrees of success. As with any "unorthodox" method of treatment, use it with caution.

Rashes. To treat a skin rash effectively, first determine what is causing it. This determination may be difficult even in the best of situations. Observe the following rules to treat rashes:

- If it is moist, keep it dry.
- If it is dry, keep it moist.
- Do not scratch it.

Use a compress of vinegar or tannic acid derived from tea or from boiling acorns or the bark of a hardwood tree to dry weeping rashes. Keep dry rashes moist by rubbing a small amount of rendered animal fat or grease on the affected area.

Remember, treat rashes as open wounds and clean and dress them daily. There are many substances available to survivors in the wild or in captivity for use as antiseptics to treat wounds:

- *Iodine tablets.* Use 5 to 15 tablets in a liter of water to produce a good rinse for wounds during healing.
- *Garlic.* Rub it on a wound or boil it to extract the oils and use the water to rinse the affected area.
- *Salt water.* Use 2 to 3 tablespoons per liter of water to kill bacteria.
- *Bee honey.* Use it straight or dissolved in water.
- *Sphagnum moss.* Found in boggy areas worldwide, it is a natural source of iodine. Use as a dressing.

Again, use noncommercially prepared materials with caution.

Frostbite. This injury results from frozen tissues. Light frostbite involves only the skin that takes on a dull, whitish pallor. Deep frostbite extends to a depth below the skin. The tissues become solid and immovable. Your feet, hands, and exposed facial areas are particularly vulnerable to frostbite.

When with others, prevent frostbite by using the buddy system. Check your buddy's face often and make sure that he checks yours. If you are alone, periodically cover your nose and lower part of your face with your mittens.

Do not try to thaw the affected areas by placing them close to an open flame. Gently rub them in luke-warm water. Dry the part and place it next to your skin to warm it at body temperature.

Trench Foot. This condition results from many hours or days of exposure to wet or damp conditions at a temperature just above freezing. The nerves and muscles sustain the main damage, but gangrene can occur. In extreme cases the flesh dies and it may become necessary to have the foot or leg amputated. The best prevention is to keep your feet dry. Carry extra socks with you in a waterproof packet. Dry wet socks against your body. Wash your feet daily and put on dry socks.

Burns. The following field treatment for burns relieves the pain somewhat, seems to help speed healing, and offers some protection against infection:

- First, stop the burning process. Put out the fire by removing clothing, dousing with water or sand, or by rolling on the ground. Cool the burning skin with ice or water. For burns caused by white phosphorous, pick out the white phosphorous with tweezers; do not douse with water.
- Soak dressings or clean rags for 10 minutes in a boiling tannic acid solution (obtained from tea, inner bark of hardwood trees, or acorns boiled in water).
- Cool the dressings or clean rags and apply over burns.
- Rest as an open wound.
- Replace fluid loss.
- Maintain airway.
- Treat for shock.
- Consider using morphine, unless the burns are near the face.

ENVIRONMENTAL INJURIES

Heatstroke, hypothermia, diarrhea, and intestinal parasites are environmental injuries you could face.

Heatstroke. The breakdown of the body's heat regulatory system (body temperature more than 40.5 degrees C [105 degrees F]) causes a heatstroke. Other heat injuries, such as cramps or dehydration, do not always precede a heatstroke. Signs and symptoms of heatstroke are—

- Swollen, beet-red face.
- Reddened whites of eyes.
- Victim not sweating.
- Unconsciousness or delirium, which can cause pallor, a bluish color to lips and nail beds (cyanosis), and cool skin.

Note: By this time the victim is in severe shock. Cool the victim as rapidly as possible. Cool him by dipping him in a cool stream. If one is not available, douse the victim with urine, water, or at the very least, apply cool wet compresses to all the joints, especially the neck, armpits, and crotch. Be sure to wet the victim's head. Heat loss through the scalp is great. Administer IVs and provide drinking fluids. You may fan the individual.

Expect, during cooling—

- Vomiting.
- Diarrhea.

- Struggling.
- Shivering.
- Shouting.
- Prolonged unconsciousness.
- Rebound heatstroke within 48 hours.
- Cardiac arrest; be ready to perform CPR.

Note: Treat for dehydration with lightly salted water.

Hypothermia. Defined as the body's failure to maintain a temperature of 36 degrees C (97 degrees F). Exposure to cool or cold temperature over a short or long time can cause hypothermia. Dehydration and lack of food and rest predispose the survivor to hypothermia.

Unlike heatstroke, you must gradually warm the hypothermia victim. Get the victim into dry clothing. Replace lost fluids, and warm him.

Diarrhea. A common, debilitating ailment caused by a change of water and food, drinking contaminated water, eating spoiled food, becoming fatigued, and using dirty dishes. You can avoid most of these causes by practicing preventive medicine. If you get diarrhea, however, and do not have anti-diarrheal medicine, one of the following treatments may be effective:

- Limit your intake of fluids for 24 hours.
- Drink one cup of a strong tea solution every 2 hours until the diarrhea slows or stops. The tannic acid in the tea helps to control the diarrhea. Boil the inner bark of a hardwood tree for 2 hours or more to release the tannic acid.
- Make a solution of one handful of ground chalk, charcoal, or dried bones and treated water. If you have some apple pomace or the rind of citrus fruit, add an equal portion to the mixture to make it more effective. Take 2 tablespoons of the solution every 2 hours until the diarrhea slows or stops.

Intestinal Parasites. You can usually avoid worm infestations and other intestinal parasites if you take preventive measures. For example, never go barefoot. The most effective way to prevent intestinal parasites is to avoid uncooked meat and raw vegetables contaminated by raw sewage or human waste used as a fertilizer. However, should you become infested and lack proper medicine, you can use home remedies. Keep in mind that these home remedies work on the principle of changing the environment of the gastrointestinal tract. The following are home remedies you could use:

- *Salt water.* Dissolve 4 tablespoons of salt in 1 liter of water and drink. Do not repeat this treatment.
- *Tobacco.* Eat 1 to 1.5 cigarettes. The nicotine in the cigarette will kill or stun the worms long enough for your system to pass them. If the infestation is severe, repeat the treatment in 24 to 48 hours, but no sooner.
- *Kerosene.* Drink 2 tablespoons of kerosene but no more. If necessary, you can repeat this treatment in 24 to 48 hours. Be careful not to inhale the fumes. They may cause lung irritation.
- *Hot peppers.* Peppers are effective only if they are a steady part of your diet. You can eat them raw or put them in soups or rice and meat dishes. They create an environment that is prohibitive to parasitic attachment.

HERBAL MEDICINES

Our modern wonder drugs, laboratories, and equipment have obscured more primitive types of medicine involving determination, commonsense, and a few simple treatments. In many areas of the world, however, the people still depend on local "witch doctors" or healers to cure their ailments. Many of the herbs (plants) and treatments they use areas effective as the most modern medications available. In fact, many modern medications come from refined herbs.

WARNING
Use herbal medicines with extreme care, however, and only when you lack or have limited medical supplies. Some herbal medicines are dangerous and may cause further damage or even death.

CHAPTER 1

Fundamental Criteria for First Aid

Soldiers may have to depend upon their first aid knowledge and skills to save themselves or other soldiers. They may be able to save a life, prevent permanent disability, and reduce long periods of hospitalization by knowing what to do, what not to do, and when to seek medical assistance. Anything soldiers can do to keep others in good fighting condition is part of the primary mission to fight or to support the weapons system. Most injured or ill soldiers are able to return to their units to fight and/or support primarily because they are given appropriate and timely first aid followed by the best medical care possible. Therefore, all soldiers must remember the basics:

- Check for BREATHING: Lack of oxygen intake (through a compromised airway or inadequate breathing) can lead to brain damage or death in very few minutes.
- Check for BLEEDING: Life cannot continue without an adequate volume of blood to carry oxygen to tissues.
- Check for SHOCK: Unless shock is prevented or treated, death may result even though the injury would not otherwise be fatal.

SECTION I. EVALUATE CASUALTY

1-1. Casualty Evaluation (081-831-1000). The time may come when you must instantly apply your knowledge of lifesaving and first aid measures, possibly under combat or other adverse conditions. Any soldier observing an unconscious and/or ill, injured, or wounded person must carefully and skillfully evaluate him to determine the first aid measures required to prevent further injury or death. He should seek help from medical personnel as soon as possible, but must NOT interrupt his evaluation or treatment of the casualty. A second person may be sent to find medical help. One of the cardinal principles of treating a casualty is that the initial rescuer must continue the evaluation and treatment, as the tactical situation permits, until he is relieved by another individual. If, during any part of the evaluation, the casualty exhibits the conditions for which the soldier is checking, the soldier must stop the evaluation and immediately administer first aid. Ina chemical environment, the soldier should not evaluate the casualty until the casualty has been masked and given the antidote. After providing first aid, the soldier must proceed with the evaluation and continue to monitor the casualty for further medical complications until relieved by medical personnel. Learn the following procedures well. You may become that soldier who will have to give first aid some day.

✍ NOTE

A casualty in shock after suffering a heart attack, chest wound, or breathing difficulty, may breathe easier in a sitting position. If this is the case, allow him to sit upright, but monitor carefully in case his condition worsens.

WARNING

Again, remember, if there are any signs of chemical or biological agent poisoning, you should immediately mask the casualty. If it is nerve agent poisoning, administer the antidote, using the casualty's injector/ampules. See task 081-831-1031, Administer First Aid to a Nerve Agent Casualty (Buddy Aid).

a. Step ONE. Check the casualty for responsiveness by gently shaking or tapping him while calmly asking, "Are you okay?" Watch for response. If the casualty does not respond, go to step TWO. See Chapter 2, paragraph 2-5 for more information. If the casualty responds, continue with the evaluation.

    (1) If the casualty is conscious, ask him where he feels different than usual or where it hurts. Ask him to identify the locations of pain if he can, or to identify the area in which there is no feeling.

    (2) If the casualty is conscious but is choking and cannot talk, stop the evaluation and begin treatment. See task 081-831-1003, Clear an Object from the Throat of a Conscious Casualty. Also see Chapter 2, paragraph 2-13 for specific details on opening the airway.

WARNING

If a broken neck or back is suspected, do not move the casualty unless to save his life. Movement may cause permanent paralysis or death.

b. Step TWO. Check for breathing. See Chapter 2, paragraph 2-5c for procedure.

    (1) If the casualty is breathing, proceed to step FOUR.

    (2) If the casualty is not breathing, stop the evaluation and begin treatment (attempt to ventilate). See task 081-831-1042, Perform Mouth-to-Mouth Resuscitation. If an airway obstruction is apparent, clear the airway obstruction, then ventilate.

    (3) After successfully clearing the casualty's airway, proceed to step THREE.

c. Step THREE. Check for pulse. If pulse is present, and the casualty is breathing, proceed to step FOUR.

    (1) If pulse is present, but the casualty is still not breathing, start rescue breathing. See Chapter 2, paragraphs 2-6, and 2-7 for specific methods.

    *(2) If pulse is not found, seek medically trained personnel for help.

d. Step FOUR. Check for bleeding. Look for spurts of blood or blood-soaked clothes. Also check for both entry and exit wounds. If the casualty is bleeding from an open wound, stop the evaluation and begin first aid treatment in accordance with the following tasks, as appropriate:

    (1) Arm or leg wound–Task 081-831-1016, Put on a Field or Pressure Dressing. See Chapter 2, paragraphs 2-15, 2-17, 2-18, and 2-19.

    (2) Partial or complete amputation–Task 081-831-1017, Put on a Tourniquet. See Chapter 2, paragraph 2-20.

    (3) Open head wound–Task 081-831-1033, Apply a Dressing to an Open Head Wound. See Chapter 3, Section I.

    (4) Open abdominal wound–Task 081-831-1025, Apply a Dressing to an Open Abdominal Wound. See Chapter 3, paragraph 3-12.

    (5) Open chest wound–Task 081-831-1026, Apply a Dressing to an Open Chest Wound. See Chapter 3, paragraphs 3-9 and 3-10.

WARNING

In a chemically contaminated area, do not expose the wound(s).

e. Step FIVE. Check for shock. If signs/symptoms of shock are present, stop the evaluation and begin treatment immediately. The following are nine signs and/or symptoms of shock.

(1) Sweaty but cool skin (clammy skin).

(2) Paleness of skin.

(3) Restlessness or nervousness.

(4) Thirst.

(5) Loss of blood (bleeding).

(6) Confusion (does not seem aware of surroundings).

(7) Faster than normal breathing rate.

(8) Blotchy or bluish skin, especially around the mouth.

(9) Nausea and/or vomiting.

WARNING

Leg fractures must be splinted before elevating the legs as a treatment for shock.

See Chapter 2, Section III for specific information regarding the causes and effects, signs/symptoms, and the treatment/prevention of shock.

f. Step SIX. Check for fractures (Chapter 4).

(1) Check for the following signs/symptoms of a back or neck injury and treat as necessary.

- Pain or tenderness of the neck or back area.
- Cuts or bruises in the neck or back area.
- Inability of a casualty to move (paralysis or numbness).
  - Ask about ability to move (paralysis).
  - Touch the casualty's arms and legs and ask whether he can feel your hand (numbness).
- Unusual body or limb position.

WARNING

Unless there is immediate life threatening danger, do not move a casualty who has a suspected back or neck injury. Movement may cause permanent paralysis or death.

(2) Immobilize any casualty suspected of having a neck or back injury by doing the following

- Tell the casualty not to move.
- If a *back injury* is suspected, place padding (rolled or folded to conform to the shape of the arch) under the natural arch of the casualty's back. For example, a blanket may be used as padding.
- If a *neck injury* is suspected, place a roll of cloth under the casualty's neck and put weighted boots (filled with dirt, sand and so forth) or rocks on both sides of his head.

(3) Check the casualty's arms and legs for open or closed fractures.

Check for open fractures.

Look for bleeding.

Look for bone sticking through the skin.

Check for closed fractures.
Look for swelling.
Look for discoloration.
Look for deformity.
Look for unusual body position.

*(4) Stop the evaluation and begin treatment if a fracture to an arm or leg is suspected. See Task 081-831-1034, Splint a Suspected Fracture, Chapter 4, paragraphs 4-4 through 4-7.

(5) Check for signs/symptoms of fractures of other body areas (for example, shoulder or hip) and treat as necessary.

g. Step SEVEN. Check for burns. Look carefully for reddened, blistered, or charred skin, also check for singed clothing. If burns are found, stop the evaluation and begin treatment (Chapter 3, paragraph 3-14). See task 081-831-1007, *Give First Aid for Burns.*

h. Step EIGHT. Check for possible head injury.

(1) Look for the following signs and symptoms
Unequal pupils.
Fluid from the ear(s), nose, mouth, or injury site.
Slurred speech.
Confusion.
Sleepiness.
Loss of memory or consciousness.
Staggering in walking.
Headache.
Dizziness.
Vomiting and/or nausea.
Paralysis.
Convulsions or twitches.

(2) If a head injury is suspected, continue to watch for signs which would require performance of mouth-to-mouth resuscitation, treatment for shock, or control of bleeding and seek medical aid. See Chapter 3, Section I for specific indications of head injury and treatment. See task 081-831-1033, Apply a Dressing to an Open Head Wound.

1-2. Medical Assistance (081-831-1000). When a non-medically trained soldier comes upon an unconscious and/or injured soldier, he must accurately evaluate the casualty to determine the first aid measures needed to prevent further injury or death. He should seek medical assistance as soon as possible, but he MUST NOT interrupt treatment. To interrupt treatment may cause more harm than good to the casualty. A second person may be sent to find medical help. If, during any part of the evaluation, the casualty exhibits the conditions for which the soldier is checking, the soldier must stop the evaluation and immediately administer first aid. Remember that in a chemical environment, the soldier should not evaluate the casualty until the casualty has been masked and given the antidote. After performing first aid, the soldier must proceed with the evaluation and continue to monitor the casualty for development of conditions which may require the performance of necessary basic life saving measures, such as clearing the airway, mouth-to-mouth resuscitation, preventing shock, and/or bleeding control. He should continue to monitor until relieved by medical personnel.

SECTION II. UNDERSTAND VITAL BODY FUNCTIONS

1-3. Respiration and Blood Circulation. Respiration (inhalation and exhalation) and blood circulation are vital body functions. Interruption of either of these two functions need not be fatal IF appropriate first aid measures are correctly applied.

a. Respiration. When a person inhales, oxygen is taken into the body and when he exhales, carbon dioxide is expelled from the body–this is respiration. Respiration involves the—
   - Airway (nose, mouth, throat, voice box, windpipe, and bronchial tree). The canal through which air passes to and from the lungs.
   - Lungs (two elastic organs made up of thousands of tiny air spaces and covered by an airtight membrane).
   - Chest cage (formed by the muscle-connected ribs which join the spine in back and the breastbone in front). The top part of the chest cage is closed by the structure of the neck, and the bottom part is separated from the abdominal cavity by a large dome-shaped muscle called the diaphragm (Figure 1-1). The diaphragm and rib muscles, which are under the control of the respiratory center in the brain, automatically contract and relax. Contraction increases and relaxation decreases the size of the chest cage.

When the chest cage increases and then decreases, the air pressure in the lungs is first less and then more than the atmospheric pressure, thus causing the air to rush in and out of the lungs to equalize the pressure. This cycle of inhaling and exhaling is repeated about 12 to 18 times per minute.

b. Blood Circulation. The heart and the blood vessels (arteries, veins, and capillaries) circulate blood through the body tissues. The heart is divided into two separate halves, each acting as a pump. The left side pumps oxygenated blood (bright red) through the arteries into the capillaries; nutrients and oxygen pass from the blood through the walls of the capillaries into the cells. At the same time waste products and carbon dioxide enter the capillaries. From the capillaries the oxygen poor blood is carried through the veins to the right side of the heart and then into the lungs where it

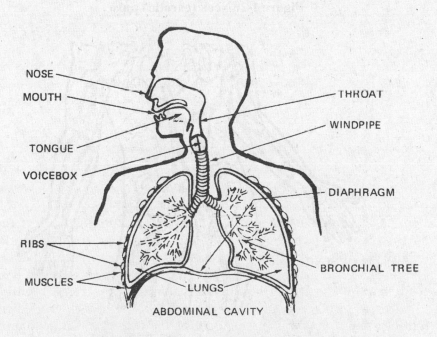

Figure 1-1: Airway, lungs, and chest cage.

expels carbon dioxide and picks up oxygen, Blood in the veins is dark red because of its low oxygen content. Blood does not flow through the veins in spurts as it does through the arteries.

(1) Heartbeat. The heart functions as a pump to circulate the blood continuously through the blood vessels to all parts of the body. It contracts, forcing the blood from its chambers; then it relaxes, permitting its chambers to refill with blood. The rhythmical cycle of contraction and relaxation is called the heartbeat. The normal heartbeat is from 60 to 80 beats per minute.

(2) Pulse. The heartbeat causes a rhythmical expansion and contraction of the arteries as it forces blood through them. This cycle of expansion and contraction can be felt (monitored) at various body points and is called the pulse. The common points for checking the pulse are at the side of the neck (carotid), the groin (femoral), the wrist (radial), and the ankle (posterial tibial).

  (a) Neck (carotid) pulse. To check the neck (carotid) pulse, feel for a pulse on the side of the casualty's neck closest to you by placing the tips of your first two fingers beside his Adam's apple (Figure 1-2).

  (b) Groin (femoral) pulse. To check the groin (femoral) pulse, press the tips of two fingers into the middle of the groin (Figure 1-3).

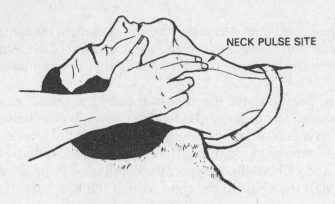

Figure 1-2: Neck (carotid) pulse.

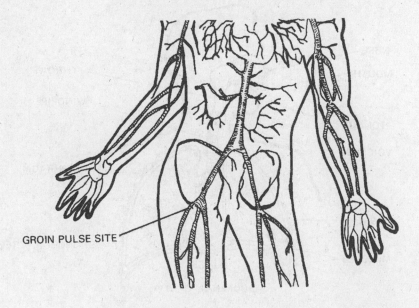

Figure 1-3: Groin (femoral) pulse.

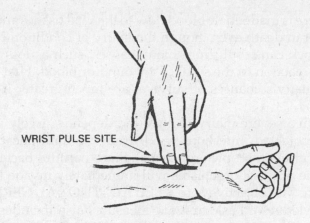

Figure 1-4: Wrist (radial) pulse.

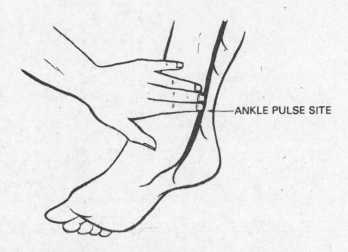

Figure 1-5: Ankle (posterial tibial) pulse.

(c) Wrist (radial) pulse. To check the wrist (radial) pulse, place your first two fingers on the thumb side of the casualty's wrist (Figure 1-4).

(d) Ankle (posterial tibial) pulse. To check the ankle (posterial tibial) pulse, place your first two fingers on the inside of the ankle (Figure 1-5).

✍ **NOTE**

DO NOT use your thumb to check a casualty's pulse because you may confuse your pulse beat with that of the casualty.

1-4. Adverse Conditions

a. *Lack of Oxygen.* Human life cannot exist without a continuous intake of oxygen. Lack of oxygen rapidly leads to death. First aid involves knowing how to OPEN THE AIRWAY AND RESTORE BREATHING AND HEARTBEAT (Chapter 2, Section I).

b. *Bleeding.* Human life cannot continue without an adequate volume of blood to carry oxygen to the tissues. An important first aid measure is to STOP THE BLEEDING to prevent loss of blood (Chapter 2, Section II).

c. *Shock.* Shock means there is inadequate blood flow to the vital tissues and organs. Shock that remains uncorrected may result in death even though the injury or condition causing the shock would not otherwise be fatal. Shock can result from many causes, such as loss of blood, loss of fluid from deep burns, pain, and reaction to the sight of a wound or blood. First aid includes PREVENTING SHOCK, since the casualty's chances of survival are much greater if he does not develop shock (Chapter 2, Section III).

d. *Infection.* Recovery from a severe injury or a wound depends largely upon how well the injury or wound was initially protected. Infections result from the multiplication and growth (spread) of germs (bacteria: harmful microscopic organisms). Since harmful bacteria are in the air and on the skin and clothing, some of these organisms will immediately invade (contaminate) a break in the skin or an open wound. The objective is to KEEP ADDITIONAL GERMS OUT OF THE WOUND. A good working knowledge of basic first aid measures also includes knowing how to dress the wound to avoid infection or additional contamination (Chapters 2 and 3).

CHAPTER 2

Basic Measures for First Aid

Several conditions which require immediate attention are an inadequate airway, lack of breathing or lack of heartbeat, and excessive loss of blood. A casualty without a clear airway or who is not breathing may die from lack of oxygen. Excessive loss of blood may lead to shock, and shock can lead to death; therefore, you must act immediately to control the loss of blood. All wounds are considered to be contaminated, since infection-producing organisms (germs) are always present on the skin, on clothing, and in the air. Any missile or instrument causing the wound pushes or carries the germs into the wound. Infection results as these organisms multiply. That a wound is contaminated does not lessen the importance of protecting it from further contamination. You must dress and bandage a wound as soon as possible to prevent further contamination. It is also important that you attend to any airway, breathing, or bleeding problem IMMEDIATELY because these problems may become life-threatening.

SECTION I. OPEN THE AIRWAY AND RESTORE BREATHING

*2-1. Breathing Process. All living things must have oxygen to live. Through the breathing process, the lungs draw oxygen from the air and put it into the blood. The heart pumps the blood through the body to be used by the living cells which require a constant supply of oxygen. Some cells are more dependent on a constant supply of oxygen than others. Cells of the brain may die within 4 to 6 minutes without oxygen. Once these cells die, they are lost forever since they DO NOT regenerate. This could result in permanent brain damage, paralysis, or death.

2-2. Assessment (Evaluation) Phase (081-831-1000 and 081-831-1042)
    a. Check for responsiveness (Figure 2-1A)—establish whether the casualty is conscious by gently shaking him and asking, "Are you O.K.?"
    b. Call for help (Figure 2-1B).
    c. Position the unconscious casualty so that he is lying on his back and on a firm surface (Figure 2-1C) (081-831-1042).

>
> **WARNING (081-831-1042)**
> If the casualty is lying on his chest (prone position), cautiously roll the casualty as a unit so that his body does not twist (which may further complicate a neck, back or spinal injury).

    (1) Straighten the casualty's legs. Take the casualty's arm that is nearest to you and move it so that it is straight and above his head. Repeat procedure for the other arm.
    (2) Kneel beside the casualty with your knees near his shoulders (leave space to roll his body) (Figure 2-1B). Place one hand behind his head and neck for support. With your other hand, grasp the casualty under his far arm (Figure 2-1C).
    (3) Roll the casualty toward you using a steady and even pull. His head and neck should stay in line with his back.
    (4) Return the casualty's arms to his sides. Straighten his legs. Reposition yourself so that you are now kneeling at the level of the casualty's shoulders. However, if a neck injury is suspected, and the jaw-thrust will be used, kneel at the casualty's head, looking toward his feet.

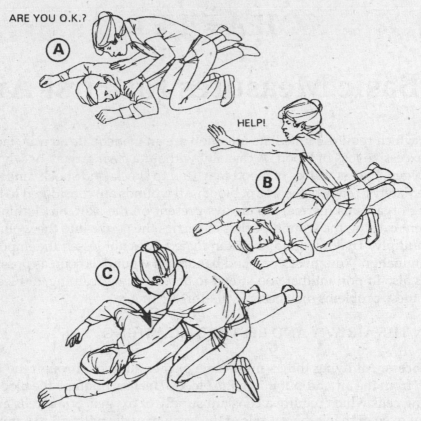

SOURCE: Copyright. American Heart Association. *Instructor's Manual for Basic Life Support*. Dallas: American Heart Association, 1987.

Figure 2-1: Responsiveness checked.

2-3. Opening the Airway—Unconscious and Not Breathing Casualty (081-831-1042). *The tongue is the single most common cause of an airway obstruction (Figure 2-2). In most cases, the airway can be cleared by simply using the head-tilt/chin-lift technique. This action pulls the tongue away from the air passage in the throat (Figure 2-3).

    a. Step ONE (081-331-1042). Call for help and then position the casualty. Move (roll) the casualty onto his back (Figure 2-1C above).

SOURCE: Copyright. American Heart Association. *Instructor's Manual for Basic Life Support*. Dallas: American Heart Association, 1987.

Figure 2-2: Airway blocked by tongue.

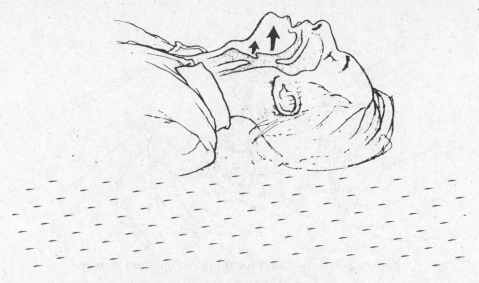

Figure 2-3: Airway opened (cleared).

> ⚠ **CAUTION**
> Take care in moving a casualty with a suspected neck or back injury. Moving an injured neck or back may permanently injure the spine.

> ✍ **NOTE (081-831-1042)**
> If foreign material or vomitus is visible in the mouth, it should be removed, but do not spend an excessive amount of time doing so.

b. Step TWO (081-831-1042). Open the airway using the jaw-thrust or head-tilt/chin-lift technique.

> ✍ **NOTE**
> The head-tilt/chin-lift is an important procedure in opening the airway; however, use extreme care because excess force in performing this maneuver may cause further spinal injury. In a casualty with a suspected neck injury or severe head trauma, the safest approach to opening the airway is the jaw-thrust technique because in most cases it can be accomplished without extending the neck.

(1) Perform the jaw-thrust technique. The jaw-thrust maybe accomplished by the rescuer grasping the angles of the casualty's lower jaw and lifting with both hands, one on each side, displacing the jaw forward and up (Figure 2-4). The rescuer's elbows should rest on the surface on which the casualty is lying. If the lips close, the lower lip can be retracted with the thumb. If mouth-to-mouth breathing is necessary, close the nostrils by placing your cheek tightly against them. The head should be carefully supported without tilting it backwards or turning it from side to side. If this is unsuccessful, the head should be tilted back very slightly. The jaw-thrust is the safest first approach to opening the airway of a casualty who has a suspected neck injury because in most cases it can be accomplished without extending the neck.

(2) Perform the head-tilt/chin-lift technique (081-831-1042). Place one hand on the casualty's forehead and apply firm, backward pressure with the palm to tilt the head back. Place the fingertips

Figure 2-4: Jaw-thrust technique of opening airway.

of the other hand under the bony part of the lower jaw and lift, bringing the chin forward. The thumb should not be used to lift the chin (Figure 2-5).

✍ **NOTE**

The fingers should not press deeply into the soft tissue under the chin because the airway may be obstructed.

---

c. Step THREE. Check for breathing (while maintaining an airway). After establishing an open airway, it is important to maintain that airway in an open position. Often the act of just opening and maintaining the airway will allow the casualty to breathe properly. Once the rescuer uses one of the techniques to open the airway (jaw-thrust or head-tilt/chin-lift), he should maintain that head position to keep the airway open. Failure to maintain the open airway will prevent the casualty from receiving an adequate supply of oxygen. Therefore, while maintaining an open airway, the rescuer should check for breathing by observing the casualty's chest and performing the following actions within 3 to 5 seconds:

(1) LOOK for the chest to rise and fall.

(2) LISTEN for air escaping during exhalation by placing your ear near the casualty's mouth.

Figure 2-5: Head-tilt/chin-lift technique of opening airway.

(3) FEEL for the flow of air on your cheek (see Figure 2-6),

(4) If the casualty does not resume breathing, give mouth-to-mouth resuscitation.

> ✍ **NOTE**
>
> If the casualty resumes breathing, monitor and maintain the open airway. If he continues to breathe, he should be transported to a medical treatment facility.

2-4. Rescue Breathing (Artificial Respiration)

a. If the casualty does not promptly resume adequate spontaneous breathing after the airway is open, rescue breathing (artificial respiration) must be started. Be calm! Think and act quickly! The sooner you begin rescue breathing, the more likely you are to restore the casualty's breathing. If you are in doubt whether the casualty is breathing, give artificial respiration, since it can do no harm to a person who is breathing. If the casualty is breathing, you can feel and see his chest move. Also, if the casualty is breathing, you can feel and hear air being expelled by putting your hand or ear close to his mouth and nose.

b. There are several methods of administering rescue breathing. The mouth-to-mouth method is preferred; however, it cannot be used in all situations. If the casualty has a severe jaw fracture or mouth wound or his jaws are tightly closed by spasms, use the mouth-to-nose method.

2-5. Preliminary Steps—All Rescue Breathing Methods (081-831-1042)

a. Step ONE. Establish unresponsiveness. Call for help. Turn or position the casualty.

b. Step TWO. Open the airway.

c. Step THREE. Check for breathing by placing your ear over the casualty's mouth and nose, and looking toward his chest:

(1) **Look** for rise and fall of the casualty's chest (Figure 2-6).

(2) **Listen** for sounds of breathing.

Figure 2-6:

(3) **Feel** for breath on the side of your face. If the chest does not rise and fall and no air is exhaled, then the casualty is breathless (not breathing). (This evaluation procedure should take only 3 to 5 seconds. Perform rescue breathing if the casualty is not breathing.

> ✎ **NOTE**
> Although the rescuer may notice that the casualty is making respiratory efforts, the airway may still be obstructed and opening the airway may be all that is needed. If the casualty resumes breathing, the rescuer should continue to help maintain an open airway.

2-6. Mouth-to-Mouth Method (081-831-1042). In this method of rescue breathing, you inflate the casualty's lungs with air from your lungs. This can be accomplished by blowing air into the person's mouth. The mouth-to-mouth rescue breathing method is performed as follows:

a. Preliminary Steps.
   (1) Step ONE (081-831-1042). If the casualty is not breathing, place your hand on his forehead, and pinch his nostrils together with the thumb and index finger of this same hand. Let this same hand exert pressure on his forehead to maintain the backward head-tilt and maintain an open airway. With your other hand, keep your fingertips on the bony part of the lower jaw near the chin and lift (Figure 2-7).

> ✎ **NOTE**
> If you suspect the casualty has a neck injury and you are using the jaw-thrust technique, close the nostrils by placing your cheek tightly against them.

   (2) Step TWO (081-831-1042).Take a deep breath and place your mouth (in an airtight seal) around the casualty's mouth (Figure 2-8). (If the injured person is small, cover both his nose and mouth with your mouth, sealing your lips against the skin of his face.)
   (3) Step THREE (081-831-1042). Blow two full breaths into the casualty's mouth (1 to 1 1/2 seconds per breath), taking a breath of fresh air each time before you blow. Watch out of the corner of

Figure 2-7: Head-tilt/chin-lift.

Figure 2-8: Rescue breathing.

your eye for the casualty's chest to rise. If the chest rises, sufficient air is getting into the casualty's lungs. Therefore, proceed as described in step FOUR below. If the chest does not rise, do the following (a, b, and c below) and then attempt to ventilate again.

(a) Take corrective action immediately by reestablishing the airway. Make sure that air is not leaking from around your mouth or out of the casualty's pinched nose.

(b) Reattempt to ventilate.

(c) If chest still does not rise, take the necessary action to open an obstructed airway (paragraph 2-14).

✍ **NOTE**

If the initial attempt to ventilate the casualty is unsuccessful, reposition the casualty's head and repeat rescue breathing. Improper chin and head positioning is the most common cause of difficulty with ventilation. If the casualty cannot be ventilated after repositioning the head, proceed with foreign-body airway obstruction maneuvers (see Open an Obstructed Airway, paragraph 2-14).

---

(4) Step FOUR (081-831-1042). After giving two breaths which cause the chest to rise, attempt to locate a pulse on the casualty. Feel for a pulse on the side of the casualty's neck closest to you by placing the first two fingers (index and middle fingers) of your hand on the groove beside the casualty's Adam's apple (carotid pulse) (Figure 2-9). (Your thumb should not be used for pulse

Figure 2-9: Placement of fingers to detect pulse.

taking because you may confuse your pulse beat with that of the casualty.) Maintain the airway by keeping your other hand on the casualty's forehead. Allow 5 to 10 seconds to determine if there is a pulse.

(a) If a pulse is found and the casualty is breathing—STOP and allow the casualty to breathe on his own. If possible, keep him warm and comfortable.

(b) If a pulse is found and the casualty is not breathing, continue rescue breathing.

* (c) If a pulse is not found, seek medically trained personnel for help.

b. Rescue Breathing (mouth-to-mouth resuscitation) (081-831-1042). Rescue breathing (mouth-to-mouth or mouth-to-nose resuscitation) is performed at the rate of about one breath every 5 seconds (12 breaths per minute) with rechecks for pulse and breathing after every 12 breaths. Rechecks can be accomplished in 3 to 5 seconds. See steps ONE through SEVEN (below) for specifics.

✍ **NOTE**

Seek help (medical aid), if not done previously.

---

(1) Step ONE. If the casualty is not breathing, pinch his nostrils together with the thumb and index finger of the hand on his forehead and let this same hand exert pressure on the forehead to maintain the backward head-tilt (Figure 2-7).

(2) Step TWO. Take a deep breath and place your mouth (in an airtight seal) around the casualty's mouth (Figure 2-8).

(3) Step THREE. Blow a quick breath into the casualty's mouth forcefully to cause his chest to rise. If the casualty's chest rises, sufficient air is getting into his lungs.

(4) Step FOUR. When the casualty's chest rises, remove your mouth from his mouth and listen for the return of air from his lungs (exhalation).

(5) Step FIVE. Repeat this procedure (mouth-to-mouth resuscitation) at a rate of one breath every 5 seconds to achieve 12 breaths per minute. Use the following count: "one, one-thousand; two, one-thousand; three, one-thousand; four, one-thousand; BREATH; one, one-thousand;" and so forth. To achieve a rate of one breath every 5 seconds, the breath must be given on the fifth count.

* (6) Step SIX. Feel for a pulse after every 12th breath. This check should take about 3 to 5 seconds. If a pulse beat is not found, seek medically trained personnel for help.

* (7) Step SEVEN. Continue rescue breathing until the casualty starts to breathe on his own, until you are relieved by another person, or until you are too tired to continue. Monitor pulse and return of spontaneous breathing after every few minutes of rescue breathing. If spontaneous breathing returns, monitor the casualty closely. The casualty should then be transported to a medical treatment facility. Maintain an open airway and be prepared to resume rescue breathing, if necessary.

2-7. Mouth-to-Nose Method. Use this method if you cannot perform mouth-to-mouth rescue breathing because the casualty has a severe jaw fracture or mouth wound or his jaws are tightly closed by spasms. The mouth-to-nose method is performed in the same way as the mouth-to-mouth method except that you blow into the nose while you hold the lips closed with one hand at the chin. You then remove your mouth to allow the casualty to exhale passively. It may be necessary to separate the casualty's lips to allow the air to escape during exhalation.

* 2-8. Heartbeat. If a casualty's heart stops beating, you must immediately seek medically trained personnel for help. SECONDS COUNT! Stoppage of the heart is soon followed by cessation of respiration unless it has occurred first. Be calm! Think and act! When a casualty's heart has stopped, there is no pulse at all; the person is unconscious and limp, and the pupils of his eyes are open wide. When evaluating a casualty or when performing the preliminary steps of rescue breathing, feel for a pulse. If you DO NOT detect a pulse, immediately seek medically trained personnel.

Note: The U.S. Army deleted paragraphs 2-9, 2-10, and 2-11 of this manual as part of a revision.

2-12. Airway Obstructions. In order for oxygen from the air to flow to and from the lungs, the upper airway must be unobstructed.

   a. Upper airway obstructions often occur because—
     (1) The casualty's tongue falls back into his throat while he is unconscious as a result of injury, cardiopulmonary arrest, and so forth. (The tongue falls back and obstructs, it is not swallowed.)
     (2) Foreign bodies become lodged in the throat. These obstructions usually occur while eating (meat most commonly causes obstructions). Choking on food is associated with—
      • Attempting to swallow large pieces of poorly chewed food.
      • Drinking alcohol.
      • Slipping dentures.
     (3) The contents of the stomach are regurgitated and may block the airway.
     (4) Blood clots may form as a result of head and facial injuries.
   b. Upper airway obstructions may be prevented by taking the following precautions:
     (1) Cut food into small pieces and take care to chew slowly and thoroughly.
     (2) Avoid laughing and talking when chewing and swallowing.
     (3) Restrict alcohol while eating meals.
     (4) Keep food and foreign objects from children while they walk, run, or play.
     (5) Consider the correct positioning/maintenance of the open airway for the injured or unconscious casualty.
   c. Upper airway obstruction may cause either *partial* or *complete* airway blockage.
    * (1) *Partial airway obstruction.* The casualty may still have an air exchange. A good air exchange means that the casualty can cough forcefully, though he may be wheezing between coughs. You, the rescuer, should not interfere, and should encourage the casualty to cough up the object on his own. A poor air exchange may be indicated by weak coughing with a high pitched noise between coughs. Additionally, the casualty may show signs of shock (for example, paleness of the skin, bluish or grayish tint around the lips or fingernail beds) indicating a need for oxygen. You should assist the casualty and treat him as though he had a complete obstruction.
     (2) *Complete airway obstruction.* A complete obstruction (no air exchange) is indicated if the casualty cannot speak, breathe, or cough at all. He may be clutching his neck and moving erratically. In

an unconscious casualty a complete obstruction is also indicated if after opening his airway you cannot ventilate him.

2-13. Opening the Obstructed Airway-Conscious Casualty (081-831-1003). Clearing a conscious casualty's airway obstruction can be performed with the casualty either standing or sitting, and by following a relatively simple procedure.

> **WARNING**
> Once an obstructed airway occurs, the brain will develop an oxygen deficiency resulting in unconsciousness. Death will follow rapidly if prompt action is not taken.

a. Step ONE. Ask the casualty if he can speak or if he is choking. Check for the universal choking sign (Figure 2-18).
b. Step TWO. If the casualty can speak, encourage him to attempt to cough; the casualty still has a good air exchange. If he is able to speak or cough effectively, DO NOT interfere with his attempts to expel the obstruction.
c. Step THREE. Listen for high pitched sounds when the casualty breathes or coughs (poor air exchange). If there is poor air exchange or no breathing, CALL for HELP and immediately deliver manual thrusts (either an abdominal or chest thrust).

✍ **NOTE**

The manual thrust with the hands centered between the waist, and the rib cage is called an abdominal thrust (or Heimlich maneuver). The chest thrust (the hands are centered in the middle of the breastbone) is used only for an individual in the advanced stages of pregnancy, in the markedly obese casualty, or if there is a significant abdominal wound.

• Apply ABDOMINAL THRUSTS using the procedures below:
  ○ Stand behind the casualty and wrap your arms around his waist. Make a fist with one hand and grasp it with the other. The thumb side of your fist should be against the casualty's abdomen, in the midline and slightly above the casualty's navel, but well below the tip of the breastbone (Figure 2-19).
  ○ Press the fists into the abdomen with a quick backward and upward thrust (Figure 2-20).
  ○ Each thrust should be a separate and distinct movement.

Figure 2-18: Universal sign of choking.

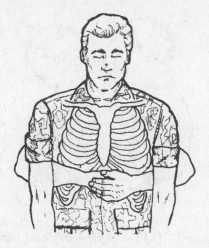

Figure 2-19: Anatomical view of abdominal thrust procedure.

Figure 2-20: Profile view of abdominal thrust.

> ✍ ***NOTE***
> Continue performing abdominal thrusts until the obstruction is expelled or the casualty becomes unconscious.

- ○ If the casualty becomes unconscious, call for help as you proceed with steps to open the airway and perform rescue breathing (See task 081-831-1042, Perform Mouth-to-Mouth Resuscitation.)
- • Applying CHEST THRUSTS. An alternate technique to the abdominal thrust is the chest thrust. This technique is useful when the casualty has an abdominal wound, when the casualty is pregnant, or when the casualty is so large that you cannot wrap your arms around the abdomen. TO apply chest thrusts with casualty sitting or standing:
- ○ Stand behind the casualty and wrap your arms around his chest with your arms under his armpits.
- ○ Make a fist with one hand and place the thumb side of the fist in the middle of the breastbone (take care to avoid the tip of the breastbone and the margins of the ribs).
- ○ Grasp the fist with the other hand and exert thrusts (Figure 2-21).
- ○ Each thrust should be delivered slowly, distinctly, and with the intent of relieving the obstruction.

Figure 2-21: Profile view of chest thrust.

- ○ Perform chest thrusts until the obstruction is expelled or the casualty becomes unconscious.
- ○ If the casualty becomes unconscious, call for help as you proceed with steps to open the airway and perform rescue breathing. (See task 081-831-1042, Perform Mouth-to-Mouth Resuscitation.)

2-14. Open an Obstructed Airway—Casualty Lying or Unconscious (081-831-1042). The following procedures are used to expel an airway obstruction in a casualty who is lying down, who becomes unconscious, or is found unconscious (the cause unknown):

- If a casualty who is choking becomes unconscious, call for help, open the airway, perform a finger sweep, and attempt rescue breathing (paragraphs 2-2 through 2-4). If you still cannot administer rescue breathing due to an airway blockage, then remove the airway obstruction using the procedures in steps *a* through *e* below.
- If a casualty is unconscious when you find him (the cause unknown), assess or evaluate the situation, call for help, position the casualty on his back, open the airway, establish breathlessness, and attempt to perform rescue breathing (paragraphs 2-2 through 2-8).

a. Open the airway and attempt rescue breathing. (See task 081-831-1042, Perform Mouth-to-Mouth Resuscitation.)

b. If still unable to ventilate the casualty, perform 6 to 10 manual (abdominal or chest) thrusts. (Note that the abdominal thrusts are used when casualty does not have abdominal wounds; is not pregnant or extremely overweight.) To perform the abdominal thrusts:
   (1) Kneel astride the casualty's thighs (Figure 2-22).
   (2) Place the heel of one hand against the casualty's abdomen (in the midline slightly above the navel but well below the tip of the breastbone). Place your other hand on top of the first one. Point your fingers toward the casualty's head.
   (3) Press into the casualty's abdomen with a quick, forward and upward thrust. You can use your body weight to perform the maneuver. Deliver each thrust slowly and distinctly.
   (4) Repeat the sequence of abdominal thrusts, finger sweep, and rescue breathing (attempt to ventilate) as long as necessary to remove the object from the obstructed airway. See paragraph d below.
   (5) If the casualty's chest rises, proceed to feeling for pulse.

c. Apply chest thrusts. (Note that the chest thrust technique is an alternate method that is used when the casualty has an abdominal wound, when the casualty is so large that you cannot wrap your arms around the abdomen, or when the casualty is pregnant.) To perform the chest thrusts:

Figure 2-22: Abdominal thrust on unconscious casualty.

(1) Place the unconscious casualty on his back, face up, and open his mouth. Kneel close to the side of the casualty's body.
- Locate the lower edge of the casualty's ribs with your fingers. Run the fingers up along the rib cage to the notch (Figure 2-23A).
- Place the middle finger on the notch and the index finger next to the middle finger on the lower edge of the breastbone. Place the heel of the other hand on the lower half of the breastbone next to the two fingers (Figure 2-23B).
- Remove the fingers from the notch and place that hand on top of the positioned hand on the breastbone, extending or interlocking the fingers (Figure 2-23C).
- Straighten and lock your elbows with your shoulders directly above your hands without bending the elbows, rocking, or allowing the shoulders to sag. Apply enough pressure to

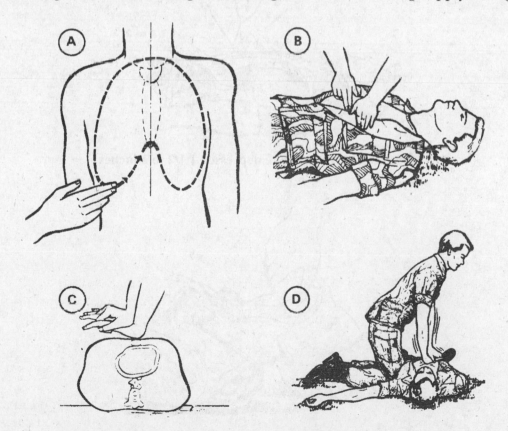

Figure 2-23: Hand placement for chest thrust (Illustrated A-D).

depress the breastbone 1 1/2 to 2 inches, then release the pressure completely (Figure 2-23D). Do this 6 to 10 times. Each thrust should be delivered slowly and distinctly. See Figure 2-24 for another view of the breastbone being depressed.

    (2) Repeat the sequence of chest thrust, finger sweep, and rescue breathing as long as necessary to clear the object from the obstructed airway. See paragraph d below.

    (3) If the casualty's chest rises, proceed to feeling for his pulse.

d. Finger Sweep. If you still cannot administer rescue breathing due to an airway obstruction, then remove the airway obstruction using the procedures in steps (1) and (2) below.

    (1) Place the casualty on his back, face up, turn the unconscious casualty as a unit, and call out for help.

    (2) Perform finger sweep, keep casualty face up, use tongue-jaw lift to open mouth.

- Open the casualty's mouth by grasping both his tongue and lower jaw between your thumb and fingers and lifting (tongue-jaw lift) (Figure 2-25). If you are unable to open his mouth, cross your fingers and thumb (crossed-finger method) and push his teeth apart (Figure 2-26) by pressing your thumb against his upper teeth and pressing your finger against his lower teeth.

- Insert the index finger of the other hand down along the inside of his cheek to the base of the tongue. Use a hooking motion from the side of the mouth toward the center to dislodge the foreign body (Figure 2-27).

WARNING

Take care not to force the object deeper into the airway by pushing it with the finger.

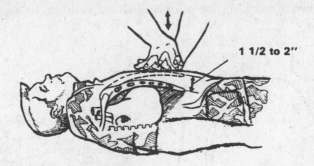

Figure 2-24: Breastbone depressed 1 1/2 to 2 inches

Figure 2-25: Opening casualty's mouth (tongue-jaw lift).

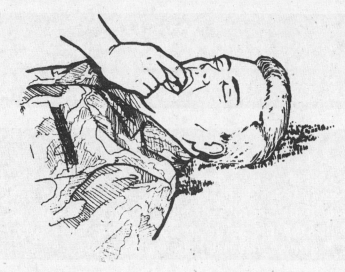

Figure 2-26: Opening casualty's mouth (crossed-finger method).

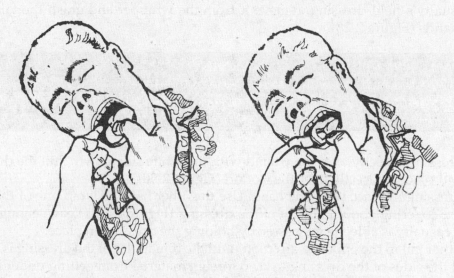

Figure 2-27: Using finger to dislodge foreign body.

SECTION II. STOP THE BLEEDING AND PROTECT THE WOUND

2-15. Clothing (081-831-1016). In evaluating the casualty for location, type, and size of the wound or injury, cut or tear his clothing and carefully expose the entire area of the wound. This procedure is necessary to avoid further contamination. Clothing stuck to the wound should be left in place to avoid further injury. DO NOT touch the wound; keep it as clean as possible.

WARNING (081-831-1016)
DO NOT REMOVE protective clothing in a chemical environment. Apply dressings over the protective clothing.

2-16. Entrance and Exit Wounds. Before applying the dressing, carefully examine the casualty to determine if there is more than one wound. A missile may have entered at one point and exited at another point. The EXIT wound is usually LARGER than the entrance wound.

> **WARNING**
>
> Casualty should be continually monitored for development of conditions which may require the performance of necessary basic lifesaving measures, such as clearing the airway and mouth-to-mouth resuscitation. All open (or penetrating) wounds should be checked for a point of entry and exit and treated accordingly.

> **WARNING**
>
> If the missile lodges in the body (fails to exit), DO NOT attempt to remove it or probe the wound. Apply a dressing. If there is an object extending from (impaled in) the wound, DO NOT remove the object. Apply a dressing around the object and use additional improvised bulky materials dressings (use the cleanest material available) to build up the area around the object. Apply a supporting bandage over the bulky materials to hold them in place.

2-17. Field Dressing (081-831-1016)

a. Use the casualty's field dressing; remove it from the wrapper and grasp the tails of the dressing with both hands (Figure 2-28).

> **WARNING**
>
> DO NOT touch the white (sterile) side of the dressing, and DO NOT allow the white (sterile) side of the dressing to come in contact with any surface other than the wound.

b. Hold the dressing directly over the wound with the white side down. Pull the dressing open (Figure 2-29) and place it directly over the wound (Figure 2-30).

c. Hold the dressing in place with one hand. Use the other hand to wrap one of the tails around the injured part, covering about one-half of the dressing (Figure 2-31). Leave enough of the tail for a knot. If the casualty is able, he may assist by holding the dressing in place.

d. Wrap the other tail in the opposite direction until the remainder of the dressing is covered. The tails should seal the sides of the dressing to keep foreign material from getting under it.

e. Tie the tails into a non-slip knot over the outer edge of the dressing (Figure 2-32). DO NOT TIE THE KNOT OVER THE WOUND. In order to allow blood to flow to the rest of an injured limb, tie the dressing firmly enough to prevent it from slipping but without causing a tourniquet-like effect; that is, the skin beyond the injury becomes cool, blue, or numb.

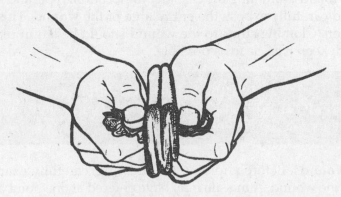

Figure 2-28: Grasping tails of dressing with both hands.

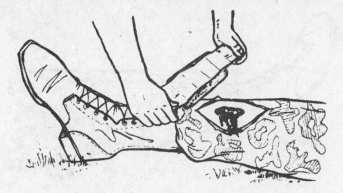

Figure 2-29: Pulling dressing open.

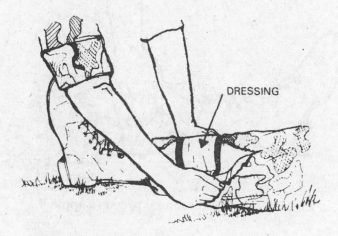

DRESSING

Figure 2-30: Placing dressing directly on wound.

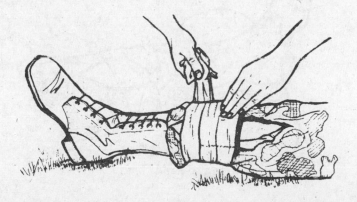

Figure 2-31: Wrapping tail of dressing around injured part.

2-18. Manual Pressure (081-831-1016)

a. If bleeding continues after applying the sterile field dressing, direct manual pressure may be used to help control bleeding. Apply such pressure by placing a hand on the dressing and exerting firm pressure for 5 to 10 minutes (Figure 2-33). The casualty may be asked to do this himself if he is conscious and can follow instructions.

b. Elevate an injured limb slightly above the level of the heart to reduce the bleeding (Figure 2-34).

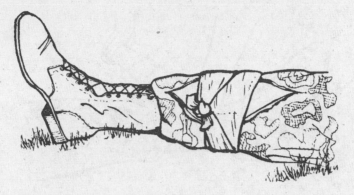

Figure 2-32: Tails tied into nonslip knot.

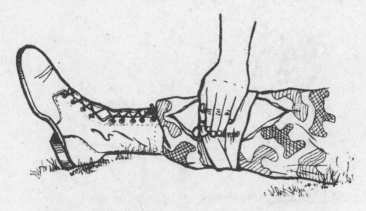

Figure 2-33: Direct manual pressure applied.

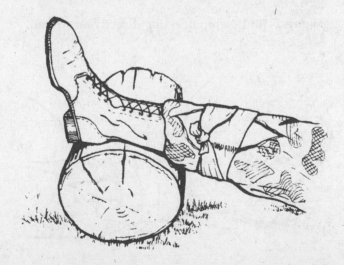

Figure 2-34: Injured limb elevated.

WARNING

DO NOT elevate a suspected fractured limb unless it has been properly splinted. (To splint a fracture before elevating, see task 081-831-1034, Splint a Suspected Fracture.)

c. If the bleeding stops, check and treat for shock. If the bleeding continues, apply a pressure dressing.

2-19. Pressure Dressing (081-831-1016). Pressure dressings aid in blood clotting and compress the open blood vessel. If bleeding continues after the application of a field dressing, manual pressure, and elevation, then a pressure dressing must be applied as follows:

a. Place a wad of padding on top of the field dressing, directly over the wound (Figure 2-35). Keep injured extremity elevated.

> **✍ NOTE**
>
> Improvised bandages may be made from strips of cloth. These strips may be made from T-shirts, socks, or other garments.

b. Place an improvised dressing (or cravat, if available) over the wad of padding (Figure 2-36). Wrap the ends tightly around the injured limb, covering the previously placed field dressing (Figure 2-37).

c. Tie the ends together in a non-slip knot, directly over the wound site (Figure 2-38). DO NOT tie so tightly that it has a tourniquet-like effect. If bleeding continues and all other measures have failed, or if the limb is severed, then apply a tourniquet. Use the tourniquet as a LAST RESORT. When the bleeding stops, check and treat for shock.

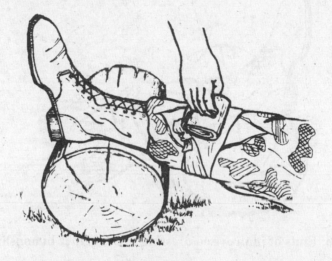

Figure 2-35: Wad of padding on top of field dressing.

Figure 2-36: Improvised dressing over wad of padding.

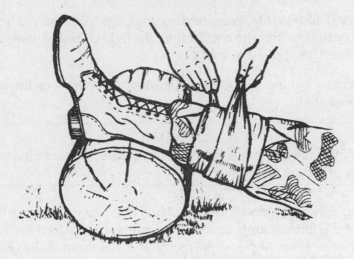

Figure 2-37: Ends of improvised dressing wrapped tightly around limb.

Figure 2-38: Ends of improvised dressing tied together in nonslip knot.

✍ **NOTE**
Wounded extremities should be checked periodically for adequate circulation. The dressing must be loosened if the extremity becomes cool, blue or gray, or numb.

✍ ***NOTE**
If bleeding continues and all other measures have failed (dressing and covering wound, applying direct manual pressure, elevating limb above heart level, and applying pressure dressing maintaining limb elevation), then apply digital pressure. See Appendix E for appropriate pressure points.

2-20. Tourniquet (081-831-1017). A tourniquet is a constricting band placed around an arm or leg to control bleeding. A soldier whose arm or leg has been completely amputated may not be bleeding when first discovered, but a tourniquet should be applied anyway. This absence of bleeding is due to the body's normal defenses (contraction of blood vessels) as a result of the amputation, but after a period of time bleeding will start as the blood vessels relax. Bleeding from a major artery of the thigh, lower leg, or arm and bleeding from multiple arteries (which occurs in a traumatic amputation) may prove to be beyond control by manual pressure. If the pressure dressing under firm hand pressure becomes soaked with blood and the wound continues to bleed, apply a tourniquet.

WARNING

Casualty should be continually monitored for development of conditions which may require the performance of necessary basic life-saving measures, such as: clearing the airway, performing mouth-to-mouth resuscitation, preventing shock, and/or bleeding control. All open (or penetrating) wounds should be checked for a point of entry or exit and treated accordingly.

* The tourniquet should not be used unless a pressure dressing has failed to stop the bleeding or an arm or leg has been cut off. On occasion, tourniquets have injured blood vessels and nerves. If left in place too long, a tourniquet can cause loss of an arm or leg. Once applied, it must stay in place, and the casualty must be taken to the nearest medical treatment facility as soon as possible. DO NOT loosen or release a tourniquet after it has been applied and the bleeding has stopped.

a. Improvising a Tourniquet (081-831-1017). In the absence of a specially designed tourniquet, a tourniquet may be made from a strong, pliable material, such as gauze or muslin bandages, clothing, or kerchiefs. An improvised tourniquet is used with a rigid stick-like object. To minimize skin damage, ensure that the improvised tourniquet is at least 2 inches wide.

WARNING

The tourniquet must be easily identified or easily seen.

WARNING

DO NOT use wire or shoestring for a tourniquet band.

WARNING

A tourniquet is only used on arm(s) or leg(s) where there is danger of loss of casualty's life.

b. Placing the Improvised Tourniquet (081-831-1017).
   (1) Place the tourniquet around the limb, between the wound and the body trunk (or between the wound and the heart). Place the tourniquet 2 to 4 inches from the edge of the wound site (Figure 2-39). Never place it directly over a wound or fracture or directly on a joint (wrist, elbow, or knee). For wounds just below a joint, place the tourniquet just above and as close to the joint as possible.
   (2) The tourniquet should have padding underneath. If possible, place the tourniquet over the smoothed sleeve or trouser leg to prevent the skin from being pinched or twisted. If the tourniquet is long enough, wrap it around the limb several times, keeping the material as flat as possible. Damaging the skin may deprive the surgeon of skin required to cover an amputation. Protection of the skin also reduces pain.

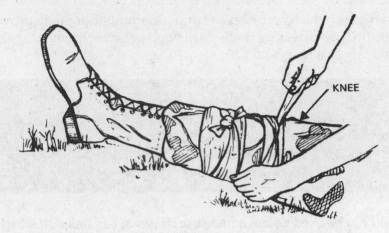

Figure 2-39: Tourniquet 2 to 4 inches above wound.

c. Applying the Tourniquet (081-831-1017).
   (1) Tie a half-knot. (A half-knot is the same as the first part of tying a shoe lace.)
   (2) Place a stick (or similar rigid object) on top of the half-knot (Figure 2-40).
   (3) Tie a full knot over the stick (Figure 2-41).
   (4) Twist the stick (Figure 2-42) until the tourniquet is tight around the limb and/or the bright red bleeding has stopped. In the case of amputation, dark oozing blood may continue for a short time. This is the blood trapped in the area between the wound and tourniquet.

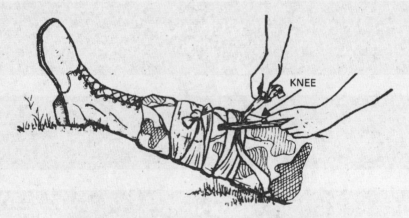

Figure 2-40: Rigid object on top of half-knot.

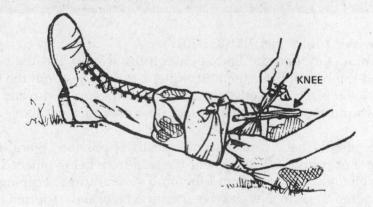

Figure 2-41: Full knot over rigid object.

TWIST THE STICK

ALIGN THE STICK LENGTHWISE WITH THE LIMB

Figure 2-42: Stick twisted.

(5) Fasten the tourniquet to the limb by looping the free ends of the tourniquet over the ends of the stick. Then bring the ends around the limb to prevent the stick from loosening. Tie them together under the limb (Figure 2-43A and B).

✍ **NOTE (081-831-1017)**

Other methods of securing the stick may be used as long as the stick does not unwind and no further injury results.

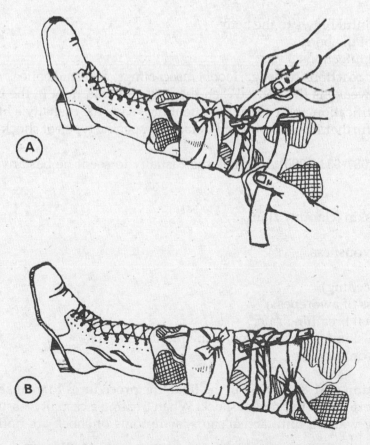

Figure 2-43: Free ends looped (Illustrated A and B).

✍ **NOTE**

If possible, save and transport any severed (amputated) limbs or body parts with (but out of sight of) the casualty.

---

(6) DO NOT cover the tourniquet—you should leave it in full view. If the limb is missing (total amputation), apply a dressing to the stump.

(7) Mark the casualty's forehead, if possible, with a "T" to indicate a tourniquet has been applied. If necessary, use the casualty's blood to make this mark.

(8) Check and treat for shock.

(9) Seek medical aid.

⚠ CAUTION (081-831-1017)

Do not loosen or release the tourniquet once it has been applied because it could enhance the probability of shock.

SECTION III. CHECK AND TREAT FOR SHOCK

2-21. Causes and Effects

a. Shock may be caused by severe or minor trauma to the body. It usually is the result of—
- Significant loss of blood.
- Heart failure.
- Dehydration.
- Severe and painful blows to the body.
- Severe burns of the body.
- Severe wound infections.
- Severe allergic reactions to drugs, foods, insect stings, and snakebites.

b. Shock stuns and weakens the body. When the normal blood flow in the body is upset, death can result. Early identification and proper treatment may save the casualty's life.

c. See FM 8-230 for further information and details on specific types of shock and treatment.

2-22. Signs/Symptoms (081-831-1000). Examine the casualty to see if he has any of the following signs/symptoms:

- Sweaty but cool skin (clammy skin).
- Paleness of skin.
- Restlessness, nervousness.
- Thirst.
- Loss of blood (bleeding).
- Confusion (or loss of awareness).
- Faster-than-normal breathing rate.
- Blotchy or bluish skin (especially around the mouth and lips).
- Nausea and/or vomiting.

2-23. Treatment/Prevention (081-831-1005). In the field, the procedures to treat shock are identical to procedures that would be performed to prevent shock. When treating a casualty, assume that shock is present or will occur shortly. By waiting until actual signs/symptoms of shock are noticeable, the rescuer may jeopardize the casualty's life.

a. Position the Casualty. (DO NOT move the casualty or his limbs if suspected fractures have not been splinted. See Chapter 4 for details.)

(1) Move the casualty to cover, if cover is available and the situation permits.

(2) Lay the casualty on his back.

✍ **NOTE**

A casualty in shock after suffering a heart attack, chest wound, or breathing difficulty, may breathe easier in a sitting position. If this is the case, allow him to sit upright, but monitor carefully in case his condition worsens.

---

(3) Elevate the casualty's feet higher than the level of his heart. Use a stable object (a box, field pack, or rolled up clothing) so that his feet will not slip off (Figure 2-44).

WARNING
DO NOT elevate legs if the casualty has an unsplinted broken leg, head injury, or abdominal injury. (See task 081-831-1034, Splint a Suspected Fracture, and task 081-831-1025, Apply a Dressing to an Open Abdominal Wound.)

WARNING (081-831-1005)
Check casualty for leg fracture(s) and splint, if necessary, before elevating his feet. For a casualty with an abdominal wound, place knees in an upright (flexed) position.

(4) Loosen clothing at the neck, waist, or wherever it may be binding.

⚠ **CAUTION (081-831-1005)**
DO NOT LOOSEN OR REMOVE protective clothing in a chemical environment.

(5) Prevent chilling or overheating. The key is to maintain body temperature. In cold weather, place a blanket or other like item over him to keep him warm and under him to prevent chilling (Figure 2-45). However, if a tourniquet has been applied, leave it exposed (if possible). In hot weather, place the casualty in the shade and avoid excessive covering.

(6) Calm the casualty. Throughout the entire procedure of treating and caring for a casualty, the rescuer should reassure the casualty and keep him calm. This can be done by being authoritative (taking charge) and by showing self-confidence. Assure the casualty that you are there to help him.

(7) Seek medical aid.

Figure 2-44: Clothing loosened and feet elevated.

Figure 2-45: Body temperature maintained.

Figure 2-46: Casualty's head turned to side.

b. Food and/or Drink. During the treatment/prevention of shock, DO NOT give the casualty any food or drink. If you must leave the casualty or if he is unconscious, turn his head to the side to prevent him from choking should he vomit (Figure 2-46).

c. Evaluate Casualty. If necessary, continue with the casualty's evaluation.

CHAPTER 3

First Aid for Special Wounds

* Basic lifesaving steps are discussed in Chapters 1 and 2: clear the airway/restore breathing, stop the bleeding, protect the wound, and treat/prevent shock. They apply to first aid measures for all injuries. Certain types of wounds and burns will require special precautions and procedures when applying these measures. This chapter discusses first aid procedures for special wounds of the head, face, and neck; chest and stomach wounds; and burns. It also discusses the techniques for applying dressings and bandages to specific parts of the body.

SECTION I. GIVE PROPER FIRST AID FOR HEAD INJURIES

3-1. Head Injuries. A head injury may consist of one or a combination of the following conditions: a concussion, a cut or bruise of the scalp, or a fracture of the skull with injury to the brain and the blood vessels of the scalp. The damage can range from a minor cut on the scalp to a severe brain injury which rapidly causes death. Most head injuries lie somewhere between the two extremes. Usually, serious skull fractures and brain injuries occur together; however, it is possible to receive a serious brain injury without a skull fracture. The brain is a very delicate organ; when it is injured, the casualty may vomit, become sleepy, suffer paralysis, or lose consciousness and slip into a coma. All severe head injuries are potentially life threatening. For recovery and return to normal function, casualties require proper first aid as a vital first step.

3-2. Signs/Symptoms (081-831-1000). A head injury may be open or closed. In open injuries, there is a visible wound and, at times, the brain may actually be seen. In closed injuries, no visible injury is seen, but the casualty may experience the same signs and symptoms. Either closed or open head injuries can be life-threatening if the injury has been severe enough; thus, if you suspect a head injury, evaluate the casualty for the following:

- Current or recent unconsciousness (loss of consciousness).
- Nausea or vomiting.
- Convulsions or twitches (involuntary jerking and shaking).
- Slurred speech.
- Confusion.
- Sleepiness (drowsiness).
- Loss of memory (does casualty know his own name, where he is, and so forth).
- Clear or bloody fluid leaking from nose or ears.
- Staggering in walking.
- Dizziness.
- A change in pulse rate.
- Breathing problems.
- Eye (vision) problems, such as unequal pupils.
- Paralysis.
- Headache.
- Black eyes.
- Bleeding from scalp/head area.
- Deformity of the head.

3-3. General First Aid Measures (081-831-1000)

a. General Considerations. The casualty with a head injury (or suspected head injury) should be continually monitored for the development of conditions which may require the performance of the necessary basic lifesaving measures, therefore be prepared to—

- Clear the airway (and be prepared to perform the basic lifesaving measures). Treat as a suspected neck/spinal injury until proven otherwise. (See Chapter 4 for more information.)
- Place a dressing over the wounded area. DO NOT attempt to clean the wound.
- Seek medical aid.
- Keep the casualty warm.
- DO NOT attempt to remove a protruding object from the head.
- DO NOT give the casualty anything to eat or drink.

b. Care of the Unconscious Casualty. If a casualty is unconscious as the result of a head injury, he is not able to defend himself. He may lose his sensitivity to pain or ability to cough up blood or mucus that may be plugging his airway. An unconscious casualty must be evaluated for breathing difficulties, uncontrollable bleeding, and spinal injury.

(1) Breathing. The brain requires a constant supply of oxygen. A bluish (or in an individual with dark skin—grayish) color of skin around the lips and nail beds indicates that the casualty is not receiving enough air (oxygen). Immediate action must be taken to clear the airway, to position the casualty on his side, or to give artificial respiration. Be prepared to give artificial respiration if breathing should stop.

(2) Bleeding. Bleeding from a head injury usually comes from blood vessels within the scalp. Bleeding can also develop inside the skull or within the brain. In most instances bleeding from the head can be controlled by proper application of the field first aid dressing.

> ⚠ CAUTION (081-831-1033)
> DO NOT attempt to put unnecessary pressure on the wound or attempt to push any brain matter back into the head (skull). DO NOT apply a pressure dressing.

(3) Spinal injury. A person that has an injury above the collar bone or a head injury resulting in an unconscious state should be suspected of having a neck or head injury with spinal cord damage. Spinal cord injury may be indicated by a lack of responses to stimuli, stomach distention (enlargement), or penile erection.

(a) Lack of responses to stimuli. Starting with the feet, use a sharp pointed object–a sharp stick or something similar, and prick the casualty lightly while observing his face. If the casualty blinks or frowns, this indicates that he has feeling and may not have an injury to the spinal cord. If you observe no response in the casualty's reflexes after pricking upwards toward the chest region, you must use extreme caution and treat the casualty for an injured spinal cord.

(b) Stomach distention (enlargement). Observe the casualty's chest and stomach. If the stomach is distended (enlarged) when the casualty takes a breath and the chest moves slightly, the casualty may have a spinal injury and must be treated accordingly.

(c) Penile erection. A male casualty may have a penile erection, an indication of a spinal injury.

> ⚠ CAUTION
> Remember to suspect any casualty who has a severe head injury or who is unconscious as possibly having a broken neck or a spinal cord injury! It is better to treat conservatively and assume that the neck/spinal cord is injured rather than to chance further injuring the casualty. Consider this when you position the casualty. See Chapter 4, paragraph 4-9 for treatment procedures of spinal column injuries.

c. Concussion. If an individual receives a heavy blow to the head or face, he may suffer a brain concussion, which is an injury to the brain that involves a temporary loss of some or all of the brain's ability to function. For example, the casualty may not breathe properly for a short period of time, or he may become confused and stagger when he attempts to walk. A concussion may only last for a short period of time. However, if a casualty is suspected of having suffered a concussion, he must be seen by a physician as soon as conditions permit.

d. Convulsions. Convulsions (seizures/involuntary jerking) may occur after a mild head injury. When a casualty is convulsing, protect him from hurting himself. Take the following measures:
(1) Ease him to the ground.
(2) Support his head and neck.
(3) Maintain his airway.
(4) Call for assistance.
(5) Treat the casualty's wounds and evacuate him immediately.

e. Brain Damage. In severe head injuries where brain tissue is protruding, leave the wound alone; carefully place a first aid dressing over the tissue. DO NOT remove or disturb any foreign matter that may be in the wound. Position the casualty so that his head is higher than his body. Keep him warm and seek medical aid immediately.

✍ NOTE
- DO NOT forcefully hold the arms and legs if they are jerking because this can lead to broken bones.
- DO NOT force anything between the casualty's teeth especially if they are tightly clenched because this may obstruct the casualty's airway.
- Maintain the casualty's airway if necessary.

3-4. Dressings and Bandages (081-831-1000 and 081-831-1033)
* a. Evaluate the Casualty (081-831-1000). Be prepared to perform lifesaving measures. The basic lifesaving measures may include clearing the airway, rescue breathing, treatment for shock, and/or bleeding control.

b. Check Level of Consciousness/Responsiveness (081-831-1033). With a head injury, an important area to evaluate is the casualty's level of consciousness and responsiveness. Ask the casualty questions such as—
- "What is your name?" (Person)
- "Where are you?" (Place)
- "What day/month/year is it?" (Time)

Any incorrect responses, inability to answer, or changes in responses should be reported to medical personnel. Check the casualty's level of consciousness every 15 minutes and note any changes from earlier observations.

c. Position the Casualty (081-831-1033).

WARNING (081-831-1033)
DO NOT move the casualty if you suspect he has sustained a neck, spine, or severe head injury (which produces any signs or symptoms other than minor bleeding). See task 081-831-1000, *Evaluate the Casualty.*

- If the casualty is conscious or has a minor (superficial) scalp wound:
  ○ Have the casualty sit up (unless other injuries prohibit or he is unable); OR

- If the casualty is lying down and is not accumulating fluids or drainage in his throat, elevate his head slightly; OR
- If the casualty is bleeding from or into his mouth or throat, turn his head to the side or position him on his side so that the airway will be clear. Avoid pressure on the wound or place him on his side–opposite the site of the injury (Figure 3-1).
- If the casualty is unconscious or has a severe head injury, then suspect and treat him as having a potential neck or spinal injury, immobilize and DO NOT move the casualty.

✍ **NOTE**

If the casualty is choking and/or vomiting or is bleeding from or into his mouth (thus compromising his airway), position him on his side so that his airway will be clear. Avoid pressure on the wound; place him on his side opposite the side of the injury.

 WARNING (081-831-1033)

If it is necessary to turn a casualty with a suspected neck/spine injury, roll the casualty gently onto his side, keeping the head, neck, and body aligned while providing support for the head and neck. DO NOT roll the casualty by yourself but seek assistance. Move him only if absolutely necessary, otherwise keep the casualty immobilized to prevent further damage to the neck/spine.

d. Expose the Wound (081-831-1033).
- Remove the casualty's helmet (if necessary).
- In a chemical environment:
  - If mask and/or hood is not breached, apply no dressing to the head wound casualty. If the "all clear" has not been given, DO NOT remove the casualty's mask to attend the head wound: OR
  - If mask and/or hood have been breached and the "all clear" has not been given, try to repair the breach with tape and apply no dressing; OR
  - If mask and/or hood have been breached and the "all clear" has been given the mask can be removed and a dressing applied.

WARNING

DO NOT attempt to clean the wound, or remove a protruding object.

Figure 3-1: Casualty lying on side opposite injury.

> **✍ NOTE**
> If there is an object extending from the wound, DO NOT remove the object. Improvise bulky dressings from the cleanest material available and place these dressings around the protruding object for support after applying the field dressing.

> **✍ NOTE**
> Always use the casualty's field dressing, not your own!

e. Apply a Dressing to a Wound of the Forehead/Back of Head (081-831-1033). To apply a dressing to a wound of the forehead or back of the head—
   (1) Remove the dressing from the wrapper.
   (2) Grasp the tails of the dressing in both hands.
   (3) Hold the dressing (white side down) directly over the wound. DO NOT touch the white (sterile) side of the dressing or allow anything except the wound to come in contact with the white side.
   (4) Place it directly over the wound.
   (5) Hold it in place with one hand. If the casualty is able, he may assist.
   (6) Wrap the first tail horizontally around the head; ensure the tail covers the dressing (Figure 3-2).
   (7) Hold the first tail in place and wrap the second tail in the opposite direction, covering the dressing (Figure 3-3).
   (8) Tie a non slip knot and secure the tails at the side of the head, making sure they DO NOT cover the eyes or ears (Figure 3-4).

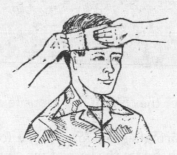

Figure 3-2: First tail of dressing wrapped horizontally around head.

Figure 3-3: Second tail wrapped in opposite direction.

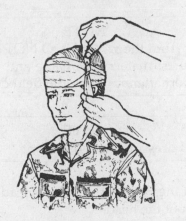

Figure 3-4: Tails tied in nonslip knot at side of head.

f. Apply a Dressing to a Wound on Top of the Head (081-831-1033). To apply a dressing to a wound on top of the head–
    (1) Remove the dressing from the wrapper.
    (2) Grasp the tails of the dressing in both hands.
    (3) Hold it (white side down) directly over the wound.
    (4) Place it over the wound (Figure 3-5).
    (5) Hold it in place with one hand. If the casualty is able, he may assist.
    (6) Wrap one tail down under the chin (Figure 3-6), up in front of the ear, over the dressing, and in front of the other ear.

 WARNING
(Make sure the tails remain wide and close to the front of the chin to avoid choking the casualty.)

    (7) Wrap the remaining tail under the chin in the opposite direction and up the side of the face to meet the first tail (Figure 3-7).
    (8) Cross the tails (Figure 3-8), bringing one around the forehead (above the eyebrows) and the other around the back of the head (at the base of the skull) to a point just above and in front of the opposite ear, and tie them using a non slip knot (Figure 3-9).

Figure 3-5: Dressing placed over wound.

Figure 3-6: One tail of dressing wrapped under chin.

Figure 3-7: Remaining tail wrapped under chin in opposite direction.

Figure 3-8: Tails of dressing crossed with one around forehead.

g. Apply a Triangular Bandage to the Head. To apply a triangular bandage to the head–
   (1) Turn the base (longest side) of the bandage up and center its base on center of the forehead, letting the point (apex) fall on the back of the neck (Figure 3-10 A).
   (2) Take the ends behind the head and cross the ends over the apex.
   (3) Take them over the forehead and tie them (Figure 3-10 B).
   (4) Tuck the apex behind the crossed part of the bandage and/or secure it with a safety pin, if available (Figure 3-10 C).

Figure 3-9: Tails tied in nonslip knot (in front of and above ear).

Figure 3-10: Triangular bandage applied to head (Illustrated A thru C).

Figure 3-11: Cravat bandage applied to head (Illustrated A thru C).

h. **Apply a Cravat Bandage to the Head.** To apply a cravat bandage to the head–
   (1) Place the middle of the bandage over the dressing (Figure 3-11 A).
   (2) Cross the two ends of the bandage in opposite directions completely around the head (Figure 3-11 B).
   (3) Tie the ends over the dressing (Figure 3-11 C).

SECTION II. GIVE PROPER FIRST AID FOR FACE AND NECK INJURIES

3-5. Face Injuries. Soft tissue injuries of the face and scalp are common. Abrasions (scrapes) of the skin cause no serious problems. Contusions (injury without a break in the skin) usually cause swelling. A contusion of the scalp looks and feels like a lump. Laceration (cut) and avulsion (torn away tissue) injuries are also common. Avulsions are frequently caused when a sharp blow separates the scalp from the skull beneath it. Because the face and scalp are richly supplied with blood vessels (arteries and veins), wounds of these areas usually bleed heavily.

3-6. Neck Injuries. Neck injuries may result in heavy bleeding. Apply manual pressure above and below the injury and attempt to control the bleeding. Apply a dressing. Always evaluate the casualty for a possible neck fracture/spinal cord injury; if suspected, seek medical treatment immediately.

✎ * NOTE

Establish and maintain the airway in cases of facial or neck injuries. If a neck fracture or spinal cord injury is suspected, immobilize or stabilize casualty. See Chapter 4 for further information on treatment of spinal injuries.

3-7. Procedure. When a casualty has a face or neck injury, perform the measures below.
   a. Step ONE. Clear the airway. Be prepared to perform any of the basic lifesaving steps. Clear the casualty's airway (mouth) with your fingers, remove any blood, mucus, pieces of broken teeth or bone, or bits of flesh, as well as any dentures.
   b. Step TWO. Control any bleeding, especially bleeding that obstructs the airway. Do this by applying direct pressure over a first aid dressing or by applying pressure at specific pressure points on the face, scalp, or temple. (See Appendix E for further information on pressure points.) If the casualty is bleeding from the mouth, position him as indicated (c below) and apply manual pressure.

⚠ CAUTION

Take care not to apply too much pressure to the scalp if a skull fracture is suspected.

   c. Step THREE. Position the casualty. If the casualty is bleeding from the mouth (or has other drainage, such as mucus, vomitus, or so forth) and is conscious, place him in a comfortable sitting position and have him lean forward with his head tilted slightly down to permit free drainage (Figure 3-12). DO NOT use the sitting position if—

Figure 3-12: Casualty leaning forward to permit drainage.

- It would be harmful to the casualty because of other injuries.
- The casualty is unconscious, in which case, place him on his side (Figure 3-13). If there is a suspected injury to the neck or spine, immobilize the head before turning the casualty on his side.

⚠ **CAUTION**

If you suspect the casualty has a neck/spinal injury, then immobilize his head/neck and treat him as outlined in Chapter 4.

d. Step FOUR. Perform other measures.
(1) Apply dressings/bandages to specific areas of the face.
(2) Check for missing teeth and pieces of tissue. Check for detached teeth in the airway. Place detached teeth, pieces of ear or nose on a field dressing and send them along with the casualty to the medical facility. Detached teeth should be kept damp.
(3) Treat for shock and seek medical treatment IMMEDIATELY.

3-8. Dressings and Bandages (081-831-1033)

a. Eye Injuries. The eye is a vital sensory organ, and blindness is a severe physical handicap. Timely first aid of the eye not only relieves pain but also helps prevent shock, permanent eye injury, and possible loss of vision. Because the eye is very sensitive, any injury can be easily aggravated if it is improperly handled. Injuries of the eye may be quite severe. Cuts of the eyelids can appear to be very serious, but if the eyeball is not involved, a person's vision usually will not be damaged. However, lacerations (cuts) of the eyeball can cause permanent damage or loss of sight.
(1) Lacerated/torn eyelids. Lacerated eyelids may bleed heavily, but bleeding usually stops quickly. Cover the injured eye with a sterile dressing. DO NOT put pressure on the wound because you may injure the eyeball. Handle torn eyelids very carefully to prevent further injury. Place any detached pieces of the eyelid on a clean bandage or dressing and immediately send them with the casualty to the medical facility.
(2) Lacerated eyeball (injury to the globe). Lacerations or cuts to the eyeball may cause serious and permanent eye damage. Cover the injury with a loose sterile dressing. DO NOT put pressure on the eyeball because additional damage may occur. An important point to remember is that when one eyeball is injured, you should immobilize both eyes. This is done by applying a bandage to both eyes. Because the eyes move together, covering both will lessen the chances of further damage to the injured eye.

⚠ **CAUTION**

DO NOT apply pressure when there is a possible laceration of the eyeball. The eyeball contains fluid. Pressure applied over the eye will force the fluid out, resulting in/permanent injury. APPLY PROTECTIVE DRESSING WITHOUT ADDED PRESSURE.

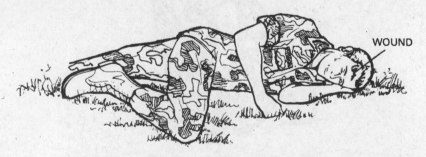

WOUND

Figure 3-13: Casualty lying on side.

(3) Extruded eyeballs. Soldiers may encounter casualties with severe eye injuries that include an extruded eyeball (eyeball out-of-socket). In such instances you should gently cover the extruded eye with a loose moistened dressing and also cover the unaffected eye. DO NOT bind or exert pressure on the injured eye while applying a loose dressing. Keep the casualty quiet, place him on his back, treat for shock (make warm and comfortable), and evacuate him immediately.

(4) Burns of the eyes. Chemical burns, thermal (heat) burns, and light burns can affect the eyes.

   (a) Chemical burns. Injuries from chemical burns require immediate first aid. Chemical burns are caused mainly by acids or alkalies. The first aid is to flush the eye(s) immediately with large amounts of water for at least 5 to 20 minutes, or as long as necessary to flush out the chemical. If the burn is an acid burn, you should flush the eye for at least 5 to 10 minutes. If the burn is an alkali burn, you should flush the eye for at least 20 minutes. After the eye has been flushed, apply a bandage over the eyes and evacuate the casualty immediately.

   (b) Thermal burns. When an individual suffers burns of the face from a fire, the eyes will close quickly due to extreme heat. This reaction is a natural reflex to protect the eyeballs; however, the eyelids remain exposed and are frequently burned. If a casualty receives burns of the eyelids/face, DO NOT apply a dressing; DO NOT TOUCH; seek medical treatment immediately.

   (c) Light burns. Exposure to intense light can burn an individual. Infrared rays, eclipse light (if the casualty has looked directly at the sun), or laser burns cause injuries of the exposed eyeball. Ultraviolet rays from arc welding can cause a superficial burn to the surface of the eye. These injuries are generally not painful but may cause permanent damage to the eyes. Immediate first aid is usually not required. Loosely bandaging the eyes may make the casualty more comfortable and protect his eyes from further injury caused by exposure to other bright lights or sunlight.

> ⚠ CAUTION
>
> In certain instances both eyes are usually bandaged; but, in hazardous surroundings leave the uninjured eye uncovered so that the casualty may be able to see.

b. Side-of-Head or Cheek Wound (081-831-1033). Facial injuries to the side of the head or the cheek may bleed profusely (Figure 3-14). Prompt action is necessary to ensure that the airway remains open and also to control the bleeding. It may be necessary to apply a dressing. To apply a dressing—

(1) Remove the dressing from its wrapper.

(2) Grasp the tails in both hands.

(3) Hold the dressing directly over the wound with the white side down and place it directly on the wound (Figure 3-15 A).

Figure 3-14: Side of head or cheek wound.

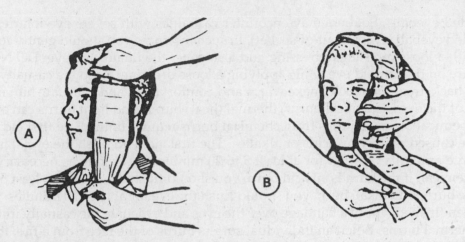

Figure 3-15: Dressing placed directly on wound. Top tail wrapped over top of head, down in front of ear, and under chin (Illustrated A and B).

(4) Hold the dressing in place with one hand (the casualty may assist if able). Wrap the top tail over the top of the head and bring it down in front of the ear (on the side opposite the wound), under the chin (Figure 3-15 B) and up over the dressing to a point just above the ear (on the wound side).

> ✍ **NOTE**
>
> When possible, avoid covering the casualty's ear with the dressing, as this will decrease his ability to hear.

(5) Bring the second tail under the chin, up in front of the ear (on the side opposite the wound), and over the head to meet the other tail (on the wound side) (Figure 3-16).

(6) Cross the two tails (on the wound side) (Figure 3-17) and bring one end across the forehead (above the eyebrows) to a point just in front of the opposite ear (on the uninjured side).

(7) Wrap the other tail around the back of the head (at the base of the skull), and tie the two ends just in front of the ear on the uninjured side with a non slip knot (Figure 3-18).

c. Ear Injuries. Lacerated (cut) or avulsed (torn) ear tissue may not, in itself, be a serious injury. Bleeding, or the drainage of fluids from the ear canal, however, may be a sign of a head injury, such as a skull fracture. DO NOT attempt to stop the flow from the inner ear canal nor put anything into the ear canal to block it. Instead, you should cover the ear lightly with a dressing. For minor cuts or wounds to the external ear, apply a cravat bandage as follows:

Figure 3-16: Bringing second tail under the chin.

Figure 3-17: crossing the tails on the side of the wound.

Figure 3-18: Tying the tails of the dressing in a nonslip knot.

Figure 3-19: Applying cravat bandage to ear (Illustrated A thru C).

(1) Place the middle of the bandage over the ear (Figure 3-19 A).

(2) Cross the ends, wrap them in opposite directions around the head, and tie them (Figures 3-19 B and 3-19 C).

(3) If possible, place some dressing material between the back of the ear and the side of the head to avoid crushing the ear against the head with the bandage.

d. Nose Injuries. Nose injuries generally produce bleeding. The bleeding may be controlled by placing an ice pack over the nose, or pinching the nostrils together. The bleeding may also be controlled by placing torn gauze (rolled) between the upper teeth and the lip.

> ⚠ **CAUTION**
>
> DO NOT attempt to remove objects inhaled in the nose. An untrained person who removes such an object could worsen the casualty's condition and cause permanent injury.

e. **Jaw Injuries.** Before applying a bandage to a casualty's jaw, remove all foreign material from the casualty's mouth. If the casualty is unconscious, check for obstructions in the airway. When applying the bandage, allow the jaw enough freedom to permit passage of air and drainage from the mouth.

(1) Apply bandages attached to field first aid dressing to the jaw. After dressing the wound, apply the bandages using the same technique illustrated in Figures 3-5 through 3-8.

> ✍ **NOTE**
>
> The dressing and bandaging procedure outlined for the jaw serves a twofold purpose. In addition to stopping the bleeding and protecting the wound, it also immobilizes a fractured jaw.

(2) Apply a cravat bandage to the jaw.
(a) Place the bandage under the chin and carry its ends upward. Adjust the bandage to make one end longer than the other (Figure 3-20 A).
(b) Take the longer end over the top of the head to meet the short end at the temple and cross the ends over (Figure 3-20 B).
(c) Take the ends in opposite directions to the other side of the head and tie them over the part of the bandage that was applied first (Figure 3-20 C).

> ✍ **NOTE**
>
> The cravat bandage technique is used to immobilize a fractured jaw or to maintain a sterile dressing that does not have tail bandages attached.

SECTION III. GIVE PROPER FIRST AID FOR CHEST AND ABDOMINAL WOUNDS AND BURN INJURIES

3-9. Chest Wounds (081-831-1026). Chest injuries may be caused by accidents, bullet or missile wounds, stab wounds, or falls. These injuries can be serious and may cause death quickly if proper treatment is

Figure 3-20: Applying cravat bandage to jew (Illustrated A thru C).

not given. A casualty with a chest injury may complain of pain in the chest or shoulder area; he may have difficulty with his breathing. His chest may not rise normally when he breathes. The injury may cause the casualty to cough up blood and to have a rapid or a weak heartbeat. A casualty with an open chest wound has a punctured chest wall. The sucking sound heard when he breathes is caused by air leaking into his chest cavity. This particular type of wound is dangerous and will collapse the injured lung (Figure 3-21). Breathing becomes difficult for the casualty because the wound is open. The soldier's life may depend upon how quickly you make the wound airtight.

3-10. Chest Wound(s) Procedure (081-831-1026)

* a. Evaluate the Casualty (081-831-1000). Be prepared to perform lifesaving measures. The basic life-saving measures may include clearing the airway, rescue breathing, treatment for shock, and/or bleeding control.
b. Expose the Wound. If appropriate, cut or remove the casualty's clothing to expose the entire area of the wound. Remember, DO NOT remove clothing that is stuck to the wound because additional injury may result. DO NOT attempt to clean the wound.

✍ **NOTE**

Examine the casualty to see if there is an entry and/or exit wound. If there are two wounds (entry, exit), perform the same procedure for both wounds. Treat the more serious (heavier bleeding, larger) wound first. It may be necessary to improvise a dressing for the second wound by using strips of cloth, such as a torn T-shirt, or whatever material is available. Also, listen for sucking sounds to determine if the chest wall is punctured.

⚠ **CAUTION**

If there is an object extending from (impaled in) the wound, DO NOT remove the object. Apply a dressing around the object and use additional improvised bulky materials/dressings (use the cleanest materials available) to build up the area around the object. Apply a supporting bandage over the bulky materials to hold them in place.

⚠ **CAUTION (081-831-1026)**

DO NOT REMOVE protective clothing in a chemical environment. Apply dressings over the protective clothing.

---

c. Open the Casualty's Field Dressing Plastic Wrapper. The plastic wrapper is used with the field dressing to create an airtight seal. If a plastic wrapper is not available, or if an additional wound needs to be treated; cellophane, foil, the casualty's poncho, or similar material maybe used. The

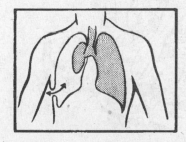

Figure 3-21: Collapsed lung.

covering should be wide enough to extend 2 inches or more beyond the edges of the wound in all directions.

(1) Tear open one end of the casualty's plastic wrapper covering the field dressing. Be careful not to destroy the wrapper and DO NOT touch the inside of the wrapper.

(2) Remove the inner packet (field dressing).

(3) Complete tearing open the empty plastic wrapper using as much of the wrapper as possible to create a flat surface.

d. Place the Wrapper Over the Wound (081-831-1026). Place the inside surface of the plastic wrapper directly over the wound when the casualty exhales and hold it in place (Figure 3-22). The casualty may hold the plastic wrapper in place if he is able.

e. Apply the Dressing to the Wound (081-831-1026).

(1) Use your free hand and shake open the field dressing (Figure 3-23).

(2) Place the white side of the dressing on the plastic wrapper covering the wound (Figure 3-24).

✍ **NOTE (081-831-1026)**

Use the casualty's field dressing, not your own.

---

(3) Have the casualty breathe normally.

Figure 3-22: Open chest wound sealed with plastic wrapper.

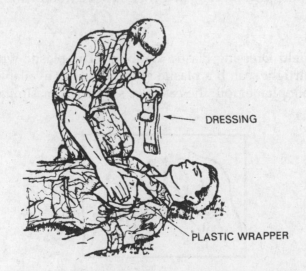

DRESSING

PLASTIC WRAPPER

Figure 3-23: Shaking open the field dressing.

Figure 3-24: Field dressing placed on plastic wrapper.

(4) While maintaining pressure on the dressing, grasp one tail of the field dressing with the other hand and wrap it around the casualty's back.

(5) Wrap the other tail in the opposite direction, bringing both tails over the dressing (Figure 3-25).

(6) Tie the tails into a non slip knot in the center of the dressing after the casualty exhales and before he inhales. This will aid in maintaining pressure on the bandage after it has been tied (Figure 3-26). Tie the dressing firmly enough to secure the dressing without interfering with the casualty's breathing.

✍ **NOTE (081-831-1026)**

When practical, apply direct manual pressure over the dressing for 5 to 10 minutes to help control the bleeding.

Figure 3-25: Tails of field dressing wrapped around casualty in opposite direction.

Figure 3-26: Tails of dressing tied into nonslip knot over center of dressing.

f. Position the Casualty (081-831-1026). Position the casualty on his injured side or in a sitting position, whichever makes breathing easier (Figure 3-27).

g. Seek Medical Aid. Contact medical personnel.

> *** WARNING**
>
> Even if an airtight dressing has been placed properly, air may still enter the chest cavity without having means to escape. This causes a life-threatening condition called tension pneumothorax. If the casualty's condition (for example, difficulty breathing, shortness of breath, restlessness, or grayness of skin in a dark-skinned individual [or blueness in an individual with light skin]) worsens after placing the dressing, quickly lift or remove, then replace the airtight dressing.

3-11. Abdominal Wounds. The most serious abdominal wound is one in which an object penetrates the abdominal wall and pierces internal organs or large blood vessels. In these instances, bleeding may be severe and death can occur rapidly.

3-12. Abdominal Wound(s) Procedure (081-831-1025)

a. Evaluate the Casualty. Be prepared to perform basic lifesaving measures. It is necessary to check for both entry and exit wounds. If there are two wounds (entry and exit), treat the wound that appears more serious first (for example, the heavier bleeding, protruding organs, larger wound, and so forth). It may be necessary to improvise dressings for the second wound by using strips of cloth, a T-shirt, or the cleanest material available.

b. Position the Casualty. Place and maintain the casualty on his back with his knees in an upright (flexed) position (Figure 3-28). The knees-up position helps relieve pain, assists in the treatment of shock, prevents further exposure of the bowel (intestines) or abdominal organs, and helps relieve abdominal pressure by allowing the abdominal muscles to relax.

c. Expose the Wound.

(1) Remove the casualty's loose clothing to expose the wound. However, DO NOT attempt to remove clothing that is stuck to the wound; it may cause further injury. Thus, remove any loose clothing from the wound but leave in place the clothing that is stuck.

> ⚠ **CAUTION (081-831-1000 and 081-831-1025)**
>
> DO NOT REMOVE protective clothing in a chemical environment. Apply dressings over the protective clothing.

(2) Gently pick up any organs which may be on the ground. Do this with a clean, dry dressing or with the cleanest available material. Place the organs on top of the casualty's abdomen (Figure 3-29).

Figure 3-27: Casualty positioned (lying) on injured side.

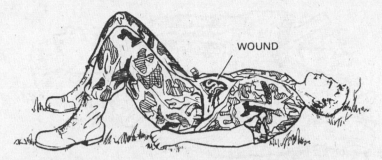

PLACE CASUALTY ON BACK TO PREVENT FURTHER EXPOSURE OF THE
BOWEL UNLESS OTHER WOUNDS PREVENT SUCH ACTION. FLEX
CASUALTY'S KNEES TO RELAX ABDOMINAL MUSCLES AND ANY INTERNAL
PRESSURE.

Figure 3-28: Casualty positioned (lying) on back with knees (flexed) up.

BEFORE APPLYING DRESSINGS, CAREFULLY PLACE PROTRUDING ORGANS
NEAR THE WOUND TO PROTECT THEM AND CONTROL CONTAMINATION.

Figure 3-29: Protruding organs placed near wound.

✍ NOTE (081-831-1025)

DO NOT probe, clean, or try to remove any foreign object from the abdomen.
DO NOT touch with bare hands any exposed organs.
DO NOT push organs back inside the body.

---

d. Apply the Field Dressing. Use the casualty's field dressing not your own. If the field dressing is not large enough to cover the entire wound, the plastic wrapper from the dressing may be used to cover the wound first (placing the field dressing on top). Open the plastic wrapper carefully without touching the inner surface, if possible. If necessary other improvised dressings may be made from clothing, blankets, or the cleanest materials available because the field dressing and/or wrapper may not be large enough to cover the entire wound.

WARNING

If there is an object extending from the wound, DO NOT remove it. Place as much of the wrapper over the wound as possible without dislodging or moving the object. DO NOT place the wrapper over the object.

(1) Grasp the tails in both hands.
(2) Hold the dressing with the white, or cleanest, side down directly over the wound.
(3) Pull the dressing open and place it directly over the wound (Figure 3-30). If the casualty is able, he may hold the dressing in place.

IF THE DRESSING WRAPPER IS LARGE ENOUGH TO EXTEND WELL BEYOND
THE PROTRUDING BOWEL. THE STERILE SIDE OF THE DRESSING WRAPPER
CAN BE PLACED DIRECTLY OVER THE WOUND, WITH THE FIELD DRESSING
ON THE TOP.

Figure 3-30: Dressing placed directly over the wound.

(4) Hold the dressing in place with one hand and use the other hand to wrap one of the tails around the body.
(5) Wrap the other tail in the opposite direction until the dressing is completely covered. Leave enough of the tail for a knot.
(6) Loosely tie the tails with a non slip knot at the casualty's side (Figure 3-31).

> **WARNING**
> When dressing is applied, DO NOT put pressure on the wound or exposed internal parts, because pressure could cause further injury (vomiting, ruptured intestines, and so forth). Therefore, tie the dressing ties (tails) loosely at casualty's side, not directly over the dressing.

(7) Tie the dressing firmly enough to prevent slipping without applying pressure to the wound site (Figure 3-32).

Field dressings can be covered with improvised reinforcement material (cravats, strips of torn T-shirt, or other cloth), if available, for additional support and protection. Tie improvised bandage on the opposite side of the dressing ties firmly enough to prevent slipping but without applying additional pressure to the wound.

Figure 3-31: Dressing applied and tails tied with a nonslip knot.

Figure 3-32: Field dressing covered with improvised material and loosely tied.

⚠ **CAUTION (081-831-1025)**
DO NOT give casualties with abdominal wounds food or water (moistening the lips is allowed).

    e. Seek Medical Aid. Notify medical personnel.

3-13. Burn Injuries. Burns often cause extreme pain, scarring, or even death. Proper treatment will minimize further injury of the burned area. Before administering the proper first aid, you must be able to recognize the type of burn to be treated. There are four types of burns: (1) thermal burns caused by fire, hot objects, hot liquids, and gases or by nuclear blast or fire ball; (2) electrical burns caused by electrical wires, current, or lightning; (3) chemical burns caused by contact with wet or dry chemicals or white phosphorus (WP)—from marking rounds and grenades; and (4) laser burns.

3-14. First Aid for Burns (081-831-1007)

    a. Eliminate the Source of the Burn. The source of the burn must be eliminated before any evaluation or treatment of the casualty can occur.
        (1) Remove the casualty quickly and cover the thermal burn with any large non synthetic material, such as a field jacket. Roll the casualty on the ground to smother (put out) the flames (Figure 3-33).

⚠ **CAUTION**
Synthetic materials, such as nylon, may melt and cause further injury.

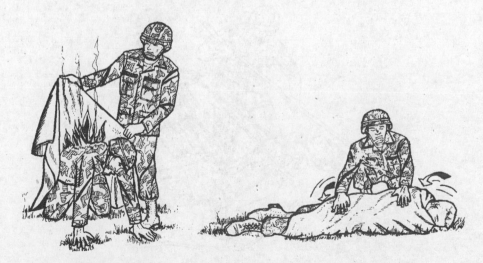

Figure 3-33: Casualty covered and rolled on ground.

(2) Remove the electrical burn casualty from the electrical source by turning off the electrical current. DO NOT attempt to turn off the electricity if the source is not close by. Speed is critical, so DO NOT waste unnecessary time. If the electricity cannot be turned off, wrap any nonconductive material (dry rope, dry clothing, dry wood, and so forth) around the casualty's back and shoulders and drag the casualty away from the electrical source (Figure 3-34). DO NOT make body-to-body contact with the casualty or touch any wires because you could also become an electrical burn casualty.

WARNING
High voltage electrical burns may cause temporary unconsciousness, difficulties in breathing, or difficulties with the heart (heartbeat).

(3) Remove the chemical from the burned casualty. Remove liquid chemicals by flushing with as much water as possible. If water is not available, use any nonflammable fluid to flush chemicals off the casualty. Remove dry chemicals by brushing off loose particles (DO NOT use the bare surface of your hand because you could become a chemical burn casualty) and then flush with large amounts of water, if available. If large amounts of water are not available, then NO water should be applied because small amounts of water applied to a dry chemical burn may cause a chemical reaction. When white phosphorous strikes the skin, smother with water, a wet cloth, or wet mud. Keep white phosphorous covered with a wet material to exclude air which will prevent the particles from burning.

WARNING
Small amounts of water applied to a dry chemical burn may cause a chemical reaction, transforming the dry chemical into an active burning substance.

(4) Remove the laser burn casualty from the source. (NOTE: Lasers produce a narrow amplified beam of light. The word laser means Light Amplification by Stimulated Emission of Radiation and sources include range finders, weapons/guidance, communication systems, and weapons simulations such as MILES.) When removing the casualty from the laser beam source, be careful

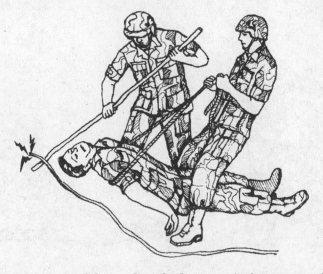

Figure 3-34: Casualty removed from electrical source (using nonconductive material).

not to enter the beam or you may become a casualty. Never look directly at the beam source and if possible, wear appropriate eye protection.

✍ **NOTE**

After the casualty is removed from the source of the burn, he should be evaluated for conditions requiring basic lifesaving measures (Evaluate the Casualty).

---

b. Expose the Burn. Cut and gently lift away any clothing covering the burned area, without pulling clothing over the burns. Leave in place any clothing that is stuck to the burns. If the casualty's hands or wrists have been burned, remove jewelry if possible without causing further injury (rings, watches, and so forth) and place in his pockets. This prevents the necessity to cut off jewelry since swelling usually occurs as a result of a burn.

⚠ **CAUTION (081-831-1007)**
- DO NOT lift or cut away clothing if in a chemical environment. Apply the dressing directly over the casualty's protective clothing.
- DO NOT attempt to decontaminate skin where blisters have formed.

---

c. Apply a Field Dressing to the Burn.
  (1) Grasp the tails of the casualty's dressing in both hands.
  (2) Hold the dressing directly over the wound with the white (sterile) side down, pull the dressing open, and place it directly over the wound. If the casualty is able, he may hold the dressing in place.
  (3) Hold the dressing in place with one hand and use the other hand to wrap one of the tails around the limbs or the body.
  (4) Wrap the other tail in the opposite direction until the dressing is completely covered.
  (5) Tie the tails into a knot over the outer edge of the dressing. The dressing should be applied lightly over the burn. Ensure that dressing is applied firmly enough to prevent it from slipping.

✍ **NOTE**

Use the cleanest improvised dressing material available if a field dressing is not available or if it is not large enough for the entire wound.

---

d. Take the Following Precautions (081-831-1007):
  - DO NOT place the dressing over the face or genital area.
  - DO NOT break the blisters.
  - DO NOT apply grease or ointments to the burns.
    For electrical burns, check for both an entry and exit burn from the passage of electricity through the body. Exit burns may appear on any area of the body despite location of entry burn.
  - For burns caused by wet or dry chemicals, flush the burns with large amounts of water and cover with a dry dressing.
  - For burns caused by white phosphorus (WP), flush the area with water, then cover with a wet material, dressing, or mud to exclude the air and keep the WP particles from burning.
  - For laser burns, apply a field dressing.
  - If the casualty is conscious and not nauseated, give him small amounts of water.
e. Seek Medical Aid. Notify medical personnel.

SECTION IV. APPLY PROPER BANDAGES TO UPPER AND LOWER EXTREMITIES

3-15. Shoulder Bandage

a. To apply bandages attached to the field first aid dressing–
   (1) Take one bandage across the chest and the other across the back and under the arm opposite the injured shoulder.
   (2) Tie the ends with a non slip knot (Figure 3-35).
b. To apply a cravat bandage to the shoulder or armpit–
   (1) Make an extended cravat bandage by using two triangular bandages (Figure 3-36 A); place the end of the first triangular bandage along the base of the second one (Figure 3-36 B).
   (2) Fold the two bandages into a single extended bandage (Figure 3-36 C).
   (3) Fold the extended bandage into a single cravat bandage (Figure 3-36 D). After folding, secure the thicker part (overlap) with two or more safety pins (Figure 3-36 E).

Figure 3-35: Shoulder bandage.

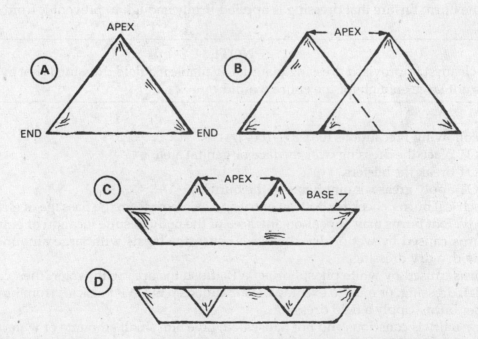

Figure 3-36: Extended cravat bandage applied to shoulder (or armpit) (Illustrated A thru H).

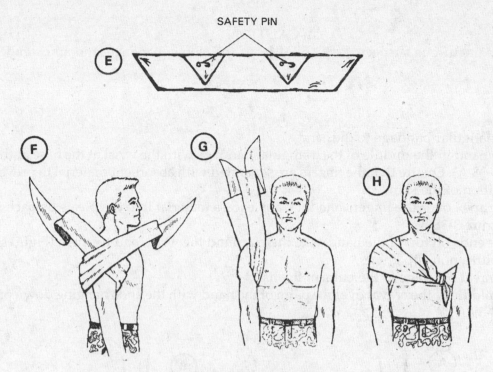

Figure 3-36: *(Continued)*

(4) Place the middle of the cravat bandage under the armpit so that the front end is longer than the back end and safety pins are on the outside (Figure 3-36 F).

(5) Cross the ends on top of the shoulder (Figure 3-36 G).

(6) Take one end across the back and under the arm on the opposite side and the other end across the chest. Tie the ends (Figure 3-36 H).

Be sure to place sufficient wadding in the armpit. DO NOT tie the cravat bandage too tightly. Avoid compressing the major blood vessels in the armpit.

3-16. Elbow Bandage. To apply a cravat bandage to the elbow–

a. Bend the arm at the elbow and place the middle of the cravat at the point of the elbow bringing the ends upward (Figure 3-37 A).

b. Bring the ends across, extending both downward (Figure 3-37 B).

c. Take both ends around the arm and tie them with a non slip knot at the front of the elbow (Figure 3-37 C).

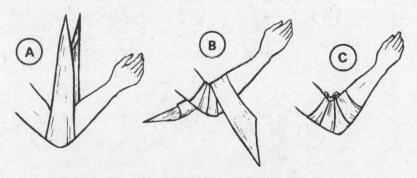

Figure 3-37: Elbow bandage (Illustrated A thru C).

⚠ **CAUTION**

If an elbow fracture is suspected, DO NOT bend the elbow; bandage it in an extended position.

3-17. Hand Bandage

a. To apply a triangular bandage to the hand–
   (1) Place the hand in the middle of the triangular bandage with the wrist at the base of the bandage (Figure 3-38 A). Ensure that the fingers are separated with absorbent material to prevent chafing and irritation of the skin.
   (2) Place the apex over the fingers and tuck any excess material into the pleats on each side of the hand (Figure 3-38 B).
   (3) Cross the ends on top of the hand, take them around the wrist, and tie them (Figures 3-38 C, D, and E) with a non slip knot.

b. To apply a cravat bandage to the palm of the hand–
   (1) Lay the middle of the cravat over the palm of the hand with the ends hanging down on each side (Figure 3-39 A).

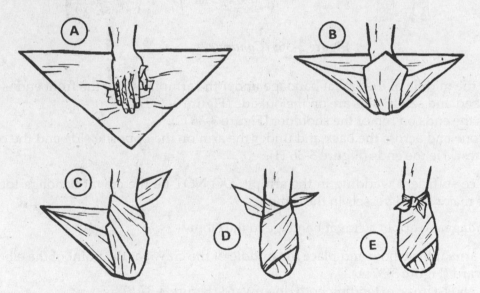

Figure 3-38: Triangular bandage applied to hand (Illustrated A thru E).

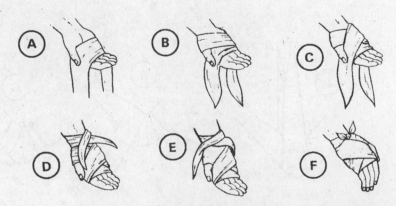

Figure 3-39: Cravat bandage applied to palm of hand (Illustrated A thru F).

(2) Take the end of the cravat at the little finger across the back of the hand, extending it upward over the base of the thumb; then bring it downward across the palm (Figure 3-39 B).

(3) Take the thumb end across the back of the hand, over the palm, and through the hollow between the thumb and palm (Figure 3-39 C).

(4) Take the ends to the back of the hand and cross them; then bring them up over the wrist and cross them again (Figure 3-39 D).

(5) Bring both ends down and tie them with a non slip knot on top of the wrist (Figure 3-39 E and F).

3-18. Leg (Upper and Lower) Bandage. To apply a cravat bandage to the leg–

a. Place the center of the cravat over the dressing (Figure 3-40 A).

b. Take one end around and up the leg in a spiral motion and the other end around and down the leg in a spiral motion, overlapping part of each preceding turn (Figure 3-40 B).

c. Bring both ends together and tie them (Figure 3-40 C) with a non slip knot.

3-19. Knee Bandage. To apply a cravat bandage to the knee as illustrated in Figure 3-41, use the same technique applied in bandaging the elbow. The same caution for the elbow also applies to the knee.

3-20. Foot Bandage. To apply a triangular bandage to the foot–

a. Place the foot in the middle of the triangular bandage with the heel well forward of the base (Figure 3-42 A). Ensure that the toes are separated with absorbent material to prevent chafing and irritation of the skin.

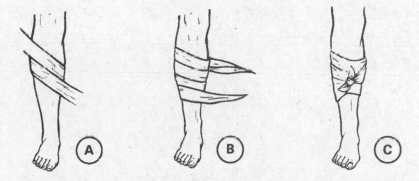

Figure 3-40: Cravat bandage applied to leg (Illustrated A thru C).

Figure 3-41: Cravat bandage applied to knee (Illustrated A thru C).

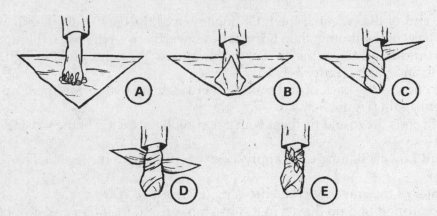

Figure 3-42: Triangular bandage applied to foot (Illustrated A thru E).

b. Place the apex over the top of the foot and tuck any excess material into the pleats on each side of the foot (Figure 3-42 B).

c. Cross the ends on top of the foot, take them around the ankle, and tie them at the front of the ankle (Figure 3-42 C, D, and E).

CHAPTER 4

First Aid for Fractures

A fracture is any break in the continuity of a bone. Fractures can cause total disability or in some cases death. On the other hand, they can most often be treated so there is complete recovery. A great deal depends upon the first aid the individual receives before he is moved. First aid includes immobilizing the fractured part in addition to applying lifesaving measures. The basic splinting principle is to immobilize the joints above and below any fracture.

4-1. Kinds of Fractures. See figure 4-1 for detailed illustration.

    a. Closed Fracture. A closed fracture is a broken bone that does not break the overlying skin. Tissue beneath the skin may be damaged. A dislocation is when a joint, such as a knee, ankle, or shoulder, is not in proper position. A sprain is when the connecting tissues of the joints have been torn. Dislocations and sprains should be treated as closed fractures.

    b. Open Fracture. An open fracture is a broken bone that breaks (pierces) the overlying skin. The broken bone may come through the skin, or a missile such as a bullet or shell fragment may go through the flesh and break the bone. An open fracture is contaminated and subject to infection.

4-2. Signs/Symptoms of Fractures (081-831-1000). Indications of a fracture are deformity, tenderness, swelling, pain, inability to move the injured part, protruding bone, bleeding, or discolored skin at the injury site. A sharp pain when the individual attempts to move the part is also a sign of a fracture. DO NOT encourage the casualty to move the injured part in order to identify a fracture since such movement could cause further damage to surrounding tissues and promote shock. If you are not sure whether a bone is fractured, treat the injury as a fracture.

4-3. Purposes of Immobilizing Fractures. A fracture is immobilized to prevent the sharp edges of the bone from moving and cutting tissue, muscle, blood vessels, and nerves. This reduces pain and helps prevent or control shock. In a closed fracture, immobilization keeps bone fragments from causing an open wound and prevents contamination and possible infection. Splint to immobilize.

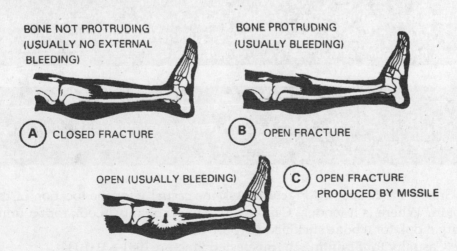

Figure 4-1: Kinds of fractures (Illustrated A thru C).

4-4. Splints, Padding, Bandages, Slings, and Swathes (081-831-1034)

  a. Splints. Splints may be improvised from such items as boards, poles, sticks, tree limbs, rolled magazines, rolled newspapers, or cardboard. If nothing is available for a splint, the chest wall can be used to immobilize a fractured arm and the uninjured leg can be used to immobilize (to some extent) the fractured leg.

  b. Padding. Padding may be improvised from such items as a jacket, blanket, poncho, shelter half, or leafy vegetation.

  c. Bandages. Bandages may be improvised from belts, rifle slings, bandoleers, kerchiefs, or strips torn from clothing or blankets. Narrow materials such as wire or cord should not be used to secure a splint in place.

  d. Slings. A sling is a bandage (or improvised material such as a piece of cloth, a belt, and so forth) suspended from the neck to support an upper extremity. Also, slings may be improvised by using the tail of a coat or shirt, and pieces torn from such items as clothing and blankets. The triangular bandage is ideal for this purpose. Remember that the casualty's hand should be higher than his elbow, and the sling should be applied so that the supporting pressure is on the uninjured side.

  e. Swathes. Swathes are any bands (pieces of cloth, pistol belts, and so forth) that are used to further immobilize a splinted fracture. Triangular and cravat bandages are often used as or referred to as swathe bandages. The purpose of the swathe is to immobilize. Therefore, the swathe bandage is placed above and/or below the fracture—not over it.

4-5. Procedures for Splinting Suspected Fractures (081-831-1034). Before beginning first aid treatment for a fracture, gather whatever splinting materials are available. Materials may consist of splints, such as wooden boards, branches, or poles. Other splinting materials include padding, improvised cravats, and/ or bandages. Ensure that splints are long enough to immobilize the joint above and below the suspected fracture. If possible, use at least four ties (two above and two below the fracture) to secure the splints. The ties should be non slip knots and should be tied away from the body on the splint.

  * a. Evaluate the Casualty (081-831-1000). Be prepared to perform any necessary lifesaving measures. Monitor the casualty for development of conditions which may require you to perform necessary basic lifesaving measures. These measures include clearing the airway, rescue breathing, preventing shock, and/or bleeding control.

> **WARNING (081-831-1000)**
> Unless there is immediate life-threatening danger, such as a fire or an explosion, DO NOT move the casualty with a suspected back or neck injury. Improper movement may cause permanent paralysis or death.

> **WARNING (081-831-1000)**
> In a chemical environment, DO NOT remove any protective clothing. Apply the dressing/splint over the clothing.

  b. Locate the Site of the Suspected Fracture. Ask the casualty for the location of the injury. Does he have any pain? Where is it tender? Can he move the extremity? Look for an unnatural position of the extremity. Look for a bone sticking out (protruding).

  c. Prepare the Casualty for Splinting the Suspected Fracture (081-831-1034).

    (1) Reassure the casualty. Tell him that you will be taking care of him and that medical aid is on the way.

    (2) Loosen any tight or binding clothing.

(3) Remove all the jewelry from the casualty and place it in the casualty's pocket. Tell the casualty you are doing this because if the jewelry is not removed at this time and swelling occurs later, further bodily injury can occur.

> **✍ NOTE**
> Boots should not be removed from the casualty unless they are needed to stabilize a neck injury, or there is actual bleeding from the foot.

d. Gather Splinting Materials (081-831-1034). If standard splinting materials (splints, padding, cravats, and so forth) are not available, gather improvised materials. Splints can be improvised from wooden boards, tree branches, poles, rolled newspapers or magazines. Splints should be long enough to reach beyond the joints above and below the suspected fracture site. Improvised padding, such as a jacket, blanket, poncho, shelter half, or leafy vegetation may be used. A cravat can be improvised from a piece of cloth, a large bandage, a shirt, or a towel. Also, to immobilize a suspected fracture of an arm or a leg, parts of the casualty's body may be used. For example, the chest wall may be used to immobilize an arm; and the uninjured leg may be used to immobilize the injured leg.

> **✍ NOTE**
> If splinting material is not available and suspected fracture CANNOT be splinted, then swathes, or a combination of swathes and slings can be used to immobilize an extremity.

e. Pad the Splints (081-831-1034). Pad the splints where they touch any bony part of the body, such as the elbow, wrist, knee, ankle, crotch, or armpit area. Padding prevents excessive pressure to the area.

f. Check the Circulation Below the Site of the Injury (081-831-1034).

(1) Note any pale, white, or bluish-gray color of the skin which may indicate impaired circulation. Circulation can also be checked by depressing the toe/fingernail beds and observing how quickly the color returns. A slower return of pink color to the injured side when compared with the uninjured side indicates a problem with circulation. Depressing the toe/fingernail beds is a method to use to check the circulation in a dark-skinned casualty.

(2) Check the temperature of the injured extremity. Use your hand to compare the temperature of the injured side with the uninjured side of the body. The body area below the injury may be colder to the touch indicating poor circulation.

(3) Question the casualty about the presence of numbness, tightness, cold, or tingling sensations.

WARNING

Casualties with fractures to the extremities may show impaired circulation, such as numbness, tingling, cold and/or pale to blue skin. These casualties should be evacuated by medical personnel and treated as soon as possible. Prompt medical treatment may prevent possible loss of the limb.

WARNING

If it is an open fracture (skin is broken; bone(s) may be sticking out), DO NOT ATTEMPT TO PUSH BONE(S) BACK UNDER THE SKIN. Apply a field dressing to protect the area. See Task 081-831-1016, Put on a Field or Pressure Dressing.

g. Apply the Splint in Place (081-831-1034).

(1) Splint the fracture(s) in the position found. DO NOT attempt to reposition or straighten the injury. If it is an open fracture, stop the bleeding and protect the wound. (See Chapter 2, Section II, for

detailed information.) Cover all wounds with field dressings before applying a splint. Remember to use the casualty's field dressing, not your own. If bones are protruding (sticking out), DO NOT attempt to push them back under the skin. Apply dressings to protect the area.

(2) Place one splint on each side of the arm or leg. Make sure that the splints reach, if possible, beyond the joints above and below the fracture.

(3) Tie the splints. Secure each splint in place above and below the fracture site with improvised (or actual) cravats. Improvised cravats, such as strips of cloth, belts, or whatever else you have, may be used. With minimal motion to the injured areas, place and tie the splints with the bandages. Push cravats through and under the natural body curvatures (spaces), and then gently position improvised cravats and tie in place. Use non slip knots. Tie all knots on the splint away from the casualty (Figure 4-2). DO NOT tie cravats directly over suspected fracture/dislocation site.

h. Check the Splint for Tightness (081-831-1034).

(1) Check to be sure that bandages are tight enough to securely hold splinting materials in place, but not so tight that circulation is impaired.

(2) Recheck the circulation after application of the splint. Check the skin color and temperature. This is to ensure that the bandages holding the splint in place have not been tied too tightly. A finger tip check can be made by inserting the tip of the finger between the wrapped tails and the skin.

(3) Make any adjustment without allowing the splint to become ineffective.

i. Apply a Sling if Applicable (081-831-1034). An improvised sling may be made from any available non stretching piece of cloth, such as a fatigue shirt or trouser, poncho, or shelter half. Slings may also be improvised using the tail of a coat, belt, or a piece of cloth from a blanket or some clothing. See Figure 4-3 for an illustration of a shirt tail used for support. A pistol belt or trouser belt also may be used for support (Figure 4-4). A sling should place the supporting pressure on the casualty's uninjured side. The supported arm should have the hand positioned slightly higher than the elbow.

Figure 4-3: Shirt tail used for support.

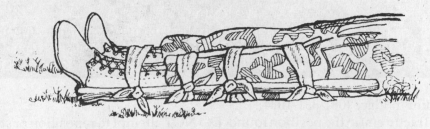

Figure 4-2: Nonslip knots tied away from casualty.

Figure 4-4: Belt Used for support.

(1) Insert the splinted arm in the center of the sling (Figure 4-5).
(2) Bring the ends of the sling up and tie them at the side (or hollow) of the neck on the uninjured side (Figure 4-6).
(3) Twist and tuck the corner of the sling at the elbow (Figure 4-7).

j. Apply a Swathe if Applicable (081-831-1034). You may use any large piece of cloth, such as a soldier's belt or pistol belt, to improvise a swathe. A swathe is any band (a piece of cloth) or wrapping used to further immobilize a fracture. When splints are unavailable, swathes, or a combination of swathes and slings can be used to immobilize an extremity.

 WARNING (081-831-1034)
The swathe should not be placed directly on top of the injury, but positioned above and/or below the fracture site.

(1) Apply swathes to the injured arm by wrapping the swathe over the injured arm, around the casualty's back and under the arm on the uninjured side. Tie the ends on the uninjured side (Figure 4-8).

Figure 4-5: Arm inserted in center of improvised sling.

Figure 4-6: Ends of improvised sling tied to side of neck.

Figure 4-7: Corner of sling twisted and tucked at elbow.

Figure 4-8: Arm immobilized with strip of clothing.

(2) A swathe is applied to an injured leg by wrapping the swathe(s) around both legs and securing it on the uninjured side.

k. Seek Medical Aid. Notify medical personnel, watch closely for development of life-threatening conditions, and if necessary, continue to evaluate the casualty.

4-6. Upper Extremity Fractures (081-831-1034). Figures 4-9 through 4-16 show how to apply slings, splints, and cravats (swathes) to immobilize and support fractures of the upper extremities. Although the padding

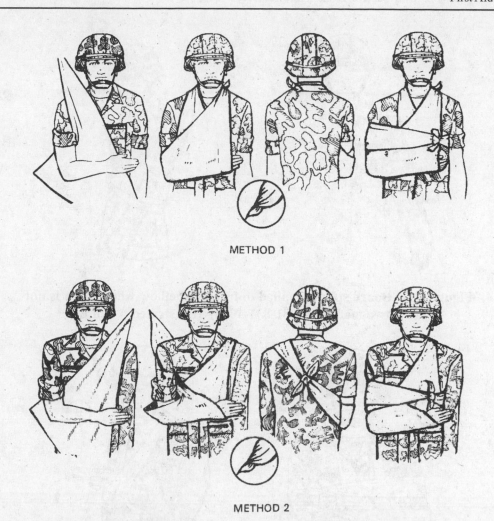

METHOD 1

METHOD 2

Figure 4-9: Application of triangular bandage to form sling (two methods) (Illustrated A and B).

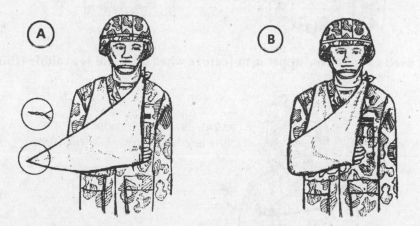

Figure 4-10: Completing sling sequence by twisting and tucking the corner of the sling at the elbow .

is not visible in some of the illustrations, it is always preferable to apply padding along the injured part for the length of the splint and especially where it touches any bony parts of the body.

4-7. Lower Extremity Fractures (081-831-1034). Figures 4-17 through 4-22 show how to apply splints to immobilize fractures of the lower extremities. Although padding is not visible in some of the figures, it

Figure 4-11: Board splints applied to fractured elbow when elbow is not bent (two methods) (081-831-1034) (Illustrated A and B).

Figure 4-12: Chest wall used as splint for upper arm frature when no splint is availble (Illustrated A and B).

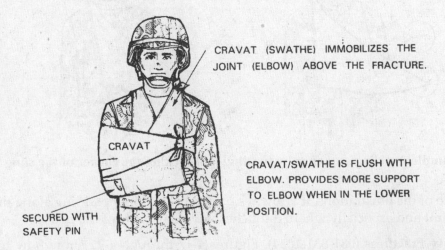

CRAVAT (SWATHE) IMMOBILIZES THE
JOINT (ELBOW) ABOVE THE FRACTURE.

CRAVAT

SECURED WITH
SAFETY PIN

CRAVAT/SWATHE IS FLUSH WITH
ELBOW. PROVIDES MORE SUPPORT
TO ELBOW WHEN IN THE LOWER
POSITION.

Figure 4-13: Chest wall, sling, and cravat used to immobilize fractured elbow when elbow is bent.

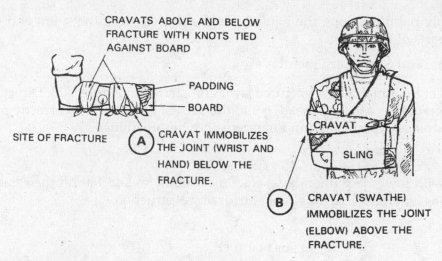

Figure 4-14: Board splint applied to fractured forearm (Illustrated A and B).

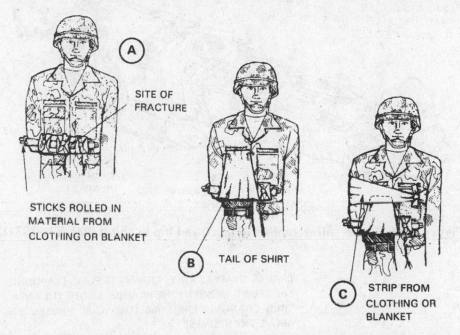

Figure 4-15: Fractured forearm or wrist splinted with sticks and supported with tail of shirt and strips of material (Illustrated A thru C).

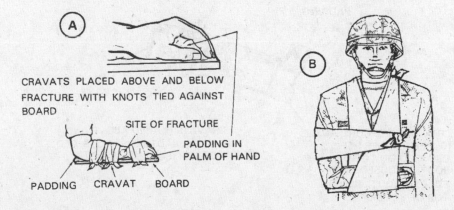

Figure 4-16: Board splint applied to fractured wrist and hand (Illustrated A thru C).

is preferable to apply padding along the injured part for the length of the splint and especially where it touches any bony parts of the body.

4-8. Jaw, Collarbone, and Shoulder Fractures

a. Apply a cravat to immobilize a fractured jaw as illustrated in Figure 4-23. Direct all bandaging support to the top of the casualty's head, not to the back of his neck. If incorrectly placed, the bandage will pull the casualty's jaw back and interfere with his breathing.

> ⚠ **CAUTION**
> Casualties with lower jaw (mandible) fractures cannot be laid flat on their backs because facial muscles will relax and may cause an airway obstruction.

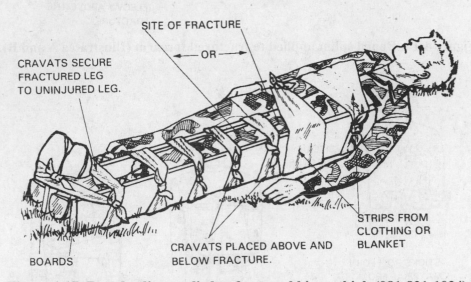

Figure 4-17: Board splint applied to fractured hip or thigh (081-831-1034).

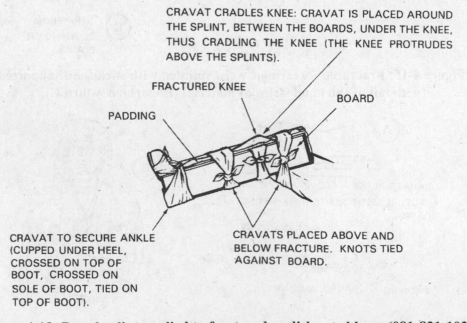

Figure 4-18: Board splint applied to fractured or dislocated knee (081-831-1034).

all

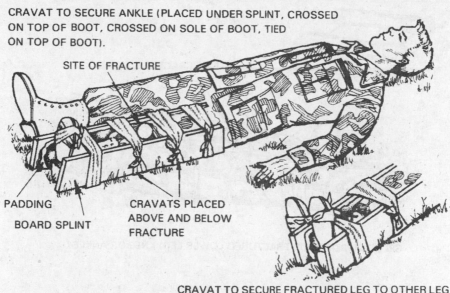

CRAVAT TO SECURE ANKLE (PLACED UNDER SPLINT, CROSSED ON TOP OF BOOT, CROSSED ON SOLE OF BOOT, TIED ON TOP OF BOOT).

SITE OF FRACTURE

PADDING

BOARD SPLINT

CRAVATS PLACED ABOVE AND BELOW FRACTURE

CRAVAT TO SECURE FRACTURED LEG TO OTHER LEG (IF MORE SUPPORT IS NEEDED).

Figure 4-19: Board splint applied to fractured lower leg or ankle.

Figure 4-20: Improvised splint applied to fractured lower leg or ankle.

b. Apply two belts, a sling, and a cravat to immobilize a fractured collarbone, as illustrated in Figure 4-24.

c. Apply a sling and a cravat to immobilize a fractured or dislocated shoulder, using the technique illustrated in Figure 4-25.

4-9. Spinal Column Fractures (081-831-1000). It is often impossible to be sure a casualty has a fractured spinal column. Be suspicious of any back injury, especially if the casualty has fallen or if his back has been sharply struck or bent. If a casualty has received such an injury and does not have feeling in his legs or cannot move them, you can be reasonably sure that he has a severe back injury which should be treated as a fracture. Remember, if the spine is fractured, bending it can cause the sharp bone fragments to bruise or cut the spinal cord and result in permanent paralysis (Figure 4-26A). The spinal column must maintain a swayback position to remove pressure from the spinal cord.

a. If the Casualty Is Not to Be Transported (081-831-1000) Until Medical Personnel Arrive—

- Caution him not to move. Ask him if he is in pain or if he is unable to move any part of his body.
- Leave him in the position in which he is found. DO NOT move any part of his body.
- Slip a blanket, if he is lying face up, or material of similar size, under the arch of his back to support the spinal column in a swayback position (Figure 4-26 B). If he is lying face down, DO NOT put anything under any part of his body.

b. If the Casualty Must Be Transported to A Safe Location Before Medical Personnel Arrive—

- And if the casualty is in a face-up position, transport him by litter or use a firm substitute, such as a wide board or a flat door longer than his height. Loosely tie the casualty's wrists together

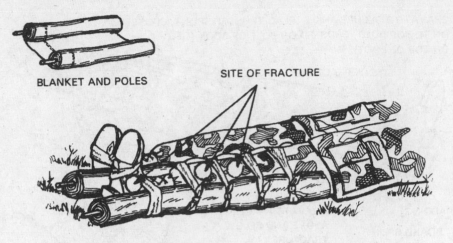

BLANKET AND POLES

SITE OF FRACTURE

SPLINT APPLIED FOR FRACTURED LOWER LEG, KNEE OR ANKLE

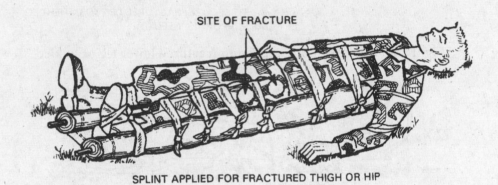

SITE OF FRACTURE

SPLINT APPLIED FOR FRACTURED THIGH OR HIP

Figure 4-21: Poles rolled in a blanket and used as splints applied to fractured lower extremity.

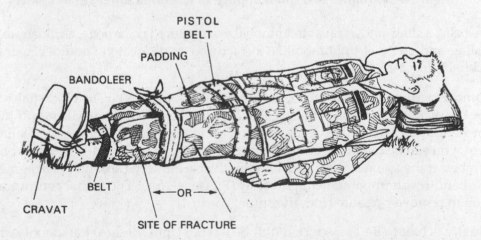

PISTOL BELT

PADDING

BANDOLEER

BELT

CRAVAT

OR

SITE OF FRACTURE

Figure 4-22: Uninjured leg used as splint for fractured leg (anatomical splint).

over his waistline, using a cravat or a strip of cloth. Tie his feet together to prevent the accidental dropping or shifting of his legs. Lay a folded blanket across the litter where the arch of his back is to be placed. Using a four-man team (Figure 4-27), place the casualty on the litter without bending his spinal column or his neck.

○   The number two, three, and four men position themselves on one side of the casualty; all kneel on one knee along the side of the casualty. The number one man positions himself to

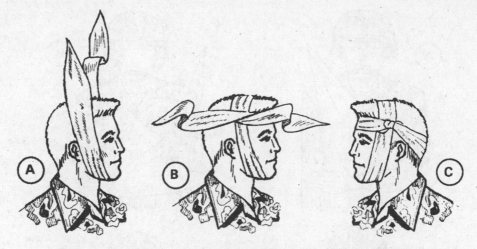

Figure 4-23: Fractured jaw immobilized (Illustrated A thru C)

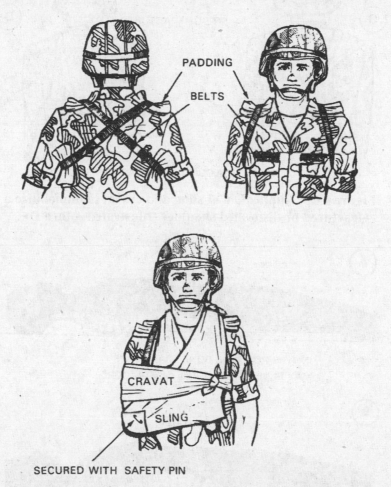

PADDING

BELTS

CRAVAT

SLING

SECURED WITH SAFETY PIN

Figure 4-24: Application of belts, sling, and cravat to immobilize a collarbone.

the opposite side of the casualty. The number two, three, and four men gently place their hands under the casualty. The number one man on the opposite side places his hands under the injured part to assist.

o   When all four men are in position to lift, the number two man commands, "PREPARE TO LIFT" and then, "LIFT." All men, in unison, gently lift the casualty about 8 inches. Once

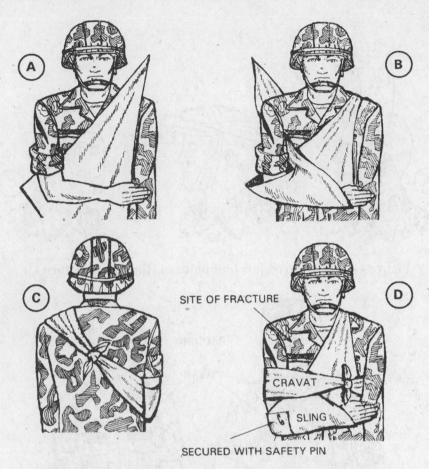

Figure 4-25: Application of sling and cravat to immobilize a fractured or dislocated shoulder (Illustrated A thru D) .

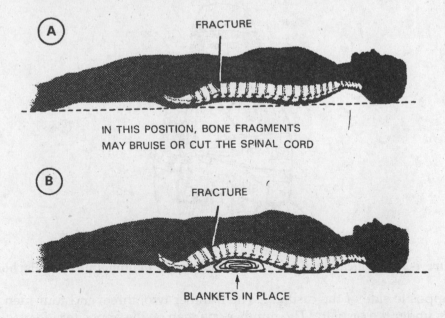

Figure 4-26: Spinal column must maintain a swayback position (Illustrated A and B).

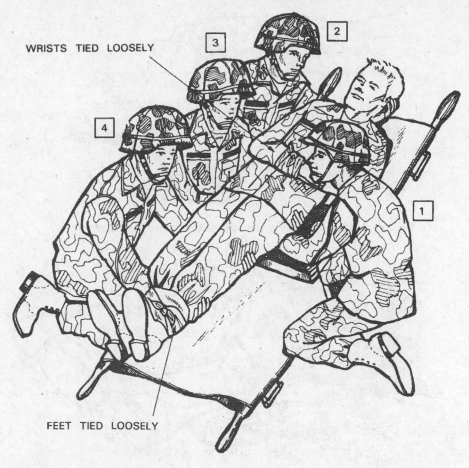

WRISTS TIED LOOSELY

FEET TIED LOOSELY

Figure 4-27: Placing face-up casualty with fractured back onto litter.

the casualty is lifted, the number one man recovers and slides the litter under the casualty, ensuring that the blanket is in proper position. The number one man then returns to his original lift position (Figure 4-27).

○ When the number two man commands, "LOWER CASUALTY," all men, in unison, gently lower the casualty onto the litter.

• And if the casualty is in a face-down position, he must be transported in this same position. The four-man team lifts him onto a regular or improvised litter, keeping the spinal column in a swayback position. If a regular litter is used, first place a folded blanket on the litter at the point where the chest will be placed.

4-10. Neck Fractures (081-831-1000). A fractured neck is extremely dangerous. Bone fragments may bruise or cut the spinal cord just as they might in a fractured back.

a. If the Casualty Is Not to Be Transported (081-831-1000) Until Medical Personnel Arrive—
   • Caution him not to move. Moving may cause death.
   • Leave the casualty in the position in which he is found. If his neck/head is in an abnormal position, immediately immobilize the neck/head. Use the procedure stated below.
      ○ Keep the casualty's head still, if he is lying face up, raise his shoulders slightly, and slip a roll of cloth that has the bulk of a bath towel under his neck (Figure 4-28). The roll should be thick enough to arch his neck only slightly, leaving the back of his head on the ground. DO NOT bend his neck or head forward. DO NOT raise or twist his head. Immobilize the casualty's head (Figure 4-29). Do this by padding heavy objects such as rocks or his boots

Figure 4-28: Casualty with roll of cloth (bulk) under neck.

Figure 4-29: Immobilization of fractured neck.

and placing them on each side of his head. If it is necessary to use boots, first fill them with stones, gravel, sand, or dirt and tie them tightly at the top. If necessary, stuff pieces of material in the top of the boots to secure the contents.

○ DO NOT move the casualty if he is lying face down. Immobilize the head/neck by padding heavy objects and placing them on each side of his head. DO NOT put a roll of cloth under the neck. DO NOT bend the neck or head, nor roll the casualty onto his back.

b. If the Casualty Must be Prepared for Transportation Before Medical Personnel Arrive—

• And he has a fractured neck, at least two persons are needed because the casualty's head and trunk must be moved in unison.

• The two persons must work in close coordination (Figure 4-30) to avoid bending the neck.

• Place a wide board lengthwise beside the casualty. It should extend at least 4 inches beyond the casualty's head and feet (Figure 4-30 A).

• If the casualty is lying face up, the number one man steadies the casualty's head and neck between his hands. At the same time the number two man positions one foot and one knee against the board to prevent it from slipping, grasps the casualty underneath his shoulder and hip, and gently slides him onto the board (Figure 4-30 B).

• If the casualty is lying face down, the number one man steadies the casualty's head and neck between his hands, while the number two man gently rolls the casualty over onto the board (Figure 4-30 C).

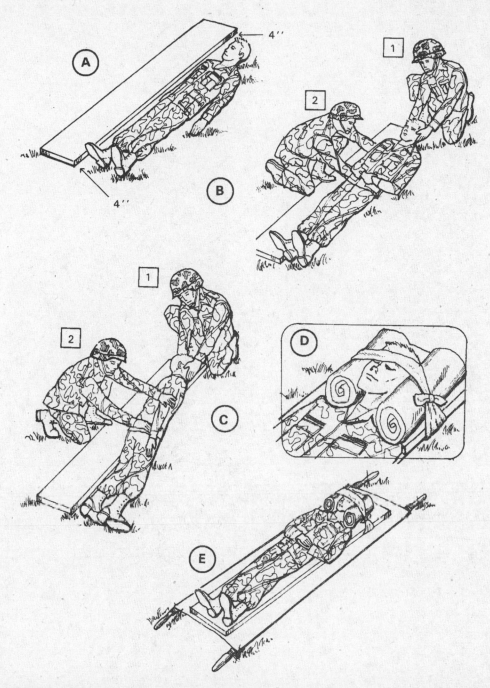

Figure 4-30: Preparing casualty with fractured neck for transportation (Illustrated A thru E).

- The number one man continues to steady the casualty's head and neck. The number two man simultaneously raises the casualty's shoulders slightly, places padding under his neck, and immobilizes the casualty's head (Figures 4-30 D, and E). The head may be immobilized with the casualty's boots, with stones rolled in pieces of blanket, or with other material.
- Secure any improvised supports in position with a cravat or strip of cloth extended across the casualty's forehead and under the board (Figure 4-30 D).
- Lift the board onto a litter or blanket in order to transport the casualty (Figure 4-30 E).

CHAPTER 5

First Aid for Climatic Injuries

It is desirable, but not always possible, for an individual's body to become adjusted (acclimatized) to an environment. Physical condition determines the time adjustment, and trying to rush it is ineffective. Even those individuals in good physical condition need time before working or training in extremes of hot or cold weather. Climate-related injuries are usually preventable; prevention is both an individual and leadership responsibility. Several factors contribute to health and well-being in any environment: diet, sleep/rest, exercise, and suitable clothing. These factors are particularly important in extremes of weather. Diet, especially, should be suited to an individual's needs in a particular climate. A special diet undertaken for any purpose should be done so with appropriate supervision. This will ensure that the individual is getting a properly balanced diet suited to both climate and personal needs, whether for weight reduction or other purposes. The wearing of specialized protective gear or clothing will sometimes add to the problem of adjusting to a particular climate. Therefore, soldiers should exercise caution and judgment in adding or removing specialized protective gear or clothing.

5-1. Heat Injuries (081-831-1008). Heat injuries are environmental injuries that may result when a soldier is exposed to extreme heat, such as from the sun or from high temperatures. Prevention depends on availability and consumption of adequate amounts of water. Prevention also depends on proper clothing and appropriate activity levels. Acclimatization and protection from undue heat exposure are also very important. Identification of high risk personnel (basic trainees, troops with previous history of heat injury, and overweight soldiers) helps both the leadership and the individual prevent and cope with climatic conditions. Instruction on living and working ingot climates also contributes toward prevention.

> ✍ **NOTE**
> Salt tablets should not be used in the prevention of heat injury. Usually, eating field rations or liberal salting of the garrison diet will provide enough salt to replace what is lost through sweating in hot weather.

   a. Diet. A balanced diet usually provides enough salt even in hot weather. But when people are on reducing or other diets, salt may need to come from other sources. DO NOT use salt tablets to supplement a diet. Anyone on a special diet (for whatever purpose) should obtain professional help to work out a properly balanced diet.

   b. Clothing.

     (1) The type and amount of clothing and equipment a soldier wears and the way he wears it also affect the body and its adjustment to the environment. Clothing protects the body from radiant heat. However, excessive or tight-fitting clothing, web equipment, and packs reduce ventilation needed to cool the body. During halts, rest stops, and other periods when such items are not needed, they should be removed, mission permitting.

     (2) The individual protective equipment (IPE) protects the soldier from chemical and biological agents. The equipment provides a barrier between him and a toxic environment. However, a serious problem associated with the chemical overgarment is heat stress. The body normally maintains a heat balance, but when the overgarment is worn the body sometimes does not function properly. Overheating may occur rapidly. Therefore, strict adherence to mission oriented protective posture (MOPP) levels directed by your commander is important. This will keep

those heat related injuries caused by wearing the IPE to a minimum. See FM 3-4 for further information on MOPP.

c. Prevention. The ideal fluid replacement is water. The availability of sufficient water during work or training in hot weather is very important. The body, which depends on water to help cool itself, can lose more than a quart of water per hour through sweat. Lost fluids must be replaced quickly. Therefore, during these work or training periods, you should drink at least one canteen full of water every hour. In extremely hot climates or extreme temperatures, drink at least a full canteen of water every half hour, if possible. In such hot climates, the body depends mainly upon sweating to keep it cool, and water intake must be maintained to allow sweating to continue. Also, keep in mind that a person who has suffered one heat injury is likely to suffer another. Before a heat injury casualty returns to work, he should have recovered well enough not to risk a recurrence. Other conditions which may increase heat stress and cause heat injury include infections, fever, recent illness or injury, overweight, dehydration, exertion, fatigue, heavy meals, and alcohol. In all this, note that salt tablets should not be used as a preventive measure.

d. Categories. Heat injury can be divided into three categories: heat cramps, heat exhaustion, and heatstroke.

e. First Aid. Recognize and give first aid for heat injuries.

WARNING

Casualty should be continually monitored for development of conditions which may require the performance of necessary basic lifesaving measures, such as: clearing the airway, performing mouth-to-mouth resuscitation, preventing shock, and/or bleeding control.

⚠ ***CAUTION**

DO NOT use salt solution in first aid procedures for heat injuries.

(1) Check the casualty for signs and symptoms of heat cramps (081-831-1008).
- Signs/Symptoms. Heat cramps are caused by an imbalance of chemicals (called electrolytes) in the body as a result of excessive sweating. This condition causes the casualty to exhibit:
  ○ Muscle cramps in the extremities (arms and legs).
  ○ Muscle cramps of the abdomen.
  ○ Heavy (excessive) sweating (wet skin).
  ○ Thirst.
- Treatment.
  ○ Move the casualty to a cool or shady area (or improvise shade).
  ○ Loosen his clothing (if not in a chemical environment).
  ○ Have him slowly drink at least one canteen full of cool water.
  ○ Seek medical aid should cramps continue.

WARNING

DO NOT loosen the casualty's clothing if in a chemical environment. 160-065 0-94-3

(2) Check the casualty for signs and symptoms of heat exhaustion (081-831-1008).
- Signs/Symptoms which occur often. Heat exhaustion is caused by loss of water through sweating without adequate fluid replacement. It can occur in an otherwise fit individual

who is involved in tremendous physical exertion in any hot environment. The signs and symptoms are similar to those which develop when a person goes into a state of shock.
- Heavy (excessive) sweating with pale, moist, cool skin.
- Headache.
- Weakness.
- Dizziness.
- Loss of appetite.

- Signs/Symptoms which occur **sometimes.**
  - Heat cramps.
  - Nausea—with or without vomiting.
  - Urge to defecate.
  - Chills (gooseflesh).
  - Rapid breathing.
  - Tingling of hands and/or feet.
  - Confusion.
- Treatment.
  - Move the casualty to a cool or shady area (or improvise shade).
  - Loosen or remove his clothing and boots (unless in a chemical environment). Pour water on him and fan him (unless in a chemical environment).
  - Have him slowly drink at least one canteen full of cool water. Elevate his legs.
  - If possible, the casualty should not participate in strenuous activity for the remainder of the day.
  - Monitor the casualty until the symptoms are gone, or medical aid arrives.

(3) Check the casualty for signs and symptoms of heatstroke (sometimes called "sunstroke") (081-831-1008).

WARNING

Heatstroke must be considered a medical emergency which may result in death if treatment is delayed.

- Signs/Symptoms. A casualty suffering from heatstroke has usually worked in a very hot, humid environment for a prolonged time. It is caused by failure of the body's cooling mechanisms. Inadequate sweating is a factor. The casualty's skin is red (flushed), hot, and dry. He may experience weakness, dizziness, confusion, headaches, seizures, nausea (stomach pains), and his respiration and pulse may be rapid and weak. Unconsciousness and collapse may occur suddenly.
- Treatment. Cool casualty immediately by—
  - Moving him to a cool or shaded area (or improvise shade).
  - Loosening or removing his clothing (except in a chemical environment).
  - *Spraying or pouring water on him; fanning him to permit a coolant effect of evaporation.
  - Massaging his extremities and skin which increases the blood flow to those body areas, thus aiding the cooling process.
  - Elevating his legs.
  - Having him slowly drink at least one canteen full of water if he is conscious.

✍ **NOTE**

Start cooling casualty immediately. Continue cooling while awaiting transportation and during the evacuation.

- Medical aid. Seek medical aid because the casualty should be transported to a medical treatment facility as soon as possible. Do not interrupt cooling process or lifesaving measures to seek help.
- Casualty should be continually monitored for development of conditions which may require the performance of necessary basic lifesaving measures, such as clearing the airway, mouth-to-mouth resuscitation, preventing shock, and/or bleeding control.

f. Table. See Table 5-1 for further information.

5-2. Cold Injuries (081-831-1009). Cold injuries are most likely to occur when an unprepared individual is exposed to winter temperatures. They can occur even with proper planning and equipment. The cold weather and the type of combat operation in which the individual is involved impact on whether he is likely to be injured and to what extent. His clothing, his physical condition, and his mental makeup also are determining factors. However, cold injuries can usually be prevented. Well-disciplined and well-trained individuals can be protected even in the most adverse circumstances. They and their leaders must know the hazards of exposure to the cold. They must know the importance of personal hygiene, exercise, care of the feet and hands, and the use of protective clothing.

a. Contributing Factors.
  (1) Weather. Temperature, humidity, precipitation, and wind modify the loss of body heat. Low temperatures and low relative humidity-dry cold—promote frostbite. Higher temperatures, together with moisture, promote immersion syndrome. Wind chill accelerates the loss of body heat and may aggravate cold injuries. These principles and risks apply equally to both men and women.
  (2) Type of combat operation. Defense, delaying, observation-post, and sentinel duties do create to a greater extent—fear, fatigue, dehydration, and lack of nutrition. These factors further increase the soldier's vulnerability to cold injury. Also, a soldier is more likely to receive a cold injury if he is—
    - Often in contact with the ground.

Table 5-1: Sun or Heat Injuries (081-831-1008).

INJURIES	SIGNS/SYMPTOMS	FIRST AID*
Heat cramps	The casualty experiences muscle cramps of arms, legs, and/or stomach. The casualty may also have heavy sweating (wet skin) and extreme thirst.	1. Move the casualty to a shady area or improvise shade and loosen his clothing.+ 2. Give him large amounts of cool water slowly. 3. Monitor the casualty and give him more water as tolerated. 4. Seek medical aid if the cramps continue.
Heat exhaustion	The casualty *often* experiences profuse (heavy) sweating with pale, moist, cool skin; headache, weakness, dizziness, and/or loss of appetite.	1. Move the casualty to a cool, shady area or improvise shade and loosen/remove his clothing.+ 2. Pour water on him and fan him to permit coolant effect of evaporation. 3. Have him slowly drink at least one canteen full of water.

(continued)

Table 5-1: *(Continued)*

Heat exhaustion *Continued.*	The casualty *sometimes* experiences heat cramps, nausea (with or without vomiting), urge to defecate, chills (gooseflesh), rapid breathing, confusion, and tingling of the hands and/or feet.	4. Elevate the casualty's legs. 5. Seek medical aid if symptoms continue; monitor the casualty until the symptoms are gone or medical aid arrives.
Heatstroke# **(sunstroke)**	The casualty stops sweating (red [flushed] hot, dry skin). He first may experience headache, dizziness, nausea, fast pulse and respiration, seizures, and mental confusion. He may collapse and suddenly become unconscious. *THIS IS A MEDICAL EMERGENCY.*	1. Move the casualty to a cool, shady area or improvise shade and loosen or remove his clothing, remove the outer garments and protective clothing if the situation permits.+ ★2. Start cooling the casualty immediately. Spray or pour water on him. Fan him. Massage his extremities and skin. 3. Elevate his legs. 4. If conscious, have him slowly drink at least one canteen full of water. 5. SEEK MEDICAL AID. CONTINUE COOLING WHILE AWAITING TRANSPORT AND DURING EVACUATION. EVACUATE AS SOON AS POSSIBLE. PERFORM ANY NECESSARY LIFESAVING MEASURES.

*The *first aid procedure* for heat related injuries caused by wearing *individual protective equipment* is to move the casualty to a clean area and give him water to drink.

+When in a chemical environment, DO NOT loosen/remove the casualty's clothing.

#Can be fatal if not treated promptly and correctly.

- Immobile for long periods, such as while riding in a crowded vehicle.
- Standing in water, such as in a foxhole.
- Out in the cold for days without being warmed.
- Deprived of an adequate diet and rest.
- Not able to take care of his personal hygiene.

(3) Clothing. The soldier should wear several layers of loose clothing. He should dress as lightly as possible consistent with the weather to reduce the danger of excessive perspiration and subsequent chilling. It is better for the body to be slightly cold and generating heat than excessively

warm and sweltering toward dehydration. He should remove a layer or two of clothing before doing any hard work. He should replace the clothing when work is completed. Most cold injuries result from soldiers having too few clothes available when the weather suddenly turns colder. Wet gloves, shoes, socks, or any other wet clothing add to the cold injury process.

⚠ **CAUTION**

In a chemical environment DO NOT take off protective chemical gear.

---

(4) Physical makeup. Physical fatigue contributes to apathy, which leads to inactivity, personal neglect, carelessness, and reduced heat production. In turn, these increase the risk of cold injury. Soldiers with prior cold injuries have a higher-than-normal risk of subsequent cold injury, not necessarily involving the part previously injured.

(5) Psychological factor. Mental fatigue and fear reduces the body's ability to rewarm itself and thus increases the incidence of cold injury. The feelings of isolation imposed by the environment are also stressful. Depressed and/or unresponsive soldiers are also vulnerable because they are less active. These soldiers tend to be careless about precautionary measures, especially warming activities, when cold injury is a threat.

b. Signs/Symptoms. Once a soldier becomes familiar with the factors that contribute to cold injury, he must learn to recognize cold injury signs/symptoms.

(1) Many soldiers suffer cold injury without realizing what is happening to them. They may be cold and generally uncomfortable. These soldiers often do not notice the injured part because it is already numb from the cold.

(2) Superficial cold injury usually can be detected by numbness, tingling, or "pins and needles" sensations. These signs/symptoms often can be relieved simply by loosening boots or other clothing and by exercising to improve circulation. In more serious cases involving deep cold injury, the soldier often is not aware that there is a problem until the affected part feels like a stump or block of wood.

(3) Outward signs of cold injury include discoloration of the skin at the site of injury. In light-skinned persons, the skin first reddens and then becomes pale or waxy white. In dark-skinned persons, grayness in the skin is usually evident. An injured foot or hand feels cold to the touch. Swelling may be an indication of deep injury. Also note that blisters may occur after rewarming the affected parts. Soldiers should work in pairs—buddy teams—to check each other for signs of discoloration and other symptoms. Leaders should also be alert for signs of cold injuries.

c. Treatment Considerations. First aid for cold injuries depends on whether they are superficial or deep. Cases of superficial cold injury can be adequately treated by warming the affected part using body heat. For example, this can be done by covering cheeks with hands, putting fingertips under armpits, or placing feet under the clothing of a buddy next to his belly. The injured part should NOT be massaged, exposed to a fire or stove, rubbed with snow, slapped, chafed, or soaked in cold water. Walking on injured feet should be avoided. Deep cold injury (frostbite) is very serious and requires more aggressive first aid to avoid or to minimize the loss of parts of the fingers, toes, hands, or feet. The sequence for treating cold injuries depends on whether the condition is life-threatening. That is, PRIORITY is given to removing the casualty from the cold. Other-than-cold injuries are treated either simultaneously while waiting for evacuation to a medical treatment facility or while en route to the facility.

✎ **NOTE**

The injured soldier should be evacuated at once to a place where the affected part can be rewarmed under medical supervision.

---

d. Conditions Caused by Cold. Conditions caused by cold are chilblain, immersion syndrome (immersion foot/trench foot), frostbite, snow blindness, dehydration, and hypothermia.

(1) Chilblain.

- Signs/Symptoms. Chilblain is caused by repeated prolonged exposure of bare skin at temperatures from 60°F, to 32°F, or 20°F for acclimated, dry, unwashed skin. The area may be acutely swollen, red, tender, and hot with itchy skin. There may be no loss of skin tissue in untreated cases but continued exposure may lead to infected, ulcerated, or bleeding lesions.

- Treatment. Within minutes, the area usually responds to locally applied body heat. Rewarm the affected part by applying firm steady pressure with your hands, or placing the affected part under your arms or against the stomach of a buddy. DO NOT rub or massage affected areas. Medical personnel should evaluate the injury, because signs and symptoms of tissue damage may be slow to appear.

- Prevention. Prevention of chilblain depends on basic cold injury prevention methods. Caring for and wearing the uniform properly and staying dry (as far as conditions permit) are of immediate importance.

(2) Immersion syndrome (immersion foot/trench foot). Immersion foot and trench foot are injuries that result from fairly long exposure of the feet to wet conditions at temperatures from approximately 50° to 32°F. Inactive feet in damp or wet socks and boots, or tightly laced boots which impair circulation are even more susceptible to injury. This injury can be very serious; it can lead to loss of toes or parts of the feet. If exposure of the feet has been prolonged and severe, the feet may swell so much that pressure closes the blood vessels and cuts off circulation. Should an immersion injury occur, dry the feet thoroughly; and evacuate the casualty to a medical treatment facility by the fastest means possible.

- Signs/Symptoms. At first, the parts of the affected foot are cold and painless, the pulse is weak, and numbness may be present. Second, the parts may feel hot, and burning and shooting pains may begin. In later stages, the skin is pale with a bluish cast and the pulse decreases. Other signs/symptoms that may follow are blistering, swelling, redness, heat, hemorrhages (bleeding), and gangrene.

- Treatment. Treatment is required for all stages of immersion syndrome injury. Rewarm the injured part gradually by exposing it to warm air. DO NOT massage it. DO NOT moisten the skin and DO NOT apply heat or ice. Protect it from trauma and secondary infections. Dry, loose clothing or several layers of warm coverings are preferable to extreme heat. Under no circumstances should the injured part be exposed to an open fire. Elevate the injured part to relieve the swelling. Evacuate the casualty to a medical treatment facility as soon as possible. When the part is rewarmed, the casualty often feels a burning sensation and pain. Symptoms may persist for days or weeks even after rewarming.

- Prevention. Immersion syndrome can be prevented by good hygienic care of the feet and avoiding moist conditions for prolonged periods. Changing socks at least daily (depending on environmental conditions) is also a preventive measure. Wet socks can be air dried, then can be placed inside the shirt to warm them prior to putting them on.

(3) Frostbite. Frostbite is the injury of tissue caused from exposure to cold, usually below 32°F depending on the wind chill factor, duration of exposure, and adequacy of protection. Individuals with a history of cold injury are likely to be more easily affected for an indefinite period. The body parts most easily frostbitten are the cheeks, nose, ears, chin, forehead, wrists, hands, and feet. Proper treatment and management depend upon accurate diagnosis. Frostbite may involve only the skin (superficial), or it may extend to a depth below the skin (deep). Deep frostbite is very serious and requires more aggressive first aid to avoid or to minimize the loss of parts of the fingers, toes, hands, or feet.

WARNING

Casualty should be continually monitored for development of conditions which may require the performance of necessary basic lifesaving measures, such as clearing the airway, performing mouth-to-mouth resuscitation, preventing shock, and/or bleeding control.

- Progressive signs/symptoms (081-831-1009).
  - Loss of sensation, or numb feeling in any part of the body.
  - Sudden blanching (whitening) of the skin of the affected part, followed by a momentary "tingling" sensation.
  - Redness of skin in light-skinned soldiers; grayish coloring in dark-skinned individuals.
  - Blister.
  - Swelling or tender areas.
  - Loss of previous sensation of pain in affected area.
  - Pale, yellowish, waxy-looking skin.
  - Frozen tissue that feels solid (or wooden) to the touch.

⚠ **CAUTION**

Deep frostbite is a very serious injury and requires immediate first aid and subsequent medical treatment to avoid or minimize loss of body parts.

- Treatment (081-831-1009).
  - Face, ears, and nose. Cover the casualty's affected area with his and/or your bare hands until sensation and color return.
  - Hands. Open the casualty's field jacket and shirt. (In a chemical environment never remove the clothing.) Place the affected hands under the casualty's armpits. Close the field jacket and shirt to prevent additional exposure.
  - Feet. Remove the casualty's boots and socks if he does not need to walk any further to receive additional treatment. (Thawing the casualty's feet and forcing him to walk on them will cause additional pain/injury.) Place the affected feet under clothing and against the body of another soldier.

WARNING (081-831-1009)

DO NOT attempt to thaw the casualty's feet or other seriously frozen areas if he will be required to walk or travel to receive further treatment. The casualty should avoid walking, if possible, because there is less danger in walking while the feet are frozen than after they have been thawed. Thawing in the field increases the possibilities of infection, gangrene, or other injury.

✍ **NOTE**

Thawing may occur spontaneously during transportation to the medical facility; this cannot be avoided since the body in general must be kept warm.

In all of the above areas, ensure that the casualty is kept warm and that he is covered (to avoid further injury). Seek medical treatment as soon as possible. Reassure the casualty, protect the affected area from further injury by covering it lightly with a blanket or any dry clothing, and seek shelter out of the wind. Remove/minimize constricting clothing and increase insulation. Ensure that the casualty exercises as much as possible, avoiding trauma to the injured part, and is prepared for pain when thawing occurs. Protect the frostbitten part from additional injury. DO NOT rub the injured part with snow or apply cold water soaks. DO NOT warm the part by massage or exposure to open fire because the frozen part may be burned due to the lack of feeling. DO NOT use ointments or other medications. DO NOT manipulate the part in any way to increase circulation. DO NOT allow the casualty to use alcohol or tobacco because this reduces the body's resistance to cold. Remember, when freezing extends to a depth below the skin, it involves a much more serious injury. Extra care is required to reduce or avoid the chances of losing all or part of the toes or feet. This also applies to the fingers and hands.

- Prevention. Prevention of frostbite or any cold injury depends on adequate nutrition, hot meals and warm fluids. Other cold injury preventive factors are proper clothing and maintenance of general body temperature. Fatigue, dehydration, tobacco, and alcoholic beverages should be avoided.
  - Sufficient clothing must be worn for protection against cold and wind. Layers of clothing that can be removed and replaced as needed are the most effective. Every effort must be made to keep clothing and body as dry as possible. This includes avoiding any excessive perspiration by removing and replacing layers of clothing. Socks should be changed whenever the feet become moist or wet. Clothing and equipment should be properly fitted to avoid any interference with blood circulation. Improper blood circulation reduces the amount of heat that reaches the extremities. Tight fitting socks, shoes, and hand wear are especially hazardous in very cold climates. The face needs extra protection against high winds, and the ears need massaging from time to time to maintain circulation. Hands may be used to massage and warm the face. By using the buddy system, individuals can watch each other's face for signs of frostbite to detect it early and keep tissue damage to a minimum. A mask or headgear tunneled in front of the face guards against direct wind injury. Fingers and toes should be exercised to keep them warm and to detect any numbness. Wearing windproof leather gloves or mittens and avoiding kerosene, gasoline, or alcohol on the skin are also preventive measures. Cold metal should not be touched with bare skin; doing so could result in severe skin damage.
  - Adequate clothing and shelter are also necessary during periods of inactivity.

(4) Snow blindness. Snow blindness is the effect that glare from an ice field or snowfield has on the eyes. It is more likely to occur in hazy, cloudy weather than when the sun is shining. Glare from the sun will cause an individual to instinctively protect his eyes. However, in cloudy weather, he may be overconfident and expose his eyes longer than when the threat is more obvious. He may also neglect precautions such as the use of protective eyewear. Waiting until discomfort (pain) is felt before using protective eyewear is dangerous because a deep burn of the eyes may already have occurred.

- Signs/Symptoms. Symptoms of snow blindness are a sensation of grit in the eyes with pain in and over the eyes, made worse by eyeball movement. Other signs/symptoms are watering, redness, headache, and increased pain on exposure to light. The same condition that causes snow blindness can cause snow burn of skin, lips, and eyelids. If a snow burn is neglected, the result is the same as a sunburn.
- Treatment. First aid measures consist of blindfolding or covering the eyes with a dark cloth which stops painful eye movement. Complete rest is desirable. If further exposure to light is not preventable, the eyes should be protected with dark bandages or the darkest glasses available. Once unprotected exposure to sunlight stops, the condition usually heals in a few days without permanent damage. The casualty should be evacuated to the nearest medical facility.
- Prevention. Putting on protective eye wear is essential not only to prevent injury, but to prevent further injury if any has occurred. When protective eye wear is not available, an emergency pair can be made from a piece of wood or cardboard cut and shaped to the width of the face. Cut slits for the eyes and attach strings to hold the improvised glasses in place.

Slits are made at the point of vision to allow just enough space to see and reduce the risk of injury. Blackening the eyelids and face around the eyes absorbs some of the harmful rays.

(5) Dehydration. Dehydration occurs when the body loses too much fluid, salt, and minerals. A certain amount of body fluid is lost through normal body processes. A normal daily intake of food and liquids replaces these losses. When individuals are engaged in any strenuous exercises or activities, an excessive amount of fluid and salt is lost through sweat. This excessive loss creates an imbalance of fluids, and dehydration occurs when fluid and salt are not replaced. It is very important to know that it can be prevented if troops are instructed in its causes, symptoms, and preventive measures. The danger of dehydration is as prevalent in cold regions as it is in hot regions. In hot weather the individual is aware of his body losing fluids and salt. He can see, taste, and feel the sweat as it runs down his face, gets into his eyes, and on his lips and tongue, and drips from his body. In cold weather, however, it is extremely difficult to realize that this condition exists. The danger of dehydration in cold weather operations is a serious problem. In cold climates, sweat evaporates so rapidly or is absorbed so thoroughly by layers of heavy clothing that it is rarely visible on the skin. Dehydration also occurs during cold weather operations because drinking is inconvenient. Dehydration will weaken or incapacitate a casualty for a few hours, or sometimes, several days. Because rest is an important part of the recovery process, casualties must take care that limited movement during their recuperative period does not enhance the risk of becoming a cold weather casualty.

- Signs/Symptoms. The symptoms of cold weather dehydration are similar to those encountered in heat exhaustion. The mouth, tongue, and throat become parched and dry, and swallowing becomes difficult. The casualty may have nausea with or without vomiting along with extreme dizziness and fainting. The casualty may also feel generally tired and weak and may experience muscle cramps (especially in the legs). Focusing eyes may also become difficult.

- Treatment. The casualty should be kept warm and his clothes should be loosened to allow proper circulation. Shelter from wind and cold will aid in this treatment. Fluid replacement, rest, and prompt medical treatment are critical. Medical personnel will determine the need for salt replacement.

- Prevention. These general preventive measures apply for both hot and cold weather. Sufficient additional liquids should be consumed to offset excessive body losses of these elements. The amount should vary according to the individual and the type of work he is doing (light, heavy, or very strenuous). Rest is equally important as a preventive measure. Each individual must realize that any work that must be done while bundled in several layers of clothing is extremely exhausting. This is especially true of any movement by foot, regardless of the distance.

(6) Hypothermia (general cooling). In intense cold a soldier may become both mentally and physically numb, thus neglecting essential tasks or requiring more time and effort to achieve them. Under some conditions (particularly cold water immersion), even a soldier in excellent physical condition may die in a matter of minutes. The destructive influence of cold on the body is called hypothermia. This means bodies lose heat faster than they can produce it. Frostbite may occur without hypothermia when extremities do not receive sufficient heat from central body stores. The reason for this is inadequate circulation and/or inadequate insulation. Nonetheless, hypothermia and frostbite may occur at the same time with exposure to below-freezing temperatures. An example of this is an avalanche accident. Hypothermia may occur from exposure to temperatures above freezing, especially from immersion in cold water, wet-cold conditions, or from the effect of wind. Physical exhaustion and insufficient food intake may also increase the risk of hypothermia. Excessive use of alcohol leading to unconsciousness in a cold environment can also result in hypothermia. General cooling of the entire body to a temperature below 95°F is caused by continued exposure to low or rapidly dropping temperatures, cold moisture, snow, or ice. Fatigue, poor physical condition, dehydration, faulty blood circulation, alcohol or other

drug intoxication, trauma, and immersion can cause hypothermia. Remember, cold affects the body systems slowly and almost without notice. Soldiers exposed to low temperatures for extended periods may suffer ill effects even if they are well protected by clothing.

- Signs/Symptoms. As the body cools, there are several stages of progressive discomfort and impairment. A sign/symptom that is noticed immediately is shivering. Shivering is an attempt by the body to generate heat. The pulse is faint or very difficult to detect. People with temperatures around 90°F may be drowsy and mentally slow. Their ability to move may be hampered, stiff, and uncoordinated, but they may be able to function minimally. Their speech may be slurred. As the body temperature drops further, shock becomes evident as the person's eyes assume a glassy state, breathing becomes slow and shallow, and the pulse becomes weaker or absent. The person becomes very stiff and uncoordinated. Unconsciousness may follow quickly. As the body temperature drops even lower, the extremities freeze, and a deep (or core) body temperature (below 85°F) increases the risk of irregular heart action. This irregular heart action or heart standstill can result in sudden death.

- Treatment. Except in cases of the most severe hypothermia (marked by coma or unconsciousness, a weak pulse, and a body temperature of approximately 90°F or below), the treatment for hypothermia is directed towards rewarming the body evenly and without delay. Provide heat by using a hot water bottle, electric blanket, campfire, or another soldier's body heat. Always call or send for help as soon as possible and protect the casualty immediately with dry clothing or a sleeping bag. Then, move him to a warm place. Evaluate other injuries and treat them. Treatment can be given while the casualty is waiting evacuation or while he is en route. In the case of an accidental breakthrough into ice water, or other hypothermic accident, strip the casualty of wet clothing immediately and bundle him into a sleeping bag. Mouth-to-mouth resuscitation should be started at once if the casualty's breathing has stopped or is irregular or shallow. Warm liquids may be given gradually but must not be forced on an unconscious or semiconscious person because he may choke. The casualty should be transported on a litter because the exertion of walking may aggravate circulation problems. A physician should immediately treat any hypothermia casualty. Hypothermia is life-threatening until normal body temperature has been restored. The treatment of a casualty with severe hypothermia is based upon the following principles: stabilize the temperature, attempt to avoid further heat loss, handle the casualty gently, and evacuate as soon as possible to the nearest medical treatment facility! Rewarming a severely hypothermic casualty is extremely dangerous in the field due to the great possibility of such complications as rewarming shock and disturbances in the rhythm of the heartbeat.

⚠ *CAUTION

Hypothermia is a MEDICAL EMERGENCY! Prompt medical treatment is necessary. Casualties with hypothermic complications should be transported to a medical treatment facility immediately.

⚠ CAUTION

The casualty is unable to generate his own body heat. Therefore, merely placing him in a blanket or sleeping bag is not sufficient.

- Prevention. Prevention of hypothermia consists of all actions that will avoid rapid and uncontrollable loss of body heat. Individuals should be properly equipped and properly dressed (as appropriate for conditions and exposure). Proper diet, sufficient rest, and general principles apply. Ice thickness must be tested before river or lake crossings. Anyone departing a fixed base by aircraft, ground vehicle, or foot must carry sufficient protective clothing and food reserves to survive during unexpected weather changes or other unforeseen emergencies. Traveling alone is never safe. Expected itinerary and arrival time should be left with responsible parties before any departure in severe weather. Anyone living in cold regions should learn how to build expedient shelters from available materials including snow.

e. Table. See Table 5-2 for further information.

Table 5-2: Cold and Wet Injuries (081-831-1009).

INJURIES	SIGNS/SYMPTOMS	FIRST AID
Chilblain	Red, swollen, hot, tender, itching skin. Continued exposure may lead to infected (ulcerated or bleeding) skin lesions.	1. Area usually responds to locally applied rewarming (body heat). 2. DO NOT rub or massage area. 3. Seek medical treatment.
Immersion foot/ Trench foot	Affected parts are cold, numb, and painless. Parts may then be hot, with burning and shooting pains. Advanced stage: skin pale with bluish cast; pulse decreases; blistering, swelling, heat, hemorrhages, and gangrene may follow.	1. Gradual rewarming by exposure to warm air. 2. DO NOT massage or moisten skin. 3. Protect affected parts from trauma. 4. Dry feet thoroughly, avoid walking. 5. Seek medical treatment.
Frostbite	Loss of sensation, or numb feeling in any part of the body. Sudden blanching (whitening) of the skin of the affected part, followed by a momentary "tingling" sensation. Redness of skin in light-skinned soldiers; grayish coloring in dark-skinned individuals. Blisters. Swelling or tender areas. Loss of previous sensation of pain in affected area. Pale,	1. Warm the area at the first sign of frostbite, using firm, steady pressure of hand, underarm or abdomen. 2. Face, ears, nose—cover area with hands (casualty's own or buddy's). 3. Hand(s)—open field jacket and place casualty's hand(s) against body, then close jacket to prevent heat loss. 4. Feet—casualty's boots/socks removed and exposed feet placed under clothing and against body of another soldier. 5. *Warning:* Do not attempt to thaw the casualty's feet or other seriously frozen areas if he will be required

(continued)

Table 5-2: *(Continued)*

Frostbite *Continued.*	yellowish, waxy-looking skin. Frozen tissue that feels solid (or wooden) to the touch.	to walk or travel to a medical center in order to receive additional treatment. The possibility of injury from walking is less when the feet are frozen than after they have been thawed. (However, if possible, avoid walking.) Thawing in the field increases the possibility of infection, gangrene, or injury. 6. Loosen or remove constricting clothing and remove any jewelry. 7. Increase insulation (cover with blanket or other dry material). Ensure casualty exercises as much as possible, avoiding trauma to injured part.
Snow Blindness	Eyes may feel scratchy. Watering, redness, headache, and increased pain with exposure to light can occur.	1. Cover the eyes with a dark cloth. 2. Seek medical treatment.
Dehydration	Similar to heat exhaustion. See Table 5-1.	1. Keep warm, loosen clothes. 2. Casualty needs fluid replacement, rest, and prompt medical treatment.
Hypothermia	Casualty is cold. Shivering stops. Core temperature is low. Consciousness may be altered. Uncoordinated movements may occur. Shock and coma may result as	*Mild Hypothermia* 1. Rewarm body evenly and without delay. (Need to provide heat source; casualty's body unable to generate heat). 2. Keep dry, protect from elements.

(continued)

Table 5-2: *(Continued)*

INJURIES	SIGNS/SYMPTOMS	FIRST AID
Hypothermia *Continued.*	body temperature drops.	3. Warm liquids may be given gradually (to conscious casualties only). ★ 4. Seek medical treatment immediately! *Severe Hypothermia* 1. Stabilize the temperature. 2. Attempt to avoid further heat loss. 3. Handle the casualty gently. 4. Evacuate to the nearest medical treatment facility as soon as possible.

★**CAUTION:** Hypothermia is a *MEDICAL EMERGENCY!* Prompt medical treatment is necessary.

CHAPTER 6

First Aid for Bites and Stings

Snakebites, insect bites, or stings can cause intense pain and/or swelling. If not treated promptly and correctly, they can cause serious illness or death. The severity of a snakebite depends upon: whether the snake is poisonous or nonpoisonous, the type of snake, the location of the bite, and the amount of venom injected. Bites from humans and other animals, such as dogs, cats, bats, raccoons, and rats can cause severe bruises and infection, and tears or lacerations of tissue. Awareness of the potential sources of injuries can reduce or prevent them from occurring. Knowledge and prompt application of first aid measures can lessen the severity of injuries from bites and stings and keep the soldier from becoming a serious casualty.

6-1. Types of Snakes

a. Nonpoisonous Snakes. There are approximately 130 different varieties of nonpoisonous snakes in the United States. They have oval-shaped heads and round eyes. Unlike poisonous snakes, discussed below, nonpoisonous snakes do not have fangs with which to inject venom. See Figure 6-1 for characteristics of a nonpoisonous snake.

b. Poisonous Snakes. Poisonous snakes are found throughout the world, primarily in tropical to moderate climates. Within the United States, there are four kinds: rattlesnakes, copperheads, water moccasins (cottonmouth), and coral snakes. Poisonous snakes in other parts of the world include sea snakes, the fer-de-lance, the bushmaster, and the tropical rattlesnake in tropical Central America; the Malayan pit viper in the tropical Far East; the cobra in Africa and Asia; the mamba (or black mamba) in Central and Southern Africa; and the krait in India and Southeast Asia. See Figure 6-2 for characteristics of a poisonous pit viper.

Figure 6-1: Characteristics of nonpoisonous snake.

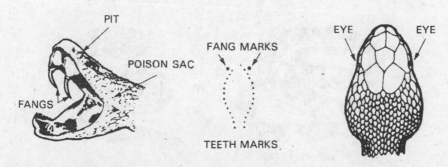

Figure 6-2: Characteristics of poisonous pit viper.

c. Pit Vipers (Poisonous). See Figure 6-3 for illustrations.

(1) Rattlesnakes, bushmasters, copperheads, fer-de-lance, Malayan pit vipers, and water moccasins (cottonmouth) are called pit vipers because of the small, deep pits between the nostrils and eyes on each side of the head (Figure 6-2). In addition to their long, hollow fangs, these snakes have other identifying features: thick bodies, slit-like pupils of the eyes, and flat, almost triangular-shaped heads. Color markings and other identifying characteristics, such as rattles or a noticeable white interior of the mouth (cottonmouth), also help distinguish these poisonous snakes. Further identification is provided by examining the bite pattern of the wound for signs of fang entry. Occasionally there will be only one fang mark, as in the case of a bite on a finger or toe where there is no room for both fangs, or when the snake has broken off a fang.

(2) The casualty's condition provides the best information about the seriousness of the situation, or how much time has passed since the bite occurred. Pit viper bites are characterized by severe burning pain. Discoloration and swelling around the fang marks usually begins within 5 to 10 minutes after the bite. If only minimal swelling occurs within 30 minutes, the bite will almost certainly have been from a nonpoisonous snake or possibly from a poisonous snake which did not inject venom. The venom destroys blood cells, causing a general discoloration of the skin. This reaction is followed by blisters and numbness in the affected area. Other signs which can occur are weakness, rapid pulse, nausea, shortness of breath, vomiting, and shock.

d. Corals, Cobras, Kraits, and Mambas. Corals, cobra, kraits, and mambas all belong to the same group even though they are found in different parts of the world. All four inject their venom through short, grooved fangs, leaving a characteristic bite pattern. See Figure 6-4 for illustration of a cobra snake.

(1) The small coral snake, found in the Southeastern United States, is brightly colored with bands of red, yellow (or almost white), and black completely encircling the body (Figure 6-5). Other nonpoisonous snakes have the same coloring, but on the coral snake found in the United States, the red ring always touches the yellow ring. To know the difference between a harmless snake and the coral snake found in the United States, remember the following:

"Red on yellow will kill a fellow. Red on black, venom will lack."

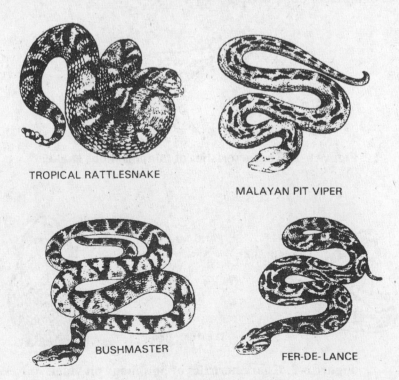

TROPICAL RATTLESNAKE

MALAYAN PIT VIPER

BUSHMASTER

FER-DE-LANCE

Figure 6-3: Poisonous snakes.

Figure 6-4: Cobra snake.

Figure 6-5: Coral snake.

(2) The venom of corals, cobras, kraits, and mambas produces symptoms different from those of pit vipers. Because there is only minimal pain and swelling, many people believe that the bite is not serious. Delayed reactions in the nervous system normally occur between 1 to 7 hours after the bite. Symptoms include blurred vision, drooping eyelids, slurred speech, drowsiness, and increased salivation and sweating. Nausea, vomiting, shock, respiratory difficulty, paralysis, convulsions, and coma will usually develop if the bite is not treated promptly.

e. Sea Snakes. Sea snakes (Figure 6-6) are found in the warm water areas of the Pacific and Indian oceans, along the coasts, and at the mouths of some larger rivers. Their venom is VERY poisonous, but their fangs are only 1/4 inch long. The first aid outlined for land snakes also applies to sea snakes.

6-2. Snakebites. If a soldier should accidentally step on or otherwise disturb a snake, it will attempt to strike. Chances of this happening while traveling along trails or waterways are remote if a soldier is alert and careful. Poisonous snakes DO NOT always inject venom when they bite or strike a person. However, all snakes may carry tetanus (lockjaw); anyone bitten by a snake, whether poisonous or nonpoisonous, should immediately seek medical attention. Poison is injected from the venom sacs through grooved or hollow fangs. Depending on the species, these fangs are either long or short. Pit vipers have long hollow fangs. These fangs are folded against the roof of the mouth and extend when the snake strikes. This allows them to strike quickly and then withdraw. Cobras, coral snakes, kraits, mambas, and sea snakes have short, grooved fangs. These snakes are less effective in their attempts to bite, since they must chew after

Figure 6-6: Sea snake.

striking to inject enough venom (poison) to be effective. See Figure 6-7 for characteristics of a poisonous snakebite. In the event you are bitten, attempt to identify and/or kill the snake. Take it to medical personnel for inspection/identification. This provides valuable information to medical personnel who deal with snakebites. TREAT ALL SNAKEBITES AS POISONOUS.

a. Venoms. The venoms of different snakes cause different effects. Pit viper venoms (hemotoxins) destroy tissue and blood cells. Cobras, adders, and coral snakes inject powerful venoms (neurotoxins) which affect the central nervous system, causing respiratory paralysis. Water moccasins and sea snakes have venom that is both hemotoxic and neurotoxic.

b. Identification. The identification of poisonous snakes is very important since medical treatment will be different for each type of venom. Unless it can be positively identified the snake should be killed and saved. When this is not possible or when doing so is a serious threat to others, identification may sometimes be difficult since many venomous snakes resemble harmless varieties. When dealing with snakebite problems in foreign countries, seek advice, professional or otherwise, which may help identify species in the particular area of operations.

*c. First Aid. Get the casualty to a medical treatment facility as soon as possible and with minimum movement. Until evacuation or treatment is possible, have the casualty lie quietly and not move anymore than necessary. The casualty should not smoke, eat, nor drink any fluids. If the casualty

Figure 6-7: Characteristics of poisonous snake bite.

has been bitten on an extremity, DO NOT elevate the limb; keep the extremity level with the body. Keep the casualty comfortable and reassure him. If the casualty is alone when bitten, he should go to the medical facility himself rather than wait for someone to find him. Unless the snake has been positively identified, attempt to kill it and send it with the casualty. Be sure that retrieving the snake does not endanger anyone or delay transporting the casualty.

* (1) If the bite is on an arm or leg, place a constricting band (narrow cravat [swathe], or narrow gauze bandage) one to two finger widths above and below the bite (Figure 6-8). However, if only one constricting band is available, place that band on the extremity between the bite site and the casualty's heart. If the bite is on the hand or foot, place a single band above the wrist or ankle. The band should be tight enough to stop the flow of blood near the skin, but not tight enough to interfere with circulation. In other words, it should not have a tourniquet-like affect. If no swelling is seen, place the bands about one inch from either side of the bite. If swelling is present, put the bands on the unswollen part at the edge of the swelling. If the swelling extends beyond the band, move the band to the new edge of the swelling. (If possible, leave the old band on, place a new one at the new edge of the swelling, and then remove and save the old one in case the process has to be repeated.) If possible, place an ice bag over the area of the bite. DO NOT wrap the limb in ice or put ice directly on the skin. Cool the bite area—do not freeze it. DO NOT stop to look for ice if it will delay evacuation and medical treatment.

⚠ CAUTION

DO NOT attempt to cut open the bite nor suck out the venom. If the venom should seep through any damaged or lacerated tissues in your mouth, you could immediately lose consciousness or even die.

---

(2) If the bite is located on an arm or leg, immobilize it at a level below the heart. DO NOT elevate an arm or leg even with or above the level of the heart.

⚠ CAUTION

When a splint is used to immobilize the arm or leg, take EXTREME care to ensure the splinting is done properly and does not bind. Watch it closely and adjust it if any changes in swelling occur.

---

(3) When possible, clean the area of the bite with soap and water. DO NOT use ointments of any kind.
(4) NEVER give the casualty food, alcohol, stimulants (coffee or tea), drugs, or tobacco.
(5) Remove rings, watches, or other jewelry from the affected limb.

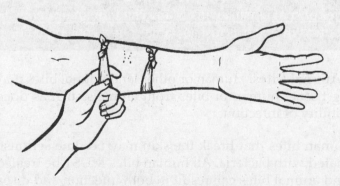

Figure 6-8: Constricting band.

✍ **NOTE**

It may be possible, in some cases, for an aidman who is specially trained and is authorized to carry and use antivenin to administer it. The use of antivenin presents special risks, and only those with specialized training should attempt to use it!

---

d. Prevention. Except for a few species, snakes tend to be shy or passive. Unless they are injured, trapped, or disturbed, snakes usually avoid contact with humans. The harmless species are often more prone to attack. All species of snakes are usually aggressive during their breeding season.

(1) Land snakes. Many snakes are active during the period from twilight to daylight. Avoid walking as much as possible during this time.

- Keep your hands off rock ledges where snakes are likely to be sunning.
- Look around carefully before sitting down, particularly if in deep grass among rocks.
- Attempt to camp on clean, level ground. Avoid camping near piles of brush, rocks, or other debris.
- Sleep on camping cots or anything that will keep you off the ground. Avoid sleeping on the ground if at all possible.
- Check the other side of a large rock before stepping over it. When looking under any rock, pull it toward you as you turn it over so that it will shield you in case a snake is beneath it.
- Try to walk only in open areas. Avoid walking close to rock walls or similar areas where snakes may be hiding.
- Determine when possible what species of snakes are likely to be found in an area which you are about to enter.
- Hike with another person. Avoid hiking alone in a snake-infested area. If bitten, it is important to have at least one companion to perform lifesaving first aid measures and to kill the snake. Providing the snake to medical personnel will facilitate both identification and treatment.
- Handle freshly killed venomous snakes only with a long tool or stick. Snakes can inflict fatal bites by reflex action even after death.
- Wear heavy boots and clothing for some protection from snakebite. Keep this in mind when exposed to hazardous conditions.
- Eliminate conditions under which snakes thrive: brush, piles of trash, rocks, or logs and dense undergrowth. Controlling their food (rodents, small animals) as much as possible is also good prevention.

(2) Sea snakes. Sea snakes may be seen in large numbers but are not known to bite unless handled. Be aware of the areas where they are most likely to appear and be especially alert when swimming in these areas. Avoid swimming alone whenever possible.

WARNING

All species of snakes can swim. Many can remain under water for long periods. A bite sustained in water is just as dangerous as one on land.

6-3. Human and Other Animal Bites. Human or other land animal bites may cause lacerations or bruises. In addition to damaging tissue, human or bites from animals such as dogs, cats, bats, raccoons, or rats always present the possibility of infection.

a. Human Bites. Human bites that break the skin may become seriously infected since the mouth is heavily contaminated with bacteria. All human bites MUST be treated by medical personnel.

b. Animal Bites. Land animal bites can result in both infection and disease. Tetanus, rabies, and various types of fevers can follow an untreated animal bite. Because of these possible complications, the animal causing the bite should, if possible, be captured or killed (without damaging its head) so that competent authorities can identify and test the animal to determine if it is carrying diseases.

c. First Aid.

(1) Cleanse the wound thoroughly with soap or detergent solution.

(2) Flush it well with water.

(3) Cover it with a sterile dressing.

(4) Immobilize an injured arm or leg.

(5) Transport the casualty immediately to a medical treatment facility.

✍ **NOTE**

If unable to capture or kill the animal, provide medical personnel with any information possible that will help identify it. Information of this type will aid in appropriate treatment.

6-4. Marine (Sea) Animals. With the exception of sharks and barracuda, most marine animals will not deliberately attack. The most frequent injuries from marine animals are wounds by biting, stinging, or puncturing. Wounds inflicted by marine animals can be very painful, but are rarely fatal.

a. Sharks, Barracuda, and Alligators. Wounds from these marine animals can involve major trauma as a result of bites and lacerations. Bites from large marine animals are potentially the most life threatening of all injuries from marine animals. Major wounds from these animals can be treated by controlling the bleeding, preventing shock, giving basic life support, splinting the injury, and by securing prompt medical aid.

b. Turtles, Moray Eels, and Corals. These animals normally inflict minor wounds. Treat by cleansing the wound(s) thoroughly and by splinting if necessary.

c. Jellyfish, Portuguese men-of-war, Anemones, and Others. This group of marine animals inflict injury by means of stinging cells in their tentacles. Contact with the tentacles produces burning pain with a rash and small hemorrhages on the skin. Shock, muscular cramping, nausea, vomiting, and respiratory distress may also occur. Gently remove the clinging tentacles with a towel and wash or treat the area. Use diluted ammonia or alcohol, meat tenderizer, and talcum powder. If symptoms become severe or persist, seek medical aid.

d. Spiny Fish, Urchins, Stingrays, and Cone Shells. These animals inject their venom by puncturing with their spines. General signs and symptoms include swelling, nausea, vomiting, generalized cramps, diarrhea, muscular paralysis, and shock. Deaths are rare. Treatment consists of soaking the wounds in hot water (when available) for 30 to 60 minutes. This inactivates the heat sensitive toxin. In addition, further first aid measures (controlling bleeding, applying a dressing, and so forth) should be carried out as necessary.

⚠ **CAUTION**

Be careful not to scald the casualty with water that is too hot because the pain of the wound will mask the normal reaction to heat.

6-5. Insect Bites/Stings. An insect bite or sting can cause great pain, allergic reaction, inflammation, and infection. If not treated correctly, some bites/stings may cause serious illness or even death. When an allergic reaction is not involved, first aid is a simple process. In any case, medical personnel should examine the casualty at the earliest possible time. It is important to properly identify the spider, bee, or creature that caused the bite/sting, especially in cases of allergic reaction when death is a possibility.

a. Types of Insects. The insects found throughout the world that can produce a bite or sting are too numerous to mention in detail. Commonly encountered stinging or biting insects include brown recluse spiders (Figure 6-9), black widow spiders (Figure 6-10), tarantulas (Figure 6-11), scorpions (Figure 6-12), urticating caterpillars, bees, wasps, centipedes, conenose beetles (kissing bugs), ants, and wheel bugs. Upon being reassigned, especially to overseas areas, take the time to become acquainted with the types of insects to avoid.

b. Signs/Symptoms. Discussed in paragraphs (1) and (2) below are the most common effects of insect bites/stings. They can occur alone or in combination with the others.

(1) Less serious. Commonly seen signs/symptoms are pain, irritation, swelling, heat, redness, and itching. Hives or wheals (raised areas of the skin that itch) may occur. These are the least

Figure 6-9: Brown recluse spider.

Figure 6-10: Black widow spider.

Figure 6-11: Tarantula.

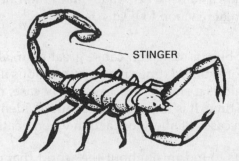

Figure 6-12: Scorpion.

severe of the allergic reactions that commonly occur from insect bites/stings. They are usually dangerous only if they affect the air passages (mouth, throat, nose, and so forth), which could interfere with breathing. The bites/stings of bees, wasps, ants, mosquitoes, fleas, and ticks are usually not serious and normally produce mild and localized symptoms. A tarantula's bite is usually no worse than that of a bee sting. Scorpions are rare and their stings (except for a specific species found only in the Southwest desert) are painful but usually not dangerous.

(2) Serious. Emergency allergic or hypersensitive reactions sometimes result from the stings of bees, wasps, and ants. Many people are allergic to the venom of these particular insects. Bites or stings from these insects may produce more serious reactions, to include generalized itching and hives, weakness, anxiety, headache, breathing difficulties, nausea, vomiting, and diarrhea. Very serious allergic reactions (called anaphylactic shock) can lead to complete collapse, shock, and even death. Spider bites (particularly from the black widow and brown recluse spiders) can be serious also. Venom from the black widow spider affects the nervous system. This venom can cause muscle cramps, a rigid, nontender abdomen, breathing difficulties, sweating, nausea and vomiting. The brown recluse spider generally produces local rather than system-wide problems; however, local tissue damage around the bite can be severe and can lead to an ulcer and even gangrene.

c. First Aid. There are certain principles that apply regardless of what caused the bite/sting. Some of these are:

- If there is a stinger present, for example, from a bee, remove the stinger by scraping the skin's surface with a fingernail or knife. DO NOT squeeze the sac attached to the stinger because it may inject more venom.
- Wash the area of the bite/sting with soap and water (alcohol or an antiseptic may also be used) to help reduce the chances of an infection and remove traces of venom.
- Remove jewelry from bitten extremities because swelling is common and may occur.
- In most cases of insect bites the reaction will be mild and localized. Use ice or cold compresses (if available) on the site of the bite/sting. This will help reduce swelling, ease the pain, and slow the absorption of venom. Meat tenderizer (to neutralize the venom) or calamine lotion (to reduce itching) may be applied locally. If necessary, seek medical aid.
- In more serious reactions (severe and rapid swelling, allergic symptoms, and so forth) treat the bite/sting like you would treat a snakebite; that is, apply constricting bands above and below the site. See paragraph 6-2c(1) above for details and illustration (Figure 6-8) of a constricting band.
- * Be prepared to perform basic lifesaving measures, such as rescue breathing. Reassure the casualty and keep him calm.
- In serious reactions, attempt to capture the insect for positive identification; however, be careful not to become a casualty yourself.
- If the reaction or symptoms appear serious, seek medical aid immediately.

> ⚠ *CAUTION
> Insect bites/stings may cause anaphylactic shock (a shock caused by a severe allergic reaction). This is a life-threatening event and a MEDICAL EMERGENCY! Be prepared to immediately transport the casualty to a medical facility.

✎ **NOTE**
Be aware that some allergic or hypersensitive individuals may carry identification (such as a MEDIC ALERT tag) or emergency insect bite treatment kits. If the casualty is having an allergic reaction and has such a kit, administer the medication in the kit according to the instructions which accompany the kit.

d. Prevention. Some prevention principles are:
- Apply insect repellent to all exposed skin, such as the ankles to prevent insects from creeping between uniform and boots. Also apply the insect repellent to the shoulder blades where the shirt fits tight enough that mosquitoes bite through. DO NOT apply insect repellent to the eyes.
- Reapply repellent, every 2 hours during strenuous activity and soon after stream crossings.
- Blouse the uniform inside the boots to further reduce risk.
- Wash yourself daily if the tactical situation permits. Pay particular attention to the groin and armpits.
- Use the buddy system. Check each other for insect bites.
- Wash your uniform at least weekly.

6-6. Table. See Table 6-1 for information on bites and stings.

Table 6-1: Bites and Stings.

TYPES	FIRST AID

Snakebite	1. Move the casualty away from the snake.
	2. Remove all rings and bracelets from the affected extremity.
	3. Reassure the casualty and keep him quiet.
	4. Place ice or freeze pack, if available, over the area of the bite.
	5. Apply constricting band(s) 1-2 finger widths from the bite. One should be able to insert a finger between the band and the skin.
	  • Arm or leg bite—place one band above and one band below the bite site.
	  • Hand or foot bite—place one band above the wrist or ankle.
	6. Immobilize the affected part in a position below the level of the heart.
	7. Kill the snake (if possible, without damaging its head or endangering yourself) and send it with the casualty.
	8. Seek medical aid immediately.

Brown Recluse Spider or Black Widow Spider Bite	1. Keep the casualty quiet. 2. Wash the area. 3. Apply ice or freeze pack, if available. 4. Seek medical aid.

(continued)

Table 6-1: *(Continued)*

TYPES	FIRST AID
Tarantula Bite or Scorpion Sting or Ant Bites	1. Wash the area. 2. Apply ice or freeze pack, if available. 3. Apply baking soda, calamine lotion, or meat tenderizer to bite site to relieve pain and itching. 4. If site of bite(s) or sting(s) is on the face, neck (possible airway problems), or genital area, or if local reaction seems severe, or if the sting is by the dangerous type of scorpion found in the Southwest desert, keep the casualty quiet as possible and seek immediate medical aid.
Bee Stings	1. If the stinger is present, remove by scraping with a knife or fingernail. DO NOT squeeze venom sac on stinger; more venom may be injected. 2. Wash the area. 3. Apply ice or freeze pack, if available. ★ 4. If allergic signs/symptoms appear, be prepared to seek immediate medical aid.

CHAPTER 7

First Aid in Toxic Environments

American forces have not been exposed to high levels of toxic substances on the battlefield since World War I. In future conflicts and wars we can expect the use of such agents. Chemical weapons will degrade unit effectiveness rapidly by forcing troops to wear hot protective clothing and by creating confusion and fear. Through training in protective procedures and first aid, units can maintain their effectiveness on the integrated battlefield.

SECTION I. INDIVIDUAL PROTECTION AND FIRST AID EQUIPMENT FOR TOXIC SUBSTANCES

7-1. Toxic Substances

a. Gasoline, chlorine, and pesticides are examples of common toxic substances. They may exist as solids, liquids, or gases depending upon temperature and pressure. Gasoline, for example, is a vaporizable liquid; chlorine is a gas; and Warfarin, a pesticide, is a solid. Some substances are more injurious to the body than others when they are inhaled or eaten or when they contact the skin or eyes. Whether they are solids, liquids, or gases (vapors and aerosols included), they may irritate, inflame, blister, burn, freeze, or destroy tissue such as that associated with the respiratory tract or the eyes. They may also be absorbed into the bloodstream, disturbing one or several of the body's major functions.

b. You may come in contact with toxic substances in combat or in everyday activities. Ordinarily, brief exposures to common household toxic substances, such as disinfectants and bleach solutions, do not cause injuries. Exposure to toxic chemical agents in warfare, even for a few seconds, could result in death, injury, or incapacitation. Remember that toxic substances employed by an enemy could persist for hours or days. To survive and operate effectively in a toxic environment, you must be prepared to protect yourself from the effects of chemical agents and to provide first aid to yourself and to others.

7-2. Protective and First Aid Equipment.
You are issued equipment for protection and first aid treatment in a toxic environment. You must know how to use the items described in *a* through *e*. It is equally important that you know when to use them. Use your protective clothing and equipment when you are ordered to and when you are under a nuclear, biological, or chemical (NBC) attack. Also, use your protective clothing and equipment when you enter an area where NBC agents have been employed.

a. Field Protective Mask With Protective Hood. Your field protective mask is the most important piece of protective equipment. You are given special training in its use and care.

b. Field Protective Clothing. Each soldier is authorized three sets of the following field protective clothing:
  • Overgarment ensemble (shirt and trousers), chemical protective.
  • Footwear cover (overboots), chemical protective.
  • Glove set, chemical protective.

c. Nerve Agent Pyridostigmine Pretreatment (NAPP). You will be issued a blister pack of pretreatment tablets when your commander directs. When ordered to take the pretreatment you must take one tablet every eight hours. This must be taken prior to exposure to nerve agents, since it may take several hours to develop adequate blood levels.

> ✍ **NOTE**
>
> Normally, one set of protective clothing is used in acclimatization training that uses various mission-oriented protective posture (MOPP) levels.

    d. **M258A1 Skin Decontamination Kit.** The M258A1 Skin Decontamination (decon) Kit contains three each of the following:
  * DECON-1 packets containing wipes (pads) moistened with decon solution.
  * DECON-2 packets containing dry wipes (pads) previously moistened with decon solution and sealed glass ampules. Ampules are crushed to moisten pads.

> **WARNING**
>
> The decon solution contained in both DECON-1 and DECON-2 packets is a poison and caustic hazard and can permanently damage the eyes. Keep wipes out of the eyes, mouth, and open wounds. Use WATER to wash toxic agent out of eyes and wounds and seek medical aid.

    e. **Nerve Agent Antidote Kit, Mark I (NAAK MKI).** Each soldier is authorized to carry three Nerve Agent Antidote Kits, Mark I, to treat nerve agent poisoning. When NAPP has been taken several hours (but no greater than 8 hours) prior to exposure, the NAAK MKI treatment of nerve agent poisoning is much more effective.

SECTION II. CHEMICAL-BIOLOGICAL AGENTS

7-3. Classification

    a. Chemical agents may be classified according to the primary physiological effects they produce, such as nerve, blister, blood, choking, vomiting, and incapacitating agents.

    b. Biological agents may be classified according to the effect they have on man. These include blockers, inhibitors, hybrids, and membrane active compounds. These agents are found in living organisms such as fungi, bacteria and viruses.

> **WARNING**
>
> Ingesting water or food contaminated with nerve, blister, and other chemical agents and with some biological agents can be fatal. NEVER consume water or food which is suspected of being contaminated until it has been tested and found safe for consumption.

7-4. Conditions for Masking Without Order or Alarm. Once an attack with a chemical or biological agent is detected or suspected, or information is available that such an agent is about to be used, you must STOP breathing and mask immediately. DO NOT WAIT to receive an order or alarm under the following circumstances:

* Your position is hit by artillery or mortar fire, missiles, rockets, smokes, mists, aerial sprays, bombs, or bomblets.
* Smoke from an unknown source is present or approaching.

- A suspicious odor, liquid, or solid is present.
- A toxic chemical or biological attack is present.
- You are entering an area known or suspected of being contaminated.
- During any motor march, once chemical warfare has begun.
- When casualties are being received from an area where chemical or biological agents have reportedly been used.
- You have one or more of the following symptoms:
  - An unexplained runny nose.
  - A feeling of choking or tightness in the chest or throat.
  - Dimness of vision.
  - Irritation of the eyes.
  - Difficulty in or increased rate of breathing without obvious reason.
  - Sudden feeling of depression.
  - Dread, anxiety, restlessness.
  - Dizziness or light-headedness.
  - Slurred speech.
- Unexplained laughter or unusual behavior is noted in others.
- Numerous unexplained ill personnel.
- Buddies suddenly collapsing without evident cause.
- Animals or birds exhibiting unusual behavior and/or sudden unexplained death.

7-5. First Aid for a Chemical Attack (081-831-1030 and 081-831-1031). Your field protective mask gives protection against chemical as well as biological agents. Previous practice enables you to mask in 9 seconds or less or to put on your mask with hood within 15 seconds.

a. Step ONE (081-831-1030 and 081-831-1031). Stop breathing. Don your mask, seat it properly, clear and check your mask, and resume breathing. Give the alarm, and continue the mission. Keep your mask on until the "all clear" signal has been given.

✍ NOTE

Keep your mask on until the area is no longer hazardous and you are told to unmask.

---

b. Step TWO (081-831-1030). If symptoms of nerve agent poisoning (paragraph 7-7) appear, immediately give yourself a nerve agent antidote. You should have taken NAPP several hours prior to exposure which will enhance the action of the nerve agent antidote.

⚠ CAUTION

Do not inject a nerve agent antidote until you are sure you need it.

---

c. Step THREE (081-831-1031). If your eyes and face become contaminated, you must immediately try to get under cover. You need this shelter to prevent further contamination while performing decon procedures on areas of the head. If no overhead cover is available, throw your poncho or shelter half over your head before beginning the decon process. Then you should put on the remaining protective clothing. (See Appendix F for decon procedure.) If vomiting occurs, the mask should be lifted momentarily and drained—while the eyes are closed and the breath is held—and replaced, cleared, and sealed.
d. Step FOUR. If nerve agents are used, mission permitting, watch for persons needing nerve agent antidotes and immediately follow procedures outlined in paragraph 7-8 b.
e. Step FIVE. When your mission permits, decon your clothing and equipment.

SECTION III. NERVE AGENTS

7-6. Background Information

a. Nerve agents are among the deadliest of chemical agents. They can be delivered by artillery shell, mortar shell, rocket, missile, landmine, and aircraft bomb, spray, or bomblet. Nerve agents enter the body by inhalation, by ingestion, and through the skin. Depending on the route of entry and the amount, nerve agents can produce injury or death within minutes. Nerve agents also can achieve their effects with small amounts. Nerve agents are absorbed rapidly, and the effects are felt immediately upon entry into the body. You will be issued three Nerve Agent Antidote Kits, Mark I. Each kit consists of one atropine autoinjector and one pralidoxime chloride (2 PAM Cl) autoinjector (also called injectors) (Figure 7-1).

b. When you have the signs and symptoms of nerve agent poisoning, you should immediately put on the protective mask and then inject yourself with one set of the Nerve Agent Antidote Kit, Mark I. You should inject yourself in the outside (lateral) thigh muscle or if you are thin, in the upper outer (lateral) part of the buttocks.

c. Also, you may come upon an unconscious chemical agent casualty who will be unable to care for himself and who will require your aid. You should be able to successfully—

(1) Mask him if he is unmasked.

(2) Inject him, if necessary, with all his autoinjectors.

(3) Decontaminate his skin.

(4) Seek medical aid.

7-7. Signs/Symptoms of Nerve Agent Poisoning (081-831-1030 and 081-831-1031).

The symptoms of nerve agent poisoning are grouped as MILD—those which you recognize and for which you can perform self-aid, and SEVERE—those which require buddy aid.

a. MILD Symptoms (081-831-1030).
- Unexplained runny nose.
- Unexplained sudden headache.
- Sudden drooling.
- Difficulty seeing (blurred vision).
- Tightness in the chest or difficulty in breathing.
- Localized sweating and twitching (as a result of small amount of nerve agent on skin).
- Stomach cramps.
- Nausea.

b. SEVERE Signs/Symptoms (081-831-1031).
- Strange or confused behavior.
- Wheezing, difficulty in breathing, and coughing.

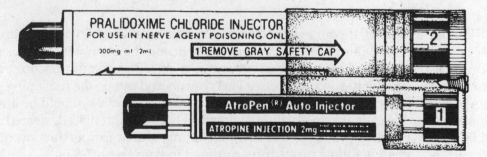

Figure 7-1: Nerve Agent Antidote Kit, Mark I.

- Severely pinpointed pupils.
- Red eyes with tearing (if agent gets into the eyes).
- Vomiting.
- Severe muscular twitching and general weakness.
- Loss of bladder/bowel control.
- Convulsions.
- Unconsciousness.
- Stoppage of breathing.

7-8. First Aid for Nerve Agent Poisoning (081-831-1030) and (081-831-1031). The injection site for administering the Nerve Agent Antidote Kit, Mark I (see Figure 7-1), is normally in the outer thigh muscle (see Figure 7-2). It is important that the injections be given into a large muscle area. If the individual is thinly-built, then the injections must be administered into the upper outer quarter (quadrant) of the buttocks (see Figure 7-3). This avoids injury to the thigh bone.

> **WARNING**
> There is a nerve that crosses the buttocks, so it is important to inject only into the upper outer quadrant (see Figure 7-3). This will avoid injuring this nerve. Hitting the nerve can cause paralysis.

a. Self-Aid (081-831-1030).
   (1) Immediately put on your protective mask after identifying any of the signs/symptoms of nerve agent poisoning (paragraph 7-7).

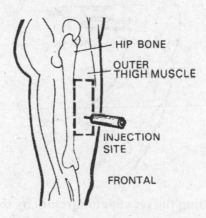

Figure 7-2: Thigh injection site.

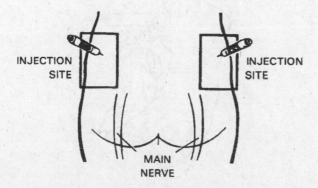

Figure 7-3: Buttocks injections site.

(2) Remove one set of the Nerve Agent Antidote Kit, Mark I.

(3) With your non dominant hand, hold the autoinjectors by the plastic clip so that the larger auto-injector is on top and both are positioned in front of you at eye level (see Figure 7-4).

(4) With the other hand, check the injection site (thigh or buttocks) for buttons or objects in pockets which may interfere with the injections.

(5) Grasp the atropine (smaller) autoinjector with the thumb and first two fingers (see Figure 7-5).

⚠ CAUTION

DO NOT cover/hold the green (needle) end with your hand or fingers—you might accidentally inject yourself.

(6) Pull the injector out of the clip with a smooth motion (see Figure 7-6).

WARNING
The injector is now armed. DO NOT touch the green (needle) end.

(7) Form a fist around the autoinjector. BE CAREFUL NOT TO INJECT YOURSELF IN THE HAND!

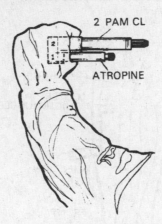

Figure 7-4: Holding the set of autoinjectors by the plastic clip.

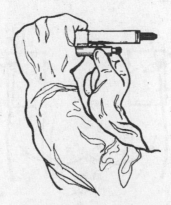

Figure 7-5: Grasping the atropine autoinjector between the thumb and first two fingers of the hand.

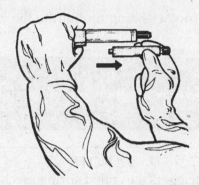

Figure 7-6: Removing the atropine autoinjector from the clip.

(8) Position the green end of the atropine autoinjector against the injection site (thigh or buttocks):
   (a) On the outer thigh muscle (see Figure 7-7).
       OR
   (b) On the upper outer portion of the buttocks (see Figure 7-8).

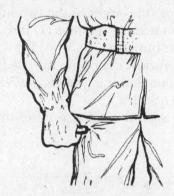

Figure 7-7: Thigh injection site for self-aid.

Figure 7-8: Buttocks injection site for self-aid.

(9) Apply firm, even pressure (not a jabbing motion) to the injector until it pushes the needle into your thigh (or buttocks).

WARNING
Using a jabbing motion may result in an improper injection or injury to the thigh or buttocks.

✍ NOTE
Firm pressure automatically triggers the coiled spring mechanism. This plunges the needle through the clothing into the muscle and injects the fluid into the muscle tissue.

---

(10) Hold the injector firmly in place for at least ten seconds. The ten seconds can be estimated by counting "one thousand and one, one thousand and two," and so forth.

(11) Carefully remove the autoinjector.

(12) Place the used atropine injector between the little finger and the ring finger of the hand holding the remaining autoinjector and the clip (see Figure 7-9). WATCH OUT FOR THE NEEDLE!

(13) Pull the 2 PAM C1 autoinjector (the larger of the two injectors) out of the clip (see Figure 7-10) and inject yourself in the same manner as steps (7) through (11) above, holding the black (needle) end against your thigh (or buttocks).

(14) Drop the empty injector clip without dropping the used autoinjectors.

(15) Attach the used injectors to your clothing (see Figure 7-11). Be careful NOT to tear your protective gloves/clothing with the needles.

   (a) Push the needle of each injector (one at a time) through one of the pocket flaps of your protective overgarment.

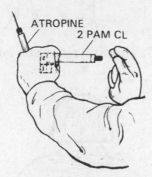

Figure 7-9: Used atropine autoinjector placed between the little finger and ring finger.

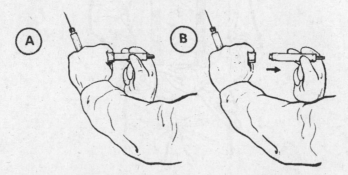

Figure 7-10: Removing the 2 PAM Cl autoinjector.

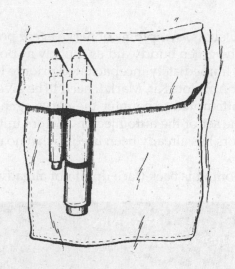

Figure 7-11: One set of used autoinjectors attached to pocket flap.

(b) Bend each needle to form a hook.

WARNING

It is important to keep track of all used autoinjectors so that medical personnel can determine how much antidote has been given and the proper follow-up treatment can be provided, if needed.

---

(16) Massage the injection site if time permits.

WARNING

If within 5 to 10 minutes after administering the first set of injections, your heart begins to beat rapidly and your mouth becomes very dry, DO NOT give yourself another set of injections. You have already received enough antidote to overcome the dangerous effects of the nerve agent. If you are able to walk without assistance (ambulate), know who you are and where you are, you WILL NOT need the second set of injections. (If not needed, giving yourself a second set of injections may create a nerve agent antidote overdose, which could cause incapacitation.) If, however, you continue to have symptoms of nerve agent poisoning for 10 to 15 minutes after receiving one set of injections, seek a buddy to check your symptoms. If your buddy agrees that your symptoms are worsening, administer the second set of injections.

> ✍ **NOTE (081-831-1030)**
> While waiting between sets (injections), you should decon your skin, if necessary, and put on the remaining protective clothing.

b. Buddy aid (081-831-1031).

A soldier exhibiting SEVERE signs/symptoms of nerve agent poisoning will not be able to care for himself and must therefore be given buddy aid as quickly as possible. Buddy aid will be required when a soldier is totally and immediately incapacitated prior to being able to apply self-aid, and all three sets of his Nerve Agent Antidote Kit, Mark I, need to be given by a buddy. Buddy aid may also be required after a soldier attempted to counter the nerve agent by self-aid but became incapacitated after giving himself one set of the autoinjectors. Before initiating buddy aid, a buddy should determine if one set of injectors has already been used so that no more than three sets of the antidote are administered.

(1) Move (roll) the casualty onto his back (face up) if not already in that position.

> **WARNING**
> Avoid unnecessary movement of the casualty so as to keep from spreading the contamination.

(2) Remove the casualty's protective mask from the carrier.
(3) Position yourself above the casualty's head, facing his feet.

> **WARNING**
> Squat, DO NOT kneel, when masking a chemical agent casualty. Kneeling may force the chemical agent into or through your protective clothing, which will greatly reduce the effectiveness of the clothing.

(4) Place the protective mask on the casualty.
(5) Have the casualty clear the mask.
(6) Check for a complete mask seal by covering the inlet valves. If properly sealed the mask will collapse.

> ✍ **NOTE**
> If the casualty is unable to follow instructions, is unconscious, or is not breathing, he will not be able to perform steps (5) or (6). It may, therefore, be impossible to determine if the mask is sealed. But you should still try to check for a good seal by placing your hands over the valves.

(7) Pull the protective hood over the head, neck, and shoulders of the casualty.
(8) Position yourself near the casualty's thigh.
(9) Remove one set of the casualty's autoinjectors.

> ✍ **NOTE (081-831-1031)**
> Use the CASUALTY'S autoinjectors. DO NOT use YOUR autoinjectors for buddy aid; if you do, you may not have any antidote if/when needed for self-aid.

(10) With your nondominant hand, hold the set of autoinjectors by the plastic clip so that the larger autoinjector is on top and both are positioned in front of you at eye level (see Figure 7-4).

(11) With the other hand, check the injection site (thigh or buttocks) for buttons or objects in pockets which may interfere with the injections.

(12) Grasp the atropine (smaller) autoinjector with the thumb and first two fingers (see Figure 7-5).

⚠ CAUTION

DO NOT cover/hold the green (needle) end with your hand or fingers–you may accidentally inject yourself.

(13) Pull the injector out of the clip with a smooth motion (see Figure 7-6).

WARNING
The injector is now armed. DO NOT touch the green (needle) end.

(14) Form a fist around the autoinjector. BE CAREFUL NOT TO INJECT YOURSELF IN THE HAND.

WARNING
Holding or covering the needle (green) end of the autoinjector may result in accidentally injecting yourself.

(15) Position the green end of the atropine autoinjector against the injection site (thigh or buttocks):
   (a) On the casualty's outer thigh muscle (see Figure 7-12).

✍ NOTE

The injections are normally given in the casualty's thigh.

Figure 7-12: Injecting the casualty's thigh.

WARNING

If this is the injection site used, be careful not to inject him close to the hip, knee, or thigh bone.

OR

(b) On the upper outer portion of the casualty's buttocks (see Figure 7-13).

✍ **NOTE**

If the casualty is thinly built, reposition him onto his side or stomach and inject the antidote into his buttocks.

WARNING

Inject the antidote only into the upper outer portion of his buttocks (see Figure 7-13). This avoids hitting the nerve that crosses the buttocks. Hitting this nerve can cause paralysis.

(16) Apply firm, even pressure (not a jabbing motion) to the injector to activate the needle. This causes the needle to penetrate both the casualty's clothing and muscle.

WARNING

Using a jabbing motion may result in an improper injection or injury to the thigh or buttocks.

(17) Hold the injector firmly in place for at least ten seconds. The ten seconds can be estimated by counting "one thousand and one, one thousand and two," and so forth.
(18) Carefully remove the autoinjector.

Figure 7-13: Injecting the casualty's buttocks.

(19) Place the used autoinjector between the little finger and ring finger of the hand holding the remaining autoinjector and the clip (see Figure 7-9). WATCH OUT FOR THE NEEDLE!

(20) Pull the 2 PAM Cl autoinjector (the larger of the two injectors) out of the clip (see Figure 7-10) and inject the casualty in the same manner as steps (9) through (19) above, holding the black (needle) end against the casualty's thigh (or buttocks).

(21) Drop the clip without dropping the used autoinjectors.

(22) Carefully lay the used injectors on the casualty's chest (if he is lying on his back), or on his back (if he is lying on his stomach), pointing the needles toward his head.

(23) Repeat the above procedure immediately (steps 9 through 22), using the second and third set of autoinjectors.

(24) Attach the three sets of used autoinjectors to the casualty's clothing (see Figure 7-14). Be careful NOT to tear either your or the casualty's protective clothing/gloves with the needles.

   (a) Push the needle of each injector (one at a time) through one of the pocket flaps of his protective overgarment.

   (b) Bend each needle to form a hook.

> **WARNING**
> It is important to keep track of all used autoinjectors so that medical personnel will be able to determine how much antidote has been given and the proper follow-up/treatment can be provided, if needed.

(25) Massage the area if time permits.

SECTION IV. OTHER AGENTS

7-9. Blister Agents. Blister agents (vesicants) include mustard (HD), nitrogen mustards(HN), lewisite (L), and other arsenicals, mixtures of mustards and arsenical, and phosgene oxime (CX). Blister agents act on the eyes, mucous membranes, lungs, and skin. They burn and blister the skin or any other body parts they contact. Even relatively low doses may cause serious injury. Blister agents damage the respiratory tract (nose, sinuses and windpipe) when inhaled and cause vomiting and diarrhea when absorbed. Lewisite and phosgene oxime cause immediate pain on contact. However, mustard agents are deceptive and there is little or no pain at the time of exposure. Thus, in some cases, signs of injury may not appear for several hours after exposure.

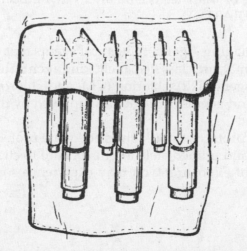

Figure 7-14: Three sets of used autoinjectors attached to pocket flap.

a. Protective Measures. Your protective mask with hood and protective overgarments provide you protection against blister agents. If it is known or suspected that blister agents are being used, STOP BREATHING, put on your mask and all your protective overgarments.

> ⚠ **CAUTION**
>
> Large drops of liquid vesicants on the protective overgarment ensemble may penetrate it if allowed to stand for an extended period. Remove large drops as soon as possible.

b. Signs/Symptoms of Blister Agent Poisoning.
   (1) Immediate and intense pain upon contact (lewisite and phosgene oxime). No initial pain upon contact with mustard.
   (2) Inflammation and blisters (burns)–tissue destruction. The severity of a chemical burn is directly related to the concentration of the agent and the duration of contact with the skin. The longer the agent is in contact with the tissue, the more serious the injury will be.
   (3) Vomiting and diarrhea. Exposure to high concentrations of vesicants may cause vomiting and or diarrhea.
   (4) Death. The blister agent vapors absorbed during ordinary field exposure will probably not cause enough internal body (systemic) damage to result in death. However, death may occur from prolonged exposure to high concentrations of vapor or from extensive liquid contamination over wide areas of the skin, particularly when decon is neglected or delayed.
c. First Aid Measures.
   (1) Use your M258A1 decon kit to decon your skin and use water to flush contaminated eyes. Decontamination of vesicants must be done immediately (within 1 minute is best).
   (2) If blisters form, cover them loosely with a field dressing and secure the dressing.

> ⚠ **CAUTION**
>
> Blisters are actually burns. DO NOT attempt to decon the skin where blisters have formed.

   (3) If you receive blisters over a wide area of the body, you are considered seriously burned. SEEK MEDICAL AID IMMEDIATELY.
   (4) If vomiting occurs, the mask should be lifted momentarily and drained—while the eyes are closed and the breath is held—and replaced, cleared, and sealed.
   (5) Remember, if vomiting or diarrhea occurs after having been exposed to blister agents, SEEK MEDICAL AID IMMEDIATELY.

7-10. Choking Agents (Lung-Damaging Agents). Chemical agents that attack lung tissue, primarily causing fluid buildup (pulmonary edema), are classified as choking agents (lung-damaging agents). This group includes phosgene (CG), diaphosgene (DP), chlorine (CL), and chloropicrin (PS). Of these four agents, phosgene is the most dangerous and is more likely to be employed by the enemy in future conflict.

a. Protective Measures. Your protective mask gives adequate protection against choking agents.
b. Signs/Symptoms. During and immediately after exposure to choking agents (depending on agent concentration and length of exposure), you may experience some or all of the following signs/symptoms:
   • Tears (lacrimation).
   • Dry throat.
   • Coughing.
   • Choking.

- Tightness of chest.
- Nausea and vomiting.
- Headaches.

c. First Aid Measures.

(1) If you come in contact with phosgene, your eyes become irritated, or a cigarette becomes tasteless or offensive, STOP BREATHING and put on your mask immediately.

(2) If vomiting occurs, the mask should be lifted momentarily and drained—while the eyes are closed and the breath is held—replaced, cleared, and sealed.

(3) Seek medical assistance if any of the above signs/symptoms occur.

⚠ NOTE

If you have no difficulty breathing, do not feel nauseated, and have no more than the usual shortness of breath on exertion, then you inhaled only a minimum amount of the agent. You may continue normal duties.

---

d. Death. With ordinary field exposure to choking agents, death will probably not occur. However, prolonged exposure to high concentrations of the vapor and neglect or delay in masking can be fatal.

7-11. Blood Agents. Blood agents interfere with proper oxygen utilization in the body. Hydrogen cyanide (AC) and cyanogen chloride (CK) are the primary agents in this group.

a. Protective Measures. Your protective mask with a fresh filter gives adequate protection against field concentrations of blood agent vapor. The protective overgarment as well as the mask are needed when exposed to liquid hydrogen cyanide.

b. Signs/Symptoms. During and immediately after exposure to blood agents (depending on agent concentration and length of exposure), you may experience some or all of the following signs/symptoms:

- Eye irritation.
- Nose and throat irritation.
- Sudden stimulation of breathing.
- Nausea.
- Coughing.
- Tightness of chest.
- Headache.
- Unconsciousness.

c. First Aid Measures.

(1) Hydrogen cyanide. During any chemical attack, if you get a sudden stimulation of breathing or notice an odor like bitter almonds, PUT ON YOUR MASK IMMEDIATELY. Speed is absolutely essential since this agent acts so rapidly that within a few seconds its effects will make it impossible for individuals to put on their mask by themselves. Stop breathing until the mask is on, if at all possible. This may be very difficult since the agent strongly stimulates respiration.

(2) Cyanogen chloride. PUT ON YOUR MASK IMMEDIATELY if you experience any irritation of the eyes, nose, or throat.

d. Medical Assistance. If you suspect that you have been exposed to blood agents, seek medical assistance immediately.

7-12. Incapacitating Agents. Generally speaking, an incapacitating agent is any compound which can interfere with your performance. The agent affects the central nervous system and produces muscular weakness and abnormal behavior. It is likely that such agents will be disseminated by smoke-producing munitions or aerosols, thus making breathing their means of entry into the body. The protective mask is, therefore, essential.

a. There is no special first aid to relieve the symptoms of incapacitating agents. Supportive first aid and physical restraint may be indicated. If the casualty is stuporous or comatose, be sure that respiration is unobstructed; then turn him on his stomach with his head to one side (in case vomiting should occur). Complete cleansing of the skin with soap and water should be done as soon as possible; or, the M258A1 Skin Decontamination Kit can be used if washing is impossible. Remove weapons and other potentially harmful items from the possession of individuals who are suspected of having these symptoms. Harmful items include cigarettes, matches, medications, and small items which might be swallowed accidentally. Delirious persons have been known to attempt to eat items bearing only a superficial resemblance to food.

b. Anticholinergic drugs (BZ - type) may produce alarming dryness and coating of the lips and tongue; however, there is usually no danger of immediate dehydration. Fluids should be given sparingly, if at all, because of the danger of vomiting and because of the likelihood of temporary urinary retention due to paralysis of bladder muscles. An important medical consideration is the possibility of heatstroke caused by the stoppage of sweating. If the environmental temperature is above 78° F, and the situation permits, remove excessive clothing from the casualty and dampen him to allow evaporative cooling and to prevent dehydration. If he does not readily improve, apply first aid measures for heat stroke and seek medical attention.

7-13. Incendiaries. Incendiaries can be grouped as white phosphorus, thickened fuel, metal, and oil and metal. You must learn to protect yourself against these incendiaries.

a. White phosphorus (WP) is used primarily as a smoke producer but can be used for its incendiary effect to ignite field expedients and combustible materials. The burns from WP are usually multiple, deep, and variable in size. When particles of WP get on the skin or clothing, they continue to burn until deprived of air. They also have a tendency to stick to a surface and must be brushed off or picked out.

(1) If burning particles of phosphorus strike and stick to your clothing, quickly take off the contaminated clothing before the phosphorus burns through to the skin.

(2) If burning phosphorus strikes your skin, smother the flame by submerging yourself in water or by dousing the WP with water from your canteen or any other source. Urine, a wet cloth, or mud can also be used.

> ✍ **NOTE**
> Since WP is poisonous to the system, DO NOT use grease or oil to smother the flame. The WP will be absorbed into the body with the grease or oil.

(3) Keep the WP particles covered with wet material to exclude air until you can remove them or get them removed from your skin.

(4) Remove the WP particles from the skin by brushing them with a wet cloth and by picking them out with a knife, bayonet, stick, or other available object.

(5) Report to a medical facility for treatment as soon as your mission permits.

b. Thickened fuel mixtures (napalm) have a tendency to cling to clothing and body surfaces, thereby producing prolonged exposure and severe burns. The first aid for these burns is the same as for other heat burns. The heat and irritating gases given off by these combustible mixtures may cause lung damage, which must be treated by a medical officer.

c. Metal incendiaries pose special problems. Thermite and thermate particles on the skin should be immediately cooled with water and then removed. Even though thermate particles have their own oxygen supply and continue to burn under water, it helps to cool them with water. The first aid for these burns is the same as for other heat burns. Particles of magnesium on the skin burn quickly and deeply. Like other metal incendiaries, they must be removed. Ordinarily, the complete removal of these particles should be done by trained personnel at a medical treatment facility, using local anesthesia. Immediate medical treatment is required.

d. Oil and metal incendiaries have much the same effect on contact with the skin and clothing as those discussed (b and c above). Appropriate first aid measures for burns are described in Chapter 3.

7-14. First Aid for Biological Agents. We are concerned with victims of biological attacks and with treating symptoms after the soldier becomes ill. However, we are more concerned with preventive medicine and hygienic measures taken before the attack. By accomplishing a few simple tasks we can minimize their effects.

a. Immunizations. In the military we are accustomed to keeping inoculations up to date. To prepare for biological defense, every effort must be taken to keep immunizations current. Based on enemy capabilities and the geographic location of our operations, additional immunizations may be required.

b. Food and Drink. Only approved food and water should be consumed. In a suspected biological warfare environment, efforts in monitoring food and water supplies must be increased. Properly treated water and properly cooked food will destroy most biological agents.

c. Sanitation Measures.
(1) Maintain high standards of personal hygiene. This will reduce the possibility of catching and spreading infectious diseases.
(2) Avoid physical fatigue. Physical fatigue lowers the body's resistance to disease. This, of course, is complemented by good physical fitness.
(3) Stay out of quarantined areas.
(4) Report sickness promptly. This ensures timely medical treatment and, more importantly, early diagnosis of the disease.

d. Medical Treatment of Casualties. Once a disease is identified, standard medical treatment commences. This may be in the form of first aid or treatment at a medical facility, depending on the seriousness of the disease. Epidemics of serious diseases may require augmentation of field medical facilities.

7-15. Toxins. Toxins are alleged to have been used in recent conflicts. Witnesses and victims have described the agent as toxic rain (or yellow rain) because it was reported to have been released from aircraft as a yellow powder or liquid that covered the ground, structures, vegetation, and people.

a. Protective Measures. Individual protective measures normally associated with persistent chemical agents will provide protection against toxins. Measures include the use of the protective mask with hood, and the overgarment ensemble with gloves and overboots (mission-oriented protective posture level-4 [MOPP 4]).

b. Signs/Symptoms. The occurrence of the symptoms from toxins may appear in a period of a few minutes to several hours depending on the particular toxin, the individual susceptibility, and the amount of toxin inhaled, ingested, or deposited on the skin. Symptoms from toxins usually involve the nervous system but are often preceded by less prominent symptoms, such as nausea, vomiting, diarrhea, cramps, or burning distress of the stomach region. Typical neurological symptoms often develop rapidly in severe cases, for example, visual disturbances, inability to swallow, speech difficulty, muscle coordination, and sensory abnormalities (numbness of mouth, throat, or extremities). Yellow rain (mycotoxins) also may have hemorrhagic symptoms which could include any/all of the following:
• Dizziness.
• Severe itching or tingling of the skin.

- Formation of multiple, small, hard blisters.
- Coughing up blood.
- Shock (which could result in death).

c. First Aid Measures. Upon recognition of an attack employing toxins or the onset (start) of symptoms listed above, you must immediately take the following actions:

(1) Step ONE. STOP BREATHING, put on your protective mask with hood, then resume breathing. Next, put on your protective clothing.

(2) Step TWO. Should severe itching of the face become unbearable, quickly—
- Loosen the cap on your canteen.
- Remove your helmet. Take and hold a deep breath and remove your mask.
- While holding your breath, close your eyes and flush your face with generous amounts of water.

> ⚠ **CAUTION**
>
> DO NOT rub or scratch your eyes. Try not to let the water run onto your clothing or protective overgarments.

- Put your protective mask back on, seat it properly, clear it, and check it for seal; then resume breathing.
- Put your helmet back on.

> ✎ **NOTE**
>
> The effectiveness of the M258A1 Skin Decon Kit for biological agent decon is unknown at this time; however, flushing the skin with large amounts of water will reduce the effectiveness of the toxins.

(3) Step THREE. If vomiting occurs, the mask should be lifted momentarily and drained—while the eyes are closed and the breath is held—and replaced, cleared, and sealed.

d. Medical Assistance. If you suspect that you have been exposed to toxins, you should seek medical assistance immediately.

7-16. Radiological. There is no direct first aid for radiological casualties. These casualties are treated for their apparent conventional symptoms and injuries.

APPENDIX A

First Aid Case and Kits, Dressings, and Bandages

A-1. First Aid Case with Field Dressings and Bandages. Every soldier is issued a first aid case (Figure A-1A) with a field first aid dressing encased in a plastic wrapper (Figure A-1B). He carries it at all times for his use. The field first aid dressing is a standard sterile (germ free) compress or pad with bandages attached (Figure A-1C). This dressing is used to cover the wound, to protect against further contamination, and to stop bleeding (pressure dressing). When a soldier administers first aid to another person, he must remember to use the wounded person's dressing; he may need his own later. The soldier must check his first aid case regularly and replace any used or missing dressing. The field first aid dressing may normally be obtained through the medical unit's assigned medical platoon or section.

A-2. General Purpose First Aid Kits. General purpose first aid kits listed in paragraph A-3 are also listed in CTA 8-100. These kits are carried on Army vehicles, aircraft, and boats for use by the operators, crew, and passengers. Individuals designated by unit standing operating procedures (SOP) to be responsible for the kits are required to check them regularly and replace all items used, or replace the entire kit when necessary. The general purpose kit and its contents can be obtained through the unit supply system.

✍ **NOTE**
Periodically check the dressings (for holes or tears in the package) and the medicines (for expiration date) that are in the first aid kits. If necessary, replace defective or outdated items.

A-3. Contents of First Aid Case and Kits. The following items are listed in the Common Table of Allowances (CTA) as indicated below. However, it is necessary to see referenced CTA for stock numbers.

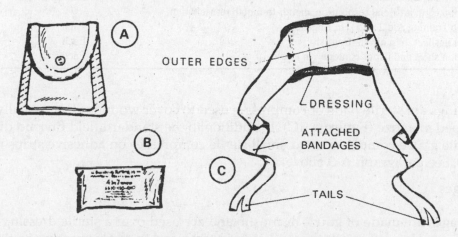

Figure A-1: Field first aid case and dressing.

CTA	Nomenclature	Unit of Issue	Quantity
a. 50-900	CASE FIELD FIRST AID DRESSING	each	1
8-100	Dressing, first aid field, individual troop, white, 4 by 7 inches	Each	1
b. 8-100	FIRST AID KIT, general purpose (Rigid Case)	Each	1
	Contents: Case, medical instrument and supply set, plastic, rigid, size A, 7½ inches long by 4½ inches wide by 2 3/4 inches high	Each	1
	Ammonia inhalation solution, aromatic, ampules, 1/3 ml, 10s	Package	1
	Povidone-iodine solution, USP: 10%,½ fl oz, 50s	Box	1/50
	Dressing, first aid, field, individual troop, camouflaged, 4 by 7 inches	Each	3
	Compress and bandage, camouflaged, 2 by 2 inches, 4s	Package	1
	Bandage, gauze, compressed, camouflaged, 3 inches by 6 yards	Each	2
	Bandage, muslin, compressed, camouflaged, 37 by 37 by 52 inches	Each	1
	Gauze, petrolatum, 3 by 36 inches, 3s	Package	1
	Adhesive tape, surgical, 1 inch by 1 1/2 yards, 100s	Package	3/100
	Bandage, adhesive, 3/4 by 3 inches,	Box	18/300
	Blade, surgical preparation razor, straight, single edge, 5s	Package	1
	First aid kit, eye dressing	Each	1
	Instruction card, artificial respiration, mouth-to-mouth resuscitation (Graphic Training Aid 21-45) (in English)	Each	1
	Instruction sheet, first aid (in English)	Each	1
	Instruction sheet and list of contents (in English)	Each	1
c. 8-100	FIRST AID KIT, general purpose (panel-mounted)	Each	1
	Contents: Case, medical instrument and supply set, nylon, nonrigid, No. 2, 7 1/2 inches long by 4 3/8 inches wide by 4 1/2 inches high	Each	1
In Upper Pocket	Ammonia Inhalation Solution aromatic, ampules, 1/3 ml, 10s	Package	1
	Compress and bandage, camouflaged, 2 by 2 inches, 4s	Package	1
	Bandage, muslin, compressed, camouflaged, 37 by 37 by 52 inches	Each	1
	Gauze, petrolatum, 3 by 36 inches, 12s	Package	3/12
	Blade, surgical preparation razor, straight, single edge, 5s	Package	1
In Lower Pocket	Pad, Povidone-Iodine, 100s	Box	10/100
	Dressing, first aid, field, individual troop, camouflaged, 4 by 6 inches	Each	3
	Bandage, gauze, compressed, camouflaged, 3 inches by 6 yards	Each	2
	Adhesive tape, surgical, 1 inch by 1½ yards, 100s	Package	3/100
	Bandage, adhesive, 3/4 by 3 inches, 300s	Box	18/300
	First aid kit, eye dressing	Each	1
	Instruction card, artificial respiration, mouth-to-mouth resuscitation (Graphic Training Aid 21-45) (in English)	Each	1
	Instruction sheet, first aid (in English)	Each	1
	Instruction sheet and list of contents (in English)	Each	1

A-4. Dressings. Dressings are sterile pads or compresses used to cover wounds. They usually are made of gauze or cotton wrapped in gauze (Figure A-1C). In addition to the standard field first aid dressing, other dressings such as sterile gauze compresses and small sterile compresses on adhesive strips may be available under CTA 8-100. See paragraph A-3 above.

A-5. Standard Bandages

a. Standard bandages are made of gauze or muslin and are used over a sterile dressing to secure the dressing in place, to close off its edge from dirt and germs, and to create pressure on the wound and control bleeding. A bandage can also support an injured part or secure a splint.

b. Tailed bandages may be attached to the dressing as indicated on the field first aid dressing (Figure A-1C).

A-6. Triangular and Cravat (Swathe) Bandages

a. Triangular and cravat (or swathe) bandages (Figure A-2) are fashioned from a triangular piece of muslin (37 by 37 by 52 inches) provided in the general purpose first aid kit. If it is folded into a strip, it is called a cravat. Two safety pins are packaged with each bandage. These bandages are valuable in an emergency since they are easily applied.

b. To improvise a triangular bandage, cut a square of available material, slightly larger than 3 feet by 3 feet, and FOLD it DIAGONALLY. If two bandages are needed, cut the material along the DIAGONAL FOLD.

c. A cravat can be improvised from such common items as T-shirts, other shirts, bed linens, trouser legs, scarfs, or any other item made of pliable and durable material that can be folded, torn, or cut to the desired size.

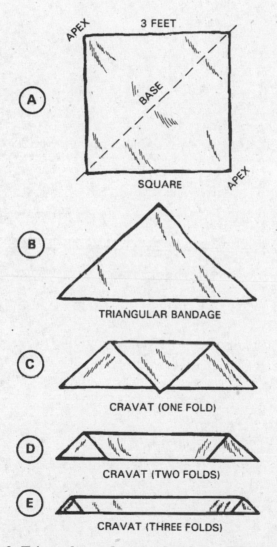

Figure A-2: Triangular and cravat bandages (Illustrated A thru E).

APPENDIX B

Rescue and Transportation Procedures

B-1. General. A basic principle of first aid is to treat the casualty before moving him. However, adverse situations or conditions may jeopardize the lives of both the rescuer and the casualty if this is done. It may be necessary first to rescue the casualty before first aid can be effectively or safely given. The life and/or the well-being of the casualty will depend as much upon the manner in which he is rescued and transported as it will upon the treatment he receives. Rescue actions must be done quickly and safely. Careless or rough handling of the casualty during rescue operations can aggravate his injuries and possibly cause death.

B-2. Principles of Rescue Operations

a. When faced with the necessity of rescuing a casualty who is threatened by hostile action, fire, water, or any other immediate hazard, DO NOT take action without first determining the extent of the hazard and your ability to handle the situation. DO NOT become a casualty.

b. The rescuer must evaluate the situation and analyze the factors involved. This evaluation involves three major steps:
  - Identify the task.
  - Evaluate circumstances of the rescue.
  - Plan the action.

B-3. Task (Rescue) Identification. First determine if a rescue attempt is actually needed. It is a waste of time, equipment, and personnel to rescue someone not in need of rescuing. It is also a waste to look for someone who is not lost or needlessly risk the lives of the rescuer(s). In planning a rescue, attempt to obtain the following information:

  - Who, what, where, when, why, and how the situation happened?
  - How many casualties are involved and the nature of their injuries?
  - What is the tactical situation?
  - What are the terrain features and the location of the casualties?
  - Will there be adequate assistance available to aid in the rescue/evacuation?
  - Can treatment be provided at the scene; will the casualties require movement to a safer location?
  - What equipment will be required for the rescue operation?
  - Will decon procedures and equipment be required for casualties, rescue personnel and rescue equipment?

B-4. Circumstances of the Rescue

a. After identifying the job (task) required, you must relate to the circumstances under which you must work. Do you need additional people, security, medical, or special rescue equipment? Are there circumstances such as mountain rescue or aircraft accidents that may require specialized skills? What is the weather like? Is the terrain hazardous? How much time is available?

b. The time element will sometimes cause a rescuer to compromise planning stages and/or treatment which can be given. A realistic estimate of time available must be made as quickly as possible to determine action time remaining. The key elements are the casualty's condition and the environment.

c. Mass casualties are to be expected on the modern battlefield. All problems or complexities of rescue are now multiplied by the number of casualties encountered. In this case, time becomes the critical element.

B-5. Plan of Action

a. The casualty's ability to endure is of primary importance in estimating the time available. Age and physical condition will differ from casualty to casualty. Therefore, to determine the time available, you will have to consider—
   - Endurance time of the casualty.
   - Type of situation.
   - Personnel and/or equipment availability.
   - Weather.
   - Terrain.

b. In respect to terrain, you must consider altitude and visibility. In some cases, the casualty may be of assistance because he knows more about the particular terrain or situation than you do. Maximum use of secure/reliable trails or roads is essential.

c. When taking weather into account, ensure that blankets and/or rain gear are available. Even a mild rain can complicate a normally simple rescue. In high altitudes and/or extreme cold and gusting winds, the time available is critically shortened.

d. High altitudes and gusting winds minimize the ability of fixed-wing or rotary wing aircraft to assist in operations. Rotary wing aircraft may be available to remove casualties from cliffs or inaccessible sites. These same aircraft can also transport the casualties to a medical treatment facility in a comparatively short time. Aircraft, though vital elements of search, rescue or evacuation, cannot be used in all situations. For this reason, do not rely entirely on their presence. Reliance on aircraft or specialized equipment is a poor substitute for careful planning.

B-6. Mass Casualties.

In situations where there are multiple casualties, an orderly rescue may involve some additional planning. To facilitate a mass casualty rescue or evacuation, recognize separate stages.

- First Stage. Remove those personnel who are not trapped among debris or who can be evacuated easily.
- Second Stage. Remove those personnel who may be trapped by debris but require only the equipment on hand and a minimum amount of time.
- Third Stage. Remove the remaining personnel who are trapped in extremely difficult or time-consuming situations, such as under large amounts of debris or behind walls.
- Fourth Stage. Remove the dead.

B-7. Proper Handling of Casualties

a. You may have saved the casualty's life through the application of appropriate first aid measures. However, his life can be lost through rough handling or careless transportation procedures. Before you attempt to move the casualty—
   - Evaluate the type and extent of his injury.
   - Ensure that dressings over wounds are adequately reinforced.
   - Ensure that fractured bones are properly immobilized and supported to prevent them from cutting through muscle, blood vessels, and skin. Based upon your evaluation of the type and extent of the casualty's injury and your knowledge of the various manual carries, you must select the best possible method of manual transportation. If the casualty is conscious, tell him how he is to be transported. This will help allay his fear of movement and gain his cooperation and confidence.

b. Buddy aid for chemical agent casualties includes those actions required to prevent an incapacitated casualty from receiving additional injury from the effects of chemical hazards. If a casualty is physically unable to decontaminate himself or administer the proper chemical agent antidote, the casualty's buddy assists him and assumes responsibility for his care. Buddy aid includes—

- Administering the proper chemical agent antidote.
- Decontaminating the incapacitated casualty's exposed skin.
- Ensuring that his protective ensemble remains correctly emplaced.
- Maintaining respiration.
- Controlling bleeding.
- Providing other standard first aid measures.
- Transporting the casualty out of the contaminated area.

B-8. Transportation of Casualties

a. Transportation of the sick and wounded is the responsibility of medical personnel who have been provided special training and equipment. Therefore, unless a good reason for you to transport a casualty arises, wait for some means of medical evacuation to be provided. When the situation is urgent and you are unable to obtain medical assistance or know that no medical evacuation facilities are available, you will have to transport the casualty. For this reason, you must know how to transport him without increasing the seriousness of his condition.

b. Transporting a casualty by litter is safer and more comfortable for him than by manual means; it is also easier for you. Manual transportation, however, may be the only feasible method because of the terrain or the combat situation; or it may be necessary to save a life. In these situations, the casualty should be transferred to a litter as soon as one can be made available or improvised.

B-9. Manual Carries (081-831-1040 and 081-831-1041). Casualties carried by manual means must be carefully and correctly handled, otherwise their injuries may become more serious or possibly fatal. Situation permitting, evacuation or transport of a casualty should be organized and unhurried. Each movement should be performed as deliberately and gently as possible. Casualties should not be moved before the type and extent of injuries are evaluated and the required emergency medical treatment is given. The exception to this occurs when the situation dictates immediate movement for safety purposes (for example, it may be necessary to remove a casualty from a burning vehicle); that is, the situation dictates that the urgency of casualty movement outweighs the need to administer emergency medical treatment. Manual carries are tiring for the bearer(s) and involve the risk of increasing the severity of the casualty's injury. In some instances, however, they are essential to save the casualty's life. Although manual carries are accomplished by one or two bearers, the two-man carries are used whenever possible. They provide more comfort to the casualty, are less likely to aggravate his injuries, and are also less tiring for the bearers, thus enabling them to carry him farther. The distance a casualty can be carried depends on many factors, such as—

- Strength and endurance of the bearer(s).
- Weight of the casualty.
- Nature of the casualty's injury.
- Obstacles encountered during transport.

a. One-man Carries (081-831-1040).

(1) Fireman's carry (081-831-1040). The fireman's carry (Figure B-1) is one of the easiest ways for one person to carry another. After an unconscious or disabled casualty has been properly positioned, he is raised from the ground. An alternate method for raising him from the ground is illustrated (Figure B-1 I). However, it should be used only when the bearer believes it to be safer for the casualty because of the location of his wounds. When the alternate method is used, take care to

(A) KNEEL AT THE CASUALTY'S UNINJURED SIDE. PLACE HIS ARMS ABOVE HIS HEAD AND CROSS HIS ANKLE FARTHER FROM YOU OVER THE ONE CLOSER TO YOU. PLACE ONE OF YOUR HANDS ON THE SHOULDER FARTHER FROM YOU AND YOUR OTHER HAND IN THE AREA OF HIS HIP OR THIGH.

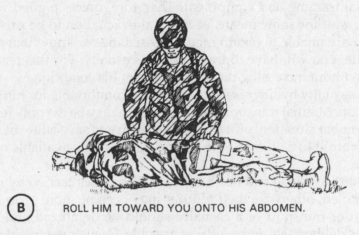

(B) ROLL HIM TOWARD YOU ONTO HIS ABDOMEN.

Figure B-1: Fireman's carry (Illustrated A thru N).

prevent the casualty's head from snapping back and causing a neck injury. The steps for raising a casualty from the ground for the fireman's carry are also used in other one-man carries.

> ✍ **NOTE**
> The alternate method of raising the casualty from the ground should be used only when the bearer believes it to be safer for the casualty because of the location of his wounds. When the alternate method is used, take care to prevent the casualty's head from snapping back and causing a neck injury.

(2) Support carry (081-831-1040). In the support carry (Figure B-2), the casualty must be able to walk or at least hop on one leg, using the bearer as a crutch. This carry can be used to assist him as far as he is able to walk or hop.

(3) Arms carry (081-831-1040). The arms carry is used when the casualty is unable to walk. This carry (Figure B-3) is useful when carrying a casualty for a short distance and when placing him on a litter.

(4) Saddleback carry (081-831-1040). Only a conscious casualty can be transported by the saddleback carry (Figure B-4), because he must be able to hold onto the bearer's neck.

(5) Pack-strap carry (081-831-1040). This carry is used when only a moderate distance will be traveled. In this carry (Figure B-5), the casualty's weight rests high on the bearer's back. To eliminate

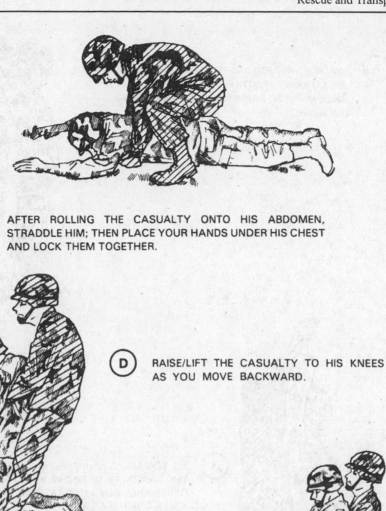

C AFTER ROLLING THE CASUALTY ONTO HIS ABDOMEN, STRADDLE HIM; THEN PLACE YOUR HANDS UNDER HIS CHEST AND LOCK THEM TOGETHER.

D RAISE/LIFT THE CASUALTY TO HIS KNEES AS YOU MOVE BACKWARD.

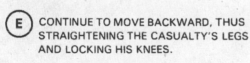

E CONTINUE TO MOVE BACKWARD, THUS STRAIGHTENING THE CASUALTY'S LEGS AND LOCKING HIS KNEES.

Figure B-1: *(Continued)*

the possibility of injury to the casualty's arms, the bearer must hold the casualty's arms in a palms-down position.

(6) Pistol-belt carry (081-831-1040). The pistol-belt carry (Figure B-6) is the best one-man carry when the distance to be traveled is long. The casualty is securely supported by a belt upon the shoulders of the bearer. The hands of both the bearer and the casualty are left free for carrying a weapon or equipment, climbing banks, or surmounting obstacles. With his hands free and the casualty secured in place, the bearer is also able to creep through shrubs and under low hanging branches.

F WALK FORWARD, BRINGING THE CASUALTY TO
A STANDING POSITION BUT TILTED SLIGHTLY
BACKWARD TO PREVENT HIS KNEES FROM
BUCKLING.

G AS YOU MAINTAIN CONSTANT SUPPORT OF
THE CASUALTY WITH ONE ARM, FREE YOUR
OTHER ARM, QUICKLY GRASP HIS WRIST,
AND RAISE HIS ARM HIGH.

Figure B-1: *(Continued)*

(7) Pistol-belt drag (081-831-1040). The pistol-belt drag (Figure B-7) and other drags are generally used for short distances. In this drag the casualty is on his back. The pistol-belt drag is useful in combat. The bearer and the casualty can remain closer to the ground in this drag than in any other.

(8) Neck drag (081-831-1040). The neck drag (Figure B-8) is useful in combat because the bearer can transport the casualty when he creeps behind a low wall or shrubbery, under a vehicle, or through a culvert. This drag is used only if the casualty does not have a broken/fractured arm. In this drag the casualty is on his back. If the casualty is unconscious, protect his head from the ground.

(9) Cradle drop drag (081-831-1040). The cradle drop drag (Figure B-9) is effective in moving a casualty up or down steps. In this drag the casualty is lying down.

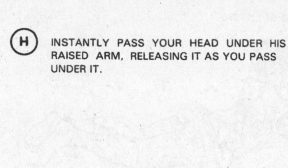

(H) INSTANTLY PASS YOUR HEAD UNDER HIS
RAISED ARM, RELEASING IT AS YOU PASS
UNDER IT.

(I) MOVE SWIFTLY TO FACE THE CASUALTY AND SECURE
YOUR ARMS AROUND HIS WAIST. IMMEDIATELY PLACE
YOUR FOOT BETWEEN HIS FEET AND SPREAD THEM
(APPROXIMATELY 6 TO 8 INCHES APART).

Figure B-1: *(Continued)*

b. Two-man Carries (081-831-1041).

    (1) Two-man support carry (081-831-1041). The two-man support carry (Figure B-10) can be used in transporting both conscious or unconscious casualties. If the casualty is taller than the bearers, it may be necessary for the bearers to lift the casualty's legs and let them rest on their forearms.

    (2) Two-man arms carry (081-831-1041). The two-man arms carry (Figure B-11) is useful in carrying a casualty for a moderate distance. It is also useful for placing him on a litter. To lessen fatigue, the bearers should carry him high and as close to their chests as possible. In extreme emergencies when there is no time to obtain a board, this manual carry is the safest one for transporting a casualty with a back/neck injury. Use two additional bearers to keep his head and legs in alignment with his body.

    (3) Two-man fore-and-aft carry (081-831-1041). The fore-and- aft carry (Figure B-12) is a most useful two-man carry for transporting a casualty for a long distance. The taller of the two bearers

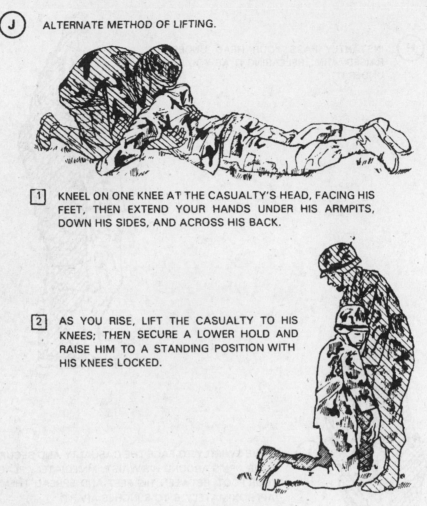

J   ALTERNATE METHOD OF LIFTING.

1 KNEEL ON ONE KNEE AT THE CASUALTY'S HEAD, FACING HIS FEET, THEN EXTEND YOUR HANDS UNDER HIS ARMPITS, DOWN HIS SIDES, AND ACROSS HIS BACK.

2 AS YOU RISE, LIFT THE CASUALTY TO HIS KNEES; THEN SECURE A LOWER HOLD AND RAISE HIM TO A STANDING POSITION WITH HIS KNEES LOCKED.

Figure B-1: *(Continued)*

should position himself at the casualty's head. By altering this carry so that both bearers face the casualty, it is also useful for placing him on a litter.

(4) Two-hand seat carry (081-831-1041). The two-hand seat carry (Figure B-13) is used in carrying a casualty for a short distance and in placing him on a litter.

(5) Four-hand seat carry (081-831-1041). Only a conscious casualty can be transported with the four-hand seat carry (Figure B-14) because he must help support himself by placing his arms around the bearers' shoulders. This carry is especially useful in transporting the casualty with a head or foot injury and is used when the distance to be traveled is moderate. It is also useful for placing a casualty on a litter.

B-10. Improvised Litters (Figures B-15 through B-17) (081-831-1041). Two men can support or carry a casualty without equipment for only short distances. By using available materials to improvise equipment, the casualty can be transported greater distances by two or more rescuers.

a. There are times when a casualty may have to be moved and a standard litter is not available. The distance may be too great for manual carries or the casualty may have an injury, such as a fractured neck, back, hip, or thigh that would be aggravated by manual transportation. In these situations, litters can be improvised from certain materials at hand. Improvised litters are emergency measures

3 SECURE YOUR ARMS AROUND THE CASUALTY'S WAIST, WITH HIS BODY TILTED SLIGHTLY BACKWARD TO PREVENT HIS KNEES FROM BUCKLING. PLACE YOUR FOOT BETWEEN HIS FEET AND SPREAD THEM (ABOUT 6 TO 8 INCHES APART).

(K) GRASP THE CASUALTY'S WRIST AND RAISE HIS ARM HIGH OVER YOUR HEAD.

Figure B-1: *(Continued)*

and must be replaced by standard litters at the first opportunity to ensure the comfort and safety of the casualty.

b. Many different types of litters can be improvised, depending upon the materials available. Satisfactory litters can be made by securing poles inside such items as blankets, ponchos, shelter halves, tarpaulins, jackets, shirts, sacks, bags, and bed tickings (fabric covers of mattresses). Poles can be improvised from strong branches, tent supports, skis, and other like items. Most flat-surface objects of suitable size can also be used as litters. Such objects include boards, doors, window shutters, benches, ladders, cots, and poles tied together. If possible, these objects should be padded.

(L) STOOP/BEND DOWN AND PULL THE
CASUALTY'S ARM OVER AND DOWN
YOUR SHOULDER, THUS BRINGING
HIS BODY ACROSS YOUR SHOULDERS.
AT THE SAME TIME, PASS YOUR ARM
BETWEEN HIS LEGS.

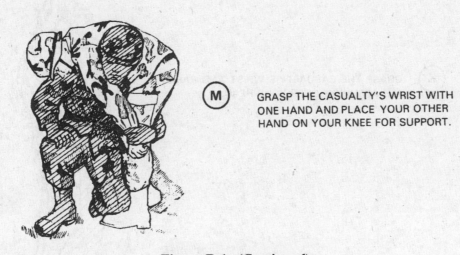

(M) GRASP THE CASUALTY'S WRIST WITH
ONE HAND AND PLACE YOUR OTHER
HAND ON YOUR KNEE FOR SUPPORT.

Figure B-1: *(Continued)*

c. If no poles can be obtained, a large item such as a blanket can be rolled from both sides toward the center. The rolls then can be used to obtain a firm grip when carrying the casualty. If a poncho is used, make sure the hood is up and under the casualty and is not dragging on the ground.

d. The important thing to remember is that an improvised litter must be well constructed to avoid the risk of dropping or further injuring the casualty.

e. Improvised litters may be used when the distance may be too long (far) for manual carries or the casualty has an injury which may be aggravated by manual transportation.

f. Any of the appropriate carries may be used to place a casualty on a litter. These carries are:
   • The one-man arms carry (Figure B-3).
   • The two-man arms carry (Figure B-11).
   • The two-man fore-and-aft carry (Figure B-12).

RISE WITH THE CASUALTY CORRECTLY POSITIONED.
YOUR OTHER HAND IS FREE FOR USE AS NEEDED.

Figure B-1: *(Continued)*

- The two-hand seat carry (Figure B-13).
- The four-hand seat carry (Figure B-14).

WARNING
Unless there is an immediate life-threatening situation (such as fire, explosion), DO NOT move the casualty with a suspected back or neck injury. Seek medical personnel for guidance on how to transport.

g. Either two or four soldiers (head/foot) may be used to lift a litter. To lift the litter, follow the procedure below.
  (1) Raise the litter at the same time as the other carriers/bearers.
  (2) Keep the casualty as level as possible.

✍ NOTE
Use caution when transporting on a sloping incline/hill.

RAISE THE CASUALTY TO A STANDING POSITION FROM GROUND AS IN FIREMAN'S CARRY. GRASP THE CASUALTY'S WRIST AND DRAW HIS ARM AROUND YOUR NECK. PLACE YOUR ARM AROUND HIS WAIST.

(THE CASUALTY IS THUS ABLE TO WALK, USING YOU AS A CRUTCH.)

Figure B-2: Support carry.

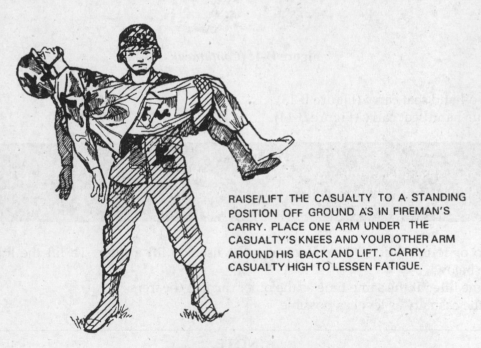

RAISE/LIFT THE CASUALTY TO A STANDING POSITION OFF GROUND AS IN FIREMAN'S CARRY. PLACE ONE ARM UNDER THE CASUALTY'S KNEES AND YOUR OTHER ARM AROUND HIS BACK AND LIFT. CARRY CASUALTY HIGH TO LESSEN FATIGUE.

Figure B-3: Arms carry.

RAISE CASUALTY TO UPRIGHT POSITION AS IN FIREMAN'S
CARRY. SUPPORT CASUALTY BY PLACING AN ARM AROUND
HIS WAIST AND MOVE IN FRONT OF HIM (YOUR BACK TO HIM).
HAVE CASUALTY ENCIRCLE HIS ARMS AROUND YOUR NECK.
STOOP, RAISE HIM UPON YOUR BACK, AND CLASP YOUR
HANDS TOGETHER BENEATH HIS THIGHS IF POSSIBLE.

Figure B-4: Saddleback carry.

(A) LIFT CASUALTY FROM GROUND TO A STANDING POSITION AS IN FIREMAN'S CARRY. SUPPORTING THE CASUALTY WITH YOUR ARMS AROUND HIM, GRASP HIS WRIST CLOSER TO YOU AND PLACE HIS ARM OVER YOUR HEAD AND ACROSS YOUR SHOULDER. MOVE IN FRONT OF HIM WHILE SUPPORTING HIS WEIGHT AGAINST YOUR BACK. GRASP HIS OTHER WRIST, AND PLACE THIS ARM OVER YOUR SHOULDER.

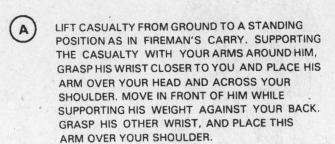

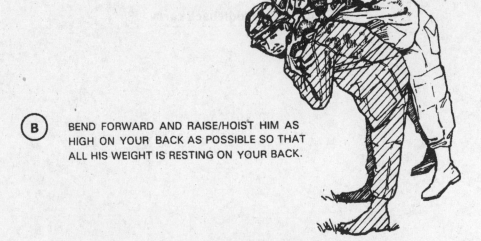

(B) BEND FORWARD AND RAISE/HOIST HIM AS HIGH ON YOUR BACK AS POSSIBLE SO THAT ALL HIS WEIGHT IS RESTING ON YOUR BACK.

Figure B-5: Pack-strap carry (Illustrated A and B).

A LINK TWO PISTOL BELTS (OR THREE, IF NECESSARY) TOGETHER TO FORM A SLING. (IF PISTOL BELTS ARE NOT AVAILABLE FOR USE, OTHER ITEMS, SUCH AS ONE RIFLE SLING, TWO CRAVAT BANDAGES, TWO LITTER STRAPS, OR ANY SUITABLE MATERIAL WHICH WILL NOT CUT OR BIND THE CASUALTY, MAY BE USED.) PLACE THIS SLING UNDER THE CASUALTY'S THIGHS AND LOWER BACK SO THAT A LOOP EXTENDS FROM EACH SIDE.

B LIE FACE UP BETWEEN THE CASUALTY'S OUTSTRETCHED LEGS. THRUST YOUR ARMS THROUGH THE LOOPS, GRASP HIS HAND AND TROUSER LEG ON HIS INJURED SIDE.

Figure B-6: Pistol-belt carry (Illustrated A thru F).

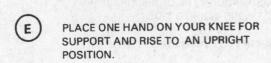 ROLL TOWARD THE CASUALTY'S UNINJURED SIDE ONTO YOUR ABDOMEN, BRINGING HIM ONTO YOUR BACK. ADJUST SLING AS NECESSARY.

(D) RISE TO A KNEELING POSITION. THE BELT WILL HOLD THE CASUALTY IN PLACE.

(E) PLACE ONE HAND ON YOUR KNEE FOR SUPPORT AND RISE TO AN UPRIGHT POSITION.

Figure B-6: *(Continued)*

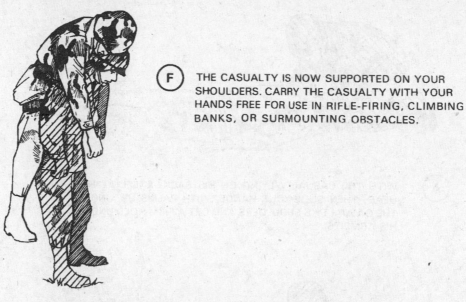

F THE CASUALTY IS NOW SUPPORTED ON YOUR SHOULDERS. CARRY THE CASUALTY WITH YOUR HANDS FREE FOR USE IN RIFLE-FIRING, CLIMBING BANKS, OR SURMOUNTING OBSTACLES.

Figure B-6: Pistol-belt carry (Illustrated A thru F).

Figure B-7: Pistol-belt drag.

TIE THE CASUALTY'S HANDS TOGETHER AT THE WRISTS. IF CASUALTY IS CONSCIOUS, HE MAY CLASP HIS HANDS TOGETHER AROUND YOUR NECK. STRADDLE THE CASUALTY IN A KNEELING FACE-TO-FACE POSITION. LOOP THE CASUALTY'S TIED HANDS OVER/AROUND YOUR NECK. CRAWL FORWARD, LOOKING FORWARD, DRAGGING THE CASUALTY WITH YOU. IF THE CASUALTY IS UNCONSCIOUS, PROTECT HIS HEAD FROM THE GROUND.

Figure B-8: Neck drag.

A WITH THE CASUALTY LYING ON HIS BACK, KNEEL AT HIS HEAD. THEN SLIDE YOUR HANDS, WITH PALMS UP, UNDER THE CASUALTY'S SHOULDERS AND GET A FIRM HOLD UNDER HIS ARMPITS.

B PARTIALLY RISE, SUPPORTING THE CASUALTY'S HEAD ON ONE OF YOUR FOREARMS. (YOU MAY BRING YOUR ELBOWS TOGETHER AND LET THE CASUALTY'S HEAD REST ON BOTH OF YOUR FOREARMS.)

Figure B-9: Cradle drop drag (Illustrated A thru D).

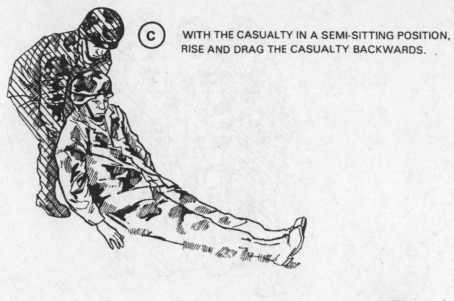

(C) WITH THE CASUALTY IN A SEMI-SITTING POSITION, RISE AND DRAG THE CASUALTY BACKWARDS.

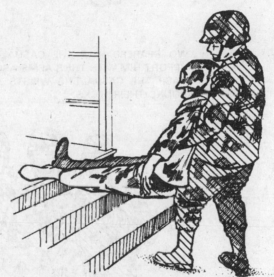

(D) THEN BACK DOWN THE STEPS, SUPPORTING THE CASUALTY'S HEAD AND BODY AND LETTING HIS HIPS AND LEGS DROP FROM STEP TO STEP. IF THE CASUALTY NEEDS TO BE MOVED *UP* THE STEPS, THEN YOU SHOULD BACK UP THE STEPS, USING THE SAME PROCEDURE.

Figure B-9: *(Continued)*

A TWO BEARERS HELP THE CASUALTY TO HIS FEET AND SUPPORT HIM WITH THEIR ARMS AROUND HIS WAIST. THEY GRASP THE CASUALTY'S WRISTS AND DRAW HIS ARMS AROUND THEIR NECKS.

B IF A CASUALTY IS TALLER THAN THE BEARERS, IT MAY BE NECESSARY FOR THE BEARERS TO LIFT HIS LEGS AND LET THEM REST ON THEIR FOREARMS.

Figure B-10: Two-man support carry (Illustrated A and B).

A TWO BEARERS KNEEL AT ONE SIDE OF THE CASUALTY AND PLACE THEIR ARMS BENEATH THE CASUALTY'S BACK (SHOULDERS), WAIST, HIPS, AND KNEES.

B THE BEARERS LIFT THE CASUALTY AS THEY RISE TO THEIR KNEES.

NOTE

Keeping the casualty's body level will prevent unnecessary movement and further injury.

Figure B-11: Two-man arms carry (Illustrated A thru D).

C AS THE BEARERS RISE TO THEIR FEET, THEY TURN THE
   CASUALTY TOWARD THEIR CHESTS.

D THEY CARRY HIM HIGH TO LESSEN FATIGUE.

Figure B-11: *(Continued)*

A   THE SHORTER BEARER SPREADS THE CASUALTY'S LEGS, KNEELS BETWEEN THE LEGS WITH HIS BACK TO THE CASUALTY, AND POSITIONS HIS HANDS BEHIND THE CASUALTY'S KNEES. THE OTHER (TALLER) BEARER KNEELS AT THE CASUALTY'S HEAD, SLIDES HIS HANDS UNDER THE ARMS AND ACROSS, AND LOCKS HIS HANDS TOGETHER.

NOTE

The taller of the two bearers should position himself at the casualty's head.

Figure B-12: Two-man fore-and-aft carry (Illustrated A thru C).

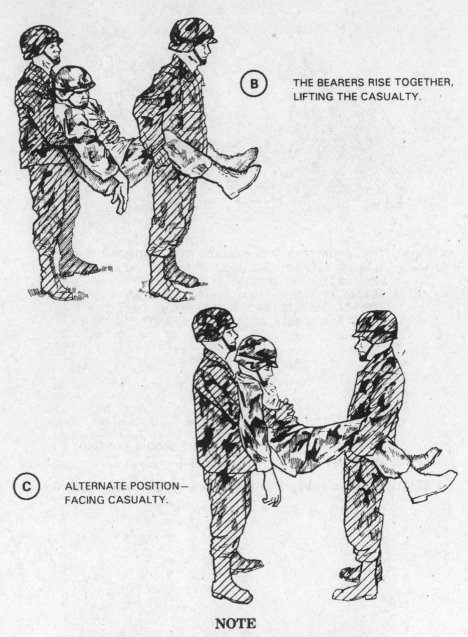

B THE BEARERS RISE TOGETHER, LIFTING THE CASUALTY.

C ALTERNATE POSITION— FACING CASUALTY.

NOTE

By altering the carry so that both bearers face the casualty, it is also useful for placing him on a litter.

Figure B-12: *(Continued)*

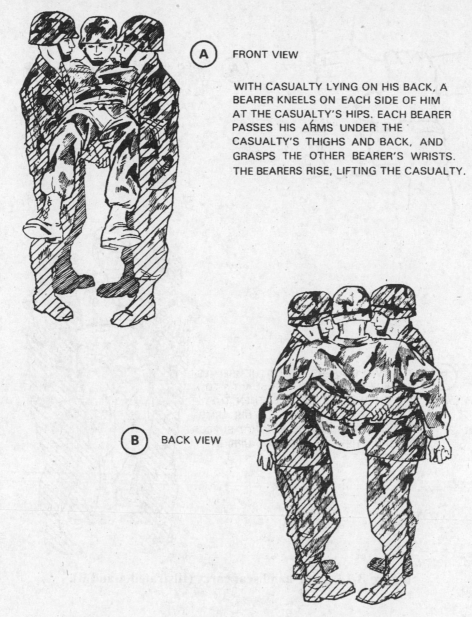

(A) FRONT VIEW

WITH CASUALTY LYING ON HIS BACK, A
BEARER KNEELS ON EACH SIDE OF HIM
AT THE CASUALTY'S HIPS. EACH BEARER
PASSES HIS ARMS UNDER THE
CASUALTY'S THIGHS AND BACK, AND
GRASPS THE OTHER BEARER'S WRISTS.
THE BEARERS RISE, LIFTING THE CASUALTY.

(B) BACK VIEW

Figure B-13: Two-hand seat carry (Illustrated A and B).

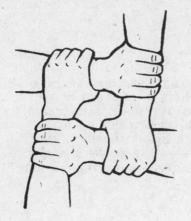

A EACH BEARER GRASPS ONE OF HIS WRISTS AND ONE OF THE OTHER BEARER'S WRISTS, THUS FORMING A PACKSADDLE.

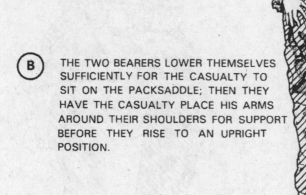

B THE TWO BEARERS LOWER THEMSELVES SUFFICIENTLY FOR THE CASUALTY TO SIT ON THE PACKSADDLE; THEN THEY HAVE THE CASUALTY PLACE HIS ARMS AROUND THEIR SHOULDERS FOR SUPPORT BEFORE THEY RISE TO AN UPRIGHT POSITION.

Figure B-14: Four-hand seat carry (Illstrated A and B).

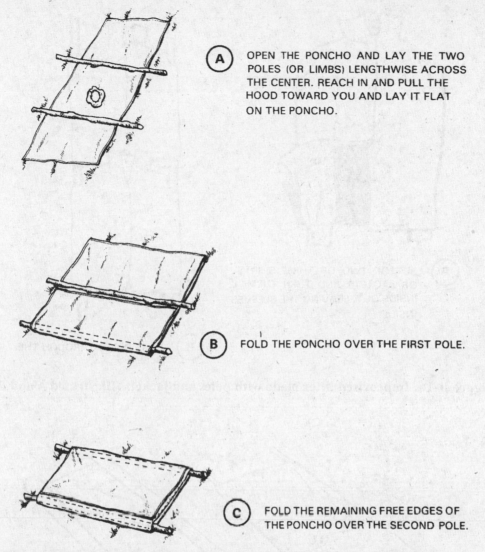

A OPEN THE PONCHO AND LAY THE TWO POLES (OR LIMBS) LENGTHWISE ACROSS THE CENTER. REACH IN AND PULL THE HOOD TOWARD YOU AND LAY IT FLAT ON THE PONCHO.

B FOLD THE PONCHO OVER THE FIRST POLE.

C FOLD THE REMAINING FREE EDGES OF THE PONCHO OVER THE SECOND POLE.

Figure B-15: Improvised litter with poncho and poles (Illustrated A thru C).

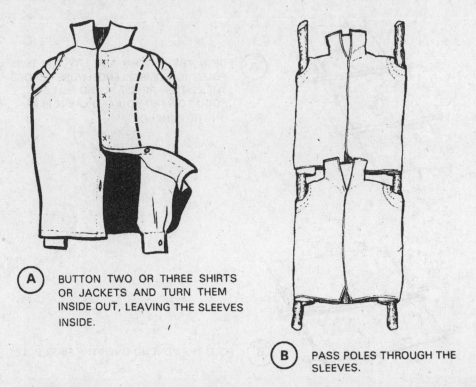

(A) BUTTON TWO OR THREE SHIRTS OR JACKETS AND TURN THEM INSIDE OUT, LEAVING THE SLEEVES INSIDE.

(B) PASS POLES THROUGH THE SLEEVES.

Figure B-16: Improvised litter made with poles and jackets (Illustrated A and B).

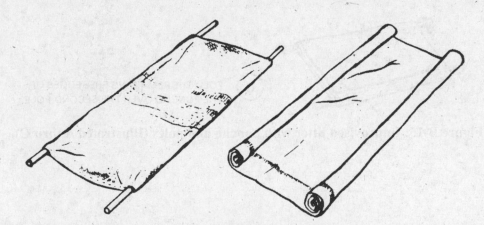

Figure B-17: Improvised litters made by inserting poles through sacks or by rolling blanket.

APPENDIX C

Common Problems/Conditions

SECTION I. HEALTH MAINTENANCE

C-1. General. History has often demonstrated that the course of battle is influenced more by the health of the troops than by strategy or tactics. Health is largely a personal responsibility. Correct cleanliness habits, regular exercise, and good nutrition have much control over a person's wellbeing. Good health does not just happen; it comes with conscious effort and good habits. This appendix outlines some basic principles that promote good health.

C-2. Personal Hygiene

    a. Because of the close living quarters frequently found in an Army environment, personal hygiene is extremely important. Disease or illness can spread and rapidly affect an entire group.

    b. Uncleanliness or disagreeable odors affect the morale of workmates. A daily bath or shower assists in preventing body odor and is necessary to maintain cleanliness. A bath or shower also aids in preventing common skin diseases. Medicated powders and deodorants help keep the skin dry. Special care of the feet is also important. You should wash your feet daily and keep them dry.

C-3. Diarrhea and Dysentery

    a. Poor sanitation can contribute to conditions which may result in diarrhea and dysentery (a medical term applied to a number of intestinal disorders characterized by stomach pain and diarrhea with passage of mucus and blood). Medical personnel can advise regarding the cause and degree of illness. Remember, however, that intestinal diseases are usually spread through contact with infectious organisms which can be spread in human waste, by flies and other insects, or in improperly prepared or disinfected food and water supplies.

    b. Keep in mind the following principles that will assist you in preventing diarrhea and/or dysentery.

    (1) Fill your canteen with treated water at every chance. When treated water is not available you must disinfect the water in your canteen by boiling it or using either iodine tablets or chlorine ampules. Iodine tablets or chlorine ampules can be obtained through your unit supply channels or field sanitation team.

    (a) To treat (disinfect) water by boiling, bring water to a rolling boil in your canteen cup for 5 to 10 minutes. In an emergency, boiling water for even 15 seconds will help. Allow the water to cool before drinking.

    (b) To treat water with iodine—

      • Remove the cap from your canteen and fill the canteen with the cleanest water available.

      • Put one tablet in clear water or two tablets in very cold or cloudy water. Double amounts if using a two quart canteen.

      • Replace the cap, wait 5 minutes, then shake the canteen. Loosen the cap and tip the canteen over to allow leakage around the canteen threads. Tighten the cap and wait an additional 25 minutes before drinking.

(c) To treat water with chlorine—
- Remove the cap from your canteen and fill your canteen with the cleanest water available.
- Mix one ampule of chlorine with one-half canteen cup of water, stir the mixture with a mess kit spoon until the contents are dissolved. Take care not to cut your hands when breaking open the glass ampule.
- Pour one canteen capful of the chlorine solution into your one quart canteen of water.
- Replace the cap and shake the canteen. Loosen the cap and tip the canteen over to allow leakage around the threads. Tighten the cap and wait 30 minutes before drinking.

(2) DO NOT buy food, drinks, or ice from civilian vendors unless approved by medical personnel.

(3) Wash your hands for at least 30 seconds after using the latrine or before touching food.

(4) Wash your mess kit in a mess kit laundry or with treated water.

(5) Food waste should be disposed of properly (covered container, plastic bags or buried) to prevent flies from using it as a breeding area.

C-4. Dental Hygiene

a. Care of the mouth and teeth by daily use of a toothbrush and dental floss after meals is essential. This care may prevent gum disease, infection, and tooth decay.

b. One of the major causes of tooth decay and gum disease is plaque. Plaque is an almost invisible film of decomposed food particles and millions of living bacteria. To prevent dental diseases, you must effectively remove this destructive plaque.

C-5. Drug (Substance) Abuse

a. Drug abuse is a serious problem in the military. It affects combat readiness, job performance, and the health of military personnel and their families. More specifically, drug abuse affects the individual. It costs millions of dollars in lost time and productivity.

b. The reasons for drug abuse are as different as the people who abuse the use of them. Generally, people seem to take drugs to change the way they feel. They may want to feel better or to feel happier. They may want to escape from pain, stress, or frustration. Some may want to forget. Some may want to be accepted or to be sociable. Some people take drugs to escape boredom; some take drugs because they are curious. Peer pressure can also be a very strong reason to use drugs.

c. People often feel better about themselves when they use drugs or alcohol, but the effects do not last. Drugs never solve problems; they just postpone or compound them. People who abuse alcohol or drugs to solve one problem run the risk of continued drug use that creates new problems and makes old problems worse.

d. Drug abuse is very serious and may cause serious health problems. Drug abuse may cause mental incapacitation and even cause death.

C-6. Sexually Transmitted Diseases. Sexually transmitted diseases (STD) formerly known as venereal diseases are caused by organisms normally transmitted through sexual intercourse. Individuals should use a prophylactic (condom) during sexual intercourse unless they have sex only within marriage or with one, steady non infected person of the opposite sex. Another good habit is to wash the sexual parts and urinate immediately after sexual intercourse. Some serious STDs include nonspecific urethritis (chlamydia), gonorrhea, syphilis and Hepatitis B and the Acquired Immunodeficiency Syndrome (AIDS). Prevention of one type of STD through responsible sex, protects both partners from all STDs. Seek the best medical attention if any discharge or blisters are found on your sexual parts.

a. Acquired Immunodeficiency Syndrome (AIDS). AIDS is the end disease stage of the HIV infection. The HIV infection is contagious, but it cannot be spread in the same manner as a common cold, measles, or chicken pox. AIDS is contagious, however, in the same way that sexually transmitted diseases, such as syphilis and gonorrhea, are contagious. AIDS can also be spread through the sharing of intravenous drug needles and syringes used for injecting illicit drugs.

b. High Risk Group. Today those practicing high risk behavior who become infected with the AIDS virus are found mainly among homosexual and bisexual persons and intravenous drug users. Heterosexual transmission is expected to account for an increasing proportion of those who become infected with the AIDS virus in the future.

(1) AIDS caused by virus. The letters A-I-D-S stand for Acquired Immunodeficiency Syndrome. When a person is sick with AIDS, he is in the final stages of a series of health problems caused by a virus (germ) that can be passed from one person to another chiefly during sexual contact or through the sharing of intravenous drug needles and syringes used for "shooting" drugs. Scientists have named the AIDS virus "HIV." The HIV attacks a person's immune system and damages his ability to fight other disease. Without a functioning immune system to ward off other germs, he now becomes vulnerable to becoming infected by bacteria, protozoa, fungi, and other viruses and malignancies, which may cause life-threatening illness, such as pneumonia, meningitis, and cancer.

(2) No known cure. There is presently no cure for AIDS. There is presently no vaccine to prevent AIDS.

(3) Virus invades blood stream. When the AIDS virus enters the blood stream, it begins to attack certain white blood cells (T-Lymphocytes). Substances called antibodies are produced by the body. These antibodies can be detected in the blood by a simple test, usually two weeks to three months after infection. Even before the antibody test is positive, the victim can pass the virus to others.

(4) Signs and Symptoms.
   • Some people remain apparently well after infection with the AIDS virus. They may have no physically apparent symptom of illness. However, if proper precautions are not used with sexual contacts and/or intravenous drug use, these infected individuals can spread the virus to others.
   • The AIDS virus may also attack the nervous system and cause delayed damage to the brain. This damage may take years to develop and the symptoms may show up as memory loss, indifference, loss of coordination, partial paralysis, or mental disorder. These symptoms may occur alone, or with other symptoms mentioned earlier.

(5) AIDS: the present situation. The number of people estimated to be infected with the AIDS virus in the United States is over 1.5 million as of April 1988. In certain parts of central Africa 50% of the sexually active population is infected with HIV. The number of persons known to have AIDS in the United States to date is over 55,000; of these, about half have died of the disease. There is no cure. The others will soon die from their disease. Most scientists predict that all HIV infected persons will develop AIDS sooner or later, if they don't die of other causes first.

(6) Sex between men. Men who have sexual relations with other men are especially at risk. About 70% of AIDS victims throughout the country are male homosexuals and bisexuals. This percentage probably will decline as heterosexual transmission increases. Infection results from a sexual relationship with an infected person.

(7) Multiple partners. The risk of infection increases according to the number of sexual partners one has, male or female. The more partners you have, the greater the risk of becoming infected with the AIDS virus.

(8) How exposed. Although the AIDS virus is found in several body fluids, a person acquires the virus during sexual contact with an infected person's blood or semen and possibly vaginal secretions. The virus then enters a person's blood stream through their rectum, vagina or penis. Small (unseen by the naked eye) tears in the surface lining of the vagina or rectum may occur during insertion of the penis, fingers, or other objects, thus opening an avenue for entrance of the virus directly into the blood stream.

(9) Prevention of sexual transmission—know your partner. Couples who maintain mutually faithful monogamous relationships (only one continuing sexual partner) are protected from AIDS through sexual transmission. If you have been faithful for at least five years and your partner has been faithful too, neither of you is at risk.

(10) Mother can infect newborn. If a woman is infected with the AIDS virus and becomes pregnant, she has about a 50% chance of passing the AIDS virus to her unborn child.

(11) Summary. AIDS affects certain groups of the population. Homosexual and bisexual persons who have had sexual contact with other homosexual or bisexual persons as well as those who "shoot" street drugs are at greatest risk of exposure, infections and eventual death. Sexual partners of these high risk individuals are at risk, as well as any children born to women who carry the virus. Heterosexual persons are increasingly at risk.

(12) Donating blood. Donating blood is not risky at all. You cannot get AIDS by donating blood.

(13) Receiving blood. High risk persons and every blood donation is now tested for the presence of antibodies to the AIDS virus. Blood that shows exposure to the AIDS virus by the presence of antibodies is not used either for transfusion or for the manufacture of blood products. Blood banks are as safe as current technology can make them. Because antibodies do not form immediately after exposure to the virus, a newly infected person may unknowingly donate blood after becoming infected but before his antibody test becomes positive.

(14) Testing of military personnel. You may wonder why the Department of Defense currently tests its uniformed services personnel for presence of the AIDS virus antibody. The military feels this procedure is necessary because the uniformed services act as their own blood bank in a combat situation. They also need to protect new recruits (who unknowingly may be AIDS virus carriers) from receiving live virus vaccines. HIV antibody positive soldiers may not be assigned overseas (includes Alaska and Hawaii). They must be rechecked every six months to determine if the disease has become worse. If the disease has progressed, they are discharged from the Army (policy per AR 600-110). This regulation requires that all soldiers receive annual education classes on AIDS.

SECTION II. FIRST AID FOR COMMON PROBLEMS

C-7. Heat Rash (or Prickly Heat)

a. Description. Heat rash is a skin rash caused by the blockage of the sweat glands because of hot, humid weather or because of fever. It appears as a rash of patches of tiny reddish pinpoints that itch.

b. First Aid. Wear clothing that is light and loose and/or uncover the affected area. Use skin powders or lotion.

C-8. Contact Poisoning (Skin Rashes)

a. General.
(1) Poison Ivy grows as a small plant (vine or shrub) and has three glossy leaflets (Figure C-1).
(2) Poison Oak grows in shrub or vine form; and has clusters of three leaflets with wavy edges (Figure C-2).
(3) Poison Sumac grows as a shrub or small tree. Leaflets grow opposite each other with one at tip (Figure C-3).
b. Signs/Symptoms.
- Redness.
- Swelling.
- Itching.

Figure C-1: Poison ivy.

Figure C-2: Western poison oak.

Figure C-3: Poison sumac.

- Rashes or blisters.
- Burning sensation.
- General headaches and fever.

✍ **NOTE**

Secondary infection may occur when blisters break.

---

c. First Aid.
   (1) Expose the affected area: remove clothing and jewelry.
   (2) Cleanse affected area with soap and water.
   (3) Apply rubbing alcohol, if available, to the affected areas.
   (4) Apply calamine lotion (helps relieve itching and burning).
   (5) Avoid dressing the affected area.
   (6) Seek medical help, evacuate if necessary. (If rash is severe, or on face or genitals, seek medical help.)

C-9. Care of the Feet. Proper foot care is essential for all soldiers in order to maintain their optimal health and physical fitness. To reduce the possibilities of serious foot trouble, observe the following rules:

a. Foot hygiene is important. Wash and dry feet thoroughly, especially between the toes. Soldiers who perspire freely should apply powder lightly and evenly twice a day.

b. Properly fitted shoes/boots should be the only ones issued. There should be no binding or pressure spots.

c. Clean, properly fitting socks should be changed and washed daily. Avoid socks with holes or poorly darned areas; they may cause blisters.

d. Attend promptly to common medical problems such as blisters, ingrown toenails, and fungus infections (like athlete's foot).

e. Foot marches are a severe test for the feet. Use only properly fitted footgear and socks. Footgear should be completely broken-in. DO NOT break-in new footgear on a long march. Any blisters, sores, and so forth, should be treated promptly. Keep the feet as dry as possible on the march; carry extra socks and change if feet get wet (socks can be dried by putting them under your shirt, around your waist or hanging on a rack). Inspect feet during rest breaks. Bring persistent complaints to the attention of medical personnel.

* C-10. Blisters. Blisters are a common problem caused by friction. They may appear on such areas as the toes, heels, or the palm of the hand (anywhere friction may occur). Unless treated promptly and correctly, they may become infected. PREVENTION is the best solution to AVOID blisters and subsequent infection. For example, ensure boots are prepared properly for a good fit, whenever possible always keep feet clean and dry; and, wear clean socks that also fit properly. Gloves should be worn whenever extensive manual work is done.

✍ NOTE

Keep blisters clean. Care should be taken to keep the feet as clean as possible at all times. Use soap and water for cleansing. Painful blisters and/or signs of infection, such as redness, throbbing, drainage, and so forth, are reasons for seeking medical treatment. Seek medical treatment only from qualified medical personnel.

Digital Pressure

APPLY DIGITAL PRESSURE

Digital pressure (also often called "pressure points") is an alternate method to control bleeding. This method uses pressure from the fingers, thumbs, or hands to press at the site or point where a main artery supplying the wounded area lies near the skin surface or over bone (Figure D-1). This pressure may help shut off or slow down the flow of blood from the heart to the wound and is used in combination with direct pressure and elevation. It may help in instances where bleeding is not easily controlled, where a pressure dressing has not yet been applied, or where pressure dressings are not readily available.

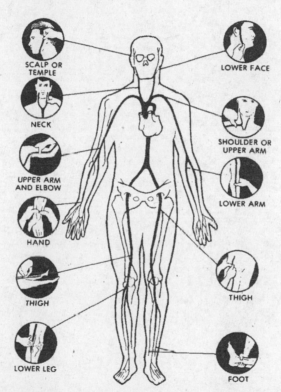

If blood is spurting from wound (artery), press at the point or site where main artery supplying the wounded area lies near skin surface or over bone as shown. This pressure shuts off or slows down the flow of blood from the heart to the wound until a pressure dressing can be unwrapped and applied. You will know you have located the artery when you feel a pulse.

Figure D-1: Digital pressure (pressure with fingers, thumbs or hands).

APPENDIX E

Decontamination Procedures

APPLY DIGITAL PRESSURE

E-1. Protective Measures and Handling of Casualties

a. Depending on the theater of operations, guidance issued may dictate the assumption of a minimum mission-oriented protective posture (MOPP) level. However, a full protective posture (MOPP 4) level will be assumed immediately when the alarm or command is given. (MOPP 4 level consists of wearing the protective overgarment, mask, hood, gloves, and overboots.) If individuals find themselves alone without adequate guidance, they should mask and assume the MOPP 4 level under any of the following conditions.

  (1) Their position is hit by a concentration of artillery, mortar, rocket fire, or by aircraft bombs if chemical agents have been used or the threat of their use is significant.

  (2) Their position is under attack by aircraft spray.

  (3) Smoke or mist of an unknown source is present or approaching.

  (4) A suspicious odor or a suspicious liquid is present.

  (5) A toxic chemical or biological attack is suspected.

  (6) They are entering an area known to be or suspected of being contaminated with a toxic chemical or biological agent.

  (7) During any motor march, once chemical warfare has been initiated.

  (8) When casualties are being received from an area where chemical agents have reportedly been used.

  (9) They have one or more of the following signs/symptoms:

    (a) An unexplained sudden runny nose.

    (b) A feeling of choking or tightness in the chest or throat.

    (c) Blurring of vision and difficulty in focusing the eyes on close objects.

    (d) Irritation of the eyes (could be caused by the presence of several toxic chemical agents).

    (e) Unexplained difficulty in breathing or increased rate of breathing.

    (f) Sudden feeling of depression.

    (g) Dread, anxiety, restlessness.

    (h) Dizziness or light-headedness.

    (i) Slurred speech.

  (10) Unexplained laughter or unusual behavior noted in others.

  (11) Buddies suddenly collapsing without evident cause.

b. Stop breathing, don the protective mask, seat it properly, clear it, and check it for seal; then resume breathing. The mask should be worn until unmasking procedures indicate no chemical agent is in the air and the "all clear" signal is given. If vomiting occurs, the mask should be lifted momentarily and drained—while the eyes are closed and the breath is held—and replaced, cleared, and sealed.

c. Casualties contaminated with a chemical agent may endanger unprotected personnel. Handlers of these casualties must wear a protective mask, protective gloves, and chemical protective clothing until the casualty's contaminated clothing has been removed. The battalion aid station should be established upwind from the most heavily contaminated areas, if it is expected that troops will remain in the area six hours or more. Collective protective shelters must be used to adequately manage casualties on the integrated battlefield. Casualties must be undressed and decontaminated, as required, in an area equipped for the removal of contaminated clothing and equipment prior to entering collective protection. Contaminated clothing and equipment should be placed in airtight containers or plastic bags, if available, or removed to a designated dump site downwind from the aid station.

F-2. Personal Decontamination. Following contamination of the skin or eyes with vesicants (mustards, lewisite, and so forth) or nerve agents, personal decontamination must be carried out immediately. This is because chemical agents are effective at very small concentrations and within a very few minutes after exposure, decontamination is marginally effective. Decontamination consists of either removal and/or neutralization of the agent. Decontamination after absorption occurs may serve little or no purpose. Soldiers will decontaminate themselves unless they are incapacitated. For soldiers who cannot decontaminate themselves, the nearest able person should assist them as the situation permits.

✍ **NOTE**

In a cyanide only environment, there would be no need for decontamination.

---

a. Eyes. Following contamination of the eyes with any chemical agent, the agent must be removed instantly. In most cases, identity of the agent will not be known immediately. Individuals who suspect contamination of their eyes or face must quickly obtain overhead shelter to protect themselves while performing the following decontamination process:

(1) Remove and open your canteen.
(2) Take a deep breath and hold it.
(3) Remove the mask.
(4) Flush or irrigate the eye, or eyes, immediately with large amounts of water. To flush the eyes with water from a canteen (or other container of uncontaminated water), tilt the head to one side, open the eyelids as wide as possible, and pour water slowly into the eye so that it will run off the side of the face to avoid spreading the contamination. This irrigation must be carried out despite the presence of toxic vapors in the atmosphere. Hold your breath and keep your mouth closed during this procedure to prevent contamination and absorption through the mucous membranes. Chemical residue flushed from the eyes should be neutralized along the flush path.

WARNING

DO NOT use the fingers or gloved hands for holding the eyelids apart. Instead, open the eyes as wide as possible and pour the water as indicated above.

(5) Replace, clear, and check your mask. Then resume breathing.
(6) If contamination was picked up while flushing the eyes, then decontaminate the face. Follow procedure outlined in paragraph b (2) *(a)* through *(ae)* below.

b. Skin (Hands, Face, Neck, Ears, and Other Exposed Areas). The M258A1 Skin Decontamination Kit (Figure F-1) is provided individuals for performing emergency decontamination of their skin (and selected small equipment, such as the protective gloves, mask, hood, and individual weapon).

(1) Description of the M258A1 kit. The M258A1 kit measures 1 3/4 by 2 3/4 by 4 inches and weighs 0.2 pounds. Each kit contains six packets: three DECON-1 packets and three DECON-2 packets. DECON-1 packet contains a pad premoistened with hydroxyethane 72%, phenol 10%, sodium hydroxide 5%, and ammonia 0.2%, and the remainder water. DECON-2 packet contains a pad impregnated with chloramine B and sealed glass ampules filled with hydroxyethane 45%, zinc

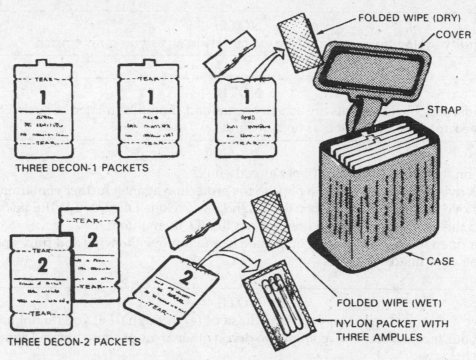

THREE DECON-1 PACKETS

THREE DECON-2 PACKETS

Figure F-1: M258A1 Skin Decontamination Kit.

chloride 5%, and the remainder water. The case fits into the pocket on the outside rear of the M17 series protective mask carrier or in an inside pocket of the carrier for the M24 and M25 series protective mask. The case can also be attached to the web belt or on the D ring of the protective mask carrier.

(2) Use of the M258A1 kit. It should be noted that the procedures outlined in paragraphs (*a*) through (*ue*) below were not intended to replace or supplant those contained in STP 21-1-SMCT but, rather, to expand on the doctrine of skin decontamination.

WARNING
The ingredients of the DECON-1 and DECON-2 packets of the M258A1 kit are poisonous and caustic and can permanently damage the eyes. KEEP PADS OUT OF THE EYES, MOUTH, AND OPEN WOUNDS. Use water to wash the toxic agent out of the eyes or wounds, except in the case of mustard. Mustard may be removed by thorough immediate wiping.

WARNING
The complete decon (WIPES 1 and 2) of the face must be done as quickly as possible–3 minutes or less.

WARNING
DO NOT attempt to decontaminate the face or neck before putting on a protective mask.

✍ NOTE

Use the buddy system to decontaminate exposed skin areas you cannot reach.

✍ NOTE

Blisters caused by blister agents are actually burns and should be treated as such. Blisters which have ruptured are treated as open wounds.

(a) Put on the protective mask (if not already on).
(b) Seek overhead cover or use a poncho for protection against further contamination.
(c) Remove the M258A1 kit. Open the kit and remove one DECON-1 WIPE packet by its tab.
(d) Fold the packet on the solid line marked BEND, then unfold it.
(e) Tear open the packet quickly at the notch, and remove the wipe and fully open it.
(f) Wipe your hands.

✍ NOTE

If you have a chemical agent on your face, do steps (g) through (t). If you do not have an agent on your face, do step (m), continue to decon other areas of contaminated skin, then go to step (n).

✍ NOTE

You must hold your breath while doing steps (g) through (l). If you need to breathe before you finish, reseal your mask, clear it and check it, then continue.

(g) Hold your breath, close your eyes, and lift the hood and mask from your chin.
(h) Scrub up and down from ear to ear.
   1. Start at an ear.
   2. Scrub across the face to the corner of the nose.
   3. Scrub an extra stroke at the corner of the nose.
   4. Scrub across the nose and tip of the nose to the corner of the nose.
   5. Scrub an extra stroke at the corner of the nose.
   6. Scrub across the face to the other ear.
(i) Scrub up and down from the ear to the end of the jawbone.
   1. Begin where step (h) ended.
   2. Scrub across the cheek to the corner of the mouth.
   3. Scrub an extra stroke at the corner of the mouth.
   4. Scrub across the closed mouth to the center of the upper lip.
   5. Scrub an extra stroke above the upper lip.
   6. Scrub across the closed mouth to the corner of the mouth.
   7. Scrub an extra stroke at the corner of the mouth.
   8. Scrub across the cheek to the end of the jawbone.
(j) Scrub up and down from one end of the jawbone to the other end of the jawbone.
   1. Begin where step (i) ended.
   2. Scrub across and under the jaw to the chin, cupping the chin.
   3. Scrub an extra stroke at the cleft of the chin.
   4. Scrub across and under the jaw to the end of the jawbone.
(k) Quickly wipe the inside of the mask which touches the face.

(l) Reseal, clear, and check the mask. Resume breathing.

(m) Using the same DECON-1 WIPE, scrub the neck and the ears.

(n) Rewipe the hands.

(o) Drop the wipe to the ground.

(p) Remove one DECON-2 WIPE packet, and crush the encased glass ampules between the thumb and fingers. DO NOT KNEAD.

(q) Fold the packet on the solid line marked CRUSH AND BEND, then unfold it.

(r) Tear open the packet quickly at the notch and remove the wipe.

(s) Fully open the wipe. Let the encased crushed glass ampules fall to the ground.

(t) Wipe your hands.

✍ NOTE

If you have an agent on your face, do steps (u) through (ae). If you do not have an agent on your face, do step (aa), continue to decon other areas of contaminated skin, then go to step (ab).

✍ NOTE

You must hold your breath while doing steps (u) through (z). If you need to breathe before you finish, reseal your mask, clear it and check it, then continue.

(u) Hold your breath, close your eyes, and lift the hood and mask away from your chin.

(v) Scrub up and down from ear to ear.
  1. Start at an ear.
  2. Scrub across the face to the corner of the nose.
  3. Scrub an extra stroke at the corner of the nose.
  4. Scrub across the nose and tip of the nose to the corner of the nose.
  5. Scrub an extra stroke at the corner of the nose.
  6. Scrub across the face to the other ear.

(w) Scrub up and down from the ear to the end of the jawbone.
  1. Begin where step (v) ended.
  2. Scrub across the cheek to the corner of the mouth.
  3. Scrub an extra stroke at the corner of the mouth.
  4. Scrub across the closed mouth to the center of the upper lip.
  5. Scrub an extra stroke above the upper lip.
  6. Scrub across the closed mouth to the corner of the mouth.
  7. Scrub an extra stroke at the corner of the mouth.
  8. Scrub across the cheek to the end of the jawbone.

(x) Scrub up and down from one end of the jawbone to the other end of the jawbone.
  1. Begin where step (w) ended.
  2. Scrub across and under the jaw to the chin, cupping the chin.
  3. Scrub an extra stroke at the cleft of the chin.
  4. Scrub across and under the jaw to the end of the jawbone.

(y) Quickly wipe the inside of the mask which touches the face.

(z) Reseal, clear, and check the mask. Resume breathing.

(aa) Using the same DECON-2 WIPE, scrub the neck and ears.

(ab) Rewipe the hands.

(ac) Drop the wipe to the ground.

    (ad) Put on the protective gloves and any other protective clothing, as appropriate. Fasten the hood straps and neck cord.

    (ae) Bury the decontaminating packet and other items dropped on the ground, if circumstances permit.

  c. Clothing and Equipment. Although the M258A1 may be used for decontamination of selected items of individual clothing and equipment (for example, the soldier's individual weapon), there is insufficient capability to do more than emergency spot decontamination. The M258A1 is not used to decontaminate the protective overgarment. The protective overgarment does not require immediate decontamination since the charcoal layer is a decontaminating device; however, it must be exchanged. The Individual Equipment Decontamination Kit (DKIE), M280 (similar in configuration to the M258A1), is used to decontaminate equipment such as the weapon, helmet, and other gear that is carried by the individual.

E-3. Casualty Decontamination. Contaminated casualties entering the medical treatment system are decontaminated through a decentralized process. This is initially started through self-aid and buddy aid procedures. Later, units should further decontaminate the casualty before evacuation. Casualty decontamination stations are established at the field medical treatment facility to further decontaminate these individuals (clothing removal and spot decontamination, as required) prior to treatment and evacuation. These stations are manned by nonmedical members of the supported unit under supervision of medical personnel. There are insufficient medical personnel to both decontaminate and treat casualties. The medical personnel must be available for treatment of the casualties during and after decontamination by nonmedical personnel. Decontamination is accomplished as quickly as possible to facilitate medical treatment, prevent the casualty from absorbing additional agent, and reduce the spread of chemical contamination.

PART III
Shelters

Introduction

A shelter can protect you from the sun, insects, wind, rain, snow, hot or cold temperatures, and enemy observation. It can give you a feeling of well-being. It can help you maintain your will to survive. In some areas, your need for shelter may take precedence over your need for food and possibly even your need for water. For example, prolonged exposure to cold can cause excessive fatigue and weakness (exhaustion). An exhausted person may develop a "passive" outlook, thereby losing the will to survive. The most common error in making a shelter is to make it too large. A shelter must be large enough to protect you. It must also be small enough to contain your body heat, especially in cold climates.

SHELTER SITE SELECTION

When you are in a survival situation and realize that shelter is a high priority, start looking for shelter as soon as possible. As you do so, remember what you will need at the site. Two requisites are—

- It must contain material to make the type of shelter you need.
- It must be large enough and level enough for you to lie down comfortably.

When you consider these requisites, however, you cannot ignore your tactical situation or your safety. You must also consider whether the site—

- Provides concealment from enemy observation.
- Has camouflaged escape routes.
- Is suitable for signaling, if necessary.
- Provides protection against wild animals and rocks and dead trees that might fall.
- Is free from insects, reptiles, and poisonous plants.

You must also remember the problems that could arise in your environment. For instance—

- Avoid flash flood areas in foothills.
- Avoid avalanche or rockslide areas in mountainous terrain.
- Avoid sites near bodies of water that are below the high water mark.

In some areas, the season of the year has a strong bearing on the site you select. Ideal sites for a shelter differ in winter and summer. During cold winter months you will want a site that will protect you from the cold and wind, but will have a source of fuel and water. During summer months in the same area you will want a source of water, but you will want the site to be almost insect free.

When considering shelter site selection, use the word BLISS as a guide.

B—Blend in with the surroundings.
L—Low silhouette.
I—Irregular shape.

S—Small.
S—Secluded location.

TYPES OF SHELTERS

When looking for a shelter site, keep in mind the type of shelter (protection) you need. However, you must also consider—

- How much time and effort you need to build the shelter.
- If the shelter will adequately protect you from the elements (sun, wind, rain, snow).
- If you have the tools to build it. If not, can you make improvised tools?
- If you have the type and amount of materials needed to build it.

To answer these questions, you need to know how to make various types of shelters and what materials you need to make them.

Poncho Lean-To. It takes only a short time and minimal equipment to build this lean-to (Figure I-1). You need a poncho, 2 to 3 meters of rope or parachute suspension line, three stakes about 30 centimeters long, and two trees or two poles 2 to 3 meters apart. Before selecting the trees you will use or the location of your poles, check the wind direction. Ensure that the back of your lean-to will be into the wind.

To make the lean-to—

- Tie off the hood of the poncho. Pull the drawstring tight, roll the hood long ways, fold it into thirds, and tie it off with the drawstring.
- Cut the rope in half. On one long side of the poncho, tie half of the rope to the corner grommet. Tie the other half to the other corner grommet.
- Attach a drip stick (about a 10-centimeter stick) to each rope about 2.5 centimeters from the grommet. These drip sticks will keep rainwater from running down the ropes into the lean-to. Tying strings (about 10 centimeters long) to each grommet along the poncho's top edge will allow the water to run to and down the line without dripping into the shelter.
- Tie the ropes about waist high on the trees (uprights). Use a round turn and two half hitches with a quick-release knot.
- Spread the poncho and anchor it to the ground, putting sharpened sticks through the grommets and into the ground.

Figure I-1: Poncho lean-to.

If you plan to use the lean-to for more than one night, or you expect rain, make a center support for the lean-to. Make this support with a line. Attach one end of the line to the poncho hood and the other end to an overhanging branch. Make sure there is no slack in the line.

Another method is to place a stick upright under the center of the lean-to. This method, however, will restrict your space and movements in the shelter.

For additional protection from wind and rain, place some brush, your rucksack, or other equipment at the sides of the lean-to.

To reduce heat loss to the ground, place some type of insulating material, such as leaves or pine needles, inside your lean-to.

Note: When at rest, you lose as much as 80 percent of your body heat to the ground.

To increase your security from enemy observation, lower the lean-to's silhouette by making two changes. First, secure the support lines to the trees at knee height (not at waist height) using two knee-high sticks in the two center grommets (sides of lean-to). Second, angle the poncho to the ground, securing it with sharpened sticks, as above.

Poncho Tent. This tent (Figure I-2) provides a low silhouette. It also protects you from the elements on two sides. It has, however, less usable space and observation area than a lean-to, decreasing your reaction time to enemy detection. To make this tent, you need a poncho, two 1.5- to 2.5-meter ropes, six sharpened sticks about 30 centimeters long, and two trees 2 to 3 meters apart.

To make the tent—

- Tie off the poncho hood in the same way as the poncho lean-to.
- Tie a 1.5- to 2.5-meter rope to the center grommet on each side of the poncho.
- Tie the other ends of these ropes at about knee height to two trees 2 to 3 meters apart and stretch the poncho tight.
- Draw one side of the poncho tight and secure it to the ground pushing sharpened sticks through the grommets.
- Follow the same procedure on the other side.

If you need a center support, use the same methods as for the poncho lean-to. Another center support is an A-frame set outside but over the center of the tent (Figure I-3). Use two 90- to 120-centimeter-long sticks, one with a forked end, to form the A-frame. Tie the hood's drawstring to the A-frame to support the center of the tent.

Figure I-2: Poncho tent using overhanging branch.

Figure I-3: Poncho tent with A-frame.

Three-Pole Parachute Tepee. If you have a parachute and three poles and the tactical situation allows, make a parachute tepee. It is easy and takes very little time to make this tepee. It provides protection from the elements and can act as a signaling device by enhancing a small amount of light from a fire or candle. It is large enough to hold several people and their equipment and to allow sleeping, cooking, and storing firewood.

You can make this tepee using parts of or a whole personnel main or reserve parachute canopy. If using a standard personnel parachute, you need three poles 3.5 to 4.5 meters long and about 5 centimeters in diameter.

To make this tepee (Figure I-4)—

- Lay the poles on the ground and lash them together at one end.
- Stand the framework up and spread the poles to form a tripod.
- For more support, place additional poles against the tripod. Five or six additional poles work best, but do not lash them to the tripod.
- Determine the wind direction and locate the entrance 90 degrees or more from the mean wind direction.
- Lay out the parachute on the "backside" of the tripod and locate the bridle loop (nylon web loop) at the top (apex) of the canopy.
- Place the bridle loop over the top of a free-standing pole. Then place the pole back up against the tripod so that the canopy's apex is at the same height as the lashing on the three poles.
- Wrap the canopy around one side of the tripod. The canopy should be of double thickness, as you are wrapping an entire parachute. You need only wrap half of the tripod, as the remainder of the canopy will encircle the tripod in the opposite direction.
- Construct the entrance by wrapping the folded edges of the canopy around two free-standing poles. You can then place the poles side by side to close the tepee's entrance.
- Place all extra canopy underneath the tepee poles and inside to create a floor for the shelter.
- Leave a 30- to 50-centimeter opening at the top for ventilation if you intend to have a fire inside the tepee.

One-Pole Parachute Tepee. You need a 14-gore section (normally) of canopy, stakes, a stout center pole, and inner core and needle to construct this tepee. You cut the suspension lines except for 40- to 45-centimeter lengths at the canopy's lower lateral band.

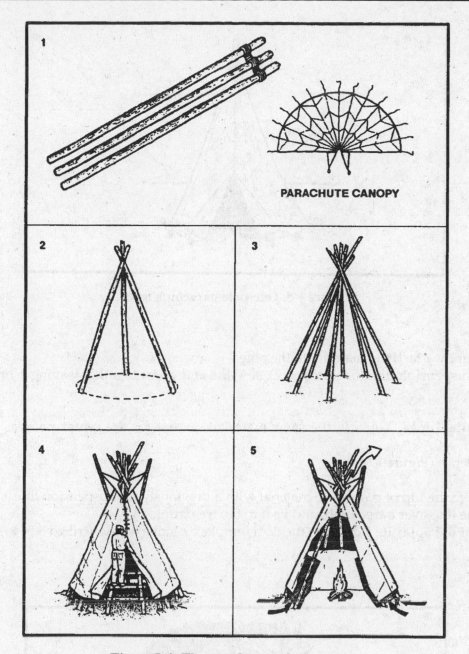

Figure I-4: Three-pole parachute tepee.

To make this tepee (Figure I-5)—

- Select a shelter site and scribe a circle about 4 meters in diameter on the ground.
- Stake the parachute material to the ground using the lines remaining at the lower lateral band.
- After deciding where to place the shelter door, emplace a stake and tie the first line (from the lower lateral band) securely to it.
- Stretch the parachute material taut to the next line, emplace a stake on the scribed line, and tie the line to it.
- Continue the staking process until you have tied all the lines.
- Loosely attach the top of the parachute material to the center pole with a suspension line you previously cut and, through trial and error, determine the point at which the parachute material will be pulled tight once the center pole is upright.

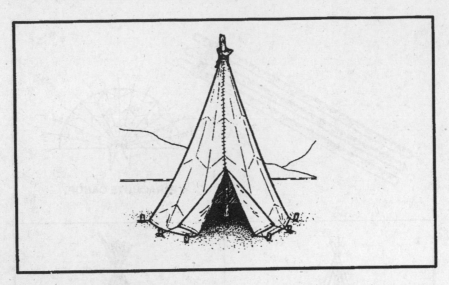

Figure I-5: One-pole parachute tepee.

- Then securely attach the material to the pole.
- Using a suspension line (or inner core), sew the end gores together leaving 1 or 1.2 meters for a door.

No-Pole Parachute Tepee. You use the same materials, except for the center pole, as for the one-pole parachute tepee.

To make this tepee (Figure I-6)—

- Tie a line to the top of parachute material with a previously cut suspension line.
- Throw the line over a tree limb, and tie it to the tree trunk.
- Starting at the opposite side from the door, emplace a stake on the scribed 3.5- to 4.3-meter circle.

Figure I-6: No-pole parachute tepee.

- Tie the first line on the lower lateral band.
- Continue emplacing the stakes and tying the lines to them.
- After staking down the material, unfasten the line tied to the tree trunk, tighten the tepee material by pulling on this line, and tie it securely to the tree trunk.

One-Man Shelter. A one-man shelter you can easily make using a parachute, a tree and three poles. One pole should be about 4.5 meters long and the other two about 3 meters long.

To make this shelter (Figure I-7)—

- Secure the 4.5-meter pole to the tree at about waist height.
- Lay the two 3-meter poles on the ground on either side of and in the same direction as the 4.5-meter pole.
- Lay the folded canopy over the 4.5 meter pole so that about the same amount of material hangs on both sides.
- Tuck the excess material under the 3-meter poles, and spread it on the ground inside to serve as a floor.
- Stake down or put a spreader between the two 3-meter poles at the shelter's entrance so they will not slide inward.
- Use any excess material to cover the entrance.
- The parachute cloth makes this shelter wind resistant, and the shelter is small enough that it is easily warmed. A candle, used carefully, can keep the inside temperature comfortable. This shelter is unsatisfactory, however, when snow is falling as even a light snowfall will cave it in.

Parachute Hammock. You can make a hammock using 6 to 8 gores of parachute canopy and two trees about 4.5 meters apart (Figure I-8).

Field-Expedient Lean-To. If you are in a wooded area and have enough natural materials, you can make a field-expedient lean-to (Figure I-9) without the aid of tools or with only a knife. It takes longer to make this type of shelter than it does to make other types, but it will protect you from the elements.

You will need two trees (or upright poles) about 2 meters apart; one pole about 2 meters long and 2.5 centimeters in diameter; five to eight poles about 3 meters long and 2.5 centimeters in diameter for beams; cord or vines for securing the horizontal support to the trees; and other poles, saplings, or vines to crisscross the beams.

Figure I-7: One-man shelter.

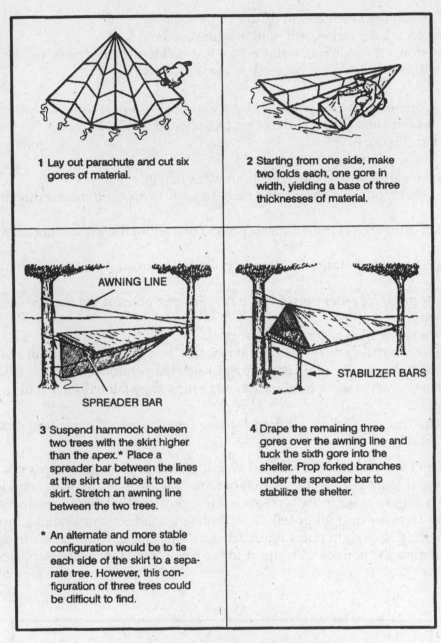

Figure I-8: Parachute hammock.

To make this lean-to—

- Tie the 2-meter pole to the two trees at waist to chest height. This is the horizontal support. If a standing tree is not available, construct a biped using Y-shaped sticks or two tripods.
- Place one end of the beams (3-meter poles) on one side of the horizontal support. As with all lean-to type shelters, be sure to place the lean-to's backside into the wind.
- Crisscross saplings or vines on the beams.
- Cover the framework with brush, leaves, pine needles, or grass, starting at the bottom and working your way up like shingling.
- Place straw, leaves, pine needles, or grass inside the shelter for bedding.

Figure I-9: Field-expedient lean-to and fire reflector.

In cold weather, add to your lean-to's comfort by building a fire reflector wall (Figure I-9). Drive four 1.5-meter-long stakes into the ground to support the wall. Stack green logs on top of one another between the support stakes. Form two rows of stacked logs to create an inner space within the wall that you can fill with dirt. This action not only strengthens the wall but makes it more heat reflective. Bind the top of the support stakes so that the green logs and dirt will stay in place.

With just a little more effort you can have a drying rack. Cut a few 2-centimeter-diameter poles (length depends on the distance between the lean-to's horizontal support and the top of the fire reflector wall). Lay one end of the poles on the lean-to support and the other end on top of the reflector wall. Place and tie into place smaller sticks across these poles. You now have a place to dry clothes, meat, or fish.

Swamp Bed. In a marsh or swamp, or any area with standing water or continually wet ground, the swamp bed (Figure I-10) keeps you out of the water. When selecting such a site, consider the weather, wind, tides, and available materials.

To make a swamp bed—

- Look for four trees clustered in a rectangle, or cut four poles (bamboo is ideal) and drive them firmly into the ground so they form a rectangle. They should be far enough apart and strong enough to support your height and weight, to include equipment.
- Cut two poles that span the width of the rectangle. They, too, must be strong enough to support your weight.
- Secure these two poles to the trees (or poles). Be sure they are high enough above the ground or water to allow for tides and high water.
- Cut additional poles that span the rectangle's length. Lay them across the two side poles, and secure them.
- Cover the top of the bed frame with broad leaves or grass to form a soft sleeping surface.
- Build a fire pad by laying clay, silt, or mud on one corner of the swamp bed and allow it to dry.

Another shelter designed to get you above and out of the water or wet ground uses the same rectangular configuration as the swamp bed. You very simply lay sticks and branches lengthwise on the

Figure I-10: Swamp bed.

inside of the trees (or poles) until there is enough material to raise the sleeping surface above the water level.

Natural Shelters. Do not overlook natural formations that provide shelter. Examples are caves, rocky crevices, clumps of bushes, small depressions, large rocks on leeward sides of hills, large trees with low-hanging limbs, and fallen trees with thick branches. However, when selecting a natural formation—

- Stay away from low ground such as ravines, narrow valleys, or creek beds. Low areas collect the heavy cold air at night and are therefore colder than the surrounding high ground. Thick, brushy, low ground also harbors more insects.
- Check for poisonous snakes, ticks, mites, scorpions, and stinging ants.
- Look for loose rocks, dead limbs, coconuts, or other natural growth that could fall on your shelter.

Debris Hut. For warmth and ease of construction, this shelter is one of the best.
   When shelter is essential to survival, build this shelter.
   To make a debris hut (Figure I-11)—

- Build it by making a tripod with two short stakes and a long ridgepole or by placing one end of a long ridgepole on top of a sturdy base.
- Secure the ridgepole (pole running the length of the shelter) using the tripod method or by anchoring it to a tree at about waist height.
- Prop large sticks along both sides of the ridgepole to create a wedge-shaped ribbing effect. Ensure the ribbing is wide enough to accommodate your body and steep enough to shed moisture.
- Place finer sticks and brush crosswise on the ribbing. These form a latticework that will keep the insulating material (grass, pine needles, leaves) from falling through the ribbing into the sleeping area.
- Add light, dry, if possible, soft debris over the ribbing until the insulating material is at least 1 meter thick—the thicker the better.
- Place a 30-centimeter layer of insulating material inside the shelter.
- At the entrance, pile insulating material that you can drag to you once inside the shelter to close the entrance or build a door.
- As a final step in constructing this shelter, add shingling material or branches on top of the debris layer to prevent the insulating material from blowing away in a storm.

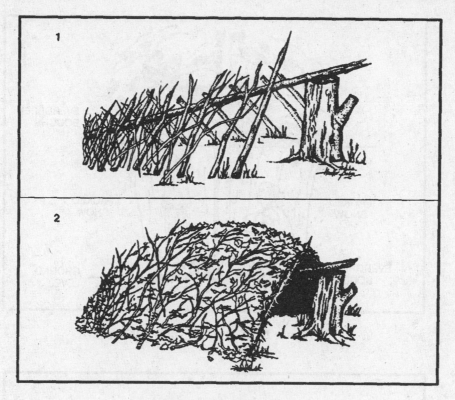

Figure I-11: Debris hut.

Tree-Pit Snow Shelter. If you are in a cold, snow-covered area where evergreen trees grow and you have a digging tool, you can make a tree-pit shelter (Figure I-12).

To make this shelter—

- Find a tree with bushy branches that provides overhead cover.
- Dig out the snow around the tree trunk until you reach the depth and diameter you desire, or until you reach the ground.
- Pack the snow around the top and the inside of the hole to provide support.
- Find and cut other evergreen boughs. Place them over the top of the pit to give you additional overhead cover. Place evergreen boughs in the bottom of the pit for insulation.

See Chapter 3 for other arctic or cold weather shelters.

Beach Shade Shelter. This shelter protects you from the sun, wind, rain, and heat. It is easy to make using natural materials.

To make this shelter (Figure I-13)—

- Find and collect driftwood or other natural material to use as support beams and as a digging tool.
- Select a site that is above the high water mark.
- Scrape or dig out a trench running north to south so that it receives the least amount of sunlight. Make the trench long and wide enough for you to lie down comfortably.
- Mound soil on three sides of the trench. The higher the mound, the more space inside the shelter.
- Lay support beams (driftwood or other natural material) that span the trench on top of the mound to form the framework for a roof.

Figure I-12: Tree-pit snow shelter.

Figure I-13: Beach shade shelter.

- Enlarge the shelter's entrance by digging out more sand in front of it.
- Use natural materials such as grass or leaves to form a bed inside the shelter.

Desert Shelters. In an arid environment, consider the time, effort, and material needed to make a shelter. If you have material such as a poncho, canvas, or a parachute, use it along with such terrain features as rock outcropping, mounds of sand, or a depression between dunes or rocks to make your shelter.

Using rock outcroppings—

- Anchor one end of your poncho (canvas, parachute, or other material) on the edge of the outcrop using rocks or other weights.
- Extend and anchor the other end of the poncho so it provides the best possible shade.

In a sandy area—

- Build a mound of sand or use the side of a sand dune for one side of the shelter.
- Anchor one end of the material on top of the mound using sand or other weights.
- Extend and anchor the other end of the material so it provides the best possible shade.

Note: If you have enough material, fold it in half and form a 30-centimeter to 45-centimeter airspace between the two halves. This airspace will reduce the temperature under the shelter.

A below ground shelter (Figure I-14) can reduce the midday heat as much as 16 to 22 degrees C (30 to 40 degrees F). Building it, however, requires more time and effort than for other shelters. Since your physical effort will make you sweat more and increase dehydration, construct it before the heat of the day.

To make this shelter—

- Find a low spot or depression between dunes or rocks. If necessary, dig a trench 45 to 60 centimeters deep and long and wide enough for you to lie in comfortably.
- Pile the sand you take from the trench to form a mound around three sides.
- On the open end of the trench, dig out more sand so you can get in and out of your shelter easily.
- Cover the trench with your material.
- Secure the material in place using sand, rocks, or other weights.

If you have extra material, you can further decrease the midday temperature in the trench by securing the material 30 to 45 centimeters above the other cover. This layering of the material will reduce the inside temperature 11 to 22 degrees C (20 to 40 degrees F).

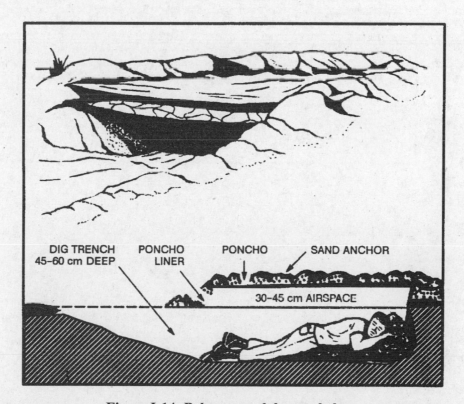

Figure I-14: Belowground desert shelter.

Another type of below ground shade shelter is of similar construction, except all sides are open to air currents and circulation. For maximum protection, you need a minimum of two layers of parachute material (Figure I-15). White is the best color to reflect heat; the innermost layer should be of darker material.

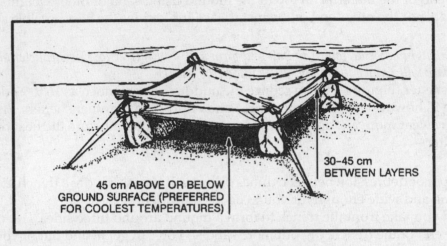

Figure I-15: Open desert shelter.

CHAPTER 1

Planning Positions

This chapter highlights basic survivability knowledge required for planning fighting and protective positions. Included are descriptions of the various directly and indirectly fired weapons and their multiple penetration capabilities and effects on the positions. Both natural and man-made materials available to construct the positions are identified and ranked according to their protection potential. Positions are then categorized and briefly described. Construction methods, including the use of hand tools as well as explosives, and special overall construction considerations such as camouflage and concealment, are also presented.

WEAPONS EFFECTS

A fighting position is a place on the battlefield from which troops engage the enemy with direct and indirect fire weapons. The positions provide necessary protection for personnel, yet allow for fields of fire and maneuver. A protective position protects the personnel and/or material not directly involved with fighting the enemy from attack or environmental extremes. In order to develop plans for fighting and protective positions, five types of weapons, their effects, and their survivability considerations are presented. Air-delivered weapons such as ATGMs, laser-guided missiles, mines, and large bombs require similar survivability considerations.

Direct Fire. Direct fire projectiles are primarily designed to strike a target with a velocity high enough to achieve penetration. The chemical energy projectile uses some form of chemical heat and blast to achieve penetration. It detonates either at impact or when maximum penetration is achieved. Chemical energy projectiles carrying impact-detonated or delayed detonation high-explosive charges are used mainly for direct fire from systems with high accuracy and consistently good target acquisition ability. Tanks, antitank weapons, and automatic cannons usually use these types of projectiles. The kinetic energy projectile uses high velocity and mass (momentum) to penetrate its target. Currently, the hypervelocity projectile causes the most concern in survivability position design. The materials used must dissipate the projectile's energy and thus prevent total penetration. Shielding against direct fire projectiles should initially stop or deform the projectiles in order to prevent or limit penetration. Direct fire projectiles are further divided into the categories of ball and tracer, armor piercing and armor piercing incendiary, and high explosive (HE) rounds.

Ball and Tracer. Ball and tracer rounds are normally of a relatively small caliber (5.56 to 14.5 millimeters (mm)) and are fired from pistols, rifles, and machine guns. The round's projectile penetrates soft targets on impact at a high velocity. The penetration depends directly on the projectile's velocity, weight, and angle at which it hits.

Armor Piercing and Armor Piercing Incendiary. Armor piercing and armor piercing incendiary rounds are designed to penetrate armor plate and other types of homogeneous steel. Armor piercing projectiles have a special jacket encasing a hard core or penetrating rod which is designed to penetrate when fired with high accuracy at an angle very close to the perpendicular of the target. Incendiary projectiles are used principally to penetrate a target and ignite its contents. They are used effectively against fuel supplies and storage areas.

High Explosive. High explosive rounds include high explosive antitank (HEAT) rounds, recoilless rifle rounds, and antitank rockets. They are designed to detonate a shaped charge on impact. At detonation, an extremely high velocity molten jet is formed. This jet perforates large thicknesses of high-density material, continues along its path, and sets fuel and ammunition on fire. The HEAT rounds generally range in size from 60 to 120 mm.

Survivability Considerations. Direct fire survivability considerations include oblique impact, or impact of projectiles at other than a perpendicular angle to the structure, which increases the apparent thickness of the structure and decreases the possibility of penetration. The potential for ricochet off a structure increases as the angle of impact from the perpendicular increases. Designers of protective structures should select the proper material and design exposed surfaces with the maximum angle from the perpendicular to the direction of fire. Also, a low structure silhouette design makes a structure harder to engage with direct fire.

Indirect Fire. Indirect fire projectiles used against fighting and protective positions include mortar and artillery shells and rockets which cause blast and fragmentation damage to affected structures.

Blast. Blast, caused by the detonation of the explosive charge, creates a shock wave which knocks apart walls or roof structures. Contact bursts cause excavation cave-in from ground shock, or structure collapse. Overhead bursts can buckle or destroy the roof,

Blasts from high explosive shells or rockets can occur in three ways:

- Overhead burst (fragmentation from an artillery airburst shell).
- Contact burst (blast from an artillery shell exploding on impact).
- Delay fuze burst (blast from an artillery shell designed to detonate after penetration into a target).

The severity of the blast effects increases as the distance from the structure to the point of impact decreases. Delay fuze bursts are the greatest threat to covered structures. Repeated surface or delay fuze bursts further degrade fighting and protective positions by the cratering effect and soil discharge. Indirect fire blast effects also cause concussions. The shock from a high explosive round detonation causes headaches, nosebleeds, and spinal and brain concussions.

Fragmentation. Fragmentation occurs when the projectile disintegrates, producing a mass of high-speed steel fragments which can perforate and become imbedded in fighting and protective positions. The pattern or distribution of fragments greatly affects the design of fighting and protective positions. Airburst of artillery shells provides the greatest unrestricted distribution of fragments. Fragments created by surface and delay bursts are restricted by obstructions on the ground.

Survivability Considerations. Indirect fire survivability from fragmentation requires shielding similar to that needed for direct fire penetration.

Nuclear. Nuclear weapons effects are classified as residual and initial. Residual effects (such as fallout) are primarily of long-term concern. However, they may seriously alter the operational plans in the immediate battle area. Figure 1-1 shows how the energy released by detonation of a tactical nuclear explosion is divided. Initial effects occur in the immediate area shortly after detonation and are the most tactically significant since they cause personnel casualties and material damage within the immediate time span of any operation. The principal initial casualty-producing effects are blast, thermal radiation (burning), and nuclear radiation. Other initial effects, such as electromagnetic pulse (EMP) and transient radiation effects on electronics (TREE), affect electrical and electronic equipment.

Blast. Blast from nuclear bursts overturns and crushes equipment, collapses lungs, ruptures eardrums, hurls debris and personnel, and collapses positions and structures.

Thermal Radiation. Thermal radiation sets fire to combustible materials, and causes flash blindness or burns in the eyes, as well as personnel casualties from skin burns.

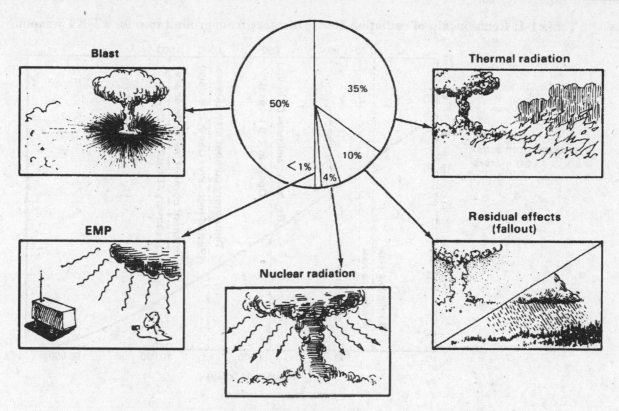

Figure 1-1: Energy distribution of tactical nuclear weapons.

Nuclear Radiation. Nuclear radiation damages cells throughout the body. This radiation damage may cause the headaches, nausea, vomiting, and diarrhea generally called "radiation sickness." The severity of radiation sickness depends on the extent of initial exposure. Table 1-1 shows the relationship between dose of nuclear radiation and distance from ground zero for a 1-kiloton weapon. Once the dose is known, initial radiation effects on personnel are determined from Table 1-2. Radiation in the body is cumulative.

Nuclear radiation is the dominant casualty-producing effect of low-yield tactical nuclear weapons. But other initial effects may produce significant damage and/or casualties depending on the weapon type, yield, burst conditions, and the degree of personnel and equipment protection. Figure 1-2 shows tactical radii of effects for nominal 1-kiloton and 10-kiloton weapons.

Electromagnetic Pulse. Electromagnetic pulse (EMP) damages electrical and electronic equipment. It occurs at distances from the burst where other nuclear weapons effects produce little or no damage, and it lasts for less than a second after the burst. The pulse also damages vulnerable electrical and electronic equipment at ranges up to 5 kilometers for a 10-kiloton surface burst, and hundreds of kilometers for a similar high-altitude burst.

Survivability Considerations. Nuclear weapons survivability includes dispersion of protective positions within a suspected target area. Deep-covered positions will minimize the danger from blast and thermal radiation. Personnel should habitually wear complete uniforms with hands, face, and neck covered. Nuclear radiation is minimized by avoiding the radioactive fallout area or remaining in deep-covered protective positions. Examples of expedient protective positions against initial nuclear effects are shown on Figure 1-3. Additionally, buttoned-up armor vehicles offer limited protection from nuclear radiation. Removal of antennae and placement of critical electrical equipment into protective positions will reduce the adverse effects of EMP and TREE.

Table 1-1: Relationship of radiation dose to distance from ground zero for a 1-KT weapon.

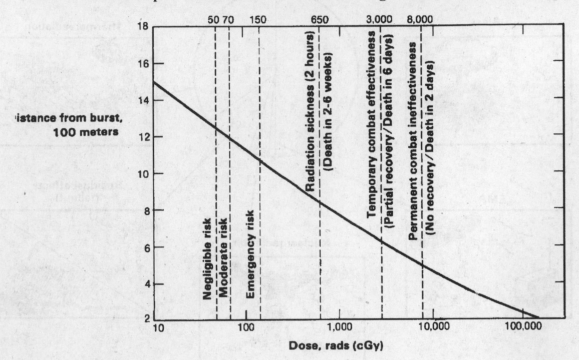

Chemical. Toxic chemical agents are primarily designed for use against personnel and to contaminate terrain and material. Agents do not destroy material and structures, but make them unusable for periods of time because of chemical contaminant absorption. The duration of chemical agent effectiveness depends on—

- Weather conditions.
- Dispersion methods.
- Terrain conditions.
- Physical properties.
- Quantity used.
- Type used (nerve, blood, or blister).

Part II of this book provides chemical agent details and characteristics. Since the vapor of toxic chemical agents is heavier than air, it naturally tends to drift to the lowest corners or sections of a structure. Thus, low, unenclosed fighting and protective positions trap chemical vapors or agents. Because chemical agents saturate an area, access to positions without airlock entrance ways is limited during and after an attack, since every entering or exiting soldier brings contamination inside.

Survivability Considerations. Survivability of chemical effects includes overhead cover of any design that delays penetration of chemical vapors and biological aerosols, thereby providing additional masking time and protection against direct liquid contamination. Packing materials and covers are used to protect sensitive equipment. Proper use of protective clothing and equipment, along with simply avoiding the contaminated area, aids greatly in chemical survivability.

Special Purpose. Fuel-air munitions and flamethrowers are considered special-purpose weapons. Fuel-air munitions disperse fuel into the atmosphere forming a fuel-air mixture that is detonated. The fuel is usually contained in a metal canister and is dispersed by detonation of a central burster charge carried within the canister. Upon proper dispersion, the fuel-air mixture is detonated. Peak pressures created within the detonated cloud reach 300 pounds per square inch (psi). Fuel-air munitions create large area loading on a

Table 1-2: Initial Radiation Effects on Personnel.

Early Symptoms*

Dose rads (cGy)	Percent of Personnel	Time to Effect	Combat Effectiveness of Personnel	Fatalities
0 to 70	<5% of personnel require hospitalization		Full	None
150	5%	≤6 hours	Effectiveness reduced depending on task. Some hospitalization required.	None
650	100%	≤2 hours	Symptoms continue intermittently for next few days. Effectiveness reduced significantly for second to sixth day. Hospitalizaton required.	More than 50% in about 16 days
2,000 to 3,000	100%	≤5 minutes	Immediate, temporary incapacitation for 30 to 40 minutes, followed by recovery period during which efficiency is impaired. No operational capability.	100% in about 7 days
8,000	100%	≤5 minutes	Immediate, permanent incapacitation for personnel performing physically demanding tasks. No period of latent "recovery."	100% in 1 to 2 days
18,000	100%	Immediate	Permanent incapacitation for personnel performing even undemanding tasks. No operational capability.	100% within 24 hours

* Symptoms include vomiting, diarrhea, "dry heaving," nausea, lethargy, depression, and mental disorientation. At lower dose levels, incapacitation is a simple slow down in performance rate due to a loss of physical mobility and/or mental disorientation. At the high dose levels, shock and coma are sometimes the "early" symptoms.

structure as compared to localized loadings caused by an equal weight high explosive charge. High temperatures ignite flammable materials. Flamethrowers and napalm produce intense heat and noxious gases which can neutralize accessible positions. The intense flame may also exhaust the oxygen content of inside air causing respiratory injuries to occupants shielded from the flaming fuel. Flame is effective in penetrating protective positions.

Survivability Considerations. Survivability of special purpose weapons effects includes covered positions with relatively small apertures and closable entrance areas which provide protection from napalm and flamethrowers. Deep-supported tunnels and positions provide protection from other fuel-air munitions and explosives.

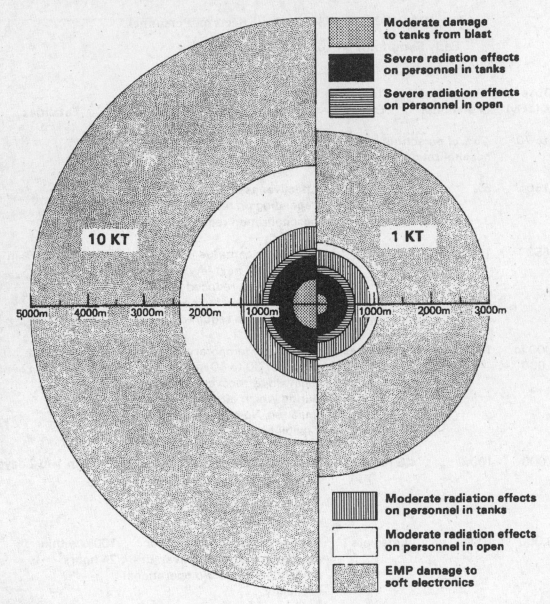

Figure 1-2: Tactical radii of effects of 1-KT and 10-KT fission weapons from low airburst.

CONSTRUCTION MATERIALS

Before designing fighting and protective positions, it is important to know how the previously-described weapons affect and interact with various materials that are fired upon. The materials used in fighting and protective position construction act as either shielding for the protected equipment and personnel, structural components to hold the shielding in place, or both.

Shielding Materials. Shielding provides protection against penetration of both projectiles and fragments, nuclear and thermal radiation, and the effects of fire and chemical agents. Various materials and amounts of materials provide varying degrees of shielding. Some of the more commonly used materials and the effects of both projectile and fragment penetration in these materials, as well as nuclear and thermal radiation suppression, are discussed in the following paragraphs. (Incendiary and chemical effects are generalized from the previous discussion of weapons effects.) The following three tables contain shielding requirements of various materials to protect against direct hits by direct fire projectiles (Table 1-3), direct

Figure 1-3: Examples of expedient protective positions against initial nuclear effects.

Table 1-3: Material thickness in inches, required to protect against direct hits by direct fire projectiles

Material	Small Caliber and Machine Gun (7.62-mm) Fire* at 100 yd	Antitank Rifle (76-mm) Fire at 100 yd	20-mm Antitank Fire at 200 yd	37-mm Antitank Fire at 400 yd	50-mm Antitank Fire at 400 yd	75-mm Direct Fire at 500 to 1,000 yd	Remarks
Solid walls**							
Brick masonry	18	24	30	60	-	-	None
Concrete, not reinforced***	12	18	24	42	48	54	Plain formed-concrete walls
Concrete, reinforced	6	12	18	36	42	48	Structurally reinforced with steel
Stone masonry	12	18	30	42	54	60	Values are guides only
Timber	36	60	-	-	-	-	Values are guides only
Wood	24	36	48	-	-	-	Values are guides only
Walls of loose material between boards**							
Brick rubble	12	24	30	60	72	-	None
Clay, dry	36	48	-	-	-	-	Add 100% to thickness if wet
Gravel/small crushed rock	12	24	30	60	72	-	None
Loam, dry	24	36	48	-	-	-	Add 50% to thickness if wet
Sand, dry	12	24	30	60	72	-	Add 100% to thickness if wet
Sandbags, filled with							
Brick rubble	20	30	30	60	70	-	None
Clay, dry	40	60	-	-	-	-	Add 100% to thickness if wet
Gravel/small crushed rock	20	30	30	60	70	-	None
Loam, dry	30	50	60	-	-	-	Add 50% to thickness if wet
Sand, dry	20	30	30	60	70	-	Add 100% to thickness if wet
Parapets of							
Clay	42	60	-	-	-	-	Add 100% to thickness if wet
Loam	36	48	60	-	-	-	Add 50% to thickness if wet
Sand	24	36	48	-	-	-	Add 100% to thickness if wet
Snow and Ice							
Frozen snow	80	80	-	-	-	-	None
Frozen soil	24	24	-	-	-	-	None
Icecrete (ice + aggregate)	18	18	-	-	-	-	None
Tamped snow	72	72	-	-	-	-	None
Unpacked snow	180	180	-	-	-	-	None

* One burst of five shots.

** Thicknesses to nearest ½ ft.

*** 3,000 psi concrete.

Note: Except where indicated, protective thicknesses are for a single shot only. Where weapons place five or six direct fire projectiles in the same area, the required protective thickness is approximately twice that indicated. Where no values are given, material is not recommended.

fire high explosive (HE) shaped charges (Table 1-4), and indirect fire fragmentation and blast (Table 1-5). Table 1-6 lists nuclear protection factors associated with earth cover and sandbags.

Soil. Direct fire and indirect fire fragmentation penetration in soil or other similar granular material is based on three considerations: for materials of the same density, the finer the grain the greater the penetration; penetration decreases with increase in density; and penetration increases with increasing water content. Nuclear and thermal radiation protection of soil is governed by the following:

- The more earth cover, the better the shielding. Each layer of sandbags filled with sand or clay reduces transmitted radiation by 50 percent.
- Sand or compacted clay provides better radiation shielding than other soils which are less dense.
- Damp or wet earth or sand provides better protection than dry material.
- Sandbags protected by a top layer of earth survive thermal radiation better than exposed bags. Exposed bags may burn, spill their contents, and become susceptible to the blast wave.

Steel. Steel is the most commonly used material for protection against direct and indirect fire fragmentation. Steel is also more likely to deform a projectile as it penetrates, and is much less likely to span than concrete. Steel plates, only 1/6 the thickness of concrete, afford equal protection against nondeforming projectiles of small and intermediate calibers. Because of its high density, steel is five times more effective in initial radiation suppression than an equal thickness of concrete. It is also effective against thermal radiation, although it transmits heat rapidly. Many field expedient types of steel are usable for shielding. Steel landing mats, culvert sections, and steel drums, for example, are effectively used in a structure as one of several composite materials. Expedient steel pieces are also used for individual protection against projectile and fragment penetration and nuclear radiation.

Concrete. When reinforcing steel is used in concrete, direct and indirect fire fragmentation protection is excellent. The reinforcing helps the concrete to remain intact even after excessive cracking caused by penetration. When a near-miss shell explodes, its fragments travel faster than its blast wave. If these fragments strike the exposed concrete surfaces of a protective position, they can weaken the concrete to such an extent that the blast wave destroys it. When possible, at least one layer of sandbags, placed on their short ends, or 15 inches of soil should cover all exposed concrete surfaces. An additional consequence of concrete penetration is spalling. If a projectile partially penetrates concrete shielding, particles and chunks of concrete

Table 1-4: Material thickness, in inches, required to protect against direct fire he shaped-charge.

Material	73-mm RCLR	82-mm RCLR	85-mm RPG-7	107-mm RCLR	120-mm Sagger
Aluminum	36	24	30	36	36
Concrete	36	24	30	36	36
Granite	30	18	24	30	30
Rock	36	24	24	36	36
Snow, packed	156	156	156	-	-
Soil	100	66	78	96	96
Soil, frozen	50	33	39	48	48
Steel	24	14	18	24	24
Wood, dry	100	72	90	108	108
Wood, green	60	36	48	60	66

Note: Thicknesses assume perpendicular impact.

Table 1-5: Material thickness, inches, required to protect against indirect fire fragmentation and blast exploding 50 feet away.

Material	Mortars 82-mm	120-mm	122-mm Rocket	HE Shells 122-mm	152-mm	Bombs 100-lb	250-lb	500-lb	1,000-lb
Solid Walls									
Brick masonry	4	6	6	6	8	8	10	13	17
Concrete	4	5	5	5	6	8	10	15	18
Concrete, reinforced	3	4	4	4	5	7	9	12	15
Timber	8	12	12	12	14	15	18	24	30
Walls of loose material between boards									
Brick rubble	9	12	12	12	12	18	24	28	30
Earth*	12	12	12	12	16	24	30	-	-
Gravel, small stones	9	12	12	12	12	18	24	28	30
Sandbags, filled with									
Brick rubble	10	18	18	18	20	20	20	30	40
Clay*	10	18	18	18	20	30	40	40	50
Gravel, small stones, soil	10	18	18	18	20	20	20	30	40
Sand*	8	16	16	16	18	30	30	40	40
Loose parapets of									
Clay*	12	20	20	20	30	36	48	60	-
Sand*	10	18	18	18	24	24	36	36	48
Snow									
Tamped	60	60	60	60	60	-	-	-	-
Unpacked	60	60	60	60	60	-	-	-	-

* Double values if material is saturated.
Note: Where no values are given, material is not recommended.

Table 1-6: Shielding values of earth cover and sandbags for a hypothetical 2,400-rads (cgy) free-in-air dose.

Type of Protection	Radiation Protection Factor	Resulting Dose rads
Soldier in open	None	2,400
Earth Cover		
Soldier in 4-ft-deep open position	8	300
with 6 in of earth cover	12	200
with 12 in of earth cover	24	100
with 18 in of earth cover	48	50
with 24 in of earth cover	96	25
Sand- and Clay-Filled Sandbags		
Soldier in 4-ft-deep open position	8	300
with 1 layer of sandbags (4 in)	16	150
with 2 layers of sandbags (8 in)	32	75
with 3 layers of sandbags (12 in)	64	38

often break or scab off the back of the shield at the time of impact. These particles can kill when broken loose. Concrete provides excellent protection against nuclear and thermal radiation.

Rock. Direct and indirect fire fragmentation penetration into rock depends on the rock's physical properties and the number of joints, fractures, and other irregularities contained in the rock. These irregularities weaken rock and can increase penetration. Several layers of irregularly-shaped rock can change the angle of penetration. Hard rock can cause a projectile or fragment to flatten out or break up and stop penetration. Nuclear and thermal radiation protection is limited because of undetectable voids and cracks in rocks. Generally, rock is not as effective against radiation as concrete, since the ability to provide protection depends on the rock's density.

Brick and Masonry. Direct and indirect fire fragmentation penetration into brick and masonry have the same protection limitations as rock. Nuclear and thermal radiation protection by brick and masonry is 1.5 times more effective than the protection afforded by soil. This characteristic is due to the higher compressive strength and hardness properties of brick and masonry. However, since density determines the degree of protection against initial radiation, unreinforced brick and masonry are not as good as concrete for penetration protection.

Snow and Ice. Although snow and ice are sometimes the only available materials in certain locations, they are used for shielding only. Weather could cause structures made of snow or ice to wear away or even collapse. Shielding composed of frozen materials provides protection from initial radiation, but melts if thermal radiation effects are strong enough.

Wood. Direct and indirect fire fragmentation protection using wood is limited because of its low density and relatively low compressive strengths. Greater thicknesses of wood than of soil are needed for protection from penetration. Wood is generally used as structural support for a survivability position. The low density of wood provides poor protection from nuclear and thermal radiation. Also, with its low ignition point, wood is easily destroyed by fire from thermal radiation.

Other Materials. Expedient materials include steel pickets, landing mats, steel culverts, steel drums, and steel shipping consolidated express (CONEX) containers. Chapter 4 discusses fighting and protective positions constructed with some of these materials.

Structural Components. The structure of a fighting and protective position depends on the weapon or weapon effect it is designed to defeat. All fighting and protective positions have some configuration of floor, walls, and roof designed to protect material and/or occupants, The floor, walls, and roof support the shielding discussed earlier, or may in themselves make up that shielding. These components must also resist blast and ground shock effects from detonation of high explosive rounds which place greater stress on the structure than the weight of the components and the shielding. Designers must make structural components of the positions stronger, larger, and/or more numerous in order to defeat blast and ground shock. Following is a discussion of materials used to build floors, walls, and roofs of positions.

Floors. Fighting and protective position floors are made from almost any material, but require resistance to weathering, wear, and trafficability. Soil is most often used, yet is least resistant to water damage and rutting from foot and vehicle traffic. Wood pallets, or other field-available materials are often cut to fit floor areas. Drainage sumps, shown in Figure 1-4, or drains are also installed when possible.

Walls. Walls of fighting and protective positions are of two basic types—below ground (earth or revetted earth) and above-ground. Below-ground walls are made of the in-place soil remaining after excavation of the position. This soil may need revetment or support, depending on the soil properties and depth of cut. When used to support roof structures, earth walls must support the roof at points no less than one fourth the depth of cutout from the edges of excavation, as Figure 1-5.

Above-ground walls are normally constructed for shielding from direct fire and fragments. They are usually built of revetted earth, sandbags, concrete, or other materials. When constructed to a thickness

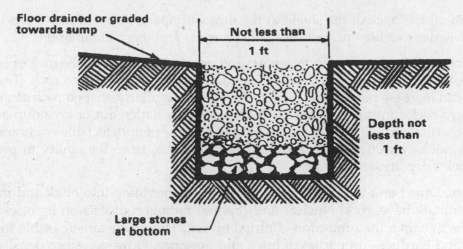

Figure 1-4: Drainage sump.

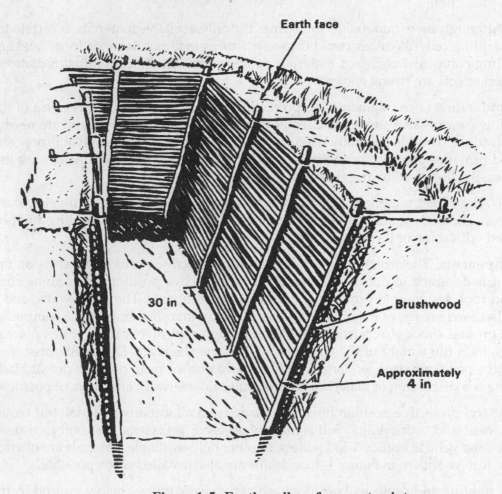

Figure 1-5: Earth wall roof support points.

adequate for shielding from direct fire and fragments, they are thick and stable enough for roof support.

Roofs. Roofs of fighting and protective positions are easily designed to support earth cover for shielding from fragments and small caliber direct fire. However, contact burst protection requires much stronger roof structures and, therefore, careful design. Roofs for support of earth cover shielding are constructed of almost any material that is usually used as beams or stringers and sheathing. Tables 1-7 and 1-8 present guidelines for wooden roof structures (for fragment shielding only). Table 1-9 converts dimensioned to round timber. Tables 1-10 and 1-11 pertain to steel pickets and landing mats for roof supports (for fragment shielding only).

When roof structures are designed to defeat contact bursts of high explosive projectiles, substantial additional roof protection is required. Table 1-13 gives basic design criteria for a roof to defeat contact bursts.

Table 1-7: Maximum span of dimensioned wood roof support for earth cover.

Thickness of Earth Cover, ft	Span Length, ft					
	2½	3	3½	4	5	6
	Wood Thickness, in					
1½	1	1	2	2	2	2
2	1	2	2	2	2	3
2½	1	2	2	2	2	3
3	2	2	2	2	3	3
3½	2	2	2	2	3	3
4	2	2	2	2	3	4

Table 1-8: Maximum span of wood stringer roof support for earth cover.

Thickness of Earth Cover, ft	Span Length, ft					
	2½	3	3½	4	5	6
	Center-to-Center Spacing, in					
1½	40	30	22	16	10	18*
2	33	22	16	12	8/20*	14*
2½	27	18	12	10	16*	10*
3	22	14	10	8/20*	14*	8*
3½	18	12	8/24*	18*	12*	8*
4	16	10	8/20*	10*	10*	7*

Note: Stringers are 2 x 4s except those marked by an asterisk (*) which are 2 by 6s.

POSITION CATEGORIES

Seven categories of fighting and protective positions or components of positions that are used together or separately are—

- Holes and simple excavations.
- Trenches.
- Tunnels.
- Earth parapets.
- Overhead cover and roof structures.

Table 1-9: Converting dimensioned timber to round timber.

4 x 4	5
6 x 6	7
6 x 8	8
8 x 8	10
8 x 10	11
10 x 10	12
10 x 12	13
12 x 12	14

*Sizes given are nominal and not rough cut timber.

Table 1-10: Maximum span of steel picket roof supports for sandbag layers.

Number of Sandbag Layers	Span Length, ft		
	3	6	9
Single-Picket Beams*	Center-to-Center Spacing, in		
2	7	7	6
5	6	5	4
10	4	4	3
15	4	3	2
20	3	3	2
Double-Picket Beams**			
2	7	7	7
5	7	7	7
10	7	6	5
15	7	5	4
20	6	5	4

* Used with open side down.
** Two pickets are welded together every 6 inches along the
span to form box beams.

Table 1-11: Maximum span of inverted landing mats (M8A1) for roof supports.

Number of Sandbag Layers	Span Length, ft
2	10
5	6½
10	5
15	4
10	3½

- Triggering screens.
- Shelters and bunkers.

Holes And Simple Excavations. Excavations, when feasible, provide good protection from direct fire and some indirect fire weapons effects. Open excavations have the advantages of—

- Providing good protection from direct fire when the occupant would otherwise be exposed.
- Permitting 360-degree observation and fire.
- Providing good protection from nuclear weapons effects.

Open excavations have the disadvantages of—

- Providing limited protection from direct fire while the occupant is firing a weapon, since frontal and side protection is negligible.
- Providing relatively no protection from fragments from overhead bursts of artillery shells. The larger the open excavation, the less the protection from artillery.
- Providing limited protection from chemical effects. In some cases, chemicals concentrate in low holes and excavations.

Trenches. Trenches provide essentially the same protection from conventional, nuclear, and chemical effects as the other excavations described, and are used almost exclusively in defensive areas. They are employed as protective positions and used to connect individual holes, weapons positions, and shelters. They provide protection and concealment for personnel moving between fighting positions or in and out of the area. They are usually open excavations, but sections are sometimes covered to provide additional protection. Trenches are difficult to camouflage and are easily detected from the air.

Trenches, like other positions, are developed progressively. As a general rule, they are excavated deeper than fighting positions to allow movement without exposure to enemy fire. It is usually necessary to provide revetment and drainage for them.

Tunnels. Tunnels are not frequently constructed in the defense of an area due to the time, effort, and technicalities involved. However, they are usually used to good advantage when the length of time an area is defended justifies the effort, and the ground lends itself to this purpose. The decision to build tunnels also depends greatly on the nature of the soil, which is usually determined by borings or similar means. Tunneling in hard rock is slow and generally impractical. Tunnels in clay or other soft soils are also impractical since builders must line them throughout to prevent collapse. Therefore, construction of tunneled defenses is usually limited to hilly terrain, steep hillsides, and favorable soils including hard chalk, soft sandstone, and other types of hard soil or soft rock.

In the tunnel system shown in Figure 1-5, the soil was generally very hard and only the entrances were timbered. The speed of excavation using hand tools varied according to the soil, and seldom exceeded 25 feet per day. In patches of hard rock, as little as 3 feet were excavated per day. Use of power tools did not significantly increase the speed of excavation. Engineer units, assisted by infantry personnel, performed the work. Tunnels of the type shown are excavated up to 30 feet below ground level. They are usually horizontal or nearly so. Entrances are strengthened against collapse under shell fire and ground shock from nuclear weapons. The first 16 1/2 feet from each entrance should have frames using 4 by 4s or larger timber supports.

Unlimbered tunnels are generally 31 1/2 feet wide and 5 to 6 1/2 feet high. Once beyond the portal or entrance, tunnels of up to this size are unlimbered if they are deep enough and the soil will stand open. Larger tunnels must have shoring. Chambers constructed in rock or extremely hard soil do not need timber supports. If timber is not used, the chamber is not wider than 6 1/2 feet; if timbers are used, the width can increase to 10 feet. The chamber is generally the same height as the tunnel, and up to 13 feet long.

Grenade traps are constructed at the bottom of straight lengths where they slope. This is done by cutting a recess about 3 1/2 feet deep in the wall facing the inclining floor of the tunnel.

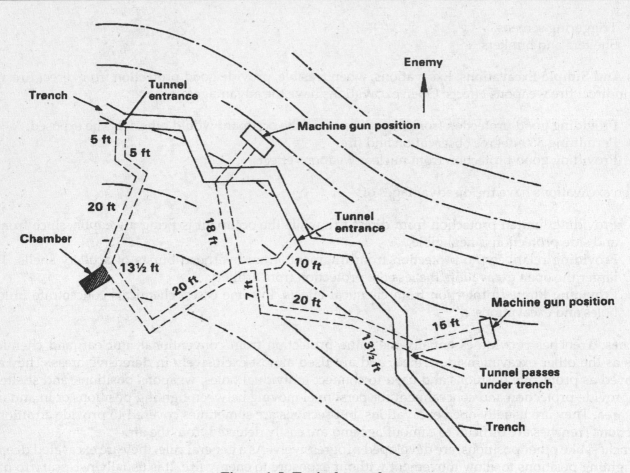

Figure 1-5: Typical tunnel system.

Much of the spoil from the excavated area requires disposal and concealment. The volume of spoil is usually estimated as one third greater than the volume of the tunnel. Tunnel entrances need concealment from enemy observation. Also, it is sometimes necessary during construction to transport spoil by hand through a trench. In cold regions, air warmer than outside air may rise from a tunnel entrance, thus revealing the position.

The danger that tunnel entrances may become blocked and trap the occupants always exists. Picks and shovels are placed in each tunnel so that trapped personnel can dig their way out. Furthermore, at least two entrances are necessary for ventilation. Whenever possible, one or more emergency exits are provided. These are usually small tunnels with entrances normally closed or concealed. A tunnel is constructed from inside the system to within a few feet of the surface so that an easy breakthrough is possible.

Earth Parapets. Excavations and trenches are usually modified to include front, rear, and side earth parapets. Parapets are constructed using spoil from the excavation or other materials carried to the site. Frontal, side, and rear parapets greatly increase the protection of occupants firing their weapons (see Figure 1-6). Thicknesses required for parapets vary according to the material's ability to deny round penetration.

Parapets are generally positioned as shown below to allow full frontal protection, thus relying on mutual support of other firing positions. Parapets are also used as a single means of protection, even in the absence of excavations.

Overhead Cover and Roof Structures. Fighting and protective positions are given overhead cover primarily to defeat indirect fire projectiles landing on or exploding above them. Defeat of an indirect fire attack on a position, then, requires that the three types of burst conditions are considered. (Note: Always place a waterproof layer over any soil cover to prevent it from gaining moisture or weathering.)

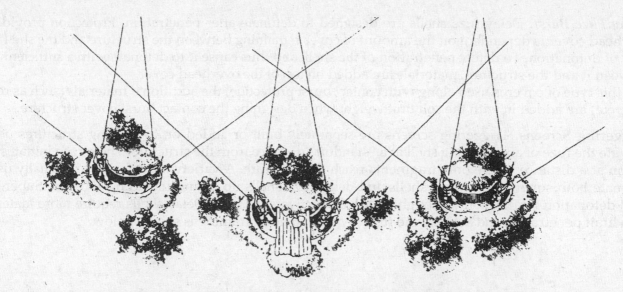

Figure 1-6: Parapets used for frontal protection relying on mutual support.

Overhead Burst (Fragments). Protection against fragments from airburst artillery is provided by a thickness of shielding required to defeat a certain size shell fragment, supported by a roof structure adequate for the dead load of the shielding. This type of roof structure is designed using the thicknesses to defeat fragment penetration given in Table 1-5. As a general guide, fragment penetration protection always requires at least 1 1/2 feet of soil cover. For example, to defeat fragments from a 120-mm mortar when available cover material is sandbags filled with soil, the cover depth required is 1 1/2 feet. Then, Table 1-8 shows that support of the 1 1/2 feet of cover (using 2 by 4 roof stringers over a 4-foot span) requires 16-inch center-to-center spacing of the 2 by 4s. This example is shown in Figure 1-7.

Contact Burst. Protection from contact burst of indirect fire HE shells requires much more cover and roof structure support than does protection from fragmentation. The type of roof structure necessary is given in Table 1-13. For example, if a position must defeat the contact burst of an 82-mm mortar, Table 1-13 provides multiple design options. If 4 by 4 stringers are positioned on 9-inch center-to-center spacings over a span of 8 feet, then 2 feet of soil (loose, gravelly sand) is required to defeat the burst.

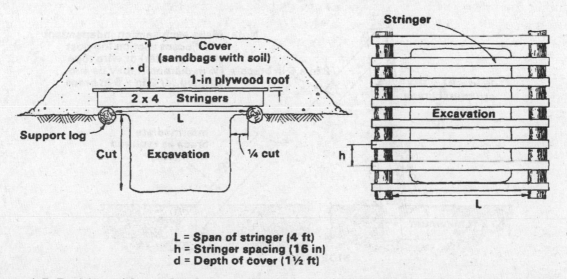

L = Span of stringer (4 ft)
h = Stringer spacing (16 in)
d = Depth of cover (1½ ft)

Figure 1-7: Position with overhead cover protection against fragments from a 120-mm mortar.

Delay Fuze Burst. Delay fuze shells are designed to detonate after penetration. Protection provided by overhead cover is dependent on the amount of cover remaining between the structure and the shell at the time of detonation. To defeat penetration of the shell, and thus cause it to detonate with a sufficient cover between it and the structure, materials are added on top of the overhead cover.

If this type of cover is used along with contact burst protection, the additional materials (such as rock or concrete) are added in with the soil unit weight when designing the contact burst cover structure.

Triggering Screens. Triggering screens are separately built or added on to existing structures used to activate the fuze of an incoming shell at a "standoff" distance from the structure. The screen initiates detonation at a distance where only fragments reach the structure. A variety of materials are usually used to detonate both super-quick fuzed shells and delay fuze shells up to and including 130 mm. Super-quick shell detonation requires only enough material to activate the fuze. Delay shells require more material to both limit penetration and activate the fuze. Typical standoff framing is shown below.

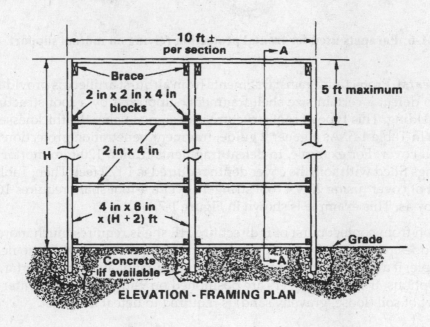

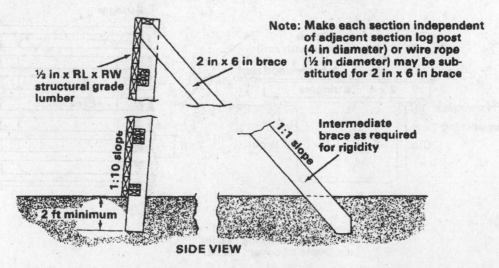

Figure 1-8: Typical standoff framing with dimensioned wood triggering screen.

Defeating Super-Quick Fuzes. Incoming shells with super-quick fuzes are defeated at a standoff distance with several types of triggering screen materials. Table 1-10 lists thicknesses of facing material required for detonating incoming shells when impacting with the triggering screen. These triggering screens detonate the incoming shell but do not defeat fragments from these shells. Protection from fragments is still necessary for a position. Table 1-11 lists required thicknesses for various materials to defeat fragments if the triggering screen is 10 feet from the structure.

Defeating Delay Fuzes. Delay fuzes are defeated by various thicknesses of protective material. Table 1-12 lists type and thickness of materials required to defeat penetration of delay fuze shells and cause their premature detonation. These materials are usually added to positions designed for contact burst protection. One method to defeat penetration and ensure premature shell detonation is to use layers of large stones. Figure 1-9 shows this added delay fuze protection on top of the contact burst protection. The rocks are placed in at least three layers on top of the required depth of cover for the expected shell size. The rock size is approximately twice the caliber of the expected shell. For example, the rock size required to defeat 82-mm mortar shell penetration is 2 x 82 mm = 164 mm (or 6 1/2 inches).

In some cases, chain link fences (shown below) also provide some standoff protection when visibility is necessary in front of the standoff and when positioned as shown in Figure 1-10. However, the fuze of some incoming shells may pass through the fence without initiating the firing mechanism.

Table 1-10: Triggering screen facing material requirements.

Material	Triggering Requirements*
Plywood, dimensioned timber	2½-in thickness
Soil in sandbags with plywood or metal facing	2-in thickness (24-gage sheet metal)
Structured steel (corrugated metal)	¼-in thickness
Tree limbs	2-in diameter
Ammunition crates	1 layer (1-in-thick wood)
Snow	3 feet

* For detonating projectiles up to and including 120-mm mortar, rocket, and artillery shells.

Table 1-11: Triggering screen material thickness, in inches, required to defeat fragments at a 10-foot standoff.

Material	Incoming Shell Size		
	82 mm	120 mm	122 mm
Soil	10	18	18
Soil, frozen	5	9	9
Sand	8	16	16
Clay	10	18	18
Steel (corrugated metal)	½	1	1
Wood (fir)	5	14	14
Concrete	2	2	3
Snow	60	80	80

Table 1-12: Required thickness, in inches, of protective material to resist penetration of different shells (Delay Fuze).

Shells	Concrete*	Rock**	Rock Size (inches)
82-mm mortar	6	20	6½
120-mm mortar	20	36	9
122-mm rocket	50	40	10
122-mm artillery	68	40	10
130-mm artillery	80	42	10½

* 3,000 psi reinforced concrete.
** Rock must be relatively strong (compressive strength of about 20,000 psi)
   and in three layers for 82 mm; four layer for others.

Note: Due to the extreme thickness required for protection, materials such as earth, sand, and clay
are not recommended.

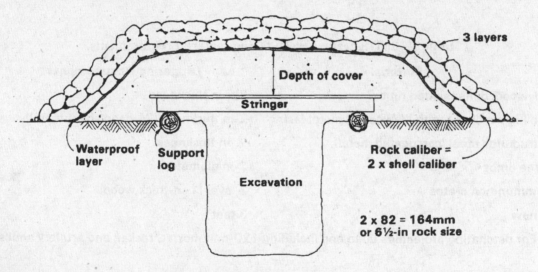

**Figure 1-9: Stone layer added to typical overhead cover to defeat the delay fuze burst
from an 82-mm mortar..**

Shelters And Bunkers. Protective shelters and fighting bunkers are usually constructed using a combination of the components of positions mentioned thus far. Protective shelters are primarily used as—

- Command posts.
- Observation posts.
- Medical aid stations.
- Supply and ammunition shelters.
- Sleeping or resting shelters.

Protective shelters are usually constructed above ground, using cavity wall revetments and earth-covered roof structures, or they are below ground using sections that are air transportable. Fighting bunkers are enlarged fighting positions designed for squad-size units or larger. They are built either above ground or below ground and are usually made of concrete. However, some are prefabricated and transported forward to the battle area by trucks or air.

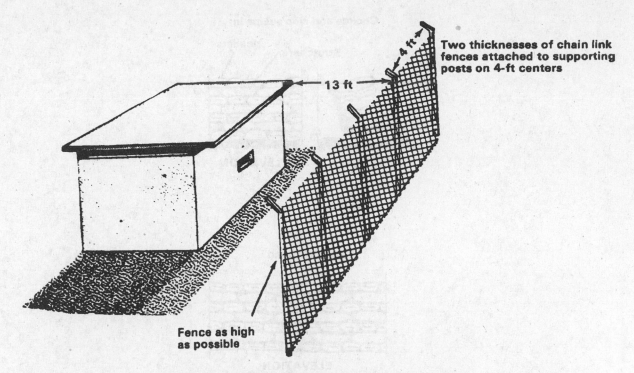

Figure 1-10: Chain link fence used for a standoff.

If shelters and bunkers are properly constructed with appropriate collective protection equipment, they can serve as protection against chemical and biological agents.

CONSTRUCTION METHODS

For individual and crew-served weapons fighting and protective position construction, hand tools are available. The individual soldier carries an entrenching tool and has access to picks, shovels, machetes, and hand carpentry tools for use in individual excavation and vertical construction work.

Earthmoving equipment and explosives are used for excavating protective positions for vehicles and supplies. Earthmoving equipment, including backhoes, bulldozers, and bucket loaders, are usually used for larger or more rapid excavation when the situation permits. Usually, these machines cannot dig out the exact shape desired or dig the amount of earth necessary. The excavation is usually then completed by hand. Descriptions and capabilities of US survivability equipment are given in appendix A.

Methods of construction include sandbagging, explosive excavation, and excavation revetments.

Sandbagging. Walls of fighting and protective positions are built of sandbags in much the same way bricks are used. Sandbags are also useful for retaining wall revetments as shown in Figure 1-11.

The sandbag is made of an acrylic fabric and is rot and weather resistant. Under all climatic conditions, the bag has a life of at least 2 years with no visible deterioration. (Some older-style cotton bags deteriorate much sooner.) The useful life of sandbags is prolonged by filling them with a mixture of dry earth and portland cement, normally in the ratio of 1 part of cement to 10 parts of dry earth. The cement sets as the bags take on moisture. A 1:6 ratio is used for sand-gravel mixtures. As an alternative, filled bags are dipped in a cement-water slurry. Each sandbag is then pounded with a flat object, such as a 2 by 4, to make the retaining wall more stable.

As a rule, sandbags are used for revetting walls or repairing trenches when the soil is very loose and requires a retaining wall. A sandbag revetment will not stand with a vertical face. The face must have a slope of 1:4, and lean against the earth it is to hold in place. The base for the revetment must stand on firm ground and dug at a slope of 4:1.

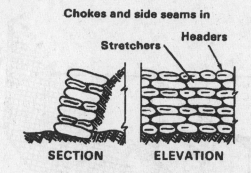

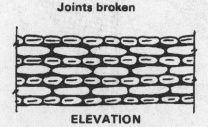

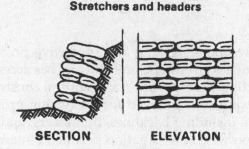

Figure 1-11: Retaining wall revetment.

The following steps are used to construct a sandbag revetment wall such as the one shown in Figure 1-11.

- The bags are filled about three-fourths full with earth or a dry soil-cement mixture and the choke cords are tied.
- The bottom corners of the bags are tucked in after filling.
- The bottom row of the revetment is constructed by placing all bags as headers. The wall is built using alternate rows of stretchers and headers with the joints broken between courses. The top row of the revetment wall consists of headers.
- Sandbags are positioned so that the planes between the layers have the same pitch as the base—at right angles to the slope of the revetment.
- All bags are placed so that side seams on stretchers and choked ends on headers are turned toward the revetted face.
- As the revetment is built, it is backfilled to shape the revetted face to this slope.

Often, the requirement for filled sandbags far exceeds the capabilities of soldiers using only shovels. If the bags are filled from a stockpile, the job is performed easier and faster by using a lumber or steel funnel as shown in Figure 1-12.

Excavation Revetments. Excavations in soil may require revetment to prevent side walls from collapsing. Several methods of excavation revetments are usually used to prevent wall collapse.

Wall Sloping. The need for revetment is sometimes avoided or postponed by sloping the walls of the excavation. In most soils, a slope of 1:3 or 1:4 is sufficient. This method is used temporarily if the soil is loose and no revetting materials are available. The ratio of 1:3, for example, will determine the slope by moving 1 foot horizontally for each 3 feet vertically. When wall sloping is used, the walls are first dug vertically and then sloped.

Facing Revetments. Facing revetments serve mainly to protect revetted surfaces from the effects of weather and occupation. It is used when soils are stable enough to sustain their own weight. This revetment consists

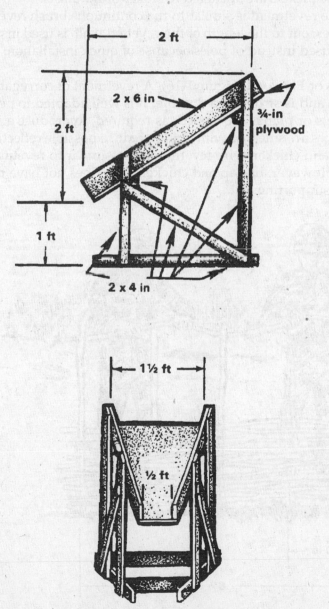

Figure 1-12: Expedient funnel for filling sandbags.

of the revetting or facing material and the supports which hold the revetting material in place. The facing material is usually much thinner than that used in a retaining wall. Facing revetments are preferable to wall sloping since less excavation is required. The top of the facing is set below ground level. The facing is constructed of brushwood hurdles, continuous brush, poles, corrugated metal, plywood, or burlap and chicken wire. The following paragraphs describe the method of constructing each type.

Brushwood Hurdle (Figure 1-13). A brushwood hurdle is a woven revetment unit usually 6 1/2 feet long and as high as the revetted wall. Pieces of brushwood about 1 inch in diameter are weaved on a framework of sharpened pickets driven into the ground at 20-inch intervals. When completed, the 6 1/2-foot lengths are carried to the position where the pickets are driven in place. The tops of the pickets are tied back to stakes or holdfasts and the ends of the hurdles are wired together.

Continuous Brush (Figure 1-14). A continuous brush revetment is constructed in place. Sharpened pickets 3 inches in diameter are driven into the bottom of the trench at 30-inch intervals and about 4 inches from the revetted earth face. The space behind the pickets is packed with small, straight brushwood laid horizontally. The tops of the pickets are anchored to stakes or holdfasts.

Pole (Figure 1-15). A pole revetment is similar to the continuous brush revetment except that a layer of small horizontal round poles, cut to the length of the revetted wall, is used instead of brushwood. If available, boards or planks are used instead of poles because of quick installation. Pickets are held in place by holdfasts or struts.

Corrugated Metal Sheets or Plywood (Figure 1-16). A revetment of corrugated metal sheets or plywood is usually installed rapidly and is strong and durable. It is well adapted to position construction because the edges and ends of sheets or planks are lapped, as required, to produce a revetment of a given height and length. All metal surfaces are smeared with mud to reduce possible reflection of thermal radiation and aid in camouflage. Burlap and chicken wire revetments are similar to revetments made from corrugated metal sheets or plywood. However, burlap and chicken wire does not have the strength or durability of plywood or sheet metal in supporting soil.

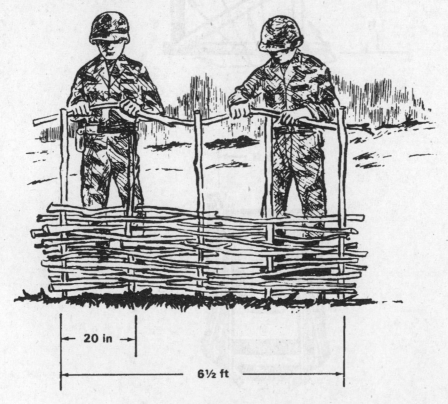

Figure 1-13: Brush wood hurdle.

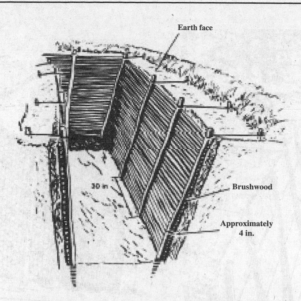

Figure 1-14: Continuous brush revetment.

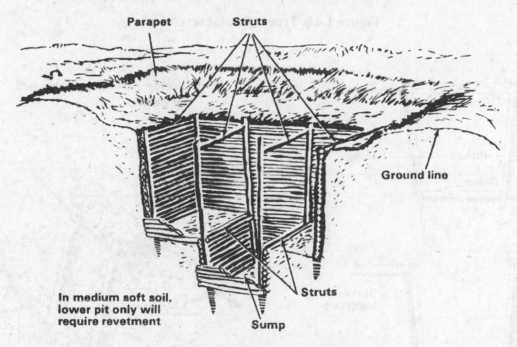

Figure 1-15: Pole revetment.

Continuous Brush (Figure 1-14). A continuous brush revetment is constructed in place. Sharpened pickets 3 inches in diameter are driven into the bottom of the trench at 30-inch intervals and about 4 inches from the revetted earth face. The space behind the pickets is packed with small, straight brushwood laid horizontally. The tops of the pickets are anchored to stakes or holdfasts.

Pole (Figure 1-15). A pole revetment is similar to the continuous brush revetment except that a layer of small horizontal round poles, cut to the length of the revetted wall, is used instead of brushwood. If available, boards or planks are used instead of poles because of quick installation. Pickets are held in place by holdfasts or struts.

Corrugated Metal Sheets or Plywood (Figure 1-16). A revetment of corrugated metal sheets or plywood is usually installed rapidly and is strong and durable. It is well adapted to position construction because

Corrugated metal sheets

Burlap and chicken wire

Figure 1-16: Types of metal revetment.

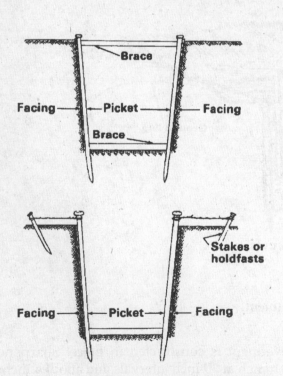

Figure 1-17: Facing revetment supported by timber frames.

Facing revetment supported by pickets

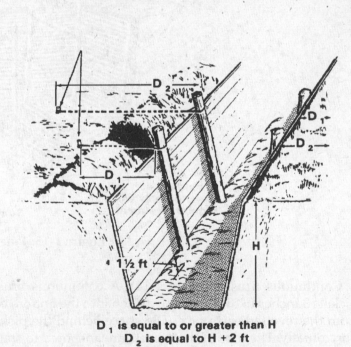

D$_1$ is equal to or greater than H
D$_2$ is equal to H + 2 ft

METHOD OF ANCHORING PICKETS

Figure 1-18: Facing revetment supported by pickets.

stake and the facing is at least equal to the height of the revetted face, with alternate anchors staggered and at least 2 feet farther back. Several strands of wire holding the pickets against the emplacement walls are placed straight and taut. A groove or channel is cut in the parapet to pass the wire through.

SPECIAL CONSTRUCTION CONSIDERATIONS

Camouflage And Concealment. The easiest and most efficient method of preventing the targeting and destruction of a position or shelter is use of proper camouflage and concealment techniques. Following are some general guidelines for position construction.

Natural concealment and good camouflage materials are used. When construction of a position begins, natural materials such as vegetation, rotting leaves, scrub brush, and snow are preserved for use as camouflage when construction is completed. If explosive excavation is used, the large area of earth spray created by detonation is camouflaged or removed by first placing tarpaulins or scrap canvas on the ground prior to charge detonation. Also, heavy equipment tracks and impressions are disguised upon completion of construction.

Fields of fire are not overcleared. In fighting position construction, clearing of fields of fire is an important activity for effective engagement of the enemy. Excessive clearing is prevented in order to reduce early enemy acquisition of the position. Procedures for clearing allow for only as much terrain modification as is needed for enemy acquisition and engagement.

Concealment from aircraft is provided. Consideration is usually given to observation from the air. Action is taken to camouflage position interiors or roofs with fresh natural materials, thus preventing contrast with the surroundings.

During construction, the position is evaluated from the enemy side. By far, the most effective means of evaluating concealment and camouflage is to check it from a suspected enemy avenue of approach.

Drainage. Positions and shelters are designed to take advantage of the natural drainage pattern of the ground. They are constructed to provide for—

- Exclusion of surface runoff.
- Disposal of direct rainfall or seepage.
- Bypassing or rerouting natural drainage channels if they are intersected by the position.

In addition to using materials that are durable and resistant to weathering and rot, positions are protected from damage due to surface runoff and direct rainfall, and are repaired quickly when erosion begins. Proper position siting can lessen the problem of surface water runoff. Surface water is excluded by excavating intercepted ditches uphill from a position or shelter. Preventing water from flowing into the excavation is easier than removing it. Positions are located to direct the runoff water into natural drainage lines. Water within a position or shelter is carried to central points by constructing longitudinal slopes in the bottom of the excavation. A very gradual slope of 1 percent is desirable.

Maintenance. If water is allowed to stand in the bottom of an excavation, the position is eventually undermined and becomes useless. Sumps and drains are kept clean of silt and refuse. Parapets around positions are kept clear and wide enough to prevent parapet soil from falling into the excavation. When wire and pickets are used to support revetment material, the pickets may become loose, especially after rain. Improvised braces are wedged across the excavation, at or near floor level, between two opposite pickets. Anchor wires are tightened by further twisting. Anchor pickets are driven in farther to hold tightened wires. Periodic inspections of sandbags are made.

Repairs. If the walls are crumbling in at the top of an excavation (ground level), soil is cut out where it is crumbling (or until firm soil is reached). Sandbags or sod blocks are used to build up the damaged area. If excavation walls are wearing away at the floor level, a plank is placed on its edge or the brushwood is shifted down. The plank is held against the excavation wall with short pickets driven into the floor. If planks are used on both sides of the excavation, a wedge is placed between the planks and earth is placed in the back of the planks. If an entire wall appears ready to collapse, the excavation is completely revetted. See Figure 1-19.

Security. In almost all instances, fighting and protective positions are prepared by teams of at least two personnel. During construction, adequate frontal and perimeter protection and observation are necessary. Additional units are sometimes required to secure an area during position construction. Unit personnel can also take turns with excavating and providing security.

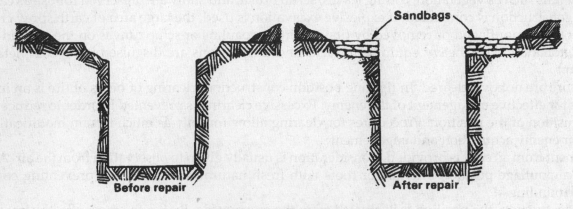

DAMAGE AT GROUND LEVEL

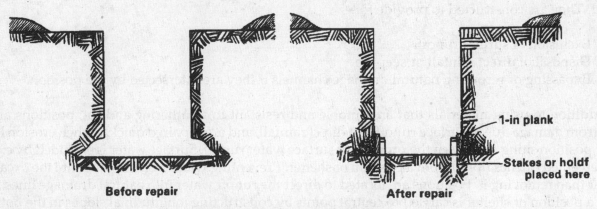

DAMAGE NEAR FLOOR LEVEL

Figure 1-19: Excavation repair.

Table 1-13: Center-to-center spacing for wood supporting soil cover to defeat contact bursts.

Nominal Stringer Size (inches)	Depth of Soil (d) (feet)	Center-to-Center Stringer Spacing (h) (inches), for Cited Span Length (L) (feet)				
		2	4	6	8	10
For Defeat of 82-mm Contact Burst						
2 x 4	2.0	3	4	4	4	3
	3.0	18	12	8	5	3
	4.0	18	14	7	4	3
2 x 6	2.0	4	7	8	8	6
	3.0	18	18	16	12	8
	4.0	18	18	18	11	7
4 x 4	2.0	7	10	10	9	7
	3.0	18	18	18	12	8
	4.0	18	18	18	10	7
4 x 8	1.5	4	5	7	8	8
	2.0	14	18	18	18	18
	3.0	18	18	18	18	18
For Defeat of 120- and 122-mm Contact Bursts						
4 x 8	2.0	-	-	-	-	-
	3.0	-	-	-	-	-
	4.0	3.5	4	5	5	6
	5.0	12	12	12	11	10
	6.0	18	18	18	16	12
6 x 6	2.0	-	-	-	-	-
	3.0	-	-	-	-	-
	4.0	-	-	5.5	6	6
	5.0	14	14	13	12	10
	6.0	18	18	18	16	12
6 x 8	2.0	-	-	-	-	-
	3.0	-	-	-	-	-
	4.0	5.5	6	8	9	10
	5.0	18	18	18	18	17
8 x 8	2.0	-	-	-	-	-
	3.0	-	-	-	-	-
	4.0	7.5	9	11	12	13
	5.0	18	18	18	18	18

(continued)

Table 1-13: *(Continued)*

For Defeat of 152-mm Contact Burst

4 x 8	4.0	-	-	-	-	3.5
	5.0	6	6	7	7	7
	6.0	17	16	14	12	10
	7.0	18	18	18	15	11
6 x 6	4.0	-	-	-	-	-
	5.0	7	8	8	8	7
	6.0	18	18	15	12	10
	7.0	18	18	18	15	11
6 x 8	3.0	-	-	-	-	-
	4.0	-	-	-	-	6
	5.0	10	11	12	12	12
	6.0	18	18	18	18	17
8 x 8	3.0	-	-	-	-	-
	4.0	-	-	-	-	8
	5.0	14	15	16	17	16
	6.0	18	18	18	18	18

Note: The maximum beam spacing listed in the above table is 18 inches. This is to preclude further design for roof material placed over the stringers to hold the earth cover. A maximum of 1 inch wood or plywood should be used over stringers to support the earth cover for 82-mm bursts; 2 inches should be used for 120-mm, 122-mm, and 152-mm bursts.

CHAPTER 2

Designing Positions

This chapter contains basic requirements which must be built into the designs of fighting and protective positions. These requirements ensure soldiers are well-protected while performing their missions. The positions are all continuously improved as time, assets, and the situation permit. The following position categories are presented: hasty and deliberate fighting position for individual soldiers; trenches connecting the positions; positions for entire units; and special designs including shelters and bunkers. The positions in each category are briefly described and accompanied by a typical design illustration. Each category is summarized providing time and equipment estimates and protection factors for each position. Complete detailed construction drawings, and time and material estimates for a variety of positions are contained in Chapter 4.

BASIC DESIGN REQUIREMENTS

Weapon Employment. While it is desirable for a fighting position to give maximum protection to personnel and equipment, primary consideration is always given to effective weapon use. In offensive combat operations, weapons are sited wherever natural or existing positions are available, or where weapon emplacement is made with minimal digging.

Cover. Positions are designed to defeat an anticipated threat. Protection against direct and indirect fire is of primary concern for position design. However, the effects of nuclear and chemical attack are taken into consideration if their use is suspected. Protection design for one type of enemy fire is not necessarily effective against another. The following three types of cover—frontal, overhead, and flank and rear—will have a direct bearing on designing and constructing positions.

Frontal. Frontal cover provides protection from small caliber direct fire. Natural frontal protection such as large trees, rocks, logs, and rubble is best because enemy detection of fighting positions becomes difficult. However, if natural frontal protection is not adequate for proper protection, dirt excavated from the position (hole) is used. Frontal cover requires the position to have the correct length so that soldiers have adequate room; the correct dirt thickness (3 feet) to stop enemy small caliber fire; the correct height for overhead protection; and, for soldiers firing to the oblique, the correct frontal distance for elbow rests and sector stakes. Protection from larger direct fire weapons (for example, tank guns) is achieved by locating the position where the enemy cannot engage it, and concealing it so pinpoint location is not possible. Almost twice as many soldiers are killed or wounded by small caliber fire when their positions do not have frontal cover.

Overhead. Overhead cover provides protection from indirect fire fragmentation. When possible, overhead cover is always constructed to enhance protection against airburst artillery shells. Overhead cover is necessary because soldiers are at least ten times more protected from indirect fire if they are in a hole with overhead cover.

Flank and Rear. Flank and rear cover ensures complete protection for fighting positions. Flank and rear cover protects soldiers against the effects of indirect fire bursts to the flanks or rear of the position, and the effects of friendly weapons located in the rear (for example, packing from discarded sabot rounds fired from tanks). Ideally, this protection is provided by natural cover. In its absence, a parapet is constructed as time and circumstances permit.

Simplicity and Economy. The position is usually uncomplicated and strong, requires as little digging as possible, and is constructed of immediately available materials.

Ingenuity. A high degree of imagination is essential to assure the best use of available materials. Many different materials existing on the battlefield and prefabricated materials found in industrial and urban areas can be used for position construction.

Progressive Development. Positions should allow for progressive development to insure flexibility, security, and protection in depth. Hasty positions are continuously improved into deliberate positions to provide maximum protection from enemy fire. Trenches or tunnels connecting fighting positions give ultimate flexibility in fighting from a battle position or strongpoint. Grenade sumps are usually dug at the bottom of a position's front wall where water collects. The sump is about 3 feet long, 1/2 foot wide, and dug at a 30-degree angle. The slant of the floor channels excess water and grenades into the sump. In larger positions, separate drainage sumps or water drains are constructed to reduce the amount of water collecting at the bottom of the position.

Camouflage and Concealment. Camouflage and concealment activities are continual during position siting preparation. If the enemy cannot locate a fighting position, then the position offers friendly forces the advantage of firing first before being detected.

INDIVIDUAL FIGHTING POSITIONS

Table 2-1 summarizes the hasty and deliberate individual fighting positions and provides time estimates, equipment requirements, and protection factors.

Hasty Positions. When time and materials are limited, troops in contact with the enemy use a hasty fighting position located behind whatever cover is available. It should provide frontal protection from direct fire while allowing fire to the front and oblique. For protection from indirect fire, a hasty fighting position is located in a depression or hole at least 1 1/2 feet deep. The following positions provide limited protection and are used when there is little or no natural cover. If the unit remains in the area, the hasty positions are further developed into deliberate positions which provide as much protection as possible.

Deliberate Positions. Deliberate fighting positions are modified hasty positions prepared during periods of relaxed enemy pressure. If the situation permits, the unit leader verifies the sectors of observation before preparing each position. Continued improvements are made to strengthen the position during the period of occupation. Small holes are dug for automatic rifle biped legs so the rifle is as close to ground level as possible. Improvements include adding overhead cover, digging trenches to adjacent positions, and maintaining camouflage.

TRENCHES

Trenches are excavated to connect individual fighting positions and weapons positions in the progressive development of a defensive area. They provide protection and concealment for personnel moving between fighting positions or in and out of the area. Trenches are usually included in the overall layout plan for the defense of a position or strongpoint. Excavating trenches involves considerable time, effort, and materials, and is only justified when an area is occupied for a long time. Trenches are usually open excavations, but covered sections provide additional protection if the overhead cover does not interfere with the fire mission of the occupying personnel. Trenches are difficult to camouflage and are easily detected, especially from the air.

Trenches, as other fighting positions, are developed progressively. They are improved by digging deeper, from a minimum of 2 feet to about 5 1/2 feet. As a general rule, deeper excavation is desired for other than fighting trenches to provide more protection or allow more headroom. Some trenches may also require widening to accommodate more traffic, including stretchers. It is usually necessary to revet

Table 2-1: Characteristics of individual fighting positions.

Type of Position	Estimated Construction Time (man-hours)	Equipment Requirements	Direct Small Caliber Fire	Indirect Fire Blast and Fragmentation (Near-Miss)*	Indirect Fire Blast and Fragmentation *(Direct Hit)	Nuclear Weapons**	Remarks
Hasty							
Crater	0.2	Hand tools	7.62mm	Better than in open - no overhead protection	None	Fair	
Skirmisher's trench	0.5	Hand tools	7.62mm	Better than in open - no overhead protection	None	Fair	
Prone position	1.0	Hand tools	7.62mm	Better than in open - no overhead protection	None	Fair	Provides all-around cover
Deliberate							
One-soldier position	3.0	Hand tools	12.7mm	Medium artillery no closer than 30 ft - no overhead protection	None	Fair	
One-soldier position with 1 1/2 ft. overhead cover	8.0	Hand tools	12.7mm	Medium artillery no closer than 30 ft	None	Good	Additional cover provides protection from direct hit small mortar blast
Two-soldier position	6.0	Hand tools	12.7mm	Medium artillery no closer than 30 ft - no overhead protection	None	Fair	
Two-Soldier position with 1 1/2 ft. overhead cover	11.0	Hand tools	12.7mm	Medium artillery no closer than 30 ft	None	Good	Additional cover provides protection from direct hit small mortar blast
LAW position	3.0	Hand tools	12.7mm	Medium artillery no closer than 30 ft - no overhead protection	None	Fair	

Note Chemical protection is assumed because of individual protective masks and clothing.

* Shell sizes are:

	Small	Medium
Mortar	82mm	120mm
Artillery	105mm	152mm

** Nuclear protection ratings are rated poor, fair, good, very good, and excellent.

Table 2-2: Shielding of m8a1 landing mats.

Percent Fragments Stopped at Cited Range

Weapon	5 ft	10 ft	20 ft	30 ft
81-mm mortar	95	98	98-100	98-100
82-mm mortar	98	98-100	98-100	98-100
4.2-in mortar	76	82	91	98
107-mm rocket	70	79	89	96
120-mm mortar	98	98-100	98-100	98-100
122-mm rocket	—	—	70	78

trenches that are more than 5 feet deep in any type of soil. In the deeper trenches, some engineer advice or assistance is usually necessary in providing adequate drainage. Two basic trenches are the crawl trench and the standard fighting trench.

UNIT POSITIONS

Survivability operations are required to support the deployment of units with branch-specific missions, or missions of extreme tactical importance. These units are required to deploy and remain in one location for a considerable amount of time to perform their mission. Thus, they may require substantial protective construction.

SPECIAL DESIGNS

Table 2-3 summarizes construction estimates and levels of protection for the fighting positions, bunkers, shelters, and protective walls presented in this section.

Fighting Positions. The following two positions are designed for use by two or more individuals armed with rifles or machine guns. Although these are beyond the construction capabilities of non-engineer troops, certain construction phases can be accomplished with little or no engineer assistance. For example, while engineer assistance may be necessary to build steel frames and cut timbers for the roof of a structure, the excavation, assembly, and installation are all within the capabilities of most units. Adequate support for overhead cover is extremely important. The support system should be strong enough to safely support the roof and soil material and survive the effects of weapon detonations.

Bunkers. Bunkers are larger fighting positions constructed for squad-size units who are required to remain in defensive positions for a longer period of time. They are built either above ground or below ground and are usually made of reinforced concrete. Because of the extensive engineer effort required to build bunkers, they are usually made during strong- point construction. If time permits, bunkers are connected to other

Table 2-3: Characteristics of special design positions.

Type of Position	Estimated Construction Time (man-hours)	Equipment Requirements	Direct Small Caliber Fire	Indirect Fire Blast and Fragmentation (Near-Miss)*	Indirect Fire Blast and Fragmentation (Direct Hit)	Nuclear Weapons**	Remarks
FIGHTING POSITIONS							
Wood-frame or steel-frame fighting position with 2½-ft overhead cover	32	Hand tools	12.7mm	Medium artillery no closer than 30 ft	Small mortar	Good	
Fabric-covered frame fighting position with 1½-ft overhead cover	16	Hand tools	12.7mm	Medium artillery no closer than 15 ft	Small mortar	Good	

Note: Chemical protection is assumed because of individual protective masks and clothing.

* Shell sizes are:

	Small	Medium
Mortar	82mm	120mm
Artillery	105mm	152mm

** Nuclear protection ratings are rated poor, fair, good, very good, and excellent.

(continued)

Table 2-3: *(Continued)*

Type of Position	Estimated Construction Time (man-hours)	Equipment Requirements	Direct Small Caliber Fire	Indirect Fire Blast and Fragmentation (Near-Miss)*	Indirect Fire Blast and Fragmentation (Direct Hit)	Nuclear Weapons**	Remarks
BUNKERS							
Corrugated metal fighting bunker with 2½-ft overhead cover	48	Hand tools, backhoe	7.62mm	Medium artillery no closer than 10 ft	Small mortar	Good	
Plywood perimeter bunker	48	Hand tools, backhoe	7.62mm	Limited protection - no overhead protection	None	Poor	
Concrete log bunker with 2½-ft overhead cover	42	Hand tools, backhoe	7.62mm	Medium artillery no closer than 10 ft	Small mortar	Good	Construction time assumes precast logs. Protection provided includes one layer of sandbags around walls
Precast concrete slab bunker with 2½-ft overhead cover	30	Hand tools, backhoe, crane	7.62mm	Medium artillery no closer than 10 ft	Small mortar,	Good	Construction time assumes prefabricated slabs. Protection provided includes one layer of sandbags around walls
Concrete arch bunker with 2½-ft overhead cover	38	Hand tools, backhoe, crane	7.62mm	Medium artillery no closer than 10 ft	Small mortar	Good	Construction time assumes prefabricated sections. Protection provided includes one layer of sandbags around walls
SHELTERS							
Two-soldier sleeping shelter with 2-ft overhead cover	10	Hand tools	7.62mm	Small mortar on contact	Small mortar	Fair	
Metal culvert shelter with 2-ft overhead cover	48	Hand tools, backhoe	7.62mm	Small mortar no closer than 5 ft	None	Fair	
Inverted metal shipping container shelter with 2-ft overhead cover	28	Hand tools, backhoe	12.7mm	Medium artillery no closer than 10 ft	Small mortar	Good	

Note: Chemical protection is assumed because of individual protective masks and clothing.

* Shell sizes are:

	Small	Medium
Mortar	82mm	120mm
Artillery	105mm	152mm

** Nuclear protection ratings are rated poor, fair, good, very good, and excellent.

(continued)

Table 2-3: (Continued)

Type of Position	Estimated Construction Time (man-hours)	Equipment Requirements	Direct Small Caliber Fire	Indirect Fire Blast and Fragmentation (Near-Miss)*	Indirect Fire Blast and Fragmentation (Direct Hit)	Nuclear Weapons**	Remarks
SHELTERS (Continued)							
Airtransportable assault with 2-ft overhead cover	60	Hand tools, backhoe	Cannot engage	Medium artillery no closer than 30 ft	Small mortar	Very good	Construction time assumes prefabricated walls and floor
Timber post buried shelter with 2½-ft overhead cover	48	Hand tools, backhoe	Cannot engage	Medium artillery no closer than 30 ft	Small mortar	Very good	
Modular timber frame shelter with 2-ft overhead cover	96	Hand tools, backhoe	Cannot engage	Medium artillery no closer than 20 ft	Small mortar	Very good	
Timber frame buried shelter with 2-ft overhead cover	84	Hand tools, backhoe	Cannot engage	Medium artillery no closer than 25 ft	Small mortar	Very good	
Aboveground cavity wall shelter with 2-ft overhead cover	700	Hand tools, backhoe, crane	12.7mm	Medium artillery no closer than 10 ft	Small mortar	Good	
Steel-frame/fabric-covered shelter with 1½-ft overhead cover	35	Hand tools, backhoe	Cannot engage	Medium artillery no closer than 10 ft	Small mortar	Very good	Construction time assumes prefabricated frame
Hardened frame/fabric shelter with 4-ft overhead cover	45	Hand tools, backhoe	Cannot engage	Medium artillery no closer than 10 ft	Medium artillery	Excellent	Shelter provides improved nuclear protection to 30 psi
Rectangular fabric/frame shelter with 1½-ft overhead cover	38	Hand tools, backhoe	Cannot engage	Medium artillery no closer than 15 ft	Medium artillery	Very good	Construction time assumes prefabricated frame
Concrete arch shelter with 4-ft overhead cover	64	Hand tools, dozer, backhoe, crane	Cannot engage	Medium artillery no closer than 5 ft	Medium artillery	Very good	Construction time assumes prefabricated arches and end walls

Note: Chemical protection is assumed because of individual protective masks and clothing.

* Shell sizes are:

	Small	Medium
Mortar	82mm	120mm
Artillery	105mm	152mm

** Nuclear protection ratings are rated poor, fair, good, very good, and excellent.

(continued)

Table 2-3: *(Continued)*

Type of Position	Estimated Construction Time (man-hours)	Equipment Requirements	Direct Small Caliber Fire	Indirect Fire Blast and Fragmentation (Near-Miss)*	Indirect Fire Blast and Fragmentation (Direct Hit)	Nuclear Weapons**	Remarks
SHELTERS (Continued)							
Metal pipe arch shelter with 4-ft overhead cover	58	Hand tools, dozer, backhoe, crane	Cannot engage	Medium artillery no closer than 5 ft	Medium artillery	Very good	Construction time assumes pre-assembled arch and end section

Type of Position	Estimated Construction Time (man-hours) per 10-ft section	Equipment Requirements	Direct Small Caliber Fire	Indirect Fire Blast and Fragmentation (Near-Miss)*	Direct Fire HEAT	Nuclear Weapons**	Remarks
PROTECTIVE WALLS							
Earth wall	3	Dozer; dump truck; scoop loader	12.7mm	Medium artillery no closer than 5 ft	120mm at wall base	Poor	
Earth wall with revetment	20	Hand tools; scoop loader	12.7mm	Medium artillery no closer than 5 ft	120mm at wall base	Poor	
Soil-cement wall	25	Hand tools; concrete mixer; crane w/concrete bucket	12.7mm	Small artillery no closer than 5 ft	82mm at wall base	Poor	Walls require forming
Soil bin wall with log revetment	35	Hand tools; scoop loader	5.45mm	Small artillery no closer than 5 ft	None	Poor	
Soil bin wall with timber revetment	30	Hand tools; scoop loader	5.45mm	Small artillery no closer than 5 ft	None	Poor	
Soil bin wall with plywood revetment	19	Hand tools; scoop loader	12.7mm	Medium artillery no closer than 5 ft	120mm at wall base	Poor	Based on plywood design. Provides nuclear blast protection for drag sensitive targets
Plywood portable wall	5	Hand tools; backhoe	5.45mm	Small mortar no closer than 5 ft	None	Poor	
Steel landing mat wall	3	Welding; crane	None	Refer to the table on page	None	Poor	M8A1 steel landing mat only

Note: Chemical protection is assumed because of individual protection masks and clothing. All walls are 5 feet high with minimum thickness as specified in construction plans.

* Shell sizes are:

	Small	Medium
Mortar	82mm	120mm
Artillery	105mm	152mm

** Nuclear protection is minimal except as noted.

(continued)

Table 2-3: *(Continued)*

Type of Position	Estimated Construction Time (man-hours) per 10-ft section	Equipment Requirements	Direct Small Caliber Fire	Indirect Fire Blast and Fragmentation (Near-Miss)*	Direct Fire HEAT	Nuclear Weapons**	Remarks
PROTECTIVE WALLS (Continued)							
Portable pre-cast concrete wall	29	Hand tools; concrete mixer; crane	7.62mm	Medium artillery no closer than 5 ft	None	Poor	One layer of sandbags on outer panel surface improves small caliber protection
Cast-in-place concrete wall wall	35	Hand tools; concrete mixer; crane w/concrete bucket	12.7mm	Small artillery no closer than 5 ft	None	Poor	One layer of sandbags on outer panel surface improves protection to include indirect fire blast and fragmentation from large artillery
Portable asphalt armor panels 2x8x4	15	Hand tools; welding; hot asphalt source	7.62mm	Small artillery no closer than 5 ft	None	Poor	

Note: Chemical protection is assumed because of individual protection masks and clothing. All walls are 5 feet high with minimum thickness as specified in construction plans.

* Shell sizes are:

	Small	Medium
Mortar	82mm	120mm
Artillery	105mm	152mm

** Nuclear protection is minimal except as noted.

fighting or supply positions by tunnels. Prefabrication of bunker assemblies affords rapid construction and placement flexibility. Bunkers offer excellent protection against direct fire and indirect fire effects and, if properly constructed with appropriate collective protection equipment, they provide protection against chemical and biological agents.

Shelters. Shelters are primarily constructed to protect soldiers, equipment, and supplies from enemy action and the weather. Shelters differ from fighting positions because there are usually no provisions for firing weapons from them. However, they are usually constructed near—or to supplement—fighting positions. When available, natural shelters such as caves, mines, or tunnels are used instead of constructing shelters. Engineers are consulted to determine suitability of caves and tunnels.

The best shelter is usually one that provides the most protection but requires the least amount of effort to construct. Shelters are frequently prepared by support troops, troops making a temporary halt due to inclement weather, and units in bivouacs, assembly areas, and rest areas. Shelters are constructed with as much overhead cover as possible. They are dispersed and limited to a maximum capacity of about 25 soldiers. Supply shelters are of any size, depending on location, time, and materials available. Large shelters require additional camouflaged entrances and exits.

All three types of shelters—below ground, aboveground, and cut-and-cover—are usually sited on reverse slopes, in woods, or in some form of natural defilade such as ravines, valleys, wadis, and other hollows or depressions in the terrain. They are not constructed in paths of natural drainage lines. All shelters require camouflage or concealment. As time permits, shelters are continuously improved.

Below ground shelters require the most construction effort but generally provide the highest level of protection from conventional, nuclear, and chemical weapons.

Cut-and-cover shelters are partially dug into the ground and backfilled on top with as thick a layer of cover material as possible. These shelters provide excellent protection from the weather and enemy action.

Above ground shelters provide the best observation and are easier to enter and exit than below ground shelters. They also require the least amount of labor to construct, but are hard to conceal and require a large amount of cover and revetting material. They provide the least amount of protection from nuclear and conventional weapons; however, they do provide protection against liquid droplets of chemical agents. Aboveground shelters are seldom used for personnel in forward combat positions unless the shelters are concealed in woods, on reverse slopes, or among buildings. Above ground shelters are used when water levels are close to the ground surface or when the ground is so hard that digging a below ground shelter is impractical.

The following shelters are suitable for a variety of uses where troops and their equipment require protection, whether performing their duties or resting.

Protective Walls. Several basic types of walls are constructed to satisfy various weather, topographical, tactical, and other military requirements. The walls range from simple ones, constructed with hand tools, to more difficult walls requiring specialized engineering and equipment capabilities.

Protection provided by the walls is restricted to stopping fragment and blast effects from near-miss explosions of mortar, rocket, or artillery shells; some direct fire protection is also provided. Overhead cover is not practical due to the size of the position surrounded by the walls. In some cases, modification of the designs shown will increase nuclear protection. The wall's effectiveness substantially increases by locating it in adequately-defended areas. The walls need close integration with other forms of protection such as dispersion, concealment, and adjacent fighting positions. The protective walls should have the minimum inside area required to perform operational duties. Further, the walls should have their height as near to the height of the equipment as practical.

CHAPTER 3

Special Operations and Situations

The two basic operations involving U.S. force deployment are combined and contingency. Combined operations are enacted in areas where U.S. forces are already established, such as NATO nations. Where few or no U.S. installations exist, usually in undeveloped regions, contingency operations are planned. In both cases, survivability missions will require intensive engineer support in all types of terrain and climate. Each environment's advantages and disadvantages are adapted to survivability planning, designing, and constructing positions. Fighting and protective positions in jungles, mountainous areas, deserts, cold regions, and urban areas require specialized knowledge, skills, techniques, and equipment. This chapter presents characteristics of five environments which impact on survivability and describes the conditions expected during combined and contingency operations.

SPECIAL TERRAIN ENVIRONMENTS

Jungles. Jungles are humid, tropic areas with a dense growth of trees and vegetation. Visibility is typically less than 100 feet, and areas are sparsely populated. Because mounted infantry and armor operations are limited in jungle areas, individual and crew-served weapons fighting position construction and use receive additional emphasis. While jungle vegetation provides excellent concealment from air and ground observation, fields of fire are difficult to establish. Vegetation does not provide adequate cover from small caliber direct fire and artillery indirect fire fragments. Adequate cover is available, though, if positions are located using the natural ravines and gullies produced by erosion from the area's high annual rainfall.

The few natural or locally-procurable materials which are available in jungle areas are usually limited to camouflage use. Position construction materials are transported to these areas and are required to be weather and rot resistant. When shelters are constructed in jungles, primary consideration is given to

Stout, pliable reeds
laid across framework

Several layers of
large fine ferns

3 ft

drainage provisions. Because of high amounts of rainfall and poor soil drainage, positions are built to allow for good, natural drainage routes. This technique not only prevents flooded positions but, because of nuclear fallout washing down from trees and vegetation, it also prevents positions from becoming radiation hot spots.

Other considerations are high water tables, dense undergrowth, and tree roots, often requiring above-ground level protective construction. A structure used in areas where groundwater is high, or where there is a low-pressure resistance soil, is the fighting position platform, depicted below. This platform provides a floating base or floor where wet or low-pressure resistance soil precludes standing or sitting. The platform is constructed of small branches or timber layered over cross-posts, thus distributing the floor load over a wider area. As shown in the following two illustrations, satisfactory rain shelters are quickly constructed using easily-procurable materials such as ponchos or natural materials.

Mountainous Areas. Characteristics of mountain ranges include rugged, poorly trafficable terrain, steep slopes, and altitudes greater than 1,600 feet. Irregular mountain terrain provides numerous places for cover and concealment. Because of rocky ground, it is difficult and often impossible to dig below ground positions; therefore, boulders and loose rocks are used in above ground construction. Irregular fields of fire and dead spaces are considered when designing and locating fighting positions in mountainous areas.

Reverse slope positions are rarely used in mountainous terrain; crest and near-crest positions on high ground are much more common. Direct fire weapon positions in mountainous areas are usually poorly concealed by large fields of fire. Indirect fire weapon positions are better protected from both direct and indirect fire when located behind steep slopes and ridges.

Another important design consideration in mountain terrain is the requirement for substantial overhead cover. The adverse effects of artillery bursts above a protective position are greatly enhanced by rock and gravel displacement or avalanche. Construction materials used for both structural and shielding components are most often indigenous rocks, boulders, and rocky soil. Often, rock formations are used as structural wall components without modification. Conventional tools are inadequate for preparing individual and crew-served weapons fighting positions in rocky terrain. Engineers assist with light equipment and tools (such as pneumatic jackhammers) delivered to mountain areas by helicopter.

In areas with rocky soil or gravel, wire cages or gabions are used as building blocks in protective walls, structural walls, and fighting positions. Gabions are constructed of lumber, plywood, wire fence, or any suitable material that forms a stackable container for soil or gravel.

The two-soldier mountain shelter is basically a hole 7 feet long, 3 1/2 feet wide, and 3 1/2 feet deep. The hole is covered with 6- to 8-inch diameter logs with evergreen branches, a shelter half, or local material such as topsoil, leaves, snow, and twigs placed on top. The floor is usually covered with evergreen twigs, a shelter half, or other expedient material. Entrances can be provided at both ends or a fire pit is sometimes dug at one end for a small fire or stove. A low earth parapet is built around the position to provide more height for the occupants.

Deserts. Deserts are extensive, arid, arid treeless, having a severe lack of rainfall and extreme daily temperature fluctuations. The terrain is sandy with boulder-strewn areas, mountains, dunes, deeply-eroded valleys, areas of rock and shale, and salt marshes. Effective natural barriers are found in steep slope rock formations. Wadis and other dried up drainage features are used extensively for protective position placement.

Designers of fighting and protective positions in desert areas must consider the lack of available natural cover and concealment. The only minimal cover available is through the use of terrain masking; therefore, positions are often completed above ground. Mountain and plateau deserts have rocky soil or "surface chalk" soil which makes digging difficult. In these areas, rocks and boulders are used for cover. Most often, parapets used in desert fighting or protective positions are undesirable because of probable enemy detection in the flat desert terrain. Deep-cut positions are also difficult to construct in soft sandy areas because of wall instability during excavations. Revetments are almost always required, unless excavations are very wide and have gently sloping sides of 45 degrees or less. Designing overhead cover is additionally important because nuclear explosions have increased fallout due to easily displaced sandy soil.

Indigenous materials are usually used in desert position construction. However, prefabricated structures and revetments for excavations, if available, are ideal. Metal culvert revetments are quickly emplaced in easily excavated sand. Sandbags and sand-filled ammunition boxes are also used for containing backsliding soil. Therefore, camouflage and concealment, as well as light and noise discipline, are important considerations during position construction. Target acquisition and observation are relatively easy in desert terrain.

Cold Regions. Cold regions of the world are characterized by deep snow, permafrost, seasonally frozen ground, frozen lakes and rivers, glaciers, and long periods of extremely cold temperatures. Digging in frozen or semifrozen ground is difficult with equipment, and virtually impossible for the soldier with an entrenching tool. When possible, positions are designed to take advantage of below ground cover. Positions are dug as deep as possible, then built up. Fighting and protective position construction in snow or frozen ground takes up to twice as long as positions in unfrozen ground. Also, positions used in cold regions are affected by wind and the possibility of thaw during warming periods. An unexpected thaw causes a severe drop in the soil strength which creates mud and drainage problems. Positions near bodies of water, such as lakes or rivers, are carefully located to prevent flooding damage during the spring melt season. Wind protection greatly decreases the effects of cold on both soldiers and equipment. The following areas offer good wind protection:

- Densely wooded areas.
- Groups of vegetation; small blocks of trees or shrubs.
- The lee side of terrain elevations. (The protected zone extends horizontally up to three times the height of the terrain elevation).
- Terrain depressions.

The three basic construction materials available in cold region terrain are snow, ice, and frozen soil. Positions are more effective when constructed with these three materials in conjunction with timber, stone, or other locally-available materials.

Snow. Dry snow is less suitable for expedient construction than wet snow because it does not pack as well. Snow piled at road edges after clearing equipment has passed densifies and begins to harden within hours after disturbance, even at very low temperatures. Snow compacted artificially, by the wind, and after a brief thaw is even more suitable for expedient shelters and protective structures. A uniform snow cover with a minimum thickness of 10 inches is sufficient for shelter from the weather and for revetment construction. Blocks of uniform size, typically 8 by 12 by 16 inches, depending upon degree of hardness and density, are cut from the snow pack with shovels, long knives (machetes), or carpenter's saws. The best practices for constructing cold weather shelters are those adopted from natives of polar regions.

The systematic overlapping block-over-seam method ensures stable construction. "Caulking" seams with loose snow ensures snug, draft-free structures. Igloo shelters in cold regions have been known to survive a whole winter. An Eskimo-style snow shelter, depicted below, easily withstands above-freezing inside temperatures, thus providing comfortable protection against wind chill and low temperatures. Snow positions are built during either freezing or thawing if the thaw is not so long or intense that significant snow melt conditions occur. Mild thaw of temperatures or 2 degrees above freezing are more favorable than below-freezing temperatures because snow conglomerates readily and assumes any shape without disintegration. Below-freezing temperatures are necessary for snow construction in order to achieve solid freezing and strength. If water is available at low temperatures, expedient protective structures are built by wetting down and shaping snow, with shovels, into the desired forms.

Ice. The initial projectile-stopping capability of ice is better than snow or frozen soil; however, under sustained fire, ice rapidly cracks and collapses. Ice structures are built in the following three ways:

Layer-by-layer freezing by water. This method produces the strongest ice but, compared to the other two methods, is more time consuming. Protective surfaces are formed by spraying water in a fine mist on a structure or fabric. The most favorable temperature for this method is –10 to –15 degrees Celsius with a moderate wind. Approximately 2 to 3 inches of ice are formed per day between these temperatures (1/5-inch of ice per degree below zero).

Freezing ice fragments into layers by adding water. This method is very effective and the most frequently used for building ice structures. The ice fragments are about 1-inch thick and prepared on nearby plots or

Eskimo-style snow shelter

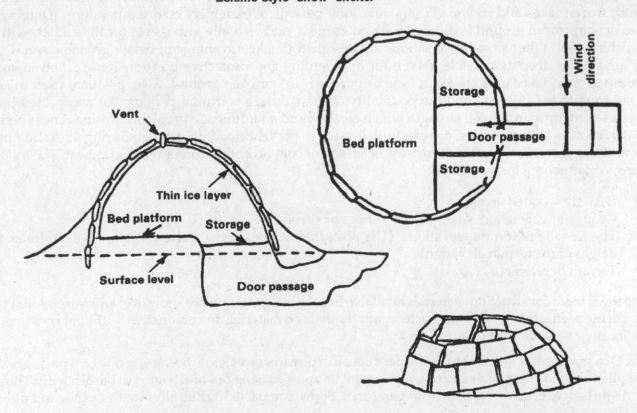

on the nearest river or water reservoir. The fragments are packed as densely as possible into a layer 8 to 12 inches thick. Water is then sprayed over the layers of ice fragments. Crushing the ice fragments weakens the ice construction. If the weather is favorable (–10 to –15 degrees Celsius with wind), a 16- to 24-inch thick ice layer is usually frozen in a day.

Laying ice blocks. This method is the quickest, but requires assets to transport the blocks from the nearest river or water reservoir to the site. Ice blocks, laid and overlapped like bricks, are of equal thickness and uniform size. To achieve good layer adhesion, the preceding layer is lightly sprayed with water before placing a new layer. Each new layer of blocks freezes onto the preceding layer before additional layers are placed.

Frozen Soil. Frozen soil is three to five times stronger than ice, and increases in strength with temperatures. Frozen soil has much better resistance to impact and explosion than to steadily-acting loads—an especially valuable feature for position construction purposes. Construction using frozen soil is performed as follows:

- Preparing blocks of frozen soil from a mixture of water and aggregate (icecrete).
- Laying prepared blocks of frozen soil.
- Freezing blocks of frozen soil together in layers.

Unfrozen soil from beneath the frozen layer is sometimes used to construct a position quickly before the soil freezes. Material made of gravel-sand-silt aggregate wetted to saturation and poured like portland cement concrete is also suitable for constructing positions. After freezing, the material has the properties of concrete. The construction methods used are analogous to those using ice. Fighting and protective positions in arctic areas are constructed both below ground and above ground.

Below ground positions. When the frost layer is one foot or less, fighting positions are usually constructed below ground, as shown. Snow packed 8 to 9 feet provides protection from sustained direct fire from small caliber weapons up to and including the Soviet 14, 5-mm KPV machine gun. When possible, unfrozen excavated soil is used to form parapets about 2-feet thick, and snow is placed on the soil for

Below ground fighting position in snow

Frontal protection

Overhead protection

Side and rear protection

camouflage and extra protection. For added frontal protection, the interior snow is reinforced with a log revetment at least 3 inches in diameter. The outer surface is reinforced with small branches to initiate bullet tumble upon impact. Bullets slow down very rapidly in snow after they begin to tumble. The wall of logs directly in front of the position safely absorbs the slowed tumbling bullet.

Overhead cover is constructed with 3 feet of packed snow placed atop a layer of 6-inch diameter logs. This protection is adequate to stop indirect fire fragmentation. A layer of small, 2-inch diameter logs is placed atop the packed snow to detonate quick fuzed shells before they become imbedded in the snow.

Above-ground positions. If the soil is frozen to a significant depth, the soldier equipped with only an entrenching tool and ax will have difficulty digging a fighting position. Under these conditions (below the tree line), snow and wood are often the only natural materials available to construct fighting positions. The fighting position is dug at least 20 inches deep, up to chest height, depending on snow conditions. Ideally, sandbags are used to revet the interior walls for added protection and to prevent cave-ins. If sandbags are not available, a lattice framework is constructed using small branches or, if time permits, a wall of 3-inch logs is built. Overhead cover, frontal protection, and side and rear parapets are built employing the same techniques described in chapter 2.

It is approximately ten times faster to build above-ground snow positions than to dig in frozen ground to obtain the same degree of protection. Fighting and protective positions constructed in cold regions are excavated with combined methods using handtools, excavation equipment, or explosives. Heavy equipment use is limited by traction and maneuverability.

Shelters. Shelters are constructed with a minimum expenditure of time and labor using available materials. They are ordinarily built on frozen ground or dug in deep snow. Shelters that are completely above ground offer protection against the weather and supplement or replace tents. Shelter sites near wooded areas are most desirable because the wood conceals the glow of fires and provides fuel for cooking and heating. Tree branches extending to the ground offer some shelter for small units or individual protective positions.

Dismounted TOW and machine gun positions in snow

A platform of plywood or timber is constructed to the rear of the frontal protection to provide a solid base from which to employ the guns. Overhead cover is usually offset from the firing position because of the difficulty of digging both the firing and protective positions together in the snow. The protective position should have at least 3 feet of packed snow as cover. The fighting position should have snow packed 8 to 9 feet thick for frontal, and at least 2 feet thick for side protection as shown. Sandbags are used to revet the interior walls for added protection and to prevent cave-ins. However, packed snow, rocks, 4-inch diameter logs, or ammunition cans filled with snow are sometimes used to complete the frontal and overhead protection, as well as side and rear parapets.

Individual fighting position in snow

Positions for individuals are constructed by placing packed snow on either side of a tree and extending the snow parapet 8 to 9 feet to the front, as illustrated. The side and rear parapets are constructed of a continuous snow mound, a minimum of 2 feet wide, and high enough to protect the soldier's head.

Snow trench with wood revetment

In deep snow, trenches and weapon positions are excavated to the dimensions outlined in chapter 4. However, unless the snow is well packed and frozen, revetment is required. In snow too shallow to permit the required depth excavation, snow walls are usually constructed, The walls are made of compacted snow, revetted, and at least 6 1/2 feet thick. The table on page 5-12 contains snow wall construction requirements.

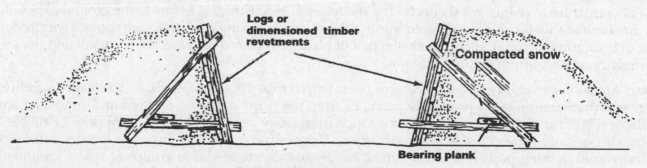

Logs or dimensioned timber revetments

Compacted snow

Bearing plank

5-6% ft between supports

Table 3-1: Snow construction for protection from grenades, small caliber fire, and HEAT projectiles.

Snow Density (lb/cu ft)	Projectiles	Muzzle Velocity	Penetration, ft		Required Minimum Thickness, ft
18.0 -25.0	Grenade frag (HE)		2.0		3.0
11.2 -13.0	5.56 mm	3,250	3.8		4.4
17.4 -23.7	5.56 mm	3,250	2.3		2.6
11.2 -13.1	7.62 mm	2,750	13.0		15.0
17.4 -23.7	7.62 mm	2,750	5.2		6.0
25.5 -28.7	7.62 mm	2,750	5.0		5.8
19.9 -24.9	12.7 mm	2,910	6.4		7.4
	14.5 mm		6.0		8.0
28.1 -31.2	70 mm HEAT	900	14.0		17.5
31.2 -34.9	70 mm liEAT	900	8.7	-10.0	13.0
27.5 -34.9	90 mm HEAT	700	9.5	-11.2	14.5

Notes: These materials degrade under sustained fire. Penetrations given for 12.7 mm or smaller are for sustained fire (30 continuous firings into a 1 by 1 foot area),

Penetration characteristics of Warsaw Pact ammunitions do nut differ significantly from US counterparts.

Figure given for HEAT weapons are for Soviet PRG-7 (70 mm) and United States M67 (90 mm) fired into machine-packed snow.

High explosive grenades produce small, high velocity fragments which stop in about 2 feet of packed snow. Effective protection from direct fire is independent of delivery method, including newer machine guns like the Soviet AGS-17 (30 mm) or United States MK 19/M75 (40 mm). Only armor penetrating rounds are effective.

Constructing winter shelters begins immediately after the halt to keep the soldiers warm. Beds of foliage, moss, straw, boards, skis, shelter halves, and ponchos are sometimes used as protection against ground dampness and cold. The entrance to the shelter, located on the side least exposed to the wind, is close to the ground and slopes up into the shelter. Openings or cracks in the shelter walls are caulked with an earth and snow mixture to reduce wind effects. The shelter itself is constructed as low to the ground as possible. Any fire built within the shelter is placed low in fire holes and cooking pits. Although snow is windproof, a layer of insulating material, such as a shelter half or blanket, is placed between the occupant and the snow to prevent body heat from melting the snow.

Urban Areas. Survivability of combat forces operating in urban areas depends on the leader's ability to locate adequate fighting and protective positions from the many apparent covered and concealed areas available. Fighting and protective positions range from hasty positions formed from piles of rubble, to deliberate positions located inside urban structures. Urban structures are the most advantageous locations for individual fighting positions. Urban structures are usually divided into groups of below ground and above-ground structures.

Wigwam shelters

This shelter is constructed easily and quickly when the ground is too hard to dig and protection is required for a short bivouac. The shelter accommodates three soldiers and provides space for cooking. About 25 evergreen saplings (2 to 3 inches in diameter, 10 feet long) are cut. The limbs are left on the saplings and are leaned against a small tree so the cut ends extend about 7 feet up the trunk. The cut ends are tied together around the tree with a tent rope, wire, or other means. The ground ends of the saplings are spaced about 1 foot apart and about 7 feet from the base of the tree. The branches on the outside of the wigwam are placed flat against the saplings. Branches on the inside are trimmed off and placed on the outside to fill in the spaces. Shelter halves wrapped around the outside make the wigwam more windproof, especially after it is covered with snow. A wigwam is also constructed by lashing the cut ends of the saplings together instead of leaning them against the tree.

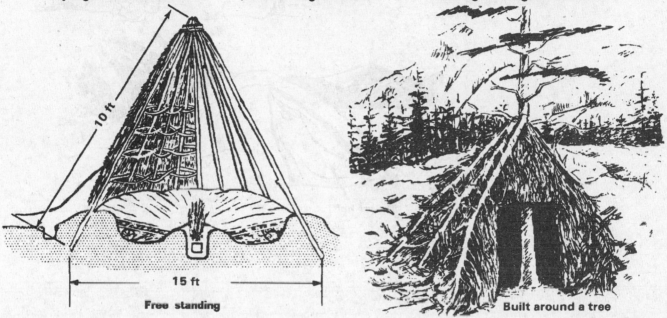

10 ft

15 ft

Free standing

Built around a tree

Below Ground Structures. A detailed knowledge of the nature and location of below ground facilities and structures is of potential value when planning survivability operations in urban terrain. Typical underground street cross sections are shown in Figure 3-13.

Sewers are separated into sanitary, storm, or combined systems. Sanitary sewers carry wastes and are normally too small for troop movement or protection. Storm sewers, however, provide rainfall removal and are often large enough to permit troop and occasional vehicle movement and protection. Except for groundwater, these sewers are dry during periods of no precipitation. During rainstorms, however, sewers fill rapidly and, though normally drained by electrical pumps, may overflow. During winter combat, snow melt may preclude daytime below ground operations. Another hazard is poor ventilation and the resultant toxic fume build-up that occurs in sewer tunnels and subways. The conditions in sewers provide an excellent breeding ground for disease, which demands proper troop hygiene and immunization.

Subways tend to run under main roadways and have the potential hazard of having electrified rails and power leads. Passageways often extend outward from underground malls or storage areas, and catacombs are sometimes encountered in older sections of cities.

Above-ground Structures. Above-ground structures in urban areas are generally of two types: frameless and framed.

Lean-to shelter

This shelter is made of the same material as the wigwam (natural saplings woven together and brush). The saplings are placed against a rock wall, a steep hillside, a deadfall, or some other existing vertical surface, on the leeward side. The ends are closed with shelter halves or evergreen branches.

Shelter half

Snow cave

Snow caves are made by burrowing into a snowdrift and fashioning a room of the desired size. This shelter gives good protection from freezing weather and a maximum amount of concealment. The entrance slopes upward for best protection against cold air penetration. Snow caves are usually built large enough for several soldiers if the consistency of the snow prevents cave-in. Two entrances are usually used while the snow is taken out of the cave; one entrance is refilled with snow when the cave is completed. Fires in snow caves are kept small to prevent melting the structure, To allow incoming fresh air, the door is not completely sealed.

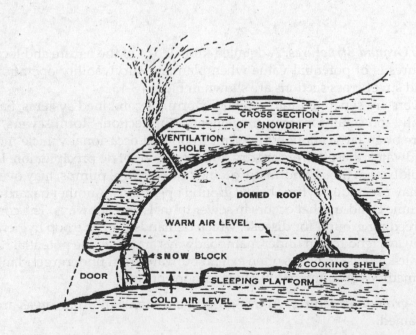

CROSS SECTION OF SNOWDRIFT

VENTILATION HOLE

DOMED ROOF

WARM AIR LEVEL

SNOW BLOCK

DOOR

SLEEPING PLATFORM

COOKING SHELF

COLD AIR LEVEL

Snow hole

The snow hole is a simple, one-soldier emergency shelter for protection against a snow storm in open, snow-covered terrain. The soldier digs a hole of body length and width with an entrenching tool or helmet. At a depth of about 3½ feet, the soldier lies down in the hole and then digs in sideways below the surface, filling the original ditch with snow that was dug out, until only a small breathing hole remains.

Snow pit

The snow pit is dug vertically with entrenching tools to form a ditch. The pit is large enough for two or three soldiers. Skis, poles, sticks, branches, shelter halves, and snow are used as roofing. The inside depth of the pit is deep enough for kneeling and reclining positions. If the snow is not deep enough, the sides of the pit can be made higher by adding snow walls. The roof should slope toward one end of the pit.

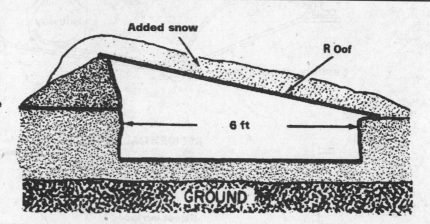

Snowhouse with snow block walls

The size and roof of a snowhouse are similar to those of a snow pit. The walls are made of snow blocks and are usually built to the soldier's height. Snow piled on the outside seals cracks and camouflages the house.

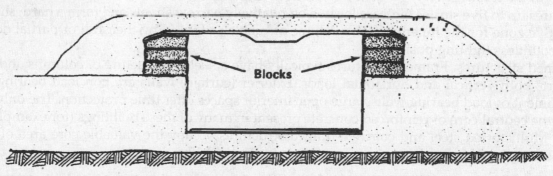

Frameless structures. In frameless structures, the mass of the exterior wall performs the principal load-bearing functions of supporting dead weight of roofs, floors, ceilings; weight of furnishings and occupants; and horizontal loads. Frameless structures are shown in Figure 3-14.

Building materials for frameless structures include mud, stone, brick, cement building blocks, and reinforced concrete. Wall thickness varies with material and building height. Frameless structures have thicker walls than framed structures, and therefore are more resistant to projectile penetration. Fighting from frameless buildings is usually restricted to the door and window areas.

Frameless buildings vary with function, age, and cost of building materials. Older institutional buildings, such as churches, are frequently made of stone. Reinforced concrete is the principal material for wall and slab structures (apartments and hotels) and for prefabricated structures used for commercial and

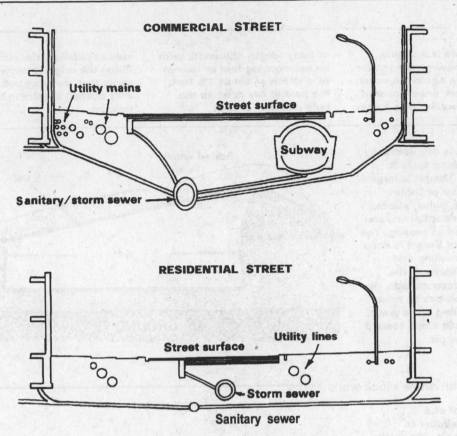

Figure 3-13: Cross sections of streets.

industrial purposes. Brick structures, the most common type of buildings, dominate the core of urban areas (except in the relatively few parts of the world where wood-framed houses are common). Close-set brick structures up to five stories high are located on relatively narrow streets and form a hard, shock-absorbing protective zone for the inner city. The volume of rubble produced by their full or partial demolition provides countless fighting positions.

Framed structures. Framed structures typically have a skeletal structure of columns and beams which supports both vertical and horizontal loads. Exterior (curtain) walls are non load-bearing. Without the impediment of load bearing walls, large open interior spaces offer little protection. The only available refuge is the central core of reinforced concrete present in many of these buildings (for example, the elevator shaft). Multistoried steel and concrete-framed structures occupy the valuable core area of most modern cities. Examples of framed structures are shown in Figure 3-15.

Material and Structural Characteristics. Urban structures, frameless and framed, fit certain material generalities. Table 3-2 converts building type and material into height/wall thicknesses. Most worldwide urban areas have more than 60 percent of their construction formed from bricks. The relationship between building height and thickness of the average brick wall is shown in Table 3-3.

Special Urban Area Positions.

Troop Protection. After urban structures are classified as either frameless or framed, and some of their material characteristics are defined, leaders evaluate them for protective soundness. The evaluation is based on troop protection available and weapon position employment requirements for cover, concealment, and routes of escape. Table 3-4 summarizes survivability requirements for troop protection.

Cover. The extent of building cover depends on the proportion of walls to windows. It is necessary to know the proportion of non-windowed wall space which might serve as protection. Frameless buildings,

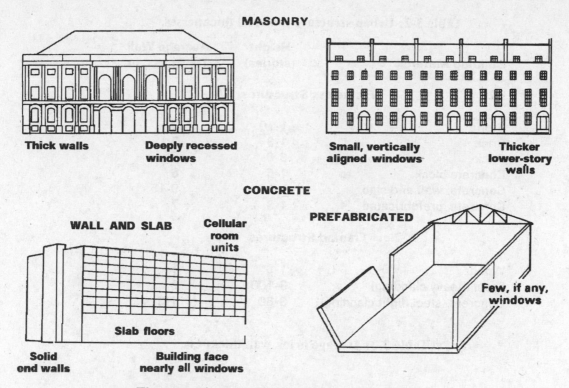

Figure 3-14: Frameless building characteristics.

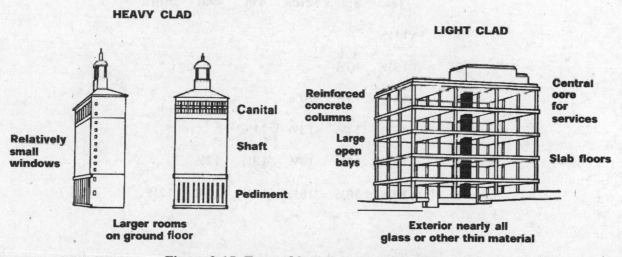

Figure 3-15: Framed building characteristics.

with their high proportion of walls to windows, afford more substantial cover than framed buildings having both a lower proportion of wall to window space and thinner (non load-bearing) walls.

Composition and thickness of both exterior and interior walls also have a significant bearing on cover assessment. Frameless buildings with their strong weight-bearing walls provide more cover than the curtain walls of framed buildings. However, interior walls of the older, heavy-clad, framed buildings are stronger than those of the new, light-clad, framed buildings. Cover within these light-clad framed buildings is very slight except in and behind their stair and elevator modules which are usually constructed of reinforced concrete. Familiarity with the location, dimension, and form of these modules is vital when assessing cover possibilities.

Table 3-2: Urban structure material thicknesses.

Building Material	Height (stories)	Average Wall Thickness, in
Frameless Structures		
Stone	1-10	30
Brick	1-3	9
Brick	3-6	15
Concrete block	1-5	8
Concrete, wall and slab	1-10	9-15
Concrete, prefabricated	1-3	7
Framed Structures		
Wood	1-5	1
Steel (heavy cladding)	3-100	5
Concrete/steel (light cladding)	3-50	1-3

Table 3-3: Average brick wall thickness.

Height (stories)	1st	2nd	3rd	4th	5th	6th
1	11½					
2	13½	10½				
3	14½	13½	10½			
4	15½	14½	13½	11½		
5	18½	15½	14½	13½	12½	
6	18½	18½	15½	14½	13½	12½

Concealment. Concealment considerations involve some of the same elements of building construction, but knowledge of the venting (window) pattern and floor plan is added.

These patterns vary with type of building construction and function. Older, heavy-clad framed buildings (such as office buildings) frequently have as full a venting pattern as possible, while hotels have only one window per room. In the newer, light-clad framed buildings, windows are sometimes used as a non-load bearing curtain wall. If the windows are all broken, no concealment possibilities exist. Another aspect of concealment—undetected movement within the building—depends on a knowledge of the floor plan and the traffic pattern within the building on each floor and from floor to floor.

Escape. In planning for escape routes, the floor plan, traffic patterns, and the relationships between building exits are considered. Possibilities range from small buildings with front street exits (posing unacceptable risks), to high-rise structures having exits on several floors, above and below ground level, and connecting with other buildings as well.

Table 3-4: Survivability requirements for troops in urban buildings.

Requirements	Building Characteristics
Cover	1. Proportion of walls to windows
	2. Wall composition and thickness
	3. Interior wall and partition composition and thickness
	4. Stair and elevator modules
Concealment	1. Proportion of walls to windows
	2. Venting pattern
	3. Floor plan (horizontal and vertical)
	4. Stair and elevator modules (framed high-rise buildings)
Escape	1. Floor plan (horizontal and vertical)
	2. Stair and elevator modules

Fighting Positions. Survivability requirements for fighting positions for individuals, machine guns, and antitank and antiaircraft weapons are summarized in Table 3-5.

Individual fighting positions. An upper floor area of a multistoried building generally provides sufficient fields of fire, although corner windows can usually encompass more area. Protection from the possibility of return fire from the streets requires that the soldier know the composition and thickness of the building's outer wall. Load bearing walls generally offer more protection than the curtain walls of framed buildings. However, the relatively thin walls of a low brick building (only two-bricks thick or 8 inches) is sometimes less effective than a 15-inch thick nonload bearing curtain wall of a high-rise framed structure.

The individual soldier is also concerned about the amount of overhead protection available. Therefore, the soldier needs to know about the properties of roof, floor, and ceiling materials. These materials vary with the type of building construction. In brick buildings, the material for the ceiling of the top floor is far lighter than that for the next floor down that performs as both ceiling and floor, and thus is capable of holding up the room's live load.

Table 3-5: Survivability requirements for fighting positions in urban buildings.

Individual positions

1. Wall composition and thickness of upper floors

2. Roof composition and thickness

3. Floor and ceiling composition and thickness

Machine gun positions

1. Wall composition and thickness

2. Local terrain

Antitank weapon positions

1. Wall composition and thickness

2. Room dimensions and volume

3. Function related interior furnishings, and so forth

4. Fields of fire (relative position of building)

5. Arming distance

6. Line-of-sight

Antiaircraft weapon positions

1. Roof composition and thickness

2. Floor plan (horizontal and vertical)

3. Line-of-sight

CHAPTER 4

Position Design Details

PRONE POSITION (HASTY)

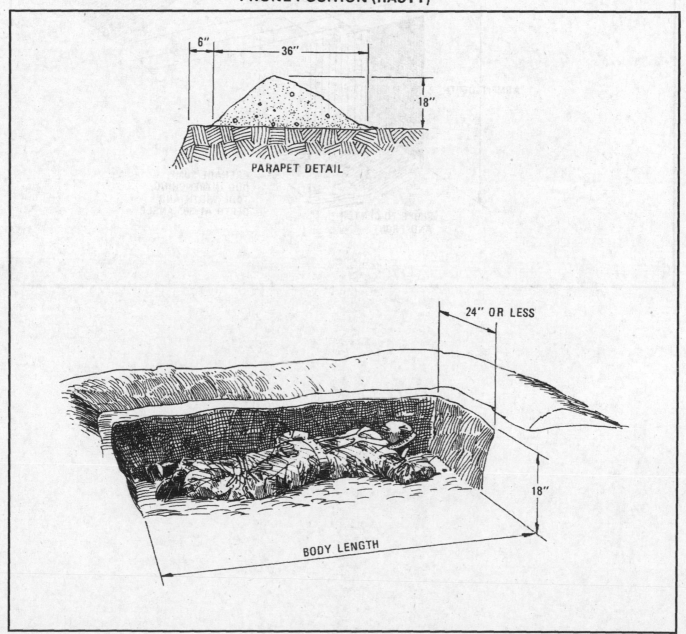

ONE-SOLDIER POSITION (DELIBERATE)

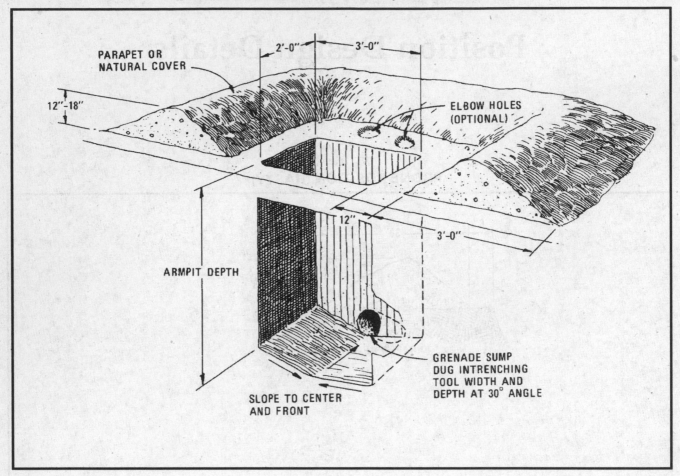

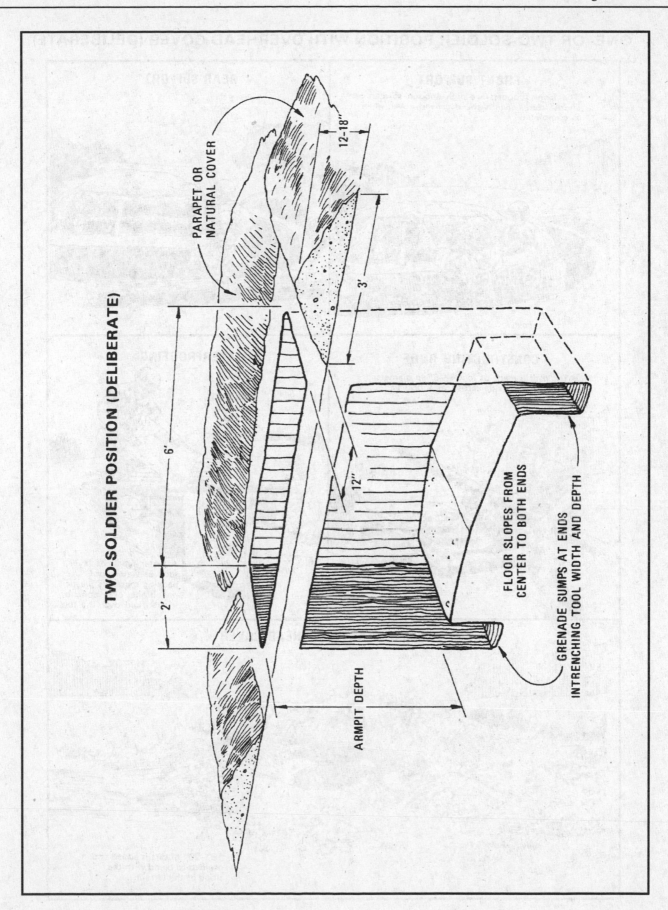

TWO-SOLDIER POSITION (DELIBERATE)

PARAPET OR NATURAL COVER

12-18"

3'

6'

12"

2'

ARMPIT DEPTH

FLOOR SLOPES FROM CENTER TO BOTH ENDS

GRENADE SUMPS AT ENDS INTRENCHING TOOL WIDTH AND DEPTH

ONE- OR TWO-SOLDIER POSITION WITH OVERHEAD COVER (DELIBERATE)

FRONT SUPPORT

The front supports are high enough so men can shoot from beneath the overhead cover when it is completed.

REAR SUPPORT

CONSTRUCTING ROOF

The roof is made of logs 4″–6″ in diameter placed side by side across the supports.

WATERPROOFING

A water-repellant layer, such as waterproof packing material, plastic membrane, or a poncho, is then laid over the logs.

CAMOUFLAGE OVERHEAD COVER

18″–20″ of dirt is added and molded to blend with the slope of the terrain.

DISMOUNTED TOW POSITION

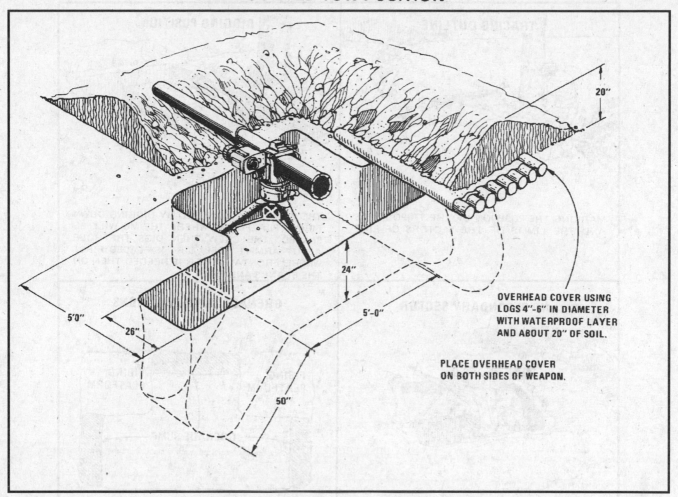

20"

24"

26"

5'0"

5'-0"

50"

OVERHEAD COVER USING
LOGS 4"-6" IN DIAMETER
WITH WATERPROOF LAYER
AND ABOUT 20" OF SOIL.

PLACE OVERHEAD COVER
ON BOTH SIDES OF WEAPON.

MACHINE GUN POSITION

TRACING OUTLINE

MARKING THE POSITION OF THE TRIPOD LEGS AND THE LIMITS OF THE SECTORS OF FIRE

DIGGING POSITION

THE WEAPON IS LOWERED BY DIGGING DOWN FIRING PLATFORMS WHERE THE MG WILL BE PLACED. THE CREW THEN DIGS THE HOLE ABOUT ARMPIT DEEP. DIRT IS PLACED FIRST WHERE FRONTAL COVER IS NEEDED THEN ON THE FLANKS AND REAR.

NO SECONDARY SECTOR

IN SOME POSITIONS, AN MG MAY NOT HAVE A SECONDARY SECTOR OF FIRE; SO, ONLY HALF OF THE POSITION IS DUG.

GRENADE SUMP LOCATIONS

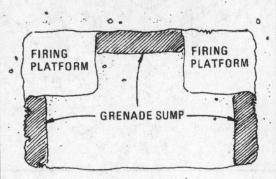

FIRING PLATFORM

FIRING PLATFORM

GRENADE SUMP

TOP VIEW

AMMO BEARER'S POSITION

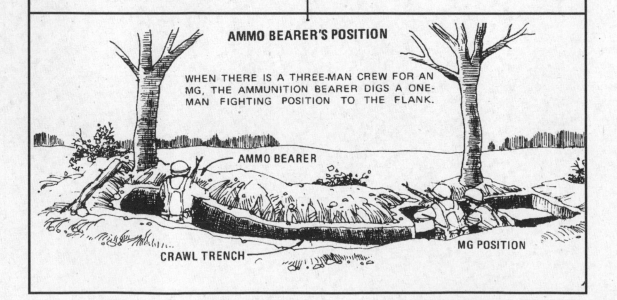

WHEN THERE IS A THREE-MAN CREW FOR AN MG, THE AMMUNITION BEARER DIGS A ONE-MAN FIGHTING POSITION TO THE FLANK.

AMMO BEARER

CRAWL TRENCH

MG POSITION

MORTAR POSITION (81MM AND 4.2-IN MORTARS)

HASTY POSITION

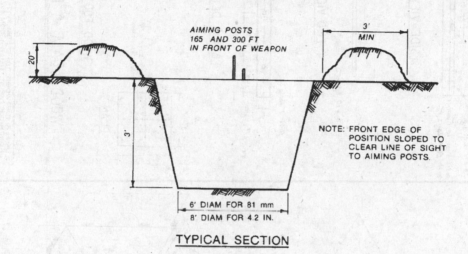

AIMING POSTS
165 AND 300 FT
IN FRONT OF WEAPON

20"

3'

3'
MIN

NOTE: FRONT EDGE OF
POSITION SLOPED TO
CLEAR LINE OF SIGHT
TO AIMING POSTS.

6' DIAM FOR 81 mm
8' DIAM FOR 4.2 IN.

TYPICAL SECTION

IMPROVED POSITION

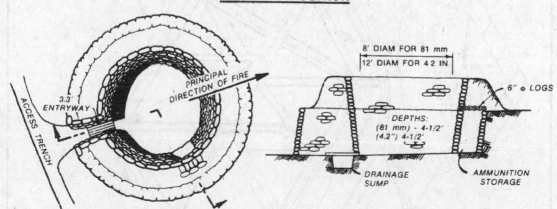

PRINCIPAL
DIRECTION OF FIRE

3.3'
ENTRYWAY

ACCESS TRENCH

8' DIAM FOR 81 mm
12' DIAM FOR 4.2 IN.

6" φ LOGS

DEPTHS:
(81 mm) - 4-1/2'
(4.2") 4-1/2'

DRAINAGE
SUMP

AMMUNITION
STORAGE

PLAN

CROSS SECTION

NOTE: TOTAL DEPTH
INCLUDES THE PARAPET.

SLOPE FLOOR TOWARD
DRAINAGE SLUMP

WOOD-FRAME FIGHTING POSITION (sheet 1 of 3)

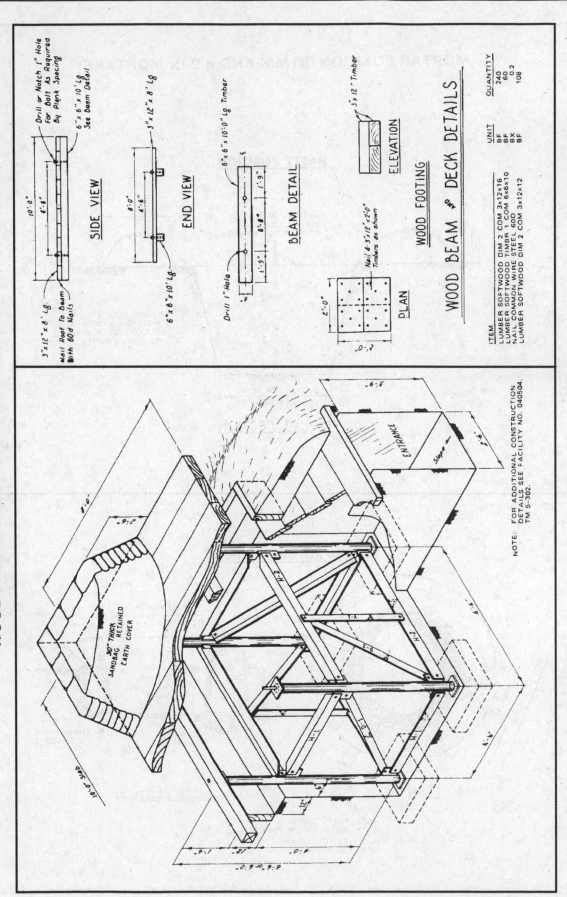

SIDE VIEW

10'-0"
6'-6"
3"x12"x 8' Lg.
6"x 6"x10'Lg.
Nail Roof To Beam With 60d Nails
Drill or Notch 1" Hole For Bolt As Required By Plank Spacing
6"x 6"x10'Lg. See Beam Detail

END VIEW

8'-0"
4'-6"
3"x12"x 8' Lg.

BEAM DETAIL

6"x 6"x10'-0"Lg Timber
1'-9"
6'-6"
1'-9"
Drill 1" Hole

WOOD FOOTING

Nail 4-3"x12"x2'-0" Timbers as shown

ELEVATION

3"x 12" Timber

PLAN

2'-0"
2'-0"

WOOD BEAM AND DECK DETAILS

ITEM	UNIT	QUANTITY
LUMBER SOFTWOOD DIM 2 COM 3x12x16	BF	240
LUMBER SOFTWOOD TIMBR 1 COM 6x6x10	BX	60
NAIL COMMON WIRE STEEL 60D	BX	0.2
LUMBER SOFTWOOD DIM 2 COM 3x12x12	BF	108

30" THICK RETAINED EARTH COVER
SANDBAG
4'-0"
6'-0"
10'-0" Slab

ENTRANCE
Slope
3'-0"
2'-0"
8'-0"
6'-6"
4'-6"
1'-0"
12"

NOTE: FOR ADDITIONAL CONSTRUCTION DETAILS SEE FACILITY NO. 040504. TM 5-302.

WOOD-FRAME FIGHTING POSITION (sheet 2 of 3)

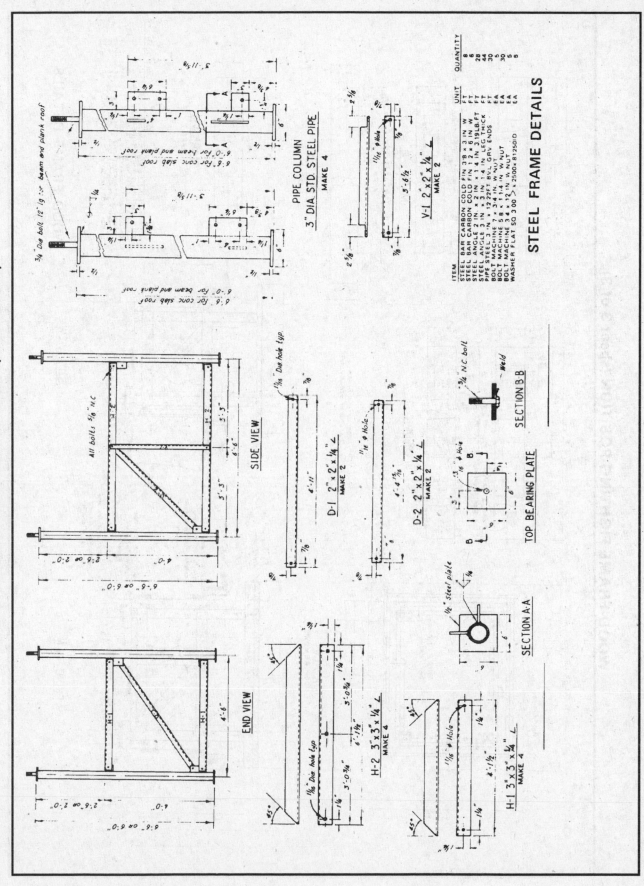

PIPE COLUMN
3" DIA STD. STEEL PIPE
MAKE 4

SIDE VIEW

END VIEW

STEEL FRAME DETAILS

ITEM	UNIT	QUANTITY
STEEL BAR CARBON COLD FIN 3/8 x 3 IN W	FT	8
STEEL BAR CARBON COLD FIN 1/2 x 6 IN W	FT	6
STEEL ANGLE 2 IN x 2 IN x 1/4 IN 3.19 LB-FT	FT	28
STEEL ANGLE 3 IN x 3 IN x 1/4 IN LEG THICK	FT	44
PIPE STEEL 3 IN x 12.22 FT BL GAL ENDS	FT	30
BOLT MACHINE 7 x 3/4 IN. W NUT	EA	5
BOLT MACHINE 5/8 x 1.1/4 IN W NUT	EA	30
BOLT MACHINE 3/4 x 12 IN W NUT	EA	5
WASHER FLAT SQ 3.00 > x 2500 x .8125 OID	EA	5

V-1 2" x 2" x 1/4" L
MAKE 2

D-1 2" x 2" x 1/4" L
MAKE 2

D-2 2" x 2" x 1/4" L
MAKE 2

SECTION B-B

TOP BEARING PLATE

SECTION A-A

H-2 3" x 3" x 1/4" L
MAKE 4

H-1 3" x 3" x 1/4" L
MAKE 4

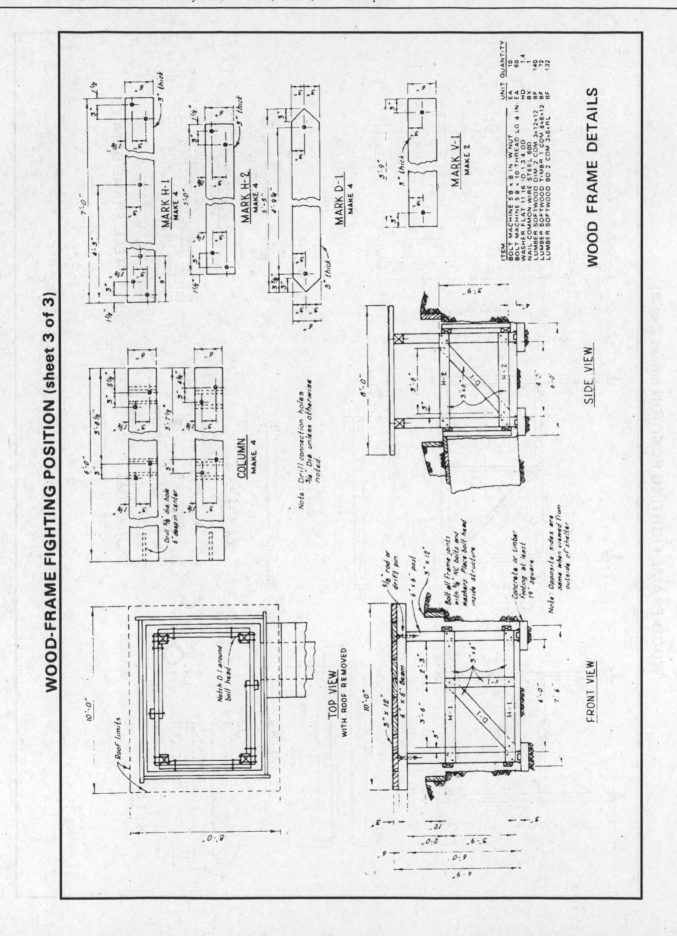

WOOD-FRAME FIGHTING POSITION (sheet 3 of 3)

WOOD FRAME DETAILS

FABRIC-COVERED FRAME POSITION (sheet 1 of 2)

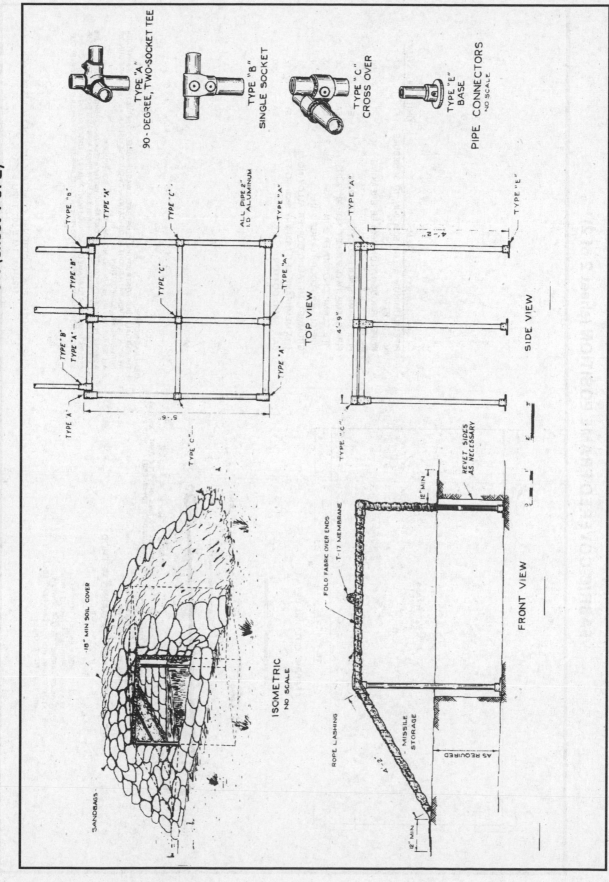

FABRIC-COVERED FRAME POSITION (sheet 2 of 2)

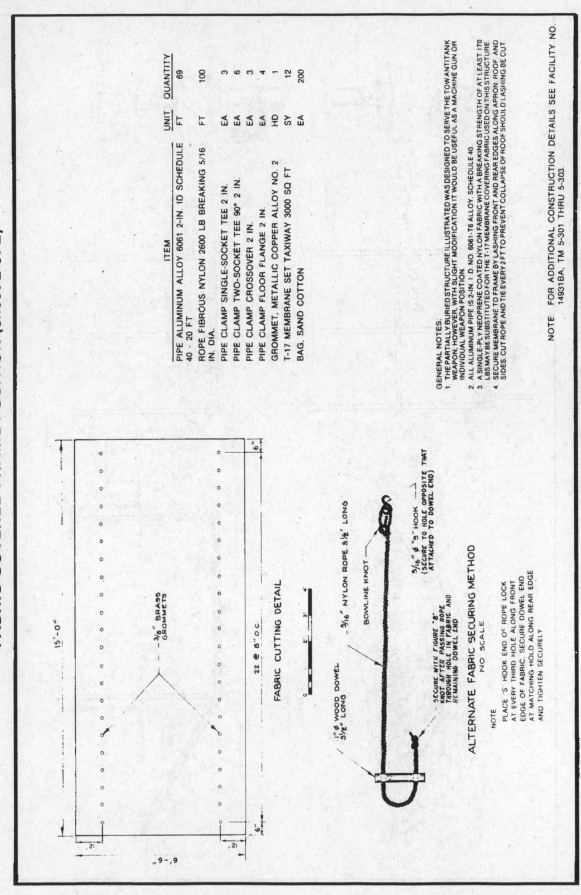

ITEM	UNIT	QUANTITY
PIPE - ALUMINUM ALLOY 6061 2-IN. ID SCHEDULE 40 - 20 FT	FT	69
ROPE FIBROUS NYLON 2600 LB BREAKING 5/16 IN. DIA.	FT	100
PIPE CLAMP SINGLE-SOCKET TEE 2 IN.	EA	3
PIPE CLAMP TWO-SOCKET TEE 90° 2 IN.	EA	6
PIPE CLAMP CROSSOVER 2 IN.	EA	3
PIPE CLAMP FLOOR FLANGE 2 IN.	EA	4
GROMMET, METALLIC COPPER ALLOY NO. 2	HD	1
T-17 MEMBRANE SET TAXIWAY 3000 SQ FT	SY	12
BAG, SAND COTTON	EA	200

GENERAL NOTES:

1. THE PARTIALLY BURIED STRUCTURE ILLUSTRATED WAS DESIGNED TO SERVE THE TOW ANTITANK WEAPON. HOWEVER, WITH SLIGHT MODIFICATION IT WOULD BE USEFUL AS A MACHINE GUN OR INDIVIDUAL WEAPON POSITION.

2. ALL ALUMINUM PIPE IS 2-IN. I.D. NO. 6061-T6 ALLOY, SCHEDULE 40.

3. A SINGLE-PLY NEOPRENE COATED NYLON FABRIC WITH A BREAKING STRENGTH OF AT LEAST 170 LBS MAY BE SUBSTITUTED FOR THE T-17 MEMBRANE COVERING FABRIC USED ON THIS STRUCTURE.

4. SECURE MEMBRANE TO FRAME BY LASHING FRONT AND REAR EDGES ALONG APRON. ROOF AND SIDES. CUT ROPE AND TIE EVERY 2 FT TO PREVENT COLLAPSE OF ROOF SHOULD LASHING BE CUT.

NOTE: FOR ADDITIONAL CONSTRUCTION DETAILS SEE FACILITY NO. 14931BA, TM 5-301 THRU 5-303.

15'-0"

3/8" BRASS GROMMETS

22 @ 8" O.C.

6"

6"

.21

.21

6'-0"

FABRIC CUTTING DETAIL

0   1'   2'   3'   4'

5/16" NYLON ROPE 5½' LONG

BOWLINE KNOT

5/16" ∅ "S" HOOK (SECURE TO HOLE OPPOSITE THAT ATTACHED TO DOWEL END)

SECURE WITH FIGURE "8" KNOT AFTER PASSING ROPE THROUGH HOLE IN FABRIC AND REMAINING DOWEL END

1" ∅ WOOD DOWEL 3½' LONG

ALTERNATE FABRIC SECURING METHOD
NO SCALE

NOTE

PLACE "S" HOOK END OF ROPE LOCK AT EVERY THIRD HOLE ALONG FRONT EDGE OF FABRIC, SECURE DOWEL END AT MATCHING HOLD ALONG REAR EDGE AND TIGHTEN SECURELY

PLYWOOD PERIMETER BUNKER

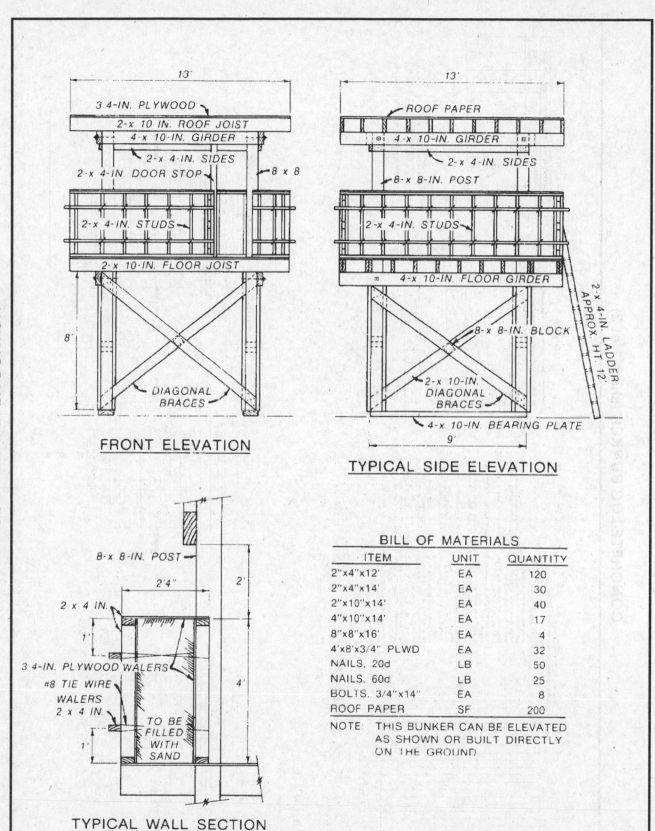

FRONT ELEVATION

- 3 4-IN. PLYWOOD
- 2- x 10 IN. ROOF JOIST
- 4- x 10-IN. GIRDER
- 2- x 4-IN. SIDES
- 2- x 4-IN. DOOR STOP
- 8 x 8
- 2- x 4-IN. STUDS
- 2- x 10-IN. FLOOR JOIST
- 8'
- DIAGONAL BRACES
- 13'

TYPICAL SIDE ELEVATION

- ROOF PAPER
- 4- x 10-IN. GIRDER
- 2- x 4-IN. SIDES
- 8- x 8-IN. POST
- 2- x 4-IN. STUDS
- 4- x 10-IN. FLOOR GIRDER
- 8- x 8-IN. BLOCK
- 2- x 10-IN. DIAGONAL BRACES
- 4- x 10-IN. BEARING PLATE
- 9'
- 13'
- 2- x 4-IN. LADDER APPROX. HT. 12'

TYPICAL WALL SECTION

- 8- x 8-IN. POST
- 2'4"
- 2'
- 2 x 4 IN.
- 1'
- 3 4-IN. PLYWOOD WALERS
- #8 TIE WIRE WALERS
- 2 x 4 IN.
- TO BE FILLED WITH SAND
- 4'
- 1'

BILL OF MATERIALS

ITEM	UNIT	QUANTITY
2"x4"x12'	EA	120
2"x4"x14'	EA	30
2"x10"x14'	EA	40
4"x10"x14'	EA	17
8"x8"x16'	EA	4
4'x8'x3/4" PLWD	EA	32
NAILS, 20d	LB	50
NAILS, 60d	LB	25
BOLTS, 3/4"x14"	EA	8
ROOF PAPER	SF	200

NOTE: THIS BUNKER CAN BE ELEVATED AS SHOWN OR BUILT DIRECTLY ON THE GROUND

CONCRETE LOG BUNKER (sheet 1 of 2)

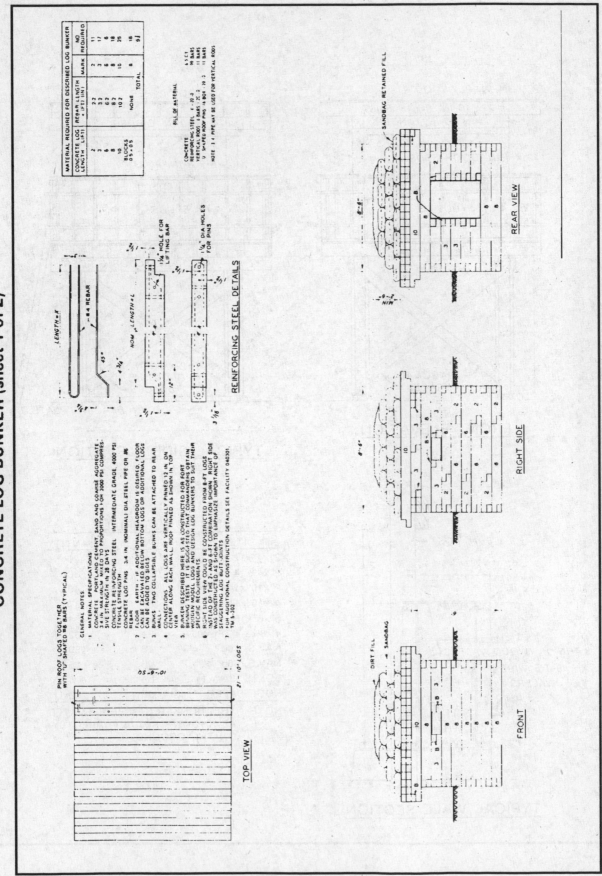

CONCRETE LOG BUNKER (sheet 2 of 2)

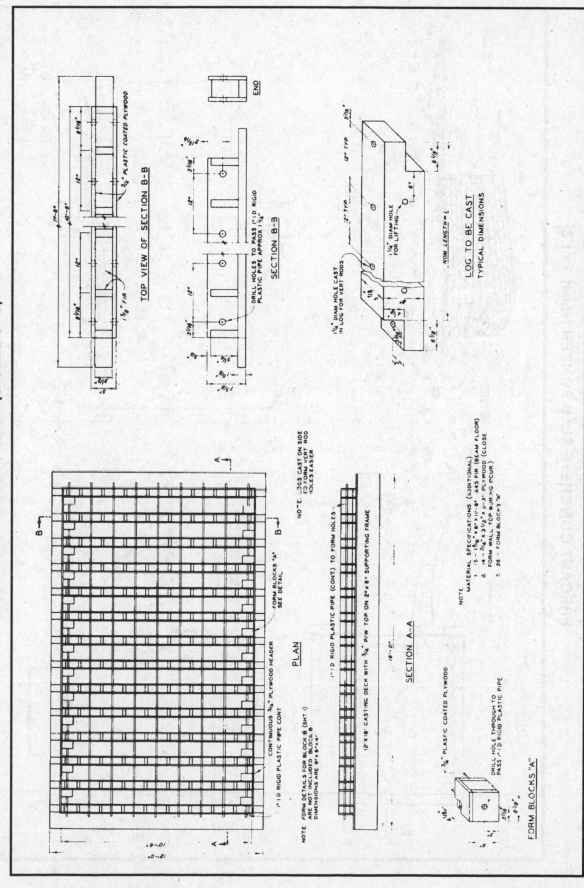

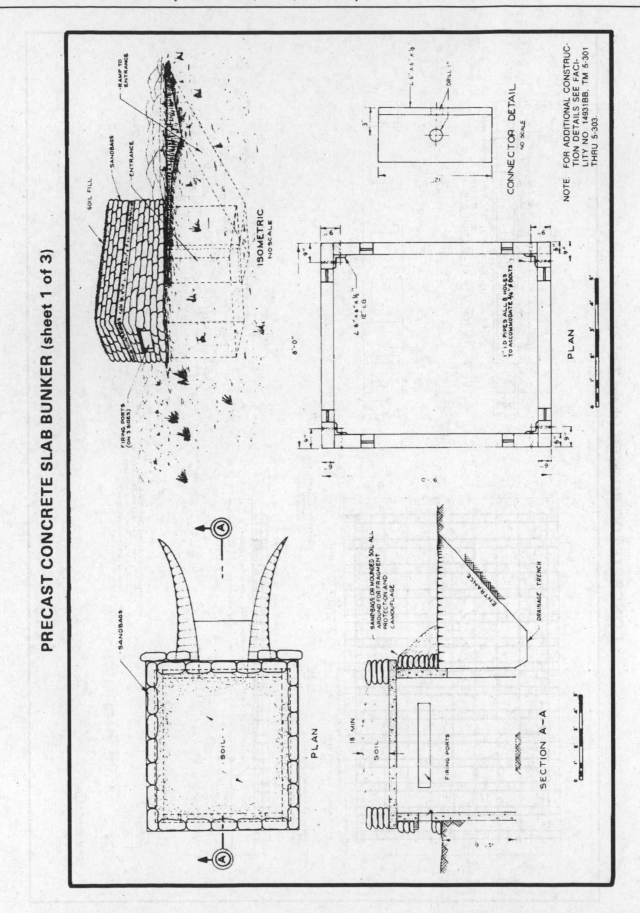

PRECAST CONCRETE SLAB BUNKER (sheet 1 of 3)

PRECAST CONCRETE SLAB BUNKER (sheet 2 of 3)

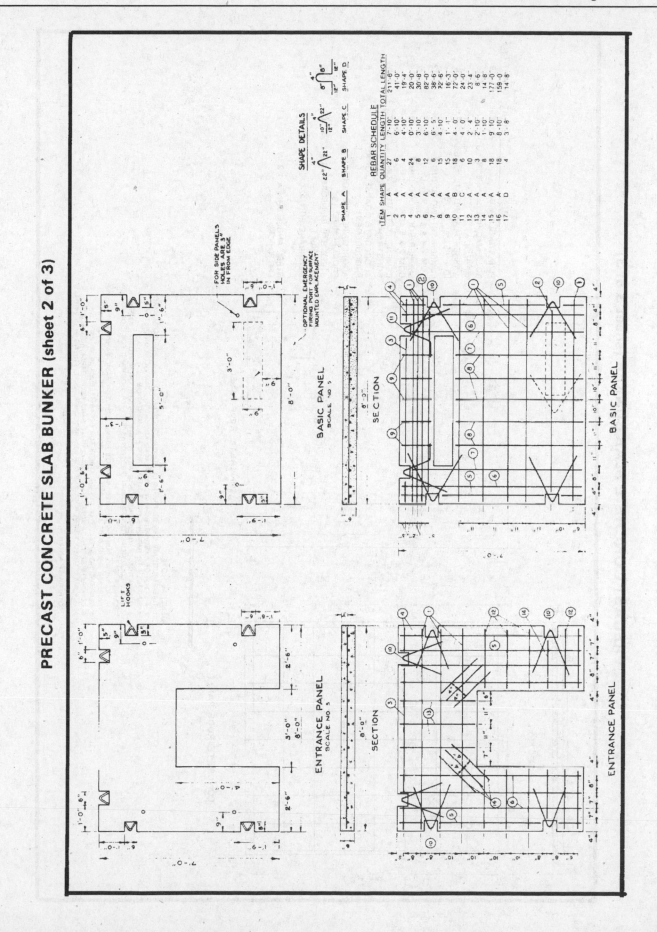

BASIC PANEL
SCALE NO 5

BASIC PANEL

SECTION

ENTRANCE PANEL
SCALE NO 5

ENTRANCE PANEL

SECTION

SHAPE DETAILS

SHAPE A   SHAPE B   SHAPE C   SHAPE D

REBAR SCHEDULE

ITEM	SHAPE	QUANTITY	LENGTH	TOTAL LENGTH
1	A	27	7'-10"	211'-6"
2	A	6	6'-10"	41'-0"
3	A	4	4'-10"	19'-4"
4	A	24	0'-10"	20'-0"
5	A	8	3'-10"	30'-8"
6	A	12	6'-10"	82'-0"
7	A	6	6'-5"	38'-6"
8	A	15	4'-10"	72'-6"
9	B	15	1'-1"	16'-3"
10	B	18	4'-0"	72'-0"
11	C	6	4'-0"	24'-0"
12	A	10	2'-4"	23'-4"
13	A	3	2'-10"	8'-6"
14	A	8	1'-10"	14'-8"
15	A'	18	9'-10"	177'-0"
16	A'	18	8'-10"	159'-0"
17	D	4	3'-8"	14'-8"

OPTIONAL EMERGENCY
FIRING PORT FOR SURFACE
MOUNTED EMPLACEMENT

FOR SIDE PANELS
HOLES ARE 3"
IN FROM EDGE

LIFT HOOKS

PRECAST CONCRETE SLAB BUNKER (sheet 3 of 3)

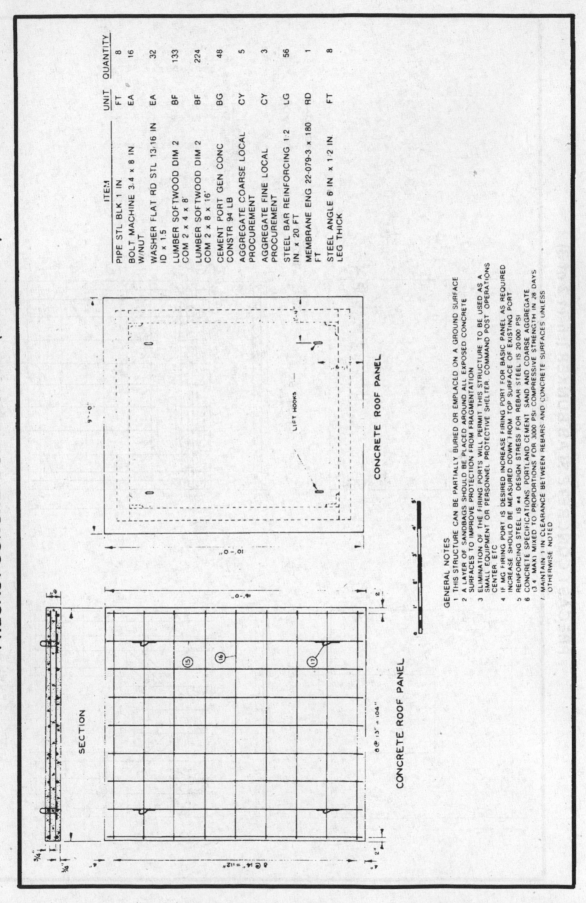

ITEM	UNIT	QUANTITY
PIPE STL BLK 1 IN	FT	8
BOLT MACHINE 3⁄4 x 8 IN W/NUT	EA	16
WASHER FLAT RD STL 13⁄16 IN ID x 1.5	EA	32
LUMBER SOFTWOOD DIM 2 COM 2 x 4 x 8'	BF	133
LUMBER SOFTWOOD DIM 2 COM 2 x 8 x 16'	BF	224
CEMENT PORT GEN CONC CONSTR 94 LB	BG	48
AGGREGATE COARSE LOCAL PROCUREMENT	CY	5
AGGREGATE FINE LOCAL PROCUREMENT	CY	3
STEEL BAR REINFORCING 1⁄2 IN. x 20 FT	LG	56
MEMBRANE ENG 22-079-3 x 180 FT	RD	1
STEEL ANGLE 6 IN x 1⁄2 IN LEG THICK	FT	8

CONCRETE ROOF PANEL

LIFT HOOKS

SECTION

CONCRETE ROOF PANEL

GENERAL NOTES

1 THIS STRUCTURE CAN BE PARTIALLY BURIED OR EMPLACED ON A GROUND SURFACE
2 A LAYER OF SANDBAGS SHOULD BE PLACED AROUND ALL EXPOSED CONCRETE SURFACES TO IMPROVE PROTECTION FROM FRAGMENTATION
3 ELIMINATION OF THE FIRING PORTS WILL PERMIT THIS STRUCTURE TO BE USED AS A SMALL EQUIPMENT OR PERSONNEL PROTECTIVE SHELTER COMMAND POST OPERATIONS CENTER ETC
4 IF MG FIRING PORT IS DESIRED INCREASE FIRING PORT FOR BASIC PANEL AS REQUIRED INCREASE SHOULD BE MEASURED DOWN 7 FROM TOP SURFACE OF EXISTING PORT
5 REINFORCING STEEL IS #4 DESIGN STRESS FOR REBAR STEEL IS 20,000 PSI
6 CONCRETE SPECIFICATIONS PORTLAND CEMENT SAND AND COARSE AGGREGATE (3.4 MAX) MIXED TO PROPORTIONS FOR 3000 PSI COMPRESSIVE STRENGTH IN 28 DAYS
7 MAINTAIN 1 IN CLEARANCE BETWEEN REBARS AND CONCRETE SURFACES UNLESS OTHERWISE NOTED

CONCRETE ARCH BUNKER

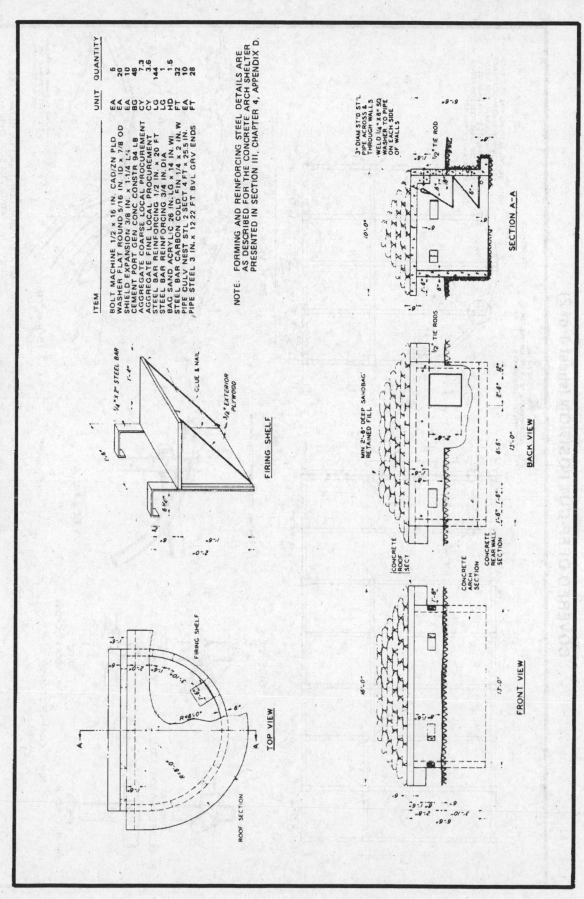

ITEM	UNIT	QUANTITY
BOLT MACHINE 1/2 x 16 IN. CAD/ZN PLD	EA	6
WASHER FLAT ROUND 5/16 IN. ID x 7/8 OD	EA	20
SHIELD EXPANSION 3/8 IN. x 1-1/4 L4	EA	10
CEMENT PORT GEN CONC CONSTR 94 LB	BG	48
AGGREGATE COARSE LOCAL PROCUREMENT	CY	7.3
AGGREGATE FINE LOCAL PROCUREMENT	CY	3.6
STEEL BAR REINFORCING 1/2 IN. x 20 FT	LG	144
STEEL BAR REINFORCING 3/4 IN. DIA.	LG	1
BAG SAND ACRYLIC 26 IN. LG x 14 IN. WI	HD	1.5
STEEL BAR CARBON COLD FIN 1/4 x 2 IN. W	FT	32
PIPE CULV NEST STL 2 SECT 4 FT x 25.5 IN.	EA	10
PIPE STEEL 3 IN. x 12 22 FT BVL GRV ENDS	FT	28

NOTE: FORMING AND REINFORCING STEEL DETAILS ARE AS DESCRIBED FOR THE CONCRETE ARCH SHELTER PRESENTED IN SECTION III, CHAPTER 4, APPENDIX D.

FIRING SHELF

TOP VIEW

BACK VIEW

SECTION A-A

FRONT VIEW

COVERED DEEP-CUT POSITION (sheet 1 of 2)

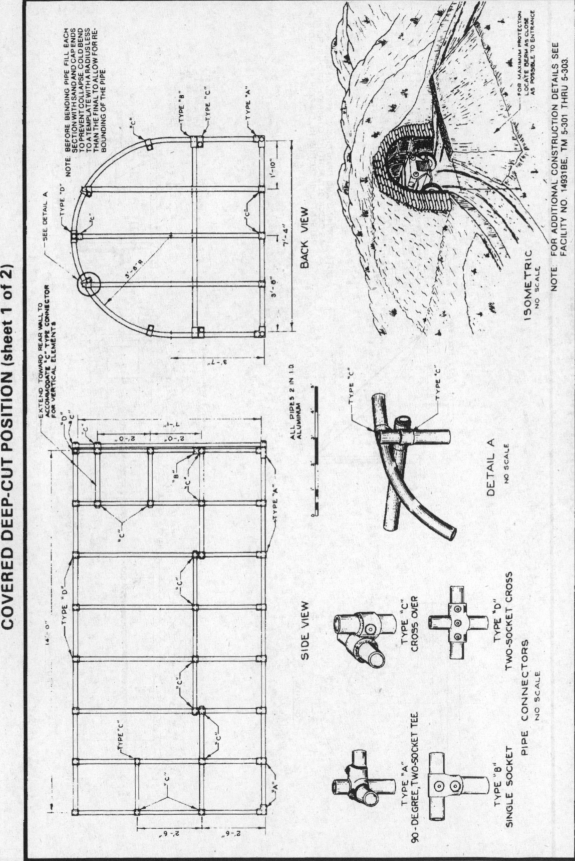

NOTE: BEFORE BENDING PIPE FILL EACH SECTION WITH SAND AND CAP ENDS TO PREVENT COLLAPSE. COLD BEND TO A TEMPLATE WITH A RADIUS LESS THAN THE FINAL TO ALLOW FOR RE-BOUNDING OF THE PIPE.

BACK VIEW

EXTEND TOWARD REAR WALL TO ACCOMMODATE "C" TYPE CONNECTOR FOR VERTICAL ELEMENTS

SEE DETAIL A

ALL PIPES 2 IN I.D. ALUMINUM

SIDE VIEW

DETAIL A
NO SCALE

TYPE "C"
CROSS OVER

TYPE "D"
TWO-SOCKET CROSS

TYPE "A"
90-DEGREE, TWO-SOCKET TEE

TYPE "B"
SINGLE SOCKET

PIPE CONNECTORS
NO SCALE

ISOMETRIC
NO SCALE

FOR MAXIMUM PROTECTION LOCATE BERM AS CLOSE AS POSSIBLE TO ENTRANCE.

NOTE: FOR ADDITIONAL CONSTRUCTION DETAILS SEE FACILITY NO. 14931BE, TM 5-301 THRU 5-303.

COVERED DEEP-CUT POSITION (sheet 2 of 2)

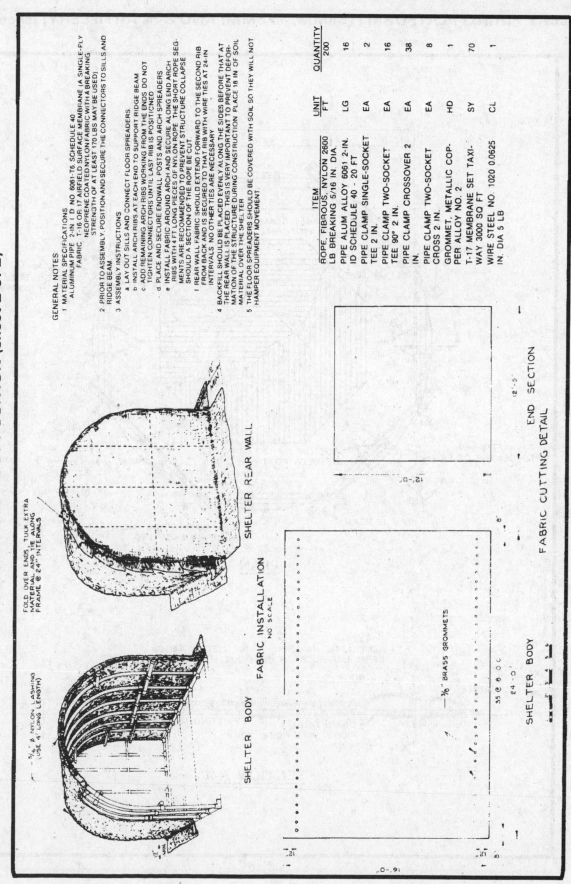

GENERAL NOTES

1. MATERIAL SPECIFICATIONS
   ALUMINUM PIPE 2-IN I D NO 6061-T6. SCHEDULE 40
   FABRIC T-16 OR 17 AIRFIELD SURFACE MEMBRANE (A SINGLE-PLY
   NEOPRENE COATED NYLON FABRIC WITH A BREAKING
   STRENGTH OF AT LEAST 170 LBS MAY BE USED)

2. PRIOR TO ASSEMBLY, POSITION AND SECURE THE CONNECTORS TO SILLS AND
   RIDGE BEAM.

3. ASSEMBLY INSTRUCTIONS
   a. LAY OUT SILLS AND CONNECT FLOOR SPREADERS
   b. INSTALL ARCH RIBS AT EACH END TO SUPPORT RIDGE BEAM
   c. ADD REMAINING ARCH RIBS WORKING FROM THE ENDS DO NOT
      TIGHTEN CONNECTORS UNTIL LAST RIB IS POSITIONED
   d. PLACE AND SECURE END WALL POSTS AND ARCH SPREADERS
   e. INSTALL FABRIC AROUND ARCH AND SECURE ALONG END ARCH
      RIBS WITH 4 FT LONG PIECES OF NYLON ROPE. THE SHORT ROPE SEG-
      MENTS ARE RECOMMENDED TO PREVENT STRUCTURE COLLAPSE
      SHOULD A SECTION OF THE ROPE BE CUT.

4. REAR WALL FABRIC SHOULD EXTEND FORWARD TO THE SECOND RIB
   FROM BACK AND IS SECURED TO THAT RIB WITH WIRE TIES AT 24-IN
   INTERVALS NO OTHER TIES ARE NECESSARY

5. BACKFILL SHOULD BE PLACED EVENLY ALONG THE SIDES BEFORE THAT AT
   THE REAR WALL IS PLACED THIS IS VERY IMPORTANT TO PREVENT DEFOR-
   MATION OF THE STRUCTURE DURING CONSTRUCTION PLACE 18 IN OF SOIL
   MATERIAL OVER THE SHELTER

6. THE FLOOR SPREADERS SHOULD BE COVERED WITH SOIL SO THEY WILL NOT
   HAMPER EQUIPMENT MOVEMENT.

ITEM	UNIT	QUANTITY
ROPE, FIBROUS, NYLON 2600 LB BREAKING 5/16 IN. DIA.	FT	200
PIPE ALUM ALLOY 6061 2-IN. ID SCHEDULE 40 - 20 FT	LG	16
PIPE CLAMP SINGLE-SOCKET TEE 2 IN.	EA	2
PIPE CLAMP TWO-SOCKET TEE 90° 2 IN.	EA	16
PIPE CLAMP CROSSOVER 2 IN.	EA	38
PIPE CLAMP TWO-SOCKET CROSS 2 IN.	EA	8
GROMMET, METALLIC COP-PER ALLOY NO. 2	HD	1
T-17 MEMBRANE SET TAXI-WAY 3000 SQ FT	SY	70
WIRE STEEL NO. 1020 0.0625 IN. DIA 5 LB	CL	1

SHELTER REAR WALL

SHELTER BODY   FABRIC INSTALLATION
NO SCALE

3/4" ø NYLON LASHING
(USE 4' LONG LENGTH)

FOLD OVER ENDS, TUCK EXTRA
MATERIAL AND TIE ALONG
FRAME @ 24" INTERVALS

END SECTION

FABRIC CUTTING DETAIL

3/8" BRASS GROMMETS

35 @ 8 OC
24'-0"

SHELTER BODY

ARTILLERY FIRING PLATFORM (155MM, 175MM, AND 8-IN ARTILLERY) (sheet 1 of 3)

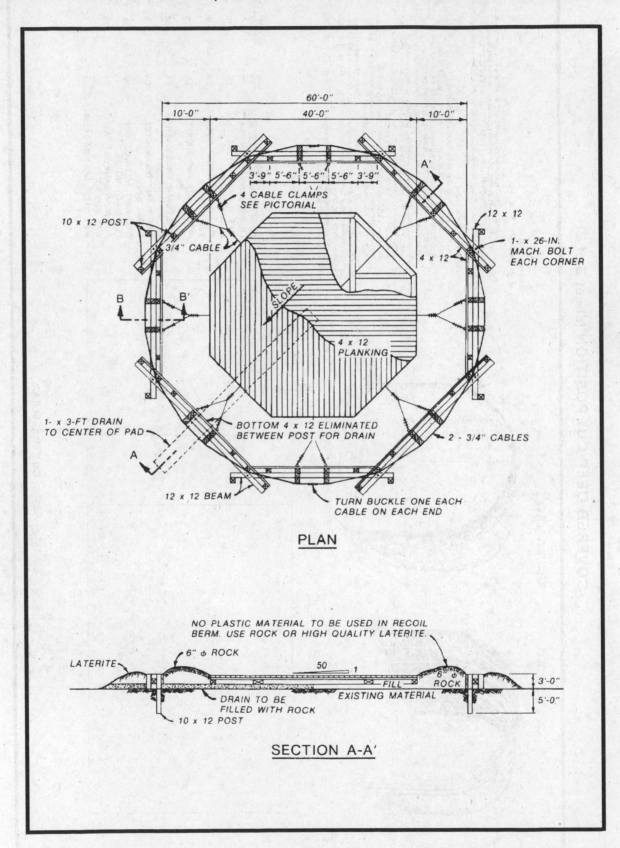

PLAN

SECTION A-A'

ARTILLERY FIRING PLATFORM (155MM, 175MM, AND 8-IN ARTILLERY) (sheet 2 of 3)

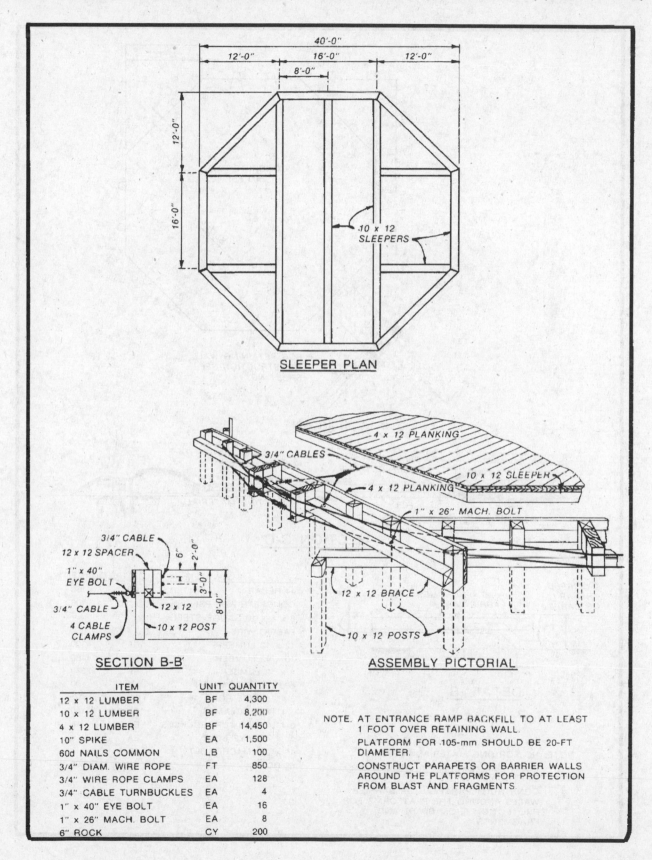

SLEEPER PLAN

SECTION B-B'

ASSEMBLY PICTORIAL

ITEM	UNIT	QUANTITY
12 x 12 LUMBER	BF	4,300
10 x 12 LUMBER	BF	8,200
4 x 12 LUMBER	BF	14,450
10" SPIKE	EA	1,500
60d NAILS COMMON	LB	100
3/4" DIAM. WIRE ROPE	FT	850
3/4" WIRE ROPE CLAMPS	EA	128
3/4" CABLE TURNBUCKLES	EA	4
1" x 40" EYE BOLT	EA	16
1" x 26" MACH. BOLT	EA	8
6" ROCK	CY	200

NOTE. AT ENTRANCE RAMP BACKFILL TO AT LEAST 1 FOOT OVER RETAINING WALL.

PLATFORM FOR 105-mm SHOULD BE 20-FT DIAMETER.

CONSTRUCT PARAPETS OR BARRIER WALLS AROUND THE PLATFORMS FOR PROTECTION FROM BLAST AND FRAGMENTS.

ARTILLERY FIRING PLATFORM (155MM, 175MM, AND 8-IN ARTILLERY) (sheet 3 of 3)

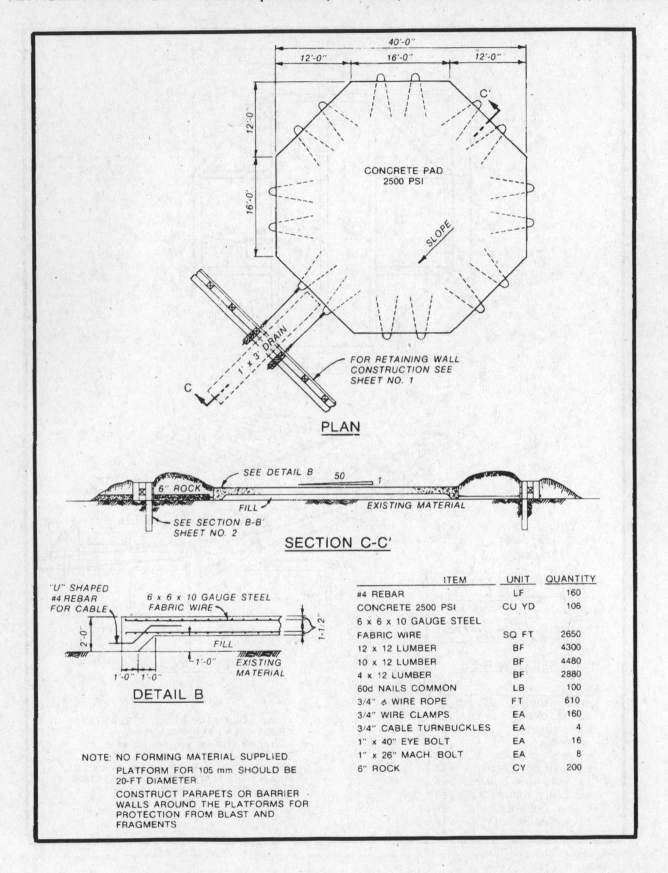

PLAN

SECTION C-C'

DETAIL B

ITEM	UNIT	QUANTITY
#4 REBAR	LF	160
CONCRETE 2500 PSI	CU YD	106
6 x 6 x 10 GAUGE STEEL FABRIC WIRE	SQ FT	2650
12 x 12 LUMBER	BF	4300
10 x 12 LUMBER	BF	4480
4 x 12 LUMBER	BF	2880
60d NAILS COMMON	LB	100
3/4" ⌀ WIRE ROPE	FT	610
3/4" WIRE CLAMPS	EA	160
3/4" CABLE TURNBUCKLES	EA	4
1" x 40" EYE BOLT	EA	16
1" x 26" MACH. BOLT	EA	8
6" ROCK	CY	200

NOTE: NO FORMING MATERIAL SUPPLIED.
PLATFORM FOR 105 mm SHOULD BE
20-FT DIAMETER.

CONSTRUCT PARAPETS OR BARRIER
WALLS AROUND THE PLATFORMS FOR
PROTECTION FROM BLAST AND
FRAGMENTS

PARAPET POSITION FOR ADA

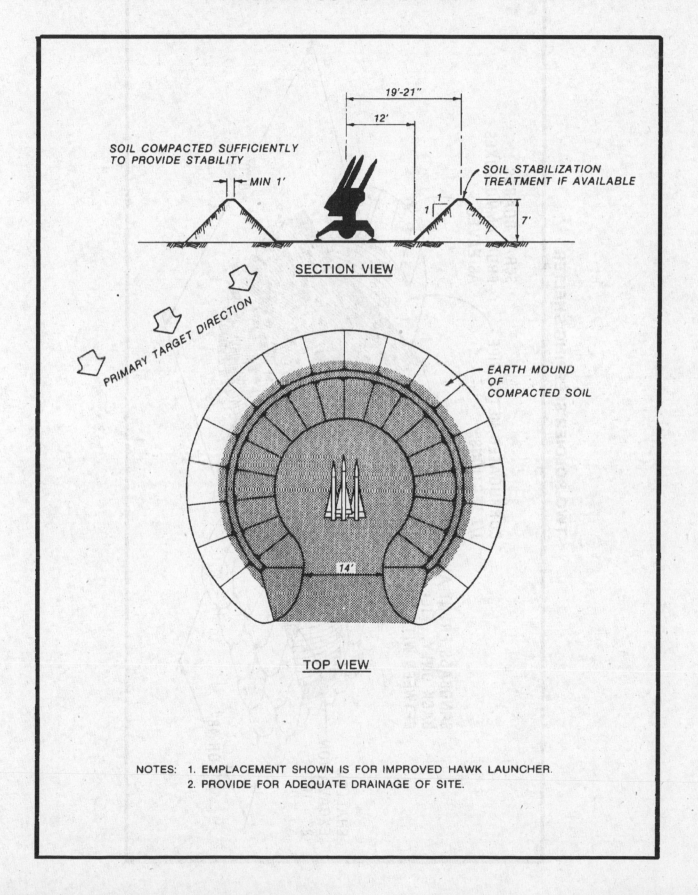

19'-21"

12'

SOIL COMPACTED SUFFICIENTLY
TO PROVIDE STABILITY

MIN 1'

SOIL STABILIZATION
TREATMENT IF AVAILABLE

1
1

1

7'

SECTION VIEW

PRIMARY TARGET DIRECTION

EARTH MOUND
OF
COMPACTED SOIL

14"

TOP VIEW

NOTES: 1. EMPLACEMENT SHOWN IS FOR IMPROVED HAWK LAUNCHER.
        2. PROVIDE FOR ADEQUATE DRAINAGE OF SITE.

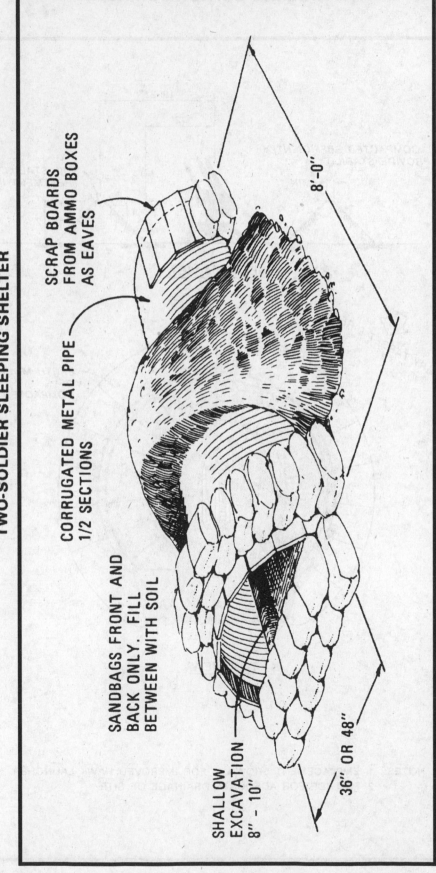

TWO-SOLDIER SLEEPING SHELTER

SCRAP BOARDS FROM AMMO BOXES AS EAVES

CORRUGATED METAL PIPE 1/2 SECTIONS

SANDBAGS FRONT AND BACK ONLY. FILL BETWEEN WITH SOIL

SHALLOW EXCAVATION 8" – 10"

8'-0"

36" OR 48"

METAL CULVERT SHELTER

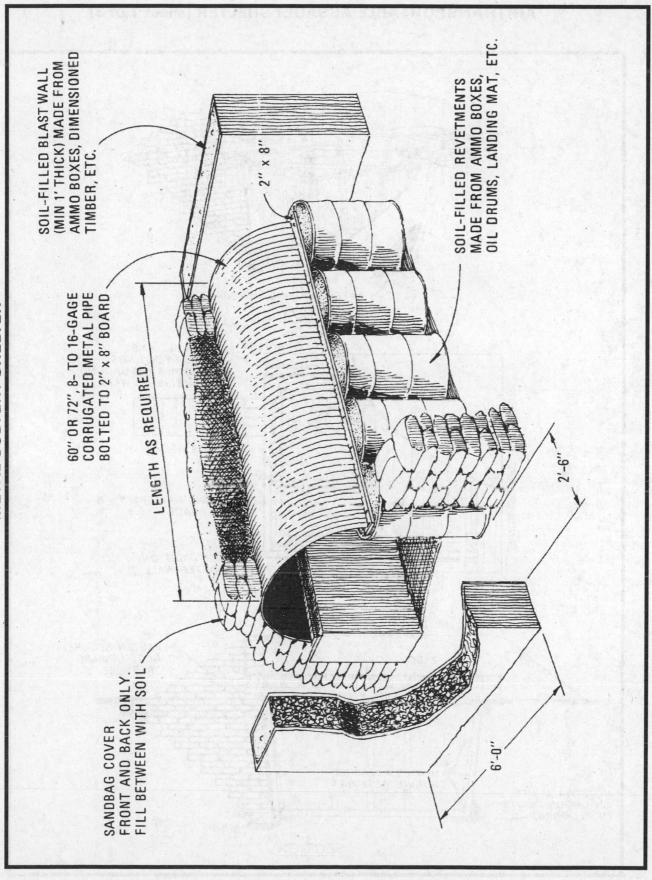

SOIL-FILLED BLAST WALL (MIN 1' THICK) MADE FROM AMMO BOXES, DIMENSIONED TIMBER, ETC.

2" x 8"

SOIL-FILLED REVETMENTS MADE FROM AMMO BOXES, OIL DRUMS, LANDING MAT, ETC.

60" OR 72", 8- TO 16-GAGE CORRUGATED METAL PIPE BOLTED TO 2" x 8" BOARD

LENGTH AS REQUIRED

2'-6"

SANDBAG COVER FRONT AND BACK ONLY. FILL BETWEEN WITH SOIL

6'-0"

AIRTRANSPORTABLE ASSAULT SHELTER (sheet 1 of 3)

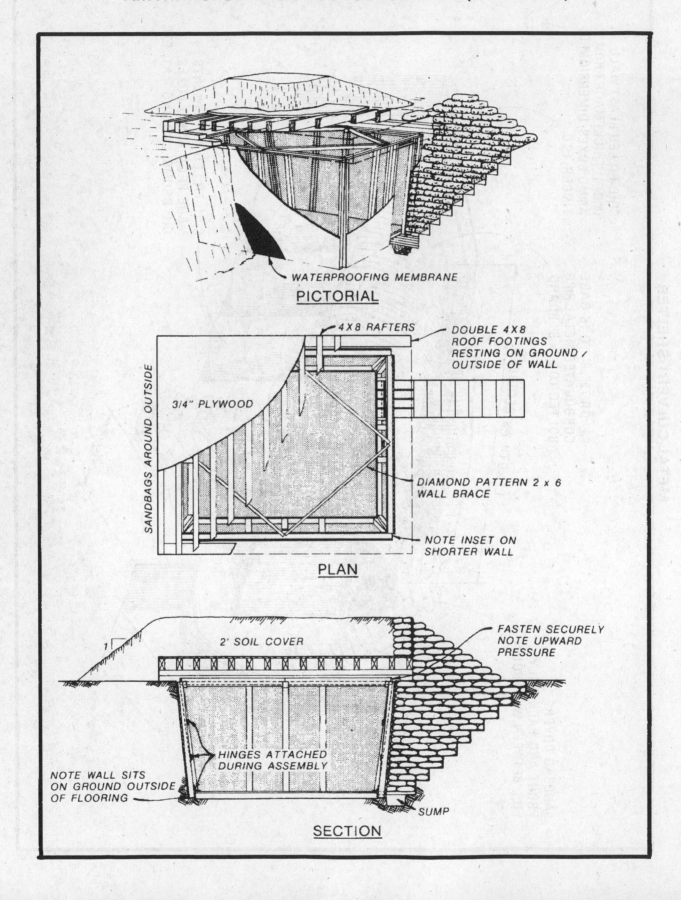

WATERPROOFING MEMBRANE

PICTORIAL

4 X 8 RAFTERS

DOUBLE 4 X 8 ROOF FOOTINGS RESTING ON GROUND / OUTSIDE OF WALL

SANDBAGS AROUND OUTSIDE

3/4" PLYWOOD

DIAMOND PATTERN 2 x 6 WALL BRACE

NOTE INSET ON SHORTER WALL

PLAN

1

2' SOIL COVER

FASTEN SECURELY NOTE UPWARD PRESSURE

HINGES ATTACHED DURING ASSEMBLY

NOTE WALL SITS ON GROUND OUTSIDE OF FLOORING

SUMP

SECTION

AIRTRANSPORTABLE ASSAULT SHELTER (sheet 2 of 3)

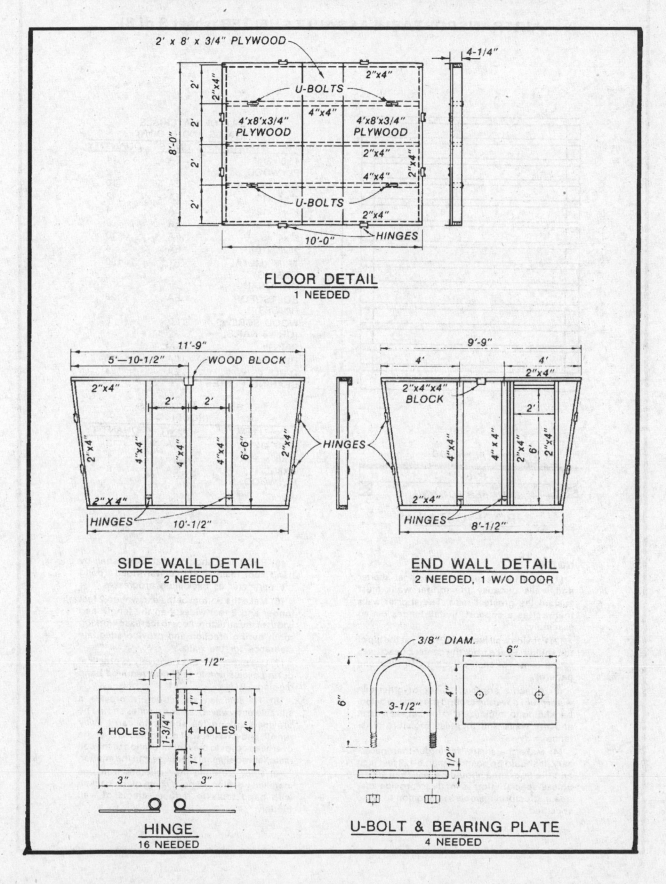

FLOOR DETAIL
1 NEEDED

SIDE WALL DETAIL
2 NEEDED

END WALL DETAIL
2 NEEDED, 1 W/O DOOR

HINGE
16 NEEDED

U-BOLT & BEARING PLATE
4 NEEDED

AIRTRANSPORTABLE ASSAULT SHELTER (sheet 3 of 3)

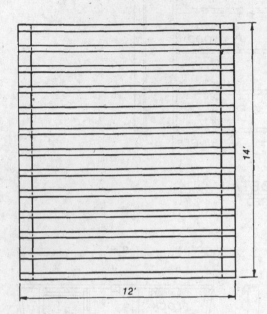

14'

12'

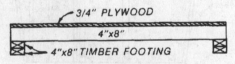

3/4" PLYWOOD
4"x8"
4"x8" TIMBER FOOTING

ROOF DETAIL

BILL OF MATERIALS (WALLS AND FLOOR)		
ITEM	UNITS	QUANTITY
4'x8'x3/4" PLYWOOD	EA	14
4"x4"x8'	EA	10
4"x4"x10'	EA	2
2"x4"x12'	EA	4
2"x4"x10'	EA	9
2"x4"x8'	EA	10
2"x6"x10'	EA	4
TRIM (METAL EDGING) OPTIONAL	FT	190
BOLTS (FOR HINGES)	EA	128
WOOD SCREWS (OR #8 NAILS)	LB	5
PAINT	GAL	1
HINGES	EA	16
U-BOLTS W/ BEARING PLATES	EA	4

BILL OF MATERIALS (ROOF)		
ITEM	UNIT	QUANTITY
4"x8"x12'	EA	13
4"x8"x14'	EA	4
4'x8'x3/4" PLYWOOD	EA	6

NOTES:

(1) Abut longer side walls against shorter end walls because the longer walls must sustain the greatest load. The shorter walls then act as a support. Install hinges during assembly.

(2) Provide wall bracing (2" x 6") at the top of the shelter. Brace from the center of each wall to the center of each adjacent wall (diamond pattern).

(3) Attach a sheet of plastic or other thin waterproof covering around the outside before backfilling to minimize friction between earth and the walls and increase moisture resistance.

(4) Make the shelter no larger than necessary. It should be no more than 6-1/2 feet high and the floor area should be less than 100 ft^2 unless special effort is made to provide adequate structural members in addition to those specified.

(5) Backfilling should be accomplished by hand labor, maintaining a uniform load around the perimeter as backfilling progresses.

(6) Make the bottom of the excavation 2 feet longer and 2 feet wider than the length and width of the structure floor to increase working room during erection and provide adequate clearance for the walls.

(7) Use explosives as extensively as practical during excavation to minimize required hand digging.

(8) To complete the structure provide a suitable entryway. Drainage ditches should be provided around the shelter to carry away runoff, and a waterproof cover placed over the overhead cover to prevent saturation of the soil material and eliminate seepage into the interior.

(9) Prior to lifting the structure from the installed position, remove some of the backfill with hand tools to reduce effects of wall friction.

TIMBER POST BURIED SHELTER

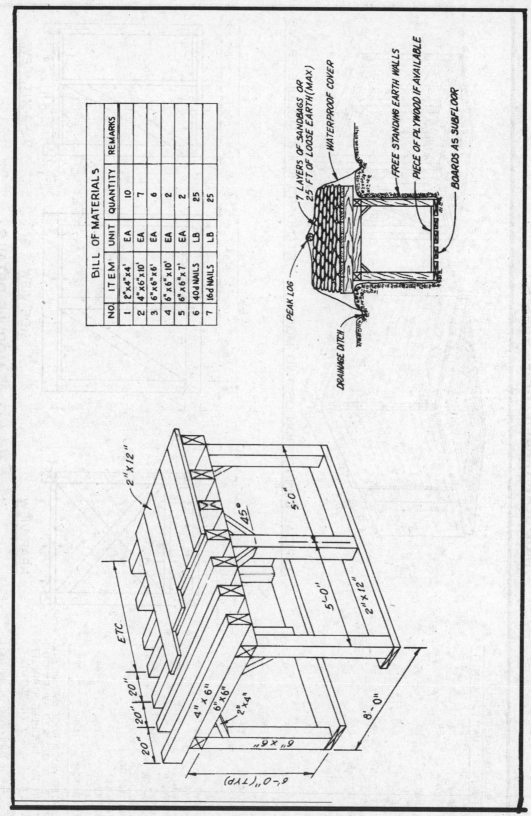

NO	ITEM	UNIT	QUANTITY	REMARKS
1	2"x4"x4'	EA	10	
2	4"x6"x10'	EA	7	
3	6"x6"x6'	EA	6	
4	6"x6"x10'	EA	2	
5	6"x6"x7'	EA	2	
6	40d NAILS	LB	25	
7	16d NAILS	LB	25	

BILL OF MATERIALS

7 LAYERS OF SANDBAGS OR 2.5 FT OF LOOSE EARTH (MAX)

WATERPROOF COVER

FREE STANDING EARTH WALLS

PIECE OF PLYWOOD IF AVAILABLE

BOARDS AS SUBFLOOR

PEAK LOG

DRAINAGE DITCH

2"X12"

45°

5'-0"

5'-0"

2"X12"

ETC

20" 20" 20"

4" X 6"

6"X 6"

2"X 4"

6" X 6"

8'-0"

8'-0" (TYP)

MODULAR TIMBER FRAME SHELTER

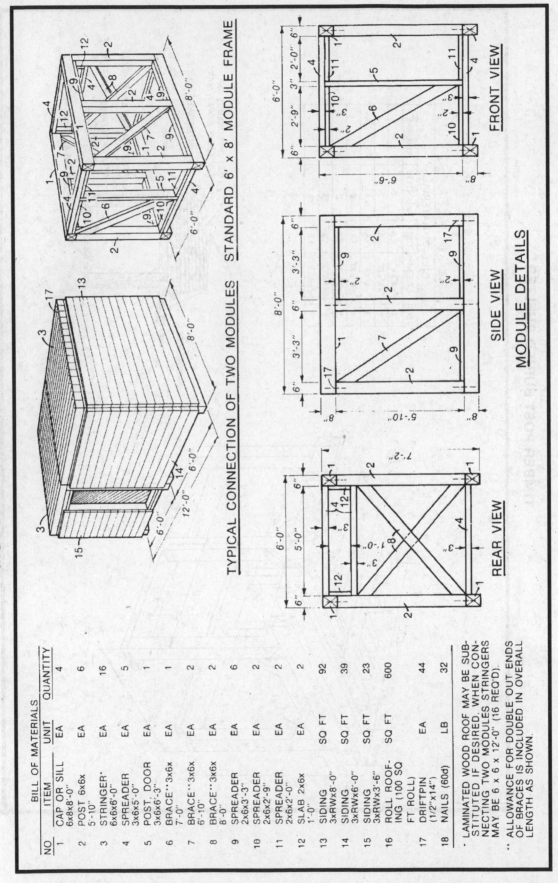

STANDARD 6' x 8' MODULE FRAME

TYPICAL CONNECTION OF TWO MODULES

FRONT VIEW

SIDE VIEW

REAR VIEW

MODULE DETAILS

BILL OF MATERIALS			
NO	ITEM	UNIT	QUANTITY
1	CAP OR SILL 6x8x8'-0"	EA	4
2	POST 6x6x 5'-10"	EA	6
3	STRINGER· 6x6x6'-0"	EA	16
4	SPREADER 3x6x5'-0"	EA	5
5	POST, DOOR 3x6x6'-3"	EA	1
6	BRACE·· 3x6x 7'-0"	EA	1
7	BRACE·· 3x6x 6'-10"	EA	2
8	BRACE·· 3x6x 8'-0"	EA	2
9	SPREADER 2x6x3'-3"	EA	6
10	SPREADER 2x6x2'-9"	EA	2
11	SPREADER 2x6x2'-0"	EA	2
12	SLAB 2x6x 1'-0"	EA	2
13	SIDING 3xRWx8'-0"	SQ FT	92
14	SIDING, 3xRWx6'-0"	SQ FT	39
15	SIDING 3xRWx3'-6"	SQ FT	23
16	ROLL ROOF- ING (100 SQ FT ROLL)	SQ FT	600
17	DRIFTPIN (1/2"x14")	EA	44
18	NAILS (60d)	LB	32

· LAMINATED WOOD ROOF MAY BE SUB-
STITUTED IF DESIRED. WHEN CON-
NECTING TWO MODULES STRINGERS
MAY BE 6 x 6 x 12'-0" (16 REQ'D).

·· ALLOWANCE FOR DOUBLE OUT ENDS
OF BRACES IS INCLUDED IN OVERALL
LENGTH AS SHOWN.

TIMBER FRAME BURIED SHELTER

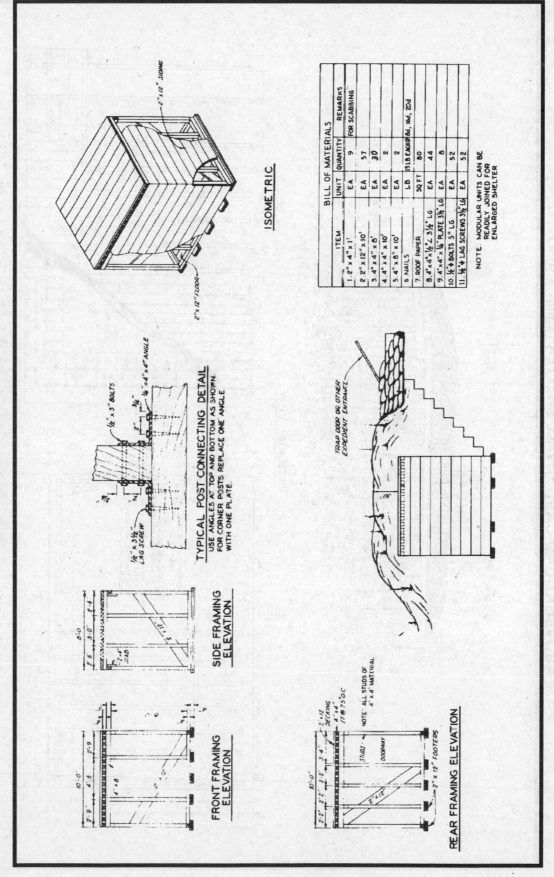

ABOVEGROUND CAVITY WALL SHELTER (sheet 1 of 2)

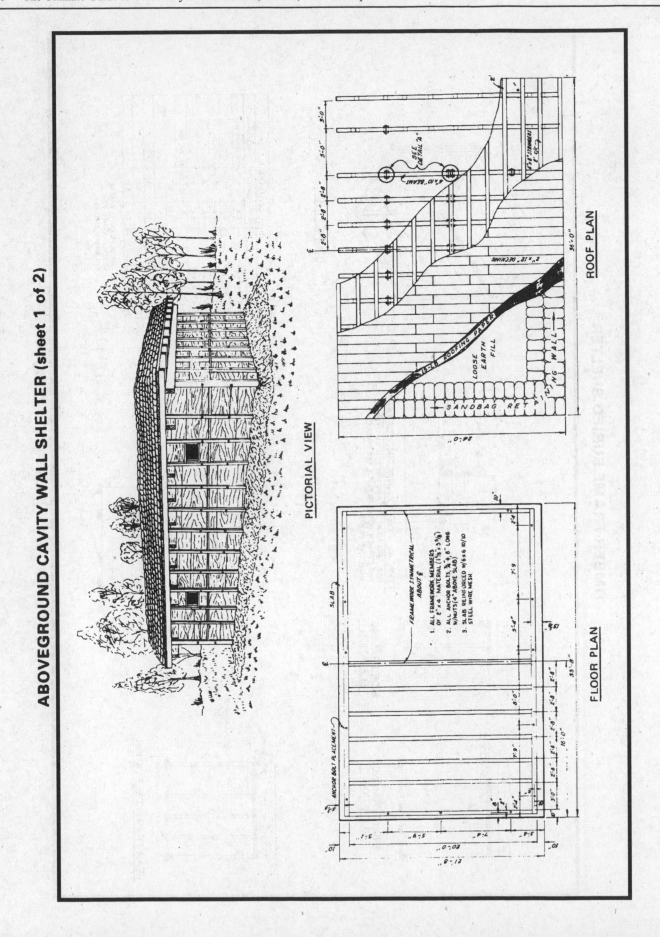

PICTORIAL VIEW

ROOF PLAN

FLOOR PLAN

ABOVEGROUND CAVITY WALL SHELTER (sheet 2 of 2)

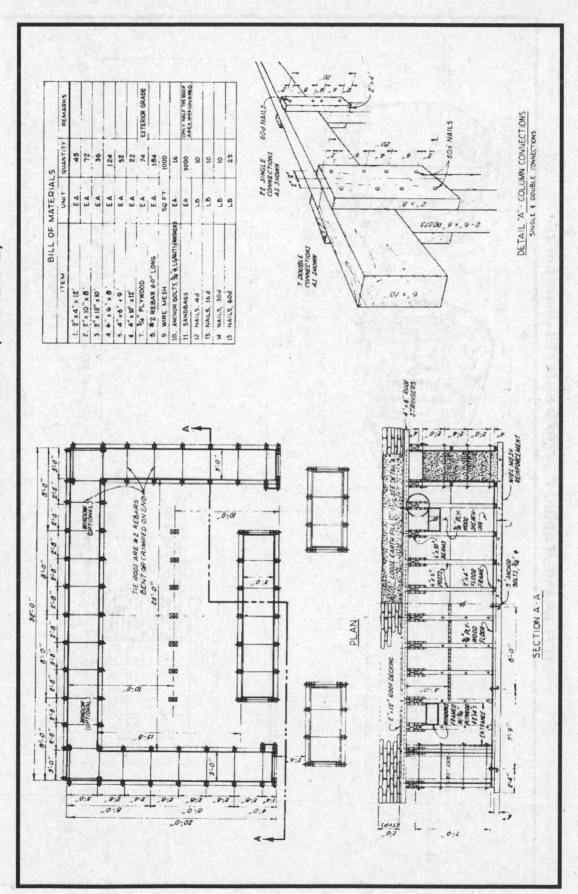

BILL OF MATERIALS

	ITEM	UNIT	QUANTITY	REMARKS
1.	2" x 4" x 12'	EA	45	
2.	2" x 10" x 8'	EA	72	
3.	2" x 12" x 10'	EA	36	
4.	6" x 6" x 8'	EA	124	
5.	4" x 6" x 9'	EA	52	
6.	6" x 10" x 12'	EA	22	
7.	3/4" PLYWOOD	EA	74	EXTERIOR GRADE
8.	#2 REBAR 60" LONG	EA	184	
9.	WIRE MESH	SQ FT	1000	ONLY HALF THE ROOF AREA WAS COVERED
10.	ANCHOR BOLTS, 3/8" dia L. NUTS & WASHERS	EA	16	
11.	SANDBAGS	EA	3000	
12.	NAILS, 4d	LB	10	
13.	NAILS, 16d	LB	10	
14.	NAILS, 30d	LB	10	
15.	NAILS, 60d	LB	25	

DETAIL "A" - COLUMN CONNECTIONS
SINGLE & DOUBLE CONNECTIONS

PLAN

SECTION A-A

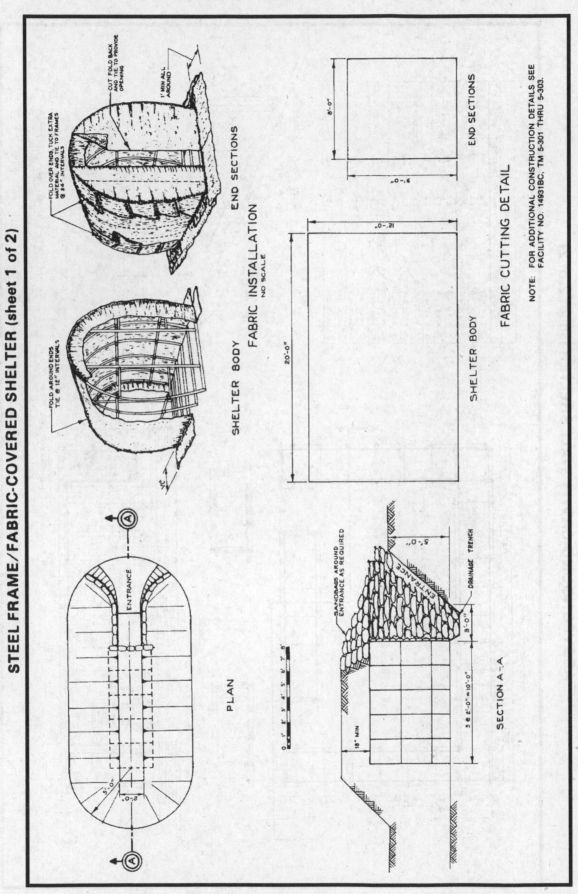

STEEL FRAME/FABRIC-COVERED SHELTER (sheet 1 of 2)

STEEL FRAME/FABRIC-COVERED SHELTER (sheet 2 of 2)

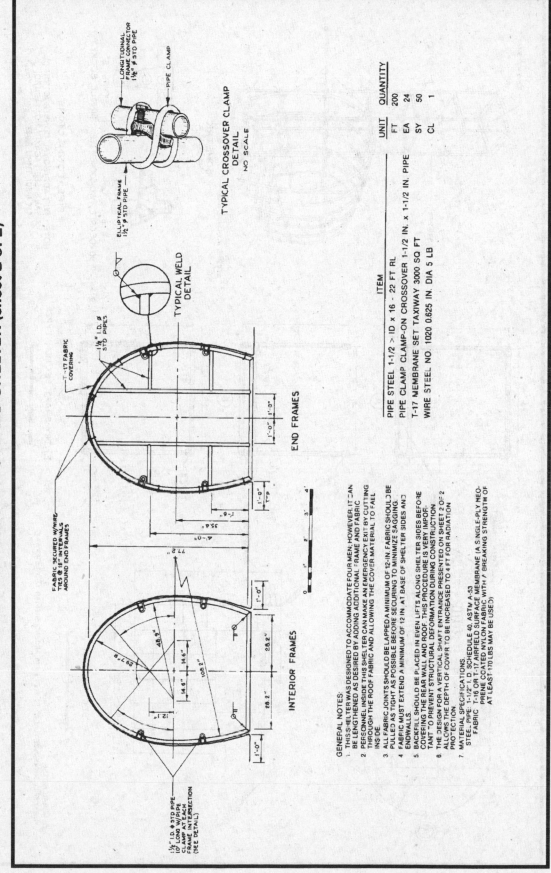

TYPICAL CROSSOVER CLAMP DETAIL
NO SCALE

END FRAMES

INTERIOR FRAMES

TYPICAL WELD DETAIL

ITEM	UNIT	QUANTITY
PIPE STEEL 1-1/2 > ID x 16 - 22 FT RL	FT	200
PIPE CLAMP CLAMP-ON CROSSOVER 1-1/2 IN. x 1-1/2 IN. PIPE	EA	24
T-17 MEMBRANE SET TAXIWAY 3000 SQ FT	SY	50
WIRE STEEL NO. 1020 0.625 IN. DIA 5 LB	CL	1

GENERAL NOTES:
1. THIS SHELTER WAS DESIGNED TO ACCOMMODATE FOUR MEN. HOWEVER, IT CAN BE LENGTHENED AS DESIRED BY ADDING ADDITIONAL FRAME AND FABRIC.
2. PERSONNEL INSIDE THIS SHELTER CAN MAKE AN EMERGENCY EXIT BY CUTTING THROUGH THE ROOF FABRIC AND ALLOWING THE COVER MATERIAL TO FALL INSIDE.
3. ALL FABRIC JOINTS SHOULD BE LAPPED A MINIMUM OF 12-IN. FABRIC SHOULD BE PULLED AS TIGHT AS POSSIBLE BEFORE SECURING TO MINIMIZE SAGGING.
4. FABRIC MUST EXTEND A MINIMUM OF 12 IN. AT BASE OF SHELTER SIDES AND ENDWALLS.
5. BACKFILL SHOULD BE PLACED IN EVEN LIFTS ALONG SHELTER SIDES BEFORE COVERING THE REAR WALL AND ROOF. THIS PROCEDURE IS VERY IMPORTANT TO PREVENT STRUCTURAL DEFORMATION DURING CONSTRUCTION.
6. THE DESIGN FOR A VERTICAL SHAFT ENTRANCE PRESENTED ON SHEET 2 OF 2 ALLOWS THE DEPTH OF COVER TO BE INCREASED TO 4 FT FOR RADIATION PROTECTION.
7. MATERIAL SPECIFICATIONS.
   STEEL, PIPE: 1-1/2" I.D. SCHEDULE 40, ASTM A-53.
   FABRIC: T-16 OR T-17 AIRFIELD SURFACE MEMBRANE (A SINGLE-PLY NEO-PRENE COATED NYLON FABRIC WITH A BREAKING STRENGTH OF AT LEAST 170 LBS MAY BE USED)

HARDENED FRAME/FABRIC SHELTER (sheet 1 of 3)

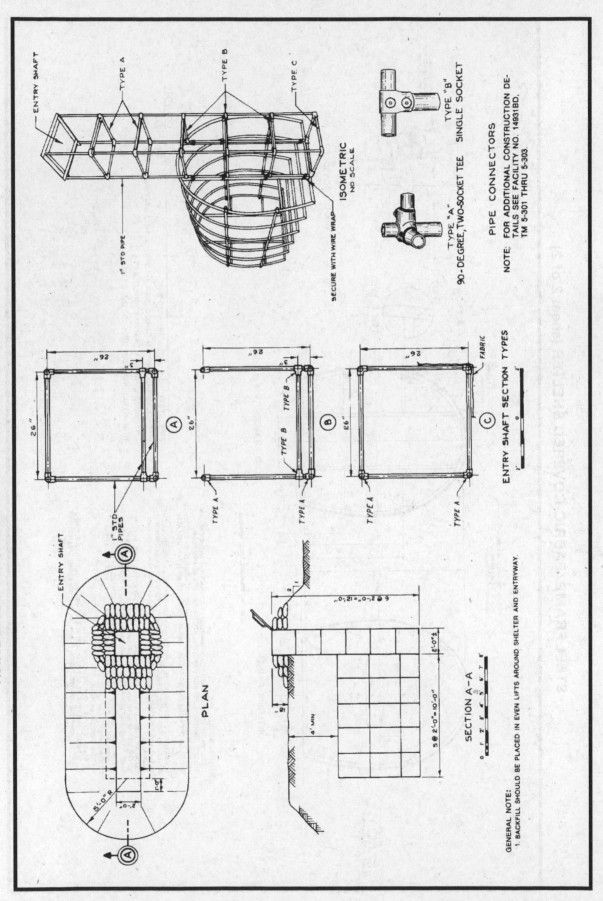

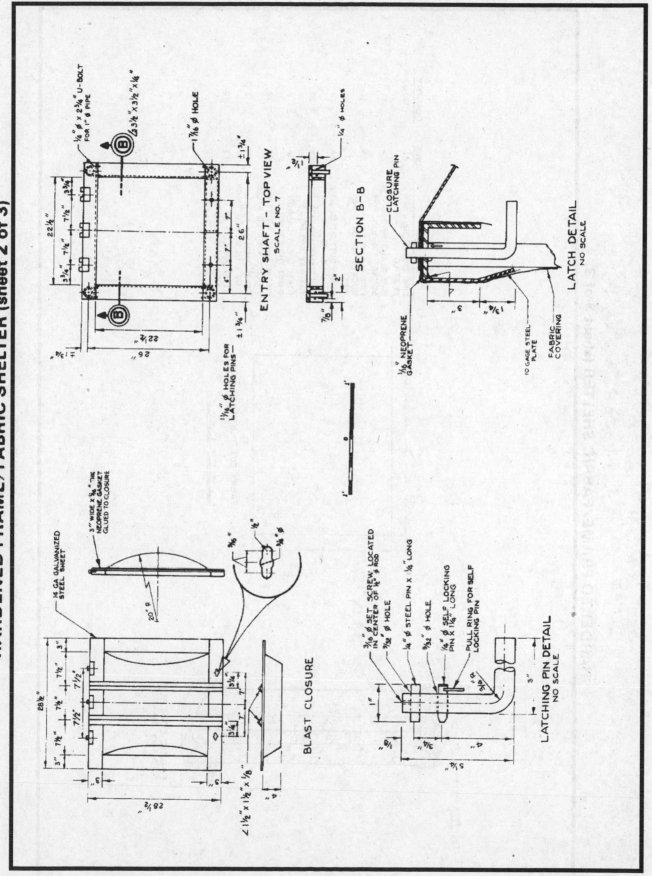

HARDENED FRAME/FABRIC SHELTER (sheet 2 of 3)

ENTRY SHAFT – TOP VIEW
SCALE NO. 7

SECTION B–B

LATCH DETAIL
NO SCALE

BLAST CLOSURE

LATCHING PIN DETAIL
NO SCALE

HARDENED FRAME/FABRIC SHELTER (sheet 3 of 3)

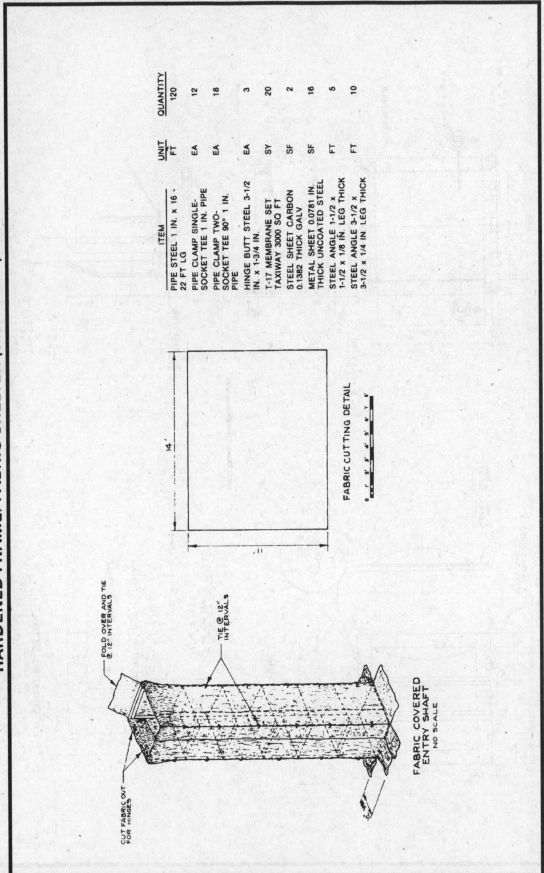

ITEM	UNIT	QUANTITY
PIPE STEEL 1 IN. x 16 - 22 FT LG	FT	120
PIPE CLAMP SINGLE- SOCKET TEE 1 IN. PIPE	EA	12
PIPE CLAMP TWO- SOCKET TEE 90° 1 IN. PIPE	EA	18
HINGE BUTT STEEL 3-1/2 IN. x 1-3/4 IN.	EA	3
T-17 MEMBRANE SET TAXIWAY 3000 SQ FT	SY	20
STEEL SHEET CARBON 0.1382 THICK GALV	SF	2
METAL SHEET 0.0781 IN. THICK UNCOATED STEEL	SF	16
STEEL ANGLE 1-1/2 x 1-1/2 x 1/8 IN. LEG THICK	FT	5
STEEL ANGLE 3-1/2 x 3-1/2 x 1/4 IN. LEG THICK	FT	10

14'

11'

FABRIC CUTTING DETAIL

FOLD OVER AND TIE @ 12" INTERVALS

TIE @ 12" INTERVALS

CUT FABRIC OUT FOR HINGES

FABRIC COVERED ENTRY SHAFT
NO SCALE

RECTANGULAR FABRIC/FRAME SHELTER (sheet 1 of 2)

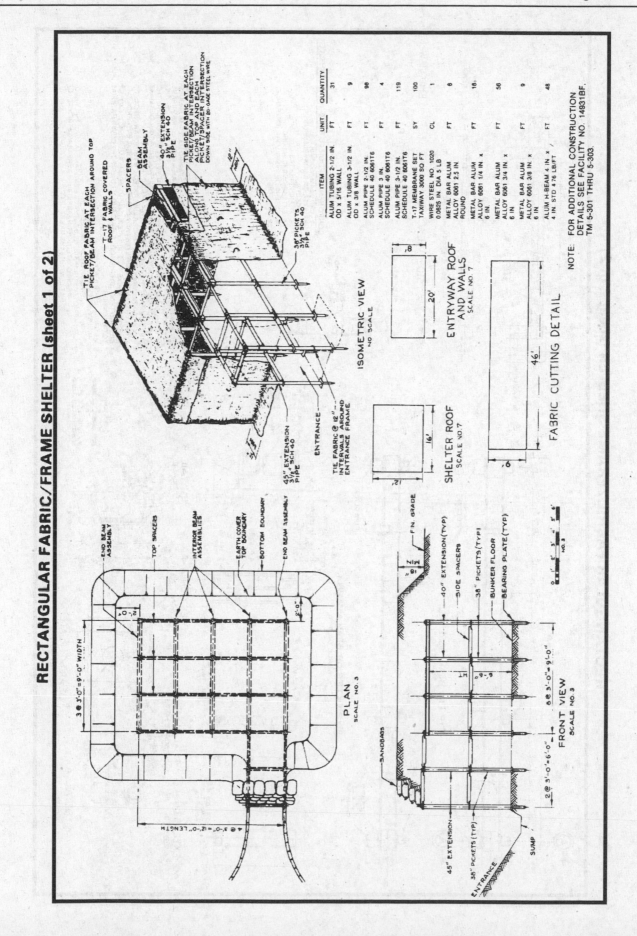

ITEM	UNIT	QUANTITY
ALUM TUBING 2-1/2 IN OD x 5/16 WALL	FT	31
ALUM TUBING 3-1/2 IN OD x 3/8 WALL	FT	9
ALUM PIPE 2-1/2 IN SCHEDULE 40 6061T6	FT	98
ALUM PIPE 3 IN SCHEDULE 40 6061T6	FT	4
ALUM PIPE 3-1/2 IN SCHEDULE 40 6061T6	FT	119
T-17 MEMBRANE SET TAXIWAY 3000 SQ FT	SY	100
WIRE STEEL NO 1020 0.0625 IN DIA 5 LB	CL	1
METAL BAR ALUM ALLOY 6061 2.5 IN ROUND	FT	6
METAL BAR ALUM ALLOY 6061 1/4 IN x 6 IN	FT	18
METAL BAR ALUM ALLOY 6061 3/4 IN x 6 IN	FT	56
METAL BAR ALUM ALLOY 6061 3/8 IN x 6 IN	FT	9
ALUM H-BEAM 4 IN x 4 IN STD 4.76 LB/FT	FT	48

NOTE: FOR ADDITIONAL CONSTRUCTION DETAILS SEE FACILITY NO. 14931BF. TM 5-301 THRU 5-303.

ISOMETRIC VIEW
NO SCALE

ENTRYWAY ROOF AND WALLS
SCALE NO. 7

SHELTER ROOF
SCALE NO. 7

FABRIC CUTTING DETAIL

PLAN
SCALE NO. 3

FRONT VIEW
SCALE NO. 3

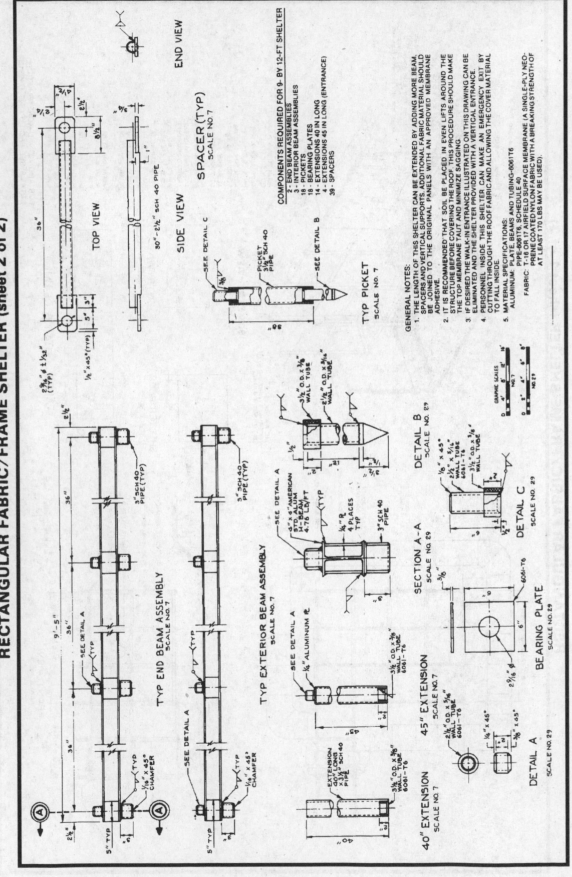

RECTANGULAR FABRIC/FRAME SHELTER (sheet 2 of 2)

CONCRETE ARCH SHELTER (sheet 1 of 3)

GENERAL NOTES

1. MATERIAL SPECIFICATIONS:
   CONCRETE: PORTLAND CEMENT, SAND & COARSE AGGREGATE 3/4" MAX, MIXED TO PROPORTIONS FOR 3000 PSI COMPRESSIVE STRENGTH IN 28 DAYS.

   CONCRETE REINFORCING STEEL: INTERMEDIATE GRADE 40,000 PSI TENSILE STRENGTH

   MULTIPLATE PIPE ARCH: 8-GAGE CORRUGATED STEEL, 6" x 2" CORRUGATIONS, GALVANIZED.

   MISCELLANEOUS FASTENERS: GRADE AS AVAILABLE.

2. DESIGN LOADING: 8' EARTH COVER AT CROWN.

3. GROUNDWATER SEEPAGE: ALL EXTERIOR SURFACES OF SHELTER & ENTRANCEWAY MUST BE COVERED WITH WATERPROOFING MEMBRANE TO PREVENT GROUND WATER SEEPAGE.

4. FLOOR: CORRUGATED-STEEL FLOOR PROVIDED.

5. OCCUPANCY: 12' x 12' BASIC SHELTER DESIGNED FOR 10 MEN ASSUMING ONE SLEEPING BUNK PER MAN.

6. EQUIPMENT PROVIDED: BASIC SHELTER PROVIDES ONLY STRUCTURAL COMPONENTS AND ASSEMBLIES. ALL EQUIPMENT MUST BE PROVIDED SEPARATELY.

7. EMERGENCY EXIT: USE ONE OF DOOR OPENINGS AS EMERGENCY EXIT.

8. BUNK INSTALLATION: TO BE SUSPENDED FROM ARCH SECTION BY ANCHOR HOOKS.

9. TRANSPORTABILITY: ARCH SECTION 4300 LB, ENDWALL 7900 LB.

10. FOR ADDITIONAL CONSTRUCTION DETAILS SEE FACILITY NO. 040101, TM 5-301 THRU 5-303.

ERECTION PROCEDURES

1. LEVEL SITE.

2. PLACE WATERPROOFING MEMBRANE ON GROUND BELOW SHELTER.

3. PLACE ARCH SECTION.

4. PLACE REAR END WALL AND BRACE TEMPORARILY.

5. PLACE AND TEMPORARILY BRACE FRONT END WALL.

6. CONNECT, TIGHTEN, TIE CABLES FROM END WALL TO END WALL (SEE CABLE DETAIL SHEET 2).

7. PLACE AND CONNECT ENTRANCEWAY TO SHELTER (ENTRANCEWAY NOT INCLUDED IN THIS SET OF DRAWINGS)

8. PLACE WATERPROOFING MEMBRANE OVER SHELTER AND ENTRANCEWAY.

9. BACKFILL SHELTER TO DESIRED EARTH COVER BEING CAREFUL TO RAISE AND PACK EARTH FILL IN UNIFORM LIFTS OF EQUAL DEPTH ON OPPOSITE SIDES OF ARCH SECTIONS AND END WALL SECTIONS.

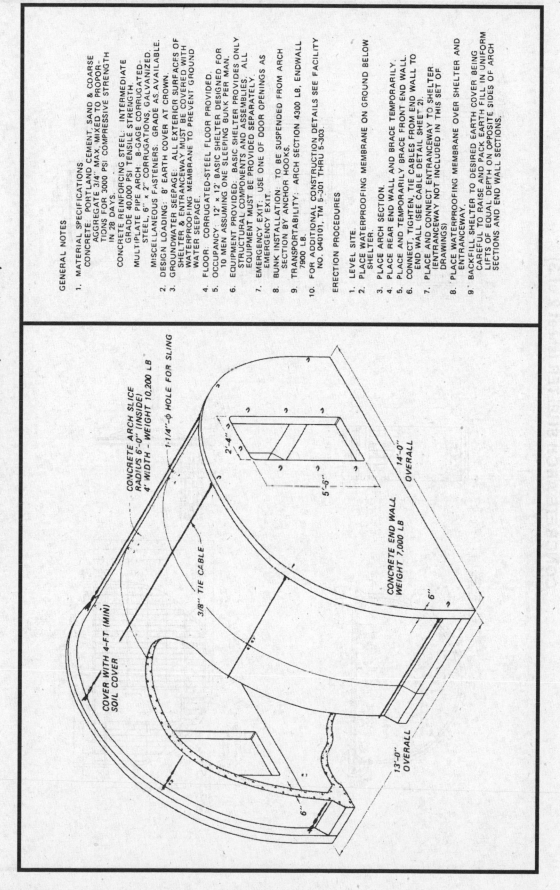

COVER WITH 4-FT (MIN) SOIL COVER

CONCRETE ARCH SLICE RADIUS 6'-0" (INSIDE) 4' WIDTH - WEIGHT 10,200 LB

1-1/4"-ø HOLE FOR SLING

3/8" TIE CABLE

CONCRETE END WALL WEIGHT 7,000 LB

13'-0" OVERALL

14'-0" OVERALL

2'-4"

5'-6"

6"

6"

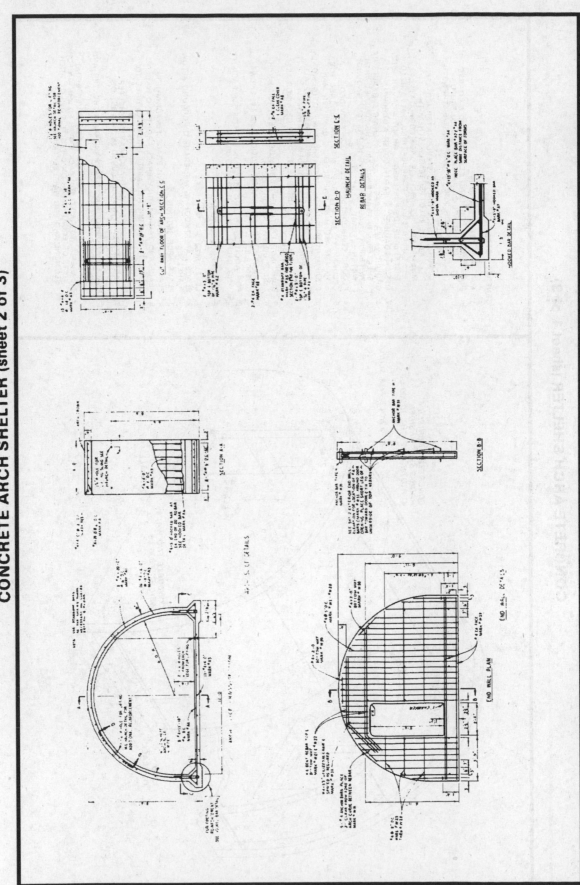

CONCRETE ARCH SHELTER (sheet 2 of 3)

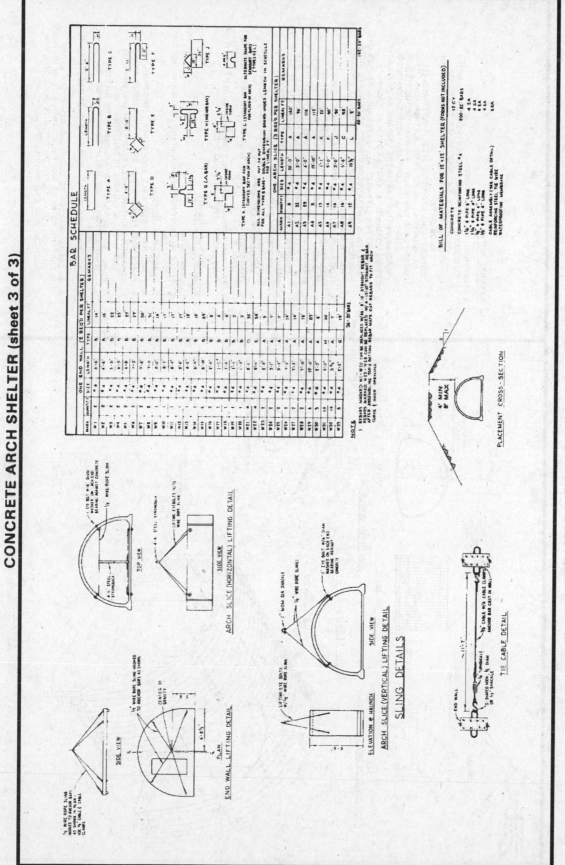

CONCRETE ARCH SHELTER (sheet 3 of 3)

METAL PIPE ARCH SHELTER (sheet 1 of 2)

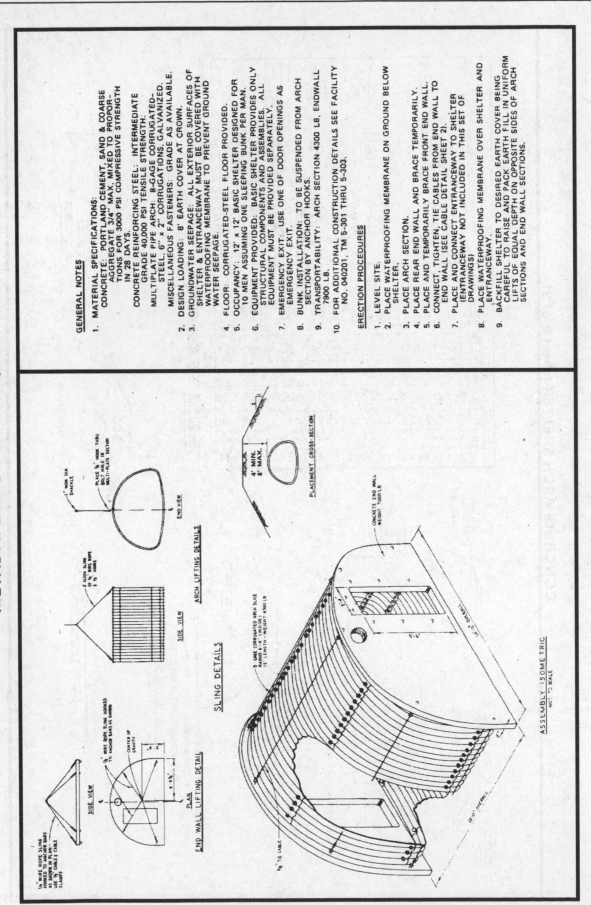

GENERAL NOTES

1. MATERIAL SPECIFICATIONS:
   CONCRETE: PORTLAND CEMENT, SAND & COARSE AGGREGATE 3/4" MAX, MIXED TO PROPORTIONS FOR 3000 PSI COMPRESSIVE STRENGTH IN 28 DAYS.
   CONCRETE REINFORCING STEEL: INTERMEDIATE GRADE 40,000 PSI TENSILE STRENGTH.
   MULTIPLATE PIPE ARCH: 8-GAGE CORRUGATED-STEEL, 6" x 2" CORRUGATIONS, GALVANIZED.
   MISCELLANEOUS FASTENERS: GRADE AS AVAILABLE.

2. DESIGN LOADING: 8' EARTH COVER AT CROWN.

3. GROUNDWATER SEEPAGE: ALL EXTERIOR SURFACES OF SHELTER & ENTRANCEWAY MUST BE COVERED WITH WATERPROOFING MEMBRANE TO PREVENT GROUND WATER SEEPAGE.

4. FLOOR: CORRUGATED-STEEL FLOOR PROVIDED.

5. OCCUPANCY: 12' x 12' BASIC SHELTER DESIGNED FOR 10 MEN ASSUMING ONE SLEEPING BUNK PER MAN.

6. EQUIPMENT PROVIDED: BASIC SHELTER PROVIDES ONLY STRUCTURAL COMPONENTS AND ASSEMBLIES. ALL EQUIPMENT MUST BE PROVIDED SEPARATELY.

7. EMERGENCY EXIT: USE ONE OF DOOR OPENINGS AS EMERGENCY EXIT.

8. BUNK INSTALLATION: TO BE SUSPENDED FROM ARCH SECTION BY ANCHOR HOOKS.

9. TRANSPORTABILITY: ARCH SECTION 4300 LB, ENDWALL 7900 LB.

10. FOR ADDITIONAL CONSTRUCTION DETAILS SEE FACILITY NO. 040201, TM 5-301 THRU 5-303.

ERECTION PROCEDURES

1. LEVEL SITE.

2. PLACE WATERPROOFING MEMBRANE ON GROUND BELOW SHELTER.

3. PLACE ARCH SECTION.

4. PLACE REAR END WALL AND BRACE TEMPORARILY.

5. PLACE AND TEMPORARILY BRACE FRONT END WALL.

6. CONNECT, TIGHTEN, TIE CABLES FROM END WALL TO END WALL (SEE CABLE DETAIL SHEET 2).

7. PLACE AND CONNECT ENTRANCEWAY TO SHELTER (ENTRANCEWAY NOT INCLUDED IN THIS SET OF DRAWINGS)

8. PLACE WATERPROOFING MEMBRANE OVER SHELTER AND ENTRANCEWAY.

9. BACKFILL SHELTER TO DESIRED EARTH COVER BEING CAREFUL TO RAISE AND PACK EARTH FILL IN UNIFORM LIFTS OF EQUAL DEPTH ON OPPOSITE SIDES OF ARCH SECTIONS AND END WALL SECTIONS.

END VIEW

ARCH LIFTING DETAILS

PLACEMENT CROSS-SECTION

SIDE VIEW

SLING DETAILS

SIDE VIEW

PLAN

END WALL LIFTING DETAIL

ASSEMBLY ISOMETRIC

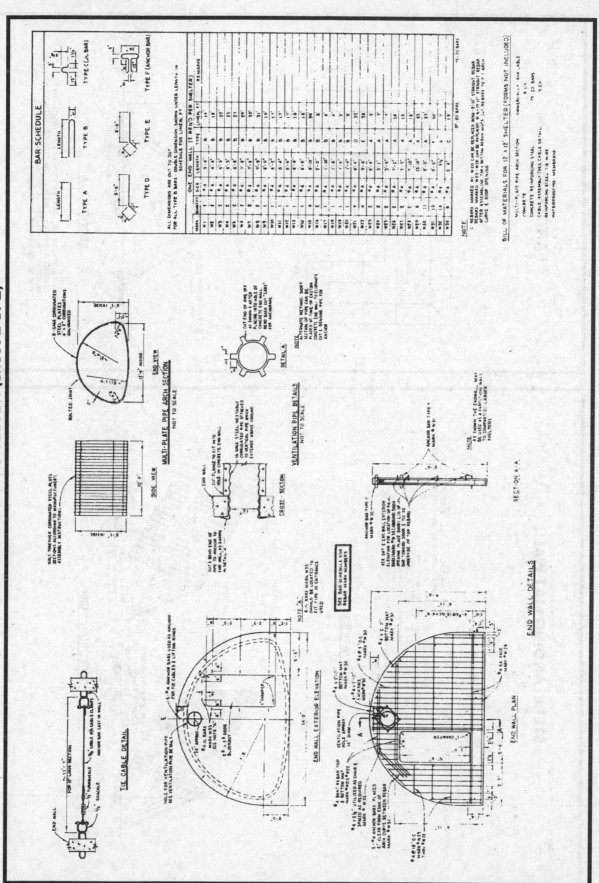

METAL PIPE ARCH SHELTER (sheet 2 of 2)

EARTH WALLS

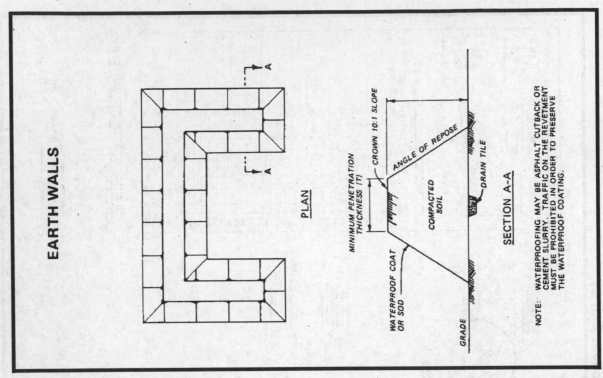

PLAN

SECTION A-A

NOTE: WATERPROOFING MAY BE ASPHALT CUTBACK OR
CEMENT SLURRY. TRAFFIC ON THE REVETMENT
MUST BE PROHIBITED IN ORDER TO PRESERVE
THE WATERPROOF COATING.

STEEL LANDING MAT WALL

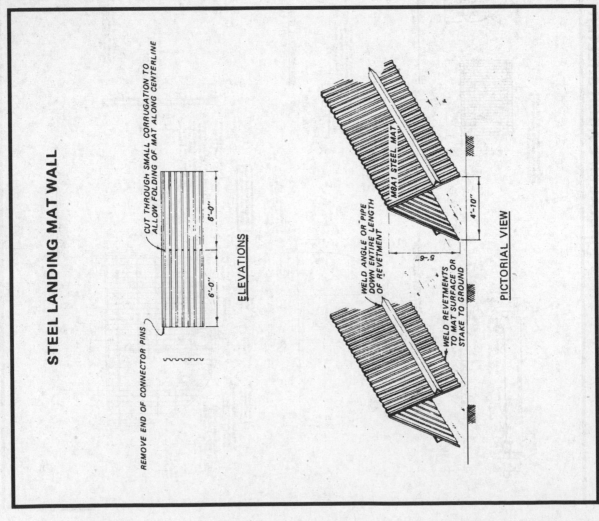

ELEVATIONS

PICTORIAL VIEW

SOIL-CEMENT WALL

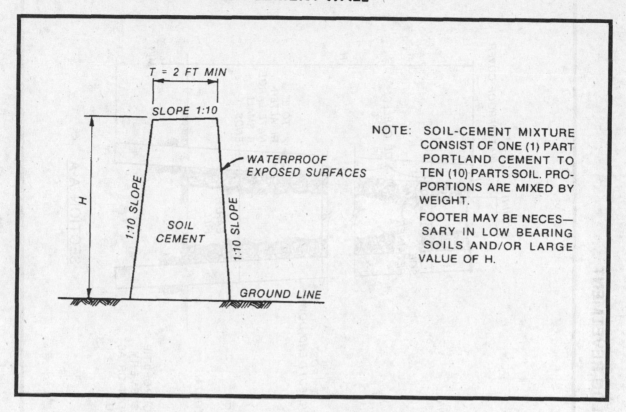

NOTE: SOIL-CEMENT MIXTURE CONSIST OF ONE (1) PART PORTLAND CEMENT TO TEN (10) PARTS SOIL. PROPORTIONS ARE MIXED BY WEIGHT.

FOOTER MAY BE NECESSARY IN LOW BEARING SOILS AND/OR LARGE VALUE OF H.

EARTH WALL WITH REVETMENT

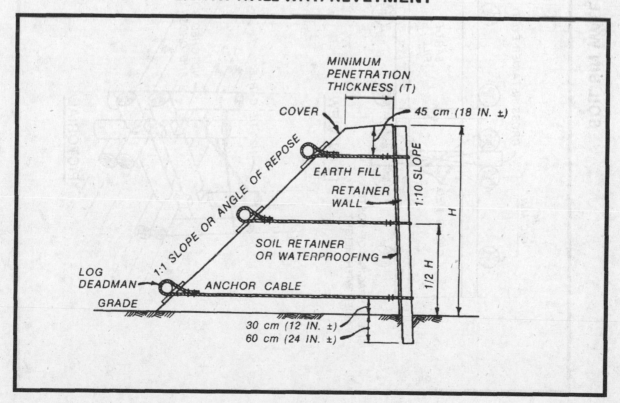

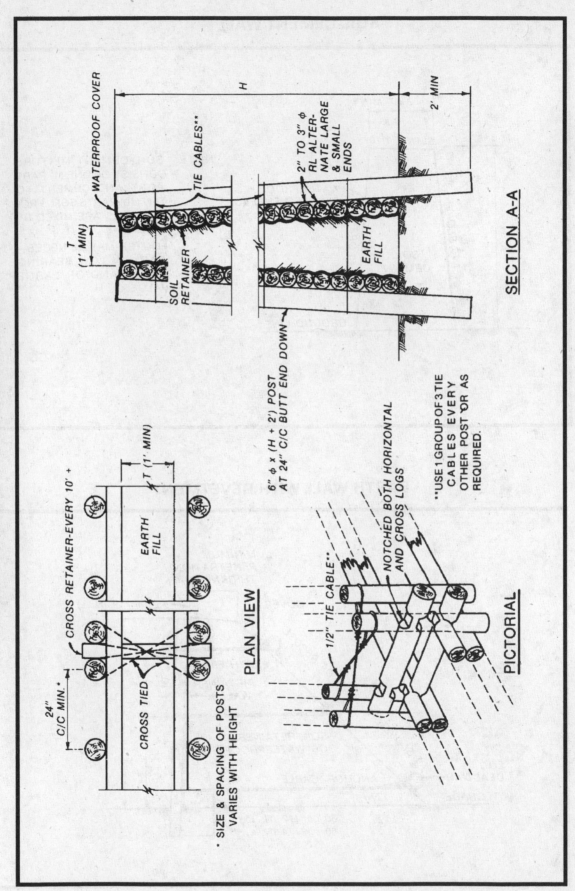

SOIL BIN WALL WITH LOG REVETMENT

WATERPROOF COVER

TIE CABLES**

2" TO 3" φ ALTER-
NATE LARGE
& SMALL ENDS

T (1' MIN)

SOIL RETAINER

EARTH FILL

6" φ x (H + 2') POST
AT 24" C/C BUTT END DOWN

SECTION A-A

CROSS RETAINER-EVERY 10' +

T (1' MIN)

EARTH FILL

24" C/C MIN.*

CROSS TIED

PLAN VIEW

* SIZE & SPACING OF POSTS
VARIES WITH HEIGHT

NOTCHED BOTH HORIZONTAL
AND CROSS LOGS

1/2" TIE CABLE**

**USE 1 GROUP OF 3 TIE
CABLES EVERY
OTHER POST OR AS
REQUIRED.

PICTORIAL

SOIL BIN WALL WITH TIMBER REVETMENT

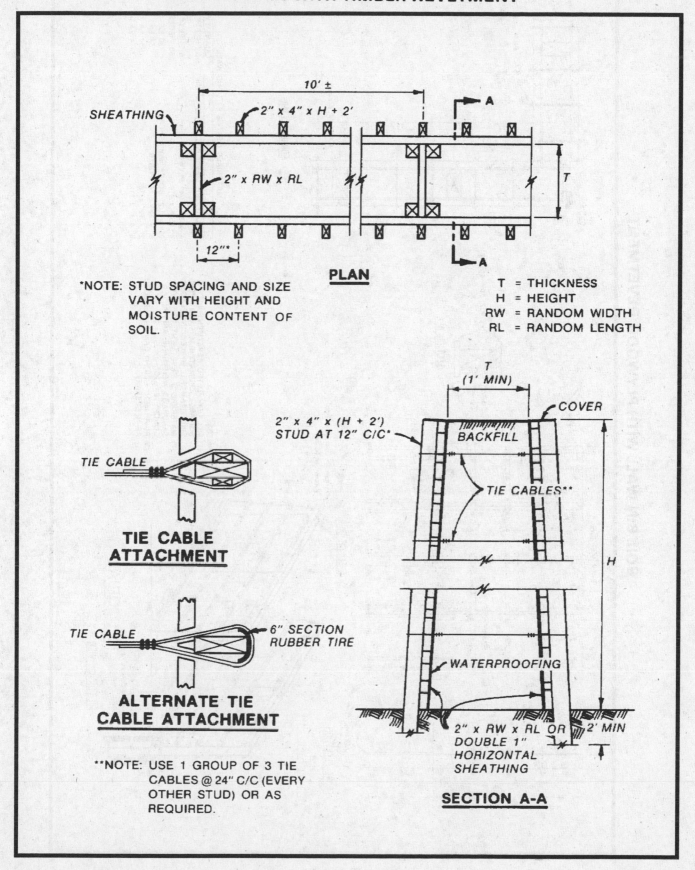

PLAN

10' ±

SHEATHING

2" x 4" x H + 2'

2" x RW x RL

12"*

T

A

*NOTE: STUD SPACING AND SIZE
VARY WITH HEIGHT AND
MOISTURE CONTENT OF
SOIL.

T = THICKNESS
H = HEIGHT
RW = RANDOM WIDTH
RL = RANDOM LENGTH

TIE CABLE

**TIE CABLE
ATTACHMENT**

TIE CABLE

6" SECTION
RUBBER TIRE

**ALTERNATE TIE
CABLE ATTACHMENT**

**NOTE: USE 1 GROUP OF 3 TIE
CABLES @ 24" C/C (EVERY
OTHER STUD) OR AS
REQUIRED.

T
(1' MIN)

COVER

2" x 4" x (H + 2')
STUD AT 12" C/C*

BACKFILL

TIE CABLES**

H

WATERPROOFING

2" x RW x RL OR
DOUBLE 1"
HORIZONTAL
SHEATHING

2' MIN

SECTION A-A

SOIL BIN WALL WITH PLYWOOD REVETMENT

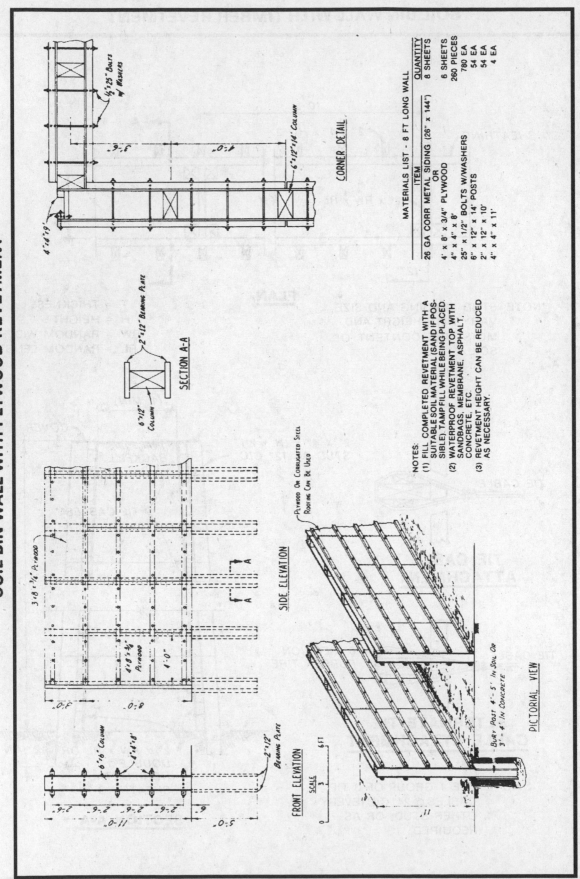

CORNER DETAIL

SECTION A-A

SIDE ELEVATION

FRONT ELEVATION

PICTORIAL VIEW

SCALE

MATERIALS LIST - 8 FT LONG WALL	
ITEM	QUANTITY
26 GA CORR METAL SIDING (26" x 144")	8 SHEETS
OR	
4' x 8' x 3/4" PLYWOOD	6 SHEETS
4" x 4" x 8'	260 PIECES
25" x 1/2" BOLTS W/WASHERS	780 EA
6" x 12" x 14' POSTS	54 EA
2" x 12" x 10'	54 EA
4" x 4" x 11'	4 EA

NOTES:

(1) FILL COMPLETED REVETMENT WITH A SUITABLE SOIL MATERIAL (SAND IF POSSIBLE) TAMP FILL WHILE BEING PLACED.

(2) WATERPROOF REVETMENT TOP WITH SANDBAGS, MEMBRANE, ASPHALT, CONCRETE, ETC.

(3) REVETMENT HEIGHT CAN BE REDUCED AS NECESSARY.

HARDENED SOIL BIN WALL WITH PLYWOOD REVETMENT (sheet 1 of 2)

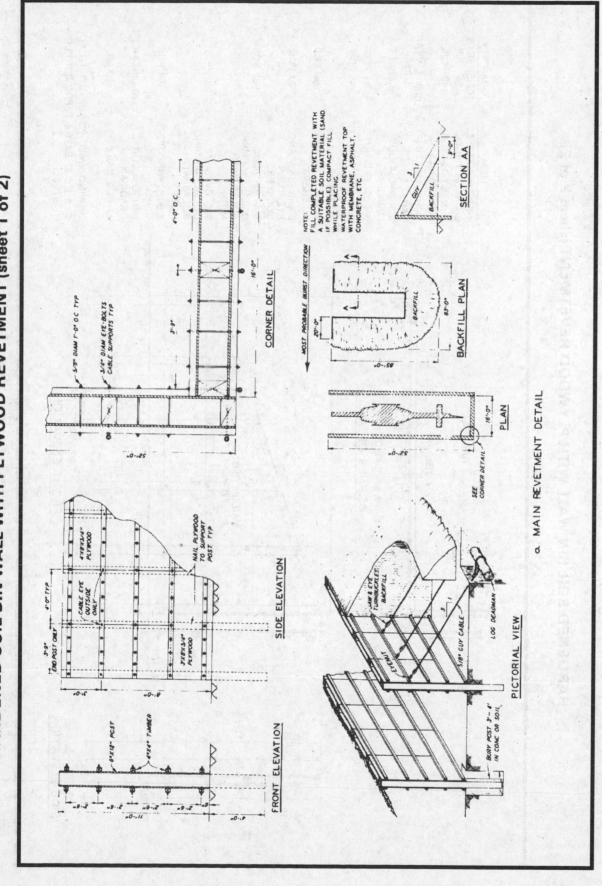

CORNER DETAIL

SIDE ELEVATION

FRONT ELEVATION

NOTE:
FILL COMPLETED REVETMENT WITH A SUITABLE SOIL MATERIAL (SAND IF POSSIBLE) COMPACT FILL WHILE PLACING.
WATERPROOF REVETMENT TOP WITH MEMBRANE, ASPHALT, CONCRETE, ETC

MOST PROBABLE BURST DIRECTION

SECTION AA

BACKFILL PLAN

PLAN

PICTORIAL VIEW

a. MAIN REVETMENT DETAIL

HARDENED SOIL BIN WALL WITH PLYWOOD REVETMENT (sheet 2 of 2)

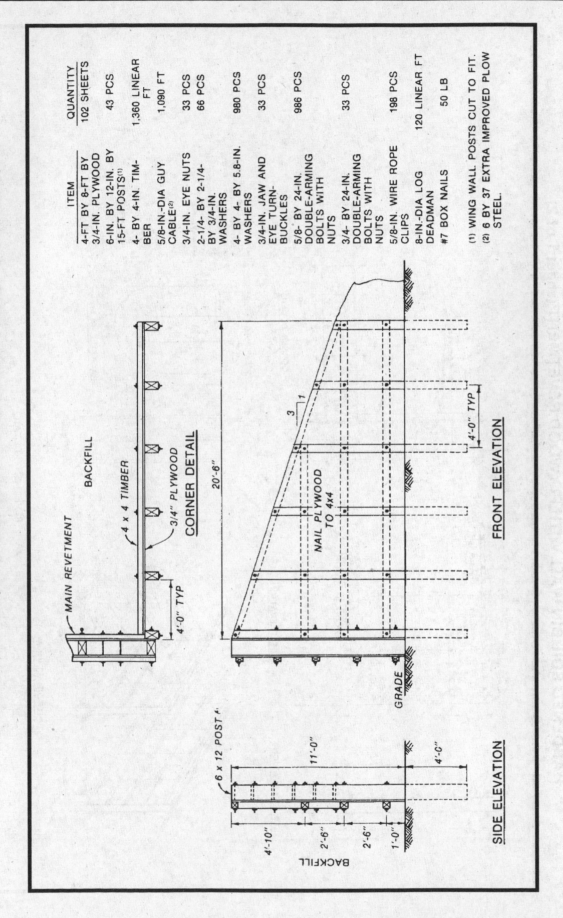

ITEM	QUANTITY
4-FT BY 8-FT BY 3/4-IN. PLYWOOD	102 SHEETS
6-IN. BY 12-IN. BY 15-FT POSTS[1]	43 PCS
4- BY 4-IN. TIMBER	1,360 LINEAR FT
5/8-IN.-DIA GUY CABLE[2]	1,090 FT
3/4-IN. EYE NUTS	33 PCS
2-1/4- BY 2-1/4- BY 3/4-IN. WASHERS	66 PCS
4- BY 4- BY 5.8-IN. WASHERS	980 PCS
3/4-IN. JAW AND EYE TURNBUCKLES	33 PCS
5/8- BY 24-IN. DOUBLE-ARMING BOLTS WITH NUTS	986 PCS
3/4- BY 24-IN. DOUBLE-ARMING BOLTS WITH NUTS	33 PCS
5/8-IN. WIRE ROPE CLIPS	198 PCS
8-IN.-DIA LOG DEADMAN	120 LINEAR FT
#7 BOX NAILS	50 LB

(1) WING WALL POSTS CUT TO FIT.
(2) 6 BY 37 EXTRA IMPROVED PLOW STEEL.

PLYWOOD (OR CORRUGATED METAL) PORTABLE WALL

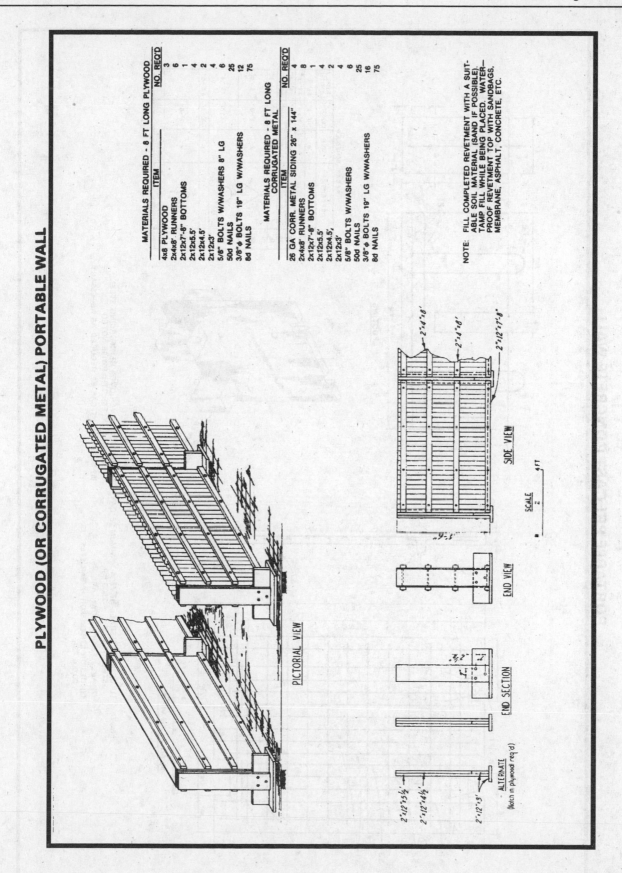

MATERIALS REQUIRED - 8 FT LONG PLYWOOD	
ITEM	NO. REQ'D
4x8 PLYWOOD	3
2x4x8' RUNNERS	6
2x12x7'-8" BOTTOMS	1
2x12x5.5'	4
2x12x4.5'	2
2x12x3'	4
5/8" BOLTS W/WASHERS 8" LG	6
50d NAILS	25
3/8"φ BOLTS 19" LG W/WASHERS	12
8d NAILS	75

MATERIALS REQUIRED - 8 FT LONG CORRUGATED METAL	
ITEM	NO. REQ'D
26 GA CORR. METAL SIDING 26" x 144"	4
2x4x8' RUNNERS	8
2x12x7'-8" BOTTOMS	1
2x12x5.5'	4
2x12x4.5'	2
2x12x3'	4
5/8" BOLTS W/WASHERS	6
50d NAILS	25
3/8"φ BOLTS 19" LG W/WASHERS	16
8d NAILS	75

NOTE: FILL COMPLETED REVETMENT WITH A SUIT-ABLE SOIL MATERIAL (SAND IF POSSIBLE). TAMP FILL WHILE BEING PLACED. WATER-PROOF REVETMENT TOP WITH SANDBAGS, MEMBRANE, ASPHALT, CONCRETE, ETC.

PICTORIAL VIEW

SIDE VIEW

END VIEW

END SECTION

ALTERNATE
(Notch in plywood req'd)

SCALE
2   4 FT

PORTABLE PRECAST CONCRETE WALL

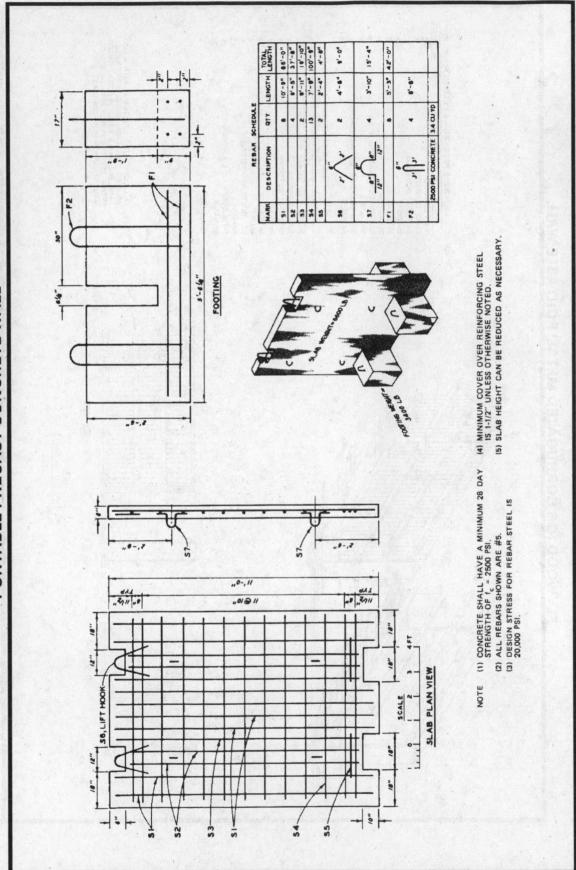

FOOTING

SLAB PLAN VIEW

REBAR SCHEDULE

MARK	DESCRIPTION	QTY	LENGTH	TOTAL LENGTH
S1		8	10'-9"	86'-0"
S2		4	9'-5"	37'-8"
S3		2	9'-11"	19'-10"
S4		13	7'-9"	100'-9"
S5		2	2'-4"	4'-8"
S6		2	4'-6"	9'-0"
S7		4	3'-10"	15'-4"
F1		8	5'-3"	42'-0"
F2		4	6'-6"	26'-0"
2500 PSI CONCRETE	3.4 CU YD			

NOTE   (1) CONCRETE SHALL HAVE A MINIMUM 28 DAY
           STRENGTH OF $f'_c = 2500$ PSI.
       (2) ALL REBARS SHOWN ARE #5.
       (3) DESIGN STRESS FOR REBAR STEEL IS
           20,000 PSI.
       (4) MINIMUM COVER OVER REINFORCING STEEL
           IS 1-1/2" UNLESS OTHERWISE NOTED.
       (5) SLAB HEIGHT CAN BE REDUCED AS NECESSARY.

CAST-IN-PLACE CONCRETE WALL

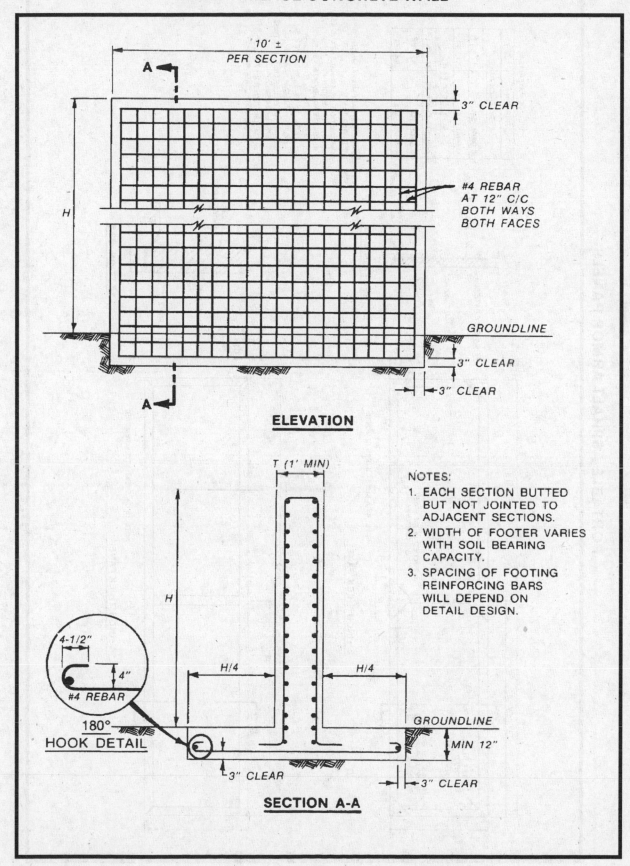

ELEVATION

SECTION A-A

HOOK DETAIL

NOTES:
1. EACH SECTION BUTTED BUT NOT JOINTED TO ADJACENT SECTIONS.
2. WIDTH OF FOOTER VARIES WITH SOIL BEARING CAPACITY.
3. SPACING OF FOOTING REINFORCING BARS WILL DEPEND ON DETAIL DESIGN.

PORTABLE ASPHALT ARMOR PANELS

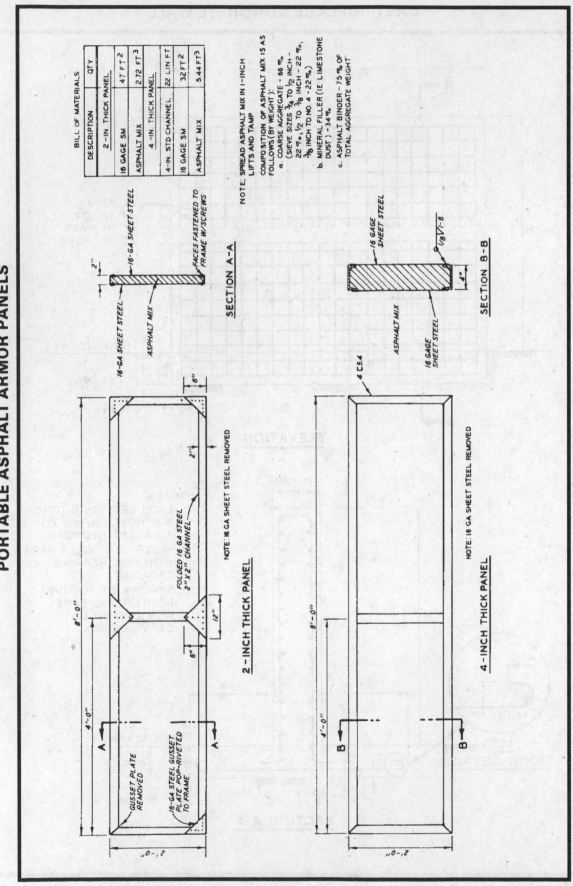

BILL OF MATERIALS	
DESCRIPTION	QTY
2-IN. THICK PANEL	
16 GAGE SM	47 FT²
ASPHALT MIX	2.72 FT³
4-IN. THICK PANEL	
4-IN STD CHANNEL	22 LIN FT
16 GAGE SM	32 FT²
ASPHALT MIX	5.44 FT³

NOTE: SPREAD ASPHALT MIX IN 1-INCH LIFTS AND TAMP

COMPOSITION OF ASPHALT MIX IS AS FOLLOWS (BY WEIGHT):
a. COARSE AGGREGATE - 66 %
   (SIEVE SIZES ¾ TO ½ INCH - 22 %, ½ TO ⅜ INCH - 22 %, ⅜ INCH TO NO. 4 - 22 %)
b. MINERAL FILLER (IE LIMESTONE DUST) - 34 %
c. ASPHALT BINDER - 75 % OF TOTAL AGGREGATE WEIGHT

16-GA SHEET STEEL
ASPHALT MIX
16-GA SHEET STEEL
FACES FASTENED TO FRAME W/SCREWS
2"

SECTION A-A

16 GAGE SHEET STEEL
⅛ √1-6
ASPHALT MIX
16 GAGE SHEET STEEL
4"

SECTION B-B

GUSSET PLATE REMOVED
16-GA STEEL GUSSET PLATE POP-RIVETED TO FRAME
FOLDED 16 GA STEEL 2"x 2" CHANNEL
NOTE: 16 GA SHEET STEEL REMOVED
8'-0"
4'-0"
2'-0"
12"
6"
2"

2-INCH THICK PANEL

4 C5.4
NOTE: 16 GA SHEET STEEL REMOVED
8'-0"
4'-0"
2'-0"

4-INCH THICK PANEL

STANDARD FIGHTING TRENCH

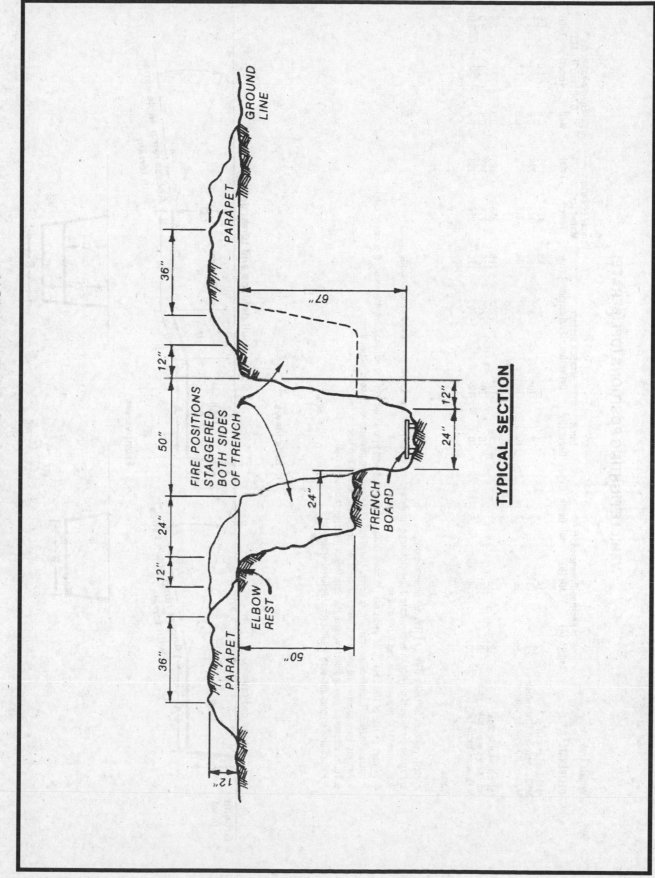

TYPICAL SECTION

VEHICLE FIGHTING POSITIONS (DELIBERATE)

Vehicle Type DELIBERATE[1]	Position Dimension, ft [2]				Weapon System		Volume of Earth Moved (cy)			Equipment Hours [4] D7 Dozer/M9 ACE		
	Length (A)	Width (B)	Hull Depth (C)[5]	Turret Depth (D)[5]	Deflection	Evaluation	Hull	Turret[6]	Total[7]	Hull	Turret[6]	Total[7]
M113 series carrier [3]	22	14	6	7½	—	+30°	69	124	193	0.6	1.0	1.6
M901 improved TOW vehicle	22	14	7	9	-10° gun	+60°	80	148	228	0.6	1.1	1.7
M2 and M3 fighting vehicle	26	16	7	10	-10° TOW	+30°	108	218	326	0.8	1.7	2.5
M1 main battle tank	32	18	5½	9	-10°	+20°	118	268	386	0.9	2.0	2.9
M60 series main battle tank	30	18	6	10	-10°	+20°	120	278	398	0.9	2.1	3.0
M48 series battle tank	30	18	6	10	-10°	+20°	120	278	398	0.9	2.1	3.0

Notes:

1. Hasty positions for tanks, IFVs, and ITVs not recommended.
2. Position dimensions provide an approximate 3-foot clearance around vehicle for movement and maintenance and do not include access ramp(s).
3. Includes M132 flamethrower and M103 Vulcan.
4. Production rate of 100 bank cubic yards per .75 hour. Divide construction time by 0.85 for rocky or hard soil, night conditions, or closed hatch operations (M9). Ripper needed if ground is frozen. Use of natural terrain features will reduce construction time.
5. All depths are approximate and will need adjustment for surrounding terrain and fields of fire.
6. Turret volume (c) plus approach volume (b). Path length (E) is approximately ½(A).
7. Hull volume (a) plus approach volume (b) plus turret volume (c).

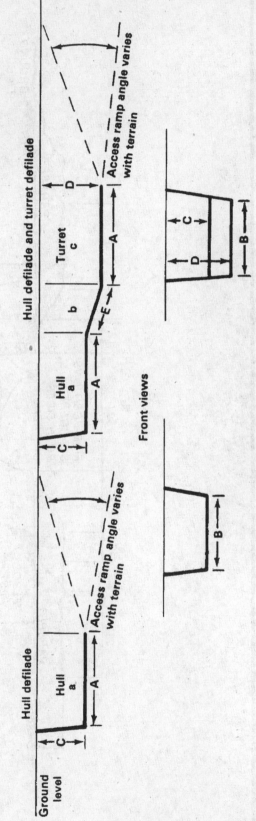

PART IV

Water, Food, Plants, Herbal Remedies, and Dangerous Plants and Animals

CHAPTER 1

Water Procurement

Water is one of your most urgent needs in a survival situation. You can't live long without it, especially in hot areas where you lose water rapidly through perspiration. Even in cold areas, you need a minimum of 2 liters of water each day to maintain efficiency.

More than three-fourths of your body is composed of fluids. Your body loses fluid as a result of heat, cold, stress, and exertion. To function effectively, you must replace the fluid your body loses. So, one of your first goals is to obtain an adequate supply of water.

WATER SOURCES

Almost any environment has water present to some degree. Table 1-1 lists possible sources of water in various environments. It also provides information on how to make the water potable.

Note: If you do not have a canteen, a cup, a can, or other type of container, improvise one from plastic or water-resistant cloth. Shape the plastic or cloth into a bowl by pleating it. Use pins or other suitable items—even your hands—to hold the pleats.

Table 1-1: Water sources in different environments.

Environment	Source of Water	Means of Obtaining and/or Making Potable	Remarks
Frigid areas	Snow and ice	Melt and purify.	**Do not eat** without melting! Eating snow and ice can reduce body temperature and will lead to more dehydration.
			Snow and ice are no purer than the water from which they come.
			Sea ice that is gray in color or opaque is salty. Do not use it without desalting it. Sea ice that is crystalline with a bluish cast has little salt in it.
At sea	Sea	Use desalter kit.	**Do not** drink seawater without desalting.
	Rain	Catch rain in tarps or in other water-holding material or containers.	If tarp or water-holding material has become encrusted with salt, wash it in the sea before using (very little salt will remain on it).
	Sea ice		See remarks above for frigid areas.

Table 1-1: *(Continued)*

Environment	Source of Water	Means of Obtaining and/or Making Potable	Remarks
Beach	Ground	Dig hole deep enough to allow water to seep in; obtain rocks, build fire, and heat rocks; drop hot rocks in water; hold cloth over hole to absorb steam; wring water from cloth.	Alternate method if a container or bark pot is available: Fill container or pot with seawater; build fire and boil water to produce steam; hold cloth over container to absorb steam; wring water from cloth.
Desert	Ground • in valleys and low areas • at foot of concave banks of dry river beads • at foot of cliffs or rock outcrops • at first depression behind first sand dune of dry desert lakes • wherever you find damp surface sand • wherever you find green vegetation	Dig holes deep enough to allow water to seep in.	In a sand dune belt, any available water will be found beneath the original valley floor at the edge of dunes.
	Cacti	Cut off the top of a barrel cactus and mash or squeeze the pulp. **CAUTION: Do not eat pulp. Place pulp in mouth, suck out juice, and discard pulp.**	Without a machete, cutting into a cactus is difficult and takes time since you must get past the long, strong spines and cut through the tough rind.

If you do not have a reliable source to replenish your water supply, stay alert for ways in which your environment can help you.

⚠ **CAUTION**

Do not substitute the fluids listed in Table 1-2 for water.

Heavy dew can provide water. Tie rags or tufts of fine grass around your ankles and walk through dew-covered grass before sunrise. As the rags or grass tufts absorb the dew, wring the water into a container. Repeat the process until you have a supply of water or until the dew is gone. Australian natives sometimes mop up as much as a liter an hour this way.

Table 1-1: *(Continued)*

Environment	Source of Water	Means of Obtaining and/or Making Potable	Remarks
Desert (continued)	Depressions or holes in rocks		Periodic rainfall may collect in pools, seep into fissures, or collect in holes in rocks.
	Fissures in rock	Insert flexible tubing and siphon water. If fissure is large enough, you can lower a container into it.	
	Porous rock	Insert flexible tubing and siphon water.	
	Condensation on metal	Use cloth to absorb water, then wring water from cloth.	Extreme temperature variations between night and day may cause condensation on metal surfaces. Following are signs to watch for in the desert to help you find water: • All trails lead to water. You should follow in the direction in which the trails converge. Signs of camps, campfire ashes, animal droppings, and trampled terrain may mark trails. • Flocks of birds will circle over water holes. Some birds fly to water holes at dawn and sunset. Their flight at these times is generally fast and close to the ground. Bird tracks or chirping sounds in the evening or early morning sometimes indicate that water is nearby.

Bees or ants going into a hole in a tree may point to a water-filled hole. Siphon the water with plastic tubing or scoop it up with an improvised dipper. You can also stuff cloth in the hole to absorb the water and then wring it from the cloth.

Water sometimes gathers in tree crotches or rock crevices. Use the above procedures to get the water. In arid areas, bird droppings around a crack in the rocks may indicate water in or near the crack.

Green bamboo thickets are an excellent source of fresh water. Water from green bamboo is clear and odorless. To get the water, bend a green bamboo stalk, tie it down, and cut off the top (Figure 1-1). The water will drip freely during the night. Old, cracked bamboo may contain water.

⚠ CAUTION

Purify the water before drinking it.

Table 1-2: The effects of substitute fluids.

Fluid	Remarks
Alcoholic beverages	Dehydrate the body and cloud judgment.
Urine	Contains harmful body wastes. Is about 2 percent salt.
Blood	Is salty and considered a food; therefore, requires additional body fluids to digest. May transmit disease.
Seawater	Is about 4 percent salt. It takes about 2 liters of body fluids to rid the body of waste from 1 liter of seawater. Therefore, by drinking seawater you deplete your body's water supply, which can cause death.

Wherever you find banana or plantain trees, you can get water. Cut down the tree, leaving about a 30-centimeter stump, and scoop out the center of the stump so that the hollow is bowl-shaped. Water from the roots will immediately start to fill the hollow. The first three fillings of water will be bitter, but succeeding fillings will be palatable. The stump (Figure 1-2) will supply water for up to four days. Be sure to cover it to keep out insects.

Some tropical vines can give you water. Cut a notch in the vine as high as you can reach, then cut the vine off close to the ground. Catch the dropping liquid in a container or in your mouth (Figure 1-3).

⚠ **CAUTION**

Do not drink the liquid if it is sticky, milky, or bitter tasting.

The milk from green (unripe) coconuts is a good thirst quencher. However, the milk from mature coconuts contains an oil that acts as a laxative. Drink in moderation only.

In the American tropics you may find large trees whose branches support air plants. These air plants may hold a considerable amount of rainwater in their overlapping, thickly growing leaves. Strain the water through a cloth to remove insects and debris.

You can get water from plants with moist pulpy centers. Cut off a section of the plant and squeeze or smash the pulp so that the moisture runs out. Catch the liquid in a container.

Figure 1-1: Water from green bamboo.

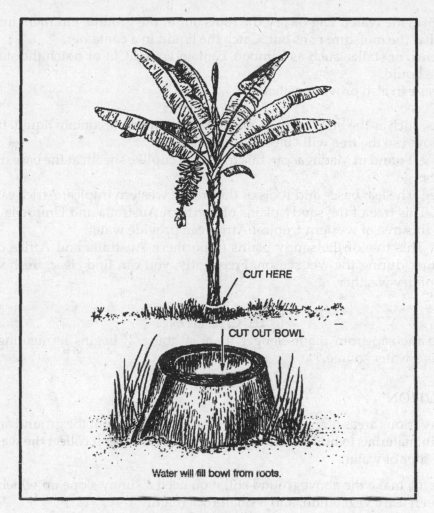

Figure 1-2: Water from plantain or banana tree stump.

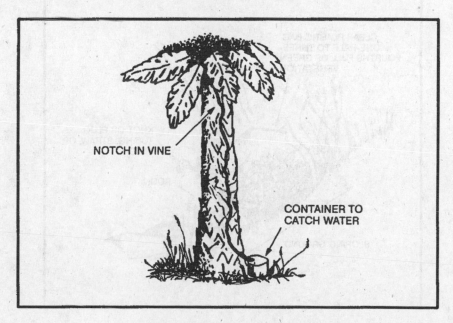

Figure 1-3: Water from a vine.

Plant roots may provide water. Dig or pry the roots out of the ground, cut them into short pieces, and smash the pulp so that the moisture runs out. Catch the liquid in a container.

Fleshy leaves, stems, or stalks, such as bamboo, contain water. Cut or notch the stalks at the base of a joint to drain out the liquid.

The following trees can also provide water:

- Palms. Palms, such as the buri, coconut, sugar, rattan, and nips, contain liquid. Bruise a lower frond and pull it down so the tree will "bleed" at the injury.
- Traveler's tree. Found in Madagascar, this tree has a cuplike sheath at the base of its leaves in which water collects.
- Umbrella tree. The leaf bases and roots of this tree of western tropical Africa can provide water.
- Baobab tree. This tree of the sandy plains of northern Australia and Umbrella tree. The leaf bases and roots of this tree of western tropical Africa can provide water.
- Baobab tree. This tree of the sandy plains of northern Australia and Africa collects water in its bottlelike trunk during the wet season. Frequently, you can find clear, fresh water in these trees after weeks of dry weather.

> ⚠ CAUTION
>
> Do not keep the sap from plants longer than 24 hours. It begins fermenting, becoming dangerous as a water source.

STILL CONSTRUCTION

You can use stills in various areas of the world. They draw moisture from the ground and from plant material. You need certain materials to build a still, and you need time to let it collect the water. It takes about 24 hours to get 0.5 to 1 liter of water.

Aboveground Still. To make the aboveground still, you need a sunny slope on which to place the still, a clear plastic bag, green leafy vegetation, and a small rock (Figure 1-4).

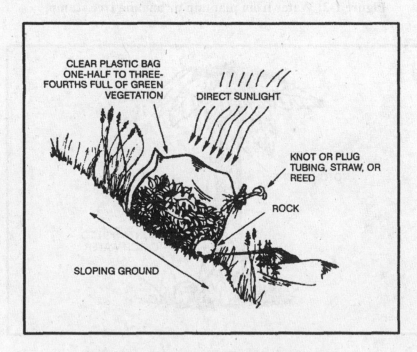

Figure 1-4: Aboveground solar water still.

To make the still—

- Fill the bag with air by turning the opening into the breeze or by "scooping" air into the bag.
- Fill the plastic bag half to three-fourths full of green leafy vegetation. Be sure to remove all hard sticks or sharp spines that might puncture the bag.

> ⚠ **CAUTION**
> Do not use poisonous vegetation. It will provide poisonous liquid.

- Place a small rock or similar item in the bag.
- Close the bag and tie the mouth securely as close to the end of the bag as possible to keep the maximum amount of air space. If you have a piece of tubing, a small straw, or a hollow reed, insert one end in the mouth of the bag before you tie it securely. Then tie off or plug the tubing so that air will not escape. This tubing will allow you to drain out condensed water without untying the bag.
- Place the bag, mouth downhill, on a slope in full sunlight. Position the mouth of the bag slightly higher than the low point in the bag.
- Settle the bag in place so that the rock works itself into the low point in the bag.

To get the condensed water from the still, loosen the tie around the bag's mouth and tip the bag so that the water collected around the rock will drain out. Then retie the mouth securely and reposition the still to allow further condensation.

Change the vegetation in the bag after extracting most of the water from it. This will ensure maximum output of water.

Belowground Still. To make a belowground still, you need a digging tool, a container, a clear plastic sheet, a drinking tube, and a rock (Figure 1-5).

Select a site where you believe the soil will contain moisture (such as a dry stream bed or a low spot where rainwater has collected). The soil at this site should be easy to dig, and sunlight must hit the site most of the day.

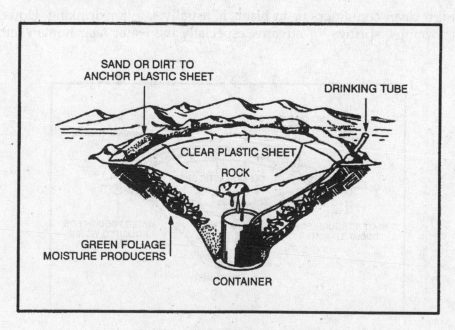

Figure 1-5: Belowground still.

To construct the still—

- Dig a bowl-shaped hole about 1 meter across and 60 centimeters deep.
- Dig a sump in the center of the hole. The sump's depth and perimeter will depend on the size of the container that you have to place in it. The bottom of the sump should allow the container to stand upright.
- Anchor the tubing to the container's bottom by forming a loose overhand knot in the tubing.
- Place the container upright in the sump.
- Extend the unanchored end of the tubing up, over, and beyond the lip of the hole.
- Place the plastic sheet over the hole, covering its edges with soil to hold it in place.
- Place a rock in the center of the plastic sheet.
- Lower the plastic sheet into the hole until it is about 40 centimeters below ground level. It now forms an inverted cone with the rock at its apex. Make sure that the cone's apex is directly over your container. Also make sure the plastic cone does not touch the sides of the hole because the earth will absorb the condensed water.
- Put more soil on the edges of the plastic to hold it securely in place and to prevent the loss of moisture.
- Plug the tube when not in use so that the moisture will not evaporate.

You can drink water without disturbing the still by using the tube as a straw.

You may want to use plants in the hole as a moisture source. If so, dig out additional soil from the sides of the hole to form a slope on which to place the plants. Then proceed as above.

If polluted water is your only moisture source, dig a small trough outside the hole about 25 centimeters from the still's lip (Figure 1-6). Dig the trough about 25 centimeters deep and 8 centimeters wide. Pour the polluted water in the trough. Be sure you do not spill any polluted water around the rim of the hole where the plastic sheet touches the soil. The trough holds the polluted water and the soil filters it as the still draws it. The water then condenses on the plastic and drains into the container. This process works extremely well when your only water source is salt water.

You will need at least three stills to meet your individual daily water intake needs.

WATER PURIFICATION

Rainwater collected in clean containers or in plants is usually safe for drinking. However, purify water from lakes, ponds, swamps, springs, or streams, especially the water near human settlements or in the

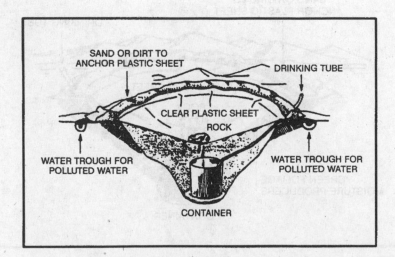

Figure 1-6: Belowground still to get potable water from polluted water.

tropics. When possible, purify all water you got from vegetation or from the ground by using iodine or chlorine, or by boiling.

Purify water by—

- Using water purification tablets. (Follow the directions provided.)
- Placing 5 drops of 2 percent tincture of iodine in a canteen full of clear water. If the canteen is full of cloudy or cold water, use 10 drops. (Let the canteen of water stand for 30 minutes before drinking.)
- Boiling water for 1 minute at sea level, adding 1 minute for each additional 300 meters above sea level, or boil for 10 minutes no matter where you are.

By drinking nonpotable water you may contract diseases or swallow organisms that can harm you. Examples of such diseases or organisms are—

- Dysentery. Severe, prolonged diarrhea with bloody stools, fever, and weakness.
- Cholera and typhoid. You may be susceptible to these diseases regardless of inoculations.
- Flukes. Stagnant, polluted water—especially in tropical areas—often contains blood flukes. If you swallow flukes, they will bore into the bloodstream, live as parasites, and cause disease.
- Leeches. If you swallow a leech, it can hook onto the throat passage or inside the nose. It will suck blood, create a wound, and move to another area. Each bleeding wound may become infected.

WATER FILTRATION DEVICES

If the water you find is also muddy, stagnant, and foul smelling, you can clear the water—

- By placing it in a container and letting it stand for 12 hours.
- By pouring it through a filtering system.

Note: These procedures only clear the water and make it more palatable. You will have to purify it.

To make a filtering system, place several centimeters or layers of filtering material such as sand, crushed rock, charcoal, or cloth in bamboo, a hollow log, or an article of clothing (Figure 1-7).

Remove the odor from water by adding charcoal from your fire. Let the water stand for 45 minutes before drinking it.

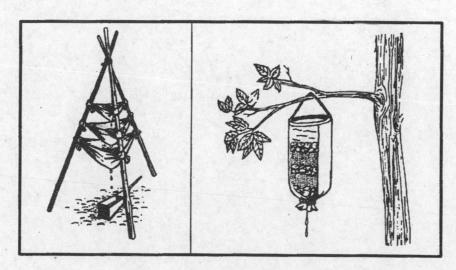

Figure 1-7: Water filtering systems.

CHAPTER 2

Food Procurement

After water, man's most urgent requirement is food. In contemplating virtually any hypothetical survival situation, the mind immediately turns to thoughts of food. Unless the situation occurs in an arid environment, even water, which is more important to maintaining body functions, will almost always follow food in our initial thoughts. The survivor must remember that the three essentials of survival—water, food, and shelter–are prioritized according to the estimate of the actual situation. This estimate must not only be timely but accurate as well. Some situations may well dictate that shelter precede both food and water.

ANIMALS FOR FOOD

Unless you have the chance to take large game, concentrate your efforts on the smaller animals, due to their abundance. The smaller animal species are also easier to prepare. You must not know all the animal species that are suitable as food. Relatively few are poisonous, and they make a smaller list to remember. What is important is to learn the habits and behavioral patterns of classes of animals. For example, animals that are excellent choices for trapping, those that inhabit a particular range and occupy a den or nest, those that have somewhat fixed feeding areas, and those that have trails leading from one area to another. Larger, herding animals, such as elk or caribou, roam vast areas and are somewhat more difficult to trap. Also, you must understand the food choices of a particular species.

You can, with relatively few exceptions, eat anything that crawls, swims, walks, or flies. The first obstacle is overcoming your natural aversion to a particular food source. Historically, people in starvation situations have resorted to eating everything imaginable for nourishment. A person who ignores an otherwise healthy food source due to a personal bias, or because he feels it is unappetizing, is risking his own survival. Although it may prove difficult at first, a survivor must eat what is available to maintain his health.

Insects. The most abundant life-form on earth, insects are easily caught. Insects provide 65 to 80 percent protein compared to 20 percent for beef. This fact makes insects an important, if not overly appetizing, food source. Insects to avoid include all adults that sting or bite, hairy or brightly colored insects, and caterpillars and insects that have a pungent odor. Also avoid spiders and common disease carriers such as ticks, flies, and mosquitoes. See Chapter 3 for more information about dangerous insects and arachnids.

Rotting logs lying on the ground are excellent places to look for a variety of insects including ants, termites, beetles, and grubs, which are beetle larvae. Do not overlook insect nests on or in the ground. Grassy areas, such as fields, are good areas to search because the insects are easily seen. Stones, boards, or other materials lying on the ground provide the insects with good nesting sites. Check these sites. Insect larvae are also edible. Insects such as beetles and grasshoppers that have a hard outer shell will have parasites. Cook them before eating. Remove any wings and barbed legs also. You can eat most insects raw. The taste varies from one species to another. Wood grubs are bland, while some species of ants store honey in their bodies, giving them a sweet taste.

You can grind a collection of insects into a paste. You can mix them with edible vegetation. You can cook them to improve their taste.

Worms. Worms (Annelidea) are an excellent protein source. Dig for them in damp humus soil or watch for them on the ground after a rain. After capturing them, drop them into clean, potable water for a few minutes. The worms will naturally purge or wash themselves out, after which you can eat them raw.

Leeches. Leeches are blood-sucking creatures with a wormlike appearance. You find them in the tropics and in temperate zones. You will certainly encounter them when swimming in infested waters or making expedient water crossings. You can find them when passing through swampy, tropical vegetation and bogs. You can also find them while cleaning food animals, such as turtles, found in fresh water. Leeches can crawl into small openings; therefore, avoid camping in their habitats when possible. Keep your trousers tucked in your boots. Check yourself frequently for leeches. Swallowed or eaten, leeches can be a great hazard. It is therefore essential to treat water from questionable sources by boiling or using chemical water treatments. Survivors have developed severe infections from wounds inside the throat or nose when sores from swallowed leeches became infected.

Crustaceans. Freshwater shrimp range in size from 0.25 centimeter up to 2.5 centimeters. They can form rather large colonies in mats of floating algae or in mud bottoms of ponds and lakes.

Crayfish are akin to marine lobsters and crabs. You can distinguish them by their hard exoskeleton and five pairs of legs, the front pair having oversized pincers. Crayfish are active at night, but you can locate them in the daytime by looking under and around stones in streams. You can also find them by looking in the soft mud near the chimney-like breathing holes of their nests. You can catch crayfish by tying bits of offal or internal organs to a string. When the crayfish grabs the bait, pull it to shore before it has a chance to release the bait.

You find saltwater lobsters, crabs, and shrimp from the surf's edge out to water 10 meters deep. Shrimp may come to a light at night where you can scoop them up with a net. You can catch lobsters and crabs with a baited trap or a baited hook. Crabs will come to bait placed at the edge of the surf, where you can trap or net them. Lobsters and crabs are nocturnal and caught best at night.

Mollusks. This class includes octopuses and freshwater and saltwater shell fish such as snails, clams, mussels, bivalves, barnacles, periwinkles, chitons, and sea urchins (Figure 2-1). You find bivalves similar to our freshwater mussel and terrestrial and aquatic snails worldwide under all water conditions.

River snails or freshwater periwinkles are plentiful in rivers, streams, and lakes of northern coniferous forests. These snails may be pencil point or globular in shape.

In fresh water, look for mollusks in the shallows, especially in water with a sandy or muddy bottom. Look for the narrow trails they leave in the mud or for the dark elliptical slit of their open valves.

Near the sea, look in the tidal pools and the wet sand. Rocks along beaches or extending as reefs into deeper water often bear clinging shellfish. Snails and limpets cling to rocks and seaweed from the low water mark upward. Large snails, called chitons, adhere tightly to rocks above the surf line.

Mussels usually form dense colonies in rock pools, on logs, or at the base of boulders.

⚠ **CAUTION**
Mussels may be poisonous in tropical zones during the summer!

Steam, boil, or bake mollusks in the shell. They make excellent stews in combination with greens and tubers.

⚠ **CAUTION**
Do not eat shellfish that are not covered by water at high tide!

See chapter 5 for more information about dangerous shellfish.

Fish. Fish represent a good source of protein and fat. They offer some distinct advantages to the survivor or evader. They are usually more abundant than mammal wildlife, and the ways to get them are silent. To be successful at catching fish, you must know their habits. For instance, fish tend to feed heavily before a storm. Fish are not likely to feed after a storm when the water is muddy and swollen. Light often attracts

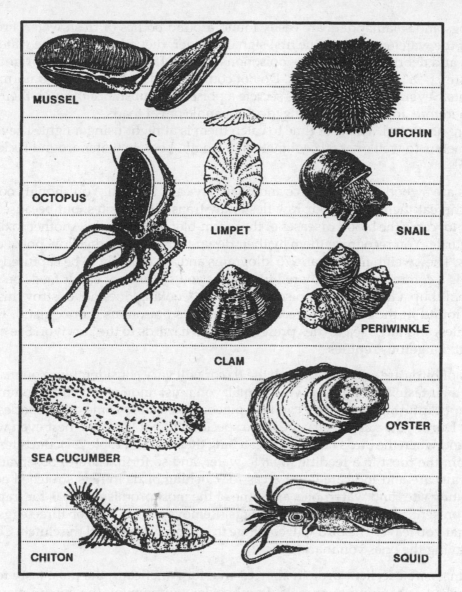

Figure 2-1: Edible Mollusks.

fish at night. When there is a heavy current, fish will rest in places where there is an eddy, such as near rocks. Fish will also gather where there are deep pools, under overhanging brush, and in and around submerged foliage, logs, or other objects that offer them shelter.

There are no poisonous freshwater fish. However, the catfish species has sharp, needlelike protrusions on its dorsal fins and barbels. These can inflict painful puncture wounds that quickly become infected.

Cook all freshwater fish to kill parasites. Also cook saltwater fish caught within a reef or within the influence of a freshwater source as a precaution. Any marine life obtained farther out in the sea will not contain parasites because of the saltwater environment. You can eat these raw.

Many fish living in reefs near shore, or in lagoons and estuaries, are poisonous to eat, though some are only seasonally dangerous. The majority are tropical fish; however, be wary of eating any unidentifiable fish wherever you are. Some predatory fish, such as barracuda and snapper, may become toxic if the fish they feed on in shallow waters are poisonous. The most poisonous types appear to have parrotlike beaks and hard shell-like skins with spines and often can inflate their bodies like balloons. However, at certain times of the year, indigenous populations consider the puffer a delicacy. See chapter 5, Dangerous Fish, Mollusks, and Freshwater Animals for more information.

Amphibians. Frogs and salamanders are easily found around bodies of fresh water. Frogs seldom move from the safety of the water's edge. At the first sign of danger, they plunge into the water and bury themselves in the mud and debris. There are few poisonous species of frogs. Avoid any brightly colored frog or one that has a distinct "X" mark on its back. Do not confuse toads with frogs. You normally find toads in drier environments. Several species of toads secrete a poisonous substance through their skin as a defense against attack. Therefore, to avoid poisoning, do not handle or eat toads.

Salamanders are nocturnal. The best time to catch them is at night using a light. They can range in size from a few centimeters to well over 60 centimeters in length. Look in water around rocks and mud banks for salamanders.

Reptiles. Reptiles are a good protein source and relatively easy to catch. You should cook them, but in an emergency, you can eat them raw. Their raw flesh may transmit parasites, but because reptiles are cold-blooded, they do not carry the blood diseases of the warm-blooded animals. Another toxic meat is the flesh of the hawksbill turtle. You recognize them by their down-turned bill and yellow polka dots on their neck and front flippers. They weigh more than 275 kilograms and are unlikely to be captured.

The box turtle is a commonly encountered turtle that you should not eat. It feeds on poisonous mushrooms and may build up a highly toxic poison in its flesh. Cooking does not destroy this toxin. Avoid the hawksbill turtle, found in the Atlantic Ocean, because of its poisonous thorax gland. Poisonous snakes, alligators, crocodiles, and large sea turtles present obvious hazards to the survivor. See chapter 4 for more information about dangerous reptiles.

Birds. All species of birds are edible, although the flavor will vary considerably. You may skin fish-eating birds to improve their taste. As with any wild animal, you must understand birds' common habits to have a realistic chance of capturing them. You can take pigeons, as well as some other species, from their roost at night by hand. During the nesting season, some species will not leave the nest even when approached. Knowing where and when the birds nest makes catching them easier (Table 2-1). Birds tend to have regular flyways going from the roost to a feeding area, to water, and so forth. Careful observation should reveal where these flyways are and indicate good areas for catching birds in nets stretched across the flyways (Figure 2-2). Roosting sites and waterholes are some of the most promising areas for trapping or snaring.

Nesting birds present another food source—eggs. Remove all but two or three eggs from the clutch, marking the on that you leave. The bird will continue to lay more eggs to fill the clutch. Continue removing the fresh eggs, leaving the ones you marked.

Mammals. Mammals are excellent protein sources and, for Americans, the most tasty food source. There are some drawbacks to obtaining mammals. In a hostile environment, the enemy may detect any traps or snares placed on land. The amount of injury an animal can inflict is in direct proportion to its size. All mammals have teeth and nearly all will bite in self-defense. Even a squirrel can inflict a serious wound and any bite presents a serious risk of infection. Also, a mother can be extremely aggressive in defense of her young. Any animal with no route of escape will fight when cornered.

All mammals are edible; however, the polar bear and bearded seal have toxic levels of vitamin A in their livers. The platypus, native to Australia and Tasmania, is an egg-laying, semiaquatic mammal that has poisonous glands. Scavenging mammals, such as the opossum, may carry diseases. Common sense tells the survivor to avoid encounters with lions, bears, and other large or dangerous animals. You should also avoid large grazing animals with horns, hooves, and great weight. Your actions may prevent unexpected meetings. Move carefully through their environment. Do not attract large predators by leaving food lying around your camp. Carefully survey the scene before entering water or forests.

Bats. Despite the legends, bats (Desmodus species) are a relatively small hazard to the survivor. There are many bat varieties worldwide, but you find the true vampire bats only in Central and South America. They are small, agile fliers that land on their sleeping victims, mostly cows and horses, to lap a blood meal after biting their victim. Their saliva contains an anticoagulant that keeps the blood slowly flowing while they feed. Only a small percentage of these bats actually carry rabies; however, avoid any sick or injured

Table 2-1: Bird nesting places.

Types of Birds	Frequent Nesting Places	Nesting Periods
Inland birds	Trees, woods, or fields	Spring and early summer in temperate and arctic regions; year round in the tropics
Cranes and herons	Mangrove swamps or high trees near water	Spring and early summer
Some species of owls	High trees	Late December through March
Ducks, geese, and swans	Tundra areas near ponds, rivers, or lakes	Spring and early summer in arctic regions
Some sea birds	Sandbars or low sand islands	Spring and early summer in temperate and arctic regions
Gulls, auks, murres, and cormorants	Steep rocky coasts	Spring and early summer in temperate and arctic regions

bat. They can carry other diseases and infections and will bite readily when handled. Taking shelter in a cave occupied by bats, however, presents the much greater hazard of inhaling powdered bat dung, or guano. Bat dung carries many organisms that can cause diseases. Eating thoroughly cooked flying foxes or other bats presents no danger from rabies and other diseases, but again, the emphasis is on thorough cooking.

TRAPS AND SNARES

For an unarmed survivor or evader, or when the sound of a rifle shot could be a problem, trapping or snaring wild game is a good alternative. Several well-placed traps have the potential to catch much more game than a man with a rifle is likely to shoot. To be effective with any type of trap or snare, you must—

- Be familiar with the species of animal you intend to catch.
- Be capable of constructing a proper trap.
- Not alarm the prey by leaving signs of your presence.

There are no catchall traps you can set for all animals. You must determine what species are in a given area and set your traps specifically with those animals in mind. Look for the following

- Runs and trails.
- Tracks.
- Droppings.
- Chewed or rubbed vegetation.
- Nesting or roosting sites.
- Feeding and watering areas.

Position your traps and snares where there is proof that animals pass through. You must determine if it is a "run" or a "trail." A trail will show signs of use by several species and will be rather distinct. A run is usually smaller and less distinct and will only contain signs of one species. You may construct a perfect

Figure 2-2: Catching birds in a net.

snare, but it will not catch anything if haphazardly placed in the woods. Animals have bedding areas, water-holes, and feeding areas with trails leading from one to another. You must place snares and traps around these areas to be effective.

For an evader in a hostile environment, trap and snare concealment is important. It is equally important, however, not to create a disturbance that will alarm the animal and cause it to avoid the trap. Therefore, if you must dig, remove all fresh dirt from the area. Most animals will instinctively avoid a pitfall-type trap. Prepare the various parts of a trap or snare away from the site, carry them in, and set them up. Such actions make it easier to avoid disturbing the local vegetation, thereby alerting the prey. Do not use freshly cut, live vegetation to construct a trap or snare. Freshly cut vegetation will "bleed" sap that has an odor the prey will be able to smell. It is an alarm signal to the animal.

You must remove or mask the human scent on and around the trap you set. Although birds do not have a developed sense of smell, nearly all mammals depend on smell even more than on sight. Even the slightest human scent on a trap will alarm the prey and cause it to avoid the area. Actually removing the scent from a trap is difficult but masking it is relatively easy. Use the fluid from the gall and urine bladders of previous kills. Do not use human urine. Mud, particularly from an area with plenty of rotting vegetation, is also good. Use it to coat your hands when handling the trap and to coat the trap when setting it. In nearly all parts of the world, animals know the smell of burned vegetation and smoke. It is only when a fire is actually burning that they become alarmed. Therefore, smoking the trap parts is an effective means to mask your scent. If one of the above techniques is not practical, and if time permits, allow a trap to weather for a few days and then set it. Do not handle a trap while it is weathering. When you position the trap, camouflage it as naturally as possible to prevent detection by the enemy and to avoid alarming the prey.

Traps or snares placed on a trail or run should use channelization. To build a channel, construct a funnel-shaped barrier extending from the sides of the trail toward the trap, with the narrowest part nearest the trap. Channelization should be inconspicuous to avoid alerting the prey. As the animal gets to the trap, it cannot turn left or right and continues into the trap. Few wild animals will back up, preferring to face the direction of travel. Channelization does not have to be an impassable barrier. You only have to make it inconvenient for the animal to go over or through the barrier. For best effect, the channelization should reduce the trail's width to just slightly wider than the targeted animal's body. Maintain this constriction at least as far back from the trap as the animal's body length, then begin the widening toward the mouth of the funnel.

Use of Bait. Baiting a trap or snare increases your chances of catching an animal. When catching fish, you must bait nearly all the devices. Success with an unbaited trap depends on its placement in a good location. A baited trap can actually draw animals to it. The bait should be something the animal knows. This bait, however, should not be so readily available in the immediate area that the animal can get it close by. For example, baiting a trap with corn in the middle of a corn field would not be likely to work. Likewise, if corn is not grown in the region, a corn-baited trap may arouse an animal's curiosity and keep it alerted while it ponders the strange food. Under such circumstances it may not go for the bait. One bait that works well on small mammals is the peanut butter from a meal, ready-to-eat (MRE) ration. Salt is also a good bait. When using such baits, scatter bits of it around the trap to give the prey a chance to sample it and develop a craving for it. The animal will then overcome some of its caution before it gets to the trap.

If you set and bait a trap for one species but another species takes the bait without being caught, try to determine what the animal was. Then set a proper trap for that animal, using the same bait.

Note: Once you have successfully trapped an animal, you will not only gain confidence in your ability, you also will have resupplied yourself with bait for several more traps.

Trap and Snare Construction. Traps and snares crush, choke, hang, or entangle the prey. A single trap or snare will commonly incorporate two or more of these principles. The mechanisms that provide power to the trap are almost always very simple. The struggling victim, the force of gravity, or a bent sapling's tension provides the power.

The heart of any trap or snare is the trigger. When planning a trap or snare, ask yourself how it should affect the prey, what is the source of power, and what will be the most efficient trigger. Your answers will help you devise a specific trap for a specific species. Traps are designed to catch and hold or to catch and kill. Snares are traps that incorporate a noose to accomplish either function.

Simple Snare. A simple snare (Figure 2-3) consists of a noose placed over a trail or den hole and attached to a firmly planted stake. If the noose is some type of cordage placed upright on a game trail, use small twigs or blades of grass to hold it up. Filaments from spider webs are excellent for holding nooses open. Make sure the noose is large enough to pass freely over the animal's head. As the animal continues to move, the noose tightens around its neck. The more the animal struggles, the tighter the noose gets. This type of snare usually does not kill the animal. If you use cordage, it may loosen enough to slip off the animal's neck. Wire is therefore the best choice for a simple snare.

Figure 2-3: Simple snare.

Drag Noose. Use a drag noose on an animal run (Figure 2-4). Place forked sticks on either side of the run and lay a sturdy crossmember across them. Tie the noose to the crossmember and hang it at a height above the animal's head. (Nooses designed to catch by the head should never be low enough for the prey to step into with a foot.) As the noose tightens around the animal's neck, the animal pulls the crossmember from the forked sticks and drags it along. The surrounding vegetation quickly catches the crossmember and the animal becomes entangled.

Twitch-Up. A twitch-up is a supple sapling, which, when bent over and secured with a triggering device, will provide power to a variety of snares. Select a hardwood sapling along the trail. A twitch-up will work much faster and with more force if you remove all the branches and foliage.

Twitch-Up Snare. A simple twitch-up snare uses two forked sticks, each with a long and short leg (Figure 2-5). Bend the twitch-up and mark the trail below it. Drive the long leg of one forked stick firmly into the ground at that point. Ensure the cut on the short leg of this stick is parallel to the ground. Tie the long leg of the remaining forked stick to a piece of cordage secured to the twitch-up. Cut the short leg so that it catches on the short leg of the other forked stick. Extend a noose over the trail. Set the trap by bending the twitch-up and engaging the short legs of the forked sticks. When an animal catches its head in the noose, it pulls the forked sticks apart, allowing the twitch-up to spring up and hang the prey.

Note: Do not use green sticks for the trigger. The sap that oozes out could glue them together.

Squirrel Pole. A squirrel pole is a long pole placed against a tree in an area showing a lot of squirrel activity (Figure 2-6). Place several wire nooses along the top and sides of the pole so that a squirrel trying to go up or down the pole will have to pass through one or more of them. Position the nooses (5 to 6 centimeters in diameter) about 2.5 centimeters off the pole. Place the top and bottom wire nooses 45 centimeters from the top and bottom of the pole to prevent the squirrel from getting its feet on a solid surface. If this happens, the squirrel will chew through the wire. Squirrels are naturally curious. After an initial period of caution, they will try to go up or down the pole and will get caught in a noose. The struggling animal will soon fall

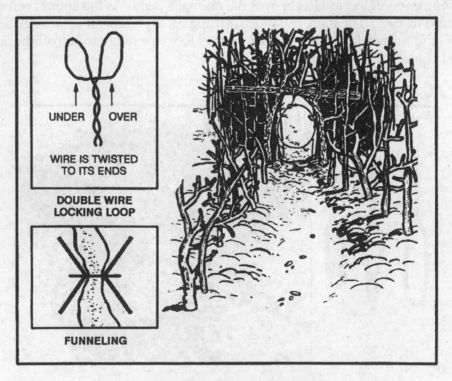

Figure 2-4: Drag noose.

Figure 2-5: Twitch-up snare.

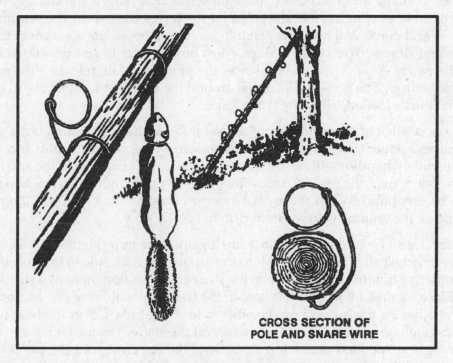

CROSS SECTION OF POLE AND SNARE WIRE

Figure 2-6: Squirrel pole.

from the pole and strangle. Other squirrels will soon follow and, in this way, you can catch several squirrels. You can emplace multiple poles to increase the catch.

Ojibwa Bird Pole. An Ojibwa bird pole is a snare used by native Americans for centuries (Figure 2-7). To be effective, place it in a relatively open area away from tall trees. For best results, pick a spot near feeding areas, dusting areas, or watering holes. Cut a pole 1.8 to 2.1 meters long and trim away all limbs and foliage. Do not use resinous wood such as pine. Sharpen the upper end to a point, then drill a small diameter hole 5 to 7.5 centimeters down from the top. Cut a small stick 10 to 15 centimeters long and shape one end so that it will almost fit into the hole. This is the perch. Plant the long pole in the ground with the pointed end up. Tie a small weight, about equal to the weight of the targeted species, to a length of cordage. Pass the free end of the cordage through the hole, and tie a slip noose that covers the perch. Tie a single

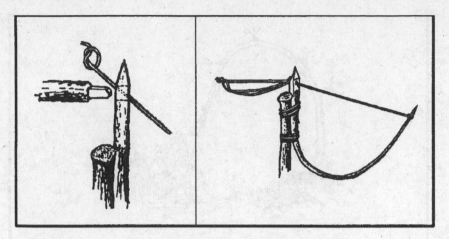

Figure 2-7: Ojibwa bird pole.

overhand knot in the cordage and place the perch against the hole. Allow the cordage to slip through the hole until the overhand knot rests against the pole and the top of the perch. The tension of the overhand knot against the pole and perch will hold the perch in position. Spread the noose over the perch, ensuring it covers the perch and drapes over on both sides. Most birds prefer to rest on something above ground and will land on the perch. As soon as the bird lands, the perch will fall, releasing the overhand knot and allowing the weight to drop. The noose will tighten around the bird's feet, capturing it. If the weight is too heavy, it will cut the bird's feet off, allowing it to escape.

Noosing Wand. A noose stick or "noosing wand" is useful for capturing roosting birds or small mammals (Figure 2-8). It requires a patient operator. This wand is more a weapon than a trap. It consists of a pole (as long as you can effectively handle) with a slip noose of wire or stiff cordage at the small end. To catch an animal, you slip the noose over the neck of a roosting bird and pull it tight. You can also place it over a den hole and hide in a nearby blind. When the animal emerges from the den, you jerk the pole to tighten the noose and thus capture the animal. Carry a stout club to kill the prey.

Treadle Spring Snare. Use a treadle snare against small game on a trail (Figure 2-9). Dig a shallow hole in the trail. Then drive a forked stick (fork down) into the ground on each side of the hole on the same side of the trail. Select two fairly straight sticks that span the two forks. Position these two sticks so that their ends engage the forks. Place several sticks over the hole in the trail by positioning one end over the lower horizontal stick and the other on the ground on the other side of the hole. Cover the hole with enough sticks so that the prey must step on at least one of them to set off the snare. Tie one end of a piece of cordage to a twitch-up or to a weight suspended over a tree limb. Bend the twitch-up or raise the suspended weight to determine where you will tie a 5 centimeter or so long trigger. Form a noose with the other end of the cordage. Route and spread the noose over the top of the sticks over the hole. Place the trigger stick against the horizontal sticks and route the cordage behind the sticks so that the tension of the power source will hold it in place. Adjust the bottom horizontal stick so that it will barely hold against the trigger. As the animal places its foot on a stick across the hole, the bottom horizontal stick moves down, releasing the trigger and

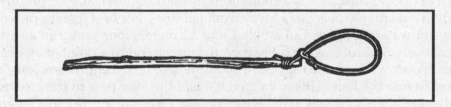

Figure 2-8: Noosing wand.

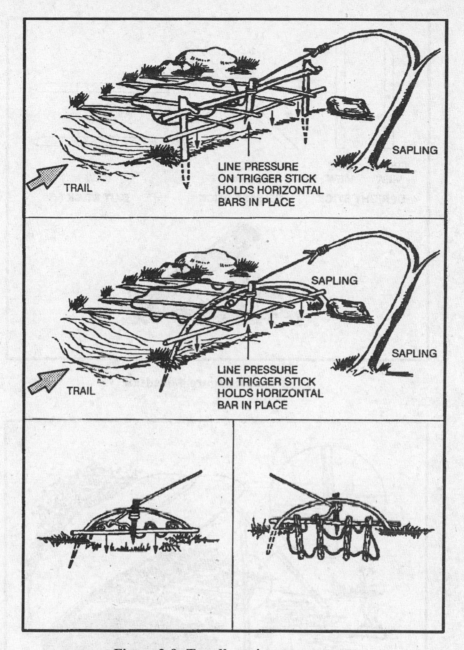

Figure 2-9: Treadle spring snare.

allowing the noose to catch the animal by the foot. Because of the disturbance on the trail, an animal will be wary. You must therefore use channelization.

Figure 4 Deadfall. The figure 4 is a trigger used to drop a weight onto a prey and crush it (Figure 2-10). The type of weight used may vary, but it should be heavy enough to kill or incapacitate the prey immediately. Construct the figure 4 using three notched sticks. These notches hold the sticks together in a figure 4 pattern when under tension. Practice making this trigger beforehand; it requires close tolerances and precise angles in its construction.

Paiute Deadfall. The Paiute deadfall is similar to the figure 4 but uses a piece of cordage and a catch stick (Figure 2-11). It has the advantage of being easier to set than the figure 4. Tie one end of a piece of cordage to the lower end of the diagonal stick. Tie the other end of the cordage to another stick about 5 centimeters

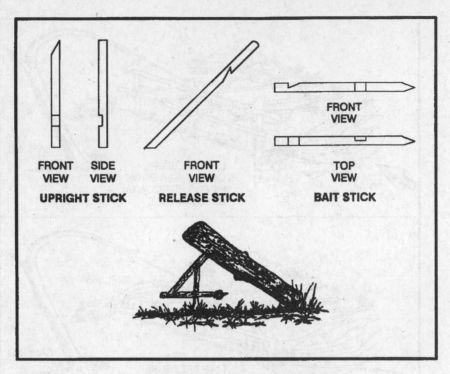

Figure 2-10: Figure 4 deadfall.

Figure 2-11: Paiute deadfall.

long. This 5-centimeter stick is the catch stick. Bring the cord halfway around the vertical stick with the catch stick at a 90-degree angle. Place the bait stick with one end against the drop weight, or a peg driven into the ground, and the other against the catch stick. When a prey disturbs the bait stick, it falls free, releasing the catch stick. As the diagonal stick flies up, the weight falls, crushing the prey.

Bow Trap. A bow trap is one of the deadliest traps. It is dangerous to man as well as animals (Figure 2-12). To construct this trap, build a bow and anchor it to the ground with pegs. Adjust the aiming point as you

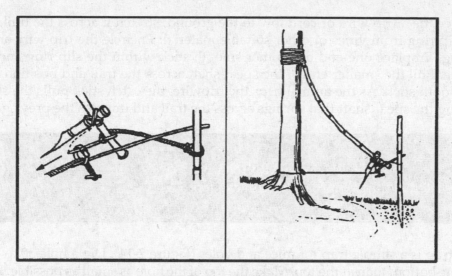

Figure 2-12: Bow trap.

anchor the bow. Lash a toggle stick to the trigger stick. Two upright sticks driven into the ground hold the trigger stick in place at a point where the toggle stick will engage the pulled bow string. Place a catch stick between the toggle stick and a stake driven into the ground. Tie a tripwire or cordage to the catch stick and route it around stakes and across the game trail where you tie it off (as in Figure 2-12). When the prey trips the trip wire, the bow looses an arrow into it. A notch in the bow serves to help aim the arrow.

WARNING
This is a lethal trap. Approach it with caution and from the rear only!

Pig Spear Shaft. To construct the pig spear shaft, select a stout pole about 2.5 meters long (Figure 2-13). At the smaller end, firmly lash several small stakes. Lash the large end tightly to a tree along the game trail. Tie a length of cordage to another tree across the trail. Tie a sturdy, smooth stick to the other end of the

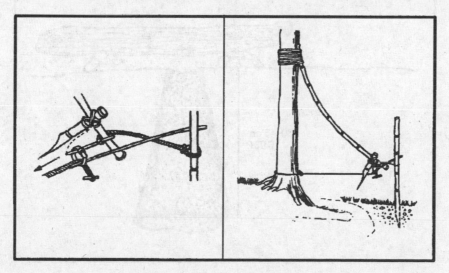

Figure 2-13: Pig spear shaft.

cord. From the first tree, tie a trip wire or cord low to the ground, stretch it across the trail, and tie it to a catch stick. Make a slip ring from vines or other suitable material. Encircle the trip wire and the smooth stick with the slip ring. Emplace one end of another smooth stick within the slip ring and its other end against the second tree. Pull the smaller end of the spear shaft across the trail and position it between the short cord and the smooth stick. As the animal trips the tripwire, the catch stick pulls the slip ring off the smooth sticks, releasing the spear shaft that springs across the trail and impales the prey against the tree.

WARNING
This is a lethal trap. Approach it with caution!

Bottle Trap. A bottle trap is a simple trap for mice and voles (Figure 2-14). Dig a hole 30 to 45 centimeters deep that is wider at the bottom than at the top. Make the top of the hole as small as possible. Place a piece of bark or wood over the hole with small stones under it to hold it up 2.5 to 5 centimeters off the ground. Mice or voles will hide under the cover to escape danger and fall into the hole. They cannot climb out because of the wall's backward slope. Use caution when checking this trap; it is an excellent hiding place for snakes.

KILLING DEVICES

There are several killing devices that you can construct to help you obtain small game to help you survive. The rabbit stick, the spear, the bow and arrow, and the sling are such devices.

Rabbit Stick. One of the simplest and most effective killing devices is a stout stick as long as your arm, from fingertip to shoulder, called a "rabbit stick." You can throw it either overhand or sidearm and with considerable force. It is very effective against small game that stops and freezes as a defense.

Spear. You can make a spear to kill small game and to fish. Jab with the spear, do not throw it. See Figure 2-20.

Bow and Arrow. A good bow is the result of many hours of work. You can construct a suitable short-term bow fairly easily. When it loses its spring or breaks, you can replace it. Select a hardwood stick about one meter long that is free of knots or limbs. Carefully scrape the large end down until it has the same pull as

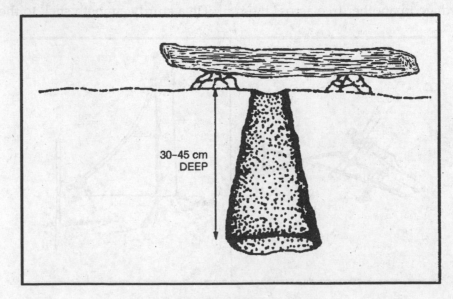

Figure 2-14: Bottle trap.

the small end. Careful examination will show the natural curve of the stick. Always scrape from the side that faces you, or the bow will break the first time you pull it. Dead, dry wood is preferable to green wood. To increase the pull, lash a second bow to the first, front to front, forming an "X" when viewed from the side. Attach the tips of the bows with cordage and only use a bowstring on one bow.

Select arrows from the straightest dry sticks available. The arrows should be about half as long as the bow. Scrape each shaft smooth all around. You will probably have to straighten the shaft. You can bend an arrow straight by heating the shaft over hot coals. Do not allow the shaft to scorch or burn. Hold the shaft straight until it cools.

You can make arrowheads from bone, glass, metal, or pieces of rock. You can also sharpen and fire harden the end of the shaft. To fire harden wood, hold it over hot coals, being careful not to burn or scorch the wood.

You must notch the ends of the arrows for the bowstring. Cut or file the notch; do not split it. Fletching (adding feathers to the notched end of an arrow) improves the arrow's flight characteristics, but is not necessary on a field-expedient arrow.

Sling. You can make a sling by tying two pieces of cordage, about sixty centimeters long, at opposite ends of a palm-sized piece of leather or cloth. Place a rock in the cloth and wrap one cord around the middle finger and hold in your palm. Hold the other cord between the forefinger and thumb. To throw the rock, spin the sling several times in a circle and release the cord between the thumb and forefinger. Practice to gain proficiency. The sling is very effective against small game. See Part V, Chapter 2 for more information about slings and arrows of outrageous fortune.

FISHING DEVICES

You can make your own fishhooks, nets and traps and use several methods to obtain fish in a survival situation.

Improvised Fishhooks. You can make field-expedient fishhooks from pins, needles, wire, small nails, or any piece of metal. You can also use wood, bone, coconut shell, thorns, flint, seashell, or tortoise shell. You can also make fish hooks from any combination of these items (Figure 2-15).

To make a wooden hook, cut a piece of hardwood about 2.5 centimeters long and about 6 millimeters in diameter to form the shank. Cut a notch in one end in which to place the point. Place the point (piece of bone, wire, nail) in the notch. Hold the point in the notch and tie securely so that it does not move out of position. This is a fairly large hook. To make smaller hooks, use smaller material.

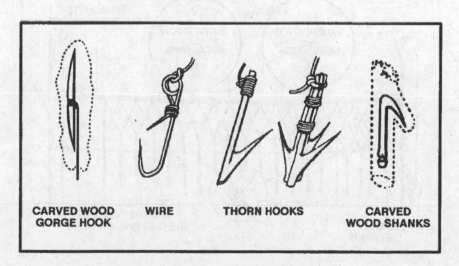

CARVED WOOD   WIRE   THORN HOOKS   CARVED
GORGE HOOK   WOOD SHANKS

Figure 2-15: Improvised fishhooks.

A gorge is a small shaft of wood, bone, metal, or other material. It is sharp on both ends and notched in the middle where you tie cordage. Bait the gorge by placing a piece of bait on it lengthwise. When the fish swallows the bait, it also swallows the gorge.

Stakeout. A stakeout is a fishing device you can use in a hostile environment (Figure 2-16). To construct a stakeout, drive two supple saplings into the bottom of the lake, pond, or stream with their tops just below the water surface. Tie a cord between them and slightly below the surface. Tie two short cords with hooks or gorges to this cord, ensuring that they cannot wrap around the poles or each other. They should also not slip along the long cord. Bait the hooks or gorges.

Gill Net. If a gill net is not available, you can make one using parachute suspension line or similar material (Figure 2-17). Remove the core lines from the suspension line and tie the easing between two trees. Attach several core lines to the easing by doubling them over and tying them with prusik knots or girth hitches. The length of the desired net and the size of the mesh determine the number of core lines used and the space between them. Starting at one end of the easing, tie the second and the third core lines together using an overhand knot. Then tie the fourth and fifth, sixth and seventh, and so on, until you reach the last core

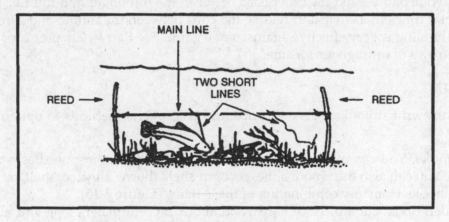

Figure 2-16: Stakeout.

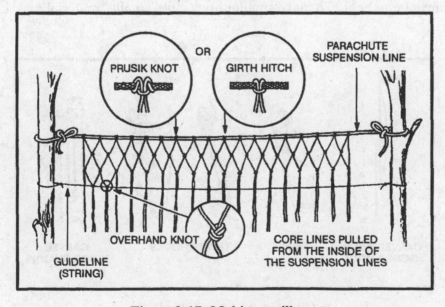

Figure 2-17: Making a gill net.

line. You should now have all core lines tied in pairs with a single core line hanging at each end. Start the second row with the first core line, tie it to the second, the third to the fourth, and so on.

To keep the rows even and to regulate the size of the mesh, tie a guideline to the trees. Position the guideline on the opposite side of the net you are working on. Move the guideline down after completing each row. The lines will always hang in pairs and you always tie a cord from one pair to a cord from an adjoining pair. Continue tying rows until the net is the desired width. Thread a suspension line easing along the bottom of the net to strengthen it. Use the gill net as shown in Figure 2-18.

Fish Traps. You may trap fish using several methods (Figure 2-19). Fish baskets are one method. You construct them by lashing several sticks together with vines into a funnel shape. You close the top, leaving a hole large enough for the fish to swim through.

You can also use traps to catch saltwater fish, as schools regularly approach the shore with the incoming tide and often move parallel to the shore. Pick a location at high tide and build the trap at low tide. On rocky shores, use natural rock pools. On coral islands, use natural pools on the surface of reefs by blocking the openings as the tide recedes. On sandy shores, use sandbars and the ditches they enclose. Build the trap as a low stone wall extending outward into the water and forming an angle with the shore.

Spearfishing. If you are near shallow water (about waist deep) where the fish are large and plentiful, you can spear them. To make a spear, cut a long, straight sapling (Figure 2-20). Sharpen the end to a point or attach a knife, jagged piece of bone, or sharpened metal. You can also make a spear by splitting the shaft a few inches down from the end and inserting a piece of wood to act as a spreader. You then sharpen the two separated halves to points. To spear fish, find an area where fish either gather or where there is a fish run. Place the spear point into the water and slowly move it toward the fish. Then, with a sudden push, impale the fish on the stream bottom. Do not try to lift the fish with the spear, as it will probably slip off and you will lose it; hold the spear with one hand and grab and hold the fish with the other. Do not throw the spear, especially if the point is a knife. You cannot afford to lose a knife in a survival situation. Be alert to the problems caused by light refraction when looking at objects in the water.

Figure 2-18: Setting a gill net in the stream.

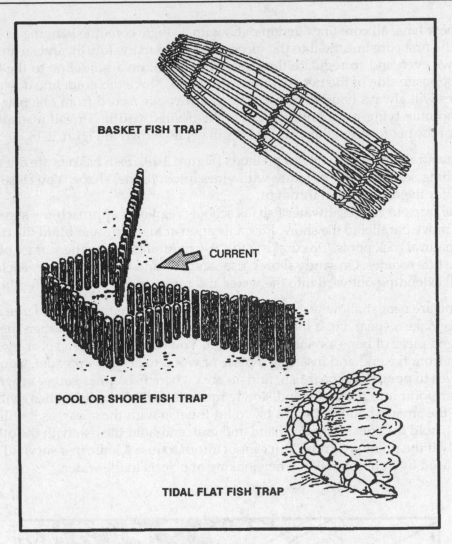

Figure 2-19: Various types of fish traps.

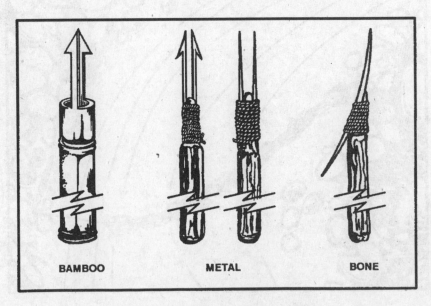

Figure 2-20: Types of spear points.

Chop Fishing. At night, in an area with a good fish density, you can use a light to attract fish. Then, armed with a machete or similar weapon, you can gather fish using the back side of the blade to strike them. Do not use the sharp side as you will cut them in two pieces and end up losing some of the fish.

Fish Poison. Another way to catch fish is by using poison. Poison works quickly. It allows you to remain concealed while it takes effect. It also enables you to catch several fish at one time. When using fish poison, be sure to gather all of the affected fish, because many dead fish floating downstream could arouse suspicion. Some plants that grow in warm regions of the world contain rotenone, a substance that stuns or kills cold-blooded animals but does not harm persons who eat the animals. The best place to use rotenone, or rotenone-producing plants, is in ponds or the headwaiters of small streams containing fish. Rotenone works quickly on fish in water 21 degrees C (70 degrees F) or above. The fish rise helplessly to the surface. It works slowly in water 10 to 21 degrees C (50 to 70 degrees F) and is ineffective in water below 10 degrees C (50 degrees F). The plants in Figure 2-21, used as indicated, will stun or kill fish:

- *Anamirta cocculus:* This woody vine grows in southern Asia and on islands of the South Pacific. Crush the bean-shaped seeds and throw them in the water.
- *Croton tiglium:* This shrub or small tree grows in waste areas on islands of the South Pacific. It bears seeds in three angled capsules. Crush the seeds and throw them into the water.
- *Barringtonia:* These large trees grow near the sea in Malaya and parts of Polynesia. They bear a fleshy one-seeded fruit. Crush the seeds and bark and throw into the water.
- *Derris eliptica:* This large genus of tropical shrubs and woody vines is the main source of commercially produced rotenone. Grind the roots into a powder and mix with water. Throw a large quantity of the mixture into the water.
- *Duboisia:* This shrub grows in Australia and bears white clusters of flowers and berrylike fruit. Crush the plants and throw them into the water.
- *Tephrosia:* This species of small shrubs, which bears beanlike pods, grows throughout the tropics. Crush or bruise bundles of leaves and stems and throw them into the water.
- *Lime:* You can get lime from commercial sources and in agricultural areas that use large quantities of it. You may produce your own by burning coral or seashells. Throw the lime into the water.
- *Nut husks:* Crush green husks from butternuts or black walnuts. Throw the husks into the water.

PREPARATION OF FISH AND GAME FOR COOKING AND STORAGE

You must know how to prepare fish and game for cooking and storage in a survival situation. Improper cleaning or storage can result in inedible fish or game.

Fish. Do not eat fish that appears spoiled. Cooking does not ensure that spoiled fish will be edible. Signs of spoilage are—

- Sunken eyes.
- Peculiar odor.
- Suspicious color. (Gills should be red to pink. Scales should be a pronounced shade of gray, not faded.)
- Dents stay in the fish's flesh after pressing it with your thumb.
- Slimy, rather than moist or wet body.
- Sharp or peppery taste.

Eating spoiled or rotten fish may cause diarrhea, nausea, cramps, vomiting, itching, paralysis, or a metallic taste in the mouth. These symptoms appear suddenly, one to six hours after eating. Induce vomiting if symptoms appear.

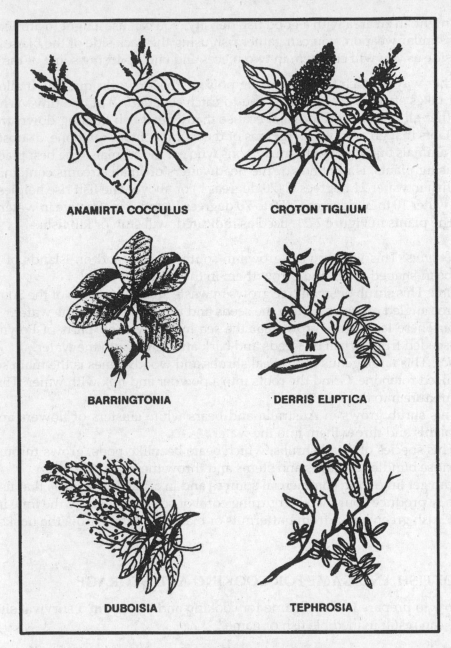

Figure 2-21: Fish-poisoning plants.

Fish spoils quickly after death, especially on a hot day. Prepare fish for eating as soon as possible after catching it. Cut out the gills and large blood vessels that lie near the spine. Gut fish that is more than 10 centimeters long. Scale or skin the fish.

You can impale a whole fish on a stick and cook it over an open fire. However, boiling the fish with the skin on is the best way to get the most food value. The fats and oil are under the skin and, by boiling, you can save the juices for broth. You can use any of the methods used to cook plant food to cook fish. Pack fish into a ball of clay and bury it in the coals of a fire until the clay hardens. Break open the clay ball to get to the cooked fish. Fish is done when the meat flakes off. If you plan to keep the fish for later, smoke or fry it. To prepare fish for smoking, cut off the head and remove the backbone.

Snakes. To skin a snake, first cut off its head and bury it. Then cut the skin down the body 15 to 20 centimeters (Figure 2-22). Peel the skin back, then grasp the skin in one hand and the body in the other and pull

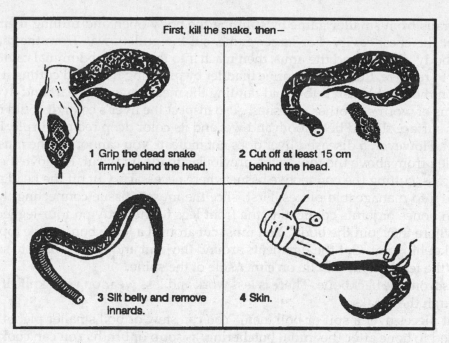

Figure 2-22: Cleaning a snake.

apart. On large, bulky snakes it may be necessary to slit the belly skin. Cook snakes in the same manner as small game. Remove the entrails and discard. Cut the snake into small sections and boil or roast it.

Birds. After killing the bird, remove its feathers by either plucking or skinning. Remember, skinning removes some of the food value. Open up the body cavity and remove its entrails, saving the craw (in seed-eating birds), heart, and liver. Cut off the feet. Cook by boiling or roasting over a spit. Before cooking scavenger birds, boil them at least 20 minutes to kill parasites.

Skinning and Butchering Game. Bleed the animal by cutting its throat. If possible, clean the carcass near a stream. Place the carcass belly up and split the hide from throat to tail, cutting around all sexual organs (Figure 2-23). Remove the musk glands at points A and B to avoid tainting the meat. For smaller mammals, cut the hide around the body and insert two fingers under the hide on both sides of the cut and pull both pieces off (Figure 2-24).

Note: When cutting the hide, insert the knife blade under the skin and turn the blade up so that only the hide gets cut. This will also prevent cutting hair and getting it on the meat.

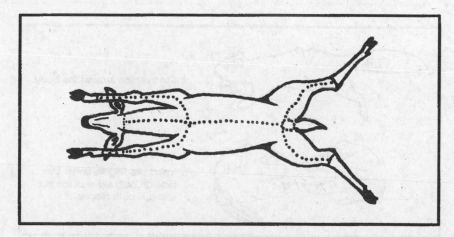

Figure 2-23: Skinning and butchering large game.

Remove the entrails from smaller game by splitting the body open and pulling them out with the fingers. Do not forget the chest cavity. For larger game, cut the gullet away from the diaphragm. Roll the entrails out of the body. Cut around the anus, then reach into the lower abdominal cavity, grasp the lower intestine, and pull to remove. Remove the urine bladder by pinching it off and cutting it below the fingers. If you spill urine on the meat, wash it to avoid tainting the meat. Save the heart and liver. Cut these open and inspect for signs of worms or other parasites. Also inspect the liver's color; it could indicate a diseased animal. The liver's surface should be smooth and wet and its color deep red or purple. If the liver appears diseased, discard it. However, a diseased liver does not indicate you cannot eat the muscle tissue.

Cut along each leg from above the foot to the previously made body cut. Remove the hide by pulling it away from the carcass, cutting the connective tissue where necessary. Cut off the head and feet.

Cut larger game into manageable pieces. First, slice the muscle tissue connecting the front legs to the body. There are no bones or joints connecting the front legs to the body on four-legged animals. Cut the hindquarters off where they join the body. You must cut around a large bone at the top of the leg and cut to the ball and socket hip joint. Cut the ligaments around the joint and bend it back to separate it. Remove the large muscles (the tenderloin) that lie on either side of the spine.

Separate the ribs from the backbone. There is less work and less wear on your knife if you break the ribs first, then cut through the breaks.

Cook large meat pieces over a spit or boil them. You can stew or boil smaller pieces, particularly those that remain attached to bone after the initial butchering, as soup or broth. You can cook body organs such as the heart, liver, pancreas, spleen, and kidneys using the same methods as for muscle meat. You can also cook and eat the brain. Cut the tongue out, skin it, boil it until tender, and eat it.

Smoking Meat. To smoke meat, prepare an enclosure around a fire (Figure 2-25). Two ponchos snapped together will work. The fire does not need to be big or hot. The intent is to produce smoke, not heat. Do not use resinous wood in the fire because its smoke will ruin the meat. Use hardwoods to produce good smoke. The wood should be somewhat green. If it is too dry, soak it. Cut the meat into thin slices, no more than 6 centimeters thick, and drape them over a framework. Make sure none of the meat touches another piece. Keep the poncho enclosure around the meat to hold the smoke and keep a close watch on the fire. Do not let the fire get too hot. Meat smoked overnight in this manner will last about 1 week. Two days of continuous smoking will preserve the meat for 2 to 4 weeks. Properly smoked meat will look like a dark, curled, brittle stick and you can eat it without further cooking. You can also use a pit to smoke meat (Figure 2-26).

Drying Meat. To preserve meat by drying, cut it into 6-millimeter strips with the grain. Hang the meat strips on a rack in a sunny location with good air flow. Keep the strips out of the reach of animals and cover them to keep blowflies off. Allow the meat to dry thoroughly before eating. Properly dried meat will have a dry, crisp texture and will not feel cool to the touch.

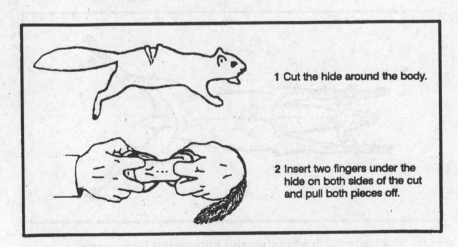

1 Cut the hide around the body.

2 Insert two fingers under the hide on both sides of the cut and pull both pieces off.

Figure 2-24: Skinning small game.

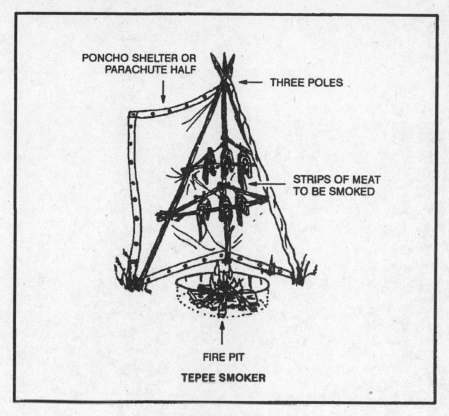

Figure 2-25: Smoking meat.

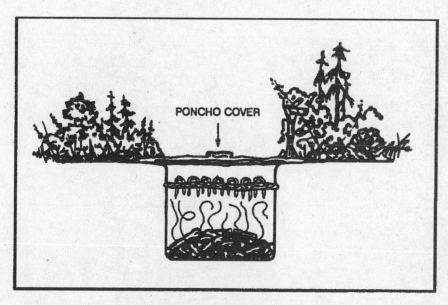

Figure 2-26: Smoking meat over a pit.

Other Preservation Methods. You can also preserve meats using the freezing or brine and salt methods.

Freezing. In cold climates, you can freeze and keep meat indefinitely. Freezing is not a means of preparing meat. You must still cook it before eating.

Brine and Salt. You can preserve meat by soaking it thoroughly in a saltwater solution. The solution must cover the meat. You can also use salt by itself. Wash off the salt before cooking.

CHAPTER 3

Dangerous Insects and Arachnids

Insects are often overlooked as a danger to the survivor. More people in the United States die each year from bee stings, and resulting anaphylactic shock, than from snake bites. A few other insects are venomous enough to kill, but often the greatest danger is the transmission of disease.

> ⚠ CAUTION
>
> Scorpions sting with their tails, causing local pain, swelling, possible incapacitation, and death.

Scorpion
Scorpionidae order

Description: Dull brown, yellow, or black. Have 7.5- to 20-centimeter-long lobsterlike pincers and jointed tail usually held over the back. There are 800 species of scorpions.

Habitat: Decaying matter, under debris, logs, and rocks. Feeds at night. Sometimes hides in boots.

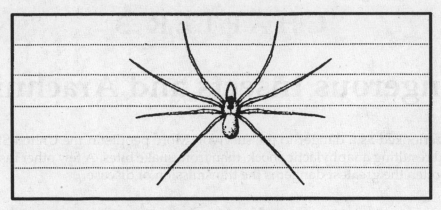

Brown house spider or brown recluse spider
Laxosceles reclusa

Description: Brown to black with obvious "fiddle" on back of head and thorax. Chunky body with long, slim legs 2.5 to 4 centimeters long.

Habitat: Under debris, rocks, and logs. In caves and dark places.

Distribution: North America.

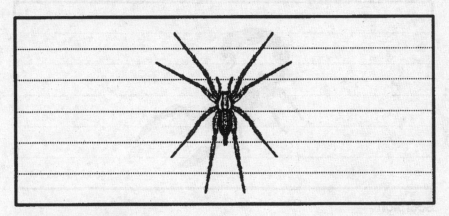

Funnelweb spider
Atrax species (A. robustus, A. formidablis)

Description: Large, brown, bulky spiders. Aggressive when disturbed.

Habitat: Woods, jungles, and brushy areas. Web has funnellike opening.

Distribution: Australia. (Other nonvenomous species worldwide.)

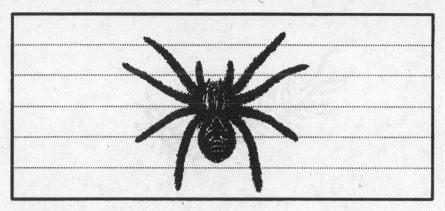

Tarantula
Theraphosidae and *Lycosa* species

Description: Very large, brown, black, reddish hairy spiders. Large fangs inflict painful bite.

Habitat: Desert areas, tropics.

Distribution: Americas, southern Europe.

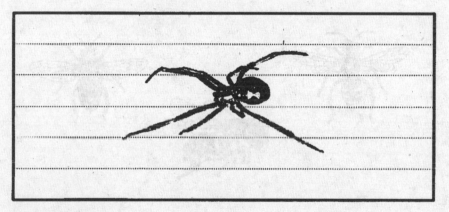

Widow spider
Latrodectus species

Description: Dark spiders with light red or orange markings on female's abdomen.

Habitat: Under logs, rocks, and debris. In shaded places.

Distribution: Varied species worldwide. Black widow in United States, red widow in Middle East, and brown widow in Australia.

Note: Females are the poisonous gender. Red widow in the Middle East is the only spider known to be deadly to man.

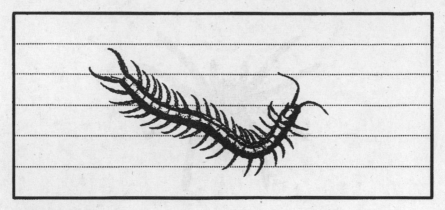

Centipede

Description: Multijointed body to 30 centimeters long. Dull orange to brown, with black point eyes at base of antennae. There are 2,800 species worldwide.

Habitat: Under bark and stones by day. Active at night.

Distribution: Worldwide.

Bee

Description: Insect with brown or black, thick, hairy bodies. Generally found in colonies. Many build wax combs.

Habitat: Hollow trees, caves, dwellings. Near water in desert areas.

Distribution: Worldwide.

Note: Bees have barbed stingers and die after stinging because their venom sac and internal organs are pulled out during the attack.

Tick

Description: Round body from size of pinhead to 2.5 centimeters. Has 8 legs and sucking mouth parts. There are 850 species worldwide.

Habitat: Mainly in forests and grasslands. Also in urban areas and farmlands.

Distribution: Worldwide.

CHAPTER 4

Poisonous Snakes and Lizards

If you fear snakes, it is probably because you are unfamiliar with them or you have wrong information about them. There is no need for you to fear snakes if you know—

- Their habits.
- How to identify the dangerous kinds.
- Precautions to take to prevent snakebite.
- What actions to take in case of snakebite.

For a man wearing shoes and trousers and living in a camp, the danger of being bitten by a poisonous snake is small compared to the hazards of malaria, cholera, dysentery, or other diseases.

Nearly all snakes avoid man if possible. Reportedly, however, a few—the king cobra of Southeast Asia, the bushmaster and tropical rattlesnake of South America, and the mamba of Africa—sometimes aggressively attack man, but even these snakes do so only occasionally. Most snakes get out of the way and are seldom seen.

WAYS TO AVOID SNAKEBITE

Snakes are widely distributed. They are found in all tropical, subtropical, and most temperate regions. Some species of snakes have specialized glands that contain a toxic venom and long hollow fangs to inject their venom.

Poisonous Snakes of the Americas

- American Copperhead (*Agkistrodon contortrix*)
- Bushmaster (*Lachesis mutus*)
- Coral snake (*Micrurus fulvius*)
- Cottonmouth (*Agkistrodon piscivorus*)
- Fer-de-lance (*Bothrops atrox*)
- Rattlesnake (*Crotalus* species)

Poisonous Snakes of Europe

- Common adder (*Vipers berus*)
- Pallas' viper (*Agkistrodon halys*)

Poisonous Snakes of Africa and Asia

- Boomslang (*Dispholidus typus*)
- Cobra (*Naja* species)
- Gaboon viper (*Bitis gabonica*)
- Green tree pit viper (*Trimeresurus gramineus*)
- Habu pit viper (*Trimeresurus flavoviridis*)
- Krait (*Bungarus caeruleus*)

- Malayan pit viper (*Callaselasma rhodostoma*)
- Mamba (*Dendraspis species*)
- Puff adder (*Bitis arietans*)
- Rhinoceros viper (*Bitis nasicornis*)
- Russell's viper (*Vipera russellii*)
- Sand viper (*Cerastes vipera*)
- Saw-scaled viper (*Echis carinatus*)
- Wagler's pit viper (*Trimeresurus wagleri*)

Poisonous Snakes of Australasia

- Death adder (*Acanthophis antarcticus*)
- Taipan (Oxyuranus scutellatus)
- Tiger snake (*Notechis scutatus*)
- Yellow-bellied sea snake (*Pelamis platurus*)

The polar regions are free of snakes due to their inhospitable environments. Other areas considered to be free of poisonous snakes are New Zealand, Cuba, Haiti, Jamaica, Puerto Rico, Ireland, Polynesia, and Hawaii.

There are no infallible rules for expedient identification of poisonous snakes in the field, because the guidelines all require close observation or manipulation of the snake's body. The best strategy is to leave all snakes alone. Where snakes are plentiful and poisonous species are present, the risk of their bites negates their food value.

Although venomous snakes use their venom to secure food, they also use it for self-defense. Human accidents occur when you don't see or hear the snake, when you step on them, or when you walk too close to them.

Follow these simple rules to reduce the chance of accidental snakebite:

- Don't sleep next to brush, tall grass, large boulders, or trees. They provide hiding places for snakes. Place your sleeping bag in a clearing. Use mosquito netting tucked well under the bag. This netting should provide a good barrier.
- Don't put your hands into dark places, such as rock crevices, heavy brush, or hollow logs, without first investigating.
- Don't step over a fallen tree. Step on the log and look to see if there is a snake resting on the other side.
- Don't walk through heavy brush or tall grass without looking down. Look where you are walking.
- Don't pick up any snake unless you are absolutely positive it is not venomous.
- Don't pick up freshly killed snakes without first severing the head. The nervous system may still be active and a dead snake can deliver a bite.

SNAKE GROUPS

Snakes dangerous to man usually fall into two groups: proteroglypha and solenoglypha. Their fangs and their venom best describe these two groups (Table 4-1).

Fangs. The proteroglypha have, in front of the upper jaw and preceding the ordinary teeth, permanently erect fangs. These fangs are called fixed fangs.

The solenoglypha have erectile fangs; that is, fangs they can raise to an erect position. These fangs are called folded fangs.

Table 4-1: Snake group characteristics.

Group	Fang Type	Venom Type
Proteroglypha	Fixed	Usually dominant neurotoxic
Solenoglypha	Folded	Usually dominant hemotoxic

Venom. The fixed-fang snakes (proteroglypha) usually have neurotoxic venoms. These venoms affect the nervous system, making the victim unable to breathe.

The folded-fang snakes (solenoglypha) usually have hemotoxic venoms. These venoms affect the circulatory system, destroying blood cells, damaging skin tissues, and causing internal hemorrhaging.

Remember, however, that most poisonous snakes have both neurotoxic and hemotoxic venom. Usually one type of venom in the snake is dominant and the other is weak.

Poisonous Versus Nonpoisonous Snakes. No single characteristic distinguishes a poisonous snake from a harmless one except the presence of poison fangs and glands. Only in dead specimens can you determine the presence of these fangs and glands without danger.

DESCRIPTIONS OF POISONOUS SNAKES

There are many different poisonous snakes throughout the world. It is unlikely you will see many except in a zoo. This manual describes only a few poisonous snakes. You should, however, be able to spot a poisonous snake if you—

- Learn about the two groups of snakes and the families in which they fall (Table 4-2).
- Examine the pictures and read the descriptions of snakes in this appendix.

Viperidae. The viperidae or true vipers usually have thick bodies and heads that are much wider than their necks (Figure 4-1). However, there are many different sizes, markings, and colorations.

This snake group has developed a highly sophisticated means for delivering venom. They have long, hollow fangs that perform like hypodermic needles. They deliver their venom deep into the wound.

The fangs of this group of snakes are movable. These snakes fold their fangs into the roof of their mouths. When they strike, their fangs come forward, stabbing the victim. The snake controls the movement of its fangs; fang movement is not automatic. The venom is usually hemotoxic. There are, however, several species that have large quantities of neurotoxic elements, thus making them even more dangerous. The vipers are responsible for many human fatalities around the world.

Crotalidae. The crotalids, or pit vipers (Figure 4-2), may be either slender or thick-bodied. Their heads are usually much wider than their necks. These snakes take their name from the deep pit located between the eye and the nostril. They are commonly brown with dark blotches, though some kinds are green.

Rattlesnakes, copperheads, cottonmouths, and several species of dangerous snakes from Central and South America, Asia, China, and India fall into the pit viper group. The pit is a highly sensitive organ capable of picking up the slightest temperature variance. Most pit vipers are nocturnal. They hunt for food at night with the aid of these specialized pits that let them locate prey in total darkness. Rattlesnakes are the only pit vipers that possess a rattle at the tip of the tail.

India has about 12 species of these snakes. You find them in trees or on the ground in all types of terrain. The tree snakes are slender; the ground snakes are heavy-bodied. All are dangerous.

China has a pit viper similar to the cottonmouth found in North America. You find it in the rocky areas of the remote mountains of South China. It reaches a length of 1.4 meters but is not vicious unless irritated.

Table 4-2: Clinical effects of snake bites.

Group	Family	Local Effects	Systemic Effects
Solenoglypha *Usually dominant* **hemotoxic** *venom affecting the circulatory system*	Viperidae *True vipers with movable front fangs*	Strong pain, swelling, necrosis	Hemorrhaging, internal organ breakdown, destroying of blood cells
	Crotalidae *Pit vipers with movable front fangs*		
	Trimeresurus		
Proteroglypha *Usually dominant* **neurotoxic** *venom affecting the nervous system*	Elapidae *Fixed front fangs*		
	Cobra	Various pains, swelling, necrosis	Respiratory collapse
	Krait	No local effects	Respiratory collapse
	Micrurus	Little or no pain; no local symptoms	Respiratory collapse
	Laticaudinae and Hydrophidae *Ocean-living with fixed front fangs*	Pain and local swelling	Respiratory collapse

Note: The venom of the Gaboon viper, the rhinoceros viper, the tropical rattlesnake, and the Mojave rattlesnake is both strongly hemotoxic and strongly neurotoxic.

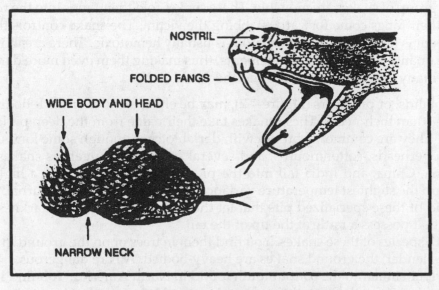

Figure 4-1: Positive identification of vipers.

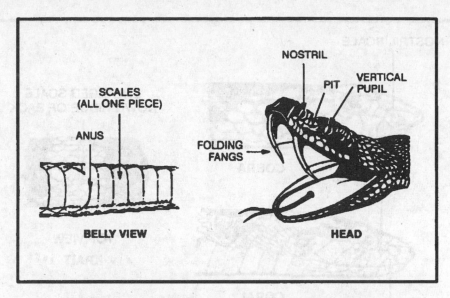

Figure 4-2: Positive identification of pit vipers.

You can also find a small pit viper, about 45 centimeters long, on the plains of eastern China. It is too small to be dangerous to a man wearing shoes.

There are about 27 species of rattlesnakes in the United States and Mexico. They vary in color and may or may not have spots or blotches. Some are small while others, such as the diamondbacks, may grow to 2.5 meters long.

There are five kinds of rattlesnakes in Central and South America, but only the tropical rattlesnake is widely distributed. The rattle on the tip of the tail is sufficient identification for a rattlesnake.

Most will try to escape without a fight when approached, but there is always a chance one will strike at a passerby. They do not always give a warning; they may strike first and rattle afterwards or not at all.

The genus Trimeresurus is a subgroup of the crotalidae. These are Asian pit vipers. These pit vipers are normally tree-loving snakes with a few species living on the ground. They basically have the same characteristics of the crotalidae—slender build and very dangerous. Their bites usually are on the upper extremities—head, neck, and shoulders. Their venom is largely hemotoxic.

Elapidae. A group of highly dangerous snakes with powerful neurotoxic venom that affects the nervous system, causing respiratory paralysis. Included in this family are coral snakes, cobras, mambas, and all the Australian venomous snakes. The coral snake is small and has caused human fatalities. The Australian death adder, tiger, taipan, and king brown snakes are among the most venomous in the world, causing many human fatalities.

Only by examining a dead snake can you positively determine if it is a cobra or a near relative (Figure 4-3). On cobras, kraits, and coral snakes, the third scale on the upper lip touches both the nostril scale and the eye. The krait also has a row of enlarged scales down its ridged back.

You can find the cobras of Africa and the Near East in almost any habitat. One kind may live in or near water, another in trees. Some are aggressive and savage. The distance a cobra can strike in a forward direction is equal to the distance its head is raised above the ground. Some cobras, however, can spit venom a distance of 3 to 3.5 meters. This venom is harmless unless it gets into your eyes; then it may cause blindness if not washed out immediately. Poking around in holes and rock piles is dangerous because of the chance of encountering a spitting cobra.

Laticaudinae and Hydrophidae. A subfamily of elapidae, these snakes are specialized in that they found a better environment in the oceans. Why they are in the oceans is not clear to science.

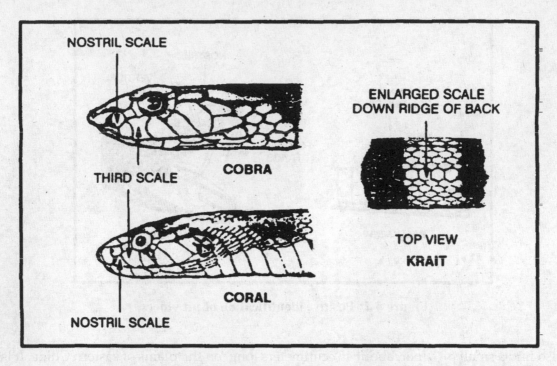

Figure 4-3: Positive identification of cobras, kraits, and coral snakes.

Sea snakes differ in appearance from other snakes in that they have an oarlike tail to aid in swimming. Some species of sea snakes have venom several times more toxic than the cobra's. Because of their marine environment, sea snakes seldom come in contact with humans. The exceptions are fishermen who capture these dangerous snakes in fish nets and scuba divers who swim in waters where sea snakes are found.

There are many species of sea snakes. They vary greatly in color and shape. Their scales distinguish them from eels that have no scales.

Sea snakes occur in salt water along the coasts throughout the Pacific. There are also sea snakes on the east coast of Africa and in the Persian Gulf. There are no sea snakes in the Atlantic Ocean.

There is no need to fear sea snakes. They have not been known to attack a man swimming. Fishermen occasionally get bit by a sea snake caught in a net. The bite is dangerous.

Colubridae. The largest group of snakes worldwide. In this family there are species that are rear-fanged; however, most are completely harmless to man. They have a venom-producing gland and enlarged, grooved rear fangs that allow venom to flow into the wound. The inefficient venom apparatus and the specialized venom is effective on cold-blooded animals (such as frogs and lizards) but not considered a threat to human life. The boomslang and the twig snake of Africa have, however, caused human deaths.

Table 4-3:

Viperidae

- Common adder
- Long-nosed adder
- Gaboon viper
- Horned desert viper
- McMahon's viper
- Mole viper
- Palestinian viper
- Puff adder

- Rhinoceros viper
- Russell's viper
- Sand viper
- Saw-scaled viper
- Ursini's viper

Elapidae

- Australian copperhead
- Common cobra

Table 4-3: *(Continued)*

- Coral snake
- Death adder
- Egyptian cobra
- Green mamba
- King cobra
- Krait
- Taipan
- Tiger snake

Crotallidae

- American copperhead
- Boomslang
- Bush viper
- Bushmaster
- Cottonmouth
- Easter diamondback rattlesnake
- Eyelash pit viper

- Fer-de-lance
- Green tre pit viper
- Habu pit viper
- Jumping ciper
- Malayan pit viper
- Mojave rattlesnake
- Pallas' viper
- Tropical rattlesnake
- Wagler's pit viper
- Western diamondback rattlesnake

Hydrophilidae

- Banded sea snake
- Yellow-bellied sea snake

LIZARDS

There is little to fear from lizards as long as you follow the same precautions as for avoiding snakebite. Usually, there are only two poisonous lizards: the Gila monster and the Mexican beaded lizard. The venom of both these lizards is neurotoxic. The two lizards are in the same family, and both are slow moving with a docile nature. The komodo dragon (Varanus komodoensis), although not poisonous, can be dangerous due to its large size. These lizards can reach lengths of 3 meters and weigh over 115 kilograms. Do not try to capture this lizard.

POISONOUS SNAKES OF THE AMERICAS

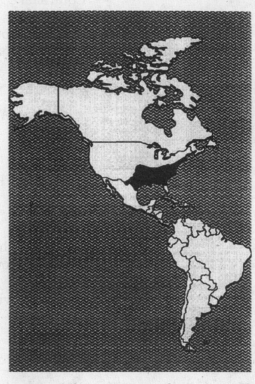

American copperhead
Agkistrodon contortrix

Description: Chestnut color dominates overall, with darker crossbands of rich browns that become narrower on top and widen at the bottom. The top of the head is a coppery color.

Characteristics: Very common over much of its range, with a natural camouflage ability to blend in the environment. Copperheads are rather quiet and inoffensive in disposition but will defend themselves vigorously. Bites occur when the snakes are stepped on or when a victim is lying next to one. A copperhead lying on a bed of dead leaves becomes invisible. Its venom is hemotoxic.

Habitat: Found in wooded and rocky areas and mountainous regions.

Length: Average 60 centimeters, maximum 120 centimeters.

Distribution: Eastern Gulf States, Texas, Arkansas, Maryland, North Florida, Illinois, Oklahoma, Kansas, Ohio, New York, Alabama, Tennessee, and Massachusetts.

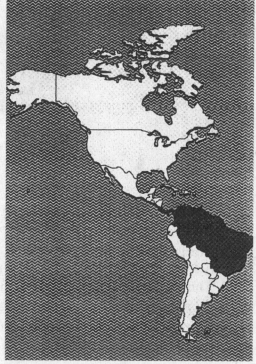

Bushmaster
Lachesis mutus

Description: The body hue is rather pale brown or pinkish, with a series of large bold dark brown or black blotches extending along the body. Its scales are extremely rough.

Characteristics: The world's largest pit viper has a bad reputation. This huge venomous snake is not common anywhere in its range. It lives in remote and isolated habitats and is largely nocturnal in its feeding habits; it seldom bites anyone, so few bites are recorded. A bite from one would indeed be very serious and fatal if medical aid was not immediately available. Usually, the bites occur in remote, dense jungles, many kilometers and several hours or even days away from medical help. Bushmaster fangs are long. In large bushmasters, they can measure 3.8 centimeters. Its venom is a powerful hemotoxin.

Habitat: Found chiefly in tropical forests in their range.

Length: Average 2.1 meters, maximum 3.7 meters.

Distribution: Nicaragua, Costa Rica, Panama, Trinidad, and Brazil.

Coral snake
Micrurus fulvius

Description: Beautifully marked with bright blacks, reds, and yellows. To identify the species, remember that when red touches yellow it is a coral snake.

Characteristics: Common over range, but secretive in its habits, therefore seldom seen. It has short fangs that are fixed in an erect position. It often chews to release its venom into a wound. Its venom is very powerful. The venom is neurotoxic, causing respiratory paralysis in the victim, who succumbs to suffocation.

Habitat: Found in a variety of habitats including wooded areas, swamps, palmetto and scrub areas. Coral snakes often venture into residential locations.

Length: Average 60 centimeters, maximum 115 centimeters.

Distribution: Southeast North Carolina, Gulf States, west central Mississippi, Florida, Florida Keys, and west to Texas. Another genus of coral snake is found in Arizona. Coral snakes are also found throughout Central and most of South America.

Cottonmouth
Agkistrodon piscivorus

Description: Colors are variable. Adults are uniformly olive brown or black. The young and subadults are strongly crossbanded with dark brown.

Characteristics: These dangerous semiaquatic snakes closely resemble harmless water snakes that have the same habitat. Therefore, it is best to leave all water snakes alone. Cottonmouths often stand their ground. An aroused cottonmouth will draw its head close to its body and open its mouth showing its white interior. Cottonmouth venom is hemotoxic and potent. Bites are prone to gangrene.

Habitat: Found in swamps, lakes, rivers, and ditches.

Length: Average 90 centimeters, maximum 1.8 meters.

Distribution: Southeast Virginia, west central Alabama, south Georgia, Illinois, east central Kentucky, south central Oklahoma, Texas, North and South Carolina, Florida, and the Florida Keys.

Eastern diamondback rattlesnake
Crotalus adamanteus

Description: Diamonds are dark brown or black, outlined by a row of cream or yellowish scales. Ground color is olive to brown.

Characteristics: The largest venomous snake in the United States. Large individual snakes can have fangs that measure 2.5 centimeters in a straight line. This species has a sullen disposition, ready to defend itself when threatened. Its venom is potent and hemotoxic, causing great pain and damage to tissue.

Habitat: Found in palmettos and scrubs, swamps, pine woods, and flatwoods. It has been observed swimming many miles out in the Gulf of Mexico, reaching some of the islands off the Florida coast.

Length: Average 1.4 meters, maximum 2.4 meters.

Distribution: Coastal areas of North Carolina, South Carolina, Louisiana, Florida, and the Florida Keys.

Eyelash pit viper
Bothrops schlegeli

Description: Identified by several spiny scales over each eye. Color is highly variable, from bright yellow over its entire body to reddish-yellow spots throughout the body.

Characteristics: Arboreal snake that seldom comes to the ground. It feels more secure in low-hanging trees where it looks for tree frogs and birds. It is a dangerous species because most of its bites occur on the upper extremities. It has an irritable disposition. It will strike with little provocation. Its venom is hemotoxic, causing severe tissue damage. Deaths have occurred from the bites of these snakes.

Habitat: Tree-loving species found in rain forests; common on plantations and in palm trees.

Length: Average 45 centimeters, maximum 75 centimeters.

Distribution: Southern Mexico, throughout Central America, Columbia, Ecuador, and Venezuela.

Fer-de-lance
Bothrops atrox

There are several closely related species in this group. All are very dangerous to man.

Description: Variable coloration, from gray to olive, brown, or reddish, with dark triangles edged with light scales. Triangles are narrow at the top and wide at the bottom.

Characteristics: This highly dangerous snake is responsible for a high mortality rate.

It has an irritable disposition, ready to strike with little provocation. The female fer-de-lance is highly prolific, producing up to 60 young born with a dangerous bite. The venom of this species is hemotoxic, painful, and hemorrhagic (causing profuse internal bleeding). The venom causes massive tissue destruction.

Habitat: Found on cultivated land and farms, often entering houses in search of rodents.

Length: Average 1.4 meters, maximum 2.4 meters.

Distribution: Southern Mexico, throughout Central and South America.

Jumping viper
Bothrops nummifer

Description: It has a stocky body. Its ground color varies from brown to gray and it has dark brown or black dorsal blotches. It has no pattern on its head.

Characteristics: It is chiefly a nocturnal snake. It comes out in the early evening hours to feed on lizards, rodents, and frogs. As the name implies, this species can strike with force as it actually leaves the ground. Its venom is hemotoxic. Humans have died from the bites inflicted by large jumping vipers. They often hide under fallen logs and piles of leaves and are difficult to see.

Habitat: Found in rain forests, on plantations, and on wooded hillsides.

Length: Average 60 centimeters, maximum 120 centimeters.

Distribution: Southern Mexico, Honduras, Guatemala, Costa Rica, Panama, and El Salvador.

Mojave rattlesnake
Crotalus scutulatus

Description: This snake's entire body is a pallid or sandy odor with darker diamond-shaped markings bordered by lighter-colored scales and black bands around the tail.

Characteristics: Although this rattlesnake is of moderate size, its bite is very serious. Its venom has quantities of neurotoxic elements that affect the central nervous system. Deaths have resulted from this snake's bite.

Habitat: Found in arid regions, deserts, and rocky hillsides from sea level to 2400-meter elevations.

Length: Average 75 centimeters, maximum 1.2 meters.

Distribution: Mojave Desert in California, Nevada, southwest Arizona, and Texas into Mexico.

Tropical rattlesnake
Crotalus terrificus

Description: Coloration is light to dark brown with a series of darker rhombs or diamonds bordered by a buff color.

Characteristics: Extremely dangerous with an irritable disposition, ready to strike with little or no warning (use of its rattle). This species has a highly toxic venom containing neurotoxic and hemotoxic components that paralyze the central nervous system and cause great damage to tissue.

Habitat: Found in sandy places, plantations, and dry hillsides.

Length: Average 1.4 meters, maximum 2.1 meters.

Distribution: Southern Mexico, Central America, and Brazil to Argentina.

Western diamondback rattlesnake
Crotalus atrox

Description: The body is a light buff color with darker brown diamond-shaped markings. The tail has heavy black and white bands.

Characteristics: This bold rattlesnake holds its ground. When coiled and rattling, it is ready to defend itself. It injects a large amount of venom when it bites, making it one of the most dangerous snakes. Its venom is hemotoxic, causing considerable pain and tissue damage.

Habitat: It is a very common snake over its range. It is found in grasslands, deserts, woodlands, and canyons.

Length: Average 1.5 meters, maximum 2 meters.

Distribution: Southeast California, Oklahoma, Texas, New Mexico, and Arizona.

POISONOUS SNAKES OF EUROPE

Common adder
Vipera berus

Description: Its color is variable. Some adult specimens are completely black while others have a dark zigzag pattern running along the back.

Characteristics: The common adder is a small true viper that has a short temper and often strikes without hesitation. Its venom is hemotoxic, destroying blood cells and causing tissue damage. Most injuries occur to campers, hikers, and field workers.

Habitat: Common adders are found in a variety of habitats, from grassy fields to rocky slopes, and on farms and cultivated lands.

Length: Average 45 centimeters, maximum 60 centimeters.

Distribution: Very common throughout most of Europe.

Long-nosed adder
Vipera ammodytes

Description: Coloration is gray, brown, or reddish with a dark brown or black zigzag pattern running the length of its back. A dark stripe is usually found behind each eye.

Characteristics: A small snake commonly found in much of its range. The term "long-nosed" comes from the projection of tiny scales located on the tip of its nose. This viper is responsible for many bites. Deaths have been recorded. Its venom is hemotoxic, causing severe pain and massive tissue damage. The rate of survival is good with medical aid.

Habitat: Open fields, cultivated lands, farms, and rocky slopes.

Length: Average 45 centimeters, maximum 90 centimeters.

Distribution: Italy, Yugoslavia, northern Albania, and Romania.

Pallas' viper
Agkistrodon halys

Description: Coloration is gray, tan, or yellow, with markings similar to those of the American copperhead.

Characteristics: This snake is timid and rarely strikes. Its venom is hemotoxic but rarely fatal.

Habitat: Found in open fields, hillsides, and farming regions.

Length: Average 45 centimeters, maximum 90 centimeters.

Distribution: Throughout southeastern Europe.

Ursini's viper
Vipera ursinii

Description: The common adder, long-nosed adder, and Ursini's viper basically have the same coloration and dorsal zigzag pattern. The exception among these adders is that the common adder and Ursini's viper lack the projection of tiny scales on the tip of the nose.

Characteristics: These little vipers have an irritable disposition. They will readily strike when approached. Their venom is hemotoxic. Although rare, deaths from the bites of these vipers have been recorded.

Habitat: Meadows, farmlands, rocky hillsides, and open, grassy fields.

Length: Average 45 centimeters, maximum 90 centimeters.

Distribution: Most of Europe, Greece, Germany, Yugoslavia, France, Italy, Hungary, Romania, Bulgaria, and Albania.

POISONOUS SNAKES OF AFRICA AND ASIA

Boomslang
Dispholidus typus

Description: Coloration varies but is generally green or brown, which makes it very hard to see in its habitat.

Characteristics: Will strike if molested. Its venom is hemotoxic; even small amounts cause severe hemorrhaging, making it dangerous to man.

Habitat: Found in forested areas. It will spend most of its time in trees or looking for chameleons and other prey in bushes.

Length: Generally less than 60 centimeters.

Distribution: Found throughout sub-Saharan Africa.

Bush viper
Atheris squamiger

Description: Often called leaf viper, its color varies from ground colors of pale green to olive, brown, or rusty brown. It uses it prehensile tail to secure itself to branches.

Characteristics: An arboreal species that often comes down to the ground to feed on small rodents. It is not aggressive, but it will defend itself when molested or touched. Its venom is hemotoxic; healthy adults rarely die from its bite.

Habitat: Found in rain forests and woodlands bordering swamps and forests. Often found in trees, low-hanging branches, or brush.

Length: Average 45 centimeters, maximum 75 centimeters.

Distrubition: Most of Africa, Angola, Cameroon, Uganda, Kenya, and Zaire.

Common cobra
Naja naja

Description: Also known as the Asiatic cobra. Usually slate gray to brown overall. The back of the hood may or may not have a pattern.

Characteristics: A very common species responsible for many deaths each year. When aroused or threatened, the cobra will lift its head off the ground and spread its hood, making it more menacing. Its venom is highly neurotoxic, causing respiratory paralysis with some tissue damage. The cobra would rather retreat if possible, but if escape is shut off, it will be a dangerous creature to deal with.

Habitat: Found in any habitat cultivated farms, swamps, open fields, and human dwelling where it searches for rodents.

Length: Average 1.2 meters, maximum 2.1 meters.

Distribution: All of Asia.

Egyptian cobra
Naja haje

Description: Yellowish, dark brown, or black uniform top with brown crossbands. Its head is sometimes black.

Characteristics: It is extremely dangerous. It is responsible for many human deaths. Once aroused or threatened, it will attack and continue the attack until it feels an escape is possible. Its venom is neurotoxic and much stronger than the common cobra. Its venom causes paralysis and death due to respiratory failure.

Habitat: Cultivated farmlands, open fields, and arid countrysides. It is often seen around homes searching for rodents.

Length: Average 1.5 meters, maximum 2.5 meters.

Distribution: Africa, Iraq, Syria, and Saudi Arabia.

Gaboon viper
Bitis gabonica

Description: Pink to brown with a vertebral series of elongated yellowish or light brown spots connected by hourglass-shaped markings on each side. It has a dark brown stripe behind each eye. This dangerous viper is almost invisible on the forest floor. A 1.8-meter-long Gaboon viper could weigh 16 kilograms.

Characteristics: The largest and heaviest of all true vipers, having a very large triangular head. It comes out in the evening to feed. Fortunately, it is not aggressive, but it will stand its ground if approached. It bites when molested or stepped on. Its fangs are enormous, often measuring 5 centimeters long. It injects a large amount of venom when it strikes. Its venom is neurotoxic and hemotoxic.

Habitat: Dense rain forests. Occasionally found in open country.

Length: Average 1.2 meters, maximum 1.8 meters.

Distribution: Most of Africa.

Green mamba
Dendraspis angusticeps

Description: Most mambas are uniformly bright green over their entire body. The black mamba, the largest of the species, is uniformly olive to black.

Characteristics: The mamba is the dreaded snake species of Africa. Treat it with great respect. It is considered one of the most dangerous snakes known. Not only is it highly venomous but it is aggressive and its victim has little chance to escape from a bite. Its venom is highly neurotoxic.

Habitat: Mambas are at home in brush, trees, and low-hanging branches looking for birds, a usual diet for this species.

Length: Average 1.8 meters, maximum 3.7 meters.

Distribution: Most of Africa.

Green tree pit viper
Trimeresurus gramineus

Description: Uniform bright or dull green with light yellow on the facial lips.

Characteristics: A small arboreal snake of some importance, though not considered a deadly species. It is a dangerous species because most of its bites occur in the head, shoulder, and neck areas. It seldom comes to the ground. It feeds on young birds, lizards, and tree frogs.

Habitat: Found in dense rain forests and plantations.

Length: Average 45 centimeters, maximum 75 centimeters.

Distribution: India, Burma, Malaya, Thailand, Laos, Cambodia, Vietnam, China, Indonesia, and Formosa.

Habu pit viper
Trimeresurus flavoviridis

Description: Light brown or olive-yellow with black markings and a yellow or greenish-white belly.

Characteristics: This snake is responsible for biting many humans and its bite could be fatal. It is an irritable species ready to defend itself. Its venom is hemotoxic, causing pain and considerable tissue damage.

Habitat: Found in a variety of habitats, ranging from lowlands to mountainous regions. Often encountered in old houses and rock walls surroundings buildings.

Length: Average 1 meter, maximum 1.5 meters.

Distribution: Okinawa and neighboring islands and Kyushu.

Horned desert viper
Cerastes cerastes

Description: Pale buff color with obscure markings and a sharp spine (scale) over each eye.

Characteristics: As with all true vipers that live in the desert, it finds refuge by burrowing in the heat of the day, coming out at night to feed. It is difficult to detect when buried; therefore, many bites result from the snake being accidentally stepped on. Its venom is hemotoxic, causing severe damage to blood cells and tissue.

Habitat: Only found in very arid places within its range.

Length: Average 45 centimeters, maximum 75 centimeters.

Distribution: Arabian Peninsula, Africa, Iran, and Iraq.

King cobra
Ophiophagus hannah

Description: Uniformly olive, brown, or green with ringlike crossbands of black.

Characteristics: Although it is the largest venomous snake in the world and it has a disposition to go with this honor, it causes relatively few bites on humans. It appears to have a degree of intelligence. It avoids attacking another venomous snake for fear of being bitten. It feeds exclusively on harmless species. The female builds a nest then deposits her eggs. Lying close by, she guards the nest and is highly aggressive toward anything that closely approaches the nest. Its venom is a powerful neurotoxin. Without medical aid, death is certain for its victims.

Habitat: Dense jungle and cultivated fields.

Length: Average 3.5 meters, maximum 5.5 meters.

Distibution: Thailand, southern China, Malaysia Peninsula, and Phillipines.

Krait
Bungarus caeruleus

Description: Black or bluish-black with white narrow crossbands and a narrow head.

Characteristics: Kraits are found only in Asia. This snake is of special concern to man. It is deadly—about 15 times more deadly than the common cobra. It is active at night and relatively passive during the day. The native people often step on kraits while walking through their habitats. The krait has a tendency to seek shelter in sleeping bags, boots, and tents. Its venom is a powerful neurotoxin that causes respiratory failure.

Habitat: Open fields, human settlements, and dense jungle.

Length: Average 90 centimeters, maximum 1.5 meters.

Distribution: India, Sri Lanka, and Pakistan.

Levant viper
Vipera lebetina

Description: Gray to pale brown with large dark brown spots on the top of the back and a " ^ " mark on top of the head.

Characteristics: This viper belongs to a large group of true vipers. Like its cousins, it is large and dangerous. Its venom is hemotoxic. Many deaths have been reported from bites of this species. It is a strong snake with an irritable disposition; it hisses loudly when ready to strike.

Habitat: Varies greatly, from farmlands to mountainous areas.

Length: Average 1 meter, maximum 1.5 meters.

Distribution: Greece, Iraq, Syria, Lebanon, Turkey, Afganistan, lower portion of the former USSR, and Saudi Arabia.

Malayan pit viper
Callaselasma rhostoma

Description: Reddish running into pink tinge toward the belly with triangular-shaped, brown markings bordered with light-colored scales. The base of the triangular-shaped markings end at the midline. It has dark brown, arrow-shaped markings on the top and each side of its head.

Characteristics: This snake has long fangs, is ill-tempered, and is responsible for many bites. Its venom is hemotoxic, destroying blood cells and tissue, but a victim's chances of survival are good with medical aid. This viper is a ground dweller that moves into many areas in search of food. The greatest danger is in stepping on the snake with bare feet.

Habitat: Rubber plantations, farms, rural villages, and rain forests.

Length: Average 60 centimeters, maximum 1 meter.

Distribution: Thailand, Laos, Cambodia, Java, Sumatra, Malaysia, Vietnam, Burma, and China.

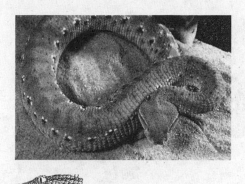

McMahon's viper
Eristicophis macmahonii

Description: Sandy buff color dominates the body with darker brown spots on the side of the body. Its nose shield is broad, aiding in burrowing.

Characteristics: Very little is known about this species. It apparently is rare or seldom seen. This viper is very irritable; it hisses, coils, and strikes at any intruder that ventures too close. Its venom is highly hemotoxic, causing great pain and tissue damage.

Habitat: Arid or semidesert. It hides during the day's sun, coming out only at night to feed on rodents.

Length: Average 45 centimeters, maximum 1 meter.

Distribution: West Pakistan and Afghanistan.

Mole viper or burrowing viper
Atracaspis microlepidota

Description: Uniformly black or dark brown with a small, narrow head.

Characteristics: A viper that does not look like one. It is small in size, and its small head does not indicate the presence of venom glands. It has a rather inoffensive disposition; however, it will quickly turn and bite if restrained or touched. Its venom is a potent hemotoxin for such a small snake. Its fangs are exceptionally long. A bite can result even when picking it up behind the head. It is best to leave this snake alone.

Habitat: Agricultural areas and arid localities

Length: Average 55 centimeters, maximum 75 centimeters

Distribution: Sudan, Ethiopia, Somaliland, Kenya, Tanganyika, Uganda, Cameroon, Niger, Congo, and Urundi.

Palestinian viper
Vipera palaestinae

Description: Olive to rusty brown with a dark V-shaped mark on the head and a brown, zigzag band along the back.

Characteristics: The Palestinian viper is closely related to the Russell's viper of Asia. Like its cousin, it is extremely dangerous. It is active and aggressive at night but fairly placid during the day. When threatened or molested, it will tighten its coils, hiss loudly, and strike quickly.

Habitat: Arid regions, but may be found around barns and stables. It has been seen entering houses in search of rodents.

Length: Average 0.8 meter, maximum 1.3 meters.

Distribution: Turkey, Syria, Palestine, Israel, Lebanon, and Jordan.

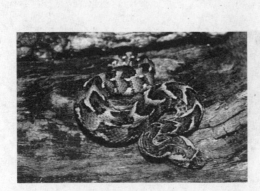

Puff adder
Bitis arietans

Description: Yellowish, light brown, or orange with chevron-shaped dark brown or black bars.

Characteristics: The puff adder is the second largest of the dangerous vipers. It is one of the most common snakes in Africa. It is largely nocturnal, hunting at night and seeking shelter during the day's heat. It is not shy when approached. It draws its head close to its coils, makes a loud hissing sound, and is quick to strike any intruder. Its venom is strongly hemotoxic, destroying bloods cells and causing extensive tissue damage.

Habitat: Arid regions to swamps and dense forests. Common around human settlements.

Length: Average 12 meters, maximum 1.8 meters.

Distribution: Most of Africa, Saudi Arabia, Iraq, Lebanon, Israel, and Jordan.

Rhinoceros viper or river jack
Bitis nasicornis

Description: Brightly colored with purplish to reddish-brown markings and black and light olive markings along the back. On its head it has a triangular marking that starts at the tip of the nose. It has a pair of long horns (scales) on the tip of its nose.

Characteristics: Its appearance is awesome; its horns and very rough scales give it a sinister look. It has an irritable disposition. It is not aggressive but will stand its ground ready to strike if disturbed. Its venom is neurotoxic and hemotoxic.

Habitat: Rain forests, along waterways, and in swamps.

Length: Average 75 centimeters, maximum 1 meter.

Distribution: Equatorial Africa.

Russell's viper
Vipera rus sellii

Description: Light brown body with three rows of dark brown or black splotches bordered with white or yellow extending its entire length.

Characteristics: This dangerous species is abundant over its entire range. It is responsible for more human fatalities than any other venomous snake. It is irritable. When threatened, it coils tightly, hisses, and strikes with such speed that its victim has little chance of escaping. Its hemotoxic venom is a powerful coagulant, damaging tissue and blood cells.

Habitat: Variable, from farmlands to dense rain forests. It is commonly found around human settlements.

Length: Average 1 meter, maximum 1.5 meters

Distribution: Sri Lanka, south China, India, Malaysian Peninsula, Java, Sumatra, Borneo, and surrounding islands.

Sand viper
Cerastes vipera

Description: Usually uniformly very pallid, with three rows of darker brown spots.

Characteristics: A very small desert dweller that can bury itself in the sand during the day's heat. It is nocturnal, coming out at night to feed on lizards and small desert rodents. It has a short temper and will strike several times. Its venom is hemotoxic.

Habitat: Restricted to desert areas.

Length: Average 45 centimeters, maximum 60 centimeters.

Distribution: Northern Sahara, Algeria, Egypt, Sudan, Nigeria, Chad, Somalia, and central Africa.

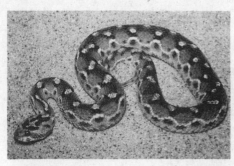

Saw-scaled viper
Echis carinatus

Description: Color is light buff with shades of brown, dull red, or gray. Its sides have a white or light-colored pattern. Its head usually has two dark stripes that start behind the eye and extend to the rear.

Characteristics: A small but extremely dangerous viper. It gets the name saw-scaled from rubbing the sides of its body together, producing a rasping sound. This ill-tempered snake will attack any intruder. Its venom is highly hemotoxic and quite potent. Many deaths are attributed to this species.

Habitat: Found in a variety of environments. It is common in rural settlements, cultivated fields, arid regions, barns, and rock walls.

Length: Average 45 centimeters, maximum 60 centimeters.

Distribution: Asia, Syria, India, Africa, Iraq, Iran, Saudi Arabia, Pakistan, Jordan, Lebanon, Sri Lanka, Algeria, Egypt, and Israel.

Wagler's pit viper or temple viper
Trimeresurus wagleri

Description: Green with white crossbands edged with blue or purple. It has two dorsal lines on both sides of its head.

Characteristics: It is also known as the temple viper because certain religious cults have placed venomous snakes in their temples. Bites are not uncommon for the species; fortunately, fatalities are very rare. It has long fangs. Its venom is hemotoxic causing cell and tissue destruction. It is an arboreal species and its bites often occur on the upper extremities.

Habitat: Dense rain forests, but often found near human settlements.

Length: Average 60 centimeters, maximum 100 centimeters.

Distribution: Malaysian Peninsula and Archipelago, Indonesia, Borneo, the Philippines. and Ryuku Islands.

POISONOUS SNAKES OF AUSTRALASIA

Australian copperhead
Denisonia superba

Description: Coloration is reddish brown to dark brown. A few from Queensland are black.

Characteristics: Rather sluggish disposition but will bite if stepped on. When angry, rears its head a few inches from the ground with its neck slightly arched. Its venom is neurotoxic.

Habitat: Swamps.

Length: Average 1.2 meters, maximum 1.8 meters.

Distribution: Tasmania, South Australia, Queensland, and Kangaroo Island.

Death adder
Acanthophis antarcticus

Description: Reddish, yellowish, or brown color with distinct dark brown crossbands. The end of its tail is black, ending in a hard spine.

Characteristics: When aroused, this highly dangerous snake will flatten its entire body, ready to strike over a short distance. It is nocturnal, hiding by day and coming out to feed at night. Although it has the appearance of a viper, it is related to the cobra family. Its venom is a powerful neurotoxin; it causes mortality in about 50 percent of the victims, even with treatment.

Habitat: Usually found in arid regions, fields, and wooded lands.

Length: Average 45 centimeters, maximum 90 centimeters.

Distribution: Australia, New Guinea, and Moluccas.

Taipan
Oxyuranus scutellatus

Description: Generally uniformly olive or dark brown, the head is somewhat darker brown.

Characteristics: Considered one of the most deadly snakes. It has an aggressive disposition. When aroused, it can display a fearsome appearance by flattening its head, raising it off the ground, waving it back and forth, and suddenly striking with such speed that the victim may receive several bites before it retreats. Its venom is a powerful neurotoxin, causing respiratory paralysis. Its victim has little chance for recovery without prompt medical aid.

Habitat: At home in a variety of habitats, it is found from the savanna forests to the inland plains.

Length: Average 1.8 meters, maximum 3.7 meters.

Distribution: Northern Australia and southern New Guinea.

Tiger snake
Notechis scutatus

Description: Olive to dark brown above with yellowish or olive belly and crossbands. The subspecies in Tasmania and Victoria is uniformly black.

Characteristics: It is the most dangerous snake in Australia. It is very common and bites many humans. It has a very potent neurotoxic venom that attacks the nervous system. When aroused, it is aggressive and attacks any intruder. It flattens its neck making a narrow band.

Habitat: Found in many habitats from arid regions to human settlements along waterways to grasslands.

Length: Average 1.2 meters, maximum 1.8 meters.

Distribution: Australia, Tasmania, Bass Strait islands, and New Guinea.

POISONOUS SEA SNAKES

Banded sea snake
Laticauda colubrina

Description: Smooth-scaled snake that is a pale shade of blue with black bands. Its oarlike tail provides propulsion in swimming.

Characteristics: Most active at night, swimming close to shore and at times entering tide pools. Its venom is a very strong neurotoxin. Its victims are usually fishermen who untangle these deadly snakes from large fish nets.

Habitat: Common in all oceans, absent in the Atlantic Ocean.

Length: Average 75 centimeters, maximum 1.2 meters.

Distribution: Coastal waters of New Guinea, Pacific islands, the Philippines, Southeast Asia, Sri Lanka, and Japan.

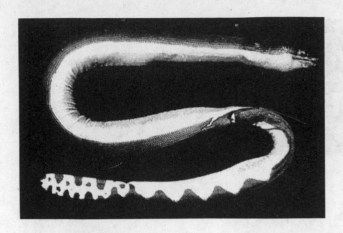

Yellow-bellied sea snake

Pelamis platurus

Description: Upper part of body is black or dark brown and lower part is bright yellow.

Characteristics: A highly venomous snake belonging to the cobra family. This snake is truly of the pelagic species—it never leaves the water to come to shore. It has an oarlike tail to aid its swimming. This species is quick to defend itself. Sea snakes do not really strike, but deliberately turn and bite if molested. A small amount of their neurotoxic venom can cause death.

Habitat: Found in all oceans except the Atlantic Ocean.

Length: Average 0.7 meter, maximum 1.1 meters.

Distribution: Throughout the Pacific Ocean from many of the Pacific islands to Hawaii and to the coast of Costa Rica and Panama.

POISONOUS LIZARDS

Gila monster
Heloderma suspectum

Description: Robust, with a large head and a heavy tail. Its body is covered with beadlike scales. It is capable of storing fat against lean times when food is scarce. Its color is striking in rich blacks laced with yellow or pinkish scales.

Characteristics: Not an aggressive lizard, but it is ready to defend itself when provoked. If approached too closely, it will turn toward the intruder with its mouth open. If it bites, it hangs on tenaciously and must be pried off. Its venom glands and grooved teeth are on its bottom jaw.

Habitat: Found in arid areas, coming out at night or early morning hours in search of small rodents and bird eggs. During the heat of the day it stays under brush or rocks.

Length: Average 30 centimeters, maximum 50 centimeters.

Distribution: Arizona, New Mexico, Utah, Nevada, northern Mexico, and extreme corner of southeast California.

Mexican beaded lizard
Heloderma horridum

Description: Less colorful than its cousin, the gila monster. It has black or pale yellow bands or is entirely black.

Characteristics: Very strong legs let this lizard crawl over rocks and dig burrows. It is short-tempered. It will turn and open its mouth in a threatening manner when molested. Its venom is hemotoxic and potentially dangerous to man.

Habitat: Found in arid or desert areas, often in rocky hillsides, coming out during evening and early morning hours.

Length: Average 60 centimeters, maximum 90 centimeters.

Distribution: Mexico through Central America.

CHAPTER 5

Dangerous Fish and Mollusks

Since fish and mollusks may be one of your major sources of food, it is wise to know which ones are dangerous to you should you catch them. Know which ones are dangerous, what the dangers of the various fish are, what precautions to take, and what to do if you are injured by one of these fish.

Fish and mollusks will present a danger in one of three ways: by attacking and biting you, by injecting toxic venom into you through its venomous spines or tentacles, and through eating fish or mollusks whose flesh is toxic.

The danger of actually encountering one of these dangerous fish is relatively small, but it is still significant. Any one of these fish can kill you. Avoid them if at all possible.

DANGERS IN RIVERS

Common sense will tell you to avoid confrontations with hippopotami, alligators, crocodiles, and other large river creatures. There are, however, a few smaller river creatures with which you should be cautious.

Electric Eel. Electric eels (*Electrophorus electricus*) may reach 2 meters in length and 20 centimeters in diameter. Avoid them. They are capable of generating up to 500 volts of electricity in certain organs in their body. They use this shock to stun prey and enemies. Normally, you find these eels in the Orinoco and Amazon River systems in South America. They seem to prefer shallow waters that are more highly oxygenated and provide more food. They are bulkier than our native eels. Their upper body is dark gray or black, with a lighter-colored underbelly.

Piranha. Piranhas (*Serrasalmo* species) are another hazard of the Orinoco and Amazon River systems, as well as the Paraguay River Basin, where they are native. These fish vary greatly in size and coloration, but usually have a combination of orange undersides and dark tops. They have white, razor-sharp teeth that are clearly visible. They may be as long as 50 centimeters. Use great care when crossing waters where they live. Blood attracts them. They are most dangerous in shallow waters during the dry season.

Turtle. Be careful when handling and capturing large freshwater turtles, such as the snapping turtles and soft-shelled turtles of North America and the matamata and other turtles of South America. All of these turtles will bite in self-defense and can amputate fingers and toes.

Platypus. The platypus or duckbill (*Ornithorhyncus anatinus*) is the only member of its family and is easily recognized. It has a long body covered with grayish, short hair, a tail like a beaver, and a bill like a duck. Growing up to 60 centimeters in length, it may appear to be a good food source, but this egg-laying mammal, the only one in the world, is very dangerous. The male has a poisonous spur on each hind foot that can inflict intensely painful wounds. You find the platypus only in Australia, mainly along mud banks on waterways.

FISH THAT ATTACK MAN

The shark is usually the first fish that comes to mind when considering fish that attack man. Other fish also fall in this category, such as the barracuda, the moray eel, and the piranha.

Sharks. Whether you are in the water or in a boat or raft, you may see many types of sea life around you. Some may be more dangerous than others. Generally, sharks are the greatest danger to you. Other animals such as whales, porpoises, and stingrays may look dangerous, but really pose little threat in the open sea.

Of the many hundreds of shark species, only about 20 species are known to attack man. The most dangerous are the great white shark, the hammerhead, the make, and the tiger shark. Other sharks known to attack man include the gray, blue, lemon, sand, nurse, bull, and oceanic white tip sharks. See Figure 5-1 for illustrations of sharks. Consider any shark longer than 1 meter dangerous.

There are sharks in all oceans and seas of the world. While many live and feed in the depths of the sea, others hunt near the surface. The sharks living near the surface are the ones you will most likely see. Their dorsal fins frequently project above the water. Sharks in the tropical and subtropical seas are far more aggressive than those in temperate waters.

All sharks are basically eating machines. Their normal diet is live animals of any type, and they will strike at injured or helpless animals. Sight, smell, or sound may guide them to their prey. Sharks have an acute sense of smell and the smell of blood in the water excites them. They are also very sensitive to any

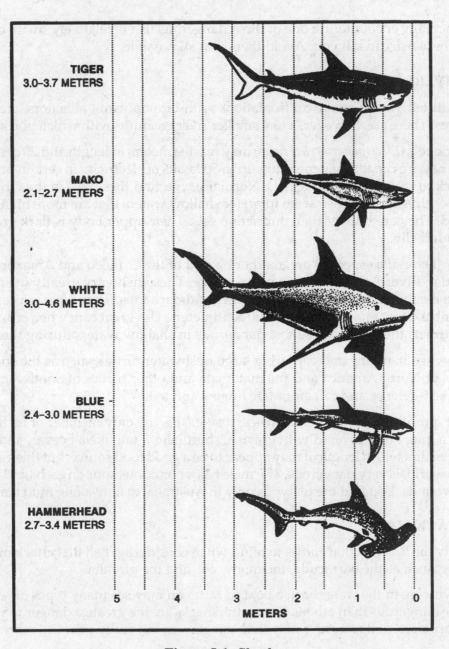

Figure 5-1: Sharks.

abnormal vibrations in the water. The struggles of a wounded animal or swimmer, underwater explosions, or even a fish struggling on a fish line will attract a shark.

Sharks can bite from almost any position; they do not have to turn on their side to bite. The jaws of some of the larger sharks are so far forward that they can bite floating objects easily without twisting to the side.

Sharks may hunt alone, but most reports of attacks cite more than one shark present. The smaller sharks tend to travel in schools and attack in mass. Whenever one of the sharks finds a victim, the other sharks will quickly join it. Sharks will eat a wounded shark as quickly as their prey.

Sharks feed at all hours of the day and night. Most reported shark contacts and attacks were during daylight, and many of these have been in the late afternoon. Some of the measures that you can take to protect yourself against sharks when you are in the water are—

- Stay with other swimmers. A group can maintain a 360-degree watch. A group can either frighten or fight off sharks better than one man.
- Always watch for sharks. Keep all your clothing on, to include your shoes. Historically, sharks have attacked the unclothed men in groups first, mainly in the feet. Clothing also protects against abrasions should the shark brush against you.
- Avoid urinating. If you must, only do so in small amounts. Let it dissipate between discharges. If you must defecate, do so in small amounts and throw it as far away from you as possible. Do the same if you must vomit.

If a shark attack is imminent while you are in the water, splash and yell just enough to keep the shark at bay. Sometimes yelling underwater or slapping the water repeatedly will scare the shark away. Conserve your strength for fighting in case the shark attacks.

If attacked, kick and strike the shark. Hit the shark on the gills or eyes if possible. If you hit the shark on the nose, you may injure your hand if it glances off and hits its teeth.

When you are in a raft and see sharks—

- Do not fish. If you have hooked a fish, let it go. Do not clean fish in the water.
- Do not throw garbage overboard.
- Do not let your arms, legs, or equipment hang in the water.
- Keep quiet and do not move around.
- Bury all dead as soon as possible. If there are many sharks in the area, conduct the burial at night.

When you are in a raft and a shark attack is imminent, hit the shark with anything you have, except your hands. You will do more damage to your hands than the shark. If you strike with an oar, be careful not to lose or break it.

If bitten by a shark, the most important measure for you to take is to stop the bleeding quickly. Blood in the water attracts sharks. Get yourself or the victim into a raft or to shore as soon as possible. If in the water, form a circle around the victim (if not alone), and stop the bleeding with a tourniquet.

Other Ferocious Fish. In salt water, other ferocious fish include the barracuda, sea bass, and moray eel (Figure 5-2). The sea bass is usually an open water fish. It is dangerous due to its large size. It can remove large pieces of flesh from a human. Barracudas and moray eels have been known to attack man and inflict vicious bites. Be careful of these two species when near reefs and in shallow water. Moray eels are very aggressive when disturbed.

VENOMOUS FISH AND INVERTEBRATES

There are several species of venomous fish and invertebrates, all of which live in salt water. All of these are capable of injecting poisonous venom through spines located in their fins, tentacles, or bites. Their venoms cause intense pain and are potentially fatal. If injured by one of these fish or invertebrates, treat the injury as for snakebite.

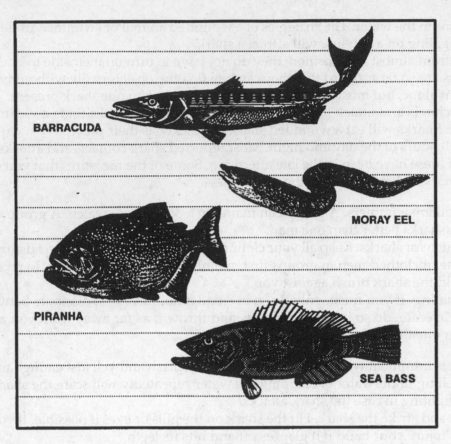

Figure 5-2: Ferocious fish.

Stingrays inhabit shallow water, especially in the tropics and in temperate regions as well. All have distinctive ray shape but coloration may make them hard to spot unless they are swimming. The venomous, barbed spines in their tails can cause severe or fatal injury. When moving about in shallow water, wear some form of footwear and shuffle your feet along the bottom, rather than picking up your feet and stepping.

Rabbitfish are found predominantly on reefs in the Pacific and Indian Oceans. They average about 30 centimeters long and have very sharp spines in their fins. The spines are venomous and can inflict intense

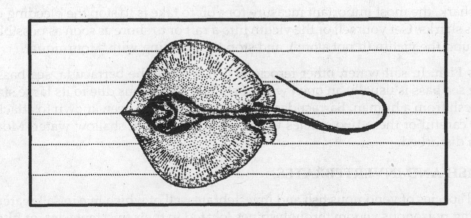

Figure 5-3: Stingray (*Dasyatidae* species).

Figure 5-4: Rabbitfish (*Siganidae* species).

pain. Rabbitfish are considered edible by native peoples where the fish are found, but deaths occur from careless handling. Seek other nonpoisonous fish to eat if possible.

Scorpion fish or Zebra fish live mainly in the reefs in the Pacific and Indian oceans, and occasionally in the Mediterranean and Aegean seas. They vary from 30 to 90 centimeters long, are unusually reddish in coloration, and have long, wavy fins and spines. They inflict an intensely painful sting.

The siganus fish is small, about 10 to 15 centimeters long, and looks much like a small tuna. It has venomous spines in its dorsal and ventral fins. These spines can inflict painful stings.

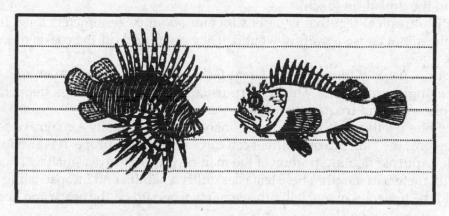

Figure 5-5: Scorpion fish or Zebra fish (*Scorpaenidae* species).

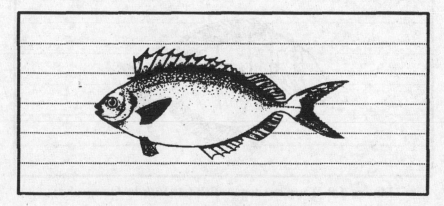

Figure 5-6: Siganus fish.

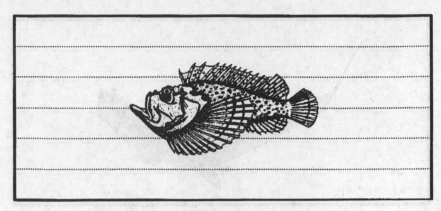

Figure 5-7: Stonefish (*Synanceja* species).

Stonefish are found in the tropical waters of the Pacific and Indian oceans. Averaging about 30 centimeters in length, their subdued colors and lumpy shape provide them with exceptional camouflage. When stepped on, the fins in the dorsal spine inflict an extremely painful and sometimes fatal wound.

Tang or surgeonfish average 20 to 25 centimeters in length, with a deep body, small mouth, and bright coloration. They have scalpellike spines on the side of the tail that cause fatal and extremely painful wounds including infection, envenomation, and blood loss.

Toadfish are found in the tropical waters of the Gulf Coast of the United States and along both coasts of South and Central America. They are between 17.5 and 25 centimeters long and have a dull color and large mouths. They bury themselves in the sand and may be easily stepped on. They have very sharp, extremely poisonous spines on the dorsal fin (back).

The weever fish is a tropical fish that is fairly slim and about 30 centimeters long. Its gills and all fins have venomous spines that cause a painful wound. They are found off the coasts of Europe, Africa, and the Mediterranean.

This small octopus is usually found on the Great Barrier Reef off eastern Australia. It is grayish-white with iridescent blue ringlike markings. This octopus usually will not bite unless stepped on or handled. Its bite is extremely poisonous and frequently lethal.

Although it resembles a jellyfish, the Portuguese man-of-war is actually a colony of sea animals. Mainly found in tropical regions, the Gulf Stream current can carry it as far as Europe. It is also found as far south as Australia. The pink or purple floating portion of the man-of-war may be as small as 15 centimeters, but the tentacles can reach 12 meters in length. These tentacles inflict a painful and incapacitating sting, but the sting is rarely fatal. Avoid the tentacles of any jellyfish, even if washed up on the beach and apparently dead.

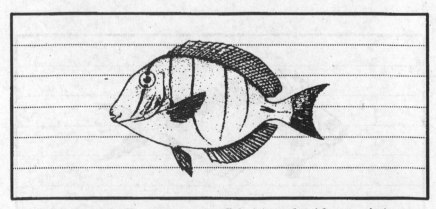

Figure 5-8: Tang or surgeonfish (*Acanthuridae* species).

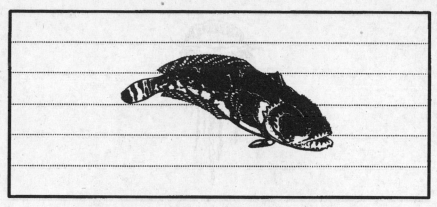

Figure 5-9: Toadfish (*Batrachoidiae* species).

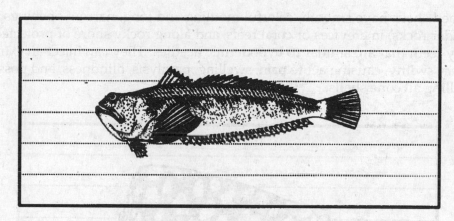

Figure 5-10: Weever fish (*Trachinidae* species).

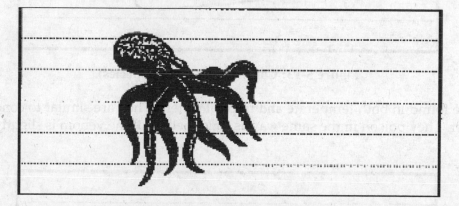

Figure 5-11: Blue-ringed octopus (*Hapalochlaena lunulata*).

Figure 5-12: Portuguese man-of-war (*Physalis* species).

These cone-shaped shells have smooth, colorful mottling and long, narrow openings in the base of the shell. They live under rocks, in crevices or coral reefs, and along rocky shore of protected bays in tropical areas. All have tiny teeth that are similar to hypodermic needles. They can inject an extremely poisonous venom that acts very swiftly, causing acute pain, swelling, paralysis, blindness, and possible death within hours. Avoid handling all cone shells.

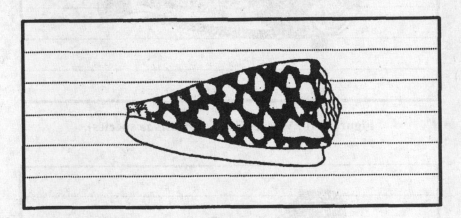

Figure 5-13: Cone shells (*Conidae* species).

These shells are found in both temperate and tropical waters. They are similar to cone shells but much thinner and longer. They poison in the same way as cone shells, but the venom is slightly less poisonous.

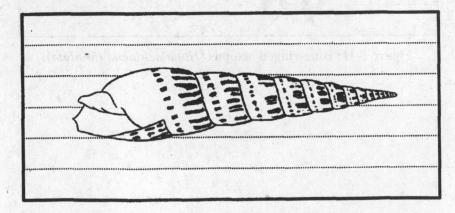

Figure 5-14: Terebra shells or Auger shells (*Terebridae* species).

These cone-shaped shells have smooth, colorful mottling and long, narrow openings in the base of the shell. They live under rocks, in crevices or coral reefs, and along rocky shore of protected bays in tropical areas. All have tiny teeth that are similar to hypodermic needles. They can inject an extremely poisonous venom that acts very swiftly, causing acute pain, swelling, paralysis, blindness, and possible death within hours. Avoid handling all cone shells.

These shells are found in both temperate and tropical waters. They are similar to cone shells but much thinner and longer. They poison in the same way as cone shells, but the venom is slightly less poisonous.

FISH WITH TOXIC FLESH

There are no simple rules to tell edible fish from those with poisonous flesh. The most common toxic fish are shown in Figure 5-15. All of these fish contain various types of poisonous substances or toxins in their flesh and are dangerous to eat. They have the following common characteristics:

- Most live in shallow water around reefs or lagoons.
- Many have boxy or round bodies with hard shell-like skins covered with bony plates or spines. They have small parrotlike mouths, small gills, and small or absent belly fins. Their names suggest their shape.

Blowfish or puffer (*Tetraodontidae* species) are more tolerant of coldwater. You find them along tropical and temperate coasts worldwide, even in some of the rivers of Southeast Asia and Africa. Stout-bodied and round, many of these fish have short spines and can inflate themselves into a ball when alarmed or agitated. Their blood, liver, and gonads are so toxic that as little as 28 milligrams (1 ounce) can be fatal. These fish vary in color and size, growing up to 75 centimeters in length.

The triggerfish (*Balistidae* species) occur in great variety, mostly in tropical seas. They are deep-bodied and compressed, resembling a seagoing pancake up to 60 centimeters in length, with large and sharp dorsal spines. Avoid them all, as many have poisonous flesh.

Although most people avoid them because of their ferocity, they occasionally eat barracuda (*Sphyraena barracuda*). These predators of mostly tropical seas can reach almost 1.5 meters in length and have attacked humans without provocation. They occasionally carry the poison ciguatera in their flesh, making them deadly if consumed.

In addition to the above fish and their characteristics, red snapper fish may carry ciguatera, a toxin that accumulates in the systems of fish that feed on tropical marine reefs.

Without specific local information, take the following precautions:

- Be very careful with fish taken from normally shallow lagoons with sandy or broken coral bottoms. Reef-feeding species predominate and some may be poisonous.
- Avoid poisonous fish on the leeward side of an island. This area of shallow water consists of patches of living corals mixed with open spaces and may extend seaward for some distance. Many different types of fish inhabit these shallow waters, some of which are poisonous.
- Do not eat fish caught in any area where the water is unnaturally discolored. This may be indicative of plankton that cause various types of toxicity in plankton-feeding fish.
- Try fishing on the windward side or in deep passages leading from the open sea to the lagoon, but be careful of currents and waves. Live coral reefs drop off sharply into deep water and form a dividing line between the suspected fish of the shallows and the desirable deepwater species. Deepwater fish are usually not poisonous. You can catch the various toxic fish even in deep water. Discard all suspected reef fish, whether caught on the ocean or the reef side.

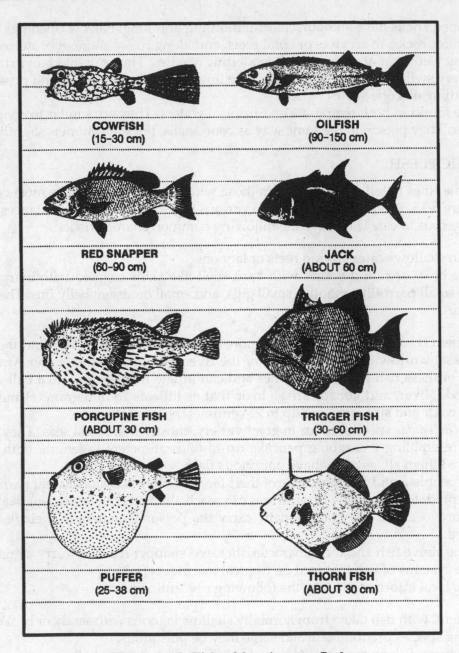

Figure 5-15: Fish with poisonous flesh.

CHAPTER 6

Survival Use of Plants

After having solved the problems of finding water, shelter, and animal food, you will have to consider the use of plants you can eat. In a survival situation you should always be on the lookout for familiar wild foods and live off the land whenever possible.

You must not count on being able to go for days without food as some sources would suggest. Even in the most static survival situation, maintaining health through a complete and nutritious diet is essential to maintaining strength and peace of mind.

Nature can provide you with food that will let you survive any ordeal, if you don't eat the wrong plant. You must therefore learn as much as possible beforehand about the flora of the region where you will be operating. Plants can provide you with medicines in a survival situation. Plants can supply you with weapons and raw materials to construct shelters and build fires. Plants can even provide you with chemicals for poisoning fish, preserving animal hides, and for camouflaging yourself and your equipment.

Note: You will find illustrations of the plants described at the end of this chapter and the end of Chapter 7.

EDIBILITY OF PLANTS

Plants are valuable sources of food because they are widely available, easily procured, and, in the proper combinations, can meet all your nutritional needs.

WARNING

The critical factor in using plants for food is to avoid accidental poisoning. Eat only those plants you can positively identify and you know are safe to eat. See Chapter 7, Poisonous Plants, for more information.

Absolutely identify plants before using them as food. Poison hemlock has killed people who mistook it for its relatives, wild carrots and wild parsnips.

At times you may find yourself in a situation for which you could not plan. In this instance you may not have had the chance to learn the plant life of the region in which you must survive. In this case you can use the Universal Edibility Test to determine which plants you can eat and those to avoid.

It is important to be able to recognize both cultivated and wild edible plants in a survival situation. Most of the information in this chapter is directed towards identifying wild plants because information relating to cultivated plants is more readily available.

Remember the following when collecting wild plants for food:

- Plants growing near homes and occupied buildings or along roadsides may have been sprayed with pesticides. Wash them thoroughly. In more highly developed countries with many automobiles, avoid roadside plants, if possible, due to contamination from exhaust emissions.
- Plants growing in contaminated water or in water containing *Giardia lamblia* and other parasites are contaminated themselves. Boil or disinfect them.
- Some plants develop extremely dangerous fungal toxins. To lessen the chance of accidental poisoning, do not eat any fruit that is starting to spoil or showing signs of mildew or fungus.

- Plants of the same species may differ in their toxic or subtoxic compounds content because of genetic or environmental factors. One example of this is the foliage of the common choke-cherry. Some chokecherry plants have high concentrations of deadly cyanide compounds while others have low concentrations or none. Horses have died from eating wilted wild cherry leaves. Avoid any weed, leaves, or seeds with an almondlike scent, a characteristic of the cyanide compounds.

- Some people are more susceptible to gastric distress (from plants) than others. If you are sensitive in this way, avoid unknown wild plants. If you are extremely sensitive to poison ivy, avoid products from this family, including any parts from sumacs, mangoes, and cashews.

- Some edible wild plants, such as acorns and water lily rhizomes, are bitter. These bitter substances, usually tannin compounds, make them unpalatable. Boiling them in several changes of water will usually remove these bitter properties.

- Many valuable wild plants have high concentrations of oxalate compounds, also known as oxalic acid. Oxalates produce a sharp burning sensation in your mouth and throat and damage the kidneys. Baking, roasting, or drying usually destroys these oxalate crystals. The corm (bulb) of the jack-in-the-pulpit is known as the "Indian turnip," but you can eat it only after removing these crystals by slow baking or by drying.

WARNING

Do not eat mushrooms in a survival situation! The only way to tell if a mushroom is edible is by positive identification. There is no room for experimentation. Symptoms of the most dangerous mushrooms affecting the central nervous system may show up after several days have passed when it is too late to reverse their effects.

Plant Identification. You identify plants, other than by memorizing particular varieties through familiarity, by using such factors as leaf shape and margin, leaf arrangements, and root structure.

The basic leaf margins (Figure 6-1) are toothed, lobed, and toothless or smooth.

These leaves may be lance-shaped, elliptical, egg-shaped, oblong, wedge-shaped, triangular, long-pointed, or top-shaped (Figure 6-2).

The basic types of leaf arrangements (Figure 6-3) are opposite, alternate, compound, simple, and basal rosette.

The basic types of root structures (Figure 6-4) are the bulb, clove, taproot, tuber, rhizome, corm, and crown. Bulbs are familiar to us as onions and, when sliced in half, will show concentric rings. Cloves are those bulblike structures that remind us of garlic and will separate into small pieces when broken apart. This characteristic separates wild onions from wild garlic. Taproots resemble carrots and may be single-rooted or branched, but usually only one plant stalk arises from each root. Tubers are like potatoes and daylilies and you will find these structures either on strings or in clusters underneath the parent plants. Rhizomes are large creeping rootstock or underground stems and many plants arise from the "eyes" of these roots. Corms are similar to bulbs but are solid when cut rather than possessing rings. A crown is the type of root structure found on plants such as asparagus and looks much like a mophead under the soil's surface.

Learn as much as possible about plants you intend to use for food and their unique characteristics. Some plants have both edible and poisonous parts. Many are edible only at certain times of the year. Others may have poisonous relatives that look very similar to the ones you can eat or use for medicine.

Universal Edibility Test. There are many plants throughout the world. Tasting or swallowing even a small portion of some can cause severe discomfort, extreme internal disorders, and even death. Therefore, if you have the slightest doubt about a plant's edibility, apply the Universal Edibility Test (Table 6-1) before eating any portion of it.

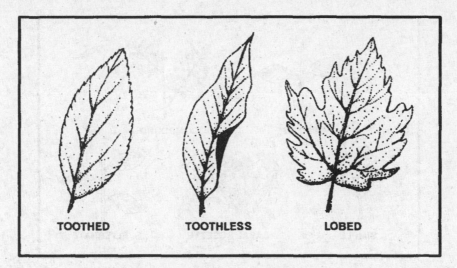

Figure 6-1: Leaf margins.

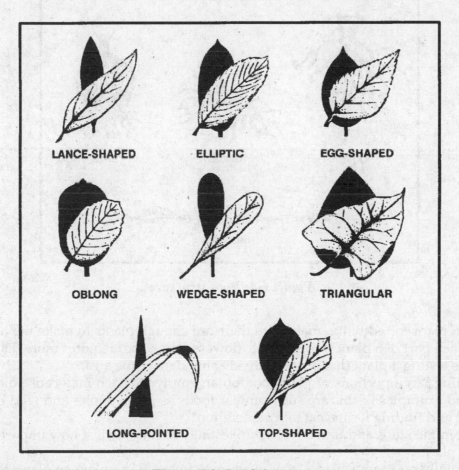

Figure 6-2: Leaf shapes.

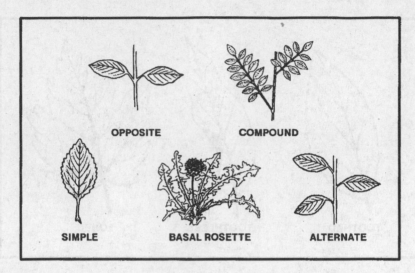

Figure 6-3: Leaf arrangements.

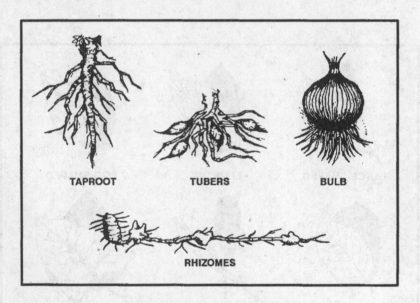

Figure 6-4: Root structures.

Before testing a plant for edibility, make sure there are enough plants to make the testing worth your time and effort. Each part of a plant (roots, leaves, flowers, and so on) requires more than 24 hours to test. Do not waste time testing a plant that is not relatively abundant in the area.

Remember, eating large portions of plant food on an empty stomach may cause diarrhea, nausea, or cramps. Two good examples of this are such familiar foods as green apples and wild onions. Even after testing plant food and finding it safe, eat it in moderation.

You can see from the steps and time involved in testing for edibility just how important it is to be able to identify edible plants.

To avoid potentially poisonous plants, stay away from any wild or unknown plants that have—

- Milky or discolored sap.
- Beans, bulbs, or seeds inside pods.
- Bitter or soapy taste.

- Spines, fine hairs, or thorns.
- Dill, carrot, parsnip, or parsleylike foliage.
- "Almond" scent in woody parts and leaves.
- Grain heads with pink, purplish, or black spurs.
- Three-leaved growth pattern.

Using the above criteria as eliminators when choosing plants for the Universal Edibility Test will cause you to avoid some edible plants. More important, these criteria will often help you avoid plants that are potentially toxic to eat or touch.

An entire encyclopedia of edible wild plants could be written, but space limits the number of plants presented here. Learn as much as possible about the plant life of the areas where you train regularly and where you expect to be traveling or working. Listed below and on the following pages are some of the most

Table 6-1: Universal Edibility Test.

1	Test only one part of a potential food plant at a time.
2	Separate the plant into its basic components—leaves, stems, roots, buds, and flowers.
3	Smell the food for strong or acid odors. Remember, smell alone does not indicate a plant is edible or inedible.
4	Do not eat for 8 hours before starting the test.
5	During the 8 hours you abstain from eating, test for contact poisoning by placing a piece of the plant part you are testing on the inside of your elbow or wrist. Usually 15 minutes is enough time to allow for a reaction.
6	During the test period, take nothing by mouth except purified water and the plant part you are testing.
7	Select a small portion of a single part and prepare it the way you plan to eat it.
8	Before placing the prepared plant part in your mouth, touch a small portion (a pinch) to the outer surface of your lip to test for burning or itching.
9	If after 3 minutes there is no reaction on your lip, place the plant part on your tongue, holding it there for 15 minutes.
10	If there is no reaction, thoroughly chew a pinch and hold it in your mouth for 15 minutes. **Do not swallow.**
11	If no burning, itching, numbing, stinging, or other irritation occurs during the 15 minutes, swallow the food.
12	Wait 8 hours. If any ill effects occur during this period, induce vomiting and drink a lot of water.
13	If no ill effects occur, eat 0.25 cup of the same plant part prepared the same way. Wait another 8 hours. If no ill effects occur, the plant part as prepared is safe for eating.

CAUTION

Test all parts of the plant for edibility, as some plants have both edible and inedible parts. Do not assume that a part that proved edible when cooked is also edible when raw. Test the part raw to ensure edibility before eating raw. The same part or plant may produce varying reactions in different individuals.

common edible and medicinal plants. Detailed descriptions and photographs of these and other common plants are at the end of this chapter.

Table 6-2: Temperate food zone plants.

TEMPERATE ZONE FOOD PLANTS
• Amaranth *(Amaranthus retrof/exus* and other species)
• Arrowroot *(Sagittaria* species)
• Asparagus *(Asparagus officinalis)*
• Beechnut *(Fagus* species)
• Blackberries *(Rubus* species)
• Blueberries *(Vaccinium* species)
• Burdock *(Arctium lappa)*
• Cattail *(Typha* species)
• Chestnut *(Castanea* species)
• Chicory *(Cichoriurn intybus)*
• Chufa *(Cyperus esculentus)*
• Dandelion *(Taraxacum officinale)*
• Daylily *(Hemerocallis fulva)*
• Nettle *(Utica* species)
• Oaks *(Quercus* species)
• Persimmon *(Diospyros virginiana)*
• Plantain *(Plantago* species)
• Pokeweed *(Phytolacca americana)*
• Prickly pear cactus *(Opuntia species)*
• Purslane *(Portulaca oleracea)*
• Sassafras *(Sassafras albidum)*
• Sheep sorrel *(Rumex acetosella)*
• Strawberries *(Fragaria* species)
• Thistle *(Cirsium* species)
• Water lily and lotus *(Nuphar, Nelumbo,* and other species)
• Wild onion and garlic *(Allium* species)
• Wild rose *(Rosa* species)
• Wood sorrel *(Oxalis* species)

Table 6-3: Tropical zone food plants.

TROPICAL ZONE FOOD PLANTS
• Bamboo *(Bambusa* and other species)
• Bananas (Musa species)
• Breadfruit *(Artocarpus incisa)*
• Cashew nut *(Anacardium occidental)*
• Coconut (Cocos *nucifera)*
• Mango *(Mangifera indica)*
• Palms (various species)
• Papaya *(Carica* species)
• Sugarcane *(Saccharum officinarum)*
• Taro *(Colocasia* species)

Table 6-4: Desert zone food plants.

DESERT ZONE FOOD PLANTS
• Acacia *(Acacia farnesiana)*
• Agave *(Agave species)*
• Cactus (various species)
• Date palm *(Phoenix dactylifera)*
• Desert amaranth *(Amaranths palmeri)*

Seaweeds. One plant you should never overlook is seaweed. It is a form of marine algae found on or near ocean shores. There are also some edible freshwater varieties. Seaweed is a valuable source of iodine, other minerals, and vitamin C. Large quantities of seaweed in an unaccustomed stomach can produce a severe laxative effect.

When gathering seaweeds for food, find living plants attached to rocks or floating free. Seaweed washed on shore any length of time may be spoiled or decayed. You can dry freshly harvested seaweeds for later use.

Its preparation for eating depends on the type of seaweed. You can dry thin and tender varieties in the sun or over a fire until crisp. Crush and add these to soups or broths. Boil thick, leathery seaweeds for a short time to soften them. Eat them as a vegetable or with other foods. You can eat some varieties raw after testing for edibility.

Table 6-5: Seaweeds.

SEAWEEDS
• Dulse *(Rhodymenia palmata)*
• Green seaweed *(Ulva lactuca)*
• Irish moss *(Chondrus crispus)*
• Kelp *(Alaria esculenta)*
• Laver *(Porphyra species)*
• Mojaban *(Sargassum fulvellum)*
• Sugar wrack *(Laminaria saccharin)*

Preparation of Plant Food. Although some plants or plant parts are edible raw, you must cook others to be edible or palatable. Edible means that a plant or food will provide you with necessary nutrients, while palatable means that it actually is pleasing to eat. Many wild plants are edible but barely palatable. It is a good idea to learn to identify, prepare, and eat wild foods.

Methods used to improve the taste of plant food include soaking, boiling, cooking, or leaching. Leaching is done by crushing the food (for example, acorns), placing it in a strainer, and pouring boiling water through it or immersing it in running water.

Boil leaves, stems, and buds until tender, changing the water, if necessary, to remove any bitterness.

Boil, bake, or roast tubers and roots. Drying helps to remove caustic oxalates from some roots like those in the Arum family.

Leach acorns in water, if necessary, to remove the bitterness. Some nuts, such as chestnuts, are good raw, but taste better roasted.

You can eat many grains and seeds raw until they mature. When hard or dry, you may have to boil or grind them into meal or flour.

The sap from many trees, such as maples, birches, walnuts, and sycamores, contains sugar. You may boil these saps down to a syrup for sweetening. It takes about 35 liters of maple sap to make one liter of maple syrup!

PLANTS FOR MEDICINE

In a survival situation you will have to use what is available. In using plants and other natural remedies, positive identification of the plants involved is as critical as in using them for food. Proper use of these plants is equally important.

Terms and Definitions

The following terms, and their definitions, are associated with medicinal plant use:

- Poultice. The name given to crushed leaves or other plant parts, possibly heated, that you apply to a wound or sore either directly or wrapped in cloth or paper.
- Infusion or tisane or tea. The preparation of medicinal herbs for internal or external application. You place a small quantity of a herb in a container, pour hot water over it, and let it steep (covered or uncovered) before use.
- Decoction. The extract of a boiled down or simmered herb leaf or root. You add herb leaf or root to water. You bring them to a sustained boil or simmer to draw their chemicals into the water. The average ratio is about 28 to 56 grams (1 to 2 ounces) of herb to 0.5 liter of water.
- Expressed juice. Liquids or saps squeezed from plant material and either applied to the wound or made into another medicine.

Many natural remedies work slower than the medicines you know. Therefore, start with smaller doses and allow more time for them to take effect. Naturally, some will act more rapidly than others.

Specific Remedies. The following remedies are for use only in a survival situation, not for routine use:

- Diarrhea. Drink tea made from the roots of blackberries and their relatives to stop diarrhea. White oak bark and other barks containing tannin are also effective. However, use them with caution when nothing else is available because of possible negative effects on the kidneys. You can also stop diarrhea by eating white clay or campfire ashes. Tea made from cowberry or cranberry or hazel leaves works too.
- Antihemorrhagics. Make medications to stop bleeding from a poultice of the puffball mushroom, from plantain leaves, or most effectively from the leaves of the common yarrow or woundwort (*Achilles millefolium*).
- Antiseptics. Use to cleanse wounds, sores, or rashes. You can make them from the expressed juice from wild onion or garlic, or expressed juice from chickweed leaves or the crushed leaves of dock. You can also make antiseptics from a decoction of burdock root, mallow leaves or roots, or white oak bark. All these medications are for external use only.
- Fevers. Treat a fever with a tea made from willow bark, an infusion of elder flowers or fruit, linden flower tea, or elm bark decoction.
- Colds and sore throats. Treat these illnesses with a decoction made from either plantain leaves or willow bark. You can also use a tea made from burdock roots, mallow or mullein flowers or roots, or mint leaves.
- Aches, pains, and sprains. Treat with externally applied poultices of dock, plantain, chickweed, willow bark, garlic, or sorrel. You can also use salves made by mixing the expressed juices of these plants in animal fat or vegetable oils.
- Itching. Relieve the itch from insect bites, sunburn, or plant poisoning rashes by applying a poultice of jewelweed (*Impatiens biflora*) or witch hazel leaves (*Hamamelis virginiana*).The jewelweed juice will help when applied to poison ivy rashes or insect stings. It works on sunburn as well as aloe vera.
- Sedatives. Get help in falling asleep by brewing a tea made from mint leaves or passionflower leaves.

- Hemorrhoids. Treat them with external washes from elm bark or oak bark tea, from the expressed juice of plantain leaves, or from a Solomon's seal root decoction.
- Constipation. Relieve constipation by drinking decoctions from dandelion leaves, rose hips, or walnut bark. Eating raw daylily flowers will also help.
- Worms or intestinal parasites. Using moderation, treat with tea made from tansy (*Tanacetum vulgare*) or from wild carrot leaves.
- Gas and cramps. Use a tea made from carrot seeds as an antiflatulent; use tea made from mint leaves to settle the stomach.
- Antifungal washes. Make a decoction of walnut leaves or oak bark or acorns to treat ringworm and athlete's foot. Apply frequently to the site, alternating with exposure to direct sunlight.

MISCELLANEOUS USE OF PLANTS

- Make dyes from various plants to color clothing or to camouflage your skin. Usually, you will have to boil the plants to get the best results. Onion skins produce yellow, walnut hulls produce brown, and pokeberries provide a purple dye.
- Make fibers and cordage from plant fibers. Most commonly used are the stems from nettles and milkweeds, yucca plants, and the inner bark of trees like the linden.
- Make fish poison by immersing walnut hulls in a small area of quiet water. This poison makes it impossible for the fish to breathe but doesn't adversely affect their edibility.
- Make tinder for starting fires from cattail fluff, cedar bark, lighter knot wood from pine trees, or hardened sap from resinous wood trees.
- Make insulation by fluffing up female cattail heads or milkweed down.
- Make insect repellents by applying the expressed juice of wild garlic or onion to the skin, by placing sassafras leaves in your shelter, or by burning or smudging cattail seed hair fibers.

Plants can be your ally as long as you use them cautiously. The key to the safe use of plants is positive identification whether you use them as food or medicine or in constructing shelters or equipment.

EDIBLE AND MEDICINAL PLANTS

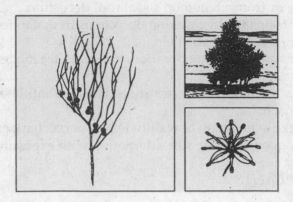

Abal
Calligonum comosum

Description: The abal is one of the few shrubby plants that exists in the shady deserts. This plant grows to about 1.2 meters, and its branches look like wisps from a broom. The stiff, green branches produce an abundance of flowers in the early spring months (March, April).

Habitat and Distribution: This plant is found in desert scrub and waste in any climatic zone. It inhabits much of the North African desert. It may also be found on the desert sands of the Middle East and as far eastward as the Rajputana desert of westen India.

Edible Parts: This plant's general appearance would not indicate its usefulness to the survivor, but while this plant is flowering in the spring, its fresh flowers can be eaten. This plant is common in the areas where it is found. An analysis of the food value of this plant has shown it to be high in sugar and nitrogenous components.

Acacia
Acacia farnesiana

Description: Acacia is a spreading, usually short tree with spines and alternate compound leaves. Its individual leaflets are small. Its flowers are ball-shaped, bright yellow, and very fragrant. Its bark is a whitish-gray color. Its fruits are dark brown and podlike.

Habitat and Distribution: Acacia grows in open, sunny areas. It is found throughout all tropical regions.

Note: There are about 500 species of acacia. These plants are especially prevalent in Africa, southern Asia, and Australia, but many species are found in the warmer and drier parts of America.

Edible Parts: Its young leaves, flowers, and pods are edible raw or cooked.

Agave
Agave species

Description: These plants have large clusters of thick, fleshy leaves borne close to the ground and surrounding a central stalk. The plants flower only once, then die. They produce a massive flower stalk.

Habitat and Distribution: Agaves prefer dry, open areas. They are found throughout Central America, the Caribbean, and parts of the western deserts of the United States and Mexico.

Edible Parts: Its flowers and flower buds are edible. Boil them before eating.

> ⚠ **CAUTION**
> The juice of some species causes dermatitis in some individuals.

Other Uses: Cut the huge flower stalk and collect the juice for drinking. Some species have very fibrous leaves. Pound the leaves and remove the fibers for weaving and making ropes. Most species have thick, sharp needles at the tips of the leaves. Use them for sewing or making hacks. The sap of some species contains a chemical that makes the sap suitable for use as a soap.

Almond
Prunus amygdalus

Description: The almond tree, which sometimes grows to 12.2 meters, looks like a peach tree. The fresh almond fruit resembles a gnarled, unripe peach and grows in clusters. The stone (the almond itself) is covered with a thick, dry, woolly skin.

Habitat and Distribution: Almonds are found in the scrub and thorn forests of the tropics, the evergreen scrub forests of temperate areas, and in desert scrub and waste in all climatic zones. The almond tree is also found in the semidesert areas of the Old World in southern Europe, the eastern Mediterranean, Iran, the Middle East, China, Madeira, the Azores, and the Canary Islands.

Edible Parts: The mature almond fruit splits open lengthwise down the side, exposing the ripe almond nut. You can easily get the dry kernel by simply cracking open the stone. Almond meats are rich in food value, like all nuts. Gather them in large quantities and shell them for further use as survival food. You could live solely on almonds for rather long periods. When you boil them, the kernel's outer covering comes off and only the white meat remains.

Amaranth
Amaranthus species

Description: These plants, which grow 90 centimeters to 150 centimeters tall, are abundant weeds in many parts of the world. All amaranth have alternate simple leaves. They may have some red color present on the stems. They bear minute, greenish flowers in dense clusters at the top of the plants. Their seeds may be brown or black in weedy species and light-colored in domestic species.

Habitat and Distribution: Look for amaranth along roadsides, in disturbed waste areas, or as weeds in crops throughout the world. Some amaranth species have been grown as a grain crop and a garden vegetable in various parts of the world, especially in South America.

Edible Parts: All parts are edible, but some may have sharp spines you should remove before eating. The young plants or the growing tips of alder plants are an excellent vegetable. Simply boil the young plants or eat them raw. Their seeds are very nutritious. Shake the tops of alder plants to get the seeds. Eat the seeds raw, boiled, ground into flour, or popped like popcorn.

Arctic willow
Salix arctica

Description: The arctic willow is a shrub that never exceeds more than 60 centimeters in height and grows in clumps that form dense mats on the tundra.

Habitat and Distribution: The arctic willow is common on tundras in North America, Europe, and Asia. You can also find it in some mountainous areas in temperate regions.

Edible Parts: You can collect the succulent, tender young shoots of the arctic willow in early spring. Strip off the outer bark of the new shoots and eat the inner portion raw. You can also peel and eat raw the young underground shoots of any of the various kinds of arctic willow. Young willow leaves are one of the richest sources of vitamin C, containing 7 to 10 times more than an orange.

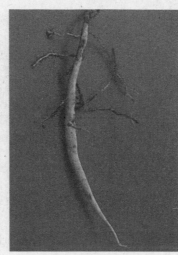

Arrowroot
Maranta and Sagittaria species

Description: The arrowroot is an aquatic plant with arrow-shaped leaves and potatolike tubers in the mud.

Habitat and Distribution: Arrowroot is found worldwide in temperate zones and the tropics. It is found in moist to wet habitats.

Edible Parts: The rootstock is a rich source of high quality starch. Boil the rootstock and eat it as a vegetable.

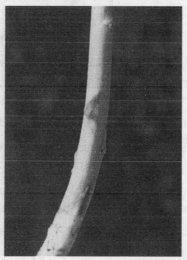

Asparagus
Asparagus officinalis

Description: The spring growth of this plant resembles a cluster of green fingers. The mature plant has fernlike, wispy foliage and red berries. Its flowers are small and greenish in color. Several species have sharp, thornlike structures.

Habitat and Distribution: Asparagus is found worldwide in temperate areas. Look for it in fields, old homesites, and fencerows.

Edible Parts: Eat the young stems before leaves form. Steam or boil them for 10 to 15 minutes before eating. Raw asparagus may cause nausea or diarrhea. The fleshy roots are a good source of starch.

Bael fruit
Aegle marmelos

Description: This is a tree that grows from 2.4 to 4.6 meters tall, with a dense spiny growth. The fruit is 5 to 10 centimeters in diameter, gray or yellowish, and full of seeds.

Habitat and Distribution: Bael fruit is found in rain forests and semi-evergreen seasonal forests of the tropics. It grows wild in India and Burma.

Edible Parts: The fruit, which ripens in December, is at its best when just turning ripe. The juice of the ripe fruit, diluted with water and mixed with a small amount of tamarind and sugar or honey, is sour but refreshing. Like other citrus fruits, it is rich in vitamin C.

Bamboo
Various species including *Bambusa, Dendrocalamus, Phyllostachys*

Description: Bamboos are woody grasses that grow up to 15 meters tall. The leaves are grasslike and the stems are the familiar bamboo used in furniture and fishing poles.

Habitat and Distribution: Look for bamboo in warm, moist regions in open or jungle country, in lowland, or on mountains. Bamboos are native to the Far East (Temperate and Tropical zones) but have bean widely planted around the world.

Edible Parts: The young shoots of almost all species are edible raw or cooked. Raw shoots have a slightly bitter taste that is removed by boiling. To prepare, remove the tough protective sheath that is coated with tawny or red hairs. The seed grain of the flowering bamboo is also edible. Boil the seeds like rice or pulverize them, mix with water, and make into cakes.

Other Uses: Use the mature bamboo to build structures or to make containers, ladles, spoons, and various other cooking utensils. Also use bamboo to make tools and weapons. You can make a strong bow by splitting the bamboo and putting several pieces together.

⚠ **CAUTION**

Green bamboo may explode in a fire. Green bamboo has an internal membrane you must remove before using it as a food or water container.

Banana and plantain
Musa species

Description: These are treelike plants with several large leaves at the top. Their flowers are borne in dense hanging clusters.

Habitat and Distribution: Look for bananas and plantains in open fields or margins of forests where they are grown as a crop. They grow in the humid tropics.

Edible Parts: Their fruits are edible raw or cooked. They may be boiled or baked. You can boil their flowers and eat them like a vegetable. You can cook and eat the rootstocks and leaf sheaths of many species. The center or "heart" or the plant is edible year-round, cooked or raw.

Other Uses: You can use the layers of the lower third of the plants to cover coals to roast food. You can also use their stumps to get water (see Chapter 6). You can use their leaves to wrap other foods for cooking or storage.

Baobab
Adansonia digitata

Description: The baobab tree may grow as high as 18 meters and may have a trunk 9 meters in diameter. The tree has short, stubby branches and a gray, thick bark. Its leaves are compound and their segments are arranged like the palm of a hand. Its flowers, which are white and several centimeters across, hang from the higher branches. Its fruit is shaped like a football, measures up to 45 centimeters long, and is covered with short dense hair.

Habitat and Distribution: These trees grow in savannas. They are found in Africa, in parts of Australia, and on the island of Madagascar.

Edible Parts: You can use the young leaves as a soup vegetable. The tender root of the young baobab tree is edible. The pulp and seeds of the fruit are also edible. Use one handful of pulp to about one cup of water for a refreshing drink. To obtain flour, roast the seeds, then grind them.

Other Uses: Drinking a mixture of pulp and water will help cure diarrhea. Often the hollow trunks are good sources of fresh water. The bark can be cut into strips and pounded to obtain a strong fiber for making rope.

Batoko plum
Flacourtia inermis

Description: This shrub or small tree has dark green, alternate, simple leaves. Its fruits are bright red and contain six or more seeds.

Habitat and Distribution: This plant is a native of the Philippines but is widely cultivated for its fruit in other areas. It can be found in clearings and at the edges of the tropical rain forests of Africa and Asia.

Edible Parts: Eat the fruit raw or cooked.

Bearberry or kinnikinnick
Arctostaphylos uvaursi

Description: This plant is a common evergreen shrub with reddish, scaly bark and thick, leathery leaves 4 centimeters long and 1 centimeter wide. It has white flowers and bright red fruits.

Habitat and Distribution: This plant is found in arctic, subarctic, and temperate regions, most often in sandy or rocky soil.

Edible Parts: Its berries are edible raw or cooked. You can make a refreshing tea from its young leaves.

Beech
Fagus species

Description: Beech trees are large (9 to 24 meters), symmetrical forest trees that have smooth, light-gray bark and dark green foliage. The character of its bark, plus its clusters of prickly seedpods, clearly distinguish the beech tree in the field.

Habitat and Distribution: This tree is found in the Temperate Zone. It grows wild in the eastern United States, Europe, Asia, and North Africa. It is found in moist areas, mainly in the forests. This tree is common throughout southeastern Europe and across temperate Asia. Beech relatives are also found in Chile, New Guinea, and New Zealand.

Edible Parts: The mature beechnuts readily fall out of the husklike seedpods. You can eat these dark brown triangular nuts by breaking the thin shell with your fingernail and removing the white, sweet kernel inside. Beechnuts are one of the most delicious of all wild nuts. They are a most useful survival food because of the kernel's high oil content. You can also use the beechnuts as a coffee substitute. Roast them so that the kernel becomes golden brown and quite hard. Then pulverize the kernel and, after boiling or steeping in hot water, you have a passable coffee substitute.

Bignay
Antidesma bunius

Description: Bignay is a shrub or small tree, 3 to 12 meters tall, with shiny, pointed leaves about 15 centimeters long. Its flowers are small, clustered, and green. It has fleshy, dark red or black fruit and a single seed. The fruit is about 1 centimeter in diameter.

Habitat and Distribution: This plant is found in rain forests and semi-evergreen seasonal forests in the tropics. It is found in open places and in secondary forests. It grows wild from the Himalayas to Ceylon and eastward through Indonesia to northern Australia. However, it may be found anywhere in the tropics in cultivated forms.

Edible Parts: The fruit is edible raw. Do not eat any other parts of the tree. In Africa, the roots are toxic. Other parts of the plant may be poisonous.

⚠ CAUTION
Eaten in large quantities, the fruit may have a laxative effect.

Blackberry, raspberry, and dewberry
Rubus species

Description: These plants have prickly stems (canes) that grow upward, arching back toward the ground. They have alternate, usually compound leaves. Their fruits may be red, black, yellow, or orange.

Habitat and Distribution: These plants grow in open, sunny areas at the margin of woods, lakes, streams, and roads throughout temperate regions. There is also an arctic raspberry.

Edible Parts: The fruits and peeled young shoots are edible. Flavor varies greatly.

Other Uses: Use the leaves to make tea. To treat diarrhea, drink a tea made by brewing the dried root bark of the blackberry bush.

Blueberry and huckleberry
Vaccinium and Gaylussacia species

Description: These shrubs vary in size from 30 centimeters to 3.7 meters tall. All have alternate, simple leaves. Their fruits may be dark blue, black, or red and have many small seeds.

Habitat and Distribution: These plants prefer open, sunny areas. They are found throughout much of the north temperate regions and at higher elevations in Central America.

Edible Parts: Their fruits are edible raw.

Breadfruit
Artocarpus incisa

Description: This tree may grow up to 9 meters tall. It has dark green, deeply divided leaves that are 75 centimeters long and 30 centimeters wide. Its fruits are large, green, ball-like structures up to 30 centimeters across when mature.

Habitat and Distribution: Look for this tree at the margins of forests and homesites in the humid tropics. It is native to the South Pacific region but has been widely planted in the West Indies and parts of Polynesia.

Edible Parts: The fruit pulp is edible raw. The fruit can be sliced, dried, and ground into flour for later use. The seeds are edible cooked.

Other Uses: The thick sap can serve as glue and caulking material. You can also use it as birdlime (to entrap small birds by smearing the sap on twigs where they usually perch).

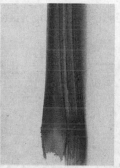

Burdock
Arctium lappa

Description: This plant has wavy-edged, arrow-shaped leaves and flower heads in burrlike clusters. It grows up to 2 meters tall, with purple or pink flowers and a large, fleshy root.

Habitat and Distribution: Burdock is found worldwide in the North Temperate Zone. Look for it in open waste areas during the spring and summer.

Edible Parts: Peel the tender leaf stalks and eat them raw or cook them like greens. The roots are also edible boiled or baked.

⚠ CAUTION
Do not confuse burdock with rhubarb that has poisonous leaves.

Other Uses: A liquid made from the roots will help to produce sweating and increase urination. Dry the root, simmer it in water, strain the liquid, and then drink the strained liquid. Use the fiber from the dried stalk to weave cordage.

Burl Palm
Corypha elata

Description: This tree may reach 18 meters in height. It has large, fan-shaped leaves up to 3 meters long and split into about 100 narrow segments. It bears flowers in huge clusters at the top of the tree. The tree dies after flowering.

Habitat and Distribution: This tree grows in coastal areas of the East Indies.

Edible Parts: The trunk contains starch that is edible raw. The very tip of the trunk is also edible raw or cooked. You can get large quantities of liquid by bruising the flowering stalk. The kernels of the nuts are edible.

> ⚠ **CAUTION**
> The seed covering may cause dermatitis in some individuals.

Other Uses: You can use the leaves as weaving material.

Canna lily
Canna indica

Description: The canna lily is a coarse perennial herb, 90 centimeters to 3 meters tall. The plant grows from a large, thick, underground rootstock that is edible. Its large leaves resemble those of the banana plant but are not so large. The flowers of wild canna lily are usually small, relatively inconspicuous, and brightly colored reds, oranges, or yellows.

Habitat and Distribution: As a wild plant, the canna lily is found in all tropical areas, especially in moist places along streams, springs, ditches, and the margins of woods. It may also be found in wet temperate, mountainous regions. It is easy to recognize because it is commonly cultivated in flower gardens in the United States.

Edible Parts: The large and much branched rootstocks are full of edible starch. The younger parts may be finely chopped and then boiled or pulverized into a meal. Mix in the young shoots of palm cabbage for flavoring.

Carob tree
Ceratonia siliqua

Description: This large tree has a spreading crown. Its leaves are compound and alternate. Its seedpods, also known as Saint John's bread, are up to 45 centimeters long and are filled with round, hard seeds and a thick pulp.

Habitat and Distribution: This tree is found throughout the Mediterranean, the Middle East, and parts of North Africa.

Edible Parts: The young tender pods are edible raw or boiled. You can pulverize the seeds in mature pods and cook as porridge.

Cashew nut
Anacardium occidentale

Description: The cashew is a spreading evergreen tree growing to a height of 12 meters, with leaves up to 20 centimeters long and 10 centimeters wide. Its flowers are yellowish-pink. Its fruit is very easy to recognize because of its peculiar structure. The fruit is thick and pear-shaped, pulpy and red or yellow when ripe. This fruit bears a hard, green, kidney-shaped nut at its tip. This nut is smooth, shiny, and green or brown according to its maturity.

Habitat and Distribution: The cashew is native to the West Indies and northern South America, but transplantation has spread it to all tropical climates. In the Old World, it has escaped from cultivation and appears to be wild at least in parts of Africa and India.

Edible Parts: The nut encloses one seed. The seed is edible when roasted. The pear-shaped fruit is juicy, sweet-acid, and astringent. It is quite safe and considered delicious by most people who eat it.

⚠ CAUTION
The green hull surrounding the nut contains a resinous irritant poison that will blister the lips and tongue like poison ivy. Heat destroys this poison when roasting the nuts.

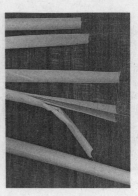

Cattail
Typha latifolia

Description: Cattails are grasslike plants with strap-shaped leaves 1 to 5 centimeters wide and growing up to 1.8 meters tall. The male flowers are borne in a dense mass above the female flowers. These last only a short time, leaving the female flowers that develop into the brown cattail. Pollen from the male flowers is often abundant and bright yellow.

Habitat and Distribution: Cattails are found throughout most of the world. Look for them in full sun areas at the margins of lakes, streams, canals, rivers, and brackish water.

Edible Parts: The young tender shoots are edible raw or cooked. The rhizome is often very tough but is a rich source of starch. Pound the rhizome to remove the starch and use as a flour. The pollen is also an exceptional source of starch. When the cattail is immature and still green, you can boil the female portion and eat it like corn on the cob.

Other Uses: The dried leaves are an excellent source of weaving material you can use to make floats and rafts. The cottony seeds make good pillow stuffing and insulation. The fluff makes excellent tinder. Dried cattails are effective insect repellents when burned.

Cereus cactus
Cereus species

Description: These cacti are tall and narrow with angled stems and numerous spines.

Habitat and Distribution: They may be found in true deserts and other dry, open, sunny areas throughout the Caribbean region, Central America, and the western United States.

Edible Parts: The fruits are edible, but some may have a laxative effect.

Other Uses: The pulp of the cactus is a good source of water. Break open the stem and scoop out the pulp.

Chestnut
Castanea sativa

Description: The European chestnut is usually a large tree, up to 18 meters in height.

Habitat and Distribution: In temperate regions, the chestnut is found in both hardwood and coniferous forests. In the tropics, it is found in semi-evergreen seasonal forests. They are found over all of middle and south Europe and across middle Asia to China and Japan. They are relatively abundant along the edge of meadows and as a forest tree. The European chestnut is one of the most common varieties. Wild chestnuts in Asia belong to the related chestnut species.

Edible Parts: Chestnuts are highly useful as survival food. Ripe nuts are usually picked in autumn, although unripe nuts picked while green may also be used for food. Perhaps the easiest way to prepare them is to roast the ripe nuts in embers. Cooked this way, they are quite tasty, and you can eat large quantities. Another way is to boil the kernels after removing the outer shell. After being boiled until fairly soft, you can mash the nuts like potatoes.

Chicory
Cichorium intybus

Description: This plant grows up to 1.8 meters tall. It has leaves clustered at the base of the stem and some leaves on the stem. The base leaves resemble those of the dandelion. The flowers are sky blue and stay open only on sunny days. Chicory has a milky juice.

Habitat and Distribution: Look for chicory in old fields, waste areas, weedy lots, and along roads. It is a native of Europe and Asia, but is also found in Africa and most of North America where it grows as a weed.

Edible Parts: All parts are edible. Eat the young leaves as a salad or boil to eat as a vegetable. Cook the roots as a vegetable. For use as a coffee substitute, roast the roots until they are dark brown and then pulverize them.

Chufa
Cyperus esculentus

Description: This very common plant has a triangular stem and grasslike leaves. It grows to a height of 20 to 60 centimeters. The mature plant has a soft furlike bloom that extends from a whorl of leaves. Tubers 1 to 2.5 centimeters in diameter grow at the ends of the roots.

Habitat and Distribution: Chufa grows in moist sandy areas throughout the world. It is often an abundant weed in cultivated fields.

Edible Parts: The tubers are edible raw, boiled, or baked. You can also grind them and use them as a coffee substitute.

Coconut
Cocos nucifera

Description: This tree has a single, narrow, tall trunk with a cluster of very large leaves at the top. Each leaf may be over 6 meters long with over 100 pairs of leaflets.

Habitat and Distribution: Coconut palms are found throughout the tropics. They are most abundant near coastal regions.

Edible Parts: The nut is a valuable source of food. The milk of the young coconut is rich in sugar and vitamins and is an excellent source of liquid. The nut meat is also nutritious but is rich in oil. To preserve the meat, spread it in the sun until it is completely dry.

Other Uses: Use coconut oil to cook and to protect metal objects from corrosion. Also use the oil to treat saltwater sores, sunburn, and dry skin. Use the oil in improvised torches. Use the tree trunk as building material and the leaves as thatch. Hollow out the large stump for use as a food container. The coconut husks are good flotation devices and the husk's fibers are used to weave ropes and other items. Use the gauzelike fibers at the leaf bases as strainers or use them to weave a bug net or to make a pad to use on wounds. The husk makes a good abrasive. Dried husk fiber is an excellent tinder. A smoldering husk helps to repel mosquitoes. Smoke caused by dripping coconut oil in a fire also repels mosquitoes. To render coconut oil, put the coconut meat in the sun, heat it over a slow fire, or boil it in a pot of water. Coconuts washed out to sea are a good source of fresh liquid for the sea survivor.

Common jujube
Ziziphus jujuba

Description: The common jujube is either a deciduous tree growing to a height of 12 meters or a large shrub, depending upon where it grows and how much water is available for growth. Its branches are usually spiny. Its reddish-brown to yellowish-green fruit is oblong to ovoid, 3 centimeters or less in diameter, smooth, and sweet in flavor, but has rather dry pulp around a comparatively large stone. Its flowers are green.

Habitat and Distribution: The jujube is found in forested areas of temperate regions and in desert scrub and waste areas worldwide. It is common in many of the tropical and subtropical areas of the Old World. In Africa, it is found mainly bordering the Mediterranean. In Asia, it is especially common in the drier parts of India and China. The jujube is also found throughout the East Indies. It can be found bordering some desert areas.

Edible Parts: The pulp, crushed in water, makes a refreshing beverage. If time permits, you can dry the ripe fruit in the sun like dates. Its fruits are high in vitamins A and C.

Cranberry
Vaccinium macrocarpon

Description: This plant has tiny leaves arranged alternately. Its stem creeps along the ground. Its fruits are red berries.

Habitat and Distribution: It only grows in open, sunny, wet areas in the colder regions of the Northern Hemisphere.

Edible Parts: The berries are very tart when eaten raw. Cook in a small amount of water and add sugar, if available, to make a jelly.

Other Uses: Cranberries may act as a diuretic. They are useful for treating urinary tract infections.

Crowberry
Empetrum nigrum

Description: This is a dwarf evergreen shrub with short needlelike leaves. It has small, shiny, black berries that remain on the bush throughout the winter.

Habitat and Distribution: Look for this plant in tundra throughout arctic regions of North America and Eurasia.

Edible Parts: The fruits are edible fresh or can be dried for later use.

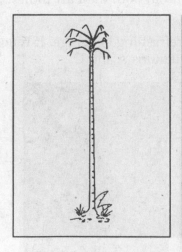

Cuipo tree
Cavanillesia platanifolia

Description: This is a very dominant and easily detected tree because it extends above the other trees. Its height ranges from 45 to 60 meters. It has leaves only at the top and is bare 11 months out of the year. It has rings on its bark that extend to the top to make is easily recognizable. Its bark is reddish or gray in color. Its roots are light reddish-brown or yellowish-brown.

Habitat and Distribution: The cuipo tree is located primarily in Central American tropical rain forests in mountainous areas.

Edible Parts: To get water from this tree, cut a piece of the root and clean the dirt and bark off one end, keeping the root horizontal. Put the clean end to your mouth or canteen and raise the other. The water from this tree tastes like potato water.

Other Uses: Use young saplings and the branches' inner bark to make rope.

Dandelion
Taraxacum officinale

Description: Dandelion leaves have a jagged edge, grow close to the ground, and are seldom more than 20 centimeters long. Its flowers are bright yellow. There are several dandelion species.

Habitat and Distribution: Dandelions grow in open, sunny locations throughout the Northern Hemisphere.

Edible Parts: All parts are edible. Eat the leaves raw or cooked. Boil the roots as a vegetable. Roots roasted and ground are a good coffee substitute. Dandelions are high in vitamins A and C and in calcium.

Other Uses: Use the white juice in the flower stems as glue.

Date palm
Phoenix dactylifera

Description: The date palm is a tall, unbranched tree with a crown of huge, compound leaves. Its fruit is yellow when ripe.

Habitat and Distribution: This tree grows in arid semitropical regions. It is native to North Africa and the Middle East but has been planted in the arid semitropics in other parts of the world.

Edible Parts: Its fruit is edible fresh but is very bitter if eaten before it is ripe. You can dry the fruits in the sun and preserve them for a long time.

Other Uses: The trunks provide valuable building material in desert regions where few other treelike plants are found. The leaves are durable and you can use them for thatching and as weaving material. The base of the leaves resembles coarse cloth that you can use for scrubbing and cleaning.

Daylily
Hemerocallis fulva

Description: This plant has unspotted, tawny blossoms that open for 1 day only. It has long, swordlike, green basal leaves. Its root is a mass of swollen and elongated tubers.

Habitat and Distribution: Daylilies are found worldwide in Tropic and Temperate Zones. They are grown as a vegetable in the Orient and as an ornamental plant elsewhere.

Edible Parts: The young green leaves are edible raw or cooked. Tubers are also edible raw or cooked. You can eat its flowers raw, but they taste better cooked. You can also fry the flowers for storage.

⚠ CAUTION
Eating excessive amounts of raw flowers may cause diarrhea.

Duchesnea or Indian strawberry
Duchesnea indica

Description: The duchesnea is a small plant that has runners and three-parted leaves. Its flowers are yellow and its fruit resembles a strawberry.

Habitat and Distribution: It is native to southern Asia but is a common weed in warmer temperate regions. Look for it in lawns, gardens, and along roads.

Edible Parts: Its fruit is edible. Eat it fresh.

Elderberry
Sambucus canadensis

Description: Elderberry is a many-stemmed shrub with opposite, compound leaves. It grows to a height of 6 meters. Its flowers are fragrant, white, and borne in large flat-topped clusters up to 30 centimeters across. Its berrylike fruits are dark blue or black when ripe.

Habitat and Distribution: This plant is found in open, usually wet areas at the margins of marshes, rivers, ditches, and lakes. It grows throughout much of eastern North America and Canada.

Edible Parts: The flowers and fruits are edible. You can make a drink by soaking the flower heads for 8 hours, discarding the flowers, and drinking the liquid.

⚠ CAUTION
All other parts of the plant are poisonous and dangerous if eaten.

Fireweed
Epilobium angustifolium

Description: This plant grows up to 1.8 meters tall. It has large, showy, pink flowers and lance-shaped leaves. Its relative, the dwarf fireweed (*Epilobium latifolium*), grows 30 to 60 centimeters tall.

Habitat and Distribution: Tall fireweed is found in open woods, on hillsides, on stream banks, and near seashores in arctic regions. It is especially abundant in burned-over areas. Dwarf fireweed is found along streams, sandbars, and lakeshores and on alpine and arctic slopes.

Edible Parts: The leaves, stems, and flowers are edible in the spring but become tough in summer. You can split open the stems of old plants and eat the pith raw.

Fishtail palm
Caryota urens

Description: Fishtail palms are large trees, at least 18 meters tall. Their leaves are unlike those of any other palm; the leaflets are irregular and toothed on the upper margins. All other palms have either fan-shaped or featherlike leaves. Its massive flowering shoot is borne at the top of the tree and hangs downward.

Habitat and Distribution: The fishtail palm is native to the tropics of India, Assam, and Burma. Several related species also exist in Southeast Asia and the Philippines. These palms are found in open hill country and jungle areas.

Edible Parts: The chief food in this palm is the starch stored in large quantities in its trunk. The juice from the fishtail palm is very nourishing and you have to drink it shortly after getting it from the palm flower shoot. Boil the juice down to get a rich sugar syrup. Use the same method as for the sugar palm to get the juice. The palm cabbage may be eaten raw or cooked.

Foxtail grass
Setaria species

Description: This weedy grass is readily recognized by the narrow, cylindrical head containing long hairs. Its grains are small, less than 6 millimeters long. The dense heads of grain often droop when ripe.

Habitat and Distribution: Look for foxtail grasses in open, sunny areas, along roads, and at the margins of fields. Some species occur in wet, marshy areas. Species of Setaria are found throughout the United States, Europe, western Asia, and tropical Africa. In some parts of the world, foxtail grasses are grown as a food crop.

Edible Parts: The grains are edible raw but are very hard and sometimes bitter. Boiling removes some of the bitterness and makes them easier to eat.

Goa bean
Psophocarpus tetragonolobus

Description: The goa bean is a climbing plant that may cover small shrubs and trees. Its bean pods are 22 centimeters long, its leaves 15 centimeters long, and its flowers are bright blue. The mature pods are 4-angled, with jagged wings on the pods.

Habitat and Distribution: This plant grows in tropical Africa, Asia, the East Indies, the Philippines, and Taiwan. This member of the bean (legume) family serves to illustrate a kind of edible bean common in the tropics of the Old World. Wild edible beans of this sort are most frequently found in clearings and around abandoned garden sites. They are more rare in forested areas.

Edible Parts: You can eat the young pods like string beans. The mature seeds are a valuable source of protein after parching or roasting them over hot coals. You can germinate the seeds (as you can many kinds of beans) in damp moss and eat the resultant sprouts. The thickened roots are edible raw. They are slightly sweet, with the firmness of an apple. You can also eat the young leaves as a vegetable, raw or steamed.

Hackberry
Celtis species

Description: Hackberry trees have smooth, gray bark that often has corky warts or ridges. The tree may reach 39 meters in height. Hackberry trees have long-pointed leaves that grow in two rows. This tree bears small, round berries that can be eaten when they are ripe and fall from the tree. The wood of the hackberry is yellowish.

Habitat and Distribution: This plant is widespread in the United States, especially in and near ponds.

Edible Parts: Its berries are edible when they are ripe and fall from the tree.

Hazelnut or wild filbert
Corylus species

Description: Hazelnuts grow on bushes 1.8 to 3.6 meters high. One species in Turkey and another in China are large trees. The nut itself grows in a very bristly husk that conspicuously contracts above the nut into a long neck. The different species vary in this respect as to size and shape.

Habitat and Distribution: Hazelnuts are found over wide areas in the United States, especially the eastern half of the country and along the Pacific coast. These nuts are also found in Europe where they are known as filberts. The hazelnut is common in Asia, especially in eastern Asia from the Himalayas to China and Japan. The hazelnut usually grows in the dense thickets along stream banks and open places. They are not plants of the dense forest.

Edible Parts: Hazelnuts ripen in the autumn when you can crack them open and eat the kernel. The dried nut is extremely delicious. The nut's high oil content makes it a good survival food. In the unripe stage, you can crack them open and eat the fresh kernel.

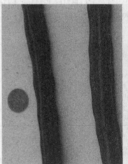

Horseradish tree
Moringa pterygosperma

Description: This tree grows from 4.5 to 14 meters tall. Its leaves have a fernlike appearance. Its flowers and long, pendulous fruits grow on the ends of the branches. Its fruit (pod) looks like a giant bean. Its 25- to 60-centimeter-long pods are triangular in cross section, with strong ribs. Its roots have a pungent odor.

Habitat and Distribution: This tree is found in the rain forests and semi-evergreen seasonal forests of the tropical regions. It is widespread in India, Southeast Asia, Africa, and Central America. Look for it in abandoned fields and gardens and at the edges of forests.

Edible Parts: The leaves are edible raw or cooked, depending on their hardness. Cut the young seedpods into short lengths and cook them like string beans or fry them. You can get oil for frying by boiling the young fruits of palms and skimming the oil off the surface of the water. You can eat the flowers as part of a salad. You can chew fresh, young seedpods to eat the pulpy and soft seeds. The roots may be ground as a substitute for seasoning similar to horseradish.

Iceland moss
Cetraria islandica

Description: This moss grows only a few inches high. Its color may be gray, white, or even reddish.

Habitat and Distribution: Look for it in open areas. It is found only in the arctic.

Edible Parts: All parts of the Iceland moss are edible. During the winter or dry season, it is dry and crunchy but softens when soaked. Boil the moss to remove the bitterness. After boiling, eat by itself or add to milk or grains as a thickening agent. Dried plants store well.

Indian potato or Eskimo potato
Claytonia species

Description: All Claytonia species are somewhat fleshy plants only a few centimeters tall, with showy flowers about 2.5 centimeters across.

Habitat and Distribution: Some species are found in rich forests where they are conspicuous before the leaves develop. Western species are found throughout most of the northern United States and in Canada.

Edible Parts: The tubers are edible but you should boil them before eating.

Juniper
Juniperus species

Description: Junipers, sometimes called cedars, are trees or shrubs with very small, scalelike leaves densely crowded around the branches. Each leaf is less than 1.2 centimeters long. All species have a distinct aroma resembling the well-known cedar. The berrylike cones are usually blue and covered with a whitish wax.

Habitat and Distribution: Look for junipers in open, dry, sunny areas throughout North America and northern Europe. Some species are found in southeastern Europe, across Asia to Japan, and in the mountains of North Africa.

Edible Parts: The berries and twigs are edible. Eat the berries raw or roast the seeds to use as a coffee substitute. Use dried and crushed berries as a seasoning for meat. Gather young twigs to make a tea.

⚠ CAUTION

Many plants may be called cedars but are not related to junipers and may be harmful. Always look for the berrylike structures, neddle leaves, and resinous, fragrant sap to be sure the plant you have is a juniper.

Lotus
Nelumbo species

Description: There are two species of lotus: one has yellow flowers and the other pink flowers. The flowers are large and showy. The leaves, which may float on or rise above the surface of the water, often reach 1.5 meters in radius. The fruit has a distinctive flattened shape and contains up to 20 hard seeds.

Habitat and Distribution: The yellow-flowered lotus is native to North America. The pink-flowered species, which is widespread in the Orient, is planted in many other areas of the world. Lotuses are found in quiet fresh water.

Edible Parts: All parts of the plant are edible raw or cooked. The underwater parts contain large quantities of starch. Dig the fleshy portions from the mud and bake or boil them. Boil the young leaves and eat them as a vegetable. The seeds have a pleasant flavor and are nutritious. Eat them raw, or parch and grind them into flour.

Malanga
Xanthosoma caracu

Description: This plant has soft, arrow-shaped leaves, up to 60 centimeters long. The leaves have no aboveground stems.

Habitat and Distribution: This plant grows widely in the Caribbean region. Look for it in open, sunny fields.

Edible Parts: The tubers are rich in starch. Cook them before eating to destroy a poison contained in all parts of the plant.

Mango
Mangifera indica

Description: This tree may reach 30 meters in height. It has alternate, simple, shiny, dark green leaves. Its flowers are small and inconspicuous. Its fruits have a large single seed. There are many cultivated varieties of mango. Some have red flesh, others yellow or orange, often with many fibers and a kerosene taste.

Habitat and Distribution: This tree grows in warm, moist regions. It is native to northern India, Burma, and western Malaysia. It is now grown throughout the tropics.

Edible Parts: The fruits are a nutritious food source. The unripe fruit can be peeled and its flesh eaten by shredding it and eating it like a salad. The ripe fruit can be peeled and eaten raw. Roasted seed kernels are edible.

⚠ CAUTION

If you are sensitive to poison ivy, avoid eating mangoes, as they cause a severe reaction in sensitive individuals.

Manioc
Manihot utillissima

Description: Manioc is a perennial shrubby plant, 1 to 3 meters tall, with jointed stems and deep green, fingerlike leaves. It has large, fleshy rootstocks.

Habitat and Distribution: Manioc is widespread in all tropical climates, particularly in moist areas. Although cultivated extensively, it may be found in abandoned gardens and growing wild in many areas.

Edible Parts: The rootstocks are full of starch and high in food value. Two kinds of manioc are known: bitter and sweet. Both are edible. The bitter type contains poisonous hydrocyanic acid. To prepare manioc, first grind the fresh manioc root into a pulp, then cook it for at least 1 hour to remove the bitter poison from the roots. Then flatten the pulp into cakes and bake as bread. Manioc cakes or flour will keep almost indefinitely if protected against insects and dampness. Wrap them in banana leaves for protection.

> ⚠ **CAUTION**
> For safety, always cook the roots of either type.

Marsh marigold
Caltha palustris

Description: This plant has rounded, dark green leaves arising from a short stem. It has bright yellow flowers.

Habitat and Distribution: This plant is found in bogs, lakes, and slow-moving streams. It is abundant in arctic and subarctic regions and in much of the eastern region of the northern United States.

Edible Parts: All parts are edible if boiled.

> ⚠ **CAUTION**
> As with all water plants, do not eat this plant raw. Raw water plants may carry dangerous organisms that are removed only by cooking.

Mulberry
Morus species

Description: This tree has alternate, simple, often lobed leaves with rough surfaces. Its fruits are blue or black and many seeded.

Habitat and Distribution: Mulberry trees are found in forests, along roadsides, and in abandoned fields in Temperate and Tropical Zones of North America, South America, Europe, Asia, and Africa.

Edible Parts: The fruit is edible raw or cooked. It can be dried for eating later.

> ⚠ **CAUTION**
>
> When eaten in quantity, mulberry fruit acts as a laxative. Green, unripe fruit can be hallucinogenic and cause extreme nausea and cramps.

Other Uses: You can shred the inner bark of the tree and use it to make twine or cord.

Nettle
Urtica and Laportea species

Description: These plants grow several feet high. They have small, inconspicuous flowers. Fine, hairlike bristles cover the stems, leafstalks, and undersides of leaves. The bristles cause a stinging sensation when they touch the skin.

Habitat and Distribution: Nettles prefer moist areas along streams or at the margins of forests. They are found throughout North America, Central America, the Caribbean, and northern Europe.

Edible Parts: Young shoots and leaves are edible. Boiling the plant for 10 to 15 minutes destroys the stinging element of the bristles. This plant is very nutritious.

Other Uses: Mature stems have a fibrous layer that you can divide into individual fibers and use to weave string or twine.

Nips palm
Nips fruticans

Description: This palm has a short, mainly underground trunk and very large, erect leaves up to 6 meters tall. The leaves are divided into leaflets. A flowering head forms on a short erect stern that rises among the palm leaves. The fruiting (seed) head is dark brown and may be 30 centimeters in diameter.

Habitat and Distribution: This palm is common on muddy shores in coastal regions throughout eastern Asia.

Edible Parts: The young flower stalk and the seeds provide a good source of water and food. Cut the flower stalk and collect the juice. The juice is rich in sugar. The seeds are hard but edible.

Other Uses: The leaves are excellent as thatch and coarse weaving material.

Oak

Quercus species

Description: Oak trees have alternate leaves and acorn fruits. There are two main groups of oaks: red and white. The red oak group has leaves with bristles and smooth bark in the upper part of the tree. Red oak acorns take 2 years to mature. The white oak group has leaves without bristles and a rough bark in the upper portion of the tree. White oak acorns mature in 1 year.

Habitat and Distribution: Oak trees are found in many habitats throughout North America, Central America, and parts of Europe and Asia.

Edible Parts: All parts are edible, but often contain large quantities of bitter substances. White oak acorns usually have a better flavor than red oak acorns. Gather and shell the acorns. Soak red oak acorns in water for 1 to 2 days to remove the bitter substance. You can speed up this process by putting wood ashes in the water in which you soak the acorns. Boil the acorns or grind them into flour and use the flour for baking. You can use acorns that you baked until very dark as a coffee substitute.

⚠ **CAUTION**

Tannic acid gives the acorns their bitter taste. Eating an excessive amount of acorns high in tannic acid can lead to kidney failure. Before eating acorns, leach out this chemical.

Other Uses: Oak wood is excellent for building or burning. Small oaks can be split and cut into long thin strips (3 to 6 millimeters thick and 1.2 centimeters wide) used to weave mats, baskets, or frameworks for packs, sleds, furniture, etc. Oak bark soaked in water produces a tanning solution used to preserve leather.

Orach
Atriplex species

Description: This plant is vinelike in growth and has arrowhead-shaped, alternate leaves up to 5 centimeters long. Young leaves maybe silver-colored. Its flowers and fruits are small and inconspicuous.

Habitat and Distribution: Orach species are entirely restricted to salty soils. They are found along North America's coasts and on the shores of alkaline lakes inland. They are also found along seashores from the Mediterranean countries to inland areas in North Africa and eastward to Turkey and central Siberia.

Edible Parts: The entire plant is edible raw or boiled.

Palmetto palm
Sabal palmetto

Description: The palmetto palm is a tall, unbranched tree with persistent leaf bases on most of the trunk. The leaves are large, simple, and palmately lobed. Its fruits are dark blue or black with a hard seed.

Habitat and Distribution: The palmetto palm is found throughout the coastal regions of the southeastern United States.

Edible Parts: The fruits are edible raw. The hard seeds may be ground into flour. The heart of the palm is a nutritious food source at any time. Cut off the top of the tree to obtain the palm heart.

Papaya or pawpaw
Carica papaya

Description: The papaya is a small tree 1.8 to 6 meters tall, with a soft, hollow trunk. When cut, the entire plant exudes a milky juice. The trunk is rough and the leaves are crowded at the trunk's apex. The fruit grows directly from the trunk, among and below the leaves. The fruit is green before ripening. When ripe, it turns yellow or remains greenish with a squashlike appearance.

Habitat and Distribution: Papaya is found in rain forests and semi-evergreen seasonal forests in tropical regions and in some temperate regions as well. Look for it in moist areas near clearings and former habitations. It is also found in open, sunny places in uninhabited jungle areas.

Edible Parts: The ripe fruit is high in vitamin C. Eat it raw or cook it like squash. Place green fruit in the sun to make it ripen quickly. Cook the young papaya leaves, flowers, and stems carefully, changing the water as for taro.

⚠ **CAUTION**
Be careful not to get the milky sap from the unripe fruit into your eyes. It will cause intense pain and temporary—sometimes even permanent—blindness.

Other Uses: Use the milky juice of the unripe fruit to tenderize tough meat. Rub the juice on the meat.

Persimmon

Diospyros virginiana and other species

Description: These trees have alternate, dark green, elliptic leaves with entire margins. The flowers are inconspicuous. The fruits are orange, have a sticky consistency, and have several seeds.

Habitat and Distribution: The persimmon is a common forest margin tree. It is wide spread in Africa, eastern North America, and the Far East.

Edible Parts: The leaves are a good source of vitamin C. The fruits are edible raw or baked. To make tea, dry the leaves and soak them in hot water. You can eat the roasted seeds.

⚠ CAUTION

Some persons are unable to digest persimmon pulp. Unripe persimmons are highly astringent and inedible.

Pincushion cactus
Mammilaria species

Description: Members of this cactus group are round, short, barrel-shaped, and without leaves. Sharp spines cover the entire plant.

Habitat and Distribution: These cacti are found throughout much of the desert regions of the western United States and parts of Central America.

Edible Parts: They are a good source of water in the desert.

Pine
Pinus species

Description: Pine trees are easily recognized by their needlelike leaves grouped in bundles. Each bundle may contain one to five needles, the number varying among species. The tree's odor and sticky sap provide a simple way to distinguish pines from similar looking trees with needlelike leaves.

Habitat and Distribution: Pines prefer open, sunny areas. They are found throughout North America, Central America, much of the Caribbean region, North Africa, the Middle East, Europe, and some places in Asia.

Edible Parts: The seeds of all species are edible. You can collect the young male cones, which grow only in the spring, as a survival food. Boil or bake the young cones. The bark of young twigs is edible. Peel off the bark of thin twigs. You can chew the juicy inner bark; it is rich in sugar and vitamins. Eat the seeds raw or cooked. Green pine needle tea is high in vitamin C.

Other Uses : Use the resin to waterproof articles. Also use it as glue. Collect the resin from the tree. If there is not enough resin on the tree, cut a notch in the bark so more sap will seep out. Put the resin in a container and heat it. The hot resin is your glue. Use it as is or add a small amount of ash dust to strengthen it. Use it immediately. You can use hardened pine resin as an emergency dental filling.

Plantain, broad and narrow leaf
Plantago species

Description: The broad leaf plantain has leaves over 2.5 centimeters across that grow close to the ground. The flowers are on a spike that rises from the middle of the cluster of leaves. The narrow leaf plantain has leaves up to 12 centimeters long and 2.5 centimeters wide, covered with hairs. The leaves form a rosette. The flowers are small and inconspicuous.

Habitat and Distribution: Look for these plants in lawns and along roads in the North Temperate Zone. This plant is a common weed throughout much of the world.

Edible Parts: The young tender leaves are edible raw. Older leaves should be cooked. Seeds are edible raw or roasted.

Other Uses: To relieve pain from wounds and sores, wash and soak the entire plant for a short time and apply it to the injured area. To treat diarrhea, drink tea made from 28 grams (1 ounce) of the plant leaves boiled in 0.5 liter of water. The seeds and seed husks act as laxatives.

Pokeweed
Phytolacca americana

Description: This plant may grow as high as 3 meters. Its leaves are elliptic and up to 1 meter in length. It produces many large clusters of purple fruits in late spring.

Habitat and Distribution: Look for this plant in open, sunny areas in forest clearings, in fields, and along roadsides in eastern North America, Central America, and the Caribbean.

Edible Parts: The young leaves and stems are edible cooked. Boil them twice, discarding the water from the first boiling. The fruits are edible if cooked.

> ⚠ **CAUTION**
>
> All parts of this plant are poisonous if eaten raw. Never eat the underground portions of the plant as these contain the highest concentrations of the poisons. Do not eat any plant over 25 centimeters tall or when red is showing in the plant.

Other Uses: Use the juice of fresh berries as a dye.

Prickly pear cactus
Opuntia species

Description: This cactus has flat, padlike stems that are green. Many round, furry dots that contain sharp-pointed hairs cover these stems.

Habitat and Distribution: This cactus is found in arid and semiarid regions and in dry, sandy areas of wetter regions throughout most of the United States and Central and South America. Some species are planted in arid and semiarid regions of other parts of the world.

Edible Parts: All parts of the plant are edible. Peel the fruits and eat them fresh or crush them to prepare a refreshing drink. Avoid the tiny, pointed hairs. Roast the seeds and grind them to a flour.

⚠ CAUTION
Avoid any prickly pear cactus like plant with milky sap.

Other Uses: The pad is a good source of water. Peel it carefully to remove all sharp hairs before putting it in your mouth. You can also use the pads to promote healing. Split them and apply the pulp to wounds.

Purslane

Portulaca oleracea

Description: This plant grows close to the ground. It is seldom more than a few centimeters tall. Its stems and leaves are fleshy and often tinged with red. It has paddle-shaped leaves, 2.5 centimeter or less long, clustered at the tips of the stems. Its flowers are yellow or pink. Its seeds are tiny and black.

Habitat and Distribution: It grows in full sun in cultivated fields, field margins, and other weedy areas throughout the world.

Edible Parts: All parts are edible. Wash and boil the plants for a tasty vegetable or eat them raw. Use the seeds as a flour substitute or eat them raw.

Rattan palm

Calamus species

Description: The rattan palm is a stout, robust climber. It has hooks on the midrib of its leaves that it uses to remain attached to trees on which it grows. Sometimes, mature stems grow to 90 meters. It has alternate, compound leaves and a whitish flower.

Habitat and Distribution: The rattan palm is found from tropical Africa through Asia to the East Indies and Australia. It grows mainly in rain forests.

Edible Parts: Rattan palms hold a considerable amount of starch in their young stem tips. You can eat them roasted or raw. In other kinds, a gelatinous pulp, either sweet or sour, surrounds the seeds. You can suck out this pulp. The palm heart is also edible raw or cooked.

Other Uses: You can obtain large amounts of potable water by cutting the ends of the long stems (see Chapter 6). The stems can be used to make baskets and fish traps.

Reed
Phragmites australis

Description: This tall, coarse grass grows to 3.5 meters tall and has gray-green leaves about 4 centimeters wide. It has large masses of brown flower branches in early summer. These rarely produce grain and become fluffy, gray masses late in the season.

Habitat and Distribution: Look for reed in any open, wet area, especially one that has been disturbed through dredging. Reed is found throughout the temperate regions of both the Northern and Southern Hemispheres.

Edible Parts: All parts of the plant are edible raw or cooked in any season. Harvest the stems as they emerge from the soil and boil them. You can also harvest them just before they produce flowers, then dry and beat them into flour. You can also dig up and boil the underground stems, but they are often tough. Seeds are edible raw or boiled, but they are rarely found.

Reindeer moss
Cladonia rangiferina

Description: Reindeer moss is a low-growing plant only a few centimeters tall. It does not flower but does produce bright red reproductive structures.

Habitat and Distribution: Look for this lichen in open, dry areas. It is very common in much of North America.

Edible Parts: The entire plant is edible but has a crunchy, brittle texture. Soak the plant in water with some wood ashes to remove the bitterness, then dry, crush, and add it to milk or to other food.

Rock tripe
Umbilicaria species

Description: This plant forms large patches with curling edges. The top of the plant is usually black. The underside is lighter in color.

Habitat and Distribution: Look on rocks and boulders for this plant. It is common throughout North America.

Edible Parts: The entire plant is edible. Scrape it off the rock and wash it to remove grit. The plant may be dry and crunchy; soak it in water until it becomes soft. Rock tripes may contain large quantities of bitter substances; soaking or boiling them in several changes of water will remove the bitterness.

⚠ **CAUTION**

There are some reports of poisoning from rock tripe, so apply the Universal Edibility Test.

Rose apple
Eugenia jambos

Description: This tree grows 3 to 9 meters high. It has opposite, simple, dark green, shiny leaves. When fresh, it has fluffy, yellowish-green flowers and red to purple egg-shaped fruit.

Habitat and Distribution: This tree is widely planted in all of the tropics. It can also be found in a semiwild state in thickets, waste places, and secondary forests.

Edible Parts: The entire fruit is edible raw or cooked.

Sago palm
Metroxylon sagu

Description: These palms are low trees, rarely over 9 meters tall, with a stout, spiny trunk. The outer rind is about 5 centimeters thick and hard as bamboo. The rind encloses a spongy inner pith containing a high proportion of starch. It has typical palmlike leaves clustered at the tip.

Habitat and Distribution: Sago palm is found in tropical rain forests. It flourishes in damp lowlands in the Malay Peninsula, New Guinea, Indonesia, the Philippines, and adjacent islands. It is found mainly in swamps and along streams, lakes, and rivers.

Edible Parts: These palms, when available, are of great use to the survivor. One trunk, cut just before it flowers, will yield enough sago to feed a person for 1 year. Obtain sago starch from nonflowering palms. To extract the edible sage, cut away the bark lengthwise from one half of the trunk, and pound the soft, whitish inner part (pith) as fine as possible. Knead the pith in water and strain it through a coarse cloth into a container. The fine, white sago will settle in the container. Once the sago settles, it is ready for use. Squeeze off the excess water and let it dry. Cook it as pancakes or oatmeal. Two kilograms of sago is the nutritional equivalent of 1.5 kilograms of rice. The upper part of the trunk's core does not yield sage, but you can roast it in lumps over a fire. You can also eat the young sago nuts and the growing shoots or palm cabbage.

Other Uses: Use the stems of tall sorghums as thatching materials.

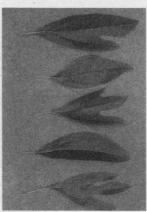

Sassafras
Sassafras albidum

Description: This shrub or small tree bears different leaves on the same plant. Some leaves will have one lobe, some two lobes, and some no lobes. The flowers, which appear in early spring, are small and yellow. The fruits are dark blue. The plant parts have a characteristics root beer smell.

Habitat and Distribution: Sassafras grows at the margins of roads and forests, usually in open, sunny areas. It is a common tree throughout eastern North America.

Edible Parts: The young twigs and leaves are edible fresh or dried. You can add dried young twigs and leaves to soups. Dig the underground portion, peel off the bark, and let it dry. Then boil it in water to prepare sassafras tea.

Other Uses: Shred the tender twigs for use as a toothbrush.

Saxaul
Haloxylon ammondendron

Description: The saxaul is found either as a small tree or as a large shrub with heavy, coarse wood and spongy, water-soaked bark. The branches of the young trees are vivid green and pendulous. The flowers are small and yellow.

Habitat and Distribution: The saxaul is found in desert and arid areas. It is found on the arid salt deserts of Central Asia, particularly in the Turkestan region and east of the Caspian Sea.

Edible Parts: The thick bark acts as a water storage organ. You can get drinking water by pressing quantities of the bark. This plant is an important source of water in the arid regions in which it grows.

Screw pine
Pandanus species

Description: The screw pine is a strange plant on stilts, or prop roots, that support the plant above-ground so that it appears more or less suspended in midair. These plants are either shrubby or treelike, 3 to 9 meters tall, with stiff leaves having sawlike edges. The fruits are large, roughened balls resembling pineapples, but without the tuft of leaves at the end.

Habitat and Distribution: The screw pine is a tropical plant that grows in rain forests and semievergreen seasonal forests. It is found mainly along seashores, although certain kinds occur inland for some distance, from Madagascar to southern Asia and the islands of the southwestern Pacific. There are about 180 types.

Edible Parts: Knock the ripe fruit to the ground to separate the fruit segments from the hard outer covering. Chew the inner fleshy part. Cook fruit that is not fully ripe in an earth oven. Before cooking, wrap the whole fruit in banana leaves, breadfruit leaves, or any other suitable thick, leathery leaves. After cooking for about 2 hours, you can chew fruit segments like ripe fruit. Green fruit is inedible.

Sea orach
Atriplex halimus

Description: The sea orach is a sparingly branched herbaceous plant with small, gray-colored leaves up to 2.5 centimeters long. Sea orach resembles Iamb's quarter, a common weed in most gardens in the United States. It produces its flowers in narrow, densely compacted spikes at the tips of its branches.

Habitat and Distribution: The sea orach is found in highly alkaline and salty areas along seashores from the Mediterranean countries to inland areas in North Africa and eastward to Turkey and central Siberia. Generally, it can be found in tropical scrub and thorn forests, steppes in temperate regions, and most desert scrub and waste areas.

Edible Parts: Its leaves are edible. In the areas where it grows, it has the healthy reputation of being one of the few native plants that can sustain man in times of want.

Sheep sorrel
Rumex acerosella

Description: These plants are seldom more than 30 centimeters tall. They have alternate leaves, often with arrowlike bases, very small flowers, and frequently reddish stems.

Habitat and Distribution: Look for these plants in old fields and other disturbed areas in North America and Europe.

Edible Parts: The plants are edible raw or cooked.

⚠ **CAUTION**
These plants contain oxalic acid that can be damaging if too many plants are eaten raw. Cooking seems to destroy the chemical.

Sorghum
Sorghum species

Description: There are many different kinds of sorghum, all of which bear grains in heads at the top of the plants. The grains are brown, white, red, or black. Sorghum is the main food crop in many parts of the world.

Habitat and Distribution: Sorghum is found worldwide, usually in warmer climates. All species are found in open, sunny areas.

Edible Parts: The grains are edible at any stage of development. When young, the grains are milky and edible raw. Boil the older grains. Sorghum is a nutritious food.

Other Uses: Use the stems of tall sorghum as building materials.

Spatterdock or yellow water lily
Nuphar species

Description: This plant has leaves up to 60 centimeters long with a triangular notch at the base. The shape of the leaves is somewhat variable. The plant's yellow flowers are 2.5 centimeter across and develop into bottle-shaped fruits. The fruits are green when ripe.

Habitat and Distribution: These plants grow throughout most of North America. They are found in quiet, fresh, shallow water (never deeper than 1.8 meters).

Edible Parts: All parts of the plant are edible. The fruits contain several dark brown seeds you can parch or roast and then grind into flour. The large rootstock contains starch. Dig it out of the mud, peel off the outside, and boil the flesh. Sometimes the rootstock contains large quantities of a very bitter compound. Boiling in several changes of water may remove the bitterness.

Sterculia
Sterculia foetida

Description: Sterculias are tall trees, rising in some instances to 30 meters. Their leaves are either undivided or palmately lobed. Their flowers are red or purple. The fruit of all sterculias is similar in aspect, with a red, segmented seedpod containing many edible black seeds.

Habitat and Distribution: There are over 100 species of sterculias distributed through all warm or tropical climates. They are mainly forest trees.

Edible Parts: The large, red pods produce a number of edible seeds. The seeds of all sterculias are edible and have a pleasant taste similar to cocoa. You can eat them like nuts, either raw or roasted.

⚠ **CAUTION**
Avoid eating large quantities. The seeds may have a laxative effect.

Strawberry
Fragaria species

Description: Strawberry is a small plant with a three-leaved growth pattern. It has small, white flowers usually produced during the spring. Its fruit is red and fleshy.

Habitat and Distribution: Strawberries are found in the North Temperate Zone and also in the high mountains of the southern Western Hemisphere. Strawberries prefer open, sunny areas. They are commonly planted.

Edible Parts: The fruit is edible fresh, cooked, or dried. Strawberries are a good source of vitamin C. You can also eat the plant's leaves or dry them and make a tea with them.

Sugarcane
Saccharum officinarum

Description: This plant grows up to 4.5 meters tall. It is a grass and has grasslike leaves. Its green or reddish stems are swollen where the leaves grow. Cultivated sugarcane seldom flowers.

Habitat and Distribution: Look for sugarcane in fields. It grows only in the tropics (throughout the world). Because it is a crop, it is often found in large numbers.

Edible Parts: The stem is an excellent source of sugar and is very nutritious. Peel the outer portion off with your teeth and eat the sugarcane raw. You can also squeeze juice out of the sugarcane.

Sugar palm
Arenga pinnata

Description: This tree grows about 15 meters high and has huge leaves up to 6 meters long. Needlelike structures stick out of the bases of the leaves. Flowers grow below the leaves and form large conspicuous clusters from which the fruits grow.

Habitat and Distribution: This palm is native to the East Indies but has been planted in many parts off the tropics. It can be found at the margins of forests.

Edible Parts: The chief use of this palm is for sugar. However, its seeds and the tip of its stems are a survival food. Bruise a young flower stalk with a stone or similar object and collect the juice as it comes out. It is an excellent source of sugar. Boil the seeds. Use the tip of the stems as a vegetable.

⚠ **CAUTION**
The flesh covering the seeds may cause dermatitis.

Other Uses: The shaggy material at the base of the leaves makes an excellent rope as it is strong and resists decay.

Sweetsop
Annona squamosa

Description: This tree is small, seldom more than 6 meters tall, and multi-branched. It has alternate, simple, elongate, dark green leaves. Its fruit is green when ripe, round in shape, and covered with protruding bumps on its surface. The fruit's flesh is white and creamy.

Habitat and Distribution: Look for sweetsop at margins of fields, near villages, and around homesites in tropical regions.

Edible Parts: The fruit flesh is edible raw.

Other Uses: You can use the finely ground seeds as an insecticide.

⚠ CAUTION
The ground seeds are extremely dangerous to the eyes.

Tamarind
Tamarindus indica

Description: The tamarind is a large, densely branched tree, up to 25 meters tall. Its has pinnate leaves (divided like a feather) with 10 to 15 pairs of leaflets.

Habitat and Distribution: The tamarind grows in the drier parts of Africa, Asia, and the Philippines. Although it is thought to be a native of Africa, it has been cultivated in India for so long that it looks like a native tree. It it also found in the American tropics, the West Indies, Central America, and tropical South America.

Edible Parts: The pulp surrounding the seeds is rich in vitamin C and is an important survival food. You can make a pleasantly acid drink by mixing the pulp with water and sugar or honey and letting the mixture mature for several days. Suck the pulp to relieve thirst. Cook the young, unripe fruits or seedpods with meat. Use the young leaves in soup. You must cook the seeds. Roast them above a fire or in ashes. Another way is to remove the seed coat and soak the seeds in salted water and grated coconut for 24 hours, then cook them. You can peel the tamarind bark and chew it.

Taro, cocoyam, elephant ears, eddo, dasheen
Colocasia and Alocasia species

Description: All plants in these groups have large leaves, sometimes up to 1.8 meters tall, that grow from a very short stem. The rootstock is thick and fleshy and filled with starch.

Habitat and Distribution: These plants grow in the humid tropics. Look for them in fields and near homesites and villages.

Edible Parts: All parts of the plant are edible when boiled or roasted. When boiling, change the water once to get rid of any poison.

> ⚠ **CAUTION**
> If eaten raw, these plants will cause a serious inflammation of the mouth and throat.

Thistle
Cirsium species

Description: This plant may grow as high as 1.5 meters. Its leaves are long-pointed, deeply lobed, and prickly.

Habitat and Distribution: Thistles grow worldwide in dry woods and fields.

Edible Parts: Peel the stalks, cut them into short sections, and boil them before eating. The roots are edible raw or cooked.

> ⚠ **CAUTION**
>
> Some thistle species are poisonous.

Other Uses: Twist the tough fibers of the stems to make a strong twine.

Ti
Cordyline terminalis

Description: The ti has unbranched stems with straplike leaves often clustered at the tip of the stem. The leaves vary in color and may be green or reddish. The flowers grow at the plant's top in large, plumelike clusters. The ti may grow up to 4.5 meters tall.

Habitat and Distribution: Look for this plant at the margins of forests or near home-sites in tropical areas. It is native to the Far East but is now widely planted in tropical areas worldwide.

Edible Parts: The roots and very tender young leaves are good survival food. Boil or bake the short, stout roots found at the base of the plant. They are a valuable source of starch. Boil the very young leaves to eat. You can use the leaves to wrap other food to cook over coals or to steam.

Other Uses: Use the leaves to cover shelters or to make a rain cloak. Cut the leaves into liners for shoes; this works especially well if you have a blister. Fashion temporary sandals from the ti leaves. The terminal leaf, if not completely unfurled, can be used as a sterile bandage. Cut the leaves into strips, then braid the strips into rope.

Tree fern
Various genera

Description: Tree ferns are tall trees with long, slender trunks that often have a very rough, barklike covering. Large, lacy leaves uncoil from the top of the trunk.

Habitat and Distribution: Tree ferns are found in wet, tropical forests.

Edible Parts: The young leaves and the soft inner portion of the trunk are edible. Boil the young leaves and eat as greens. Eat the inner portion of the trunk raw or bake it.

Tropical almond
Terminalia catappa

Description: This tree grows up to 9 meters tall. Its leaves are evergreen, leathery, 45 centimeters long, 15 centimeters wide, and very shiny. It has small, yellowish-green flowers. Its fruit is flat, 10 centimeters long, and not quite as wide. The fruit is green when ripe.

Habitat and Distribution: This tree is usually found growing near the ocean. It is a common and often abundant tree in the Caribbean and Central and South America. It is also found in the tropical rain forests of southeastern Asia, northern Australia, and Polynesia.

Edible Parts: The seed is a good source of food. Remove the fleshy, green covering and eat the seed raw or cooked.

Walnut
Juglans species

Description: Walnuts grow on very large trees, often reaching 18 meters tall. The divided leaves characterize all walnut spades. The walnut itself has a thick outer husk that must be removed to reach the hard inner shell of the nut.

Habitat and Distribution: The English walnut, in the wild state, is found from southeastern Europe across Asia to China and is abundant in the Himalayas. Several other species of walnut are found in China and Japan. The black walnut is common in the eastern United States.

Edible Parts: The nut kernel ripens in the autumn. You get the walnut meat by cracking the shell. Walnut meats are highly nutritious because of their protein and oil content.

Other Uses: You can boil walnuts and use the juice as an antifungal agent. The husks of "green" walnuts produce a dark brown dye for clothing or camouflage. Crush the husks of "green" black walnuts and sprinkle them into sluggish water or ponds for use as fish poison.

Water chestnut
Trapa natans

Description: The water chestnut is an aquatic plant that roots in the mud and has finely divided leaves that grow underwater. Its floating leaves are much larger and coarsely toothed. The fruits, borne underwater, have four sharp spines on them.

Habitat and Distribution: The water chestnut is a freshwater plant only. It is a native of Asia but has spread to many parts of the world in both temperate and tropical areas.

Edible Parts: The fruits are edible raw and cooked. The seeds are also a source of food.

Water lettuce
Ceratopteris species

Description: The leaves of water lettuce are much like lettuce and are very tender and succulent. One of the easiest ways of distinguishing water lettuce is by the little plantlets that grow from the margins of the leaves. These little plantlets grow in the shape of a rosette. Water lettuce plants often cover large areas in the regions where they are found.

Habitat and Distribution: Found in the tropics throughout the Old World in both Africa and Asia. Another kind is found in the New World tropics from Florida to South America. Water lettuce grows only in very wet places and often as a floating water plant. Look for water lettuce in still lakes, ponds, and the backwaters of rivers.

Edible Parts: Eat the fresh leaves like lettuce. Be careful not to dip the leaves in the contaminated water in which they are growing. Eat only the leaves that are well out of the water.

⚠ CAUTION
This plant has carcinogenic properties and should only be used as a last resort.

Water lily
Nymphaea odorata

Description: These plants have large, triangular leaves that float on the water's surface, large, fragrant flowers that are usually white, or red, and thick, fleshy rhizomes that grow in the mud.

Habitat and Distribution: Water lilies are found throughout much of the temperate and subtropical regions.

Edible Parts: The flowers, seeds, and rhizomes are edible raw or cooked. To prepare rhizomes for eating, peel off the corky rind. Eat raw, or slice thinly, allow to dry, and then grind into flour. Dry, parch, and grind the seeds into flour.

Other Uses: Use the liquid resulting from boiling the thickened root in water as a medicine for diarrhea and as a gargle for sore throats.

Water plantain
Alisma plantago-aquatica

Description: This plant has small, white flowers and heart-shaped leaves with pointed tips. The leaves are clustered at the base of the plant.

Habitat and Distribution: Look for this plant in fresh water and in wet, full sun areas in Temperate and Tropical Zones.

Edible Parts: The rootstocks are a good source of starch. Boil or soak them in water to remove the bitter taste.

⚠ CAUTION
To avoid parasites, always cook aquatic plants.

Wild caper
Capparis aphylla

Description: This is a thorny shrub that loses its leaves during the dry season. Its stems are gray-green and its flowers pink.

Habitat and Distribution: These shrubs form large stands in scrub and thorn forests and in desert scrub and waste. They are common throughout North Africa and the Middle East.

Edible Parts: The fruit and the buds of young shoots are edible raw.

Wild crab apple or wild apple
Malus species

Description: Most wild apples look enough like domestic apples that the survivor can easily recognize them. Wild apple varieties are much smaller than cultivated kinds; the largest kinds usually do not exceed 5 to 7.5 centimeters in diameter, and most often less. They have small, alternate, simple leaves and often have thorns. Their flowers are white or pink and their fruits reddish or yellowish.

Habitat and Distribution: They are found in the savanna regions of the tropics. In temperate areas, wild apple varieties are found mainly in forested areas. Most frequently, they are found on the edge of woods or in fields. They are found throughout the Northern Hemisphere.

Edible Parts: Prepare wild apples for eating in the same manner as cultivated kinds. Eat them fresh, when ripe, or cooked. Should you need to store food, cut the apples into thin slices and dry them. They are a good source of vitamins.

> ⚠ **CAUTION**
> Apple seeds contain cyanide compounds. Do not eat.

Wild desert gourd or colocynth
Citrullus colocynthis

Description: The wild desert gourd, a member of the watermelon family, produces an 2.4- to 3-meter-long ground-trailing vine. The perfectly round gourds are as large as an orange. They are yellow when ripe.

Habitat and Distribution: This creeping plant can be found in any climatic zone, generally in desert scrub and waste areas. It grows abundantly in the Sahara, in many Arab countries, on the southeastern coast of India, and on some of the islands of the Aegean Sea. The wild desert gourd will grow in the hottest localities.

Edible Parts: The seeds inside the ripe gourd are edible after they are completely separated from the very bitter pulp. Roast or boil the seeds—their kernels are rich in oil. The flowers are edible. The succulent stem tips can be chewed to obtain water.

Wild dock and wild sorrel
Rumex crispus and Rumex acetosella

Description: Wild dock is a stout plant with most of its leaves at the base of its stem that is commonly 15 to 30 centimeters brig. The plants usually develop from a strong, fleshy, carrotlike taproot. Its flowers are usually very small, growing in green to purplish plumelike clusters. Wild sorrel similar to the wild dock but smaller. Many of the basal leaves are arrow-shaped but smaller than those of the dock and contain a sour juice.

Habitat and Distribution: These plants can be found in almost all climatic zones of the world, in areas of high as well as low rainfall. Many kinds are found as weeds in fields, along roadsides, and in waste places.

Edible Parts: Because of tender nature of the foliage, the sorrel and the dock are useful plants, especially in desert areas. You can eat their succulent leaves fresh or slightly cooked. To take away the strong taste, change the water once or twice during cooking. This latter tip is a useful hint in preparing many kinds of wild greens.

Wild fig
Ficus species

Description: These trees have alternate, simple leaves with entire margins. Often, the leaves are dark green and shiny. All figs have a milky, sticky juice. The fruits vary in size depending on the species, but are usually yellow-brown when ripe.

Habitat and Distribution: Figs are plants of the tropics and semitropics. They grow in several different habitats, including dense forests, margins of forests, and around human settlements.

Edible Parts: The fruits are edible raw or cooked. Some figs have little flavor.

Wild gourd or luffa sponge
Luffa cylindrica

Description: The luffs sponge is widely distributed and fairly typical of a wild squash. There are several dozen kinds of wild squashes in tropical regions. Like most squashes, the luffa is a vine with leaves 7.5 to 20 centimeters across having 3 lobes. Some squashes have leaves twice this size. Luffs fruits are oblong or cylindrical, smooth, and many-seeded. Luffs flowers are bright yellow. The luffa fruit, when mature, is brown and resembles the cucumber.

Habitat and Distribution: A member of the squash family, which also includes the watermelon, cantaloupe, and cucumber, the luffa sponge is widely cultivated throughout the Tropical Zone. It may be found in a semiwild state in old clearings and abandoned gardens in rain forests and semi-evergreen seasonal forests.

Edible Parts: You can boil the young green (half-ripe) fruit and eat them as a vegetable. Adding coconut milk will improve the flavor. After ripening, the luffa sponge develops an inedible spongelike texture in the interior of the fruit. You can also eat the tender shoots, flowers, and young leaves after cooking them. Roast the mature seeds a little and eat them like peanuts.

 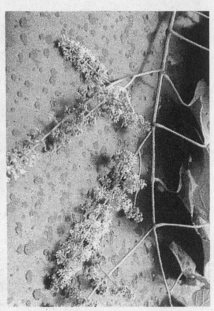

Wild grape vine
Vitis species

Description: The wild grape vine climbs with the aid of tendrils. Most grape vines produce deeply lobed leaves similar to the cultivated grape. Wild grapes grow in pyramidal, hanging bunches and are black-blue to amber, or white when ripe.

Habitat and Distribution: Wild grapes are distributed worldwide. Some kinds are found in deserts, others in temperate forests, and others in tropical areas. Wild grapes are commonly found throughout the eastern United States as well as in the southwestern desert areas. Most kinds are rampant climbers over other vegetation. The best place to look for wild grapes is on the edges of forested areas. Wild grapes are also found in Mexico. In the Old World, wild grapes are found from the Mediterranean region eastward through Asia, the East Indies, and to Australia. Africa also has several kinds of wild grapes.

Edible Parts: The ripe grape is the portion eaten. Grapes are rich in natural sugars and, for this reason, are much sought after as a source of energy-giving wild food. None are poisonous.

Other Uses: You can obtain water from severed grape vine stems. Cut off the vine at the bottom and place the cut end in a container. Make a slant-wise cut into the vine about 1.8 meters upon the hanging part. This cut will allow water to flow from the bottom end. As water diminishes in volume, make additional cuts further down the vine.

⚠ CAUTION
To avoid poisoning, do not eat grapelike fruits with only a single seed (moonseed).

Wild onion and garlic
Allium species

Description: *Allium cernuum* is an example of the many species of wild onions and garlics, all easily recognized by their distinctive odor.

Habitat and Distribution: Wild onions and garlics are found in open, sunny areas throughout the temperate regions. Cultivated varieties are found anywhere in the world.

Edible Parts: The bulbs and young leaves are edible raw or cooked. Use in soup or to flavor meat.

> ⚠ CAUTION
>
> There are several plants with onionlike bulbs that are extremely poisonous. Be certain that the plant you are using is a true onion or garlic. Do not eat bulbs with no onion smell.

Wild pistachio
Pistacia species

Description: Some kinds of pistachio trees are evergreen, while others lose their leaves during the dry season. The leaves alternate on the stem and have either three large leaves or a number of leaflets. The fruits or nuts are usually hard and dry at maturity.

Habitat and Distribution: About seven kinds of wild pistachio nuts are found in desert, or semidesert areas surrounding the Mediterranean Sea to Turkey and Afghanistan. It is generally found in evergreen scrub forests or scrub and thorn forests.

Edible Parts: You can eat the oil nut kernels after parching them over coals.

Wild rice
Zizania aquatica

Description: Wild rice is a tall grass that averages 1 to 1.5 meters in height, but may reach 4.5 meters. Its grain grows in very loose heads at the top of the plant and is dark brown or blackish when ripe.

Habitat and Distribution: Wild rice grows only in very wet areas in tropical and temperate regions.

Edible Parts: During the spring and summer, the central portion of the lower sterns and root shoots are edible. Remove the tough covering before eating. During the late summer and fall, collect the straw-covered husks. Dry and parch the husks, break them, and remove the rice. Boil or roast the rice and then beat it into flour.

Wild rose
Rosa species

Description: This shrub grows 60 centimeters to 2.5 meters high. It has alternate leaves and sharp prickles. Its flowers may be red, pink, or yellow. Its fruit, called rose hip, stays on the shrub year-round.

Habitat and Distribution: Look for wild roses in dry fields and open woods throughout the Northern Hemisphere.

Edible Parts: The flowers and buds are edible raw or boiled. In an emergency, you can peel and eat the young shoots. You can boil fresh, young leaves in water to make a tea. After the flower petals fall, eat the rose hips; the pulp is highly nutritious and an excellent source of vitamin C. Crush or grind dried rose hips to make flour.

> ⚠ CAUTION
>
> Eat only the outer portion of the fruit as the seeds of some species are quite prickly and can cause internal distress.

Wood sorrel
Oxalis species

Description: Wood sorrel resembles shamrock or four-leaf clover, with a bell-shaped pink, yellow, or white flower.

Habitat and Distribution: Wood sorrel is found in Temperate Zones worldwide, in lawns, open areas, and sunny woods.

Edible Parts: Cook the entire plant.

> ⚠ CAUTION
>
> Eat only small amounts of this plant as it contains a fairly high concentration of oxalic acid that can be harmful.

Yam
Dioscorea species

Description: These plants are vines that creep along the ground. They have alternate, heart- or arrow-shaped leaves. Their rootstock may be very large and weigh many kilograms.

Habitat and Distribution: True yams are restricted to tropical regions where they are an important food crop. Look for yams in fields, clearings, and abandoned gardens. They are found in rain forests, semi-evergreen seasonal forests, and scrub and thorn forests in the tropics. In warm temperate areas, they are found in seasonal hardwood or mixed hardwood-coniferous forests, as well as some mountainous areas.

Edible Parts: Boil the rootstock and eat it as a vegetable.

Yam bean
Pachyrhizus erosus

Description: The yam bean is a climbing plant of the bean family, with alternate, three-parted leaves and a turniplike root. The bluish or purplish flowers are pealike in shape. The plants are often so rampant that they cover the vegetation upon which they are growing.

Habitat and Distribution: The yam bean is native to the American tropics, but it was carried by man years ago to Asia and the Pacific islands. Now it is commonly cultivated in these places, and is also found growing wild in forested areas. This plant grows in wet areas of tropical regions.

Edible Parts: The tubers are about the size of a turnip and they are crisp, sweet, and juicy and have a nutty flavor. They are nourishing and at the same time quench the thirst. Eat them raw or boiled. To make flour, slice the raw tubers, let them dry in the sun, and grind into a flour that is high in starch and may be used to thicken soup.

⚠ **CAUTION**

The raw seeds are poisonous.

CHAPTER 7

Poisonous Plants

Successful use of plants in a survival situation depends on positive identification. Knowing poisonous plants is as important to a survivor as knowing edible plants. Knowing the poisonous plants will help you avoid sustaining injuries from them.

HOW PLANTS POISON

Plants generally poison by—

- Ingestion. When a person eats a part of a poisonous plant.
- Contact. When a person makes contact with a poisonous plant that causes any type of skin irritation or dermatitis.
- Absorption or inhalation. When a person either absorbs the poison through the skin or inhales it into the respiratory system.

Plant poisoning ranges from minor irritation to death. A common question asked is, "How poisonous is this plant?" It is difficult to say how poisonous plants are because–

- Some plants require contact with a large amount of the plant before noticing any adverse reaction while others will cause death with only a small amount.
- Every plant will vary in the amount of toxins it contains due to different growing conditions and slight variations in subspecies.
- Every person has a different level of resistance to toxic substances.
- Some persons may be more sensitive to a particular plant.

Some common misconceptions about poisonous plants are—

- *Watch the animals and eat what they eat.* Most of the time this statement is true, but some animals can eat plants that are poisonous to humans.
- *Boil the plant in water and any poisons will be removed.* Boiling removes many poisons, but not all.
- *Plants with a red color are poisonous.* Some plants that are red are poisonous, but not all.

The point is there is no one rule to aid in identifying poisonous plants. You must make an effort to learn as much about them as possible.

ALL ABOUT PLANTS

It is to your benefit to learn as much about plants as possible. Many poisonous plants look like their edible relatives or like other edible plants. For example, poison hemlock appears very similar to wild carrot. Certain plants are safe to eat in certain seasons or stages of growth and poisonous in other stages. For example, the leaves of the pokeweed are edible when it first starts to grow, but it soon becomes poisonous. You can eat some plants and their fruits only when they are ripe. For example, the ripe fruit of mayapple is edible, but all other parts and the green fruit are poisonous. Some plants contain both edible and poisonous parts; potatoes and tomatoes are common plant foods, but their green parts are poisonous.

Some plants become toxic after wilting. For example, when the black cherry starts to wilt, hydrocyanic acid develops. Specific preparation methods make some plants edible that are poisonous raw. You can eat the thinly sliced and thoroughly dried corms (drying may take a year) of the jack-in-the-pulpit, but they are poisonous if not thoroughly dried.

Learn to identify and use plants before a survival situation. Some sources of information about plants are pamphlets, books, films, nature trails, botanical gardens, local markets, and local natives. Gather and cross-reference information from as many sources as possible, because many sources will not contain all the information needed.

RULES FOR AVOIDING POISONOUS PLANTS

Your best policy is to be able to look at a plant and identify it with absolute certainty and to know its uses or dangers. Many times this is not possible. If you have little or no knowledge of the local vegetation, use the rules to select plants for the "Universal Edibility Test." Remember, avoid—

- All mushrooms. Mushroom identification is very difficult and must be precise, even more so than with other plants. Some mushrooms cause death very quickly. Some mushrooms have no known antidote. Two general types of mushroom poisoning are gastrointestinal and central nervous system.
- Contact with or touching plants unnecessarily.

CONTACT DERMATITIS

Contact dermatitis from plants will usually cause the most trouble in the field. The effects may be persistent, spread by scratching, and are particularly dangerous if there is contact in or around the eyes.

The principal toxin of these plants is usually an oil that gets on the skin upon contact with the plant. The oil can also get on equipment and then infect whoever touches the equipment. Never burn a contact poisonous plant because the smoke may be as harmful as the plant. There is a greater danger of being affected when overheated and sweating. The infection may be local or it may spread over the body.

Symptoms may take from a few hours to several days to appear. Signs and symptoms can include burning, reddening, itching, swelling, and blisters.

When you first contact the poisonous plants or the first symptoms appear, try to remove the oil by washing with soap and cold water. If water is not available, wipe your skin repeatedly with dirt or sand. Do not use dirt if blisters have developed. The dirt may break open the blisters and leave the body open to infection. After you have removed the oil, dry the area. You can wash with a tannic acid solution and crush and rub jewelweed on the affected area to treat plant-caused rashes. You can make tannic acid from oak bark.

Poisonous plants that cause contact dermatitis are—

- Cowhage
- Poison ivy
- Poison oak
- Poison sumac
- Rengas tree
- Trumpet vine

INGESTION POISONING

Ingestion poisoning can be very serious and could lead to death very quickly. Do not eat any plant unless you have positively identified it first. Keep a log of all plants eaten.

Signs and symptoms of ingestion poisoning can include nausea, vomiting, diarrhea, abdominal cramps, depressed heartbeat and respiration, headaches, hallucinations, dry mouth, unconsciousness, coma, and death.

If you suspect plant poisoning, try to remove the poisonous material from the victim's mouth and stomach as soon as possible. Induce vomiting by tickling the back of his throat or by giving him warm saltwater, if he is conscious. Dilute the poison by administering large quantities of water or milk, if he is conscious.

The following plants can cause ingestion poisoning if eaten:

- Castor bean
- Chinaberry
- Death camas
- Lantana
- Manchineel
- Oleander
- Pangi
- Physic nut
- Poison and water hemlocks
- Rosary pea
- Strychnine tree

POISONOUS PLANTS

Castor bean, castor-oil plant, palma Christi
Ricinus communis
Spurge (*Euphorbiaceae*) Family

Description: The castor bean is a semiwoody plant with large, alternate, starlike leaves that grows as a tree in tropical regions and as an annual in temperate regions. Its flowers are very small and inconspicuous. Its fruits grow in clusters at the tops of the plants.

⚠ CAUTION

All parts of the plant are very poisonous to eat. The seeds are large and may be mistaken for a beanlike food.

Habitat and Distribution: This plant is found in all tropical regions and has been introduced to temperate regions.

Chinaberry
Melia azedarach
Mahogany (*Meliaceae*) Family

Description: This tree has a spreading crown and grows up to 14 meters tall. It has alternate, compound leaves with toothed leaflets. Its flowers are light purple with a dark center and grow in ball-like masses. It has marble-sized fruits that are light orange when first formed but turn lighter as they become older.

> ⚠ **CAUTION**
> All parts of the tree should be considered dangerous if eaten. Its leaves are a natural insecticide and will repel insects from stored fruits and grains. Take care not to eat leaves mixed with the stored food.

Habitat and Distribution: Chinaberry is native to the Himalayas and eastern Asia but is now planted as an ornamental tree throughout the tropical and subtropical regions. It has been introduced to the southern United States and has escaped to thickets, old fields, and disturbed areas.

Cowhage, cowage, cowitch
Mucuna pruritum
Leguminosae (*Fabaceae*) Family

Description: A vinelike plant that has oval leaflets in groups of three and hairy spikes with dull purplish flowers. The seeds are brown, hairy pods.

> ⚠ **CAUTION**
> Contact with the pods and flowers causes irritation and blindness if in the eyes.

Death camas, death lily
Zigadenus species
Lily (*Liliaceae*) Family

Description: This plant arises from a bulb and may be mistaken for an onionlike plant. Its leaves are grass-like. Its flowers are six-parted and the petals have a green, heart-shaped structure on them. The flowers grow on showy stalks above the leaves.

⚠ CAUTION
All parts of this plant are very poisonous. Death camas does not have the onion smell.

Habitat and Distribution: Death camas is found in wet, open, sunny habitats, although some species favor dry, rocky slopes. They are common in parts of the western United States. Some species are found in the eastern United States and in parts of the North American western subarctic and eastern Siberia.

Lantana
Lantana camara
Vervain (*Verbenaceae*) Family

Description: Lantana is a shrublike plant that may grow up to 45 centimeters high. It has opposite, round leaves and flowers borne in flat-topped clusters. The flower color (which varies in different areas) may be white, yellow, orange, pink, or red. It has a dark blue or black berrylike fruit. A distinctive feature of all parts of this plant is its strong scent.

⚠ CAUTION
All parts of this plant are poisonous if eaten and can be fatal. This plant causes dermatitis in some individuals.

Habitat and Distribution: Lantana is grown as an ornamental in tropical and temperate areas and has escaped cultivation as a weed along roads and old fields.

Manchineel
Hippomane mancinella
Spurge (*Euphorbiaceae*) Family

Description: Manchineel is a tree reaching up to 15 meters high with alternate, shiny green leaves and spikes of small greenish flowers. Its fruits are green or greenish-yellow when ripe.

> ⚠ **CAUTION**
>
> This tree is extremely toxic. It causes severe dermatitis in most individuals after only .5 hour. Even water dripping from the leaves may cause dermatitis. The smoke from burning it irritates the eyes. No part of this plant should be considered a food.

Habitat and Distribution: The tree prefers coastal regions. Found in south Florida, the Caribbean, Central America, and northern South America.

Oleander
Nerium oleander
Dogbane (*Apocynaceae*) Family

Description: This shrub or small tree grows to about 9 meters, with alternate, very straight, dark green leaves. Its flowers may be white, yellow, red, pink, or intermediate colors. Its fruit is a brown, podlike structure with many small seeds.

> ⚠ **CAUTION**
>
> All parts of the plant are very poisonous. Do not use the wood for cooking; it gives off poisonous fumes that can poison food.

Habitat and Distribution: This native of the Mediterranean area is now grown as an ornamental in tropical and temperate regions.

Pangi
Pangium edule
Pangi Family

Description: This tree, with heart-shaped leaves in spirals, reaches a height of 18 meters. Its flowers grow in spikes and are green in color. Its large, brownish, pear-shaped fruits grow in clusters.

⚠ CAUTION

All parts are poisonous, especially the fruit.

Habitat and Distribution: Pangi trees grow in southeast Asia

Physic nut
Jatropha curcas
Spurge (*Euphoriaceae*) Family

Description: This shrub or small tree has large, 3- to 5-parted alternate leaves. It has small, greenish-yellow flowers and its yellow, apple-sized fruits contain three large seeds.

⚠ CAUTION

The seeds taste sweet but their oil is violently purgative. All parts of the physic nut are poisonous.

Habitat and Distribution: Throughout the tropics and southern United States.

Poison hemlock, fool's parsley
Conium maculatum
Parsley (*Apiaceae*) Family

Description: This biennial herb may grow to 2.5 meters high. The smooth, hollow stem may or may not be purple or red striped or mottled. Its white flowers are small and grow in small groups that tend to form flat umbels. Its long, turniplike taproot is solid.

> ⚠ **CAUTION**
> This plant is very poisonous and even a very small amount may cause death. This plant is easy to confuse with wild carrot or Queen Anne's lace, especially in its first stage of growth. Wild carrot or Queen Anne's lace has hairy leaves and stems and smells like carrot. Poison hemlock does not.

Habitat and Distribution: Poison hemlock grows in wet or moist ground like swamps, wet meadows, stream banks, and ditches. Native to Eurasia, it has been introduced to the United States and Canada.

Poison ivy and poison oak
Toxicodendron radicans and Toxicodendron diversibba
Cashew (*Anacardiacese*) Family

Description: These two plants are quite similar in appearance and will often crossbreed to make a hybrid. Both have alternate, compound leaves with three leaflets. The leaves of poison ivy are smooth or serrated. Poison oak's leaves are lobed and resemble oak leaves. Poison ivy grows as a vine along the ground or climbs by red feeder roots. Poison oak grows like a bush. The greenish-white flowers are small and inconspicuous and are followed by waxy green berries that turn waxy white or yellow, then gray.

> ⚠ **CAUTION**
> All parts, at all times of the year, can cause serious contact dermatitis.

Habitat and Distribution: Poison ivy and oak can be found in almost any habitat in North America.

Poison sumac
Toxicodendron vernix
Cashew (*Anacardiacese*) Family

Description: Poison sumac is a shrub that grows to 8.5 meters tall. It has alternate, pinnately compound leafstalks with 7 to 13 leaflets. Flowers are greenish-yellow and inconspicuous and are followed by white or pale yellow berries.

> ⚠ **CAUTION**
> All parts can cause serious contact dermatitis at all times of the year.

Habitat and Distribution: Poison sumac grows only in wet, acid swamps in North America.

Renghas tree, rengas tree, marking nut, black-varnish tree
Gluta
Cashew (*Anacardiaceae*) Family

Description: This family comprises about 48 species of trees or shrubs with alternating leaves in terminal or axillary panicles. Flowers are similar to those of poison ivy and oak.

> **CAUTION**
> Can cause contact dermatitis similar to poison ivy or poison oak.

Habitat and Distribution: India, east to Southeast Asia.

Renghas tree, rengas tree, marking nut, black-varnish tree
Gluta
Cashew (*Anacardiacese*) Family

Description: This family comprises about 48 species of trees or shrubs with alternating leaves in terminal or axillary panicles. Flowers are similar to those of poison ivy and oak.

> ⚠ **CAUTION**
> Can cause contact dermatitis similar to poison ivy and oak.

Habitat and Distribution: India, east to Southeast Asia.

Rosary pea or crab's eyes
Abrus precatorius
Leguminosae (*Fabaceae*) Family

Description: This plant is a vine with alternate compound leaves, light purple flowers, and beautiful seeds that are red and black.

> ⚠ **CAUTION**
>
> This plant is one of the most dagerous plants. One seed may contain enough poison to kill an adult.

Habitat and Distribution: This is a common weed in parts of Africa, southern Florida, Hawaii, Guam, the Caribbean, and Central and South America.

Strychnine tree
Nux vomica
Logania (*Loganiaceae*) Family

Description: The strychnine tree is a medium-sized evergreen, reaching a height of about 12 meters, with a thick, frequently crooked trunk. Its deeply veined oval leaves grow in alternate pairs. Small, loose clusters of greenish flowers appear at the ends of branches and are followed by fleshy, orange-red berries about 4 centimeters in diameter.

> ⚠ **CAUTION**
>
> The berries contain the dislike seeds that yield the poisonous substance strychnine. All parts of the plant are poisonous.

Habitat and Distribution: A native of the tropics and subtropics of southeastern Asia and Australia.

Trumpet vine or trumpet creeper
Campsis radicans
Trumpet creeper (*Bignoniaceae*) Family

Description: This woody vine may climb to 15 meters high. It has pealike fruit capsules. The leaves are pinnately compound, 7 to 11 toothed leaves per leaf stock. The trumpet-shaped flowers are orange to scarlet in color.

> ⚠ **CAUTION**
> This plant causes contact dermatitis.

Habitat and Distribution: This vine is found in wet woods and thickets throughout eastern and central North America.

Water hemlock or spotted cowbane
Cicuta maculata
Parsley (*Apiaceae*) Family

Description: This perennial herb may grow to 1.8 meters high. The stem is hollow and sectioned off like bamboo. It may or may not be purple or red striped or mottled. Its flowers are small, white, and grow in groups that tend to form flat umbels. Its roots may have hollow air chambers and, when cut, may produce drops of yellow oil.

> ⚠ **CAUTION**
> This plant is very poisonous and even a very small amount of this plant may cause death. Its roots have been mistaken for parsnips.

Habitat and Distribution: Water hemlock grows in wet or moist ground like swamps, wet meadows, stream banks, and ditches throughout the Unites States and Canada.

PART V

Firecraft, Tools, Camouflage, Tracking, Movement, and Combat Skills

CHAPTER 1

Firecraft

In many survival situations, the ability to start a fire can make the difference between living and dying. Fire can fulfill many needs. It can provide warmth and comfort. It not only cooks and preserves food, it also provides warmth in the form of heated food that saves calories our body normally uses to produce body heat. You can use fire to purify water, sterilize bandages, signal for rescue, and provide protection from animals. It can be a psychological boost by providing peace of mind and companionship. You can also use fire to produce tools and weapons.

Fire can cause problems, as well. The enemy can detect the smoke and light it produces. It can cause forest fires or destroy essential equipment. Fire can also cause burns and carbon monoxide poisoning when used in shelters.

Remember to weigh your need for fire against your need to avoid enemy detection.

BASIC FIRE PRINCIPLES

To build a fire, it helps to understand the basic principles of a fire. Fuel (in a nongaseous state) does not burn directly. When you apply heat to a fuel, it produces a gas. This gas, combined with oxygen in the air, burns.

Understanding the concept of the fire triangle is very important in correctly constructing and maintaining a fire. The three sides of the triangle represent air, heat, and fuel. If you remove any of these, the fire will go out. The correct ratio of these components is very important for a fire to burn at its greatest capability. The only way to learn this ratio is to practice.

SITE SELECTION AND PREPARATION

You will have to decide what site and arrangement to use. Before building a fire consider—

- The area (terrain and climate) in which you are operating.
- The materials and tools available. Time: how much time you have?
- Need: why do you need a fire?
- Security: how close is the enemy?

Look for a dry spot that—

- Is protected from the wind.
- Is suitably placed in relation to your shelter (if any).
- Will concentrate the heat in the direction you desire.
- Has a supply of wood or other fuel available. (See Table 1-1 for types of material you can use.)

If you are in a wooded or brush-covered area, clear the brush and scrape the surface soil from the spot you have selected. Clear a circle at least 1 meter in diameter so there is little chance of the fire spreading.

If time allows, construct a fire wall using logs or rocks. This wall will help to reflect or direct the heat where you want it (Figure 1-l). It will also reduce flying sparks and cut down on the amount of wind blowing into the fire. However, you will need enough wind to keep the fire burning.

Table 1-1: Materials for Building Fires.

Tinder	Kindling	Fuel
• Birch bark • Shredded inner bark from cedar, chestnut, red elm trees • Fine wood shavings • Dead grass, ferns, moss, fungi • Straw • Sawdust • Very fine pitchwood scrapings • Dead evergreen needles • Punk (the completely rotted portions of dead logs or trees) • Evergreen tree knots • Bird down (fine feathers) • Down seed heads (milkweed, dry cattails, bulrush, or thistle) • Fine, dried vegetable fibers • Spongy threads of dead puffball • Dead palm leaves • Skinlike membrane lining bamboo • Lint from pocket and seams • Charred cloth • Waxed paper • Outer bamboo shavings • Gunpowder • Cotton • Lint	• Small twigs • Small strips of wood • Split wood • Heavy cardboard • Pieces of wood removed from the inside of larger pieces • Wood that has been doused with highly flammable materials, such as gasoline, oil, or wax	• Dry, standing wood and dry, dead branches • Dry inside (heart) of fallen tree trunks and large branches • Green wood that is finely split • Dry grasses twisted into bunches • Peat dry enough to burn (this may be found at the top of undercut banks) • Dried animal dung • Animal fats • Coal, oil shale, or oil lying on the surface

⚠ CAUTION
Do not use wet or porous rocks as they may explode when heated.

In some situations, you may find that an underground fireplace will best meet your needs. It conceals the fire and serves well for cooking food. To make an underground fireplace or Dakota fire hole (Figure 1-2)—

- Dig a hole in the ground.
- On the upwind side of this hole, poke or dig a large connecting hole for ventilation.
- Build your fire in the hole as illustrated.

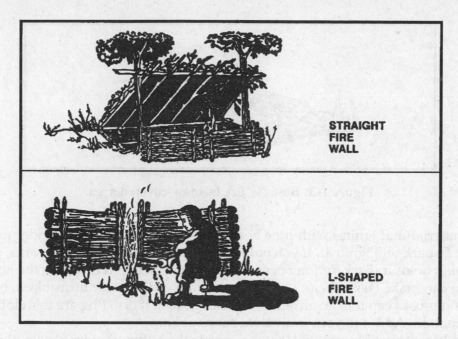

Figure 1-1: Types of fire walls.

If you are in a snow-covered area, use green logs to make a dry base for your fire (Figure 1-3). Trees with wrist-sized trunks are easily broken in extreme cold. Cut or break several green logs and lay them side by side on top of the snow. Add one or two more layers. Lay the top layer of logs opposite those below it.

FIRE MATERIAL SELECTION

You need three types of materials (Table 1-1) to build a fire—tinder, kindling, and fuel.

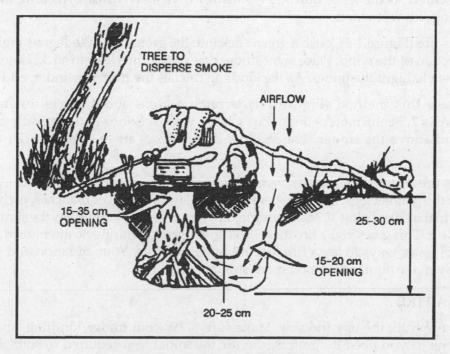

Figure 1-2: Dakota fire hole.

Figure 1-3: Base for fire in snow-covered area.

Tinder is dry material that ignites with little heat—a spark starts a fire. The tinder must be absolutely dry to be sure just a spark will ignite it. If you only have a device that generates sparks, charred cloth will be almost essential. It holds a spark for long periods, allowing you to put tinder on the hot area to generate a small flame. You can make charred cloth by heating cotton cloth until it turns black, but does not burn. Once it is black, you must keep it in an airtight container to keep it dry. Prepare this cloth well in advance of any survival situation. Add it to your individual survival kit.

Kindling is readily combustible material that you add to the burning tinder. Again, this material should be absolutely dry to ensure rapid burning. Kindling increases the fire's temperature so that it will ignite less combustible material.

Fuel is less combustible material that burns slowly and steadily once ignited.

HOW TO BUILD A FIRE

There are several methods for laying a fire, each of which has advantages. The situation you find yourself in will determine which fire to use.

Tepee. To make this fire (Figure 1-4), arrange the tinder and a few sticks of kindling in the shape of a tepee or cone. Light the center. As the tepee burns, the outside logs will fall inward, feeding the fire. This type of fire burns well even with wet wood.

Lean-To. To lay this fire (Figure 1-4), push a green stick into the ground at a 30-degree angle. Point the end of the stick in the direction of the wind. Place some tinder deep under this lean-to stick. Lean pieces of kindling against the lean-to stick. Light the tinder. As the kindling catches fire from the tinder, add more kindling.

Cross-Ditch. To use this method (Figure 1-4), scratch a cross about 30 centimeters in size in the ground. Dig the cross 7.5 centimeters deep. Put a large wad of tinder in the middle of the cross. Build a kindling pyramid above the tinder. The shallow ditch allows air to sweep under the tinder to provide a draft.

Pyramid. To lay this fire (Figure 1-4), place two small logs or branches parallel on the ground. Place a solid layer of small logs across the parallel logs. Add three or four more layers of logs or branches, each layer smaller than and at a right angle to the layer below it. Make a starter fire on top of the pyramid. As the starter fire burns, it will ignite the logs below it. This gives you a fire that burns downward, requiring no attention during the night.

There are several other ways to lay a fire that are quite effective. Your situation and the material available in the area may make another method more suitable.

HOW TO LIGHT A FIRE

Always light your fire from the upwind side. Make sure to lay your tinder, kindling, and fuel so that your fire will burn as long as you need it. Igniters provide the initial heat required to start the tinder burning. They fall into two categories: modern methods and primitive methods.

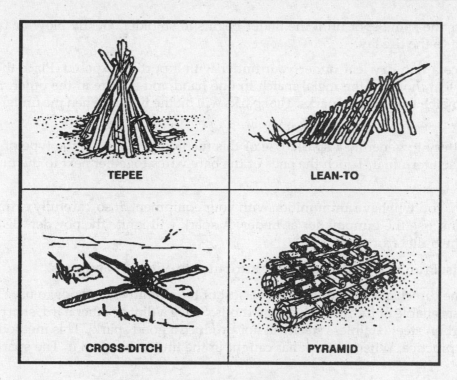

Figure 1-4: Methods for laying fires.

Modern Methods. Modern igniters use modern devices—items we normally think of to start a fire.

Matches. Make sure these matches are waterproof. Also, store them in a waterproof container along with a dependable striker pad.

Convex Lens. Use this method (Figure 1-5) only on bright, sunny days. The lens can come from binoculars, camera, telescopic sights, or magnifying glasses. Angle the lens to concentrate the sun's rays on the tinder.

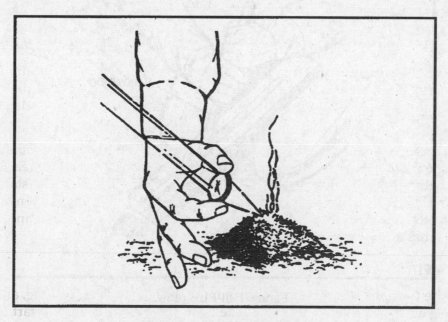

Figure 1-5: Lens method.

Hold the lens over the same spot until the tinder begins to smolder. Gently blow or fan the tinder into flame, and apply it to the fire lay.

Metal Match. Place a flat, dry leaf under your tinder with a portion exposed. Place the tip of the metal match on the dry leaf, holding the metal match in one hand and a knife in the other. Scrape your knife against the metal match to produce sparks. The sparks will hit the tinder. When the tinder starts to smolder, proceed as above.

Battery. Use a battery to generate a spark. Use of this method depends on the type of battery available. Attach a wire to each terminal. Touch the ends of the bare wires together next to the tinder so the sparks will ignite it.

Gunpowder. Often, you will have ammunition with your equipment. If so, carefully extract the bullet from the shell casing, and use the gunpowder as tinder. A spark will ignite the powder. Be extremely careful when extracting the bullet from the case.

Primitive Methods. Primitive igniters are those attributed to our early ancestors.

Flint and Steel. The direct spark method is the easiest of the primitive methods to use. The flint and steel method is the most reliable of the direct spark methods. Strike a flint or other hard, sharp-edged rock edge with a piece of carbon steel (stainless steel will not produce a good spark). This method requires a loose-jointed wrist and practice. When a spark has caught in the tinder, blow on it. The spark will spread and burst into flames.

Fire-Plow. The fire-plow (Figure 1-6) is a friction method of ignition. You rub a hardwood shaft against a softer wood base. To use this method, cut a straight groove in the base and plow the blunt tip of the shaft up and down the groove. The plowing action of the shaft pushes out small particles of wood fibers. Then, as you apply more pressure on each stroke, the friction ignites the wood particles.

Bow and Drill. The technique of starting a fire with a bow and drill (Figure 1-7) is simple, but you must exert much effort and be persistent to produce a fire. You need the following items to use this method:

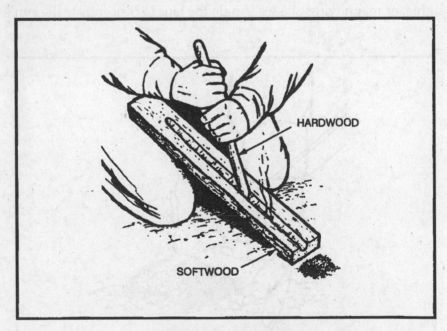

Figure 1-6: Fire-plow.

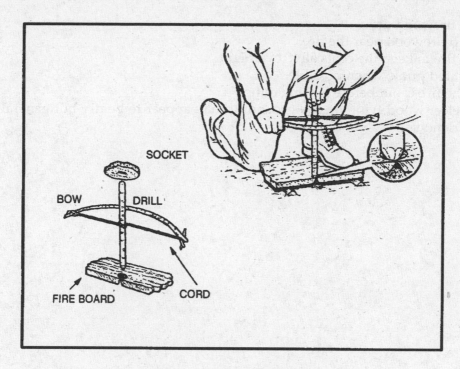

Figure 1-7: Bow and drill.

- Socket. The socket is an easily grasped stone or piece of hardwood or bone with a slight depression in one side. Use it to hold the drill in place and to apply downward pressure.
- Drill. The drill should be a straight, seasoned hardwood stick about 2 centimeters in diameter and 25 centimeters long. The top end is round and the low end blunt (to produce more friction).
- Fire board. Its size is up to you. A seasoned softwood board about 2.5 centimeters thick and 10 centimeters wide is preferable. Cut a depression about 2 centimeters from the edge on one side of the board. On the underside, make a V-shaped cut from the edge of the board to the depression.
- Bow. The bow is a resilient, green stick about 2.5 centimeters in diameter and a string. The type of wood is not important. The bowstring can be any type of cordage. You tie the bowstring from one end of the bow to the other, without any slack.

To use the bow and drill, first prepare the fire lay. Then place a bundle of tinder under the V-shaped cut in the fire board. Place one foot on the fire board. Loop the bowstring over the drill and place the drill in the precut depression on the fire board. Place the socket, held in one hand, on the top of the drill to hold it in position. Press down on the drill and saw the bow back and forth to twirl the drill (Figure 1-7). Once you have established a smooth motion, apply more downward pressure and work the bow faster. This action will grind hot black powder into the tinder, causing a spark to catch. Blow on the tinder until it ignites.

Note: Primitive fire-building methods are exhaustive and require practice to ensure success.

Helpful Hints

- Use nonaromatic seasoned hardwood for fuel, if possible.
- Collect kindling and tinder along the trail.
- Add insect repellent to the tinder.

- Keep the firewood dry.
- Dry damp firewood near the fire.
- Bank the fire to keep the coals alive overnight.
- Carry lighted punk, when possible.
- Be sure the fire is out before leaving camp.
- Do not select wood lying on the ground. It may appear to be dry but generally doesn't provide enough friction.

CHAPTER 2

Field-expedient Weapons, Tools, and Equipment

As a soldier you know the importance of proper care and use of your weapons, tools, and equipment. This is especially true of your knife. You must always keep it sharp and ready to use. A knife is your most valuable tool in a survival situation. Imagine being in a survival situation without any weapons, tools, or equipment except your knife. It could happen! You might even be without a knife. You would probably feel helpless, but with the proper knowledge and skills, you can easily improvise needed items.

In survival situations, you may have to fashion any number and type of field-expedient tools and equipment to survive. Examples of tools and equipment that could make your life much easier are ropes, rucksacks, clothes, nets, and so on.

Weapons serve a dual purpose. You use them to obtain and prepare food and to provide self-defense. A weapon can also give you a feeling of security and provide you with the ability to hunt on the move.

CLUBS

You hold clubs, you do not throw them. As a field-expedient weapon, the club does not protect you from enemy soldiers. It can, however, extend your area of defense beyond your fingertips. It also serves to increase the force of a blow without injuring yourself. There are three basic types of clubs. They are the simple, weighted, and sling club.

Simple Club. A simple club is a staff or branch. It must be short enough for you to swing easily, but long enough and strong enough for you to damage whatever you hit. Its diameter should fit comfortably in your palm, but it should not be so thin as to allow the club to break easily upon impact. A straight-grained hardwood is best if you can find it.

Weighted Club. A weighted club is any simple club with a weight on one end. The weight may be a natural weight, such as a knot on the wood, or something added, such as a stone lashed to the club.

To make a weighted club, first find a stone that has a shape that will allow you to lash it securely to the club. A stone with a slight hourglass shape works well. If you cannot find a suitably shaped stone, you must fashion a groove or channel into the stone by a technique known as pecking. By repeatedly rapping the club stone with a smaller hard stone, you can get the desired shape.

Next, find a piece of wood that is the right length for you. A straight-grained hardwood is best. The length of the wood should feel comfortable in relation to the weight of the stone. Finally, lash the stone to the handle.

There are three techniques for lashing the stone to the handle: split handle, forked branch, and wrapped handle. The technique you use will depend on the type of handle you choose. See Figure 2-1.

Sling Club. A sling club is another type of weighted club. A weight hangs 8 to 10 centimeters from the handle by a strong, flexible lashing (Figure 2-2). This type of club both extends the user's reach and multiplies the force of the blow.

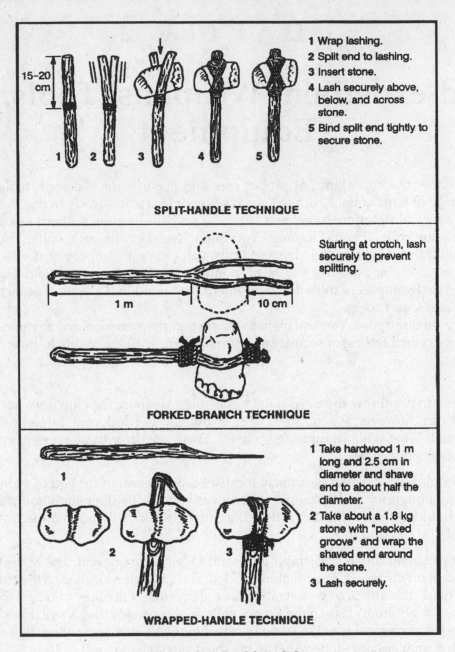

1 Wrap lashing.
2 Split end to lashing.
3 Insert stone.
4 Lash securely above, below, and across stone.
5 Bind split end tightly to secure stone.

SPLIT-HANDLE TECHNIQUE

Starting at crotch, lash securely to prevent splitting.

FORKED-BRANCH TECHNIQUE

1 Take hardwood 1 m long and 2.5 cm in diameter and shave end to about half the diameter.
2 Take about a 1.8 kg stone with "pecked groove" and wrap the shaved end around the stone.
3 Lash securely.

WRAPPED-HANDLE TECHNIQUE

Figure 2-1: Lashing clubs.

EDGED WEAPONS

Knives, spear blades, and arrow points fall under the category of edged weapons. The following paragraphs will discuss the making of such weapons.

Knives. A knife has three basic functions. It can puncture, slash or chop, and cut. A knife is also an invaluable tool used to construct other survival items. You may find yourself without a knife or you may need another type knife or a spear. To improvise you can use stone, bone, wood, or metal to make a knife or spear blade.

Stone. To make a stone knife, you will need a sharp-edged piece of stone, a chipping tool, and a flaking tool. A chipping tool is a light, blunt-edged tool used to break off small pieces of stone. A flaking tool is a pointed tool used to break off thin, flattened pieces of stone. You can make a chipping tool from wood, bone, or metal, and a flaking tool from bone, antler tines, or soft iron (Figure 2-3).

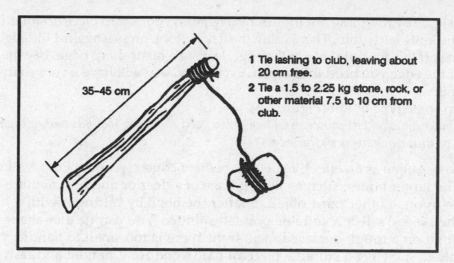

1 Tie lashing to club, leaving about 20 cm free.

2 Tie a 1.5 to 2.25 kg stone, rock, or other material 7.5 to 10 cm from club.

35–45 cm

Figure 2-2: Sling club.

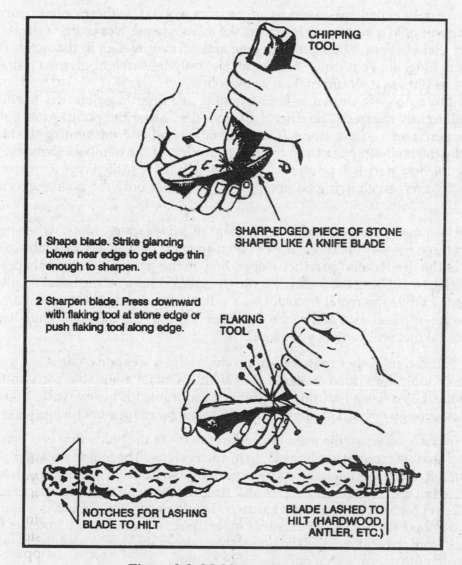

CHIPPING TOOL

SHARP-EDGED PIECE OF STONE SHAPED LIKE A KNIFE BLADE

1 Shape blade. Strike glancing blows near edge to get edge thin enough to sharpen.

2 Sharpen blade. Press downward with flaking tool at stone edge or push flaking tool along edge.

FLAKING TOOL

NOTCHES FOR LASHING BLADE TO HILT

BLADE LASHED TO HILT (HARDWOOD, ANTLER, ETC.)

Figure 2-3: Making a stone knife.

Start making the knife by roughing out the desired shape on your sharp piece of stone, using the chipping tool. Try to make the knife fairly thin. Then, using the flaking tool, press it against the edges. This action will cause flakes to come off the opposite side of the edge, leaving a razor sharp edge. Use the flaking tool along the entire length of the edge you need to sharpen. Eventually, you will have a very sharp cutting edge that you can use as a knife.

Lash the blade to some type of hilt (Figure 2-3).

Note: Stone will make an excellent puncturing tool and a good chopping tool but will not hold a fine edge. Some stones such as chert or flint can have very fine edges.

Bone. You can also use bone as an effective field-expedient edged weapon. First, you will need to select a suitable bone. The larger bones, such as the leg bone of a deer or another medium-sized animal, are best. Lay the bone upon another hard object. Shatter the bone by hitting it with a heavy object, such as a rock. From the pieces, select a suitable pointed splinter. You can further shape and sharpen this splinter by rubbing it on a rough-surfaced rock. If the piece is too small to handle, you can still use it by adding a handle to it. Select a suitable piece of hardwood for a handle and lash the bone splinter securely to it.

Note: Use the bone knife only to puncture. It will not hold an edge and it may flake or break if used differently.

Wood. You can make field-expedient edged weapons from wood. Use these only to puncture. Bamboo is the only wood that will hold a suitable edge. To make a knife using wood, first select a straight-grained piece of hardwood that is about 30 centimeters long and 2.5 centimeters in diameter. Fashion the blade about 15 centimeters long. Shave it down to a point. Use only the straight-grained portions of the wood. Do not use the core or pith, as it would make a weak point.

Harden the point by a process known as fire hardening. If a fire is possible, dry the blade portion over the fire slowly until lightly charred. The drier the wood, the harder the point. After lightly charring the blade portion, sharpen it on a coarse stone. If using bamboo and after fashioning the blade, remove any other wood to make the blade thinner from the inside portion of the bamboo. Removal is done this way because bamboo's hardest part is its outer layer. Keep as much of this layer as possible to ensure the hardest blade possible. When charring bamboo over a fire, char only the inside wood; do not char the outside.

Metal. Metal is the best material to make field-expedient edged weapons. Metal, when properly designed, can fulfill a knife's three uses—puncture, slice or chop, and cut. First, select a suitable piece of metal, one that most resembles the desired end product. Depending on the size and original shape, you can obtain a point and cutting edge by rubbing the metal on a rough-surfaced stone. If the metal is soft enough, you can hammer out one edge while the metal is cold. Use a suitable flat, hard surface as an anvil and a smaller, harder object of stone or metal as a hammer to hammer out the edge. Make a knife handle from wood, bone, or other material that will protect your hand.

Other Materials. You can use other materials to produce edged weapons. Glass is a good alternative to an edged weapon or tool, if no other material is available. Obtain a suitable piece in the same manner as described for bone. Glass has a natural edge but is less durable for heavy work. You can also sharpen plastic—if it is thick enough or hard enough—into a durable point for puncturing.

Spear Blades. To make spears, use the same procedures to make the blade that you used to make a knife blade. Then select a shaft (a straight sapling) 1.2 to1.5 meters long. The length should allow you to handle the spear easily and effectively. Attach the spear blade to the shaft using lashing. The preferred method is to split the handle, insert the blade, then wrap or lash it tightly. You can use other materials without adding a blade. Select a 1.2- to 1.5-meter long straight hardwood shaft and shave one end to a point. If possible, fire harden the point. Bamboo also makes an excellent spear. Select a piece 1.2 to 1.5 meters long. Starting 8 to 10 centimeters back from the end used as the point, shave down the end at a 45-degree angle (Figure 2-4). Remember, to sharpen the edges, shave only the inner portion.

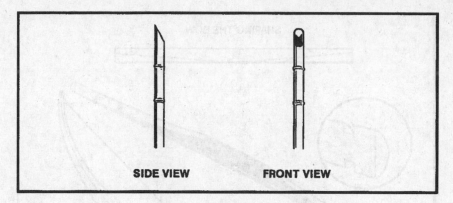

SIDE VIEW   FRONT VIEW

Figure 2-4: Bamboo spear.

Arrow Points. To make an arrow point, use the same procedures for making a stone knife blade. Chert, flint, and shell-type stones are best for arrow points. You can fashion bone like stone—by flaking. You can make an efficient arrow point using broken glass.

OTHER EXPEDIENT WEAPONS

You can make other field-expedient weapons such as the throwing stick, archery equipment, and the bola.

Throwing Stick. The throwing stick, commonly known as the rabbit stick, is very effective against small game (squirrels, chipmunks, and rabbits). The rabbit stick itself is a blunt stick, naturally curved at about a 45-degree angle. Select a stick with the desired angle from heavy hardwood such as oak. Shave off two opposite sides so that the stick is flat like a boomerang (Figure 2-5). You must practice the throwing technique for accuracy and speed. First, align the target by extending the nonthrowing arm in line with the mid to lower section of the target. Slowly and repeatedly raise the throwing arm up and back until the throwing stick crosses the back at about a 45-degree angle or is in line with the nonthrowing hip. Bring the throwing arm forward until it is just slightly above and parallel to the nonthrowing arm. This will be the throwing stick's release point. Practice slowly and repeatedly to attain accuracy.

Archery Equipment. You can make a bow and arrow (Figure 2-6) from materials available in your survival area. While it may be relatively simple to make a bow and arrow, it is not easy to use one. You must practice using it a long time to be reasonably sure that you will hit your target. Also, a field-expedient bow will not last very long before you have to make a new one. For the time and effort involved, you may well decide to use another type of field-expedient weapon.

Bola. The bola is another field-expedient weapon that is easy to make (Figure 2-7). It is especially effective for capturing running game or low-flying fowl in a flock. To use the bola, hold it by the center knot and

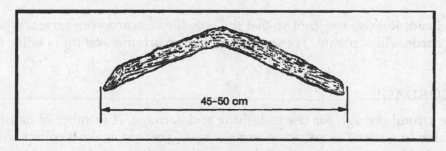

45–50 cm

Figure 2-5: Rabbit stick.

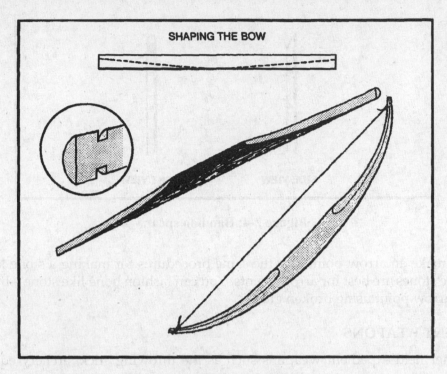

Figure 2-6: Archery equipment.

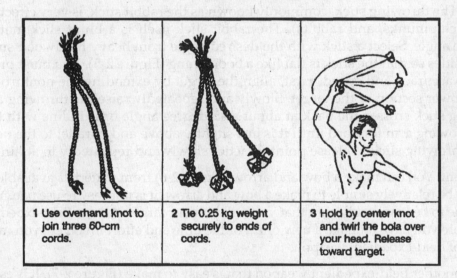

Figure 2-7: Bola.

twirl it above your head. Release the knot so that the bola flies toward your target. When you release the bola, the weighted cords will separate. These cords will wrap around and immobilize the fowl or animal that you hit.

LASHING AND CORDAGE

Many materials are strong enough for use as lashing and cordage. A number of natural and man-made materials are available in a survival situation. For example, you can make a cotton web belt much more useful by unraveling it. You can then use the string for other purposes (fishing line, thread for sewing, and lashing).

Natural Cordage Selection. Before making cordage, there are a few simple tests you can do to determine your material's suitability. First, pull on a length of the material to test for strength. Next, twist it between your fingers and roll the fibers together. If it withstands this handling and does not snap apart, tie an over-hand knot with the fibers and gently tighten. If the knot does not break, the material is usable. Figure 2-8 shows various methods of making cordage.

Lashing Material. The best natural material for lashing small objects is sinew. You can make sinew from the tendons of large game, such as deer. Remove the tendons from the game and dry them completely. Smash the dried tendons so that they separate into fibers. Moisten the fibers and twist them into a continuous strand. If you need stronger lashing material, you can braid the strands. When you use sinew for small lashings, you do not need knots as the moistened sinew is sticky and it hardens when dry.

You can shred and braid plant fibers from the inner bark of some trees to make cord. You can use the linden, elm, hickory, white oak, mulberry, chestnut, and red and white cedar trees. After you make the cord, test it to be sure it is strong enough for your purpose. You can make these materials stronger by braiding several strands together.

You can use rawhide for larger lashing jobs. Make rawhide from the skins of medium or large game. After skinning the animal, remove any excess fat and any pieces of meat from the skin. Dry the skin completely. You do not need to stretch it as long as there are no folds to trap moisture. You do not have to remove the hair from the skin. Cut the skin while it is dry. Make cuts about 6 millimeters wide. Start from the center of the hide and make one continuous circular cut, working clockwise to the hide's outer edge. Soak the rawhide for 2 to 4 hours or until it is soft. Use it wet, stretching it as much as possible while applying it. It will be strong and durable when it dries.

RUCKSACK CONSTRUCTION

The materials for constructing a rucksack or pack are almost limitless. You can use wood, bamboo, rope, plant fiber, clothing, animal skins, canvas, and many other materials to make a pack.

There are several construction techniques for rucksacks. Many are very elaborate, but those that are simple and easy are often the most readily made in a survival situation.

Horseshoe Pack. This pack is simple to make and use and relatively comfortable to carry over one shoulder. Lay available square-shaped material, such as poncho, blanket, or canvas, flat on the ground. Lay items on one edge of the material. Pad the hard items. Roll the material (with the items) toward the opposite edge and tie both ends securely. Add extra ties along the length of the bundle. You can drape the pack over one shoulder with a line connecting the two ends (Figure 2-9).

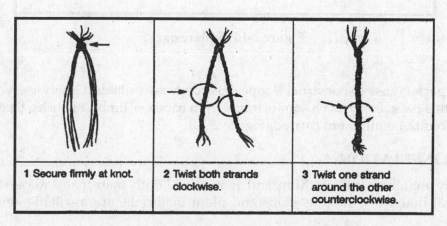

| 1 Secure firmly at knot. | 2 Twist both strands clockwise. | 3 Twist one strand around the other counterclockwise. |

Figure 2-8: Making lines from plant fibers.

Figure 2-9: Horseshoe pack.

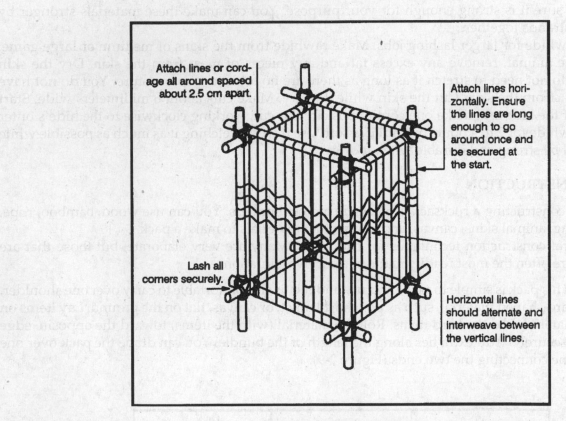

Attach lines or cord-
age all around spaced
about 2.5 cm apart.

Attach lines hori-
zontally. Ensure
the lines are long
enough to go
around once and
be secured at
the start.

Lash all
corners securely.

Horizontal lines
should alternate and
interweave between
the vertical lines.

Figure 2-10: Square pack.

Square Pack. This pack is easy to construct if rope or cordage is available. Otherwise, you must first make cordage. To make this pack, construct a square frame from bamboo, limbs, or sticks. Size will vary for each person and the amount of equipment carried (Figure 2-10).

CLOTHING AND INSULATION

You can use many materials for clothing and insulation. Both man-made materials, such as parachutes, and natural materials, such as skins and plant materials, are available and offer significant protection.

Parachute Assembly. Consider the entire parachute assembly as a resource. Use every piece of material and hardware, to include the canopy, suspension lines, connector snaps, and parachute harness. Before disassembling the parachute, consider all of your survival requirements and plan to use different portions of the parachute accordingly. For example, consider shelter requirements, need for a rucksack, and so on, in addition to clothing or insulation needs.

Animal Skins. The selection of animal skins in a survival situation will most often be limited to what you manage to trap or hunt. However, if there is an abundance of wildlife, select the hides of larger animals with heavier coats and large fat content. Do not use the skins of infected or diseased animals if at all possible. Since they live in the wild, animals are carriers of pests such as ticks, lice, and fleas. Because of these pests, use water to thoroughly clean any skin obtained from any animal. If water is not available, at least shake out the skin thoroughly. As with rawhide, lay out the skin, and remove all fat and meat. Dry the skin completely. Use the hind quarter joint areas to make shoes and mittens or socks. Wear the hide with the fur to the inside for its insulating factor.

Plant Fibers. Several plants are sources of insulation from cold. Cattail is a marshland plant found along lakes, ponds, and the backwaters of rivers. The fuzz on the tops of the stalks forms dead air spaces and makes a good down-like insulation when placed between two pieces of material. Milkweed has pollen-like seeds that act as good insulation. The husk fibers from coconuts are very good for weaving ropes and, when dried, make excellent tinder and insulation.

COOKING AND EATING UTENSILS

Many materials may be used to make equipment for the cooking, eating, and storing of food.

Bowls. Use wood, bone, horn, bark, or other similar material to make bowls. To make wooden bowls, use a hollowed out piece of wood that will hold your food and enough water to cook it in. Hang the wooden container over the fire and add hot rocks to the water and food. Remove the rocks as they cool and add more hot rocks until your food is cooked.

⚠ **CAUTION**

Do not use rocks with air pockets, such as limestone and sandstone. They may explode while heating in the fire.

You can also use this method with containers made of bark or leaves. However, these containers will burn above the waterline unless you keep them moist or keep the fire low.

A section of bamboo works very well, if you cut out a section between two sealed joints (Figure 2-11).

⚠ **CAUTION**

A sealed section of bamboo will explode if heated because of trapped air and water in the section.

Forks, Knives, and Spoons. Carve forks, knives, and spoons from nonresinous woods so that you do not get a wood resin aftertaste or do not taint the food. Nonresinous woods include oak, birch, and other hardwood trees.

Note: Do not use those trees that secrete a syrup or resinlike liquid on the bark or when cut.

Pots. You can make pots from turtle shells or wood. As described with bowls, using hot rocks in a hollowed out piece of wood is very effective. Bamboo is the best wood for making cooking containers.

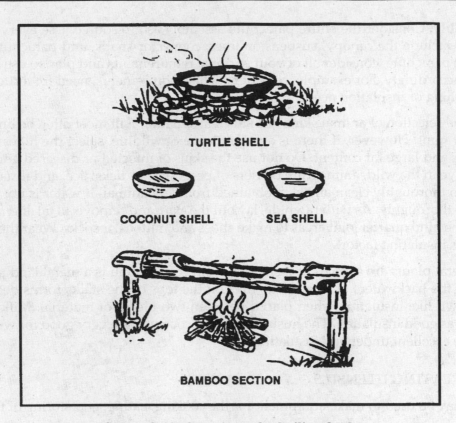

Figure 2-11: Containers for boiling food.

To use turtle shells, first thoroughly boil the upper portion of the shell. Then use it to heat food and water over a flame (Figure 2-11).

Water Bottles. Make water bottles from the stomachs of larger animals. Thoroughly flush the stomach out with water, then tie off the bottom. Leave the top open, with some means of fastening it closed.

CHAPTER 3

Hand-to-hand Combat

SECTION I: OVERVIEW

Hand-to-hand combat is an engagement between two or more persons in an empty-handed struggle or with handheld weapons such as knives, sticks, and rifles with bayonets. These fighting arts are essential military skills. Projectile weapons may be lost or broken, or they may fail to fire. When friendly and enemy forces become so intermingled that firearms and grenades are not practical, hand-to-hand combat skills become vital assets.

Purpose of Combatives Training. Today's battlefield scenarios may require silent elimination of the enemy. Unarmed combat and expedient-weapons training should not be limited to forward units. With rapid mechanized/motorized, airborne, and air assault abilities, units throughout the battle area could be faced with close-quarter or unarmed fighting situations. With low-intensity conflict scenarios and guerrilla warfare conditions, any soldier is apt to face an unarmed confrontation with the enemy, and hand-to-hand combative training can save lives. The many practical battlefield benefits of combative training are not its only advantage. It can also—

a. Contribute to individual and unit strength, flexibility, balance, and cardiorespiratory fitness.
b. Build courage, confidence, self-discipline, and esprit de corps.

Basic Principles. There are basic principles that the hand-to-hand fighter must know and apply to successfully defeat an opponent. The principles mentioned are only a few of the basic guidelines that are essential knowledge for hand-to-hand combat. There are many others, which through years of study become intuitive to a highly skilled fighter.

a. Physical Balance. Balance refers to the ability to maintain equilibrium and to remain in a stable, upright position. A hand-to-hand fighter must maintain his balance both to defend himself and to launch an effective attack. Without balance, the fighter has no stability with which to defend himself, nor does he have a base of power for an attack. The fighter must understand two aspects of balance in a struggle:

(1) How to move his body to keep or regain his own balance. A fighter develops balance through experience, but usually he keeps his feet about shoulder-width apart and his knees flexed. He lowers his center of gravity to increase stability.

(2) How to exploit weaknesses in his opponent's balance. Experience also gives the hand-to-hand fighter a sense of how to move his body in a fight to maintain his balance while exposing the enemy's weak points.

b. Mental Balance. The successful fighter must also maintain a mental balance. He must not allow fear or anger to overcome his ability to concentrate or to react instinctively in hand-to-hand combat.

c. Position. Position refers to the location of the fighter (defender) in relation to his opponent. A vital principle when being attacked is for the defender to move his body to a safe position—that is, where the attack cannot continue unless the enemy moves his whole body. To position for a counterattack, a fighter should move his whole body off the opponent's line of attack. Then, the opponent has to change his position to continue the attack. It is usually safe to move off the line of attack at a 45-degree angle, either toward the opponent or away from him, whichever is appropriate. This position affords the fighter safety and allows him to exploit weaknesses in the enemy's counterattack position. Movement to an advantageous position requires accurate timing and distance perception.

d. Timing. A fighter must be able to perceive the best time to move to an advantageous position in an attack. If he moves too soon, the enemy will anticipate his movement and adjust the attack. If the fighter moves too late, the enemy will strike him. Similarly, the fighter must launch his attack or counterattack at the critical instant when the opponent is the most vulnerable.

e. Distance. Distance is the relative distance between the positions of opponents. A fighter positions himself where distance is to his advantage. The hand-to-hand fighter must adjust his distance by changing position and developing attacks or counterattacks. He does this according to the range at which he and his opponent are engaged.

f. Momentum. Momentum is the tendency of a body in motion to continue in the direction of motion unless acted on by another force. Body mass in motion develops momentum. The greater the body mass or speed of movement, the greater the momentum. Therefore, a fighter must understand the effects of this principle and apply it to his advantage.

(1) The fighter can use his opponent's momentum to his advantage—that is, he can place the opponent in a vulnerable position by using his momentum against him.

  (a) The opponent's balance can be taken away by using his own momentum.

  (b) The opponent can be forced to extend farther than he expected, causing him to stop and change his direction of motion to continue his attack.

  (c) An opponent's momentum can be used to add power to a fighter's own attack or counterattack by combining body masses in motion.

(2) The fighter must be aware that the enemy can also take advantage of the principle of momentum. Therefore, the fighter must avoid placing himself in an awkward or vulnerable position, and he must not allow himself to extend too far.

g. Leverage. A fighter uses leverage in hand-to-hand combat by using the natural movement of his body to place his opponent in a position of unnatural movement. The fighter uses his body or parts of his body to create a natural mechanical advantage over parts of the enemy's body. He should never oppose the enemy in a direct test of strength; however, by using leverage, he can defeat a larger or stronger opponent.

SECTION II: CLOSE-RANGE COMBATIVES

In close-range combatives, two opponents have closed the gap between them so they can grab one another in hand-to-hand combat. The principles of balance, leverage, timing, and body positioning are applied. Throws and takedown techniques are used to upset the opponent's balance and to gain control of the fight by forcing him to the ground. Chokes can be applied to quickly render an opponent unconscious. The soldier should also know counters to choking techniques to protect himself. Grappling involves skillful fighting against an opponent in close-range combat so that a soldier can win through superior body movement or grappling skills. Pain can be used to disable an opponent. A soldier can use painful eye gouges and strikes to soft, vital areas to gain an advantage over his opponent.

3-1. Throws and Takedowns. Throws and takedowns enable a hand-to-hand fighter to take an opponent to the ground where he can be controlled or disabled with further techniques. Throws and takedowns make use of the principles involved in taking the opponent's balance. The fighter uses his momentum against the attacker; he also uses leverage or body position to gain an opportunity to throw the attacker.

a. It is important for a fighter to control his opponent throughout a throw to the ground to keep the opponent from countering the throw or escaping after he is thrown to the ground. One way to do this is to control the opponent's fall so that he lands on his head. It is also imperative that a fighter maintain control of his own balance when executing throws and takedowns.

b. After executing a throw or takedown and while the opponent is on the ground, the fighter must control the opponent by any means available. He can drop his weight onto exposed areas of the opponent's body, using his elbows and knees. He can control the downed opponent's limbs by stepping on them or by placing his knees and body weight on them. Joint locks, chokes, and kicks to vital areas are also good control measures. Without endangering himself, the fighter must maintain the advantage and disable his opponent after throwing him (Figures 3-1 through 3-5).

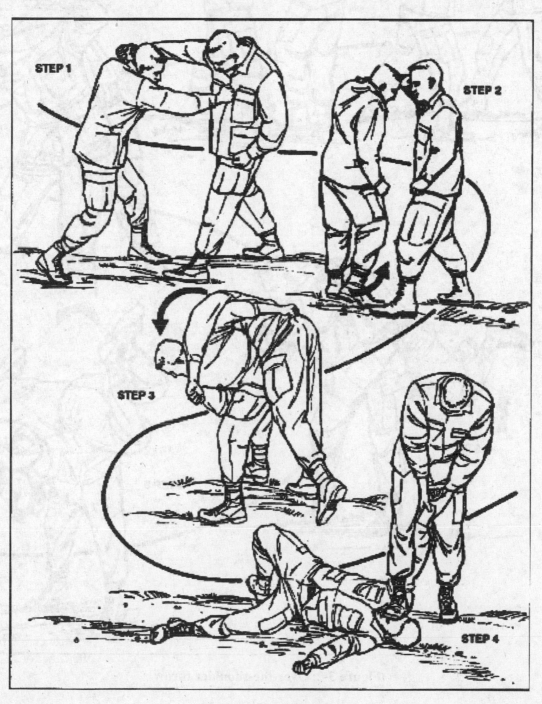

Figure 3-1: Hip throw

Figure 3-2: Over-the-shoulder throw

STEP 1

STEP 2

STEP 3

STEP 4

Figure 3-3: Throw from rear choke

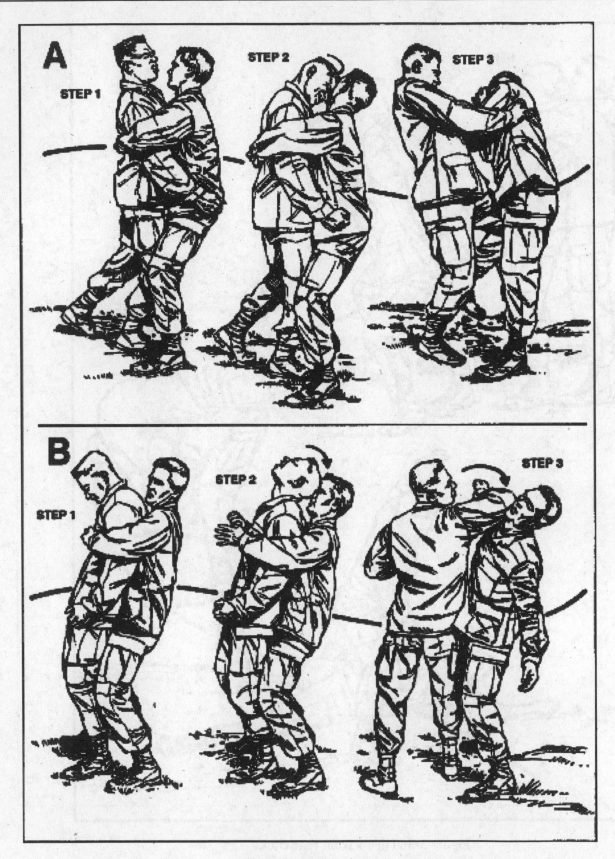

Figure 3-4: Head butt

Figure 3-5: Rear strangle takedown

NOTE: *Although the five techniques shown in Figures 3-1 through 3-5 may be done while wearing LCE—for training purposes, it is safer to conduct all throws and takedowns without any equipment.*

(1) Hip throw. The opponent throws a right punch. The defender steps in with his left foot; at the same time, he blocks the punch with his left forearm and delivers a reverse punch to the face, throat, or other vulnerable area (Figure 3-1, Step 1). (For training, deliver punches to the solar plexus.)

The defender pivots 180 degrees on the ball of his lead foot, wraps his right arm around his opponent's waist, and grasps his belt or pants (Figure 3-1, Step 2). (If opponent is wearing LCE, grasp by the pistol belt or webbing.)

The defender thrusts his hips into his opponent and maintains a grip on his opponent's right elbow. He keeps his knees shoulder-width apart and slightly bent (Figure 3-1, Step 3). He locks his knees, pulls his opponent well over his right hip, and slams him to the ground. (For training, soldier being thrown should land in a good side fall.)

By maintaining control of his opponent's arm, the defender now has the option of kicking or stomping him in the neck, face, or ribs (Figure 3-1, Step 4).

(2) Over-the-shoulder throw. The opponent lunges at the defender with a straight punch (Figure 3-2, Step 1).

The defender blocks the punch with his left forearm, pivots 180 degrees on the ball of his lead foot (Figure 3-2, Step 2), and gets well inside his opponent's right armpit with his right shoulder.

He reaches well back under his opponent's right armpit and grasps him by the collar or hair (Figure 3-2, Step 3).

The defender maintains good back-to-chest, buttock-to-groin contact, keeping his knees slightly bent and shoulder-width apart. He maintains control of his opponent's right arm by grasping the wrist or sleeve (Figure 3-2, Step 4).

The defender bends forward at the waist and holds his opponent tightly against his body. He locks his knees, thrusts his opponent over his shoulder, and slams him to the ground (Figure 3-2, Step 5). He then has the option of disabling his opponent with kicks or stomps to vital areas.

(3) Throw from rear choke. The opponent attacks the defender with a rear strangle choke. The defender quickly bends his knees and spreads his feet shoulder-width apart (Figure 3-3, Step 1). (Knees are bent quickly to put distance between you and your opponent.)

The defender reaches as far back as possible and uses his right hand to grab his opponent by the collar or hair. He then forces his chin into the vee of the opponent's arm that is around his neck. With his left hand, he grasps the opponent's clothing at the tricep and bends forward at the waist (Figure 3-3, Step 2).

The defender locks his knees and, at the same time, pulls his opponent over his shoulder and slams him to the ground (Figure 3-3, Step 3).

He then has the option of spinning around and straddling his opponent or disabling him with punches to vital areas (Figure 3-3, Step 4). (It is important to grip the opponent tightly when executing this move.)

(4) Head butt. The head butt can be applied from the front or the rear. It is repeated until the opponent either releases his grip or becomes unconscious.

 (a) The opponent grabs the defender in a bear hug from the front (A, Figure 3-4, Step 1).

   The defender uses his forehead to smash into his opponent's nose or cheek (A, Figure 3-4, Step 2) and stuns him.

   The opponent releases the defender who then follows up with a kick or knee strike to the groin (A, Figure 3-4, Step 3).

(b) The opponent grabs the defender in a bear hug from the rear (B, Figure 3-4, Step 1).

The defender cocks his head forward and smashes the back of his head into the opponent's nose or cheek area (B, Figure 3-4, Step 2).

The defender turns to face his opponent and follows up with a spinning elbow strike to the head (B, Figure 3-4, Step 3).

(5) Rear strangle takedown. The defender strikes the opponent from the rear with a forearm strike to the neck (carotid artery) (Figure 3-5, Step 1).

The defender wraps his right arm around his opponent's neck, making sure he locks the throat and windpipe in the vee formed by his elbow. He grasps his left bicep and wraps his left hand around the back of the opponent's head. He pulls his right arm in and flexes it, pushing his opponent's head forward (Figure 3-5, Step 2).

The defender kicks his legs out and back, maintains a choke on his opponent's neck, and pulls his opponent backward until his neck breaks (Figure 3-5, Step 3).

3-2. Strangulation. Strangulation is a most effective method of disabling an opponent. The throat's vulnerability is widely known and should be a primary target in close-range fighting. Your goal may be to break the opponent's neck, to crush his trachea, to block the air supply to his lungs, or to block the blood supply to his brain.

a. Strangulation by Crushing. Crushing the trachea just below the voice box is probably one of the fastest, easiest, most lethal means of strangulation. The trachea is crushed between the thumb and first two or three fingers.

b. Respiratory Strangulation. Compressing the windpipe to obstruct air flow to the lungs is most effectively applied by pressure on the cartilage of the windpipe. Unconsciousness can take place within one to two minutes. However, the technique is not always effective on a strong opponent or an opponent with a large neck. It is better to block the blood supply to weaken the opponent first.

c. Sanguineous Strangulation. Cutting off the blood supply to the brain by applying pressure to the carotid arteries results in rapid unconsciousness of the victim. The victim can be rendered unconscious within 3 to 8 seconds, and death can result within 30 to 40 seconds.

3-3. Choking Techniques. There are several choking techniques that a soldier can use to defeat his opponent in hand-to-hand combat.

a. Cross-Collar Choke. With crossed hands, the fighter reaches as far as possible around his opponent's neck and grabs his collar (Figure 3-6, Step 1). The backs of his hands should be against the neck.

The fighter keeps his elbows bent and close to the body (as in opening a tightly sealed jar), pulls outward with both hands, and chokes the sides of the opponent's neck by rotating the knuckles into the neck (Figure 3-6, Step 2). The forearm can also be used.

b. Collar Grab Choke. The fighter grabs his opponent's collar with both hands straight-on (Figure 3-7). He then rotates the knuckles inward against the neck to quickly produce a good choke. He also keeps the elbows in front and close to the body where the greatest strength is maintained.

c. Carotid Choke. The fighter grabs the sides of the opponent's throat by the muscle and sticks his thumbs into the carotids, closing them off (Figure 3-8). This is a fast and painful choke.

d. Trachea Choke. The fighter grabs the opponent's trachea (Figure 3-9) by sticking three fingers behind the voice box on one side and the thumb behind the other. He then crushes the fingers together and twists, applying pressure until the opponent is disabled.

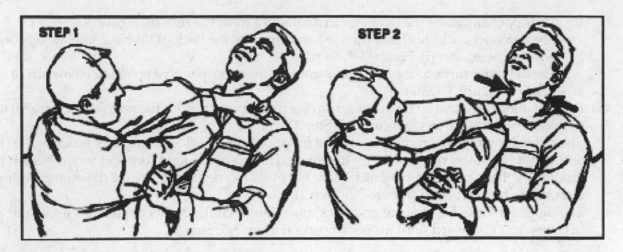

Figure 3-6: Cross-collar choke

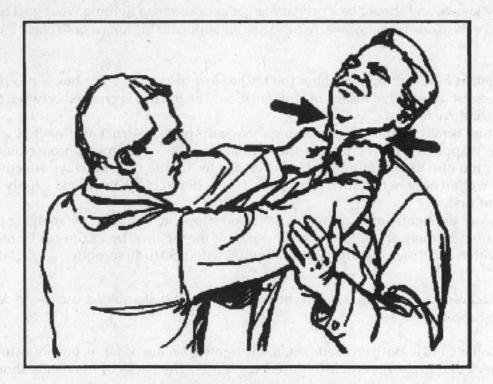

Figure 3-7: Collar grab choke

3-4. Counters to Chokes. A soldier must know how to defend against being choked. Incapacitation and unconsciousness can occur within three seconds; therefore, it is crucial for the defender to know all possible counters to chokes.

    a. Eye Gouge. The opponent attacks the defender with a frontal choke. The defender has the option of going over or under the opponent's arms. To disable the opponent, the defender inserts both

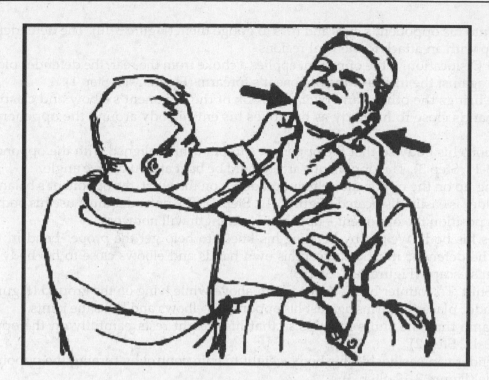

Figure 3-8: Carotid choke

Figure 3-9: Trachea choke

thumbs into his opponent's eyes and tries to gouge them (Figure 3-10). The defender is prepared to follow-up with an attack to the vital regions.

b. Shoulder Dislocation. If the opponent applies a choke from the rear, the defender places the back of his hand against the inside of the opponent's forearm (Figure 3-11, Step 1).

Then, he brings the other hand over the crook of the opponent's elbow and clasps hands, keeping his hands close to his body as he moves his entire body around the opponent (Figure 3-11, Step 2).

He positions his body so that the opponent's upper arm is aligned with the opponent's shoulders (Figure 3-11, Step 3). The opponent's arm should be bent at a 90-degree angle.

By pulling up on the opponent's elbow and down on the wrist, the opponent's balance is taken and his shoulder is easily dislocated (Figure 3-11, Step 4). The defender must use his body movement to properly position the opponent—upper body strength will not work.

He drops his body weight by bending his knees to help get the proper bend in the opponent's elbow. The defender must also keep his own hands and elbows close to his body to prevent the opponent's escape (Figure 3-11, Step 5).

c. Weight Shift. To counter being choked from above while lying on the ground (Figure 3-12, Step 1), the defender places his arms against his opponent's elbows and locks the joints.

At the same time, he shifts his hips so that his weight rests painfully on the opponent's ankle (Figure 3-12, Step 2).

The defender can easily shift his body weight to gain control by turning the opponent toward his weak side (Figure 3-12, Step 3).

Figure 3-10: Eye gouge

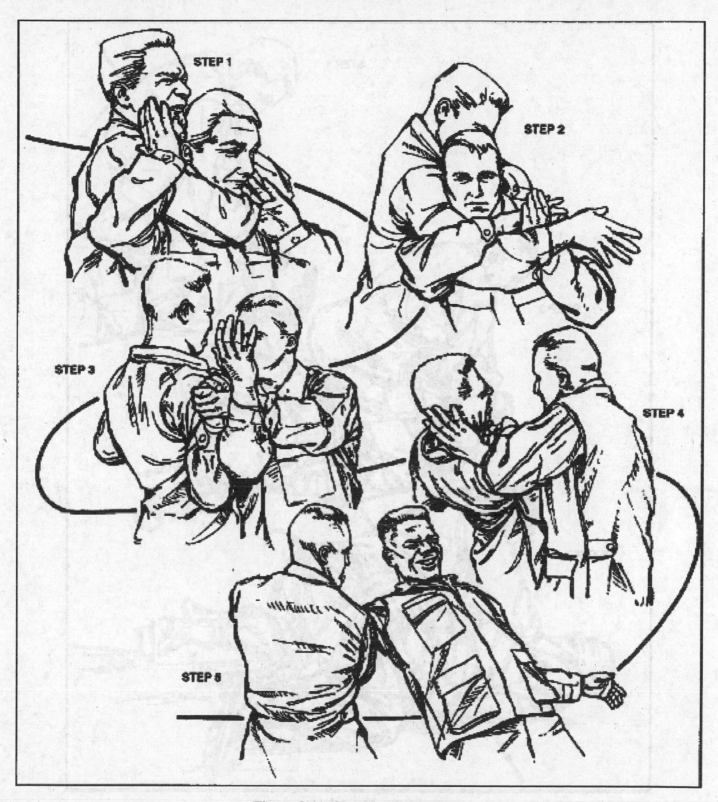

Figure 3-11: Shoulder dislocation

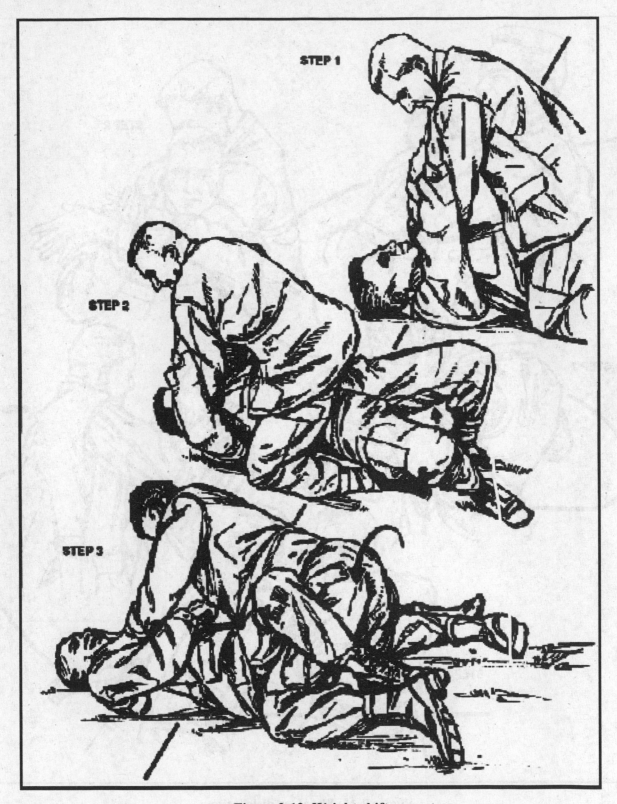

Figure 3-12: Weight shift

d. Counterstrikes to Rear Choke and Frontal Choke. As the opponent tries a rear choke (A, Figure 3- 13, Step 1), the defender can break the opponent's grip with a strong rear-elbow strike into the solar plexus (A, Figure 3-13, Step 2).

He can follow with a shin scrape down along the opponent's leg and stomp the foot (A, Figure 3-13, Step 3).

He may wish to continue by striking the groin of the opponent (A, Figure 3-13, Step 4).

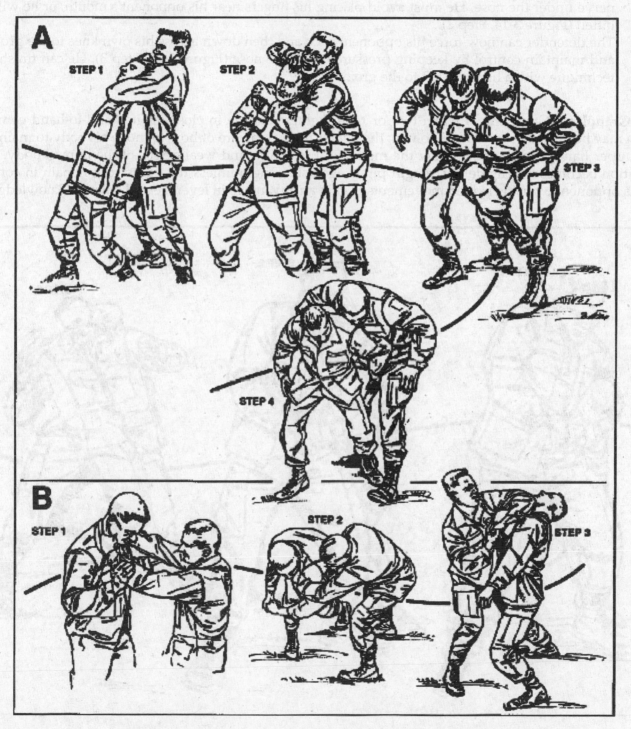

Figure 3-13: Counterstrikes to rear choke and frontal choke

As the opponent begins a frontal choke (B, Figure 3-13, Step 1), the defender turns his body and drops one arm between the opponent's arms (B, Figure 3-13, Step 2).

He sinks his body weight and drives his own hand to the ground, and then explodes upward with an elbow strike (B, Figure 3-13, Step 3) into the opponent's chin, stomach, or groin.

e. Headlock Escape. If a defender is in a headlock, he first turns his chin in toward his opponent's body to prevent choking (Figure 3-14, Step 1).

Next, he slides one hand up along the opponent's back, around to the face, and finds the sensitive nerve under the nose. He must avoid placing his fingers near his opponent's mouth, or he will be bitten (Figure 3-14, Step 2).

The defender can now force his opponent back and then down across his own knee to the ground and maintain control by keeping pressure under the nose (Figure 3-14, Step 3). He can finish the technique with a hammer fist to the groin.

3-5. Grappling. Grappling is when two or more fighters engage in close-range, hand-to-hand combat. They may be armed or unarmed. To win, the fighter must be aware of how to move his body to maintain the upper hand, and he must know the mechanical strengths and weaknesses of the human body. The situation becomes a struggle of strength pitted against strength unless the fighter can remain in control of his opponent by using skilled movements to gain an advantage in leverage and balance. Knowledge of

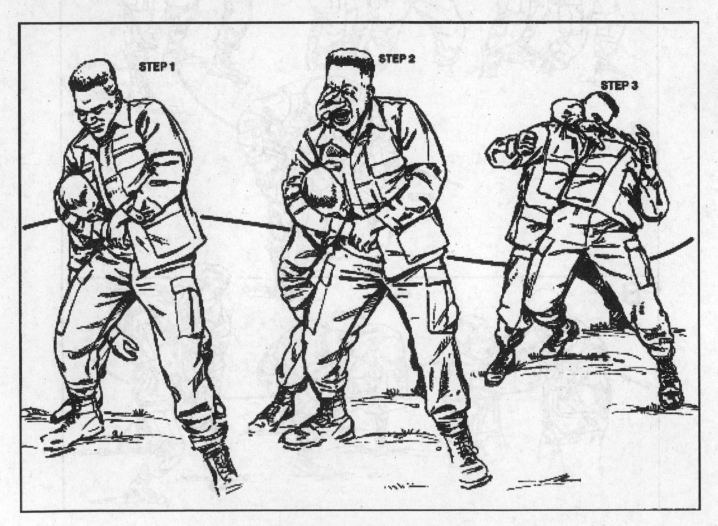

Figure 3-14: Headlock escape

the following basic movement techniques may give the fighter a way to apply and gain the advantage in grappling situations.

a. Wristlock From a Collar or Lapel Grab. When an opponent grabs the defender by the collar or by the lapel, the defender reaches up and grabs the opponent's hand (to prevent him from withdrawing it) while stepping back to pull him off balance (Figure 3-15, Step 1).
The defender peels off the opponent's grabbing hand by crushing his thumb and bending it back on itself toward the palm in a straight line (Figure 3-15, Step 2). To keep his grip on the opponent's thumb, the defender keeps his hands close to his body where his control is strongest.

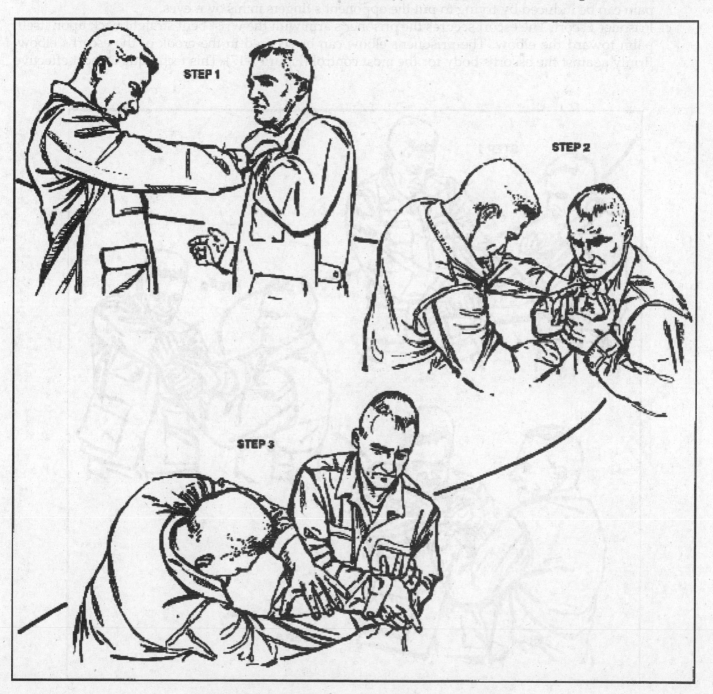

Figure 3-15: Wristlock from a collar or lapel grab

He then turns his body so that he has a wristlock on his opponent. The wristlock is produced by turning his wrist outward at a 45-degree angle and by bending it toward the elbow (Figure 3-15, Step 3). The opponent can be driven to the ground by putting his palm on the ground.

b. Wristlock From an Arm Grab. When an opponent grabs a defender's arm, the defender rotates his arm to grab the opponent's forearm (Figure 3-16, Step 1).

At the same time, he secures his other hand on the gripping hand of the opponent to prevent his escape (Figure 3-16, Step 2).

As the defender steps in toward the opponent and maintains his grip on the hand and forearm, a zee shape is formed by the opponent's arm; this is an effective wristlock (Figure 3-16, Step 3). More pain can be induced by trying to put the opponent's fingers in his own eyes.

c. Prisoner Escort. The escort secures the prisoner's arm with the wrist bent straight back upon itself, palm toward the elbow. The prisoner's elbow can be secured in the crook of the escort's elbow, firmly against the escort's body for the most control (Figure 3-17). This technique is most effective

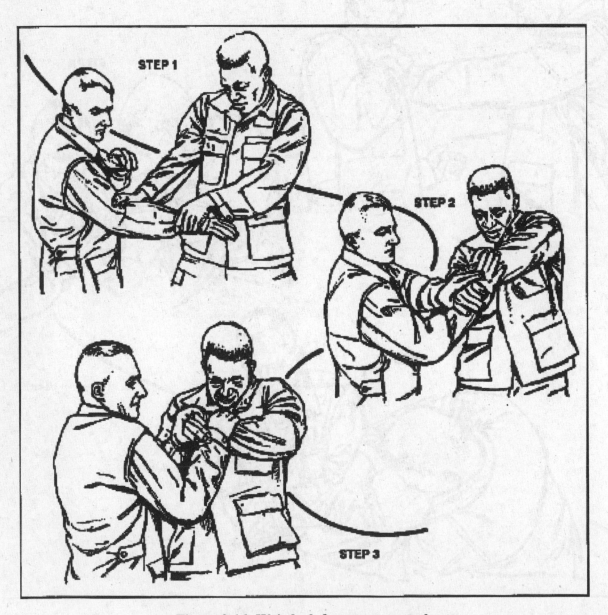

Figure 3-16: Wristlock from an arm grab

Figure 3-17: Prisoner escort

with two escorts, each holding a wrist of the prisoner. Use this technique to secure the opponent only if rope, flex cuffs, or handcuffs are unavailable.

d. Elbow Lock Against the Body. The opponent's elbow can be locked against the side of the body (Figure 3-18) by the defender. The defender turns his body to force the elbow into a position in which it was not designed to move. He can apply leverage on the opponent's wrist to gain control since the lock causes intense pain. The elbow can easily be broken to make the arm ineffective. This movement must be executed with maximum speed and force.

e. Elbow Lock Against the Knee. While grappling on the ground, a defender can gain control of the situation if he can use an elbow lock (Figure 3-19) against the opponent. He uses his knee as a fulcrum for leverage to break his opponent's arm at the elbow. Once the arm breaks, the defender must be prepared with a follow-up technique.

f. Elbow Lock Against the Shoulder. An elbow lock can be applied by locking the elbow joint against the shoulder (Figure 3-20) and pulling down on the wrist. Leverage is produced by using the shoulder as a fulcrum, by applying force, and by straightening the knees to push upward. This uses the defender's body mass and ensures more positive control. The opponent's arm must be kept straight so he cannot drive his elbow down into the defender's shoulder.

g. Shoulder Dislocation. A defender can maneuver into position to dislocate a shoulder by moving inside when an opponent launches a punch (Figure 3-21, Step 1). The defender holds his hand nearest the punching arm high to protect the head.

The defender continues to move in and places his other arm behind the punching arm (Figure 3-21, Step 2). He strikes downward into the crook of the opponent's elbow to create a bend.

Then he clasps his hands and moves to the opponent's outside until the opponent's upper arm is in alignment with his shoulders and bent 90 degrees at the elbow. As he steps, the defender pulls up on the opponent's elbow and directs the wrist downward. This motion twists the shoulder joint so it is easily dislocated and the opponent loses his balance (Figure 3-21, Step 3).

Figure 3-18: Elbow lock against the body

Figure 3-19: Elbow lock against the knee

Figure 3-20: Elbow lock against the shoulder

NOTE: The defender must keep his clasped hands close to the body and properly align the opponent's arm by maneuvering his entire body. This technique will not succeed by using upperbody strength only, the opponent will escape.

(1) Straight-arm shoulder dislocation. The shoulder can also be dislocated (Figure 3-22) by keeping the elbow straight and forcing the opponent's arm backward toward the opposite shoulder at about 45 degrees. The initial movement must take the arm down and alongside the opponent's

550 The Ultimate Guide to U.S. Army Survival Skills, Tactics, and Techniques

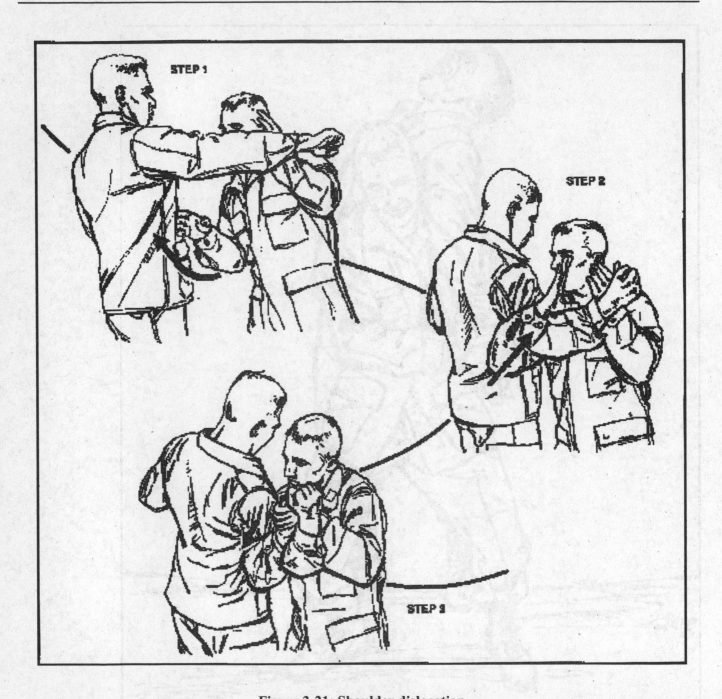

Figure 3-21: Shoulder dislocation

body. Bending the wrist toward the elbow helps to lock out the elbow. The dislocation also forces the opponent's head down-ward where a knee strike can be readily made. This dislocation technique should be practiced to get the feel of the correct direction in which to move the joint.

(2) Shoulder dislocation using the elbow. While grappling, the defender can snake his hand over the crook in the opponent's elbow and move his body to the outside, trapping one arm of the opponent against his side (Figure 3-23, Step 1).

Figure 3-22: Straight-arm shoulder dislocation

The defender can then clasp his hands in front of his body and use his body mass in motion to align the opponent's upper arm with the line between the shoulders (Figure 3-23, Step 2).

By dipping his weight and then pulling upward on the opponent's elbow, the shoulder is dislocated, and the opponent loses his balance (Figure 3-23, Step 3). If the opponent's elbow locks rather than bends to allow the shoulder dislocation, the defender can use the elbow lock to keep control.

h. Knee Lock/Break. The opponent's knee joint can be attacked to produce knee locks or breaks (Figure 3-24) by forcing the knee in a direction opposite to which it was designed to move. The knee can be attacked with the body's mass behind the defender's knee or with his entire body by falling on the opponent's knee, causing it to hyperextend.

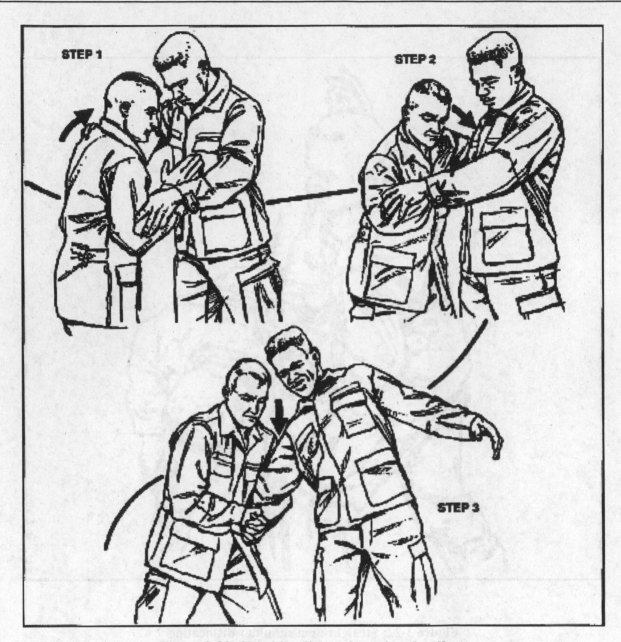

Figure 3-23: Shoulder dislocation using the elbow

Figure 3-24: Knee lock/break

CHAPTER 4

Medium-Range Combatives

In medium-range combatives, two opponents are already within touching distance. The arsenal of possible body weapons includes short punches and strikes with elbows, knees, and hands. Head butts are also effective; do not forget them during medium-range combat. A soldier uses his peripheral vision to evaluate the targets presented by the opponent and choose his target. He should be aggressive and concentrate his attack on the opponent's vital points to end the fight as soon as possible.

4-1. Vital Targets. The body is divided into three sections: high, middle, and low. Each section contains vital targets (Figure 4-1). The effects of striking these targets follow:

a. High Section. The high section includes the head and neck; it is the most dangerous target area.
   (1) Top of the head. The skull is weak where the frontal cranial bones join. A forceful strike causes trauma to the cranial cavity, resulting in unconsciousness and hemorrhage. A severe strike can result in death.
   (2) Forehead. A forceful blow can cause whiplash; a severe blow can cause cerebral hemorrhage and death.
   (3) Temple. The bones of the skull are weak at the temple, and an artery and large nerve lie close to the skin. A powerful strike can cause unconsciousness and brain concussion. If the artery is severed, the resulting massive hemorrhage compresses the brain, causing coma and or death.
   (4) Eyes. A slight jab in the eyes causes uncontrollable watering and blurred vision. A forceful jab or poke can cause temporary blindness, or the eyes can be gouged out. Death can result if the fingers penetrate through the thin bone behind the eyes and into the brain.
   (5) Ears. A strike to the ear with cupped hands can rupture the eardrum and may cause a brain concussion.
   (6) Nose. Any blow can easily break the thin bones of the nose, causing extreme pain and eye watering.
   (7) Under the nose. A blow to the nerve center, which is close to the surface under the nose, can cause great pain and watery eyes.
   (8) Jaw. A blow to the jaw can break or dislocate it. If the facial nerve is pinched against the lower jaw, one side of the face will be paralyzed.
   (9) Chin. A blow to the chin can cause paralysis, mild concussion, and unconsciousness. The jawbone acts as a lever that can transmit the force of a blow to the back of the brain where the cardiac and respiratory mechanisms are controlled.
   (10) Back of ears and base of skull. A moderate blow to the back of the ears or the base of the skull can cause unconsciousness by the jarring effect on the back of the brain. However, a powerful blow can cause a concussion or brain hemorrhage and death.
   (11) Throat. A powerful blow to the front of the throat can cause death by crushing the windpipe. A forceful blow causes extreme pain and gagging or vomiting.
   (12) Side of neck. A sharp blow to the side of the neck causes unconsciousness by shock to the carotid artery, jugular vein, and vagus nerve. For maximum effect, the blow should be focused below and slightly in front of the ear. A less powerful blow causes involuntary muscle spasms and intense pain. The side of the neck is one of the best targets to use to drop an opponent immediately or to disable him temporarily to finish him later.

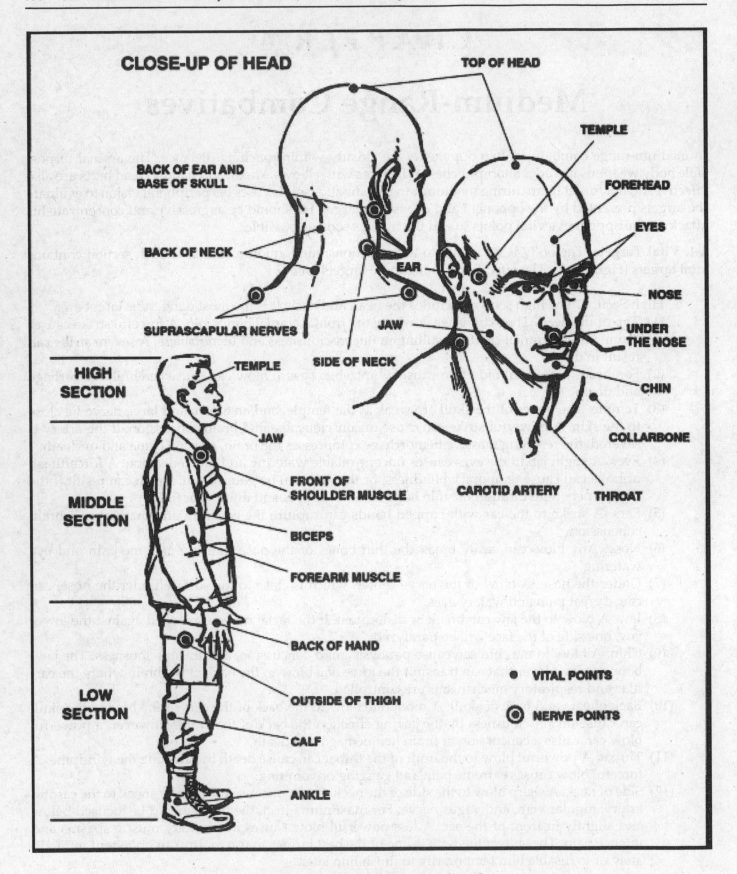

Figure 4-1: Vital targets. *(continued)*

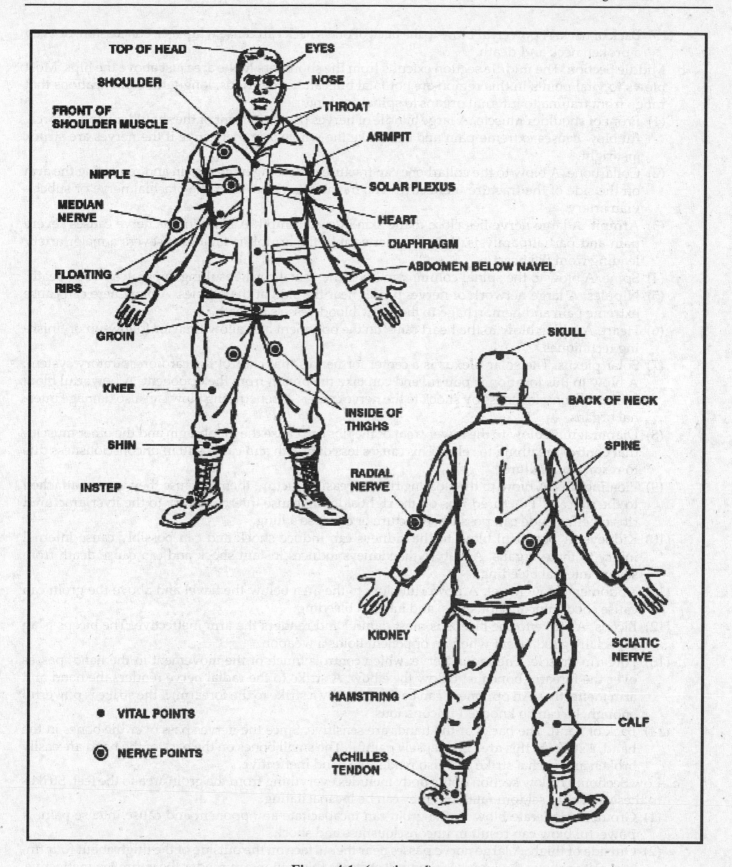

VITAL POINTS

NERVE POINTS

TOP OF HEAD
SHOULDER
FRONT OF SHOULDER MUSCLE
NIPPLE
MEDIAN NERVE
FLOATING RIBS
GROIN
KNEE
SHIN
INSTEP

EYES
NOSE
THROAT
ARMPIT
SOLAR PLEXUS
HEART
DIAPHRAGM
ABDOMEN BELOW NAVEL
SKULL
INSIDE OF THIGHS
BACK OF NECK

RADIAL NERVE
SPINE
KIDNEY
SCIATIC NERVE
HAMSTRING
CALF
ACHILLES TENDON

Figure 4-1: *(continued)*

(13) Back of neck. A powerful blow to the back of one's neck can cause whiplash, concussion, or even a broken neck and death.

b. Middle Section. The middle section extends from the shoulders to the area just above the hips. Most blows to vital points in this region are not fatal but can have serious, long-term complications that range from trauma to internal organs to spinal cord injuries.

(1) Front of shoulder muscle. A large bundle of nerves passes in front of the shoulder joint. A forceful blow causes extreme pain and can make the whole arm ineffective if the nerves are struck just right.

(2) Collarbone. A blow to the collarbone can fracture it, causing intense pain and rendering the arm on the side of the fracture ineffective. The fracture can also sever the brachial nerve or subclavian artery.

(3) Armpit. A large nerve lies close to the skin in each armpit. A blow to this nerve causes severe pain and partial paralysis. A knife inserted into the armpit is fatal as it severs a major artery leading from the heart.

(4) Spine. A blow to the spinal column can sever the spinal cord, resulting in paralysis or in death.

(5) Nipples. A large network of nerves passes near the skin at the nipples. A blow here can cause extreme pain and hemorrhage to the many blood vessels beneath.

(6) Heart. A jolting blow to the heart can stun the opponent and allow time for follow-up or finishing techniques.

(7) Solar plexus. The solar plexus is a center for nerves that control the cardiorespiratory system. A blow to this location is painful and can take the breath from the opponent. A powerful blow causes unconsciousness by shock to the nerve center. A penetrating blow can also damage internal organs.

(8) Diaphragm. A blow to the lower front of the ribs can cause the diaphragm and the other muscles that control breathing to relax. This causes loss of breath and can result in unconsciousness due to respiratory failure.

(9) Floating ribs. A blow to the floating ribs can easily fracture them because they are not attached to the rib cage. Fractured ribs on the right side can cause internal injury to the liver; fractured ribs on either side can possibly puncture or collapse a lung.

(10) Kidneys. A powerful blow to the kidneys can induce shock and can possibly cause internal injury to these organs. A stab to the kidneys induces instant shock and can cause death from severe internal bleeding.

(11) Abdomen below navel. A powerful blow to the area below the navel and above the groin can cause shock, unconsciousness, and internal bleeding.

(12) Biceps. A strike to the biceps is most painful and renders the arm ineffective. The biceps is an especially good target when an opponent holds a weapon.

(13) Forearm muscle. The radial nerve, which controls much of the movement in the hand, passes over the forearm bone just below the elbow. A strike to the radial nerve renders the hand and arm ineffective. An opponent can be disarmed by a strike to the forearm; if the strike is powerful enough, he can be knocked unconscious.

(14) Back of hand. The backs of the hands are sensitive. Since the nerves pass over the bones in the hand, a strike to this area is intensely painful. The small bones on the back of the hand are easily broken and such a strike can also render the hand ineffective.

c. Low Section. The low section of the body includes everything from the groin area to the feet. Strikes to these areas are seldom fatal, but they can be incapacitating.

(1) Groin. A moderate blow to the groin can incapacitate an opponent and cause intense pain. A powerful blow can result in unconsciousness and shock.

(2) Outside of thigh. A large nerve passes near the surface on the outside of the thigh about four fingerwidths above the knee. A powerful strike to this region can render the entire leg ineffective, causing an opponent to drop. This target is especially suitable for knee strikes and shin kicks.

(3) Inside of thigh. A large nerve passes over the bone about in the middle of the inner thigh. A blow to this area also incapacitates the leg and can cause the opponent to drop. Knee strikes and heel kicks are the weapons of choice for this target.

(4) Hamstring. A severe strike to the hamstring can cause muscle spasms and inhibit mobility. If the hamstring is cut, the leg is useless.

(5) Knee. Because the knee is a major supporting structure of the body, damage to this joint is especially detrimental to an opponent. The knee is easily dislocated when struck at an opposing angle to the joint's normal range of motion, especially when it is bearing the opponent's weight. The knee can be dislocated or hyperextended by kicks and strikes with the entire body.

(6) Calf. A powerful blow to the top of the calf causes painful muscle spasms and also inhibits mobility.

(7) Shin. A moderate blow to the shin produces great pain, especially a blow with a hard object. A powerful blow can possibly fracture the bone that supports most of the body weight.

(8) Achilles tendon. A powerful strike to the Achilles tendon on the back of the heel can cause ankle sprain and dislocation of the foot. If the tendon is torn, the opponent is incapacitated. The Achilles tendon is a good target to cut with a knife.

(9) Ankle. A blow to the ankle causes pain; if a forceful blow is delivered, the ankle can be sprained or broken.

(10) Instep. The small bones on the top of the foot are easily broken. A strike here will hinder the opponent's mobility.

4-2. Striking Principles. Effective striking with the weapons of the body to the opponent's vital points is essential for a victorious outcome in a hand-to-hand struggle. A soldier must be able to employ the principles of effective striking if he is to emerge as the survivor in a fight to the death.

a. Attitude. Proper mental attitude is of primary importance in the soldier's ability to strike an opponent. In hand-to-hand combat, the soldier must have the attitude that he will defeat the enemy and complete the mission, no matter what. In a fight to the death, the soldier must have the frame of mind to survive above all else; the prospect of losing cannot enter his mind. He must commit himself to hit the opponent continuously with whatever it takes to drive him to the ground or end his resistance. A memory aid is, "Thump him and dump him!"

b. Fluid Shock Wave. A strike should be delivered so that the target is hit and the weapon remains on the impact site for at least a tenth of a second. This imparts all of the kinetic energy of the strike into the target area, producing a fluid shock wave that travels into the affected tissue and causes maximum damage. It is imperative that all strikes to vital points and nerve motor points are delivered with this principle in mind. The memory aid is, "Hit and stick!"

c. Target Selection. Strikes should be targeted at the opponent's vital points and nerve motor points. The results of effective strikes to vital points are discussed in paragraph 4-1. Strikes to nerve motor points cause temporary mental stunning and muscle motor dysfunction to the affected areas of the body. Mental stunning results when the brain is momentarily disoriented by overstimulation from too much input—for example, a strike to a major nerve. The stunning completely disables an opponent for three to seven seconds and allows the soldier to finish off the opponent, gain total control of the situation, or make his escape. Sometimes, such a strike causes unconsciousness. A successful strike to a nerve motor center also renders the affected body part immovable by causing muscle spasms and dysfunction due to nerve overload. (Readily available nerve motor points are shown in Figure 4-1)

(1) Jugular notch pressure point. Located at the base of the neck just above the breastbone; pressure to this notch can distract and take away his balance. Pressure from fingers jabbed into the notch incurs intense pain that causes the opponent to withdraw from the pressure involuntarily.

(2) Suprascapular nerve motor point. This nerve is located where the trapezius muscle joins the side of the neck. A strike to this point causes intense pain, temporary dysfunction of the affected arm and hand, and mental stunning for three to seven seconds. The strike should be a downward knife-hand or hammerfist strike from behind.

(3) Brachial plexus origin. This nerve motor center is on the side of the neck. It is probably the most reliable place to strike someone to stun them. Any part of the hand or arm may be applied—the palm heel, back of the hand, knife hand, ridge hand, hammer fist, thumb tip, or the forearm. A proper strike to the brachial plexus origin causes—

- Intense pain.
- Complete cessation of motor activity.
- Temporary dysfunction of the affected arm.
- Mental stunning for three to seven seconds.
- Possible unconsciousness.

(4) Brachial plexus clavicle notch pressure point. This center is behind the collarbone in a hollow about halfway between the breastbone and the shoulder joint. The strike should be delivered with a small impact weapon or the tip of the thumb to create high-level mental stunning and dysfunction of the affected arm.

(5) Brachial plexus tie-in motor point. Located on the front of the shoulder joint, a strike to this point can cause the arm to be ineffective. Multiple strikes may be necessary to ensure total dysfunction of the arm and hand.

(6) Stellate ganglion. The ganglion is at the top of the pectoral muscle centered above the nipple. A severe strike to this center can cause high-level stunning, respiratory dysfunction, and possible unconsciousness. A straight punch or hammer fist should be used to cause spasms in the nerves affecting the heart and respiratory systems.

(7) Cervical vertebrae. Located at the base of the skull, a strike to this particular vertebrae can cause unconsciousness or possibly death. The harder the strike, the more likely death will occur.

(8) Radial nerve motor point. This nerve motor point is on top of the forearm just below the elbow. Strikes to this point can create dysfunction of the affected arm and hand. The radial nerve should be struck with the hammer fist or the forearm bones or with an impact weapon, if available. Striking the radial nerve can be especially useful when disarming an opponent armed with a knife or other weapon.

(9) Median nerve motor point. This nerve motor point is on the inside of the forearm at the base of the wrist, just above the heel of the hand. Striking this center produces similar effects to striking the radial nerve, although it is not as accessible as the radial nerve.

(10) Sciatic nerve. A sciatic nerve is just above each buttock, but below the belt line. A substantial strike to this nerve can disable both legs and possibly cause respiratory failure. The sciatic nerve is the largest nerve in the body besides the spinal cord. Striking it can affect the entire body, especially if an impact weapon is used.

(11) Femoral nerve. This nerve is in the center of the inside of the thigh; striking the femoral nerve can cause temporary motor dysfunction of the affected leg, high-intensity pain, and mental stunning for three to seven seconds. The knee is best to use to strike the femoral nerve.

(12) Common peroneal nerve motor point. The peroneal nerve is on the outside of the thigh about four fingers above the knee. A severe strike to this center can cause collapse of the affected leg and high intensity pain, as well as mental stunning for three to seven seconds. This highly accessible point is an effective way to drop an opponent quickly. This point should be struck with a knee, shin kick, or impact weapon.

4-3. Short Punches and Strikes. During medium-range combat, punches and strikes are usually short because of the close distance between fighters. Power is generated by using the entire body mass in motion behind all punches and strikes.

a. Hands as Weapons. A knowledge of hand-to-hand combat fighting provides the fighter another means to accomplish his mission. Hands can become deadly weapons when used by a skilled fighter.

   (1) Punch to solar plexus. The defender uses this punch for close-in fighting when the opponent rushes or tries to grab him. The defender puts his full weight and force behind the punch and strikes his opponent in the solar plexus (Figure 4-2), knocking the breath out of his lungs. The defender can then follow-up with a knee to the groin, or he can use other disabling blows to vital areas.

   (2) Thumb strike to throat. The defender uses the thumb strike to the throat (Figure 4-3) as an effective technique when an opponent is rushing him or trying to grab him. The defender thrusts his right arm and thumb out and strikes his opponent in the throat-larynx area while holding his left hand high for protection. He can follow up with a disabling blow to his opponent's vital areas.

   (3) Thumb strike to shoulder joint. The opponent rushes the defender and tries to grab him. The defender strikes the opponent's shoulder joint or upper pectoral muscle with his fist or thumb (Figure 4-4). This technique is painful and renders the opponent's arm numb. The defender then follows up with a disabling movement.

   (4) Hammer-fist strike to face. The opponent rushes the defender. The defender counters by rotating his body in the direction of his opponent and by striking him in the temple, ear, or face (Figure 4-5). The defender follows up with kicks to the groin or hand strikes to his opponent's other vital areas.

   (5) Hammer-fist strike to side of neck. The defender catches his opponent off guard, rotates at the waist to generate power, and strikes his opponent on the side of the neck (carotid artery) (Figure 4-6) with his hand clenched into a fist. This strike can cause muscle spasms at the least and may knock his opponent unconscious.

Figure 4-2: Punch to solar plexus.

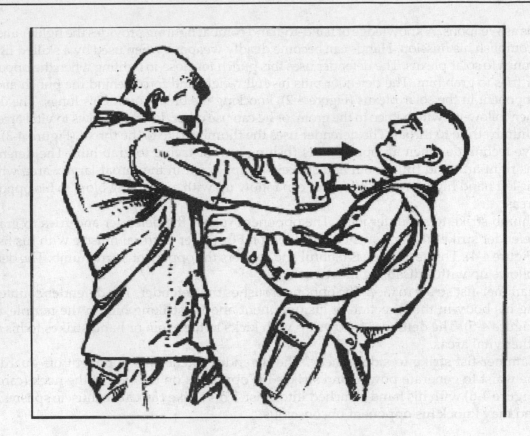

Figure 4-3: Thumb strike to throat.

Figure 4-4: Thumb strike to shoulder joint.

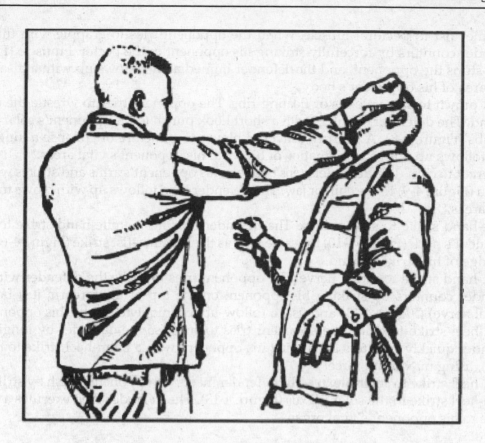

Figure 4-5: Hammer-fist strike to face.

Figure 4-6: Hammer-fist strike to neck.

(6) Hammer fist to pectoral muscle. When the opponent tries to grapple with the defender, the defender counters by forcefully striking his opponent in the pectoral muscle (Figure 4-7). This blow stuns the opponent, and the defender immediately follows up with a disabling blow to a vital area of his opponent's body.

(7) Hook punch to solar plexus or floating ribs. The opponent tries to wrestle the defender to the ground. The defender counters with a short hook punch to his opponent's solar plexus or floating ribs (Figure 4-8). A sharply delivered blow can puncture or collapse a lung. The defender then follows up with a combination of blows to his opponent's vital areas.

(8) Uppercut to chin. The defender steps between his opponent's arms and strikes with an uppercut punch (Figure 4-9) to the chin or jaw. The defender then follows up with blows to his opponent's vital areas.

(9) Knife-hand strike to side of neck. The defender executes a knife-hand strike to the side of his opponent's neck (Figure 4-10) the same way as the hammer-fist strike (Figure 4-6) except he uses the edge of his striking hand.

(10) Knife-hand strike to radial nerve. The opponent tries to strike the defender with a punch. The defender counters by striking his opponent on the top of the forearm just below the elbow (radial nerve) (Figure 4-11) and uses a follow-up technique to disable his opponent.

(11) Palm-heel strike to chin. The opponent tries to surprise the defender by lunging at him. The defender quickly counters by striking his opponent with a palm-heel strike to the chin (Figure 4- 12), using maximum force.

(12) Palm-heel strike to solar plexus. The defender meets his opponent's rush by striking him with a palm-heel strike to the solar plexus (Figure 4-13). The defender then executes a follow-up technique to his opponent's vital organs.

Figure 4-7: Hammer-fist to pectoral muscle.

Figure 4-8: Hook punch to solar plexus or floating ribs.

Figure 4-9: Uppercut to chin.

Figure 4-10: Knife-hand to side of neck.

Figure 4-11: Knife-hand strike to radial nerve.

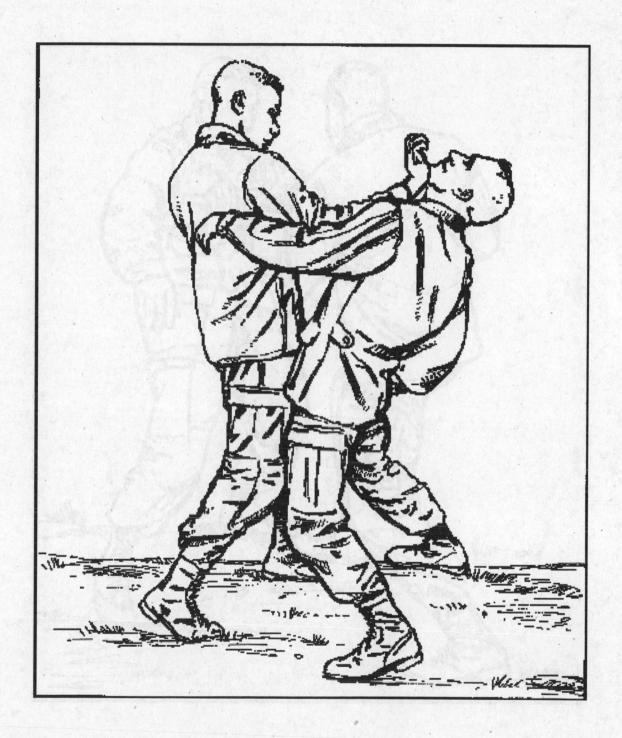

Figure 4-12: Palm heel strike to chin.

Figure 4-13: Palm-heel strike to solar plexus.

(13) Palm-heel strike to kidneys. The defender grasps his opponent from behind by the collar and pulls him off balance. He quickly follows up with a hard palm-heel strike to the opponent's kidney (Figure 4-14). The defender can then take down his opponent with a follow-up technique to the back of his knee.

Figure 4-14: Palm-heel strike to kidneys.

b. **Elbows as Weapons.** The elbows are also formidable weapons; tremendous striking power can be generated from them. The point of the elbow should be the point of impact. The elbows are strongest when kept in front of the body and in alignment with the shoulder joint; that is, never strike with the elbow out to the side of the body.

(1) Elbow strikes. When properly executed, elbow strikes (Figures 4-15 through 4-21) render an opponent ineffective. When using elbow strikes, execute them quickly, powerfully, and repetitively until the opponent is disabled.

(2) Repetitive elbow strikes. The attacker on the right throws a punch (Figure 4-22, Step 1). The defender counters with an elbow strike to the biceps (Figure 4-22, Step 2). The attacker follows with a punch from his other arm.
The defender again counters with an elbow strike to the shoulder joint (Figure 4-22, Step 3). He next strikes with an elbow from the opposite side to the throat.

Figure 4-15: Elbow strike to face.

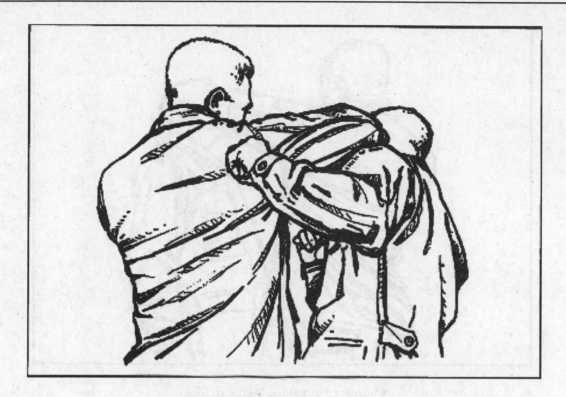

Figure 4-16: Elbow strike to temple.

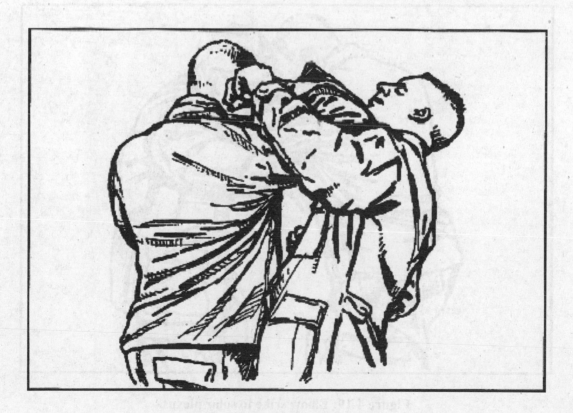

Figure 4-17: Rising elbow strike.

Figure 4-18: Elbow strike to head.

Figure 4-19: Elbow srike to solar plexus.

Figure 4-20: Elbow strike to biceps.

Figure 4-21: Elbow strike to inside of shoulder.

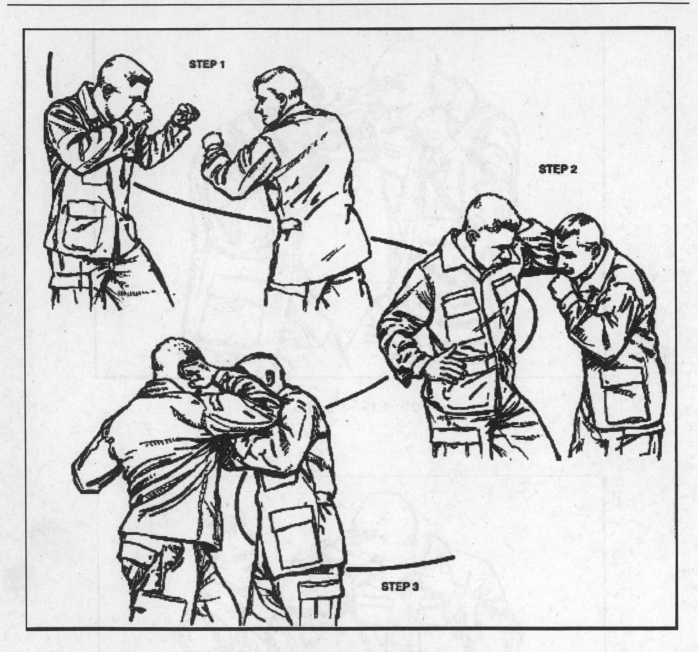

Figure 4-22: Repetitive elbow strike.

c. Knees as Weapons. When the knees are used to strike opponents, they are especially potent weapons and are hard to defend or protect against. Great power is generated by thrusting the hips in with a knee strike; however, use the point of the knee as the impact surface. All knee strikes should be executed repetitively until the opponent is disabled. The following techniques are the most effective way to overpower or disable the opponent.

(1) Front knee strike. When an opponent tries to grapple with the defender, the defender strikes his opponent in the stomach or solar plexus with his knee (Figure 4-23). This stuns the opponent and the defender can follow up with another technique.

(2) Knee strike to outside of thigh. The defender delivers a knee strike to the outside of his opponent's thigh (common peroneal nerve) (Figure 4-24). This strike causes intense pain and renders the opponent's leg ineffective.

(3) Knee strike to inside of thigh. An effective technique for close-in grappling is when the defender delivers a knee strike to the inside of his opponent's thigh (peroneal nerve) (Figure 4- 25). The defender then executes a follow-up technique to a vital point.

(4) Knee strike to groin. The knee strike to the groin is effective during close-in grappling. The defender gains control by grabbing his opponent's head, hair, ears, or shoulders and strikes him in the groin with his knee (Figure 4-26).

(5) Knee strike to face. The defender controls his opponent by grabbing behind his head with both hands and forcefully pushing his head down. At the same time, the defender brings his knee up and smashes the opponent in the face (Figure 4-27). When properly executed, the knee strike to the face is a devastating technique that can cause serious injury to the opponent.

Figure 4-23: Front knee strike.

Figure 4-24: Knee to outside of thigh.

Figure 4-25: Knee to inside of thigh.

Figure 4-26: Knee strike to groin.

Figure 4-27: Knee strike to face.

CHAPTER 5

Long-range Combatives

In long-range combatives, the distance between opponents is such that the combatants can engage one another with fully extended punches and kicks or with handheld weapons, such as rifles with fixed bayonets and clubs. As in medium-range combatives, a fighter must continuously monitor his available body weapons and opportunities for attack, as well as possible defense measures. He must know when to increase the distance from an opponent and when to close the gap. The spheres of influence that surround each fighter come into contact in long-range combatives. (See Chapter 6 for interval gaps and spheres of influence.)

SECTION I: NATURAL WEAPONS

The most dangerous natural weapons a soldier possesses are his hands and feet. This section describes natural weapon techniques of various punches, strikes, and kicks and stresses aggressive tactics with which to subdue an opponent.

5-1. Extended Arm Punches and Strikes. Extended arm punches and strikes in long-range combatives, like those in medium-range combatives, should be directed at vital points and nerve motor points. It is essential to put the entire body mass in motion behind long-range strikes. Closing the distance to the target gives the fighter an opportunity to take advantage of this principle.

a. In extended punches, the body weapon is usually the fist, although the fingers may be used—for example, eye gouging. When punching, hold the fist vertically or horizontally. Keep the wrist straight to prevent injury and use the first two knuckles in striking.

b. Another useful variation of the fist is to place the thumb on top of the vertical fist so that the tip protrudes beyond the curled index finger that supports it. The thumb strike is especially effective against soft targets. Do not fully lock out the arm when punching; keep a slight bend in the elbow to prevent hyperextension if the intended target is missed.

5-2. Kicks. Kicks during hand-to-hand combat are best directed to low targets and should be simple but effective. Combat soldiers are usually burdened with combat boots and LCE. His flexibility level is usually low during combat, and if engaged in hand-to-hand combat, he will be under high stress. He must rely on gross motor skills and kicks that do not require complicated movement or much training and practice to execute.

a. Side Knee Kick. When an opponent launches an attack—for example, with a knife (Figure 5-1, Step 1), it is most important for the defender to first move his entire body off the line of attack as the attacker moves in.

As the defender steps off at 45 degrees to the outside and toward the opponent, he strikes with a short punch to the floating ribs (Figure 5-1, Step 2).

Then the defender turns his body by rotating on the leading, outside foot and raises the knee of his kicking leg to his chest. He then drives his kick into the side of the attacker's knee with his foot turned 45 degrees outward (Figure 5-1, Step 3). This angle makes the most of the striking surface and reduces his chances of missing the target.

Figure 5-1: Side knee kick.

b. **Front Knee Kick.** As the attacker moves in, the defender immediately shifts off the line of attack and drives his kicking foot straight into the knee of the attacker (Figure 5-2). He turns his foot 45 degrees to make the most of the striking surface and to reduce the chances of missing the target. If the kick is done right, the attacker's advance will stop abruptly, and the knee joint will break.

c. **Heel Kick to Inside of Thigh.** The defender steps 45 degrees outside and toward the attacker to get off the line of attack. He is now in a position where he can drive his heel into the inside of the opponent's thigh (femoral nerve) (Figure 5-3, Steps 1 and 2). Either thigh can be targeted because the kick can still be executed if the defender moves to the inside of the opponent rather than to the outside when getting off the line of attack.

Figure 5-2: Front knee kick.

Figure 5-3: Heel kick to inside of thigh.

d. **Heel Kick to Groin.** The defender drives a heel kick into the attacker's groin (Figure 5-4) with his full body mass behind it. Since the groin is a soft target, the toe can also be used when striking it.

e. **Shin Kick.** The shin kick is a powerful kick, and it is easily performed with little training. When the legs are targeted, the kick is hard to defend against (Figure 5-5), and an opponent can be dropped by it.

Figure 5-4: Heel kick to groin.

Figure 5-5: Shin kick to legs.

The calves and common peroneal nerve (Figure 5-6) are the best striking points. The shin kick can also be used to attack the floating ribs (Figure 5-7).

f. Stepping Side Kick. A soldier starts a stepping side kick (Figure 5-8, Step 1) by stepping either behind or in front of his other foot to close the distance between him and his opponent. The movement is like that in a skip.

Figure 5-6: Shin kick to common peroneal nerve.

Figure 5-7: Shin kick to floating ribs.

Figure 5-8: Stepping side kick.

The soldier now brings the knee of his kicking foot up and thrusts out a side kick (Figure 5-8, Step 2). Tremendous power and momentum can be developed in this kick.

g. Counter to Front Kick. When the attacker tries a front kick, the defender traps the kicking foot by meeting it with his own (Figure 5-9, Step 1). The defender turns his foot 45 degrees outward to increase the likelihood of striking the opponent's kicking foot. This counter requires good timing by the defender, but not necessarily speed. Do not look at the feet; use your peripheral vision.

When an attacker tries a front kick (Figure 5-9, Step 2), the defender steps off the line of attack of the incoming foot to the outside.

As the attacker's kicking leg begins to drop, the defender kicks upward into the calf of the attacker's leg (Figure 5-9, Step 3). This kick is extremely painful and will probably render the leg ineffective. This technique does not rely on the defender's speed, but on proper timing.

Figure 5-9: Counter to front kick.

The defender can also kick to an opponent's kicking leg by moving off the line of attack to the inside and by using the heel kick to the inside of the thigh or groin (Figure 5-9, Step 4).

h. Counter to Roundhouse-Type Kick. When an opponent prepares to attack with a roundhouse-type kick (Figure 5-10, Step 1), the defender moves off the line of attack by stepping to the inside of the knee of the kicking leg.

He then turns his body to receive the momentum of the leg (Figure 5-10, Step 2). By moving to the inside of the knee, the defender lessens the power of the attacker's kicking leg. The harder the attacker kicks, the more likely he is to hyperextend his own knee against the body of the defender, but the defender will not be harmed. However, the defender must get to the inside of the knee, or

Figure 5-10: Counter to roadhouse kick.

an experienced opponent can change his roundhouse kick into a knee strike. The defender receives the energy of the kicking leg and continues turning with the momentum of the kick.

The attacker will be taken down by the defender's other leg with no effort (Figure 5-10, Step 3).

i. Kick as a Defense Against Punch. As the opponent on the left throws a punch (Figure 5-11, Step 1), the defender steps off the line of attack to the outside.

He then turns toward the opponent, brings his knee to his chest, and launches a heel kick to the outside of the opponent's thigh (Figure 5-11, Step 2). He keeps his foot turned 45 degrees to ensure striking the target and to maintain balance.

Figure 5-11: Kick as a defense against punch.

SECTION II: DEFENSIVE TECHNIQUES

A knife (or bayonet), properly employed, is a deadly weapon; however, using defensive techniques, such as maintaining separation, will greatly enhance the soldier's ability to fight and win.

5-3. Defense Against an Armed Opponent. An unarmed defender is always at a distinct disadvantage facing an armed opponent. It is imperative therefore that the unarmed defender understand and use the following principles to survive:

a. Separation. Maintain a separation of at least 10 feet plus the length of the weapon from the attacker. This distance gives the defender time to react to any attempt by the attacker to close the gap and be upon the defender. The defender should also try to place stationary objects between himself and the attacker.

b. Unarmed Defense. Unarmed defense against an armed opponent should be a last resort. If it is necessary, the defender's course of action includes:
  (1) Move the body out of the line of attack of the weapon. Step off the line of attack or redirect the attack of the weapon so that it clears the body.
  (2) Control the weapon. Maintain control of the attacking arm by securing the weapon, hand, wrist, elbow, or arm by using joint locks, if possible.

(3) Stun the attacker with an effective counterattack. Counterattack should be swift and devastating. Take the vigor out of the attacker with a low, unexpected kick, or break a locked joint of the attacking arm. Strikes to motor nerve centers are effective stuns, as are skin tearing, eye gouging, and attacking of the throat. The defender can also take away the attacker's balance.

(4) Ground the attacker. Take the attacker to the ground where the defender can continue to disarm or further disable him.

(5) Disarm the attacker. Break the attacker's locked joints. Use leverage or induce pain to disarm the attacker and finish him or to maintain physical control.

c. Precaution. Do not focus full attention on the weapon because the attacker has other body weapons to use. There may even be other attackers that you have not seen.

d. Expedient Aids. Anything available can become an expedient aid to defend against an armed attack. The kevlar helmet can be used as a shield; similarly, the LCE and shirt jacket can be used to protect the defender against a weapon. The defender can also throw dirt in the attacker's eyes as a distraction.

5-4. Angles of Attack. Any attack, regardless of the type weapon, can be directed along one of nine angles (Figure 5-12). The defense must be oriented for each angle of attack.

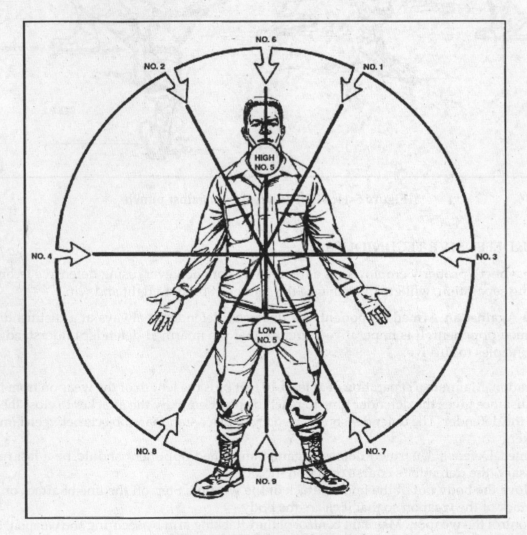

Figure 5-12: Angles of attack.

a. No. 1 Angle of Attack. A downward diagonal slash, stab, or strike toward the left side of the defender's head, neck, or torso.

b. No. 2 Angle of Attack. A downward diagonal slash, stab, or strike toward the right side of the defender's head, neck, or torso.

c. No. 3 Angle of Attack. A horizontal attack to the left side of the defender's torso in the ribs, side, or hip region.

d. No. 4 Angle of Attack. The same as No. 3 angle, but to the right side.

e. No. 5 Angle of Attack. A jabbing, lunging, or punching attack directed straight toward the defender's front.

f. No. 6 Angle of Attack. An attack directed straight down upon the defender.

g. No. 7 Angle of Attack. An upward diagonal attack toward the defender's lower-left side.

h. No. 8 Angle of Attack. An upward diagonal attack toward the defender's lower-right side.

i. No. 9 Angle of Attack. An attack directed straight up—for example, to the defender's groin.

5-5. Defense Against a Knife. When an unarmed soldier is faced with an enemy armed with a knife, he must be mentally prepared to be cut. The likelihood of being cut severely is less if the fighter is well trained in knife defense and if the principles of weapon defense are followed. A slash wound is not usually lethal or shock inducing; however, a stab wound risks injury to vital organs, arteries, and veins and may also cause instant shock or unconsciousness.

a. Types of Knife Attacks. The first line of defense against an opponent armed with a knife is to avoid close contact. The different types of knife attacks follow:

(1) Thrust. The thrust is the most common and most dangerous type of knife attack. It is a strike directed straight into the target by jabbing or lunging.

(2) Slash. The slash is a sweeping surface cut or circular slash. The wound is usually a long cut, varying from a slight surface cut to a deep gash.

(3) Flick. This attack is delivered by flicking the wrist and knife to extended limbs, inflicting numerous cuts. The flick is very distractive to the defender since he is bleeding from several cuts if the attacker is successful.

(4) Tear. The tear is a cut made by dragging the tip of the blade across the body to create a ripping type cut.

(5) Hack. The hack is delivered by using the knife to block or chop with.

(6) Butt. The butt is a strike with the knife handle.

b. Knife Defense Drills. Knife defense drills are used to familiarize soldiers with defense movement techniques for various angles of attack. For training, the soldiers should be paired off; one partner is named as the attacker and one is the defender. It is important that the attacker make his attack realistic in terms of distance and angling during training. His strikes must be accurate in hitting the defender at the intended target if the defender does not defend himself or move off the line of attack. For safety, the attacks are delivered first at one-quarter and one-half speed, and then at three-quarter speed as the defender becomes more skilled. Variations can be added by changing grips, stances, and attacks.

(1) No. 1 angle of defense—check and lift. The attacker delivers a slash along the No. 1 angle of attack. The defender meets and checks the movement with his left forearm bone, striking the inside forearm of the attacker (Figure 5-13, Step 1).

The defender's right hand immediately follows behind the strike to lift, redirect, and take control of the attacker's knife arm (Figure 5-13, Step 2).

The defender brings the attacking arm around to his right side where he can use an arm bar, wrist lock, and so forth, to disarm the attacker (Figure 5-13, Step 3).

He will have better control by keeping the knife hand as close to his body as possible (Figure 5-13, Step 4).

Figure 5-13: No. 1 angle of defense—check and lift.

(2) No. 2 angle of defense—check and ride. The attacker slashes with a No. 2 angle of attack. The defender meets the attacking arm with a strike from both forearms against the outside forearm, his bone against the attacker's muscle tissue (Figure 5-14, Step 1).

The strike checks the forward momentum of the attacking arm. The defender's right hand is then used to ride the attacking arm clear of his body (Figure 5-14, Step 2).

He redirects the attacker's energy with strength starting from the right elbow (Figure 5-14, Step 3).

(3) No. 3 angle of defense—check and lift. The attacker delivers a horizontal slash to the defender's ribs, kidneys, or hip on the left side (Figure 5-15, Step 1). The defender meets and checks the

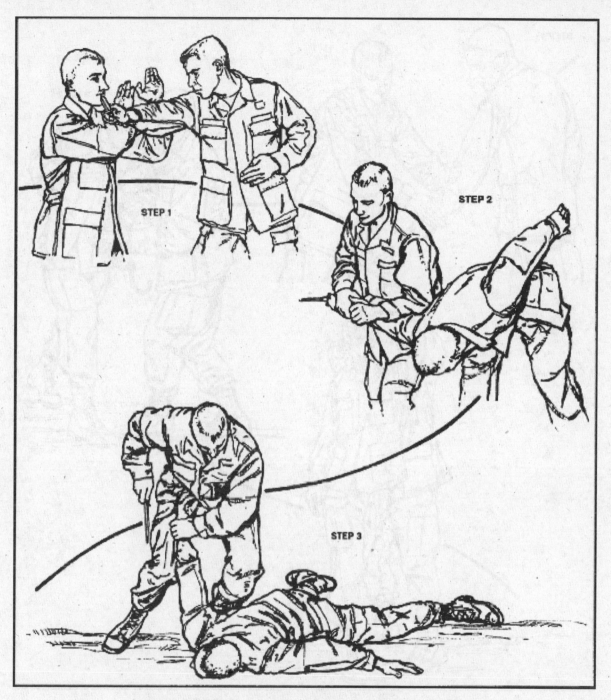

Figure 5-14: No. 2 angle of defense—check and ride.

attacking arm on the left side of his body with a downward circular motion across the front of his own body.

At the same time, he moves his body off the line of attack. He should meet the attacker's forearm with a strike forceful enough to check its momentum (Figure 5-15, Step 2). The defender then rides the energy of the attacking arm by wiping downward along the outside of his own left forearm with his right hand.

He then redirects the knife hand around to his right side where he can control or disarm the weapon (Figure 5-15, Step 3).

Figure 5-15: No. 3 angle of defense—check and lift.

(4) No. 4 angle of defense—check. The attacker slashes the defender with a backhand slashing motion to the right side at the ribs, kidneys, or hips. The defender moves his right arm in a downward circular motion and strikes the attacking arm on the outside of the body (Figure 5-16, Step 1).

At the same time, he moves off the line of attack (Figure 5-16, Step 2). The strike must be forceful enough to check the attack.

The left arm is held in a higher guard position to protect from a redirected attack or to assist in checking (Figure 5-16, Step 3).

Figure 5-16: No. 4 angle of defense—check.

The defender moves his body to a position where he can choose a proper disarming maneuver (Figure 5-16, Step 4).

(5) Low No. 5 angle of defense—parry. A lunging thrust to the stomach is made by the attacker along the No. 5 angle of attack (Figure 5-17, Step 1).

The defender moves his body off the line of attack and deflects the attacking arm by parrying with his left hand (Figure 5-17, Step 2). He deflects the attacking hand toward his right side by redirecting it with his right hand.

As he does this, the defender can strike downward with the left forearm or the wrist onto the forearm or wrist of the attacker (Figure 5-17, Step 3).

Figure 5-17: Low No. 5 angle of defense—parry.

The defender ends up in a position to lock the elbow of the attacking arm across his body if he steps off the line of attack properly (Figure 5-17, Step 4).

(6) High No. 5 angle of defense. The attacker lunges with a thrust to the face, throat, or solar plexus (Figure 5-18, Step 1).

The defender moves his body off the line of attack while parrying with either hand. He redirects the attacking arm so that the knife clears his body (Figure 5-18, Step 2).

He maintains control of the weapon hand or arm and gouges the eyes of the attacker, driving him backward and off balance (Figure 5-18, Step 3). If the attacker is much taller than the defender, it may be a more natural movement for the defender to raise his left hand to strike and

Figure 5-18: High No. 5 angle of defense.

deflect the attacking arm. He can then gouge his thumb or fingers into the jugular notch of the attacker and force him to the ground.

Still another possibility for a high No. 5 angle of attack is for the defender to move his body off the line of attack while parrying. He can then turn his body, rotate his shoulder under the elbow joint of the attacker, and lock it out (Figure 5-18, Step 4).

(7) No. 6 angle of defense. The attacker strikes straight downward onto the defender with a stab (Figure 5-19, Step 1).

The defender reacts by moving his body out of the weapon's path and by parrying or checking and redirecting the attacking arm, as the movement in the high No. 5 angle of defense (Figure 5-19, Step 2). The reactions may vary as to what is natural for the defender.

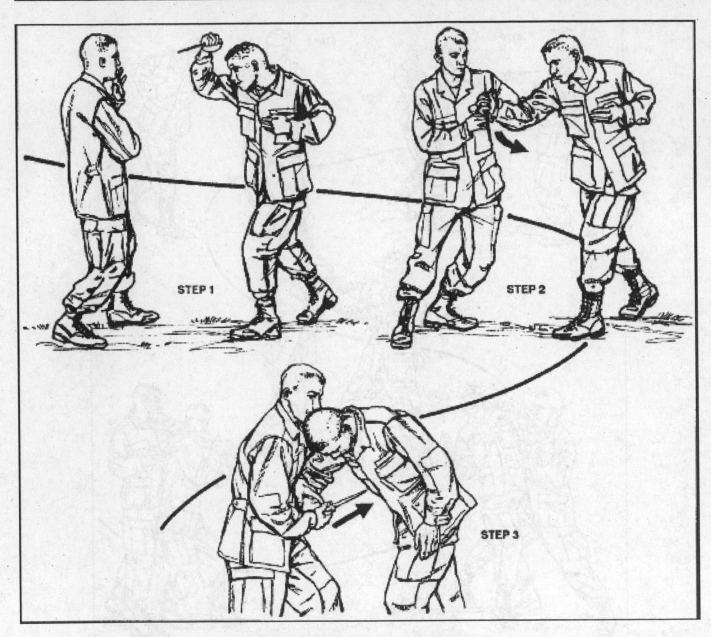

Figure 5-19: No. 6 angle of defense.

The defender then takes control of the weapon and disarms the attacker (Figure 5-19, Step 3).

c. Follow-Up Techniques. Once the instructor believes the soldiers are skilled in these basic reactions to attack, follow-up techniques may be introduced and practiced. These drills make up the defense possibilities against the various angles of attack. They also enable the soldier to apply the principles of defense against weapons and allow him to feel the movements. Through repetition, the reactions become natural, and the soldier instinctively reacts to a knife attack with the proper defense. It is important not to associate specific movements or techniques with certain types of attack. The knife fighter must rely on his knowledge of principles and his training experience in reacting to a knife attack. No two attacks or reactions will be the same; thus, memorizing techniques will not ensure a soldier's survival.

(1) Defend and clear. When the defender has performed a defensive maneuver and avoided an attack, he can push the attacker away and move out of the attacker's reach.

(2) Defend and stun. After the defender performs his first defensive maneuver to a safer position, he can deliver a stunning blow as an immediate counterattack. Strikes to motor nerve points or attacker's limbs, low kicks, and elbow strikes are especially effective stunning techniques.

(3) Defend and disarm. The defender also follows up his first defensive maneuver by maintaining control of the attacker's weapon arm, executing a stunning technique, and disarming the attacker. The stun distracts the attacker and also gives the defender some time to gain possession of the weapon and to execute his disarming technique.

5-6. Unarmed Defense Against a Rifle with Fixed Bayonet. Defense against a rifle with a fixed bayonet involves the same principles as knife defense. The soldier considers the same angles of attack and the proper response for any attack along each angle.

a. Regardless of the type weapon used by the enemy, his attack will always be along one of the nine angles of attack at any one time. The soldier must get his entire body off the line of attack by moving to a safe position. A rifle with a fixed bayonet has two weapons: a knife at one end and a butt stock at the other end. The soldier will be safe as long as he is not in a position where he can be struck by either end during the attack.

b. Usually, he is in a more advantageous position if he moves inside the length of the weapon. He can then counterattack to gain control of the situation as soon as possible. The following counterattacks can be used as defenses against a rifle with a fixed bayonet; they also provide a good basis for training.

(1) Unarmed defense against No. 1 angle of attack. The attacker prepares to slash along the No. 1 angle of attack (Figure 5-20, Step 1).

The defender waits until the last possible moment before moving so he is certain of the angle along which the attack is directed (Figure 5-20, Step 2). This way, the attacker cannot change his attack in response to movement by the defender.

When the defender is certain that the attack is committed along a specific angle (No. 1, in this case), he moves to the inside of the attacker and gouges his eyes (Figure 5-20, Step 2) while the other hand redirects and controls the weapon. He maintains control of the weapon and lunges his entire body weight into the eye gouge to drive the attacker backward and off balance. The defender now ends up with the weapon, and the attacker is in a poor recovery position (Figure 5-20, Step 3).

(2) Unarmed defense against No. 2 angle of attack. The attacker makes a diagonal slash along the No. 2 angle of attack (Figure 5-21, Step 1). Again, the defender waits until he is sure of the attack before moving.

The defender then moves to the outside of the attacker and counterattacks with a thumb jab into the right armpit (Figure 5-21, Step 2). He receives the momentum of the attacking weapon and controls it with his free hand.

He uses the attacker's momentum against him by pulling the weapon in the direction it is going with one hand and pushing with his thumb of the other hand (Figure 5-21, Step 3). The attacker is completely off balance, and the defender can gain control of the weapon.

(3) Unarmed defense against No. 3 angle of attack. The attacker directs a horizontal slash along the No. 3 angle of attack (Figure 5-22, Step 1).

The defender turns and moves to the inside of the attacker; he then strikes with his thumb into the jugular notch (Figure 5-22, Step 2).

His entire body mass is behind the thumb strike and, coupled with the incoming momentum of the attacker, the strike drives the attacker's head backward and takes his balance (Figure 5-22, Step 3).

The defender turns his body with the momentum of the weapon's attack to strip the weapon from the attacker's grip (Figure 5-22, Step 4).

Figure 5-20: Unarmed defense against No. 1 angle of attack.

(4) Unarmed defense against No. 4 angle of attack. The attack is a horizontal slash along the No. 4 angle of attack (Figure 5-23, Step 1).

The defender moves into the outside of the attacker (Figure 5-23, Step 2).

He then turns with the attack, delivering an elbow strike to the throat (Figure 5-23, Step 3). At the same time, the defender's free hand controls the weapon and pulls it from the attacker as he is knocked off balance from the elbow strike.

(5) Unarmed defense against low No. 5 angle of attack. The attacker thrusts the bayonet at the stomach of the defender (Figure 5-24, Step 1).

Figure 5-21: Unarmed defense against No. 2 angle of attack.

The defender shifts his body to the side to avoid the attack and to gouge the eyes of the attacker (Figure 5-24, Step 2).

The defender's free hand maintains control of and strips the weapon from the attacker as he is driven backward with the eye gouge (Figure 5-24, Step 3).

(6) Unarmed defense against high No. 5 angle of attack. The attacker delivers a thrust to the throat of the defender (Figure 5-25, Step 1).

The defender then shifts to the side to avoid the attack, parries the thrust, and controls the weapon with his trail hand (Figure 5-25, Step 2).

Figure 5-22: Unarmed defense against No. 3 angle of attack.

He then shifts his entire body mass forward over the lead foot, slamming a forearm strike into the attacker's throat (Figure 5-25, Step 3).

(7) Unarmed defense against No 6 angle of attack. The attacker delivers a downward stroke along the No. 6 angle of attack (Figure 5-26, Step 1).

The defender shifts to the outside to get off the line of attack and he grabs the weapon. Then, he pulls the attacker off balance by causing him to overextend himself (Figure 2-26, Step 2).

The defender shifts his weight backward and causes the attacker to fall, as he strips the weapon from him (Figure 5-26, Step 3).

5-7. Advanced Weapons Techniques and Training. For advanced training in weapons techniques, training partners should have the same skill level. Attackers can execute attacks along multiple angles of attack in combinations. The attacker must attack with a speed that offers the defender a challenge, but

Figure 5-23: Unarmed defense against No. 4 angle of attack.

does not overwhelm him. It should not be a contest to see who can win, but a training exercise for both individuals.

a. Continued training in weapons techniques will lead to the partners' ability to engage in free-response fighting or sparring—that is, the individuals become adept enough to understand the principles of weapons attacks, defense, and movements so they can respond freely when attacking or defending from any angle.

b. Instructors must closely monitor training partners to ensure that the speed and control of the individuals does not become dangerous during advanced training practice. Proper eye protection and padding should be used, when applicable. The instructor should stress the golden rule in free-response fighting—Do unto others as you would have them do unto you.

Figure 5-24: Unarmed defense against low No. 5 angle of attack.

STEP 1

STEP 2

STEP 3

Figure 5-25: Unarmed defense against high No. 5 angle of attack.

Figure 5-26: Unarmed defense against No. 6 angle of attack.

SECTION III: OFFENSIVE TECHNIQUES

At ranges of 10 meters or more in most combat situations, small arms and grenades are the weapons of choice. However, in some scenarios, today's combat soldier must engage the enemy in confined areas, such as trench clearing or room clearing where noncombatants are present or when silence is necessary. In these instances, the bayonet or knife may be the ideal weapon to dispatch the enemy. Other than the side arm, the knife is the most lethal weapon in close-quarter combat.

5-8. Bayonet/Knife. As the bayonet is an integral part of the combat soldier's equipment, it is readily available for use as a multipurpose weapon. The bayonet produces a terrifying mental effect on the enemy when in the hands of a well-trained and confident soldier. The soldier skilled in the use of the knife also increases his ability to defend against larger opponents and multiple attackers. Both these skills increase his chances of surviving and accomplishing the mission. (Although the following paragraphs say "knife," the information also applies to bayonets.)

a. Grips. The best way to hold the knife is either with the straight grip or the reverse grip.

   (1) Straight Grip. Grip the knife in the strong hand by forming a vee and by allowing the knife to fit naturally, as in gripping for a handshake. The handle should lay diagonally across the palm. Point the blade toward the enemy, usually with the cutting edge down. The cutting edge can also be held vertically or horizontally to the ground. Use the straight grip when thrusting and slashing.

   (2) Reverse Grip. Grip the knife with the blade held parallel with the forearm, cutting edge facing outward. This grip conceals the knife from the enemy's view. The reverse grip also affords the most power for lethal insertion. Use this grip for slashing, stabbing, and tearing.

b. Stances. The primary stances are the knife fighter's stance and the modified stance.

   (1) Knife fighter's stance. In this stance, the fighter stands with his feet about shoulder-width apart, dominant foot toward the rear. About 70 percent of his weight is on the front foot and 30 percent on the rear foot. He stands on the balls of both feet and holds the knife with the straight grip. The other hand is held close to his body where it is ready to use, but protected (Figure 5-27).

Figure 5-27: Stance.

(2) Modified stance. The difference in the modified stance is the knife is held close to the body with the other hand held close over the knife hand to help conceal it (Figure 5-28).

c. Range. The two primary ranges in knife fighting are long range and medium range. In long-range knife fighting, attacks consist of figure-eight slashes along the No. 1, No. 2, No. 7, and No. 8 angles of attack; horizontal slashes along the No. 3 and No. 4 angles of attack; and lunging thrusts to vital areas on the No. 5 angle of attack. Usually, the straight grip is used. In medium-range knife fighting, the reverse grip provides greater power. It is used to thrust, slash, and tear along all angles of attack.

Figure 5-28: Modified stance.

5-9. Knife-Against-Knife Sequence. The knife fighter must learn to use all available weapons of his body and not limit himself to the knife. The free hand can be used to trap the enemy's hands to create openings in his defense. The enemy's attention will be focused on the weapon; therefore, low kicks and knee strikes will seemingly come from nowhere. The knife fighter's priority of targets are the eyes, throat, abdominal region, and extended limbs. Some knife attack sequences that can be used in training to help develop soldiers' knowledge of movements, principles, and techniques in knife fighting follow:

a. Nos. 1 and 4 Angles. Two opponents assume the knife fighter's stance (Figure 5-29, Step 1).

Figure 5-29: Nos. 1 and 4 angles.

The attacker starts with a diagonal slash along the No. 1 angle of attack to the throat (Figure 5-29, Step 2).

He then follows through with a slash and continues with a horizontal slash back across the abdomen along the No. 4 angle of attack (Figure 5-29, Step 3).

He finishes the attack by using his entire body mass behind a lunging stab into the opponent's solar plexus (Figure 5-29, Step 4).

b. Nos. 5, 3, and 2 Angles. In this sequence, one opponent (attacker) starts an attack with a lunge along the No. 5 angle of attack. At the same time, the other opponent (defender) on the left moves his body off the line of attack, parries the attacking arm, and slices the biceps of his opponent (Figure 5-30, Step 1).

Figure 5-30: Nos. 5, 3, and 2 angles.

The defender slashes back across the groin along the No. 3 angle of attack (Figure 5-30, Step 2). He finishes the attacker by continuing with an upward stroke into the armpit or throat along the No. 2 angle of attack (Figure 5-30, Step 3). Throughout this sequence, the attacker's weapon hand is controlled with the defender's left hand as he attacks with his own knife hand.

c. Low No. 5 Angle. In the next sequence, the attacker on the right lunges to the stomach along a low No. 5 angle of attack.

The defender on the left moves his body off the line of attack while parrying and slashing the wrist of the attacking knife hand as he redirects the arm (Figure 5-31, Step 1).

After he slashes the wrist of his attacker, the defender continues to move around the outside and stabs the attacker's armpit (Figure 5-31, Step 2).

He retracts his knife from the armpit, continues his movement around the attacker, and slices his hamstring (Figure 5-31, Step 3).

d. Optional Low No. 5 Angle. The attacker on the right lunges to the stomach of his opponent (the defender) along the low No. 5 angle of attack. The defender moves his body off the line of attack of

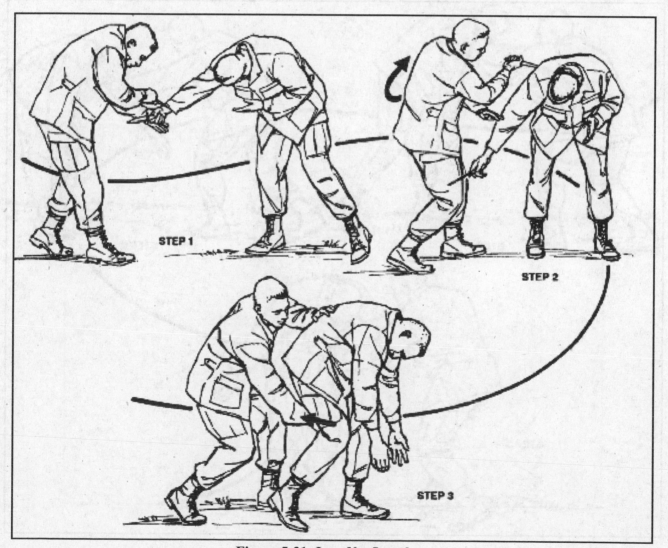

Figure 5-31: Low No. 5 angle.

the knife. Then he turns and, at the same time, delivers a slash to the attacker's throat along the No. 1 angle of attack (Figure 5-32, Step 1).

The defender immediately follows with another slash to the opposite side of the attacker's throat along the No. 2 angle of attack (Figure 5-32, Step 2).

The attacker is finished as the opponent on the left (defender) continues to slice across the abdomen with a stroke along the No. 3 angle (Figure 5-32, Step 3).

5-10. Rifle with Fixed Bayonet. The principles used in fighting with the rifle and fixed bayonet are the same as when knife fighting. Use the same angles of attack and similar body movements. The principles of timing and distance remain paramount; the main difference is the extended distance provided by the length of the weapon. It is imperative that the soldier fighting with rifle and fixed bayonet use the movement of his entire body behind all of his fighting techniques—not just upper-body strength. Unit trainers should be especially conscious of stressing full body mass in motion for power and correcting

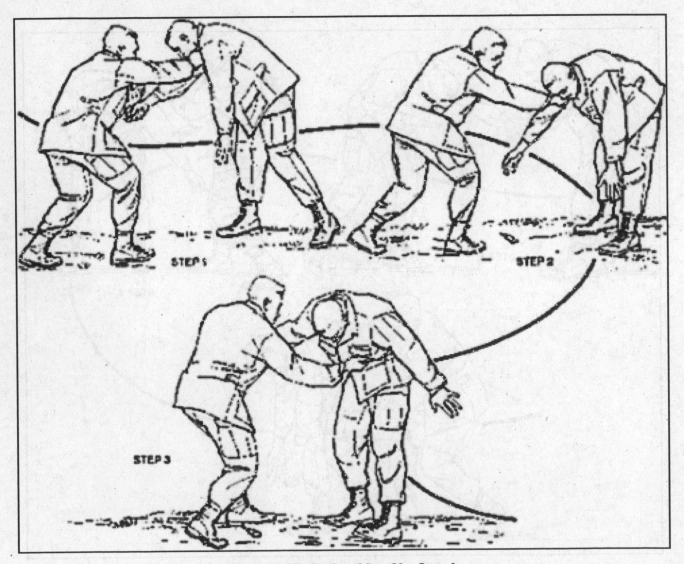

Figure 5-32: Optional low No. 5 angle.

all deficiencies during training. Whether the enemy is armed or unarmed, a soldier fighting with rifle and fixed bayonet must develop the mental attitude that he will survive the fight. He must continuously evaluate each moment in a fight to determine his advantages or options, as well as the enemy's. He should base his defenses on keeping his body moving and off the line of any attacks from his opponent. The soldier seeks openings in the enemy's defenses and starts his own attacks, using all available body weapons and angles of attack. The angles of attack with rifle and fixed bayonet are shown in Figures 5-33 through 5-39.

a. Fighting Techniques. New weapons, improved equipment, and new tactics are always being introduced; however, firepower alone will not always drive a determined enemy from his position. He will often remain in defensive emplacements until driven out by close combat. The role of the soldier, particularly in the final phase of the assault, remains relatively unchanged: His mission is to close with and disable or capture the enemy. This mission remains the ultimate goal of all individual training. The rifle with fixed bayonet is one of the final means of defeating an opponent in an assault.

(1) During infiltration missions at night or when secrecy must be maintained, the bayonet is an excellent silent weapon.

Figure 5-33: No. 1 angle of attack with rifle and fixed bayonet.

Figure 5-34: No. 2 angle of attack with rifle and fixed bayonet.

Figure 5-35: No. 3 angle of attack with rifle and fixed bayonet.

Figure 5-36: No. 4 angle of attack with rifle and fixed bayonet.

Figure 5-37: No. 5 angle of attack with rifle and fixed bayonet.

Figure 5-38: High No. 5 angle of attack with rifle and fixed bayonet.

Figure 5-39: No. 6 angle of attack with rifle and fixed bayonet.

(2) When close-in fighting determines the use of small-arms fire or grenades to be impractical, or when the situation does not permit the loading or reloading of the rifle, the bayonet is still the weapon available to the soldier.

(3) The bayonet serves as a secondary weapon should the rifle develop a stoppage.

(4) In hand-to-hand encounters, the detached bayonet may be used as a handheld weapon.

(5) The bayonet has many nonfighting uses, such as to probe for mines, to cut vegetation, and to use for other tasks where a pointed or cutting tool is needed.

b. Development. To become a successful rifle-bayonet fighter, a soldier must be physically fit and mentally alert. A well-rounded physical training program will increase his chances of survival in a bayonet encounter. Mental alertness entails being able to quickly detect and meet an opponent's attack from any direction. Aggressiveness, accuracy, balance, and speed are essential in training as well as in combat situations. These traits lead to confidence, coordination, strength, and endurance, which characterize the rifle-bayonet fighter. Differences in individual body physique may require slight changes from the described rifle-bayonet techniques. These variations will be allowed if the individual's attack is effective.

c. Principles. The bayonet is an effective weapon to be used aggressively; hesitation may mean sudden death. The soldier must attack in a relentless assault until his opponent is disabled or captured. He should be alert to take advantage of any opening. If the opponent fails to present an opening, the bayonet fighter must make one by parrying his opponent's weapon and driving his blade or rifle butt into the opponent with force.

(1) The attack should be made to a vulnerable part of the body: face, throat, chest, abdomen, or groin.

(2) In both training and combat, the rifle-bayonet fighter displays spirit by sounding off with a low and aggressive growl. This instills a feeling of confidence in his ability to close with and disable or capture the enemy.

(3) The instinctive rifle-bayonet fighting system is designed to capitalize on the natural agility and combatives movements of the soldier. It must be emphasized that precise learned movements will NOT be stressed during training.

d. Positions. The soldier holds the rifle firmly but not rigidly. He relaxes all muscles not used in a specific position; tense muscles cause fatigue and may slow him down. After proper training and thorough practice, the soldier instinctively assumes the basic positions. All positions and movements described in this manual are for right-handed men. A left-handed man, or a man who desires to learn lefthanded techniques, must use the opposite hand and foot for each phase of the movement described. All positions and movements can be executed with or without the magazine and with or without the sling attached.

(1) Attack position. This is the basic starting position (A and B, Figure 5-40) from which all attack movements originate. It generally parallels a boxer's stance. The soldier assumes this position when running or hurdling obstacles. The instructor explains and demonstrates each move.

(a) Take a step forward and to the side with your left foot so that your feet are a comfortable distance apart.

(b) Hold your body erect or bend slightly forward at the waist. Flex your knees and balance your body weight on the balls of your feet. Your right forearm is roughly parallel to the ground. Hold the left arm high, generally in front of the left shoulder. Maintain eye-to-eye contact with your opponent, watching his weapon and body through peripheral vision.

(c) Hold your rifle diagonally across your body at a sufficient distance from the body to add balance and protect you from enemy blows. Grasp the weapon in your left hand just below the upper sling swivel, and place the right hand at the small of the stock. Keep the sling facing outward and the cutting edge of the bayonet toward your opponent. The command is, ATTACK POSITION, MOVE. The instructor gives the command, and the soldiers perform the movement.

Figure 5-40: Attack position.

(2) Relaxed position. The relaxed position (Figure 5-41) gives the soldier a chance to rest during training. It also allows him to direct his attention toward the instructor as he discusses and demonstrates the positions and movements. To assume the relaxed position from the attack position, straighten the waist and knees and lower the rifle across the front of your body by extending the arms downward. The command is, RELAX. The instructor gives the command, and the soldiers perform the movement.

e. Movements. The soldier will instinctively strike at openings and become aggressive in his attack once he has learned to relax and has developed instinctive reflexes. His movements do not have to be executed in any prescribed order. He will achieve balance in his movements, be ready to strike in any direction, and keep striking until he has disabled his opponent. There are two basic movements used throughout bayonet instruction: the whirl and the crossover. These movements develop instant reaction to commands and afford the instructor maximum control of the training formation while on the training field.

(1) Whirl movement. The whirl (Figure 5-42, Steps 1, 2, and 3), properly executed, allows the rifle bayonet fighter to meet a challenge from an opponent attacking him from the rear. At the completion of a whirl, the rifle remains in the attack position. The instructor explains and demonstrates how to spin your body around by pivoting on the ball of the leading foot in the direction of the leading foot, thus facing completely about. The command is, WHIRL. The instructor gives the command, and the soldiers perform the movement.

(2) Crossover movement. While performing certain movements in rifle-bayonet training, two ranks will be moving toward each other. When the soldiers in ranks come too close to each other to

Figure 5-41: Relaxed position.

safely execute additional movements, the crossover is used to separate the ranks a safe distance apart. The instructor explains and demonstrates how to move straight forward and pass your opponent so that your right shoulder passes his right shoulder, continue moving forward about six steps, halt, and without command, execute the whirl. Remain in the attack position and wait for further commands. The command is, CROSSOVER. The instructor gives the command, and the soldiers perform the movement.

Figure 5-42: Whirl movement.

Note: Left-handed personnel cross left shoulder to left shoulder.

(3) Attack movements. There are four attack movements designed to disable or capture the opponent: thrust, butt stroke, slash, and smash. Each of these movements may be used for the initial attack or as a follow-up should the initial movement fail to find its mark. The soldiers learn these movements separately. They will learn to execute these movements in a swift and continuous series during subsequent training. During all training, the emphasis will be on conducting natural, balanced movements to effectively damage the target. Precise, learned movements will not be stressed.

(a) Thrust. The objective is to disable or capture an opponent by thrusting the bayonet blade into a vulnerable part of his body. The thrust is especially effective in areas where movement is restricted—for example, trenches, wooded areas, or built-up areas. It is also effective when an opponent is lying on the ground or in a fighting position. The instructor explains and demonstrates how to lunge forward on your leading foot without losing your balance (Figure 5-43, Step 1) and, at the same time, drive the bayonet with great force into any unguarded part of your opponent's body.

To accomplish this, grasp the rifle firmly with both hands and pull the stock in close to the right hip; partially extend the left arm, guiding the point of the bayonet in the general direction of the opponent's body (Figure 5-43, Step 2).

Quickly complete the extension of the arms and body as the leading foot strikes the ground so that the bayonet penetrates the target (Figure 5-43, Step 3).

To withdraw the bayonet, keep your feet in place, shift your body weight to the rear, and pull rearward along the same line of penetration (Figure 5-43, Step 4).

Next, assume the attack position in preparation to continue the assault (Figure 5-43, Step 5).

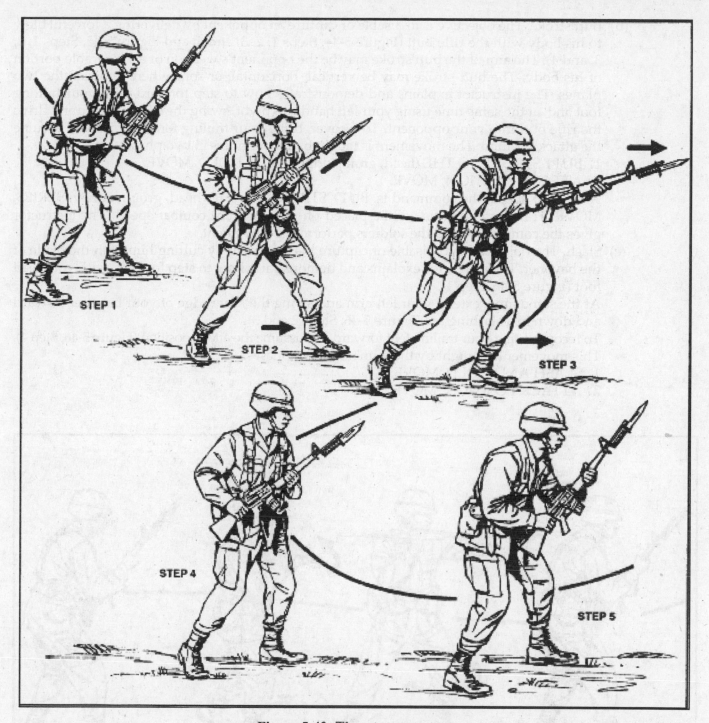

Figure 5-43: Thrust movement.

This movement is taught by the numbers in three phases:
1. THRUST AND HOLD, MOVE.
2. WITHDRAW AND HOLD, MOVE.
3. ATTACK POSITION, MOVE.

At combat speed, the command is, THRUST SERIES, MOVE. Training emphasis will be placed on movement at combat speed. The instructor gives the commands, and the soldiers perform the movements.

(b) Butt stroke. The objective is to disable or capture an opponent by delivering a forceful blow to his body with the rifle butt (Figure 5-44, Steps 1, 2, 3, and 4, and Figure 5-45, Steps 1, 2, 3, and 4). The aim of the butt stroke may be the opponent's weapon or a vulnerable portion of his body. The butt stroke may be vertical, horizontal, or somewhere between the two planes. The instructor explains and demonstrates how to step forward with your trailing foot and, at the same time using your left hand as a pivot, swing the rifle in an arc and drive the rifle butt into your opponent. To recover, bring your trailing foot forward and assume the attack position. The movement is taught by the numbers in two phases:

1. BUTT STROKE TO THE (head, groin, kidney) AND HOLD, MOVE.
2. ATTACK POSITION, MOVE.

At combat speed, the command is, BUTT STROKE TO THE (head, groin, kidney) SERIES, MOVE. Training emphasis will be placed on movement at combat speed. The instructor gives the commands, and the soldiers perform the movement.

(c) Slash. The objective is to disable or capture the opponent by cutting him with the blade of the bayonet. The instructor explains and demonstrates how to step forward with your lead foot (Figure 5-46, Step 1).

At the same time, extend your left arm and swing the knife edge of your bayonet forward and down in a slashing arc (Figure 5-46, Steps 2 and 3).

To recover, bring your trailing foot forward and assume the attack position (Figure 5-46, Step 4). This movement is taught by the number in two phases:

1. SLASH AND HOLD, MOVE.
2. ATTACK POSITION, MOVE.

Figure 5-44: Butt stroke to the head.

STEP 1

STEP 2

STEP 3

STEP 4

Figure 5-45: Butt stroke to the groin.

At combat speed, the command is, SLASH SERIES, MOVE. Training emphasis will be placed on movement at combat speed. The instructor gives the commands, and the soldiers perform the movements.

(d) Smash. The objective is to disable or capture an opponent by smashing the rifle butt into a vulnerable part of his body. The smash is often used as a follow-up to a butt stroke and is also

Figure 5-46: Slash movement.

effective in wooded areas and trenches when movement is restricted. The instructor explains and demonstrates how to push the butt of the rifle upward until horizontal (Figure 5-47, Step 1) and above the left shoulder with the bayonet pointing to the rear, sling up (Figure 5-47, Step 2). The weapon is almost horizontal to the ground at this time.

Step forward with the trailing foot, as in the butt stroke, and forcefully extend both arms, slamming the rifle butt into the opponent (Figure 5-47, Step 3).

Figure 5-47: Smash movement.

To recover, bring your trailing foot forward (Figure 5-47, Step 4) and assume the attack position (Figure 5-47, Step 5). This movement is taught by the numbers in two phases:

1. SMASH AND HOLD, MOVE.
2. ATTACK POSITION, MOVE.

At combat speed, the command is, SMASH SERIES, MOVE. Training emphasis will be placed on movement at combat speed. The instructor gives the commands, and the soldiers perform the movements.

(4) Defensive movements. At times, the soldier may lose the initiative and be forced to defend himself. He may also meet an opponent who does not present a vulnerable area to attack. Therefore, he must make an opening by initiating a parry or block movement, then follow up with a vicious attack. The follow-up attack is immediate and violent.

> ⚠ CAUTION
> TO MINIMIZE WEAPON DAMAGE WHILE USING BLOCKS AND PARRIES, LIMIT WEAPON-TO-WEAPON CONTACT TO HALF SPEED DURING TRAINING.

(a) Parry movement. The objective is to counter a thrust, throw the opponent off balance, and hit a vulnerable area of his body. Timing, speed, and judgment are essential factors in these movements. The instructor explains and demonstrates how to—
   ○ Parry right. If your opponent carries his weapon on his left hip (left-handed), you will parry it to your right. In execution, step forward with your leading foot (Figure 5-48, Step 1), strike the opponent's rifle (Figure 5-48, Step 2), deflecting it to your right (Figure 5-48, Step 3), and follow up with a thrust, slash, or butt stroke.
   ○ Parry left. If your opponent carries his weapon on his right hip (right-handed), you will parry it to your left. In execution, step forward with your leading foot (Figure 5-49, Step 1), strike the opponent's rifle (Figure 5-49, Step 2), deflecting it to your left (Figure 5-49, Step 3), and follow up with a thrust, slash, or butt stroke.
   A supplementary parry left is the follow-up attack (Figure 5-50, Steps 1, 2, 3, 4, and 5).
   ○ Recovery. Immediately return to the attack position after completing each parry and follow-up attack.
   The movement is taught by the numbers in three phases:
   1. PARRY RIGHT (OR LEFT), MOVE.
   2. THRUST MOVE.
   3. ATTACK POSITION, MOVE.
   At combat speed, the command is, PARRY RIGHT (LEFT) or PARRY (RIGHT OR LEFT) WITH FOLLOW-UP ATTACK. The instructor gives the commands, and the soldiers perform the movements.

(b) Block. When surprised by an opponent, the block is used to cut off the path of his attack by making weapon-to-weapon contact. A block must always be followed immediately with a vicious attack. The instructor explains and demonstrates how to extend your arms using the center part of your rifle as the strike area, and cut off the opponent's attack by making weapon-to-weapon contact. Strike the opponent's weapon with enough power to throw him off balance.
   ○ High block (Figure 5-51, Steps 1, 2, and 3). Extend your arms upward and forward at a 45- degree angle. This action deflects an opponent's slash movement by causing his bayonet or upper part of his rifle to strike against the center part of your rifle.
   ○ Low block (Figure 5-52, Steps 1, 2, and 3). Extend your arms downward and forward about 15 degrees from your body. This action deflects an opponent's butt stroke aimed at the groin by causing the lower part of his rifle stock to strike against the center part of your rifle.
   ○ Side block (Figure 5-53, Steps 1 and 2). Extend your arms with the left hand high and right hand low, thus holding the rifle vertical. This block is designed to stop a butt stroke aimed at your upper body or head. Push the rifle to your left to cause the butt of the opponent's rifle to strike the center portion of your rifle.

Figure 5-48: Parry right.

Figure 5-49: Parry left.

Figure 5-50: Parry left, slash, with follow-up butt stroke to kidney region.

Figure 5-51: High block against slash.

Figure 5-52: Low block against butt stroke to groin.

STEP 1

STEP 2

Figure 5-53: Side block against butt stroke.

      ○ Recovery. Counterattack each block with a thrust, butt stroke, smash, or slash.
        Blocks are taught by the numbers in two phases:
        1. HIGH (LOW) or (SIDE) BLOCK.
        2. ATTACK POSITION, MOVE.
        At combat speed, the command is the same. The instructor gives the commands, and the
        soldiers perform the movement.
  (5) Modified movements. Two attack movements have been modified to allow the rifle-bayonet
      fighter to slash or thrust an opponent without removing his hand from the pistol grip of the M16
      rifle should the situation dictate.
    (a) The modified thrust (Figure 5-54, Steps 1 and 2) is identical to the thrust (as described in
        paragraph (3)(a)) with the exception of the right hand grasping the pistol grip.
    (b) The modified slash (Figure 5-55, Steps 1, 2, 3, and 4) is identical to the slash (as described in
        paragraph (3)(c)) with the exception of the right hand grasping the pistol grip.
  (6) Follow-up movements. Follow-up movements are attack movements that naturally follow from
      the completed position of the previous movement. If the initial thrust, butt stroke, smash, or
      slash fails to make contact with the opponent's body, the soldier should instinctively follow
      up with additional movements until he has disabled or captured the opponent. It is important

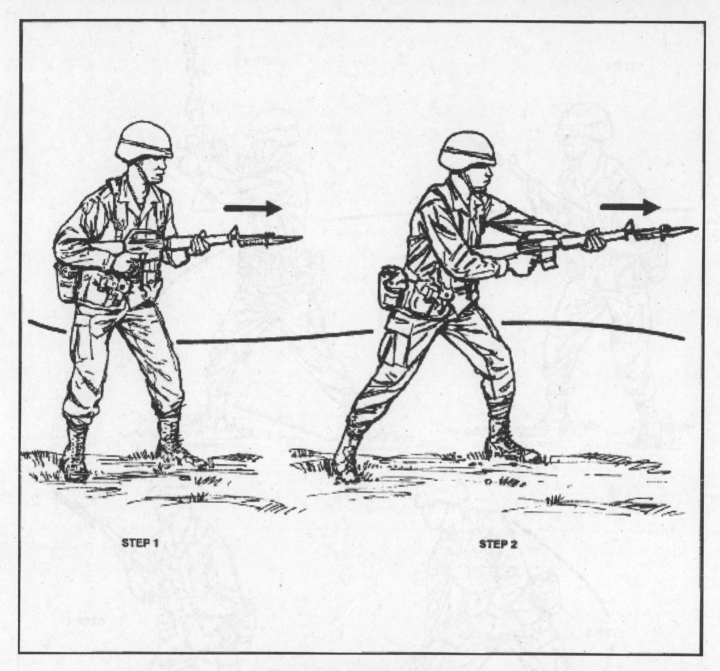

Figure 5-54: Modified thrust.

to follow up the initial attack with another aggressive action so the initiative is not lost. The instructor explains and demonstrates how instinct should govern your selection of a specific follow-up movement. For example—

- PARRY LEFT, BUTT STROKE TO THE HEAD, SMASH, SLASH, ATTACK POSITION.
- PARRY LEFT, SLASH, BUTT STROKE TO THE KIDNEY, ATTACK POSITION.
- PARRY RIGHT THRUST, BUTT STROKE TO THE GROIN, SLASH, ATTACK POSITION.
  Two examples of commands using follow-up movements are—
- PARRY LEFT (soldier executes), THRUST (soldier executes), BUTT STROKE TO THE HEAD (soldier executes), SMASH (soldier executes), SLASH (soldier executes), ATTACK POSITION (soldier assumes the attack position).

Figure 5-55: Modified slash.

- THRUST (soldier executes), THRUST (soldier executes), THRUST (soldier executes), BUTT STROKE TO THE GROIN (soldier executes), SLASH (soldier executes), ATTACK POSITION (soldier assumes the attack position).

All training will stress damage to the target and violent action, using natural movements as opposed to precise, stereotyped movements. Instinctive, aggressive action and balance are the keys to offense with the rifle and bayonet.

NOTE: For training purposes, the instructor may and should mix up the series of movements.

SECTION IV: FIELD-EXPEDIENT WEAPONS

To survive, the soldier in combat must be able to deal with any situation that develops. His ability to adapt any nearby object for use as a weapon in a win-or-die situation is limited only by his ingenuity and resourcefulness. Possible weapons, although not discussed herein, include ink pens or pencils; canteens tied to string to be swung; snap links at the end of sections of rope; kevlar helmets; sand, rocks, or liquids thrown into the enemy's eyes; or radio antennas. The following techniques demonstrate a few expedient weapons that are readily available to most soldiers for defense and counterattack against the bayonet and rifle with fixed bayonet.

5-11. Entrenching Tool. Almost all soldiers carry the entrenching tool. It is a versatile and formidable weapon when used by a soldier with some training. It can be used in its straight position—locked out and fully extended—or with its blade bent in a 90-degree configuration.

a. To use the entrenching tool against a rifle with fixed bayonet, the attacker lunges with a thrust to the stomach of the defender along a low No. 5 angle of attack (Figure 5-56, Step 1).
The defender moves just outside to avoid the lunge and meets the attacker's arm with the blade of the fully extended entrenching tool (Figure 5-56, Step 2).
The defender gashes all the way up the attacker's arm with the force of both body masses coming together. The hand gripping the entrenching tool is given natural protection from the shape of the handle. The defender continues pushing the blade of the entrenching tool up and into the throat of the attacker, driving him backward and downward (Figure 5-56, Step 3).

b. An optional use of entrenching tool against a rifle with fixed bayonet is for the attacker to lunge to the stomach of the defender (Figure 5-57, Step 1).
The defender steps to the outside of the line of attack at 45 degrees to avoid the weapon. He then turns his body and strikes downward onto the attacking arm (on the radial nerve) with the blade of the entrenching tool (Figure 5-57, Step 2).
He drops his full body weight down with the strike, and the force causes the attacker to collapse forward. The defender then strikes the point of the entrenching tool into the jugular notch, driving it deeply into the attacker (Figure 5-57, Step 3).

c. In the next two sequences, the entrenching tool is used in the bent configuration—that is, the blade is bent 90 degrees to the handle and locked into place.

(1) The attacker tries to stick the bayonet into the chest of the defender (Figure 5-58, Step 1).
When the attack comes, the defender moves his body off the line of attack by stepping to the outside. He allows his weight to shift forward and uses the blade of the entrenching tool to drag along the length of the weapon, scraping the attacker's arm and hand (Figure 5-58, Step 2). The defender's hand is protected by the handle's natural design.
He continues to move forward into the attacker, strikes the point of the blade into the jugular notch, and drives it downward (Figure 5-58, Step 3).

Figure 5-56: Entrenching tool against rifle with fixed bayonet.

Figure 5-57: Optional use of entrenching tool against rifle with fixed bayonet.

Figure 5-58: Entrenching tool in bent configuration.

(2) The attacker lunges with a fixed bayonet along the No. 5 angle of attack (Figure 5-59, Step 1). The defender then steps to the outside to move off the line of attack and turns; he strikes the point of the blade of the entrenching tool into the side of the attacker's throat (Figure 5-59, Step 2).

5-12. Three-foot Stick. Since a stick can be found almost anywhere, a soldier should know its uses as a field-expedient weapon. The stick is a versatile weapon; its capability ranges from simple prisoner control to lethal combat.

a. Use a stick about 3 feet long and grip it by placing it in the vee formed between the thumb and index finger, as in a handshake. It may also be grasped by two hands and used in an unlimited number of techniques. The stick is not held at the end, but at a comfortable distance from the butt end.

b. When striking with the stick, achieve maximum power by using the entire body weight behind each blow. The desired point of contact of the weapon is the last 2 inches at the tip of the stick. The primary targets for striking with the stick are the vital body points in Chapter 4. Effective striking points are usually the wrist, hand, knees, and other bony protuberances. Soft targets include the side of the neck, jugular notch, solar plexus, and various nerve motor points. Attack soft targets by striking or thrusting the tip of the stick into the area. Three basic methods of striking are—

(1) Thrusting. Grip the stick with both hands and thrust straight into a target with the full body mass behind it.

(2) Whipping. Hold the stick in one hand and whip it in a circular motion; use the whole body mass in motion to generate power.

(3) Snapping. Snap the stick in short, shocking blows, again with the body mass behind each strike.

c. When the attacker thrusts with a knife to the stomach of the defender with a low No. 5 angle of attack, the defender moves off the line of attack to the outside and strikes vigorously downward onto the attacking wrist, hand, or arm (Figure 5-60, Step 1).

The defender then moves forward, thrusts the tip of the stick into the jugular notch of the attacker (Figure 5-60, Step 2), and drives him to the ground with his body weight—not his upper body strength (Figure 5-60, Step 3).

d. When using a three-foot stick against a rifle with fixed bayonet, the defender grasps the stick with two hands, one at each end, as the attacker thrusts forward to the chest (Figure 5-61, Step 1).

He steps off the line of attack to the outside and redirects the weapon with the stick (Figure 5-61, Step 2).

He then strikes forward with the forearm into the attacker's throat (Figure 5-61, Step 3). The force of the two body weights coming together is devastating. The attacker's neck is trapped in the notch formed by the stick and the defender's forearm.

Using the free end of the stick as a lever, the defender steps back and uses his body weight to drive the attacker to the ground. The leverage provided by the stick against the neck creates a tremendous choke with the forearm, and the attacker loses control completely (Figure 5-61, Step 4).

5-13. Three-foot Rope. A section of rope about 3 feet long can provide a useful means of self-defense for the unarmed combat soldier in a hand-to-hand fight. Examples of field-expedient ropes are a web belt, boot laces, a portion of a 120-foot nylon rope or sling rope, or a cravat rolled up to form a rope. Hold the rope at the ends so the middle section is rigid enough to almost serve as a stick-like weapon, or the rope can be held with the middle section relaxed, and then snapped by vigorously pulling the hands apart to strike

Figure 5-59: Optional use of entrenching tool in bent configuration.

Figure 5-60: Three-foot stick against knife.

Figure 5-61: Three-foot stick against rifle with fixed bayonet.

parts of the enemy's body, such as the head or elbow joint, to cause serious damage. It can also be used to entangle limbs or weapons held by the opponent, or to strangle him.

a. When the attacker lunges with a knife to the stomach (Figure 5-62, Step 1), the defender moves off the line of attack 45 degrees to the outside.
He snaps the rope downward onto the attacking wrist, redirecting the knife (Figure 5-62, Step 2).
Then, he steps forward, allowing the rope to encircle the attacker's neck (Figure 5-62, Step 3).

Figure 5-62: Three-foot rope against knife.

He continues to turn his body and sinks his weight to drop the attacker over his hip (Figure 5-62, Step 4).

b. When the attacker thrusts with a fixed bayonet (Figure 5-63, Step 1), the defender moves off the line of attack and uses the rope to redirect the weapon (Figure 5-63, Step 2).

Then, he moves forward and encircles the attacker's throat with the rope (Figure 5-63, Step 3). He continues moving to unbalance the attacker and strangles him with the rope (Figure 5-63, Step 4).

Figure 5-63: Three-foot rope against rifle with fixed bayonet.

c. The 3-foot rope can also be a useful tool against an unarmed opponent. The defender on the left prepares for an attack by gripping the rope between his hands (Figure 5-64, Step 1).

When the opponent on the right attacks, the defender steps completely off the line of attack and raises the rope to strike the attacker's face (Figure 5-64, Step 2).

He then snaps the rope to strike the attacker either across the forehead, just under the nose, or under the chin by jerking his hands forcefully apart. The incoming momentum of the attacker against the

Figure 5-64: Three-foot rope against unarmed opponent.

rope will snap his head backward, will probably break his neck, or will at least knock him off his feet (Figure 5-64, Step 3).

5-14. Six-Foot Pole. Another field-expedient weapon that can mean the difference between life and death for a soldier in an unarmed conflict is a pole about 6 feet long. Examples of poles suitable for use are mop handles, pry bars, track tools, tent poles, and small trees or limbs cut to form a pole. A soldier skilled in the use of a pole as a weapon is a formidable opponent. The size and weight of the pole requires him to move his whole body to use it effectively. Its length gives the soldier an advantage of distance in most unarmed situations. There are two methods usually used in striking with a pole:

a. Swinging. Becoming effective in swinging the pole requires skilled body movement and practice. The greatest power is developed by striking with the last 2 inches of the pole.
b. Thrusting. The pole is thrust straight along its axis with the user's body mass firmly behind it.
   (1) An attacker tries to thrust forward with a fixed bayonet (Figure 5-65, Step 1).

Figure 5-65: Thrusting with 6-foot pole.

The defender moves his body off the line of attack; he holds the tip of the pole so that the attacker runs into it from his own momentum. He then aims for the jugular notch and anchors his body firmly in place so that the full force of the attack is felt at the attacker's throat (Figure 5-65, Step 2).

(2) The defender then shifts his entire body weight forward over his lead foot and drives the attacker off his feet (Figure 5-65, Step 3).

NOTE: During high stress, small targets, such as the throat, may be difficult to hit. Good, large targets include the solar plexus and hip/thigh joint.

CHAPTER 6

Sentry Removal

Careful planning, rehearsal, and execution are vital to the success of a mission that requires the removal of a sentry. This task may be necessary to gain access to an enemy location or to escape confinement.

6-1. Planning Considerations. A detailed schematic of the layout of the area guarded by sentries must be available. Mark known and suspected locations of all sentries. It will be necessary—

a. To learn the schedule for the changing of the guards and the checking of the posts.
b. To learn the guard's meal times. It may be best to attack a sentry soon after he has eaten when his guard is lowered. Another good time to attack the sentry is when he is going to the latrine.
c. To post continuous security.
d. To develop a contingency plan.
e. To plan infiltration and exfiltration routes.
f. To carefully select personnel to accomplish the task.
g. To carry the least equipment necessary to accomplish the mission because silence, stealth, and ease of movement are essential.
h. To conceal or dispose of killed sentries.

6-2. Rehearsals. Reproduce and rehearse the scenario of the mission as closely as possible to the execution phase.

Conduct the rehearsal on similar terrain, using sentries, the time schedule, and the contingency plan. Use all possible infiltration and exfiltration routes to determine which may be the best.

6-3. Execution. When removing a sentry, the soldier uses his stalking skills to approach the enemy undetected. He must use all available concealment and keep his silhouette as low as possible.

a. When stepping, the soldier places the ball of his lead foot down first and checks for stability and silence of the surface to be crossed. He then lightly touches the heel of his lead foot. Next, he transfers his body weight to his lead foot by shifting his body forward in a relaxed manner. With the weight on the lead foot, he can bring his rear foot forward in a similar manner.
b. When approaching the sentry, the soldier synchronizes his steps and movement with the enemy's, masking any sounds. He also uses background noises to mask his sounds. He can even follow the sentry through locked doors this way. He is always ready to strike immediately if he is discovered. He focuses his attention on the sentry's head since that is where the sentry generates all of his movement and attention. However, it is important not to stare at the enemy because he may sense the stalker's presence through a sixth sense. He focuses on the sentry's movements with his peripheral vision. He gets to within 3 or 4 feet and at the proper moment makes the kill as quickly and silently as possible.
c. The attacker's primary focus is to summon all of his mental and physical power to suddenly explode onto the target. He maintains an attitude of complete confidence throughout the execution. He must control fear and hesitation because one instant of hesitation could cause his defeat and compromise the entire mission.

6-4. Psychological Aspects. Killing a sentry is completely different than killing an enemy soldier while engaged in a firefight. It is a cold and calculated attack on a specific target. After observing a sentry for hours, watching him eat or look at his wife's photo, an attachment is made between the stalker and the sentry. Nonetheless, the stalker must accomplish his task efficiently and brutally. At such close quarters, the soldier literally feels the sentry fight for his life. The sights, sounds, and smells of this act are imprinted in the soldier's mind; it is an intensely personal experience. A soldier who has removed a sentry should be observed for signs of unusual behavior for four to seven days after the act.

6-5. Techniques. The following techniques are proven and effective ways to remove sentries. A soldier with moderate training can execute the proper technique for his situation, when he needs to.

a. Brachial Stun, Throat Cut. This technique relies on complete mental stunning to enable the soldier to cut the sentry's throat, severing the trachea and carotid arteries. Death results within 5 to 20 seconds. Some sounds are emitted from the exposed trachea, but the throat can be cut before the sentry can recover from the effect of the stunning strike and cry out. The soldier silently approaches to within striking range of the sentry (Figure 6-1, Step 1). The soldier strikes the side of the sentry's

Figure 6-1: Brachial stun, throat cut.

neck with the knife butt or a hammer fist strike (Figure 6-1, Step 2), which completely stuns the sentry for three to seven seconds. He then uses his body weight to direct the sentry's body to sink in one direction and uses his other hand to twist the sentry's head to the side, deeply cutting the throat across the front in the opposite direction (Figure 6-1, Step 3). He executes the entire length of the blade in a slicing motion. The sentry's sinking body provides most of the force—not the soldier's upper-arm strength (Figure 6-1, Step 4).

b. Kidney Stab, Throat Cut. This technique relies on a stab to the kidney (Figure 6-2, Step 1) to induce immediate shock. The kidney is relatively accessible and by inducing shock with such a stab, the soldier has the time to cut the sentry's throat. The soldier completes his stalk and stabs the kidney by pulling the sentry's balance backward and downward and inserts the knife upward against his weight. The sentry will possibly gasp at this point, but shock immediately follows. By using the sentry's body weight that is falling downward and turning, the soldier executes a cut across the front of the throat (Figure 6-2, Step 2). This completely severs the trachea and carotid arteries.

Figure 6-2: Kidney stab, throat cut.

c. Pectoral Muscle Strike, Throat Cut. The stun in this technique is produced by a vigorous strike to the stellate ganglia nerve center at the top of the pectoral muscle (Figure 6-3, Step 1). The strike is delivered downward with the attacker's body weight. Use the handle of the knife for impact. Care should be taken to avoid any equipment worn by the sentry that could obstruct the strike. Do not try this technique if the sentry is wearing a ballistic vest or bulky LCE. The sentry is unable to make a sound or move if the stun is properly delivered. The throat is then cut with a vertical stab downward into the subclavian artery at the junction of the neck and clavicle (Figure 6-3, Step 2). Death comes within 3 to 10 seconds, and the sentry is lowered to the ground.

d. Nose Pinch, Mouth Grab, Throat Cut. In this technique, completely pinch off the sentry's mouth and nose to prevent any outcry. Then cut his throat or stab his subclavian artery (Figure 6-4). The

STEP 1

STEP 2

Figure 6-3: Pectoral muscle strike, throat cut.

Figure 6-4: Nose pinch, mouth grab, throat cut.

danger with this technique is that the sentry can resist until he is killed, although he cannot make a sound.

e. Crush Larynx, Subclavian Artery Stab. Crush the sentry's larynx by inserting the thumb and two or three fingers behind his larynx, then twisting and crushing it. The subclavian artery can be stabbed at the same time with the other hand (Figure 6-5).

f. Belgian Takedown. In the Belgian take down technique, the unsuspecting sentry is knocked to the ground and kicked in the groin, inducing shock. The soldier can then kill the sentry by any proper means. Since surprise is the essential element of this technique, the soldier must use effective stalking techniques (Figure 6-6, Step 1). To initiate his attack, he grabs both of the sentry's ankles (Figure 6-6, Step 2). Then he heaves his body weight into the hips of the sentry while pulling up on the ankles. This technique slams the sentry to the ground on his face. Then, the soldier follows with a kick to the groin (Figure 6-6, Step 3).

Figure 6-5: Crush larynx, subclavian artery stab.

g. Neck Break With Sentry Helmet. The soldier can break the sentry's neck by vigorously snatching back and down on the sentry's helmet (Figure 6-7, Step 1) while forcing the sentry's body weight forward with a knee strike (Figure 6-7, Step 2). The chin strap of the helmet must be fastened for this technique to work.

h. Knockout With Helmet. The sentry's helmet is stripped from his head and used by the soldier to knock him out (Figure 6-8, Step 1). The soldier uses his free hand to stabilize the sentry during the

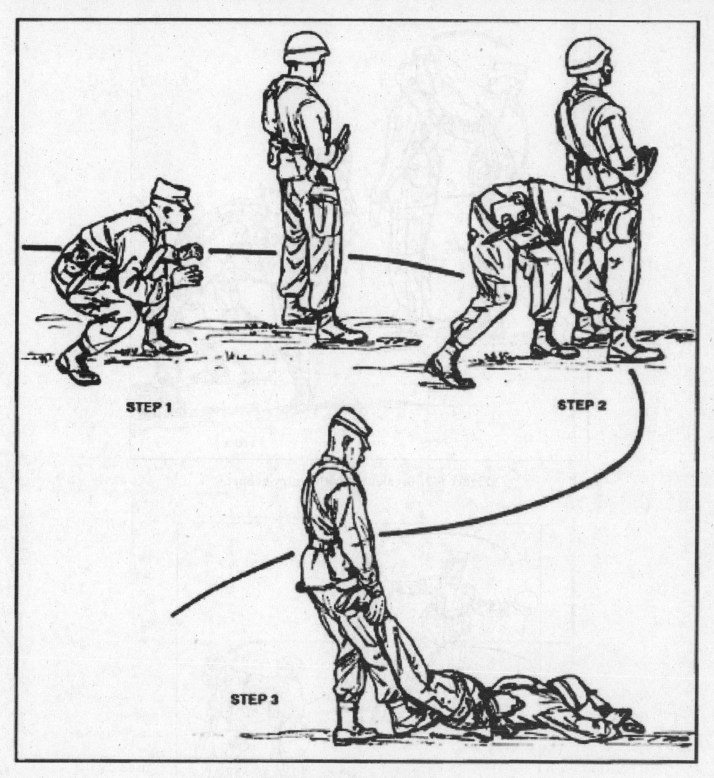

Figure 6-6: Belgian takedown.

Figure 6-7: Break neck with sentry helmet.

Figure 6-8: Knockdown with helmet.

attack. This technique can only be used when the sentry's chin strap is loose. The preferred target area for striking with the helmet is at the base of the skull or on the temple (Figure 6-8, Step 2).

i. The Garrote. In this technique, use a length of wire, cord, rope, or webbed belt to takeout a sentry. Silence is not guaranteed, but the technique is effective if the soldier is unarmed and must escape from a guarded area. The soldier carefully stalks the sentry from behind with his garrote ready (Figure 6-9, Step 1). He loops the garrote over the sentry's head across the throat (Figure 6-9, Step 2) and forcefully pulls him backward as he turns his own body to place his hips in low against the hips of the sentry. The sentry's balance is already taken at this point, and the garrote becomes crossed around the sentry's throat when the turn is made. The sentry is thrown over the soldier's shoulder and killed by strangling or breaking his neck (Figure 6-9, Step 3).

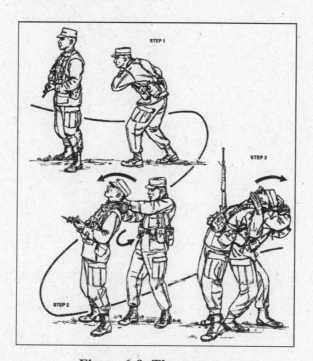

Figure 6-9: The garrote.

CHAPTER 7

Cover, Concealment, and Camouflage

GENERAL

In a survival situation where you are in hostile territory, if the enemy can see you, he can hit you with his fire. So you must be concealed from enemy observation and have cover from enemy fire.

When the terrain does not provide natural cover and concealment, you must prepare your cover and use natural and man-made materials to camouflage yourself, your equipment, and your position. This chapter provides guidance on the preparation and use of cover, concealment, and camouflage.

COVER

Cover gives protection from bullets, fragments of exploding rounds, flame, nuclear effects, and biological and chemical agents. Cover can also conceal you from enemy observation. Cover can be natural or man-made.

Natural cover includes such things as logs, trees, stumps, ravines, and hollows. Man-made cover includes such things as fighting positions, trenches, walls, rubble, and craters. Even the smallest depression or fold in the ground can give some cover. Look for and use every bit of cover the terrain offers.

TYPES OF COVER

In combat, you need protection from enemy direct and indirect fire.

To get this protection in the defense, build a fighting position (man-made cover) to add to the natural cover afforded by the terrain.

To get protection from enemy fire in the offense or when moving, use routes that put cover between you and the places where the enemy is known or thought to be. Use ravines, gullies, hills, wooded areas, walls, and other cover to keep the enemy from seeing and firing at you. Avoid open areas, and do not skyline yourself on hilltops and ridges.

CONCEALMENT

Concealment is anything that hides you from enemy observation. Concealment does not protect you from enemy fire. Do not think that you are protected from the enemy's fire just because you are concealed. Concealment, like cover, can also be natural or man-made.

Natural concealment includes such things as bushes, grass, trees, and shadows. If possible, natural concealment should not be disturbed. Man-made concealment includes such things as battle-dress uniforms, camouflage nets, face paint, and natural material that has been moved from its original location. Man-made concealment must blend into the natural concealment provided by the terrain.

Light discipline, noise discipline, movement discipline, and the use of camouflage contribute to concealment. Light discipline is controlling the use of lights at night by such things as not smoking in the open, not walking around with a flashlight on, and not using vehicle headlights. Noise discipline is taking action to deflect sounds generated by your unit (such as operating equipment) away from the enemy and, when possible, using methods to communicate that do not generate sounds (arm-and-hand signals). Movement discipline is such things as not moving about fighting positions unless necessary, and not moving on routes that lack cover and concealment. In the defense, build a well-camouflaged fighting position and avoid moving about. In the offense, conceal yourself and your equipment with camouflage and move in woods or on terrain that gives concealment. Darkness cannot hide you from enemy observation in either offense or defense. The enemy's night vision devices and other detection means let them find you in both daylight and darkness.

CAMOUFLAGE

Camouflage is anything you use to keep yourself, your equipment, and your position from looking like what they are. Both natural and man-made material can be used for camouflage.

FIGHTING POSITION WITH COVER

TROOPS MOVING ALONG A RAVINE

Change and improve your camouflage often. The time between changes and improvements depends on the weather and on the material used. Natural camouflage will often die, fade, or otherwise lose its effectiveness. Likewise, man-made camouflage may wear off or fade. When those things happen, you and your equipment or position may not blend with the surroundings. That may make it easy for the enemy to spot you.

CAMOUFLAGE CONSIDERATIONS

Movement draws attention. When you give arm-and-hand signals or walk about your position, your movement can be seen by the naked eye at long ranges. In the defense, stay low and move only when necessary. In the offense, move only on covered and concealed routes.

Positions must not be where the enemy expects to find them. Build positions on the side of a hill, away from road junctions or lone buildings, and in covered and concealed places. Avoid open areas.

Outlines and shadows may reveal your position or equipment to air or ground observers. Outlines and shadows can be broken up with camouflage. When moving, stay in the shadows when possible.

Shine may also attract the enemy's attention. In the dark, it may be a light such as a burning cigarette or flashlight. In daylight, it can be reflected light from polished surfaces such as shiny mess gear, a worn helmet, a windshield, a watch crystal and band, or exposed skin. A light, or its reflection, from a position

may help the enemy detect the position. To reduce shine, cover your skin with clothing and face paint. However, in a nuclear attack, darkly painted skin can absorb more thermal energy and may burn more readily than bare skin. Also, dull the surfaces of equipment and vehicles with paint, mud, or some type of camouflage material.

Shape is outline or form. The shape of a helmet is easily recognized. A human body is also easily recognized. Use camouflage and concealment to breakup shapes and blend them with their surroundings. Be careful not to overdo it.

The colors of your skin, uniform, and equipment may help the enemy detect you if the colors contrast with the background. For example, a green uniform will contrast with snow-covered terrain. Camouflage yourself and your equipment to blend with the surroundings.

Dispersion is the spreading of men, vehicles, and equipment over a wide area. It is usually easier for the enemy to detect soldiers when they are bunched. So, spread out. The distance between you and your fellow soldier will vary with the terrain, degree of visibility, and enemy situation. Distances will normally be set by unit leaders or by a unit's standing operating procedure (SOP).

SOLDIER IN ARCTIC CAMOUFLAGE

FIRE TEAM DISPERSED

CAMOUFLAGED SOLDIERS

HOW TO CAMOUFLAGE

Before camouflaging, study the terrain and vegetation of the area in which you are operating. Then pick and use the camouflage material that best blends with that area. When moving from one area to another, change camouflage as needed to blend with the surroundings. Take grass, leaves, brush, and other material from your location and apply it to your uniform and equipment and put face paint on your skin.

Fighting Positions. When building a fighting position, camouflage it and the dirt taken from it. Camouflage the dirt used as frontal, flank, rear, and overhead cover. Also camouflage the bottom of the hole to prevent detection from the air. If necessary, take excess dirt away from the position (to the rear).

Do not overcamouflage. Too much camouflage material may actually disclose a position. Get your camouflage material from a wide area. An area stripped of all or most of its vegetation may draw attention. Do not wait until the position is complete to camouflage it. Camouflage the position as you build.

Do not leave shiny or light-colored objects lying about. Hide mess kits, mirrors, food containers, and white underwear and towels. Do not remove your shirt in the open. Your skin may shine and be seen. Never use fires where there is a chance that the flame will be seen or the smoke will be smelled by the enemy. Also, cover up tracks and other signs of movement.

CAMOUFLAGED FIGHTING POSITION BEING IMPROVED

USING A TREE LIMB TO COVER UP A TRAIL

COLORS USED IN CAMOUFLAGE

Sand and light green for desert and dry areas

Loam and white for snow-covered terrain

Loam and light green for vegetated areas

CAMOUFLAGE MATERIAL	SKIN COLOR LIGHT OR DARK	SHINE AREAS FOREHEAD, CHEEKBONES, EARS, NOSE AND CHIN	SHADOW AREAS AROUND EYES, UNDER NOSE, AND UNDER CHIN
LOAM AND LIGHT GREEN STICK	ALL TROOPS USE IN AREAS WITH GREEN VEGETATION	USE LOAM	USE LIGHT GREEN
SAND AND LIGHT GREEN STICK	ALL TROOPS USE IN AREAS LACKING GREEN VEGETATION	USE LIGHT GREEN	USE SAND
LOAM AND WHITE	ALL TROOPS USE ONLY IN SNOW-COVERED TERRAIN	USE LOAM	USE WHITE
BURNT CORK, BARK CHARCOAL, OR LAMP BLACK	ALL TROOPS, IF CAMOUFLAGE STICKS NOT AVAILABLE	USE	DO NOT USE
LIGHT-COLOR MUD	ALL TROOPS, IF CAMOUFLAGE STICKS NOT AVAILABLE	DO NOT USE	USE

When camouflage is complete, inspect the position from the enemy's side. This should be done from about 35 meters forward of the position. Then check the camouflage periodically to see that it stays natural-looking and conceals the position. When the camouflage becomes ineffective, change and improve it.

Helmets. Camouflage your helmet with the issue helmet cover or make a cover of cloth or burlap that is colored to blend with the terrain. The cover should fit loosely with the flaps folded under the helmet or left hanging. The hanging flaps may break up the helmet outline. Leaves, grass, or sticks can also be attached to the cover. Use camouflage bands, strings, burlap strips, or rubber bands to hold those in place. If there is no material for a helmet cover, disguise and dull helmet surface with irregular patterns of paint or mud.

Uniforms. Most uniforms come already camouflaged. However, it may be necessary to add more camouflage to make the uniform blend better with the surroundings. To do this, put mud on the uniform or attach leaves, grass, or small branches to it. Too much camouflage, however, may draw attention.

When operating on snow-covered ground, wear overwhites (if issued) to help blend with the snow. If overwhites are not issued, use white cloth, such as white bedsheets, to get the same effect.

Skin. Exposed skin reflects light and may draw the enemy's attention. Even very dark skin, because of its natural oil, will reflect light. Use the following methods when applying camouflage face paint to camouflage the skin.

When applying camouflage stick to your skin, work with a buddy (in pairs) and help each other. Apply a two-color combination of camouflage stick in an irregular pattern. Paint shiny areas (forehead, cheekbones,

CAMOUFLAGED HELMETS

nose, ears, and chin) with a dark color. Paint shadow areas (around the eyes, under the nose, and under the chin) with a light color. In addition to the face, paint the exposed skin on the back of the neck, arms, and hands. Palms of hands are not normally camouflaged if arm-and-hand signals are to be used. Remove all jewelry to further reduce shine or reflection.

When camouflage sticks are not issued, use burnt cork, bark, charcoal, lamp black, or light-colored mud.

CHAPTER 8

Tracking

GENERAL

In all operations, you must be alert for signs of enemy activity. Such signs can often alert you to an enemy's presence and give your unit time to prepare for contact. The ability to track an enemy after he has broken contact also helps you regain contact with him.

TRACKER QUALITIES

Visual tracking is following the path of men or animals by the signs they leave, primarily on the ground or vegetation. Scent tracking is following men or animals by their smell.

Tracking is a precise art. You need a lot of practice to achieve and keep a high level of tracking skill. You should be familiar with the general techniques of tracking to enable you to detect the presence of a hidden enemy and to follow him, to find and avoid mines or booby- traps, and to give early warning of ambush.

With common sense and a degree of experience, you can track another person. However, you must develop the following traits and qualities:

- Be patient.
- Be able to move slowly and quietly, yet steadily, while detecting and interpreting signs.
- Avoid fast movement that may cause you to overlook signs, lose the trail, or blunder into an enemy unit.
- Be persistent and have the skill and desire to continue the mission even though signs are scarce or weather or terrain is unfavorable.
- Be determined and persistent when trying to find a trail that you have lost.
- Be observant and try to see things that are not obvious at first glance.
- Use your senses of smell and hearing to supplement your sight.
- Develop a feel for things that do not look right. It may help you regain a lost trail or discover additional signs.
- Know the enemy, his habits, equipment, and capability.

FUNDAMENTALS OF TRACKING

When tracking an enemy, you should build a picture of him in your mind. Ask yourself such questions as: How many persons am I following? How well are they trained? How are they equipped? Are they healthy? How is their morale? Do they know they are being followed?

To find the answer to such questions, use all available signs. A sign can be anything that shows you that a certain act took place at a particular place and time. For instance, a footprint tells a tracker that at a certain time a person walked on that spot.

The six fundamentals of tracking are:

- Displacement.
- Staining.

- Weathering.
- Littering.
- Camouflaging.
- Interpretation and/or immediate use intelligence.

Any sign that you find can be identified as one or more of the first five fundamentals.
In the sixth fundamental, you combine the first five and use all of them to form a picture of the enemy.

DISPLACEMENT

Displacement takes place when something is moved from its original position. An example is a footprint in soft, moist ground. The foot of the person that left the print displaced the soil, leaving an indentation in the ground. By studying the print, you can determine many facts. For example, a print that was left by a barefoot person or a person with worn or frayed footgear indicates that he may have poor equipment.

HOW TO ANALYZE FOOTPRINTS

Footprints show the following:

- The direction and rate of movement of a party.
- The number of persons in a party.
- Whether or not heavy loads are carried.
- The sex of the members of a party.
- Whether the members of a party know they are being followed.

If the footprints are deep and the pace is long, the party is moving rapidly. Very long strides and deep prints, with toe prints deeper than heel prints, indicate that the party is running. If the prints are deep, short, and widely spaced, with signs of scuffing or shuffling, a heavy load is probably being carried by the person who left the prints.

You can also determine a person's sex by studying the size and position of the footprints.

Women generally tend to be pigeon-toed, while men usually walk with their feet pointed straight ahead or slightly to the outside. Women's prints are usually smaller than men's, and their strides are usually shorter.

If a party knows that it is being followed, it may attempt to hide its tracks. Persons walking backward have a short, irregular stride. The prints have an unusually deep toe. The soil will be kicked in the direction of movement.

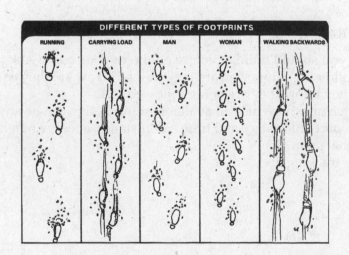

DIFFERENT TYPES OF FOOTPRINTS				
RUNNING	CARRYING LOAD	MAN	WOMAN	WALKING BACKWARDS

The last person walking in a group usually leaves the clearest footprints. Therefore, use his prints as the key set. Cut a stick the length of each key print and notch the stick to show the print width at the widest part of the sole. Study the angle of the key prints to determine the direction of march. Look for an identifying mark or feature on the prints, such as a worn or frayed part of the footwear. If the trail becomes vague or obliterated, or if the trail being followed merges with another, use the stick to help identify the key prints. That will help you stay on the trail of the group being followed.

Use the box method to count the number of persons in the group. There are two ways to use the box method—the stride as a unit of measure method and the 36-inch box method.

The stride as a unit of measure method is the most accurate of the two. Up to 18 persons can be counted using this method. Use it when the key prints can be determined. To use this method, identify a key print on a trail and draw a line from its heel across the trail. Then move forward to the key print of the opposite foot and draw a line through its instep. This should form a box with the edges of the trail forming two sides, and the drawn lines forming the other two sides. Next, count every print or partial print inside the box to determine the number of persons. Any person walking normally would have stepped in the box at least one time. Count the key prints as one.

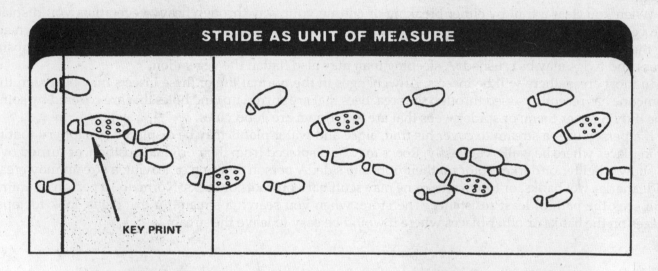

STRIDE AS UNIT OF MEASURE

KEY PRINT

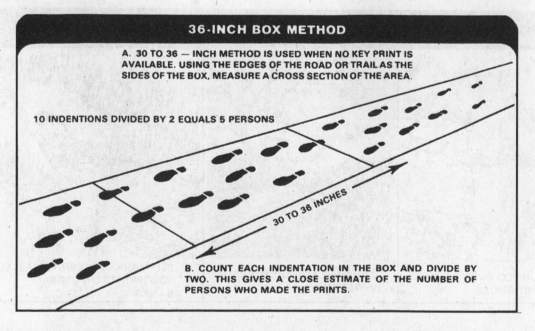

36-INCH BOX METHOD

A. 30 TO 36 — INCH METHOD IS USED WHEN NO KEY PRINT IS AVAILABLE. USING THE EDGES OF THE ROAD OR TRAIL AS THE SIDES OF THE BOX, MEASURE A CROSS SECTION OF THE AREA.

10 INDENTIONS DIVIDED BY 2 EQUALS 5 PERSONS

30 TO 36 INCHES

B. COUNT EACH INDENTATION IN THE BOX AND DIVIDE BY TWO. THIS GIVES A CLOSE ESTIMATE OF THE NUMBER OF PERSONS WHO MADE THE PRINTS.

To use the 36-inch box method, mark off a 30- to 36-inch cross section of a trail, count the prints in the box, then divide by two to determine the number of persons that used the trail. (Your M16 rifle is 39 inches long and may be used as a measuring device.)

OTHER SIGNS OF DISPLACEMENT

Footprints are only one example of displacement. Displacement occurs when anything is moved from its original position. Other examples are such things as foliage, moss, vines, sticks, or rocks that are moved from their original places; dew droplets brushed from leaves; stones and sticks that are turned over and show a different color underneath; and grass or other vegetation that is bent or broken in the direction of movement.

Bits of cloth may be torn from a uniform and left on thorns, snags, or the ground, and dirt from boots may make marks on the ground.

Another example of displacement is the movement of wild animals and birds that are flushed from their natural habitats. You may hear the cries of birds that are excited by strange movements. The movement of tall grass or brush on a windless day indicates that something is moving the vegetation from its original position.

When you clear a trail by either breaking or cutting your way through heavy vegetation, you displace the vegetation. Displacement signs can be made while you stop to rest with heavy loads. The prints made by the equipment you carry can help to identify its type. When loads are set down at a rest halt or campsite, grass and twigs may be crushed. A sleeping man may also flatten the vegetation.

In most areas, there will be insects. Any changes in the normal life of these insects may be a sign that someone has recently passed through the area. Bees that are stirred up, and holes that are covered by someone moving over them, or spider webs that are torn down are good clues.

If a person uses a stream to cover his trail, algae and water plants may be displaced in slippery footing or in places where he walks carelessly. Rocks may be displaced from their original position, or turned over to show a lighter or darker color on their opposite side. A person entering or leaving a stream may create slide marks, wet banks, or footprints, or he may scuff bark off roots or sticks. Normally, a person or animal will seek the path of least resistance. Therefore, when you search a stream for exit signs, look for open places on the banks or other places where it would be easy to leave the stream.

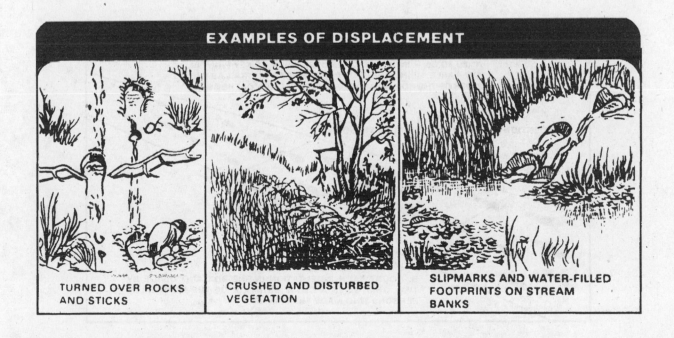

EXAMPLES OF DISPLACEMENT

TURNED OVER ROCKS AND STICKS

CRUSHED AND DISTURBED VEGETATION

SLIPMARKS AND WATER-FILLED FOOTPRINTS ON STREAM BANKS

STAINING

A good example of staining is the mark left by blood from a bleeding wound. Bloodstains often will be in the form of drops left by a wounded person. Blood signs are found on the ground and smeared on leaves or twigs.

You can determine the location of a wound on a man being followed by studying the bloodstains. If the blood seems to be dripping steadily, it probably came from a wound on his trunk. A wound in the lungs will deposit bloodstains that are pink, bubbly, frothy. A bloodstain deposited from a head wound will appear heavy, wet, and slimy, like gelatin. Abdominal wounds often mix blood with digestive juices so that the deposit will have an odor. The stains will be light in color.

Staining can also occur when a person walks over grass, stones, and shrubs with muddy boots. Thus, staining and displacement together may give evidence of movement and indicate the direction taken. Crushed leaves may stain rocky ground that is too hard for footprints.

Roots, stones, and vines may be stained by crushed leaves or berries when walked on. Yellow stains in snow may be urine marks left by personnel in the area.

In some cases, it may be hard to determine the difference between staining and displacement. Both terms can be applied to some signs. For example, water that has been muddied may indicate recent movement. The mud has been displaced and it is staining the water. Stones in streams may be stained by mud from boots. Algae can be displaced from stones in streams and can stain other stones or bark.

Water in footprints in swampy ground may be muddy if the tracks are recent. In time, however, the mud will settle and the water will clear. The clarity of the water can be used to estimate the age of the prints. Normally, the mud will clear in 1 hour. That will vary with terrain.

WEATHERING

Weather may either aid or hinder tracking. It affects signs in ways that help determine how old they are, but wind, snow, rain, and sunlight can also obliterate signs completely.

By studying the effects of weather on signs, you can determine the age of the sign. For example, when bloodstains are fresh, they may be bright red. Air and sunlight will change the appearance of blood first to a deep ruby-red color, and then to a dark brown crust when the moisture evaporates. Scuff marks on trees or bushes darken with time. Sap oozes from fresh cuts on trees but it hardens when exposed to the air.

FOOTPRINTS

Footprints are greatly affected by weather. When a foot displaces soft, moist soil to form a print, the moisture holds the edges of the print intact and sharp. As sunlight and air dry the edges of the print, small particles that were held in place by the moisture fall into the print, making the edges appear rounded. Study this process carefully to estimate the age of a print. If particles are just beginning to fall into a print, it is probably fresh. If the edges of the print are dried and crusty, the prints are probably at least an hour old. The effects of weather will vary with the terrain, so this information is furnished as a guide only.

A light rain may round out the edges of a print. Try to remember when the last rain occurred in order to put prints into a proper time frame. A heavy rain may erase all signs.

Wind also affects prints. Besides drying out a print, the wind may blow litter, sticks, or leaves into it. Try to remember the wind activity in order to help determine the age of a print. For example, you may think, "It is calm now, but the wind blew hard an hour ago. These prints have litter blown into them, so they must be over an hour old." You must be sure, however, that the litter was blown into the prints, and was not crushed into them when the prints were made.

Trails leaving streams may appear to be weathered by rain because of water running into the footprints from wet clothing or equipment. This is particularly true if a party leaves a stream in a file. From this formation, each person drips water into the prints. A wet trail slowly fading into a dry trail indicates that the trail is fresh.

WIND, SOUNDS, AND ODORS

Wind affects sounds and odors. If the wind is blowing from the direction of a trail you are following, sounds and odors are carried to you. If the wind is blowing in the same direction as the trail you are following, you must be cautious as the wind will carry your sounds toward the enemy. To find the wind direction, drop a handful of dry dirt or grass from shoulder height.

To help you decide where a sound is coming from, cup your hands behind your ears and slowly turn. When the sound is loudest, you are probably facing the origin of sound. When moving, try to keep the wind in your face.

SUN

You must also consider the effects of the sun. It is hard to look or aim directly into the sun. If possible, keep the sun at your back.

LITTERING

Poorly trained units may leave trails of litter as they move. Gum or candy wrappers, ration cans, cigarette butts, remains of fires, or human feces are unmistakable signs of recent movement.

Weather affects litter. Rain may flatten or wash litter away, or turn paper into pulp. Winds may blow litter away from its original location. Ration cans exposed to weather will rust. They first rust at the exposed edge where they were opened. Rust then moves in toward the center. Use your memory to determine the age of litter. The last rain or strong wind can be the basis of a time frame.

CAMOUFLAGE

If a party knows that you are tracking it, it will probably use camouflage to conceal its movement and to slow and confuse you. Doing so, however, will slow it down. Walking backward, brushing out trails, and moving over rocky ground or through streams are examples of camouflage that can be used to confuse you.

The party may move on hard surfaced, frequently traveled roads or try to merge with traveling civilians. Examine such routes with extreme care, because a well-defined approach that leads to the enemy will probably be mined, ambushed, or covered by snipers.

The party may try to avoid leaving a trail. Its members may wrap rags around their boots, or wear soft-soled shoes to make the edges of their footprints rounder and less distinct. The party may exit a stream in column or line to reduce the chance of leaving a well-defined exit.

If the party walks backward to leave a confusing trail, the footprints will be deepened at the toe, and the soil will be scuffed or dragged in the direction of movement.

If a trail leads across rocky or hard ground, try to work around that ground to pick up the exit trail. This process works in streams as well. On rocky ground, moss or lichens growing on the stones could be displaced by even the most careful evader. If you lose the trail, return to the last visible sign. From there, head in the direction of the party's movement. Move in ever-widening circles until you find some signs to follow.

INTERPRETATION/IMMEDIATE USE INTELLIGENCE

When reporting, do not report your interpretations as facts. Report that you have seen signs of certain things, not that those things actually exist.

Report all information quickly. The term "immediate use intelligence" includes information of the enemy that can be put to use at once to gain surprise, to keep the enemy off balance, or to keep him from escaping an area . A commander has many sources of intelligence. He puts the information from those sources together to help determine where an enemy is, what he may be planning, and where he maybe going.

Information you report gives your leader definite information on which he can act at once. For example, you may report that your leader is 30 minutes behind an enemy unit, that the enemy is moving north, and that he is now at a certain place. That gives your leader information on which he can act at once. He could then have you keep on tracking and move another unit to attack the enemy. If a trail is found that has signs of recent enemy activity, your leader can set up an ambush on it.

TRACKING TEAMS

Your unit may form tracking teams. The lead team of a moving unit can be a tracking team, or a separate unit may be a tracking team. There are many ways to organize such teams, and they can be any size. There should, however, be a leader, one or more trackers, and security for the trackers. A typical organization has three trackers, three security men, and a team leader with a radiotelephone operator (RATELO).

When a team is moving, the best tracker should be in the lead, followed by his security. The two other trackers should be on the flanks, each one followed and overmatched by his security. The leader should be where he can best control the team. The RATELO should be with the leader.

COUNTERTRACKING

In addition to knowing how to track, you must know how to counter an enemy tracker's efforts to track you. Some countertracking techniques are discussed in the following paragraphs:

- While moving from close terrain to open terrain, walk past a big tree (30 cm [12 in] in diameter or larger) toward the open area for three to five paces. Then walk backward to the forward side of the tree and make a 90-degree change of direction, passing the tree on its forward side. Step carefully and leave as little sign as possible. If this is not the direction that you want to go, change direction again about 50 meters away using the same technique. The purpose of this is to draw the enemy tracker into the open area where it is harder for him to track. That also exposes him and causes him to search the wrong area.
- When approaching a trail (about 100 meters from it), change your direction of movement and approach it at a 45-degree angle. When arriving at the trail, move along it for about 20 to 30 meters.

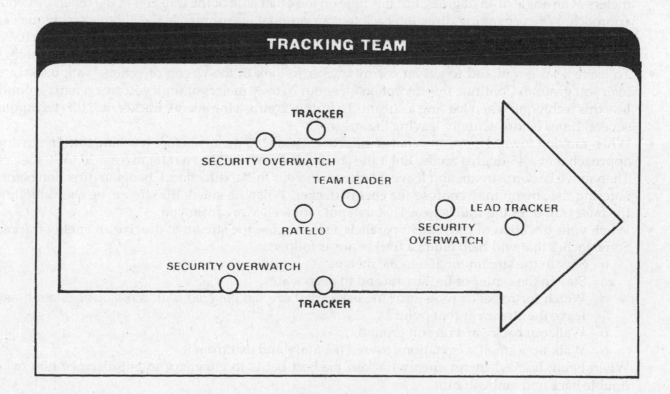

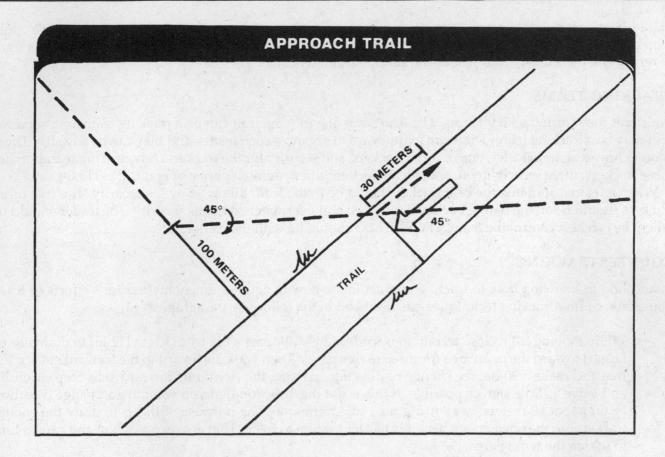

Leave several signs of your presence. Then walk backward along the trail to the point where you join edit. At that point, cross the trail and leave no sign of your leaving it. Then move about 100 meters at an angle of 45 degrees, but this time on the other side of the trail and in the reverse of your approach. When changing direction back to your original line of march, the big tree technique can be used. The purpose of this technique is to draw the enemy tracker along the easier trail. You have, by changing direction before reaching the trail, indicated that the trail is your new line of march.

- To leave a false trail and to get an enemy tracker to look in the wrong direction, walk backward over soft ground. Continue this deception for about 20 to 30 meters or until you are on hard ground. Use this technique when leaving a stream. To further confuse the enemy tracker, use this technique several times before actually leaving the stream.

- When moving toward a stream, change direction about 100 meters before reaching the stream and approach it at a 45-degree angle. Enter the stream and proceed down it for at least 20 to 30 meters. Then move back upstream and leave the stream in your initial direction. Changing direction before entering the stream may confuse the enemy tracker. When he enters the stream, he should follow the false trail until the trail is lost. That will put him well away from you.

- When your direction of movement parallels a stream, use the stream to deceive an enemy tracker. Some tactics that will help elude a tracker are as follows:
  o Stay in the stream for 100 to 200 meters.
  o Stay in the center of the stream and in deep water.
  o Watch for rocks or roots near the banks that are not covered with moss or vegetation and leave the stream at that point.
  o Walk out backward on soft ground.
  o Walk up a small, vegetation-covered tributary and exit from it.

- When being tracked by an enemy tracker, the best bet is to either try to out-distance him or to double back and ambush him.

FALSE TRAIL LEAVING STREAM

20-30 METERS

HARD GROUND

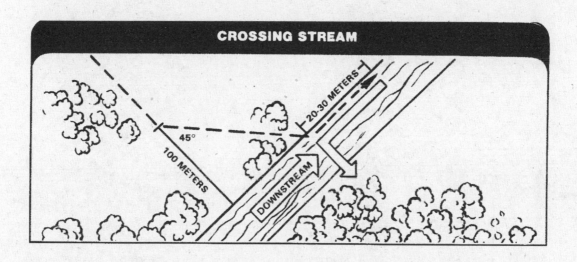

CROSSING STREAM

45°

100 METERS

20-30 METERS

DOWNSTREAM

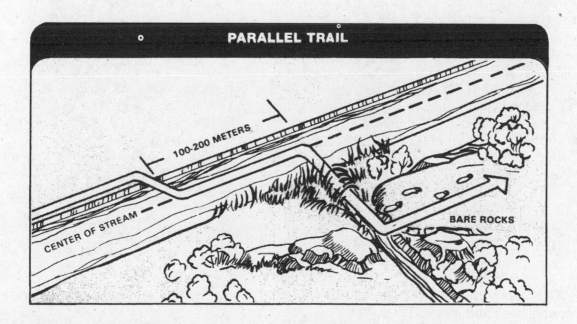

PARALLEL TRAIL

100-200 METERS

CENTER OF STREAM

BARE ROCKS

CHAPTER 9

Movement

GENERAL

Normally, you will spend more time moving than fighting. You must use proper movement techniques to avoid contact with the enemy when you are not prepared for contact.

The fundamentals of movement discussed in this chapter provide techniques that all soldiers should learn. These techniques should be practiced until they become second nature.

MOVEMENT TECHNIQUES

Your unit's ability to move depends on your movement skills and those of your fellow soldiers. Use the following techniques to avoid being seen or heard by the enemy:

- Camouflage yourself and your equipment.
- Tape your dog tags together and to the chain so they cannot slide or rattle. Tape or pad the parts of your weapon and equipment that rattle or are so loose that they may snag (the tape or padding must not interfere with the operation of the weapon or equipment).
- Jump up and down and listen for rattles.
- Wear soft, well-fitting clothes.
- Do not carry unnecessary equipment. Move from covered position to revered position (taking no longer than 3 to 5 seconds between positions).
- Stop, look, and listen before moving. Look for your next position before leaving a position.
- Look for covered and concealed routes on which to move.
- Change direction slightly from time to time when moving through tall grass.
- Stop, look, and listen when birds or animals are alarmed (the enemy may be nearby).
- Use battlefield noises, such as weapon noises, to conceal movement noises.
- Cross roads and trails at places that have the most cover and concealment (large culverts, low spots, curves, or bridges).
- Avoid steep slopes and places with loose dirt or stones.
- Avoid cleared, open areas and tops of hills and ridges.

METHODS OF MOVEMENT

In addition to walking, you may move in one of three other methods—low crawl, high crawl, or rush.

The low crawl gives you the lowest silhouette. Use it to cross places where the concealment is very low and enemy fire or observation prevents you from getting up. Keep your body flat against the ground. With your firing hand, grasp your weapon sling at the upper sling—swivel. Let the front hand guard rest on your forearm (keeping the muzzle off the ground), and let the weapon butt drag on the ground.

To move, push your arms forward and pull your firing side leg forward. Then pull with your arms and push with your leg. Continue this throughout the move.

The high crawl lets you move faster than the low crawl and still gives you a low silhouette. Use this crawl when there is good concealment but enemy fire prevents you from getting up. Keep your body

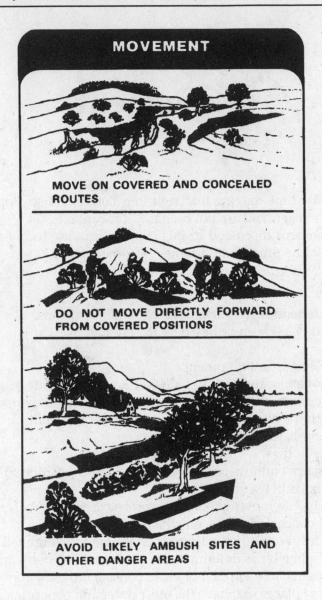

off the ground and resting on your forearms and lower legs. Cradle your weapon in your arms and keep its muzzle off the ground. Keep your knees well behind your buttocks so your body will stay low.

To move, alternately advance your right elbow and left knee, then your left elbow and right knee.

The rush is the fastest way to move from one position to another. Each rush should last from 3 to 5 seconds. The rushes are kept short to keep enemy machine gunners or riflemen from tracking you. However, do not stop and hit the ground in the open just because 5 seconds have passed. Always try to hit the ground behind some cover. Before moving, pick out your next covered and concealed position and the best route to it.

Make your move from the prone position as follows:

- Slowly raise your head and pick your next position and the route to it.
- Slowly lower your head.
- Draw your arms into your body (keeping your elbows in).
- Pull your right leg forward.

- Raise your body by straightening your arms.
- Get up quickly.
- Run to the next position.

When you are ready to stop moving, do the following:

- Plant both of your feet.
- Drop to your knees (at the same time slide a hand to the butt of your rifle).
- Fall forward, breaking the fall with the butt of the rifle.
- Go to a prone firing position.

If you have been firing from one position for some time, the enemy may have spotted you and may be waiting for you to come up from behind cover. So, before rushing forward, roll or crawl a short distance from your position. By coming up from another spot, you may fool an enemy who is aiming at one spot, waiting for you to rise.

When the route to your next position is through an open area, rush by zigzagging. If necessary, hit the ground, roll right or left, then rush again.

MOVING WITH STEALTH

Moving with stealth means moving quietly, slowly, and carefully. This requires great patience.

To move with stealth, use the following techniques:

- Hold your rifle at port arms (ready position).
- Make your footing sure and solid by keeping your body's weight on the foot on the ground while stepping.
- Raise the moving leg high to clear brush or grass.
- Gently let the moving foot down toe first, with your body's weight on the rear leg.
- Lower the heel of the moving foot after the toe is in a solid place.
- Shift your body's weight and balance to the forward foot before moving the rear foot.
- Take short steps to help maintain balance.

At night, and when moving through dense vegetation, avoid making noise. Hold your weapon with one hand, and keep the other hand forward, feeling for obstructions.

When going into a prone position, use the following techniques:

- Hold your rifle with one hand and crouch slowly.
- Feel for the ground with your freehand to make sure it is clear of mines, tripwires, and other hazards.
- Lower your knees, one at a time, until your body's weight is on both knees and your free hand.
- Shift your weight to your free hand and opposite knee.
- Raise your free leg up and back, and lower it gently to that side.
- Move the other leg into position the same way.
- Roll quietly into a prone position.

Use the following techniques when crawling:

- Crawl on your hands and knees. Hold your rifle in your firing hand.
- Use your nonfiring hand to feel for and make clear spots for your hands and knees to move to.
- Move your hands and knees to those spots, and put them down softly.

IMMEDIATE ACTIONS WHILE MOVING

This section furnishes guidance for the immediate actions you should take when reacting to enemy indirect fire and flares.

Reacting to indirect fire. If you come under indirect fire while moving, quickly look to your leader for orders. He will either tell you to run out of the impact area in a certain direction or will tell you to follow him. If you cannot see your leader, but can see other team members, follow them. If alone, or if you cannot see your leader or the other team members, run out of the area in a direction away from the incoming fire.

It is hard to move quickly on rough terrain, but the terrain may provide good cover. In such terrain, it may be best to take cover and wait for flares to burn out. After they burn out, move out of the area quickly.

Reacting to Ground Flares. The enemy puts out ground flares as warning devices. He sets them off himself or attaches tripwires to them for you to trip on and set them off. He usually puts the flares in places he can watch.

If you are caught in the light of a ground flare, move quickly out of the lighted area. The enemy will know where the ground flare is and will be ready to fire into that area. Move well away from the lighted area. While moving out of the area, look for other team members. Try to follow or join them to keep the team together.

Reacting to Aerial Flares. The enemy uses aerial flares to light up vital areas. They can be set off like ground flares; fired from hand projectors, grenade launchers, mortars, and artillery; or dropped from aircraft.

If you hear the firing of an aerial flare while you are moving, hit the ground (behind cover if possible) while the flare is rising and before it bursts and illuminates.

If moving where it is easy to blend with the background (such as in a forest) and you are caught in the light of an aerial flare, freeze in place until the flare burns out.

If you are caught in the light of an aerial flare while moving in an open area, immediately crouch low or lie down.

If you are crossing an obstacle, such as a barbed-wire fence or a wall, and get caught in the light of an aerial flare, crouch low and stay down until the flare burns out.

The sudden light of a bursting flare may temporarily blind both you and the enemy. When the enemy uses a flare to spot you, he spoils his own night vision. To protect your night vision, close one eye while the flare is burning. When the flare burns out, the eye that was closed will still have its night vision.

REACTING TO AERIAL FLARES

MOVING WITHIN A TEAM

You will usually move as a member of a team. Small teams, such as infantry fire teams, normally move in a wedge formation. Each soldier in the team has a set position in the wedge, determined by the type weapon he carries. That position, however, may be changed by the team leader to meet the situation. The normal distance between soldiers is 10 meters.

You may have to make a temporary change in the wedge formation when moving through close terrain. The soldiers in the sides of the wedge close into a single file when moving in thick brush or through a narrow pass. After passing through such an area, they should spread out, again forming the wedge. You should not wait for orders to change the formation or the interval. You should change automatically and stay in visual contact with the other team members and the team leader.

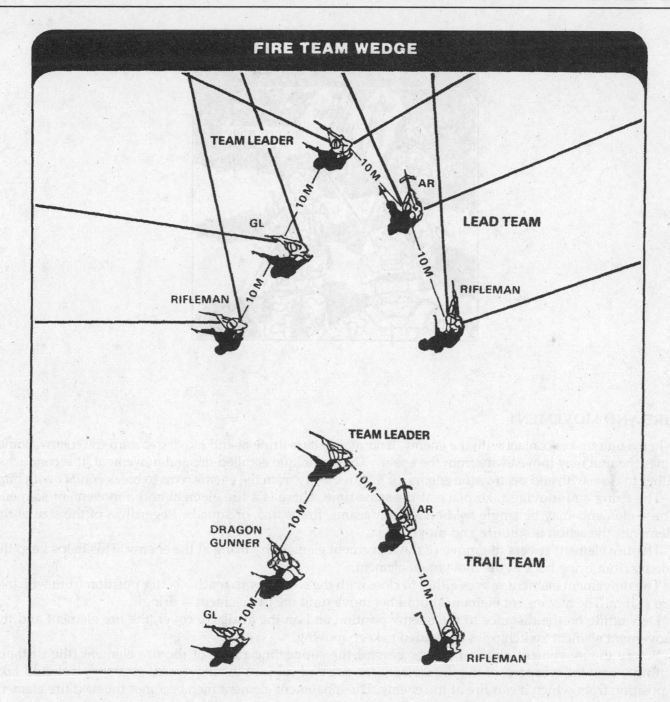

The team leader leads by setting the example. His standing order is, FOLLOW ME AND DO AS I DO. When he moves to the left, you should move to the left. When he gets down, you should get down. When he fires, you should fire.

When visibility is limited, control during movement may become difficult. Two l-inch horizontal strips of luminous tape, sewn directly on the rear of the helmet camouflage band with a l-inch space between them, are a device for night identification.

Night identification for your patrol cap could be two l-inch by 1/2-inch strips of luminous tape sewn vertically, directly on the rear of the cap. They should be centered, with the bottom edge of each tape even with the bottom edge of the cap and with a l-inch space between the two tapes.

FIRE AND MOVEMENT

When a unit makes contact with the enemy, it normally starts firing at and moving toward the enemy. Sometimes the unit may move away from the enemy. That technique is called fire and movement. It is conducted either to close with and destroy the enemy, or to move away from the enemy so as to break contact with him.

The firing and moving take place at the same time. There is a fire element and a movement element. These elements may be single soldiers, buddy teams, fire teams, or squads. Regardless of the size of the elements, the action is still fire and movement.

The fire element covers the move of the movement element by firing at the enemy. This helps keep the enemy from firing back at the movement element.

The movement element moves either to close with the enemy or to reach a better position from which to fire at him. The movement element should not move until the fire element is firing.

Depending on the distance to the enemy position and on the available cover, the fire element and the movement element switch roles as needed to keep moving.

Before the movement element moves beyond the supporting range of the fire element (the distance within which the weapons of the fire element can fire and support the movement element), it should take a position from which it can fire at the enemy. The movement element then becomes the next fire element and the fire element becomes the next movement element.

If your team makes contact, your team leader should tell you to fire or to move. He should also tell you where to fire from, what to fire at, or where to move to. When moving, use the low crawl, high crawl, or rush.

FOOT MARCH LOADS

The fighting load for a conditioned soldier should not exceed 48 pounds and the approach march load should not exceed 72 pounds. These load weights include all clothing and equipment that are worn and carried.

a. A soldier's ability to react to the enemy is reduced by the burden of his load. Load carrying causes fatigue and lack of agility, placing soldiers at a disadvantage when rapid reaction to the enemy is required. For example, the time a soldier needs to complete an obstacle course is increased from 10 to 15 percent, depending on the configuration of the load, for every 10 pounds of equipment carried. It is likely that a soldier's agility in the assault will be degraded similarly.

b. Speed of movement is as important a factor in causing exhaustion as the weight of the load carried. The chart at Figure 5-1 shows the length of time that work rates can be sustained before soldiers become exhausted and energy expenditure rates for march speeds and loads. A burst rate of energy expenditure of 900 to 1,000 calories per hour can only be sustained for 6 to 10 minutes. Fighting loads must be light so that the bursts of energy available to a soldier are used to move and to fight, rather than to carry more than the minimum fighting equipment.

c. When carrying loads during approach marches, a soldier's speed can cause a rate-of-energy expenditure of over 300 calories per hour and can erode the reserves of energy needed upon enemy contact. March speeds must be reduced when loads are heavier to stay within reasonable energy expenditure rates. Carrying awkward loads and heavy handheld items causes further degradation of march speed and agility. The distance marched in six hours decreases by about 2 km for every 10 pounds carried over 40 pounds.

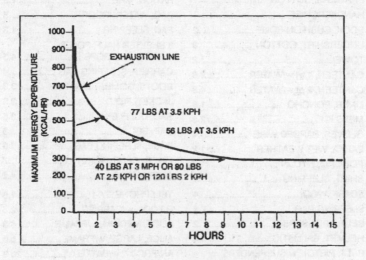

Work rate and energy expenditure.

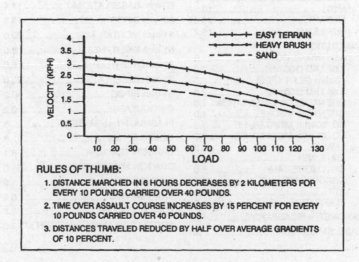

March speeds.

Battlefield stress decreases the ability of soldiers to carry their loads. Fear burns up the glycogen in the muscles required to perform physical tasks. This wartime factor is often overlooked in peacetime, but the commander must consider such a factor when establishing the load for each soldier. However, applying strong leadership to produce well-trained, highly motivated soldiers can lessen some of the effects of stress.

As the modern battlefield becomes more sophisticated, potential enemies develop better protected equipment, which could be presented as fleeting targets. Unless technological breakthroughs occur, increasingly heavy munitions and new types of target acquisition and communications equipment will be required by frontline soldiers to defeat the enemy.

a. In the future, the foot soldier's load can be decreased only by sending him into battle inadequately equipped or by providing some means of load-handling equipment to help him carry required equipment.

WEIGHT IN POUNDS (every ounce counts)

BDU	3.8	TROUSERS, WET WEATHER	1.2
DRAWERS, COTTON	.1	RATION, MRE	1.3
HANDKERCHIEF	.1	BAG, WATERPROOF	.8
SOCK, CUSHION SOLE	.2	PAD, SLEEPING	1.3
UNDERSHIRT, COTTON	.3	3 SHELTER HALF, POLES,	
TOWEL	.2	PEGS, AND ROPE	4.5
CANTEEN, 1 qt w/WATER	2.8	CARRIER, SLEEPING BAG	.4
CANTEEN, 2 qt w/WATER	4.8	BOOTS, COMBAT LEATHER	3.3
LINER, PONCHO	1.6	JACKET, FIELD	3.3
MESS KIT	2.8	BAG, DUFFLE	3.5
GLOVES, BARBED WIRE	.4	CAP, BDU	.3
PARKA, WET WEATHER	1.2	CASE, SLEEPING BAG	1.5
PONCHO, NYLON	1.3	LINER, FIELD JACKET	.7
SHIRT, SLEEPING	.7	OVERSHOES	4.2
SCARF, WOOL	.4	TELEPHONE, TA-1	1.5
SLEEPING BAG	7.1	E-TOOL, w/CARRIER	2.5
BELT, TROUSERS	.2	ALICE, MEDIUM, w/FRAME	6.3
HELMET, BALLISTIC	3.4	ALICE, LARGE, w/FRAME	6.6
BELT, PISTOL, w/SUSPENDERS		AN/PRC-77, w/BATTERY	20.8
AND FIRST-AID POUCH	1.6	M60 SPARE BARREL w/BAG	8.0
TOILET ARTICLES	2.0	60-mm MORTAR, M225	14.4
WEAPONS:		60-mm SIGHT, M64	2.5
M16	1.6	60-mm BASEPLATE, M-7	14.4
M203	10.0	60-mm BIPOD	13.2
M60 MG	23.3	81-mm MORTAR, M29	30.0
M249 SAW	15.2	81-mm SIGHT, M53	6.0
AMMUNITION:		81-mm NIGHTLIGHT	2.0
5.56 w/MAG (30 rds)	.9	81-mm BASEPLATE	25.0
7.62 LKD (100 rds)	7.0	81-mm BIPOD	40.0
40-mm (ALL TYPES)	.5	BINOCULARS	3.2
5.56 LKD (200 rds)	7.6	FLASHLIGHT, w/BATTERY	.8
GRENADE, FRAGMENTATION	1.0	COMPASS, M2	.3
GRENADE, SMOKE	1.0	DRAGON TRACKER	8.1
RD, 60-mm MORTAR, HE	3.5	DRAGON NIGHTSIGHT	34.0
RD, 81-mm MORTAR, HE	9.3	AN/PVS-5 NVG	1.9
LAW	5.2	AN/PVS-4 NVD	3.9
MINE, M21	18.0	PISTOL, CAL .45	2.5
CLAYMORE, M18	5.0	PROTECTIVE MASK, w/DECON KIT	3.0
DRAGON, MSL	25.3		
AT4	14.0		
FLARE, TRIP	1.0		
BAYONET, w/SCABBARD	1.3		
CASE, SMALL-ARMS	.9		

Weights of selected items.

b. Unless part of the load is removed from the soldier's back and carried elsewhere, all individual load weights are too heavy. Even if rucksacks are removed, key teams on the battlefield cannot fulfill their roles unless they carry excessively heavy loads. Soldiers who must carry heavy loads restrict the mobility of their units.

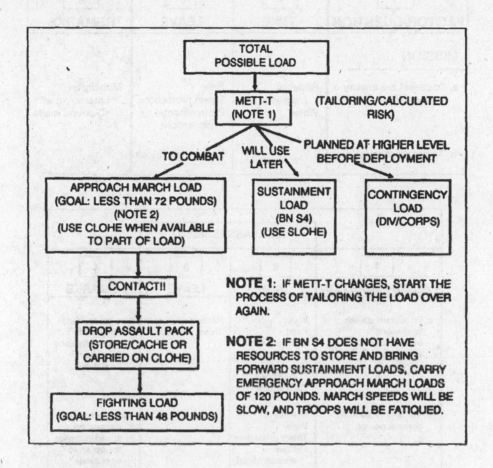

CONFLICT/ ENVIRONMENT	CLOHE (COMPANY LEVEL)	SLOHE (BATTALION LEVEL)
LOW INTENSITY		
TEMPERATE CLIMATE	1,850 LB	4,800 LB/400 CUBIC FT
COLD WET CLIMATE	1,850 LB	7,000 LB/600 CUBIC FT
MEDIUM INTENSITY		
TEMPERATE CLIMATE	2,250 LB	6,350 LB/500 CUBIC FT
COLD WET CLIMATE	2,250 LB	8,550 LB/700 CUBIC FT

RISK ASSESSMENT			
1 FACTOR/QUESTION	**2** TAKE	**3** LEAVE	**4** REMARKS
MISSION			
a. To defeat the enemy in close battle.	Reduced munitions Water	Food Threat protection Environmental protection	Mobility is paramount with 40-pound loads
b. To get there quickly.		Reserve munitions	

RISK ASSESSMENT			
1 FACTOR/QUESTION	**2** TAKE	**3** LEAVE	**4** REMARKS
c. To sustain stealth operations independent of resupply.	Water Food Environmental protection Reduced munitions Camouflage	Reserve munitions Threat protection	Maximum loads depend on speed/distance for dynamic operations.
d. To carry maximum combat power.	Munitions Water Threat protection Limited environmental protection	Food	Maximum loads depend on speed/distance for dynamic operations.
e. For static operations.	Basic load and reserve of ammunition Barrier materiel Maximum threat protection Some comfort items to achieve quality rest periods Water Food		
RESUPPLY			
a. Reliability.	Less amounts of all classes of supply		Best solution is for the commander to control his own immediate resupply transport resources.
b. On call.		Reserve munitions	

RISK ASSESSMENT			
1 **FACTOR/QUESTION**	**2** **TAKE**	**3** **LEAVE**	**4** **REMARKS**
c. Planned.		Food Environmental protection Quality rest Equipment	
DEFEAT THE THREAT			Basic load must be tailored to meet the threat.
Antipersonnel	Small-arms/ grenades/ Claymores		
Antiemplacement Antisoft vehicle Antimateriel Antitank Antiair	Grenades/66-mm Rocket/ demolitions AT4/machine gun ammunition Demolition/ grenades Dragon/AT4 Machine gun ammunition		
SURVIVE THE THREAT			Soldiers take the minimum of threat protection.
Ballistic PASGT vest protection			PAGST vest reduces casualties by 50 percent during bombardment.
NBC protection	Protective mask	MOPP suit if enemy use of chemical weapons is low	

| | RISK ASSESSMENT | | |
1 FACTOR/QUESTION	2 TAKE	3 LEAVE	4 REMARKS
Electronic Warfare	VINSON		Secure communications probably not viable below bde/bn level in light units unless COMSEC is of high priority to achieve mission.
TERRAIN			
Flat, improved road			Terrain may cause an increase of time required to conduct march; resupply cross country may be difficult.
Cross country	Water consumption increased		
Hills, improved road			
WEATHER			Energy must be maintained to fight by control of loads/march speeds.
a.Environmental Survival:			

	RISK ASSESSMENT		
1 **FACTOR/QUESTION**	**2** **TAKE**	**3** **LEAVE**	**4** **REMARKS**

Exposure	Poncho
	Extra clothing
	Limited number
	  of sleeping
	  bags
			Work rates should
			be reduced.

Heat exhaustion	Water	Threat protection	
Disease	Water purification tablets Mosquito nets		When in combat, men with excess fat can survive off natural reserves.
b. Sustenance	High-caloric food		Average of four hours quality sleep each day.
c. Quality rest	Sleeping bags/ pads		

CHAPTER 10

Field-expedient Direction Finding

In a survival situation, you will be extremely fortunate if you happen to have a map and compass. If you do have these two pieces of equipment, you will most likely be able to move toward help. If you are not proficient in using a map and compass, you must take the steps to gain this skill.

There are several methods by which you can determine direction by using the sun and the stars. These methods, however, will give you only a general direction. You can come up with a more nearly true direction if you know the terrain of the territory or country.

You must learn all you can about the terrain of the country or territory to which you or your unit may be sent, especially any prominent features or landmarks. This knowledge of the terrain together with using the methods explained below will let you come up with fairly true directions to help you navigate.

USING THE SUN AND SHADOWS

The earth's relationship to the sun can help you to determine direction on earth. The sun always rises in the east and sets in the west, but not exactly due east or due west. There is also some seasonal variation. In the northern hemisphere, the sun will be due south when at its highest point in the sky, or when an object casts no appreciable shadow. In the southern hemisphere, this same noonday sun will mark due north. In the northern hemisphere, shadows will move clockwise. Shadows will move counterclockwise in the southern hemisphere. With practice, you can use shadows to determine both direction and time of day. The shadow methods used for direction finding are the shadow-tip and watch methods.

Shadow-Tip Methods. In the first shadow-tip method, find a straight stick 1 meter long, and a level spot free of brush on which the stick will cast a definite shadow. This method is simple and accurate and consists of four steps:

Step 1. Place the stick or branch into the ground at a level spot where it will cast a distinctive shadow. Mark the shadow's tip with a stone, twig, or other means. This first shadow mark is always west—everywhere on earth.

Step 2. Wait 10 to 15 minutes until the shadow tip moves a few centimeters. Mark the shadow tip's new position in the same way as the first.

Step 3. Draw a straight line through the two marks to obtain an approximate east-west line.

Step 4. Stand with the first mark (west) to your left and the second mark to your right—you are now facing north. This fact is true everywhere on earth.

An alternate method is more accurate but requires more time. Set up your shadow stick and mark the first shadow in the morning. Use a piece of string to draw a clean arc through this mark and around the stick. At midday, the shadow will shrink and disappear. In the afternoon, it will lengthen again and at the point where it touches the arc, make a second mark. Draw a line through the two marks to get an accurate east-west line (see Figure 10-1).

The Watch Method. You can also determine direction using a common or analog watch—one that has hands. The direction will be accurate if you are using true local time, without any changes for daylight savings time. Remember, the further you are from the equator, the more accurate this method will be. If you only have a digital watch, you can overcome this obstacle. Quickly draw a watch on a circle of paper with the correct time on it and use it to determine your direction at that time.

1 Mark the shadow's tip.

2 Mark the new position and draw a line through the two marks.

3 Stand with the first mark to your left and the second mark to your right—you are now facing north.

Figure 10-1: Shadow-tip method.

In the northern hemisphere, hold the watch horizontal and point the hour hand at the sun. Bisect the angle between the hour hand and the 12 o'clock mark to get the north-south line (Figure 10-2). If there is any doubt as to which end of the line is north, remember that the sun rises in the east, sets in the west, and is due south at noon. The sun is in the east before noon and in the west after noon.

Note: If your watch is set on daylight savings time, use the midway point between the hour hand and 1 o'clock to determine the north-south line.

In the southern hemisphere, point the watch's 12 o'clock mark toward the sun and a midpoint halfway between 12 and the hour hand will give you the north-south line (Figure 10-2).

USING THE MOON

Because the moon has no light of its own, we can only see it when it reflects the sun's light. As it orbits the earth on its 28-day circuit, the shape of the reflected light varies according to its position. We say there is a new moon or no moon when it is on the opposite side of the earth from the sun. Then, as it moves away from the earth's shadow, it begins to reflect light from its right side and waxes to become a full moon before waning, or losing shape, to appear as a sliver on the left side. You can use this information to identify direction.

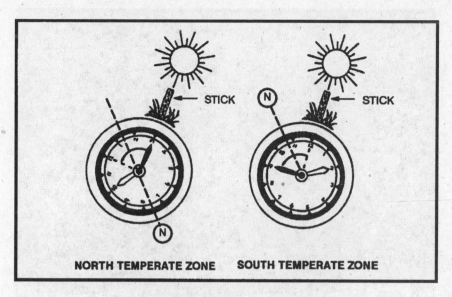

Figure 10-2: Watch method.

If the moon rises before the sun has set, the illuminated side will be the west. If the moon rises after midnight, the illuminated side will be the east. This obvious discovery provides us with a rough east-west reference during the night.

USING THE STARS

Your location in the Northern or Southern Hemisphere determines which constellation you use to determine your north or south direction.

The Northern Sky. The main constellations to learn are the Ursa Major, also known as the Big Dipper or the Plow, and Cassiopeia (Figure 10-3). Neither of these constellations ever sets. They are always visible on a clear night. Use them to locate Polaris, also known as the polestar or the North Star. The North Star forms part of the Little Dipper handle and can be confused with the Big Dipper. Prevent confusion by using both the Big Dipper and Cassiopeia together. The Big Dipper and Cassiopeia are always directly opposite each other and rotate counterclockwise around Polaris, with Polaris in the center. The Big Dipper is a seven star constellation in the shape of a dipper. The two stars forming the outer lip of this dipper are the "pointer stars" because they point to the North Star. Mentally draw a line from the outer bottom star to the outer top star of the Big Dipper's bucket. Extend this line about five times the distance between the pointer stars. You will find the North Star along this line.

Cassiopeia has five stars that form a shape like a "W" on its side. The North Star is straight out from Cassiopeia's center star.

After locating the North Star, locate the North Pole or true north by drawing an imaginary line directly to the earth.

The Southern Sky. Because there is no star bright enough to be easily recognized near the south celestial pole, a constellation known as the Southern Cross is used as a signpost to the South (Figure 10-4). The Southern Cross or Crux has five stars. Its four brightest stars form a cross that tilts to one side. The two stars that make up the cross's long axis are the pointer stars. To determine south, imagine a distance five times the distance between these stars and the point where this imaginary line ends is in the general direction of south. Look down to the horizon from this imaginary point and select a landmark to steer by. In a static survival situation, you can fix this location in daylight if you drive stakes in the ground at night to point the way.

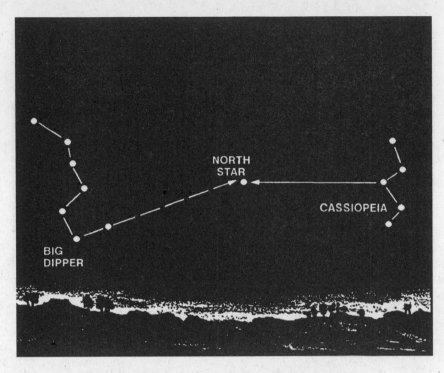

Figure 10-3: The Big Dipper and Cassiopeia.

MAKING IMPROVISED COMPASSES

You can construct improvised compasses using a piece of ferrous metal that can be needle shaped or a flat double-edged razor blade and a piece of nonmetallic string or long hair from which to suspend it. You can magnetize or polarize the metal by slowly stroking it in one direction on a piece of silk or carefully through your hair using deliberate strokes. You can also polarize metal by stroking it repeatedly at one end with a magnet. Always rub in one direction only. If you have a battery and some electric wire, you can polarize the metal electrically. The wire should be insulated. If not insulated, wrap the metal object in a single, thin strip of paper to prevent contact. The battery must be a minimum of 2 volts. Form a coil with the electric wire and touch its ends to the battery's terminals. Repeatedly insert one end of the metal object in and out of the coil. The needle will become an electromagnet. When suspended from a piece of nonmetallic string, or floated on a small piece of wood in water, it will align itself with a north-south line.

You can construct a more elaborate improvised compass using a sewing needle or thin metallic object, a nonmetallic container (for example, a plastic dip container), its lid with the center cut out and water-proofed, and the silver tip from a pen. To construct this compass, take an ordinary sewing needle and break in half. One half will form your direction pointer and the other will act as the pivot point. Push the portion used as the pivot point through the bottom center of your container; this portion should be flush on the bottom and not interfere with the lid. Attach the center of the other portion (the pointer) of the needle on the pen's silver tip using glue, tree sap, or melted plastic. Magnetize one end of the pointer and rest it on the pivot point.

OTHER MEANS OF DETERMINING DIRECTION

The old saying about using moss on a tree to indicate north is not accurate because moss grows completely around some trees. Actually, growth is more lush on the side of the tree facing the south in the Northern Hemisphere and vice versa in the Southern Hemisphere. If there are several felled trees around for comparison, look at the stumps. Growth is more vigorous on the side toward the equator and the tree growth

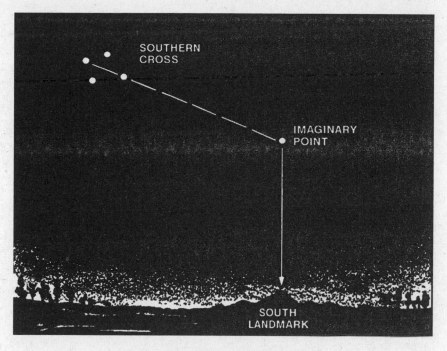

Figure 10-4: The Southern Cross.

rings will be more widely spaced. On the other hand, the tree growth rings will be closer together on the side toward the poles.

Wind direction may be helpful in some instances where there are prevailing directions and you know what they are.

Recognizing the differences between vegetation and moisture patterns on north- and south-facing slopes can aid in determining direction. In the northern hemisphere, north-facing slopes receive less sun than south-facing slopes and are therefore cooler and damper. In the summer, north-facing slopes retain patches of snow. In the winter, the trees and open areas on south-facing slopes are the first to lose their snow, and ground snow pack is shallower.

PART VI

Environment-Specific Survival

CHAPTER 1

Tropical Survival

Most people think of the tropics as a huge and forbidding tropical rain forest through which every step taken must be hacked out, and where every inch of the way is crawling with danger. Actually, over half of the land in the tropics is cultivated in some way.

A knowledge of field skills, the ability to improvise, and the application of the principles of survival will increase the prospects of survival. Do not be afraid of being alone in the jungle; fear will lead to panic. Panic will lead to exhaustion and decrease your chance of survival.

Everything in the jungle thrives, including disease germs and parasites that breed at an alarming rate. Nature will provide water, food, and plenty of materials to build shelters.

Indigenous peoples have lived for millennia by hunting and gathering. However, it will take an outsider some time to get used to the conditions and the nonstop activity of tropical survival.

TROPICAL WEATHER

High temperatures, heavy rainfall, and oppressive humidity characterize equatorial and subtropical regions, except at high altitudes. At low altitudes, temperature variation is seldom less than 10 degrees C and is often more than 35 degrees C. At altitudes over 1,500 meters, ice often forms at night. The rain has a cooling effect, but when it stops, the temperature soars.

Rainfall is heavy, often with thunder and lightning. Sudden rain beats on the tree canopy, turning trickles into raging torrents and causing rivers to rise. Just as suddenly, the rain stops. Violent storms may occur, usually toward the end of the summer months.

Hurricanes, cyclones, and typhoons develop over the sea and rush inland, causing tidal waves and devastation ashore. In choosing campsites, make sure you are above any potential flooding. Prevailing winds vary between winter and summer. The dry season has rain once a day and the monsoon has continuous rain. In Southeast Asia, winds from the Indian Ocean bring the monsoon, but it is dry when the wind blows from the landmass of China.

Tropical day and night are of equal length. Darkness falls quickly and daybreak is just as sudden.

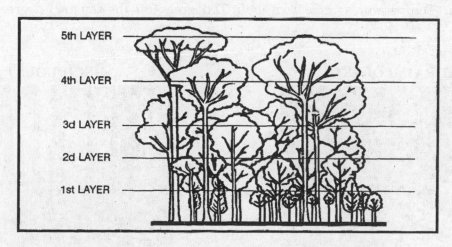

Figure 1-1: Five layers of tropical rain forest vegetation.

JUNGLE REGIONS OF THE WORLD

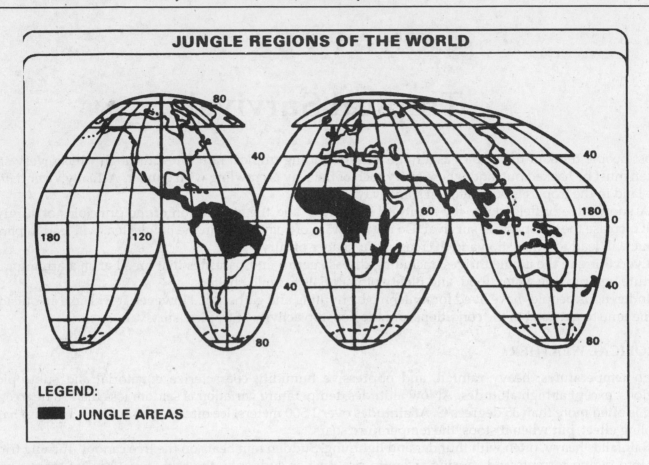

LEGEND:

■ JUNGLE AREAS

TYPES OF JUNGLES

The jungle environment includes densely forested areas, grasslands, cultivated areas, and swamps. Jungles are classified as primary or secondary jungles based on the terrain and vegetation.

Primary Jungles. These are tropical forests. Depending on the type of trees growing in these forests, primary jungles are classified either as tropical rain forests or as deciduous forests.

Tropical Rain Forests. The climate varies little in rain forests. You find these forests across the equator in the Amazon and Congo basins, parts of Indonesia, and several Pacific islands. Up to 3.5 meters of rain fall evenly throughout the year. Temperatures range from about 32 degrees C in the day to 21 degrees C at night.

TROPICAL RAIN FOREST

DECIDUOUS FOREST

There are five layers of vegetation in this jungle. Where untouched by man, jungle trees rise from buttress roots to heights of 60 meters. Below them, smaller trees produce a canopy so thick that little light reaches the jungle floor. Seedlings struggle beneath them to reach light, and masses of vines and lianas twine up to the sun. Ferns, mosses, and herbaceous plants push through a thick carpet of leaves, and a great variety of fungi grow on leaves and fallen tree trunks.

Because of the lack of light on the jungle floor, there is little undergrowth to hamper movement, but dense growth limits visibility to about 50 meters. A wet and soggy surface make vehicular traffic difficult. Foot movement is easier in tropical forests than in other types of jungle. You can easily lose your sense of direction in this jungle, and it is extremely hard for aircraft to see you.

Deciduous Forests. These are found in semitropical zones where there are both wet and dry seasons. In the wet season, trees are fully leaved; in the dry season, much of the foliage dies. Trees are generally less dense in deciduous forests than in rain forests. This allows more rain and sunlight to filter to the ground, producing thick undergrowth. In the wet season, with the trees in full leaf, observation both from the air and on the ground is limited. Movement is more difficult than in the rain forest. In the dry season, however, both observation and trafficability improve.

The characteristics of the American and African semievergreen seasonal forests correspond with those of the Asian monsoon forests. These characteristics are—

- Their trees fall into two stories of tree strata. Those in the upper story average 18 to 24 meters; those in the lower story average 7 to 13 meters.
- The diameter of the trees averages 0.5 meter.
- Their leaves fall during a seasonal drought.

Except for the sage, nipa, and coconut palms, the same edible plants grow in these areas as in the tropical rain forests.

You find these forests in portions of Columbia and Venezuela and the Amazon basin in South America; in portions of southeast coastal Kenya, Tanzania, and Mozambique in Africa; in Northeastern India, much of Burma, Thailand, Indochina, Java, and parts of other Indonesian islands in Asia.

Secondary Jungles. These are found at the edge of the rain forest and the deciduous forest, and in areas where jungles have been cleared and abandoned. Secondary jungles appear when the ground has been repeatedly exposed to sunlight. These areas are typically overgrown with weeds, grasses, thorns, ferns, canes, and shrubs. Foot movement is extremely slow and difficult. Vegetation may reach to a height of 2 meters. This will limit observation to the front to only a few meters.

SECONDARY JUNGLE

Such growth happens mainly along river banks, on jungle fringes, and where man has cleared rain forest. When abandoned, tangled masses of vegetation quickly reclaim these cultivated areas. You can often find cultivated food plants among this vegetation.

Tropical Scrub and Thorn Forests. The chief characteristics of tropical scrub and thorn forests are—

- There is a definite dry season.
- Trees are leafless during the dry season.
- The ground is bare except for a few tufted plants in bunches; grasses are uncommon.
- Plants with thorns predominate.
- Fires occur frequently.

You find tropical scrub and thorn forests on the west coast of Mexico, Yucatan peninsula, Venezuela, Brazil; on the northwest coast and central parts of Africa; and in Asia, in Turkestan and India.

Within the tropical scrub and thorn forest areas, you will find it hard to obtain food plants during the dry season. During the rainy season, plants are considerably more abundant.

COMMON JUNGLE FEATURES

Swamps. These are common to all low jungle areas where there is water and poor drainage. There are two basic types of swamps—mangrove and palm.

Mangrove Swamps. These are found in coastal areas wherever tides influence water flow. The mangrove is a shrub-like tree which grows 1 to 5 meters high. These trees have tangled root systems, both above and below the water level, which restrict movement to foot or small boats. Observation in mangrove swamps, both on the ground and from the air, is poor, and movement is extremely difficult. Sometimes, streams that you can raft form channels, but you usually must travel on foot through this swamp.

You find saltwater swamps in West Africa, Madagascar, Malaysia, the Pacific islands, Central and South America, and at the mouth of the Ganges River in India. The swamps at the mouths of the Orinoco and Amazon rivers and rivers of Guyana consist of mud and trees that offer little shade. Tides in saltwater swamps can vary as much as 12 meters.

Everything in a saltwater swamp may appear hostile to you, from leeches and insects to crocodiles and caimans. Avoid the dangerous animals in this swamp.

MANGROVE SWAMP

PALM SWAMP

Avoid this swamp altogether if you can. If there are water channels through it, you may be able to use a raft to escape.

Palm Swamps. These exist in both salt and fresh water areas. Their characteristics are masses of thorny undergrowth, reeds, grasses, and occasional short palms that reduce visibility and make travel difficult. There are often islands that dot these swamps, allowing you to get out of the water. Wildlife is abundant in these swamps. Like movement in the mangrove swamps, movement through palm swamps is mostly restricted to foot (sometimes small boats). Vehicular traffic is nearly impossible except after extensive road construction by engineers. Observation and fields-of-fire are very limited. Concealment from both air and ground observation is excellent.

Savanna. This is a broad, open jungle grassland in which trees are scarce. The thick grass is broad-bladed and grows 1 to 5 meters high. It looks like a broad, grassy meadow, and frequently has red soil. It grows scattered trees that usually appear stunted and gnarled like apple trees. Palms also occur on savannas.

Movement in the savanna is generally easier than in other types of jungle areas, especially for vehicles. The sharp-edged, dense grass and extreme heat make foot movement a slow and tiring process. Depending on the height of the grass, ground observation may vary from poor to good. Concealment from air observation is poor for both troops and vehicles. You find savannas in parts of Venezuela, Brazil, and the Guianasin South America. In Africa, you find them in the southern Sahara (north-central Cameroon and Gabon and southern Sudan), Benin, Togo, most of Nigeria, northeastern Zaire, northern Uganda, western Kenya, part of Malawi, part of Tanzania, southern Zimbabwe, Mozambique, and western Madagascar.

Bamboo. This grows in clumps of varying size in jungles throughout the tropics. Large stands of bamboo are excellent obstacles for wheeled or tracked vehicles. Troop movement through bamboo is slow, exhausting, and noisy. Troops should bypass bamboo stands if possible.

Cultivated Areas. These exist in jungles throughout the tropics and range from large, well-planned and well-managed farms and plantations to small tracts cultivated by individual farmers. There are three general types of cultivated areas—rice paddies, plantations, and small farms.

Rice Paddies. These are flat, flooded fields in which rice is grown. Flooding of the fields is controlled by a network of dikes and irrigation ditches which make movement by vehicles difficult even when the fields are dry. Concealment is poor in rice paddies. Cover is limited to the dikes, and then only from ground fire. Observation and fields of fire are excellent. Foot movement is poor when the fields are wet because soldiers must wade through water about 1/2 meter (2 feet) deep and soft mud. When the fields are dry, foot movement becomes easier. The dikes, about 2 to 3 meters tall, are the only obstacles.

SAVANNA

BAMBOO

Plantations. These are large farms or estates where tree crops, such as rubber and coconut, are grown. They are usually carefully planned and free of undergrowth (like a well-tended park). Movement through plantations is generally easy. Observation along the rows of trees is generally good. Concealment and cover can be found behind the trees, but soldiers moving down the cultivated rows are exposed.

Small Farms. These exist throughout the tropics. These small cultivated areas are usually hastily planned. After 1 or 2 years' use, they usually are abandoned, leaving behind a small open area which turns into secondary jungle. Movement through these areas may be difficult due to fallen trees and scrub brush.

Generally, observation and fields-of-fire are less restricted in cultivated areas than in uncultivated jungles. However, much of the natural cover and concealment are removed by cultivation, and troops will be more exposed in these areas.

LIFE IN THE JUNGLE

The jungle environment affects everyone. The degree to which you are trained to live and fight in harsh environments will determine your survival.

There is very little to fear from the jungle environment. Fear itself can be an enemy. Soldiers must be taught to control their fear of the jungle. A man overcome with fear is of little value in any situation. Soldiers in a jungle must learn that the most important thing is to keep their heads and calmly think out any situation.

Many of the stories written about out-of-the-way jungle places were written by writers who went there in search of adventure rather than facts. Practically without exception, these authors exaggerated or invented many of the thrilling experiences they relate. These thrillers are often a product of the author's imagination and are not facts.

Most Americans, especially those raised in cities, have lost the knack of taking care of themselves under all conditions. It would be foolish to say that, without proper training, they would be in no danger if lost in the jungles of Southeast Asia, South America, or some Pacific island. On the other hand, they would be in just as much danger if lost in the mountains of western Pennsylvania or in other undeveloped regions of our own country. The only difference would be that you are less likely to panic when lost in your homeland than abroad.

Immediate Considerations. There is less likelihood of your rescue from beneath a dense jungle canopy than in other survival situations. You will probably have to travel to reach safety.

RICE PADDIES

PLANTATIONS

SMALL FARMS

If you are the victim of an aircraft crash, the most important items to take with you from the crash site are a machete, a compass, a first aid kit, and a parachute or other material for use as mosquito netting and shelter.

Take shelter from tropical rain, sun, and insects. Malaria-carrying mosquitoes and other insects are immediate dangers, so protect yourself against bites.

Do not leave the crash area without carefully blazing or marking your route. Use your compass. Know what direction you are taking.

In the tropics, even the smallest scratch can quickly become dangerously infected. Promptly treat any wound, no matter how minor.

Effect of Climate. The discomforts of tropical climates are often exaggerated, but it is true that the heat is more persistent. In regions where the air contains a lot of moisture, the effect of the heat may seem worse than the same temperature in a dry climate. Many people experienced in jungle operations feel that the heat and discomfort in some US cities in the summertime are worse than the climate in the jungle.

Strange as it may seem, there may be more suffering from cold in the tropics than from the heat. Of course, very low temperatures do not occur, but chilly days and nights are common. In some jungles, in winter months, the nights are cold enough to require a wool blanket or poncho liner for sleeping.

Rainfall in many parts of the tropics is much greater than that in most areas of the temperate zones. Tropical downpours usually are followed by clear skies, and in most places the rains are predictable at

Precautions against malaria include:

■ Taking Dapsone and chloroquine-primaquine

■ Using insect repellent

■ Wearing clothing that covers as much of the body as possible

■ Using nets or screens at every opportunity

■ Avoiding the worst-infested areas when possible

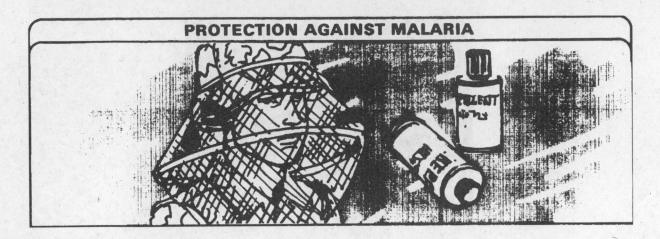

PROTECTION AGAINST MALARIA

certain times of the day. Except in those areas where rainfall may be continuous during the rainy season, there are not many days when the sun does not shine part of the time.

People who live in the tropics usually plan their activities so that they are able to stay under shelter during the rainy and hotter portions of the day. After becoming used to it, most tropical dwellers prefer the constant climate of the torrid zones to the frequent weather changes in colder climates.

Insects. Malaria-carrying mosquitoes are probably the most harmful of the tropical insects. Soldiers can contract malaria if proper precautions are not taken.

Mosquitoes are most prevalent early at night and just before dawn. Soldiers must be especially cautious at these times. Malaria is more common in populated areas than in uninhabited jungle, so soldiers must also be especially cautious when operating around villages. Mud packs applied to mosquito bites offer some relief from itching.

Wasps and bees may be common in some places, but they will rarely attack unless their nests are disturbed. When a nest is disturbed, the troops must leave the area and reassemble at the last rally point. In case of stings, mud packs are helpful. In some areas, there are tiny bees, called sweatbees, which may collect on exposed parts of the body during dry weather, especially if the body is sweating freely. They are annoying but stingless and will leave when sweating has completely stopped, or they may be scraped off with the hand.

The larger centipedes and scorpions can inflict stings which are painful but not fatal. They like dark places, so it is always advisable to shake out blankets before sleeping at night, and to make sure before dressing that they are not hidden in clothing or shoes. Spiders are commonly found in the jungle. Their bites may be painful, but are rarely serious. Ants can be dangerous to injured men lying on the ground and unable to move. Wounded soldiers should be placed in an area free of ants.

In Southeast Asian jungles, the rice-borer moth of the lowlands collects around lights in great numbers during certain seasons. It is a small, plain-colored moth with a pair of tiny black spots on the wings. It should never be brushed off roughly, as the small barbed hairs of its body may be ground into the skin. This causes a sore, much like a burn, that often takes weeks to heal.

Leeches. Leeches are common in many jungle areas, particularly throughout most of the Southwest Pacific, Southeast Asia, and the Malay Peninsula. They are found in swampy areas, streams, and moist jungle country. They are not poisonous, but their bites may become infected if not cared for properly. The small wound that they cause may provide a point of entry for the germs which cause tropical ulcers or "jungle sores." Soldiers operating in the jungle should watch for leeches on the body and brush them off before they have had time to bite. When they have taken hold, they should not be pulled off forcibly because part of the leech may remain in the skin. Leeches will release themselves if touched with insect repellent, a moist piece of tobacco, the burning end of a cigarette, a coal from a fire, or a few drops of alcohol.

Straps wrapped around the lower part of the legs ("leech straps") will prevent leeches from crawling up the legs and into the crotch area. Trousers should be securely tucked into the boots. For more information on dangerous insects, se Part IV, Chapter 3, "Dangerous Insects and Arachnids."

Snakes. A soldier in the jungle probably will see very few snakes. When he does see one, the snake most likely will be making every effort to escape.

If a soldier should accidentally step on a snake or otherwise disturb a snake, it will probably attempt to bite. The chances of this happening to soldiers traveling along trails or waterways are remote if soldiers are alert and careful. Most jungle areas pose less of a snakebite danger than do the uninhabited areas of New Mexico, Florida, or Texas. This does not mean that soldiers should be careless about the possibility of snakebites, but ordinary precautions against them are enough. Soldiers should be particularly watchful when clearing ground. *Treat all snakebites as poisonous.*

Crocodiles and Caymans. Crocodiles and caymans are meat-eating reptiles which live in tropical areas. "Crocodile-infested rivers and swamps" is a catch-phrase often found in stories about the tropics. Asian jungles certainly have their share of crocodiles, but there are few authenticated cases of crocodiles actually attacking humans. Caymans, found in South and Central America, are not likely to attack unless provoked.

Wild Animals. In Africa, where lions, leopards, and other flesh-eating animals abound, they are protected from hunters by local laws and live on large preserves. In areas where the beasts are not protected, they are shy and seldom seen. When encountered, they will attempt to escape. All large animals can be dangerous if cornered or suddenly startled at close quarters. This is especially true of females with young. In the jungles of Sumatra, Bali, Borneo, Southeast Asia, and Burma there are tigers, leopards, elephants, and buffalo. Latin America's jungles have the jaguar. Ordinarily, these will not attack a man unless they are cornered or wounded.

Certain jungle animals, such as water buffalo and elephants, have been domesticated by the local people. Soldiers should also avoid these animals. They may appear tame, but this tameness extends only to people the animals are familiar with.

Poisonous Vegetation. Another area of danger is that of poisonous plants and trees. For example, nettles, particularly tree nettles, are one of the dangerous items of vegetation. These nettles have a severe stinging that will quickly educate the victim to recognize the plant. There are ringas trees in Malaysia which affect some people in much the same way as poison oak. The poison ivy and poison sumac of the continental US can cause many of the same type troubles that may be experienced in the jungle. The danger from poisonous plants in the woods of the US eastern seaboard is similar to that of the tropics. Thorny thickets, such as rattan, should be avoided as one would avoid a blackberry patch.

Some of the dangers associated with poisonous vegetation can be avoided by keeping sleeves down and wearing gloves when practical. For more information on poisonous plants, see Part IV, Chapter 7, "Poisonous Plants."

Health and Hygiene. The climate in tropical areas and the absence of sanitation facilities increase the chance that soldiers may contract a disease. Disease is fought with good sanitation practices and preventive medicine. In past wars, diseases accounted for a significantly high percentage of casualties.

Waterborne Diseases. Water is vital in the jungle and is usually easy to find. However, water from natural sources should be considered contaminated. Water purification procedures must be taught to all soldiers. Germs of serious diseases, like dysentery, are found in impure water. Other waterborne diseases, such as blood fluke, are caused by exposure of an open sore to impure water.

Soldiers can prevent waterborne diseases by:

- Obtaining drinking water from approved engineer water points.
- Using rainwater; however, rainwater should be collected after it has been raining at least 15 to 30 minutes. This lessens the chances of impurity being washed from the jungle canopy into the water container. Even then the water should be purified.
- Insuring that all drinking water is purified.
- Not swimming or bathing in untreated water.
- Keeping the body fully clothed when crossing water obstacles.

Fungus Diseases. These diseases are caused by poor personal health practices. The jungle environment promotes fungus and bacterial diseases of the skin and warm water immersion skin diseases. Bacteria and fungi are tiny plants which multiply fast under the hot, moist conditions of the jungle. Sweat-soaked skin invites fungus attack. The following are common skin diseases that are caused by long periods of wetness of the skin:

SNAKEBITE TREATMENT

Follow these steps if bitten:

■ Remain calm, but act swiftly, and chances of survival are good. (Less than one percent of properly treated snakebites are fatal. Without treatment, the fatality rate is 10 to 15 percent.)

■ Immobilize the affected part in a position below the level of the heart.

■ Place a lightly constricting band 5 to 10 centimeters (2 to 4 inches) closer to the heart than the site of the bite. Reapply the constricting band ahead of the swelling if it moves up the arm or leg. The constricting band should be placed tightly enough to halt the flow of blood in surface vessels, but not so tight as to stop the pulse.

■ Do not attempt to cut open the bite or suck out venom.

■ Seek medical help. If possible, the snake's head with 5 to 10 centimeters (2 to 4 inches) of its body attached should be taken to the medics for identification. Identification insures use of the proper antivenom.

CROCODILES AND CAYMANS

CAYMAN

CROCODILE

ALLIGATOR

Before going into a jungle area, leaders must:

■ Make sure immunizations are current.

■ Get soldiers in top physical shape.

■ Instruct soldiers in personal hygiene.

Upon arrival in the jungle area, leaders must:

■ Allow time to adjust (acclimate) to the new environment.

■ Never limit the amount of water soldiers drink. (It is very important to replace the fluids lost through sweating.)

■ Instruct soldiers on the sources of disease. Insects cause malaria, yellow fever, and scrub typhus. Typhoid, dysentery, cholera, and hepatitis are caused by dirty food and contaminated water.

Warm Water Immersion Foot. This disease occurs usually where there are many creeks, streams, and canals to cross, with dry ground in between. The bottoms of the feet become white, wrinkled, and tender. Walking becomes painful.

Chafing. This disease occurs when soldiers must often wade through water up to their waists, and the trousers stay wet for hours. The crotch area becomes red and painful to even the lightest touch.

Most skin diseases are treated by letting the skin dry.

Heat Injuries. These result from high temperatures, high humidity, lack of air circulation, and physical exertion. All soldiers must be trained to prevent heat disorders.

To prevent these diseases, soldiers should:

■ Bathe often, and air- or sun-dry the body as often as possible.

■ Wear clean, dry, loose-fitting clothing whenever possible.

■ Not sleep in wet, dirty clothing. Soldiers should carry one dry set of clothes just for sleeping. Dirty clothing, even if wet, is put on again in the morning. This practice not only fights fungus, bacterial, and warm water immersion diseases but also prevents chills and allows soldiers to rest better.

■ Not wear underwear during wet weather. Underwear dries slower than jungle fatigues, and causes severe chafing.

■ Take off boots and massage feet as often as possible.

■ Dust feet, socks, and boots with foot powder at every chance.

■ Always carry several pairs of socks and change them frequently.

■ Keep hair cut short.

Natives. Like all other regions of the world, the jungle also has its native inhabitants. Soldiers should be aware that some of these native tribes can be hostile if not treated properly.

There may be occasions, however, when hostile tribes attack without provocation.

Food. Food of some type is always available in the jungle—in fact, there is hardly a place in the world where food cannot be secured from plants and animals. All animals, birds, reptiles, and many kinds of insects of the jungle are edible. Some animals, such as toads and salamanders, have glands on the skin which should be removed before their meat is eaten. Fruits, flowers, buds, leaves, bark, and often tubers (fleshy plant roots) may be eaten. Fruits eaten by birds and monkeys usually may be eaten by man.

There are various means of preparing and preserving food found in the jungle. Fish, for example, can be cleaned and wrapped in wild banana leaves. This bundle is then tied with string made from bark, placed on a hastily constructed wood griddle, and roasted thoroughly until done. Another method is to roast the bundle of fish underneath a pile of red-hot stones.

Other meats can be roasted in a hollow section of bamboo, about 60 centimeters (2 feet) long. Meat cooked in this manner will not spoil for three or four days if left inside the bamboo stick and sealed.

Yams, taros, yuccas, and wild bananas can be cooked in coals. They taste somewhat like potatoes. Palm hearts can make a refreshing salad, and papaya a delicious dessert. For more information on game, edible plants, and water, see Part IV.

Shelter. Jungle shelters are used to protect personnel and equipment from the harsh elements of the jungle. Shelters are necessary while sleeping, planning operations, and protecting sensitive equipment. See Part III, "Shelters" for more information.

Clothing and Equipment. Before deploying for jungle operations, troops are issued special uniforms and equipment. Some of these items are:

Jungle Fatigues. These fatigues are lighter and faster drying than standard fatigues. To provide the best ventilation, the uniform should fit loosely. It should never be starched.

Jungle Boots. These boots are lighter and faster drying than all-leather boots. Their cleated soles will maintain footing on steep, slippery slopes. The ventilating insoles should be washed in warm, soapy water when the situation allows.

HEAT INJURIES

TYPE	CAUSE	SYMPTOMS	TREATMENT
Dehydration	Dehydration is caused by the loss of too much water. About two-thirds of the human body is water. When water is not replaced as it is lost, the body becomes dried out—dehydrated.	The symptoms are sluggishness and listlessness.	The treatment is to give the victim plenty of water.
Heat Exhaustion	Heat exhaustion is caused by the loss of too much water and salt.	The symptoms are: Dizziness. Nausea. Headache. Cramps. Rapid, weak pulse. Cool, wet skin.	The treatment consists of: Moving the victim to a cool, shaded place for rest. Loosening the clothing. Elevating the feet to improve circulation. Giving the victim cool salt water (two salt tablets dissolved in a canteen of water). Natural sea water should not be used.
Heat Cramps	Heat cramps are caused by the loss of too much salt.	The symptom is painful muscle cramps which are relieved as soon as salt is replaced.	The treatment is the same as for heat exhaustion.
Heatstrokes	Heatstroke (sunstroke) is caused by a breakdown in the body's heat control mechanism. The most likely victims are those who are not acclimated to the jungle, or those who have recently had bad cases of diarrhea. Heatstroke can kill if not treated quickly.	The symptoms are: Hot, red, dry skin (most important sign). No sweating (when sweating would be expected). Very high temperature (105 to 110 degrees). Rapid pulse. Spots before eyes. Headache, nausea, dizziness, mental confusion. Sudden collapse.	Treatment consists of: Cooling the victim immediately. This is achieved by putting him in a creek or stream; pouring canteens of water over him; fanning him; and using ice, if available. Giving him cool salt water (prepared as stated earlier) if he is conscious. Rubbing his arms and legs rapidly. Evacuating him to medical aid as soon as possible.

Heat injuries are prevented by:

- Drinking plenty of water.
- Using extra salt with food and water.
- Slowing down movement.

NOTE: For more details, see FM 21-10 for field hygiene and sanitation, and FM 21-11 for first aid for soldiers.

To prevent a conflict, leaders should insure that their soldiers:

- Respect the natives' privacy and personal property
- Observe the local customs and taboos
- Do not enter a native house without being invited
- Do not pick fruits or cut trees without permission of their owners
- Treat the natives as friends

Meats that can be found in most jungles include:	The following types of fruits and nuts are common in jungle areas:		Vegetables found in most jungles include:
Wild fowl	Bananas	Wild raspberries	Taro *
Wild cattle	Coconuts		Yam *
Wild pig	Oranges and lemons	Nakarika	Yucca *
Freshwater fish *		Papaya	Hearts of palm trees
Saltwater fish	Navele nuts		
Fresh-water crawfish	Breadfruit	Mangoes	

*These items must be cooked before eating.

Insect (Mosquito) Bar. The insect (mosquito) bar or net should be used any time soldiers sleep in the jungle. Even if conditions do not allow a shelter, the bar can be hung inside the fighting position or from trees or brush. No part of the body should touch the insect net when it is hung, because mosquitoes can bite through the netting. The bar should be tucked or laid loosely, not staked down. Although this piece of equipment is very light, it can be bulky if not folded properly. It should be folded inside the poncho as tightly as possible.

JUNGLE TRAVEL, NAVIGATION AND TRACKING

Travel Through Jungle Areas. With practice, movement through thick undergrowth and jungle can be done efficiently. Always wear long sleeves to avoid cuts and scratches.

To move easily, you must develop "jungle eye," that is, you should not concentrate on the pattern of bushes and trees to your immediate front. You must focus on the jungle further out and find natural breaks in the foliage. Look through the jungle, not at it. Stop and stoop down occasionally to look along the jungle floor. This action may reveal game trails that you can follow.

Stay alert and move slowly and steadily through dense forest or jungle. Stop periodically to listen and take your bearings. Use a machete to cut through dense vegetation, but do not cut unnecessarily or you will quickly wear yourself out. If using a machete, stroke upward when cutting vines to reduce noise because sound carries long distances in the jungle. Use a stick to part the vegetation. Using a stick will also help dislodge biting ants, spiders, or snakes. Do not grasp at brush or vines when climbing slopes; they may have irritating spines or sharp thorns.

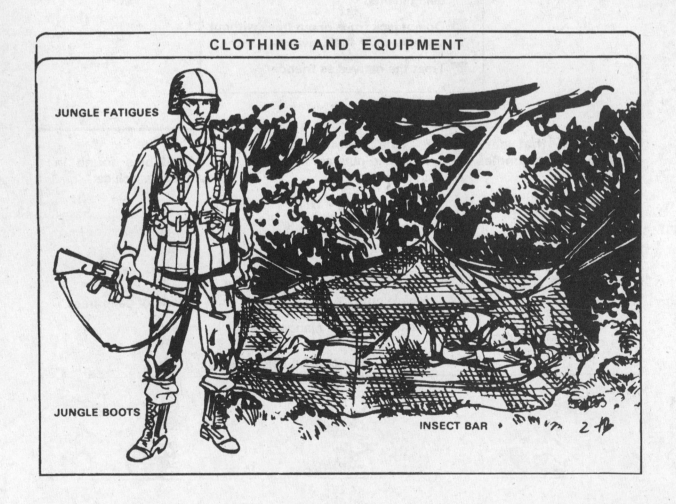

CLOTHING AND EQUIPMENT

JUNGLE FATIGUES

JUNGLE BOOTS

INSECT BAR

Many jungle and forest animals follow game trails. These trails wind and cross, but frequently lead to water or clearings. Use these trails if they lead in your desired direction of travel.

In many countries, electric and telephone lines run for miles through sparsely inhabited areas. Usually, the right-of-way is clear enough to allow easy travel. When traveling along these lines, be careful as you approach transformer and relay stations. In enemy territory, they maybe guarded.

Jungle Navigation. Navigating in the jungle can be difficult for those troops not accustomed to it. This appendix outlines techniques which have been used successfully in jungle navigation. With training and practice, troops should be able to use these techniques to navigate in even the thickest jungle.

Navigation Tools

Maps. Because of the isolation of many jungles, the rugged ground, and the presence of the canopy, topographic survey is difficult and is done mainly from the air. Therefore, although maps of jungle areas generally depict the larger features (hill, ridges, larger streams, etc.) fairly accurately, some smaller terrain features (gullies, small or intermittent streams, small swamps, etc.), which are actually on the ground, may not appear on the map. Also, many older maps are inaccurate. So, before going into the jungle, bring your maps up to date.

Compass. No one should move in the jungle without a compass. It should be tied to the clothing by a string or bootlace. The three most common methods used to follow the readings of a compass are:

Sighting along the desired azimuth. The compass man notes an object to the front (usually a tree or bush) that is on line with the proper azimuth and moves to that object. *This is not a good method in the jungle as trees and bushes tend to look very much alike.*

Holding the compass at waist level and walking in the direction of a set azimuth. This is a good method for the jungle. The compass man sets the compass for night use with the long luminous line placed over the luminous north arrow and the desired azimuth under the black index line. There is a natural tendency to drift either left or right using this method. Jungle navigators must learn their own tendencies and allow for this drift.

Sighting along the desired azimuth and guiding a man forward until he is on line with the azimuth. The unit then moves to the man and repeats the process. This is the most accurate method to use in the jungle during daylight hours, but it is slow. In this method, the compass man cannot mistake the aiming point and is free to release the compass on its string and use both hands during movement to the next aiming point.

The keys to navigation are maintaining the right direction and knowing the distance traveled. Skill with the compass (acquired through practice) takes care of the first requirement. Ways of knowing the

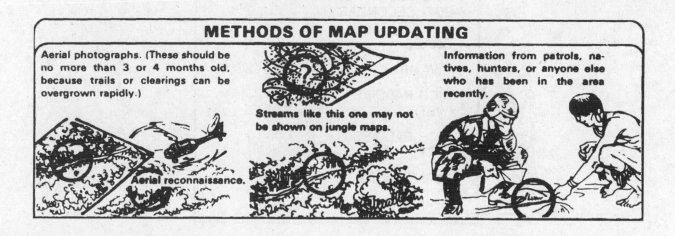

METHODS OF MAP UPDATING

Aerial photographs. (These should be no more than 3 or 4 months old, because trails or clearings can be overgrown rapidly.)

Aerial reconnaissance.

Streams like this one may not be shown on jungle maps.

Information from patrols, natives, hunters, or anyone else who has been in the area recently.

distance traveled include checking natural features with the map, knowing the rate of movement, and pacing.

Checking Features. Major recognizable features (hills, rivers, changes in the type of vegetation) should be noted as they are reached and then identified on the map. Jungle navigators must BE CAUTIOUS ABOUT TRAILS—the trail on the ground may not be the one on the map.

Rate of Movement. Speed will vary with the physical condition of the troops, the load they carry, the danger of enemy contact, and the type of jungle growth. The normal error is to overestimate the distance traveled. The following can be used as a rough guide to the maximum distance covered in 1 hour during daylight.

Pacing. In thick jungle, this is the best way of measuring distance. It is the only method which lets the soldier know how far he has traveled. With this information, he can estimate where he is at any given time. To be accurate, you must practice pacing over different types of terrain, and should make a PERSONAL PACE TABLE like the one below.

If possible, at least two men in each independent group should be compass men, and three or four should be keeping a pace count.

Location of an Objective. In open terrain, an error in navigation can be easily corrected by orienting on terrain features which are often visible from a long distance. In thick jungle, however, it is possible to be within 50 meters of a terrain feature and still not see it. Here are two methods which can aid in navigation.

Offset Method. This method is useful in reaching an objective that is not large or not on readily identifiable terrain but is on a linear feature, such as a road, stream, or ridge. The unit plans a route following an azimuth which is a few degrees to the left or right of the objective. The unit then follows the azimuth to that terrain feature. Thus, when the unit reaches the terrain feature, the members know the objective is to their right or left, and the terrain feature provides a point of reference for movement to the objective.

DAYLIGHT MOVEMENT	
TYPE TERRAIN	MAXIMUM DISTANCE (in meters per hour)
TROPICAL RAIN FOREST	1,000
DECIDUOUS FOREST, SECONDARY JUNGLE, TALL GRASS	500
SWAMPS	100 TO 300
RICE PADDIES (WET)	800
RICE PADDIES (DRY)	2,000
PLANTATIONS	2,000
TRAILS	1,500

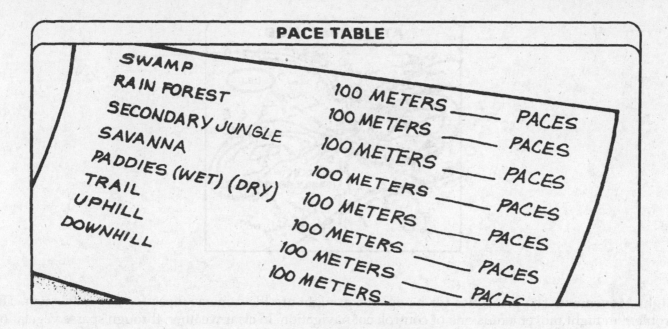

PACE TABLE

SWAMP	100 METERS	_____ PACES
RAIN FOREST	100 METERS	_____ PACES
SECONDARY JUNGLE	100 METERS	_____ PACES
SAVANNA	100 METERS	_____ PACES
PADDIES (WET) (DRY)	100 METERS	_____ PACES
TRAIL	100 METERS	_____ PACES
UPHILL	100 METERS	_____ PACES
DOWNHILL	100 METERS	_____ PACES

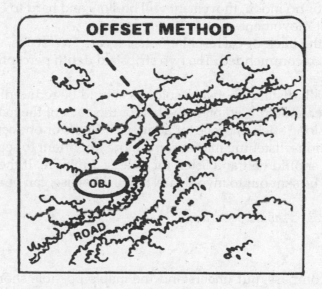

OFFSET METHOD

Attack Method. This method is used when moving to an objective not on a linear feature. An easily recognizable terrain feature is chosen as close as possible to the objective. The unit then moves to that feature. Once there, the unit follows the proper azimuth and moves the estimated distance to get to the objective.

What to do if Lost. Do not panic. Few have ever been permanently lost in the jungle, although many have taken longer to reach their destination than they should.

Disoriented navigators should try to answer these questions. (If there are other navigators in the group, they all should talk it over.)

What was the last known location?

Did you go too far and pass the objective? (Compare estimates of time and distance traveled.)

Does the terrain look the way it should? (Compare the surroundings with the map.)

What features in the area will help to fix your location? (Try to find these features.)

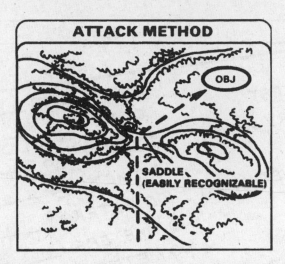

Night Movement. The principles for navigation at night are the same as those for day movement. The problem in night movement is one of control, not navigation. In clear weather, through sparse vegetation and under a bright moon, a unit can move almost as fast by night as by day. If the sky is overcast, vegetation is thick, or there is little or no moon, movement will be slow and hard to control. The following points can assist a unit during night movement.

Attach luminous tape to the back of each soldier's headgear. Two strips, side by side, each about the size of a lieutenant's bar, are recommended. The two strips aid depth perception and reduce the hypnotic effect that one strip can cause.

When there is no light at all, distance between soldiers should be reduced. When necessary to prevent breaks in contact, each soldier should hold on to the belt or the pack of the man in front of him.

The leader should carry a long stick to probe for sudden dropoffs or obstacles.

In limited visibility conditions, listening may become more important to security than observing. A unit which hears a strange noise should halt and listen for at least 1 minute. If the noise is repeated or cannot be identified, patrols should be sent out to investigate. Smell, likewise, can be an indication of enemy presence in an area.

All available night vision devices should be used.

Navigational Tips

- Trust the map and compass, but understand the map's possible shortcomings. Use the compass bezel ring, especially during night navigation.
- Break brush. Do not move on trails or roads.
- Plan the move, and use the plan.
- Do not get frustrated. If in doubt, stop and think back over the route.
- Practice leads to confidence.

Visual tracking is following the paths of men or animals by the signs they leave, primarily on the ground or vegetation. Scent tracking is following men or animals by the odors they leave.

Practice in tracking is required to achieve and maintain a high standard of skill. Because of the excellent natural concealment the jungle offers, all soldiers should be familiar with the general techniques of visual tracking to enable them to detect the presence of a concealed enemy, to follow the enemy, to locate and avoid mines or booby-traps, and to give early warning of ambush.

NAVIGATIONAL TIPS

- ■ Trust the map and compass, but understand the map's possible short-comings. Use the compass bezel ring, especially during night navigation.
- ■ Break brush. Do not move on trails or roads.
- ■ Plan the move, and use the plan.
- ■ Do not get frustrated. If in doubt, stop and think back over the route.
- ■ Practice leads to confidence.

Signs. Men or animals moving through jungle areas leave signs of their passage. Some examples of these signs are listed below.

Deception. The enemy may use any of the following methods to deceive or discourage trackers. They may, at times, mislead an experienced tracker.

TRACKING POINTS

SAVANNA

NOTE:If the grass is high, above 3 feet, trails are easy to follow because the grass is knocked down and normally stays down for several days. If the grass is short, it springs back in a shorter length of time.

■ Grass that is tramped down will point in the direction that the person or animal is traveling.

■ Grass will show a contrast in color with the surrounding undergrowth when pressed down.

■ If the grass is wet with dew, the missing dew will show a trail where a person or an animal has traveled.

■ Mud or soil from boots may appear on some of the grass.

■ If new vegetation is showing through a track, the track is old.

■ In very short grass (12 inches or less) a boot will damage the grass near the ground and a footprint can be found.

ROCKY GROUND

■ Small stones and rocks are moved aside or rolled over when walked on. The soil is also disturbed, leaving a distinct variation in color and an impression. If the soil is wet, the underside of the stones will be much darker in color than the top when moved.

■ If the stone is brittle, it will chip and crumble when walked on. A light patch will appear where the stone is broken and the chips normally remain near the broken stone.

■ Stones on a loose or soft surface are pressed into the ground when walked upon. This leaves either a ridge around the edge of the stone where it has forced the dirt out, ora hole where the stone has been pushed below the surface of the ground.

■ Where moss is growing on rocks or stones, a boot or hand will scrape off some of the moss.

TRACKING POINTS CONTINUED

PRIMARY JUNGLES

NOTE: Within rain forests and deciduous forests, there are many ways to track. This terrain includes undergrowth, live and dead leaves and trees, streams with muddy or sandy banks, and moss on the forest floor and on rocks, which makes tracking easier.

■ Disturbed leaves on the forest floor, when wet, show up a darker color when disturbed.

■ Dead leaves are brittle and will crack or break under pressure of a person walking on them. The same is true of dry twigs.

■ Where the undergrowth is thick, especially on the edges of the forest, green leaves of the bushes that have been pushed aside and twisted will show the underside of the leaf—this side is lighter in color than the upper surface. To find this sort of trail, the tracker must look through the jungle instead of directly at it.

■ Boot impressions may be left on fallen and rotting trees.

■ Marks may be left on the sides of logs lying across the path.

■ Roots running across a path may show signs that something has moved through the area.

■ Broken spiderwebs across a path indicate that something has moved through the area.

SECONDARY JUNGLE

■ Broken branches and twigs.

■ Leaves knocked off bushes and trees.

■ Branches bent in the direction of travel.

■ Footprints.

■ Tunnels made through vegetation.

■ Broken spiderwebs.

■ Pieces of clothing caught on the sharp edges of bushes.

RIVERS, STREAMS, MARSHES, AND SWAMPS

■ Footprints on the banks and in shallow water.

■ Mud stirred up and discoloring the water.

■ Rocks splashed with water in a quietly running stream.

■ Water on the ground at a point of exit.

■ Mud on grass or other vegetation near the edge of the water.

These deceptions include:

- Walking backwards. The heel mark tends to be deeper than that of the ball of the foot. The pace is shorter.

- More than one person stepping in the same tracks.

- Walking in streams.

- Splitting up into small groups.

- Walking along fallen trees or stepping from rock to rock.

- Covering tracks with leaves.

WARNING

A TRACKER SHOULD ALWAYS BE ALERT TO THE POSSIBILITY THAT THE ENEMY IS LEAVING FALSE SIGNS TO LEAD THE UNIT INTO AN AMBUSH.

MOVING ACROSS WATER OBSTACLES

Crossing Rivers and Streams. There are several expedient ways to cross rivers and streams. The ways used in any situation depends on the width and depth of the water, the speed of the current, the time and equipment available, and the friendly and enemy situation.

There is always a possibility of equipment failure. For this reason, you should be able to swim. In all water crossings several strong swimmers should be stationed either at the water's edge or, if possible, in midstream to help anyone who gets into trouble.

If you accidentally fall into the water, swim with the current to the nearer bank. Swimming against the current is dangerous because the swimmer is quickly exhausted by the force of the current.

Fording. A good site to ford a stream has these characteristics:

- Good concealment on both banks.
- Few large rocks in the river bed. (Submerged large rocks are usually slippery and make it difficult to maintain footing.)

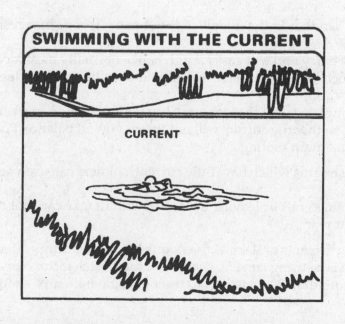

SWIMMING WITH THE CURRENT

CURRENT

SOLDIER CROSSING A STREAM

■ The standard air mattress

■ Trousers
NOTE: Trousers must be soaked in water before using.

■ Canteen safety belt

■ Poncho life belt

■ Water wings

■ Poncho brush raft

■ Australian poncho raft

■ Log rafts

- Shallow water or a sandbar in the middle of the stream. Troops may rest or regain their footing on these sandbars.
- Low banks to make entry and exit easier. High banks normally mean deep water. Deep water near the far shore is especially dangerous as the soldiers may be tired and less able to get out.

You should cross at an angle against the current. Keep your feet wide apart and drag your legs through the water, do not lift them, so that the current will not throw you off balance. Poles can be used to probe to help find deep holes and maintain footing.

Floating Aids. For deeper streams which have little current, soldiers can use a number of floating aids such as the following:

When launching any poncho raft or leaving the water with it, take care not to drag it on the ground as this will cause punctures or tears.

Rope Bridges. For crossing streams and small rivers quickly, rope bridges offer a suitable temporary system, especially when there is a strong current. Because of the stretch factor of nylon ropes, they should not be used to cross gaps of more than 20 meters. For larger gaps, manila rope should be used.

TROUSERS AS A FLOATING AID

Prepare the trousers by tying the bottoms of the legs tightly.

Enter the water to waist depth and hold the trousers behind the shoulders, with the waist open.

Bring the trousers quickly over the head and bang them onto the water in front. This action fills them with air.

Squeeze and hold the waist together, lie over the trousers, and float as if on water wings.

CANTEEN SAFETY BELT

Attach at least eight empty plastic canteens to a pistol belt (or tie them to a rope which can then be used as a belt). Insure that the caps are screwed on tightly.

PONCHO LIFE BELT

Roll green vegetation tightly inside a poncho and fold the ends over to make a watertight life belt. Roll up the life belt like a big sausage at least 8 inches in diameter and tie it. Wear it around the waist or across one shoulder and under the opposite arm like a bandoleer.

WATER WINGS

Two or more air-filled plastic bags, securely tied at the mouth, can be used as expedient water wings. Other expedients include empty water or fuel cans and ammunition canisters.

CONSTRUCTION OF PONCHO BRUSH RAFT

Use two ponchos, and tie the neck of each tightly by using the drawstring.

Spread one poncho on the ground with the hood up so that it will end up inside the raft.

Cut fresh, green brush (avoid thick branches or wood stakes) and pile it on the poncho to a height of 18 inches.

Place an X-frame made of small saplings (1 to 1 1/2 inches in diameter and 3 to 4 feet long) on the brush. Anchor this frame by tying the drawstring of the poncho to the center of the X-frame.

Pile another 18 inches of brush on top of the X-frame.

Compress the brush slightly and fold up the poncho, tying ropes or vines diagonally across the corner grommets and straight across from side grommets. The sides of the poncho should not touch.

Spread the second poncho on the ground, with hood up, next to the bundle made of the first poncho and brush. Roll the bundle over onto the center of the second poncho and tie the second poncho across the sides and diagonally across the corners. This raft will safely float 250 pounds and is very stable.

DRAWSTRING TIED TO X-FRAME

X-FRAME ON BRUSH   PONCHO TIED OVER BRUSH   COMPLETED RAFT

CONSTRUCTION OF AUSTRALIAN PONCHO RAFT

When there is not enough time to gather a lot of brush, this raft is made by using a soldier's combat equipment for bulk. Normally, two soldiers make this poncho together. It is more waterproof than the poncho brush raft but will float only about 80 pounds of weight. Two soldiers make this raft as follows.

Place one poncho on the ground with the hood facing up. Close the neck opening by tying it off with the drawstring.

Place two poles (or branches), about 1 to 1 1/2 inches in diameter and 4 feet long, in the center of the poncho about 18 inches apart.

Next, place the rucksack, and any other equipment desired, between the poles.

EQUIPMENT ON PONCHO

Snap the poncho together. Hold the snapped portion of the poncho in the air and roll it tightly down toward the equipment. Roll from the center out to both ends. At the ends, twist the poncho to form "pigtails." Fold the pigtails inward toward each other and tie them tightly together with boot laces, vines, communication wire, or other available tying material.

PONCHO ROLLED WITH PIGTAILS

Spread the second poncho on the ground, neck closed and facing up.

Place the equipment bundle formed with the first poncho, with the seam (tied pigtails) facing down, on the second poncho.

Roll and tie the second poncho in the same way as the first.

SECOND PONCHO BEING ROLLED

An empty canteen tied to one end of a rope with the other end tied to the raft helps in towing. One soldier pulls on the rope while the other pushes the raft. Place weapons on top of the raft and secure them with ropes. The weapons should be secured to the raft by the use of quick releases. The raft is now ready for the water.

SOLDIERS MOVING RAFT IN WATER

CONSTRUCTION OF LOG RAFT

Logs, either singly or lashed together, can be used to float soldiers and equipment. Be careful when selecting logs for rafts. Some jungle trees will not float. To see whether certain wood is suitable, put a wood chip from a tree in the water. If the chip sinks, so will a raft made of that wood.

DETERMINATION OF RIVER WIDTH

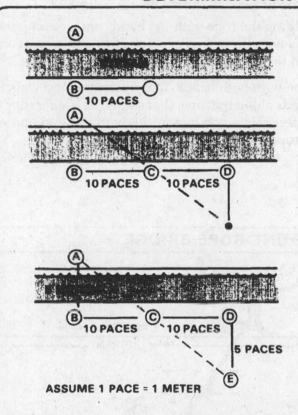

ASSUME 1 PACE = 1 METER

A method used to measure the width of a river or stream is described below.

Select a straight section of the stream.

Pick two points opposite each other (A and B).

Stand at B; turn in a direction parallel to the stream; walk off 10 paces. Mark that point as point C (B to C = 10 paces).

Continue walking in the same direction 10 more paces. Mark that point as point D (C to D = 10 paces).

Turn at a right angle away from the stream and walk until you are on line with points C and A. Mark this point as point E. Determine the distance between D and E by converting the pace count into meters. In this example, 1 pace is equal to 1 meter and the pace count is 5 paces. Therefore, the distance between D and E is 5 meters.

The distance from D to E is equal to the distance from A to B. Therefore, the width of the stream is also about 5 meters.

SWIMMER PULLING ROPE ACROSS

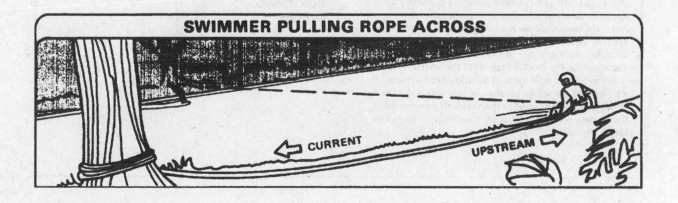

In order to erect a rope bridge, the first thing to be done is to get one end of the rope across the stream. This task can be frustrating when there is a strong current. To get the rope across, anchor one end of a rope that is at least double the width of the stream at point A. Take the other end of the line upstream as far as it will go. Then, tie a sling rope around the waist of a strong swimmer and, using a snaplink, attach the line to him. He should swim diagonally downstream to the far bank, pulling the rope across.

One-Rope Bridge. A one-rope bridge can be constructed either above water level or at water level. The leader must decide which to construct. The bridge is constructed the same regardless of the level.

Crossing Method above Water Level. Use one of the following methods.

Commando crawl. Lie on the top of the rope with the instep of the right foot hooked on the rope. Let the left leg hang to maintain balance. Pull across with the hands and arms, at the same time pushing on the rope with the right foot. (For safety, tie a rappel seat and hooks the snaplink to the rope bridge.)

Monkey crawl. Hang suspended below the rope, holding the rope with the hands and crossing the knees over the top of the rope. Pull with the hands and push with the legs. (For safety, tie a rappel seat and hook the snaplink to the rope bridge.) This is the safest and the best way to cross the one-rope bridge.

Crossing Method at Water Level. Hold onto the rope with both hands, face upstream, and walk into the water. Cross the bridge by sliding and pulling the hands along the rope. (For safety, tie a sling rope around your waist, leaving a working end of about 3 to 4 feet. Tie a bowline in the working end and attach a snaplink to the loop. Then, hook the snaplink to the rope bridge.)

To recover the rope, the last person unties the rope, ties it around his waist and, after all slack is taken up, is pulled across.

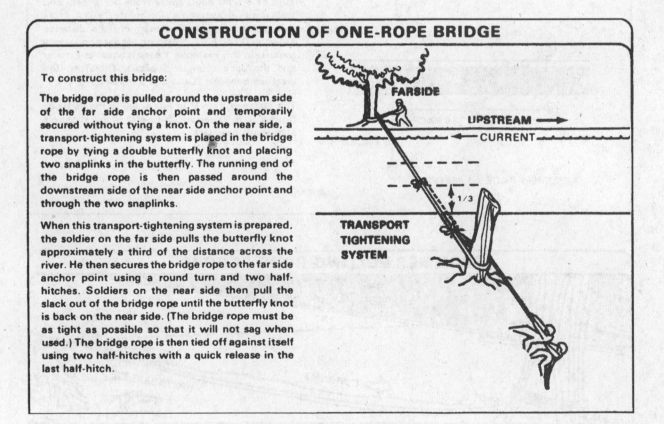

CONSTRUCTION OF ONE-ROPE BRIDGE

To construct this bridge:

The bridge rope is pulled around the upstream side of the far side anchor point and temporarily secured without tying a knot. On the near side, a transport-tightening system is placed in the bridge rope by tying a double butterfly knot and placing two snaplinks in the butterfly. The running end of the bridge rope is then passed around the downstream side of the near side anchor point and through the two snaplinks.

When this transport-tightening system is prepared, the soldier on the far side pulls the butterfly knot approximately a third of the distance across the river. He then secures the bridge rope to the far side anchor point using a round turn and two half-hitches. Soldiers on the near side then pull the slack out of the bridge rope until the butterfly knot is back on the near side. (The bridge rope must be as tight as possible so that it will not sag when used.) The bridge rope is then tied off against itself using two half-hitches with a quick release in the last half-hitch.

FARSIDE

UPSTREAM →

← CURRENT

1/3

TRANSPORT
TIGHTENING
SYSTEM

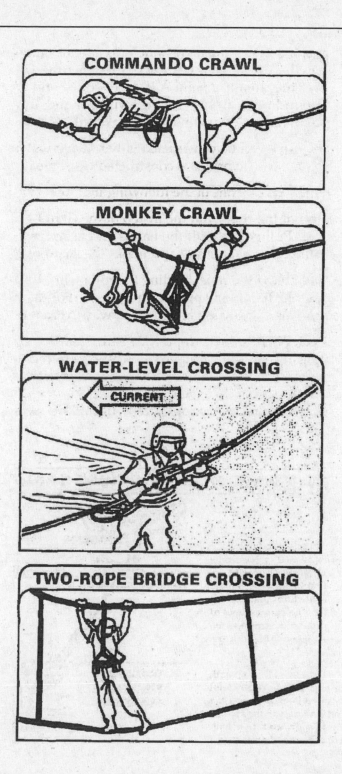

Two-rope bridge. Construction of this bridge is similar to that of the one-rope bridge, except two ropes, a hand rope, and a foot rope are used. These ropes are spaced about 1.5 meters apart vertically at the anchor points. (For added safety, make snaplink attachments to the hand and foot ropes from a rope tied around the waist. Move across the bridge using the snaplink to allow the safety rope to slide.) To keep the ropes a uniform distance apart as men cross, spreader ropes should be tied between the two ropes every 15 feet. A sling rope is used and tied to each bridge rope with a round turn and two half-hitches.

CHAPTER 2

Desert Survival

To survive and evade in arid or desert areas, you must understand and prepare for the environment you will face. You must determine your equipment needs, the tactics you will use, and how the environment will affect you and your tactics. Your survival will depend upon your knowledge of the terrain, basic climatic elements, your ability to cope with these elements, and your will to survive.

Desert terrain also varies considerably from place to place, the sole common denominator being lack of water with its consequent environmental effects, such as sparse, if any, vegetation. The basic land forms are similar to those in other parts of the world, but the topsoil has been eroded due to a combination of lack of water, heat, and wind to give deserts their characteristic barren appearance. The bedrock may be covered by a flat layer of sand, or gravel, or may have been exposed by erosion. Other common features are sand dunes, escarpments, wadis, and depressions.

It is important to realize that deserts are affected by seasons. Those in the Southern Hemisphere have summer between 21 December and 21 March. This 6-month difference from the United States makes acclimatization more difficult for persons coming from winter conditions.

TERRAIN

Key terrain in the desert is largely dependent on the restrictions to movement that are present. If the desert floor will not support wheeled vehicle traffic, the few roads and desert tracks become key terrain.

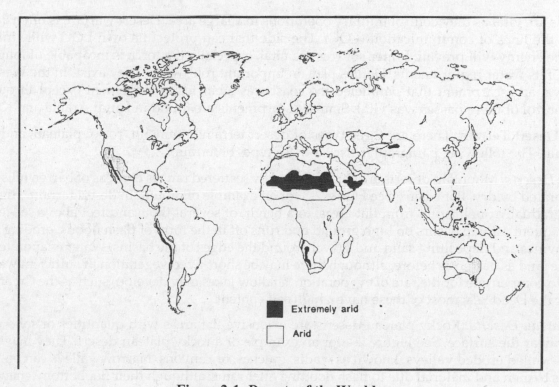

Figure 2-1: Deserts of the World.

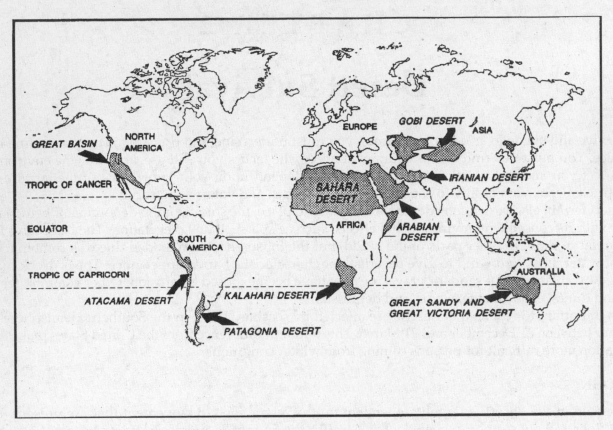

Figure 2-2: Desert locations of the world.

Crossroads are vital as they control military operations in a large area. Desert warfare is often a battle for control of the lines of communication (LOC). The side that can protect its own LOC while interdicting those of the enemy will prevail. Water sources are vital, especially if a force is incapable of long distance resupply of its water requirements. Defiles play an important role, where they exist. In the Western Desert of Libya, an escarpment that paralleled the coast was a barrier to movement except through a few passes. Control of these passes was vital. Similar escarpments are found in Saudi Arabia and Kuwait.

Types of Desert Terrain. There are three types of desert terrain: mountain, rocky plateau, and sandy or dune terrain. The following paragraphs discuss these types of terrain.

Mountain Deserts. Mountain deserts are characterized by scattered ranges or areas of barren hills or mountains, separated by dry, flat basins. See Figure 2-3 for an example of mountain desert terrain. High ground may rise gradually or abruptly from flat areas, to a height of several thousand feet above sea level. Most of the infrequent rainfall occurs on high ground and runs off in the form of flash floods, eroding deep gullies and ravines and depositing sand and gravel around the edges of the basins. Water evaporates rapidly, leaving the land as barren as before, although there may be short-lived vegetation. If sufficient water enters the basin to compensate for the rate of evaporation, shallow lakes may develop, such as the Great Salt Lake in Utah or the Dead Sea; most of these have a high salt content.

Rocky Plateau Deserts. Rocky plateau deserts are extensive flat areas with quantities of solid or broken rock at or near the surface. See Figure 2-4 for an example of a rocky plateau desert. They may be wet or dry, steep-walled eroded valleys, known as wadis, gulches, or canyons. Narrow valleys can be extremely dangerous to men and material due to flash flooding after rains; although their flat bottoms may be superficially attractive as assembly areas. The National Training Center and the Golan Heights are examples of rocky plateau deserts.

Sandy or Dune Deserts. Sandy or dune deserts are extensive flat areas covered with sand or gravel, the product of ancient deposits or modern wind erosion. "Flat" is relative in this case, as some areas may contain sand dunes that are over 1,000 feet high and 10-15 miles long; trafficability on this type of terrain will depend on windward/leeward gradients of the dunes and the texture of the sand. See Figure 2-5 for an example of a sandy desert. Other areas, however, may be totally flat for distances of 3,000 meters and beyond. Plant life may vary from none to scrub, reaching over 6 feet high. Examples of this type of desert include the ergs of the Sahara, the Empty Quarter of the Arabian desert, areas of California and New Mexico, and the Kalahari in South Africa. See Figure 2-6 for an example of a dune desert.

Salt Marshes. Salt marshes are flat, desolate areas, sometimes studded with clumps of grass but devoid of other vegetation. They occur in arid areas where rainwater has collected, evaporated, and left large deposits of alkali salts and water with a high salt concentration. The water is so salty it is undrinkable. A crust that may be 2.5 to 30 centimeters thick forms over the saltwater.

In arid areas there are salt marshes hundreds of kilometers square. These areas usually support many insects, most of which bite. Avoid salt marshes. This type of terrain is highly corrosive to boots, clothing, and skin. A good example is the Shat-el-Arab waterway along the Iran-Iraq border.

Trafficability. Roads and trails are rare in the open desert. Complex road systems beyond simple commercial links are not needed. Road systems have been used for centuries to connect centers of commerce, or important religious shrines such as Mecca and Medina in Saudi Arabia. These road systems are supplemented by routes joining oil or other mineral deposits to collection outlet points. Some surfaces, such as lava beds or salt marshes, preclude any form of routine vehicular movement, but generally ground movement is possible in all directions. Speed of movement varies depending on surface texture. Rudimentary trails are used by minor caravans and nomadic tribesmen, with wells or oases approximately every 20 to 40 miles; although there are some waterless stretches which extend over 100 miles. Trails vary in width from a few meters to over 800 meters.

Vehicle travel in mountainous desert country may be severely restricted. Available routes can be easily blinked by the enemy or by climatic conditions. Hairpin turns are common on the edges of precipitous mountain gorges, and the higher passes may be blocked by snow in the winter.

Natural Factors. The following terrain features require special considerations regarding trafficability.

Figure 2-3: Example of mountain desert terrain.

Figure 2-4: Example of rocky plateau desert terrian.

Wadis or dried water courses vary from wide, but barely perceptible depressions of soft sand, dotted with bushes, to deep, steep-sided ravines. There frequently is a passable route through the bottom of a dried wadi. Wadis can provide cover from ground observation and camouflage from visual air reconnaissance. The threat of flash floods after heavy rains poses a significant danger to troops and equipment downstream. Flooding may occur in these areas even if it is not raining in the immediate area. See Figure 2-7 for an example of a wadi.

Salt marsh (sebkha) terrain is impassable to tracks and wheels when wet. When dry it has a brittle, crusty surface, negotiable by light wheel vehicles only. Salt marshes develop at points where the water in the subsoil of the desert rose to the surface. Because of the constant evaporation in the desert, the salts carried by the water are deposited, and results in a hard, brittle crust.

Salt marshes are normally impassable, the worst type being those with a dry crust of silt on top. Marsh mud used on desert sand will, however, produce an excellent temporary road. Many desert areas have salt marshes either in the center of a drainage basin or near the sea coast. Old trails or paths may cross

Figure 2-5: Example of sandy desert terrian.

Figure 2-6: Example of dune desert terrian.

the marsh, which are visible during the dry season but not in the wet season. In the wet season trails are indicated by standing water due to the crust being too hard or too thick for it to penetrate. However, such routes should not be tried by load-carrying vehicles without prior reconnaissance and marking. Vehicles may become mired so severely as to render equipment and units combat ineffective. Heavier track-laying vehicles, like tanks, are especially susceptible to these areas, therefore reconnaissance is critical.

Man-made Factors. The ruins of earlier civilizations, scattered across the deserts of the world, often are sited along important avenues of approach and frequently dominate the only available passes in difficult terrain. Control of these positions may be imperative for any force intending to dominate the immediate area. Currently occupied dwellings have little impact on trafficability except that they are normally located near roads and trails. Apart from nomadic tribesmen who live in tents (see Figure 2-8 for an example of desert nomads), the population lives in thick-walled structures with small windows, usually built of masonry or a mud and straw (adobe) mixture. Figure 2-9 shows common man-made desert structures.

Because of exploration for and production of oil and other resources, wells, pipelines, refineries, quarries, and crushing plants may be of strategic importance in the desert. Pipelines are often raised 1 meter off the ground—where this is the case, pipelines will inhibit movement. Subsurface pipelines can also be

Figure 2-7: Example of a wadi.

Figure 2-8: Example of desert nomads.

an obstacle. In Southwest Asia, the subsurface pipelines were indicated on maps. Often they were buried at such a shallow depth that they could be damaged by heavy vehicles traversing them. Furthermore, if a pipeline is ruptured, not only is the spill of oil a consideration, but the fumes may be hazardous as well.

Agriculture in desert areas has little effect on trafficability except that canals limit surface mobility. Destruction of an irrigation system, which may be a result of military operations, could have a devastating effect on the local population and should be an important consideration in operational estimates. Figure 2-10 shows an irrigation ditch.

TEMPERATURE

The highest known ambient temperature recorded in a desert was 136 degrees Fahrenheit (58 degrees Celsius). Lower temperatures than this produced internal tank temperatures approaching 160 degrees

Figure 2-9: Common man-made desert structures.

Figure 2-10: Irrigation ditch.

Fahrenheit (71 degrees Celsius) in the Sahara Desert during the Second World War. Winter temperatures in Siberian deserts and in the Gobi reach minus 50 degrees Fahrenheit (minus 45 degrees Celsius). Low temperatures are aggravated by very strong winds producing high windchill factors. The cloudless sky of the desert permits the earth to heat during sunlit hours, yet cool to near freezing at night. In the inland Sinai, for example, day-to-night temperature fluctuations are as much as 72 degrees Fahrenheit.

Winds. Desert winds can achieve velocities of near hurricane force; dust and sand suspended within them make life intolerable, maintenance very difficult, and restrict visibility to a few meters. The Sahara "Khamseen", for example, lasts for days at a time; although it normally only occurs in the spring and summer. The deserts of Iran are equally well known for the "wind of 120 days," with sand blowing almost constantly from the north at wind velocities of up to 75 miles per hour.

Although there is no danger of a man being buried alive by a sandstorm, individuals can become separated from their units. In all deserts, rapid temperature changes invariably follow strong winds. Even without wind, the telltale clouds raised by wheels, tracks, and marching troops give away movement. Wind aggravates the problem. As the day gets warmer the wind increases and the dust signatures of vehicles may drift downwind for several hundred meters.

In the evening the wind normally settles down. In many deserts a prevailing wind blows steadily from one cardinal direction for most of the year, and eventually switches to another direction for the remaining months. The equinoctial gales raise huge sandstorms that rise to several thousand feet and may last for several days. Gales and sandstorms in the winter months can be bitterly cold. See Figure 2-11 for an example of wind erosion.

Mirages. Mirages are optical phenomena caused by the refraction of light through heated air rising from a sandy or stony surface. They occur in the interior of the desert about 10 kilometers from the coast. They make objects that are 1.5 kilometers or more away appear to move.

This mirage effect makes it difficult for you to identify an object from a distance. It also blurs distant range contours so much that you feel surrounded by a sheet of water from which elevations stand out as "islands."

The mirage effect makes it hard for a person to identify targets, estimate range, and see objects clearly. However, if you can get to high ground (3 meters or more above the desert floor), you can get above the superheated air close to the ground and overcome the mirage effect. Mirages make land navigation difficult because they obscure natural features. You can survey the area at dawn, dusk, or by moonlight when there is little likelihood of mirage.

Figure 2-11: Example of wind erosion.

Light levels in desert areas are more intense than in other geographic areas. Moonlit nights are usually crystal clear, winds die down, haze and glare disappear, and visibility is excellent. You can see lights, red flashlights, and blackout lights at great distances. Sound carries very far.

Conversely, during nights with little moonlight, visibility is extremely poor. Traveling is extremely hazardous. You must avoid getting lost, falling into ravines, or stumbling into enemy positions. Movement during such a night is practical only if you have a compass and have spent the day in a shelter, resting, observing and memorizing the terrain, and selecting your route.

Sandstorms are likely to form suddenly and stop just as suddenly. In a severe sandstorm, sand permeates everything making movement nearly impossible, not only because of limited visibility, but also because blowing sand damages moving parts of machinery.

Water. The lack of water is the most important single characteristic of the desert. The population, if any, varies directly with local water supply. A Sahara oasis may, for its size, be one of the most densely occupied places on earth (see Figure 2-12 for a typical oasis).

Desert rainfall varies from one day in the year to intermittent showers throughout the winter. Severe thunderstorms bring heavy rain, and usually far too much rain falls far too quickly to organize collection on a systematic basis. The water soon soaks into the ground and may result in flash floods. In some cases the rain binds the sand much like a beach after the tide ebbs allowing easy maneuver. However, it also turns loam into an impassable quagmire obstacle. Rainstorms tend to be localized, affecting only a few square kilometers at a time. Whenever possible, as storms approach, vehicles should move to rocky areas or high ground to avoid flash floods and becoming mired.

Permanent rivers such as the Nile, the Colorado, or the Kuiseb in the Namib Desert of Southwest Africa are fed by heavy precipitation outside the desert so the river survives despite a high evaporation rate.

Subsurface water may be so far below the surface, or so limited, that wells are normally inadequate to support any great number of people. Because potable water is absolutely vital, a large natural supply may be both tactically and strategically important. Destruction of a water supply system may become a political rather than military decision, because of its lasting effects on the resident civilian population.

Finding Water. When there is no surface water, tap into the earth's water table for ground water. Access to this table and its supply of generally pure water depends on the contour of the land and the type of soil. See Figure 2-13 for water tables.

From Rocky Soil. Look for springs and seepages. Limestone has more and larger springs than any other type rock. Because limestone is easily dissolved, caverns are readily etched in it by ground water. Look in these caverns for springs. Lava rock is a good source of seeping ground water because it is porous. Look for springs along the walls of valleys that cross the lava flow. Look for seepage where a dry canyon cuts through a layer of porous sandstone.

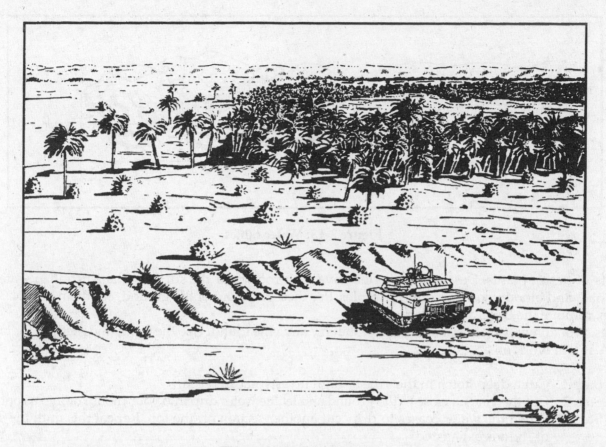

Figure 2-12: Typical oasis.

Watch for water indicators in desert environments. Some signs to look for are the direction in which certain birds fly, the location of plants, and the convergence of game trails. Asian sand grouse, crested larks, and zebra birds visit water holes at least once a day. Parrots and pigeons must live within reach of water. Cattails, greasewoods, willows, elderberry, rushes, and salt grass grow only where ground water is near the surface. Look for these signs and dig. If you do not have a bayonet or entrenching tool, dig with a flat rock or sharp stick.

Desert natives often know of lingering surface pools in low places. They cover their surface pools, so look under brush heaps or in sheltered nooks, especially in semiarid and brush country.

Places that are visibly damp, where animals have scratched, or where flies hover, indicate recent surface water. Dig in such places for water. Collect dew on clear nights by sponging it up with a handkerchief. During a heavy dew you should be able to collect about a pint an hour.

Dig in dry stream beds because water may be found under the gravel. When in snow fields, put in a water container and place it in the sun out of the wind.

From Plants. If unsuccessful in your search for ground or runoff water, or if you do not have time to purify the questionable water, a water-yielding plant may be the best source. Clear sap from many plants is easily obtained. This sap is pure and is mostly water.

Plant tissues. Many plants with fleshy leaves or stems store drinkable water. Try them wherever you find them. The barrel cactus of the southwestern United States is a possible source of water (see Figure 2-14). Use it only as a last resort and only if you have the energy to cut through the tough, spine-studded outer rind. Cut off the top of the cactus and smash the pulp within the plant. Catch the liquid in a container. Chunks may be carried as an emergency water source. A barrel cactus 3-1/2 feet high will yield about a quart of milky juice and is an exception to the rule that milky or colored sap-bearing plants should not be eaten.

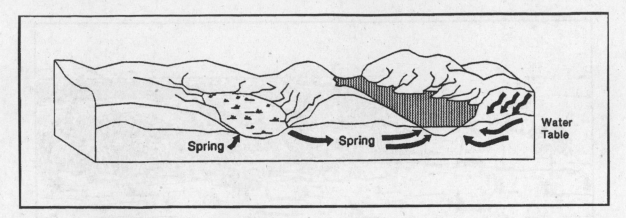

Figure 2-13: Water tables.

Roots of desert plants. Desert plants often have their roots near the surface. The Australian water tree, desert oak, and bloodwood are some examples. Pry these roots out of the ground, cut them into 24-36 inch lengths, remove the bark, and suck the water.

Vines. Not all vines yield palatable water, but try any vine found. Use the following method for tapping a vine—it will work on any species:

Step 1. Cut a deep notch in the vine as high up as you can reach.
Step 2. Cut the vine off close to the ground and let the water drip into your mouth or into a container.
Step 3. When the water ceases to drip, cut another section off the top. Repeat this until the supply of fluid is exhausted.

Palms. Burl, coconut, sugar and nipa palms contain a drinkable sugary fluid. To start the flow in coconut palm, bend the flower stalk downward and cut off the top. If a thin slice is cut off the stalk every 12 hours, you can renew the flow and collect up to a quart a day.

Interior of barrel cactus
--- watery pulp

Figure 2-14: Barrel cactus as possible source of water.

Coconut. Select green coconuts. They can be opened easily with a knife and they have more milk than ripe coconuts. The juice of a ripe coconut is extremely laxative; therefore, do not drink more than three or four cups a day.

The milk of a coconut can be obtained by piercing two eyes of the coconut with a sharp object such as a stick or a nail. To break off the outer fibrous covering of the coconut without a knife, slam the coconut forcefully on the point of a rock or protruding stump.

Survival Water Still. You can build a cheap and simple survival still that will produce drinking water in a dry desert. Basic materials for setting up this still are—

- 6-foot square sheet of clean plastic.
- A 2- to 4-quart capacity container.
- A 5-foot piece of flexible plastic tubing.

Pick an unshaded spot for the still, and dig a hole. If no shovel is available, use a stick or even your hands. The hole should be about 3 feet across for a few inches down, then slope the hole toward the bottom as shown in Figure 2-15 which depicts a cross section of a survival still. The hole should be deep enough so the point of the plastic cone will be about 18 inches below ground and will still clear the top of the container. Once the hole is properly dug, tape one end of the plastic drinking tube inside the container and center the container in the bottom of the hole. Leave the top end of the drinking tube free, lay the plastic sheet over the hole, and pile enough dirt around the edge of the plastic to hold it securely. Use a fist-size rock to weight down the center of the plastic; adjust the plastic as necessary to bring it within a couple of inches of the top of the container. Heat from the sun vaporizes the ground water. This vapor condenses under the plastic, trickles down, and drops into the container.

Vegetation. The indigenous vegetation and wildlife of a desert have physiologically adapted to the conditions of the desert environment. For example, the cacti of the American desert store moisture in enlarged stems. Some plants have drought-resistant seeds that may lie dormant for years, followed by a brief, but colorful display of growth after a rainstorm. The available vegetation is usually inadequate to provide

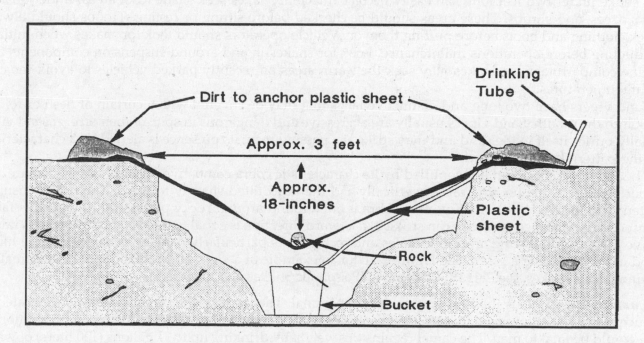

Figure 2-15: Cross-section of a survival still.

much shade, shelter, or concealment, especially from the air. Some plants, like the desert gourd, have vines which grow to 4.5 meters (15 feet). Others have wide lateral roots just below the surface to take advantage of rain and dew, while still others grow deep roots to tap subsurface water. Presence of palm trees usually indicates water within a meter of the surface, salt grass within 2 meters, cottonwood and willows up to 4 meters. In addition to indicating the presence of water, some plants are edible. For more information on edible plants, see Part IV, Chapter 6 "Survival Use of Plants."

Wildlife. Invertebrates such as ground-dwelling spiders, scorpions, and centipedes, together with insects of almost every type, are in the desert. Drawn to man as a source of moisture or food, lice, mites, and flies can be extremely unpleasant and carry diseases such as scrub typhus and dysentery. The stings of scorpions and the bites of centipedes and spiders are extremely painful, though seldom fatal. Some species of scorpion, as well as black widow and recluse spiders, can cause death. The following paragraphs describe some of the wildlife that are encountered in desert areas and the hazards they may pose to man.

Scorpions. Scorpions are prevalent in desert regions. They prefer damp locations and are particularly active at night. Scorpions are easily recognizable by their crab-like appearance, and by their long tail which ends in a sharp stinger. Adult scorpions vary from less than an inch to almost 8 inches in length. Colors range from nearly black to straw to striped. Scorpions hide in clothing, boots, or bedding, so troops should routinely shake these items before using. Although scorpion stings are rarely fatal, they can be painful.

Flies. Flies are abundant throughout desert environments. Filth-borne disease is a major health problem posed by flies. Dirt or insects in the desert can cause infection in minor cuts and scratches.

Fleas. Avoid all dogs and rats which are the major carriers of fleas. Fleas are the primary carriers of plague and murine typhus. For more information on dangerous insects, see Part IV, Chapter 3, "Dangerous Insects and Arachnids."

Reptiles. Reptiles are perhaps the most characteristic group of desert animals. Lizards and snakes occur in quantity, and crocodiles are common in some desert rivers. Lizards are normally harmless and can be ignored; although exceptions occur in North America and Saudi Arabia.

Snakes, ranging from the totally harmless to the lethal, abound in the desert. A bite from a poisonous snake under two feet long can easily become infected. Snakes seek shade (cool areas) under bushes, rocks, trees, and shrubs. These areas should be checked before sitting or resting. Troops should always check clothing and boots before putting them on. Vehicle operators should look for snakes when initially conducting before-operations maintenance. Look for snakes in and around suspension components and engine compartments as snakes may seek the warm areas on recently parked vehicles to avoid the cool night temperatures.

Sand vipers have two long and distinctive fangs that may be covered with a curtain of flesh or folded back into the mouth. Sand vipers usually are aggressive and dangerous in spite of their size. A sand viper usually buries itself in the sand and may strike at a passing man; its presence is alerted by a characteristic coiling pattern left on the sand.

The Egyptian cobra can be identified by its characteristic cobra combative posture. In this posture, the upper portion of the body is raised vertically and the head tilted sharply forward. The neck is usually flattened to form a hood. The Egyptian cobra is often found around rocky places and ruins and is fairly common. The distance the cobra can strike in a forward direction is equal to the distance the head is raised above the ground. Poking around in holes and rock piles is particularly dangerous because of the likelihood of encountering a cobra. See Figure 2-16 for an example of a viper and cobra. For more information on dangerous snakes, see Part IV, Chapter 4, "Poisonous Snakes and Lizards."

Mammals. The camel is the best known desert mammal. The urine of the camel is very concentrated to reduce water loss, allowing it to lose 30 percent of its body weight without undue distress. A proportionate loss would be fatal to man. The camel regains this weight by drinking up to 27 gallons (120 liters) of water at a time. It cannot, however, live indefinitely without water and will die of dehydration as readily as man in equivalent circumstances. Other mammals, such as gazelles, obtain most of their required water supply from the vegetation they eat and live in areas where there is no open water. Smaller animals, including rodents, conserve their moisture by burrowing underground away from the direct heat of the sun, only emerging for foraging at night. All these living things have adapted to the environment over a period of thousands of years; however, man has not made this adaptation and must carry his food and water with him and must also adapt to this severe environment.

Dogs are often found near mess facilities and tend to be in packs of 8 or 10. Dogs are carriers of rabies and should be avoided. Commanders must decide how to deal with packs of dogs; extermination and avoidance are two options. Dogs also carry fleas which may be transferred upon bodily contact. Rabies is present in most desert mammal populations. Do not take any chances of contracting fleas or rabies from any animal by adopting pets.

Rats are carriers of various parasites and gastrointestinal diseases due to their presence in unsanitary locations.

There is no reason to fear the desert environment, and it should not adversely affect the morale of a soldier/marine who is prepared for it. Lack of natural concealment has been known to induce temporary agoraphobia (fear of open spaces) in some troops new to desert conditions, but this fear normally disappears with acclimatization. Remember that there is nothing unique about either living or fighting in deserts; native tribesmen have lived in the Sahara for thousands of years. The British maintained a field army and won a campaign in the Western Desert in World War II at the far end of a 12,000-mile sea line of communication with equipment considerably inferior to that in service now. The desert is neutral, and affects both sides equally; the side whose personnel are best prepared for desert operations has a distinct advantage.

The desert is fatiguing, both physically and mentally. A high standard of discipline is essential, as a single individual's lapse may cause serious damage to his unit or to himself. Commanders must exercise a high level of leadership and train their subordinate leaders to assume greater responsibilities required by the wide dispersion of units common in desert warfare. Soldiers/marines with good leaders are more apt to accept heavy physical exertion and uncomfortable conditions. Every soldier/marine must clearly understand why he is fighting in such harsh conditions and should be kept informed of the operational situation. Ultimately, however, the maintenance of discipline will depend on individual training.

Commanders must pay special attention to the welfare of troops operating in the desert, as troops are unable to find any "comforts" except those provided by the command. Welfare is an essential factor in the maintenance of morale in a harsh environment, especially to the inexperienced. There is more to welfare

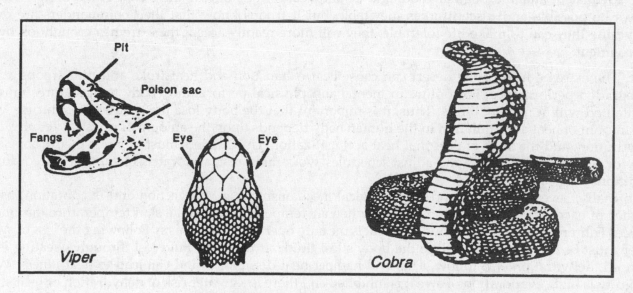

Figure 2-16: Sand viper and cobra.

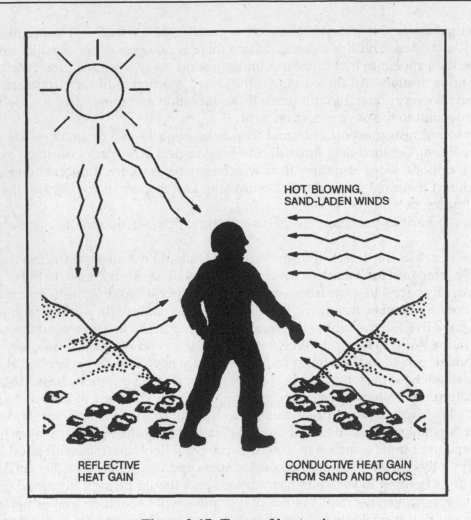

HOT, BLOWING,
SAND-LADEN WINDS

REFLECTIVE
HEAT GAIN

CONDUCTIVE HEAT GAIN
FROM SAND AND ROCKS

Figure 2-17: Types of heat gain.

than the provision of mail and clean clothing. Troops must be kept healthy and physically fit; they must have adequate, palatable, regular food, and be allowed periods of rest and sleep. These things will not always be possible and discomfort is inevitable, but if troops know that their commanders are doing everything they can to make life tolerable, they will more readily accept the extremes brought on by the environment.

Heat. The extreme heat of the desert can cause heat exhaustion and heatstroke and puts troops at risk of degraded performance. For optimum mental and physical performance, body temperatures must be maintained within narrow limits. Thus, it is important that the body lose the heat it gains during work. The amount of heat accumulation in the human body depends upon the amount of physical activity, level of hydration, and the state of personal heat acclimatization. Unit leaders must monitor their troops carefully for signs of heat distress and adjust schedules, work rates, rest, and water consumption according to conditions.

Normally, several physical and physiological mechanisms (e.g., convection and evaporation) assure transfer of excess body heat to the air. But when air temperature is above skin temperature (around 92 degrees Fahrenheit) the evaporation of sweat is the only operative mechanism. Following the loss of sweat, water must be consumed to replace the body's lost fluids. If the body fluid lost through sweating is not replaced, dehydration will follow. This will hamper heat dissipation and can lead to heat illness. When humidity is high, evaporation of sweat is inhibited and there is a greater risk of dehydration or heat stress. Consider the following to help prevent dehydration:

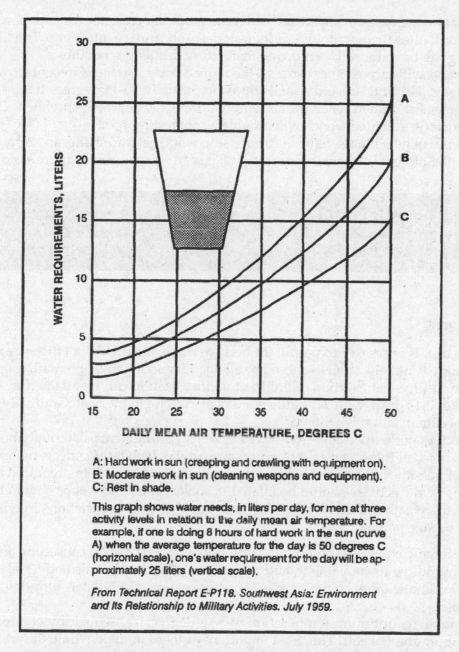

A: Hard work in sun (creeping and crawling with equipment on).
B: Moderate work in sun (cleaning weapons and equipment).
C: Rest in shade.

This graph shows water needs, in liters per day, for men at three activity levels in relation to the daily mean air temperature. For example, if one is doing 8 hours of hard work in the sun (curve A) when the average temperature for the day is 50 degrees C (horizontal scale), one's water requirement for the day will be approximately 25 liters (vertical scale).

From Technical Report E-P118. Southwest Asia: Environment and Its Relationship to Military Activities. July 1959.

Figure 2-18: Daily water requirements for three levels of activity.

- Heat, wind, and dry air combine to produce a higher individual water requirement, primarily through loss of body water as sweat. Sweat rates can be high even when the skin looks and feels dry.
- Dehydration nullifies the benefits of heat acclimatization and physical fitness, it increases the susceptibility to heat injury, reduces the capacity to work, and decreases appetite and alertness. A lack of alertness can indicate early stages of dehydration.
- Thirst is not an adequate indicator of dehydration. The soldier/marine will not sense when he is dehydrated and will fail to replace body water losses, even when drinking water is available. The universal experience in the desert is that troops exhibit "voluntary dehydration" that is, they maintain their hydration status at about 2 percent of body weight (1.5 quarts) below their ideal hydration status without any sense of thirst.

Chronic dehydration increases the incidence of several medical problems: constipation (already an issue in any field situation), piles (hemorrhoids), kidney stones, and urinary infections. The likelihood of these problems occurring can be reduced by enforcing mandatory drinking schedules.

Resting on hot sand will increase heat stress—the more a body surface is in contact with the sand, the greater the heat stress. Ground or sand in full sun is hot, usually 30-45 degrees hotter than the air, and may reach 150 degrees Fahrenheit when the air temperature is 120 degrees Fahrenheit. Cooler sand is just inches below the surface; a shaded trench will provide a cool resting spot.

At the first evidence of heat illness, have the troops stop work, get into shade, and rehydrate. Early intervention is important. Soldiers/marines who are not taken care of can become more serious casualties.

WARNING

One heat casualty is usually followed by others and is a warning that the entire unit may be at risk. This is the "Weak Link Rule." The status of the whole unit is assessed at this point.

ACCLIMATIZATION

Acclimatization to heat is necessary to permit the body to reach and maintain efficiency in its cooling process. A period of approximately 2 weeks should be allowed for acclimatization, with progressive increases in heat exposure and physical exertion. Significant acclimatization can be attained in 4-5 days, but full acclimatization takes 7-14 days, with 2-3 hours per day of exercise in the heat. Gradually increase physical activity until full acclimatization is achieved.

Acclimatization does not reduce, and may increase, water requirements. Although this strengthens heat resistance, there is no such thing as total protection against the debilitating effects of heat. Situations may arise where it is not possible for men to become fully acclimatized before being required to do heavy labor. When this happens heavy activity should be limited to cooler hours and troops should be allowed to rest frequently. Check the weather daily. Day-to-day and region-to-region variations in temperatures, wind, and humidity can be substantial.

Climatic Stress. Climatic stress on the human body in hot deserts can be caused by any combination of air temperature, humidity, air movement, and radiant heat. The body is also adversely affected by such factors as lack of acclimatization, being overweight, dehydration, alcohol consumption, lack of sleep, old age, and poor health.

The body maintains its optimum temperature of 98.6 degrees Fahrenheit by conduction/convection, radiation, and evaporation (sweat). The most important of these in the daytime desert is evaporation, as air temperature alone is probably already above skin temperature. If, however, relative humidity is high, air will not easily evaporate sweat and the cooling effect is reduced. The following paragraphs describe the effects of radiant light, wind, and sand on personnel in desert areas.

Radiant Light. Radiant light comes from all directions. The sun's rays, either direct or reflected off the ground, affect the skin and can also produce eyestrain and temporarily impaired vision. Not only does glare damage the eyes but it is very tiring; therefore, dark glasses or goggles should be worn.

Overexposure to the sun can cause sunburn. Persons with fair skin, freckled skin, ruddy complexions, or red hair are more susceptible to sunburn than others, but all personnel are susceptible to some degree. Personnel with darker complexions can also sunburn. This is difficult to monitor due to skin pigmentation, so leaders must be ever vigilant to watch for possible sunburn victims. Sunburn is characterized by painful reddened skin, and can result in blistering and lead to other forms of heat illness.

Soldier/marines should acquire a suntan in gradual stages (preferably in the early morning or late afternoon) to gain some protection against sunburn. They should not be permitted to expose bare skin to the sun for longer than five minutes on the first day, increasing exposure gradually at the rate of five minutes per day. They should be fully clothed in loose garments in all operational situations. This will also reduce sweat loss. It is important to remember that—

- The sun is as dangerous on cloudy days as it is on sunny days.
- Sunburn ointment is not designed to give complete protection against excessive exposure.
- Sunbathing or dozing in the desert sun can be fatal.

Wind. The wind can be as physically demanding as the heat, burning the face, arms, and any exposed skin with blown sand. Sand gets into eyes, nose, mouth, throat, lungs, ears, and hair, and reaches every part of the body. Even speaking and listening can be difficult. Continual exposure to blown sand is exhausting and demoralizing. Technical work spaces that are protected from dust and sand are likely to be very hot. Work/rest cycles and enforced water consumption will be required.

The combination of wind and dust or sand can cause extreme irritation to mucous membranes, chap the lips and other exposed skin surfaces, and can cause nosebleed. Cracked, chapped lips make eating difficult and cause communication problems. Irritative conjunctivitis, caused when fine particles enter the eyes, is a frequent complaint of vehicle crews, even those wearing goggles. Lip balm and skin and eye ointments must be used by all personnel. Constant wind noise is tiresome and increases soldier/marine fatigue, thus affecting alertness.

When visibility is reduced by sandstorms to the extent that military operations are impossible, soldiers/marines should not be allowed to leave their group for any purpose unless secured by lines for recovery.

The following are special considerations when performing operations in dust or sand:

- Contact lenses are very difficult to maintain in the dry dusty environment of the desert and should not be worn except by military personnel operating in air conditioned environments, under command guidance.
- Mucous membranes can be protected by breathing through a wet face cloth, snuffing small amounts of water into nostrils (native water is not safe for this purpose) or coating the nostrils with a small amount of petroleum jelly. Lips should be protected by lip balm.
- Moving vehicles create their own sandstorms and troops traveling in open vehicles should be protected.
- Scarves and bandannas can be used to protect the head and face.
- The face should be washed as often as possible. The eyelids should be cleaned daily.

High Mineral Content. All arid regions have areas where the surface soil has a high mineral content (borax, salt, alkali, and lime). Material in contact with this soil wears out quickly, and water in these areas is extremely hard and undrinkable. Wetting your uniform in such water to cool off may cause a skin rash. The Great Salt Lake area in Utah is an example of this type of mineral-laden water and soil. There is little or no plant life; therefore, shelter is hard to find. Avoid these areas if possible.

BASIC HEAT INJURY PREVENTION

The temperature of the body is regulated within very narrow limits. Too little salt causes heat cramps; too little salt and insufficient water causes heat exhaustion. Heat exhaustion will cause a general collapse of the body's cooling mechanism. This condition is heatstroke, and is potentially fatal. To avoid these illnesses, troops should maintain their physical fitness by eating adequately, drinking sufficient water, and consuming adequate salt. If soldiers/marines expend more calories than they take in, they will be more prone to heat illnesses. Since troops may lose their desire for food in hot climates, they must be encouraged to eat, with the heavier meal of the day scheduled during the cooler hours.

It is necessary to recognize heat stress symptoms quickly. When suffering from heatstroke, the most dangerous condition, there is a tendency for a soldier/marine to creep away from his comrades and attempt to hide in a shady and secluded spot; if not found and treated, he will die. When shade is required during the day, it can best be provided by tarpaulins or camouflage nets, preferably doubled to allow air circulation between layers and dampened with any surplus water.

Approximately 75 percent of the human body is fluid. All chemical activities in the body occur in a water solution, which assists in the removal of toxic body wastes and plays a vital part in the maintenance of an even body temperature. A loss of 2 quarts of body fluid (2.5 percent of body weight) decreases efficiency by 25 percent and a loss of fluid equal to 15 percent of body weight is usually fatal. The following are some considerations when operating in a desert environment:

- Consider water a tactical weapon. Reduce heat injury by forcing water consumption. Soldiers/marines in armored vehicles, MOPP gear, and in body armor need to increase their water intake.
- When possible, drink cool (50-55 degrees Fahrenheit) water.
- Drink one quart of water in the morning, at each meal, and before strenuous work. In hot climates drink at least one quart of water each hour. At higher temperatures hourly water requirements increase to over two quarts.
- Take frequent drinks since they are more often effective than drinking the same amount all at once. Larger soldiers/marines need more water.
- Replace salt loss through eating meals.
- When possible, work loads and/or duration of physical activity should be less during the first days of exposure to heat, and then should gradually be increased to follow acclimatization.
- Modify activities when conditions that increase the risk of heat injury (fatigue/loss of sleep, previous heat exhaustion, taking medication) are present.
- Take frequent rest periods in the shade, if possible. Lower the work rate and work loads as the heat condition increases.
- Perform heavy work in the cooler hours of the day such as early morning or late evening, if possible.

A description of the symptoms and treatment for heat illnesses follows:

- Heat cramps.
  - o Symptoms: Muscle cramps of arms, legs, and/or stomach. Heavy sweating (wet skin) and extreme thirst.
  - o First aid: Move soldier/marine to a shady area and loosen clothing. Slowly give large amounts of cool water. Watch the soldier/marine and continue to give him water, if he accepts it. Get medical help if cramps continue.

- Heat exhaustion.
  - o Symptoms: Heavy sweating with pale, moist, cool skin; headache, weakness, dizziness, and/or loss of appetite; heat cramps, nausea (with or without vomiting), rapid breathing, confusion, and tingling of the hands and/or feet.
  - o First aid: Move the soldier/marine to a cool, shady area and loosen/remove clothing. Pour water on the soldier/marine and fan him to increase the cooling effect. Have the soldier/marine slowly drink at least one full canteen of water. Elevate the soldier's/marine's legs. Get medical help if symptoms continue; watch the soldier/marine until the symptoms are gone or medical aid arrives.

- Heatstroke.
  - o Symptoms: Sweating stops (red, flushed, hot dry skin).
  - o First aid: Evacuate to a medical facility immediately. Move the soldier/marine to a cool, shady area and loosen or remove clothing if the situation permits. Start cooling him immediately.

Immerse him in water and fan him. Massage his extremities and skin and elevate his legs. If conscious, have the soldier/marine slowly drink one full canteen of water.

Water Supply. Maintaining safe, clean, water supplies is critical. The best containers for small quantities of water (5 gallons) are plastic water cans or coolers. Water in plastic cans will be good for up to 72 hours; storage in metal containers is safe only for 24 hours. Water trailers, if kept cool, will keep water fresh up to five days. If the air temperature exceeds 100 degrees Fahrenheit, the water temperature must be monitored. When the temperature exceeds 92 degrees Fahrenheit, the water should be changed, as bacteria will multiply. If the water is not changed the water can become a source of sickness, such as diarrhea. Ice in containers keeps water cool. If ice is put in water trailers, the ice must be removed prior to moving the trailer to prevent damage to the inner lining of the trailer.

Potable drinking water is the single most important need in the desert. Ensure nonpotable water is never mistaken for drinking water. Water that is not fit to drink but is not otherwise dangerous (it may be merely oversalinated) may be used to aid cooling. It can be used to wet clothing, for example, so the body does not use too much of its internal store of water.

Use only government-issued water containers for drinking water. Carry enough water on a vehicle to last the crew until the next planned resupply. It is wise to provide a small reserve. Carry water containers in positions that—

- Prevent vibration by clamping them firmly to the vehicle body.
- Are in the shade and benefit from an air draft.
- Are protected from puncture by shell splinters.
- Are easily dismounted in case of vehicle evacuation.

Troops must be trained not to waste water. Water that has been used for washing socks, for example, is perfectly adequate for a vehicle cooling system.

Obtain drinking water only from approved sources to avoid disease or water that may have been deliberately polluted. Be careful to guard against pollution of water sources. If rationing is in effect, water should be issued under the close supervision of officers and noncommissioned officers.

Humans cannot perform to maximum efficiency on a decreased water intake. An acclimatized soldier/marine will need as much (if not more) water as the nonacclimatized soldier/marine, as he sweats more readily. If the ration water is not sufficient, there is no alternative but to reduce physical activity or restrict it to the cooler parts of the day.

In very hot conditions it is better to drink smaller quantities of water often rather than large quantities occasionally. Drinking large quantities causes excessive sweating and may induce heat cramps. Use of alcohol lessens resistance to heat due to its dehydrating effect. As activities increase or conditions become more severe, increase water intake accordingly.

The optimum water drinking temperature is between 10 degrees Celsius and 15.5 degrees Celsius (50-60 degrees Fahrenheit). Use lister bags or even wet cloth around metal containers to help cool water.

Units performing heavy activities on a sustained basis, such as a forced march or digging in, at 80 degrees wet bulb globe temperature index, may require more than 3 gallons of drinking water per man. Any increase in the heat stress will increase this need. In high temperatures, the average soldier/marine will require 9 quarts of water per day to survive, but 5 gallons are recommended. Details on water consumption and planning factors are contained in Appendix G.

While working in high desert temperatures, a man at rest may lose as much as a pint of water per hour from sweating. In very high temperatures and low humidity, sweating is not noticeable as it evaporates so fast the skin will appear dry. Whenever possible, sweat should be retained on the skin to improve the cooling process; however, the only way to do this is to avoid direct sun on the skin. This is the most important reason why desert troops must remain fully clothed. If a soldier/marine is working, his water loss through sweating (and subsequent requirement for replenishment) increases in proportion to the amount of work done (movement). Troops will not always drink their required amount of liquid readily and will need to be encouraged or ordered to drink more than they think is necessary as the sensation of thirst is not felt until there is a body deficit of 1 to 2 quarts of water. This is particularly true during the period of acclimatization. Packets of artificial fruit flavoring encourages consumption due to the variety of pleasant tastes.

All unit leaders must understand the critical importance of maintaining the proper hydration status. Almost any contingency of military operations will act to interfere with the maintenance of hydration. Urine provides the best indicator of proper hydration. The following are considerations for proper hydration during desert operations:

- Water is the key to your health and survival. Drink before you become thirsty and drink often, When you become thirsty you will be about a "quart and a half low."
- Carry as much water as possible when away from approved sources of drinking water. Man can live longer without food than without water. Drink before you work; carry water in your belly, do not "save" it in your canteen. Learn to drink a quart or more of water at one time and drink frequently to replace sweat losses.
- Ensure troops have at least one canteen of water in reserve, and know where and when water resupply will be available.
- Carbohydrate/electrolyte beverages (e.g., Gatorade) are not required, and if used, should not be the only source of water. They are too concentrated to be used alone. Many athletes prefer to dilute these 1:1 with water. Gaseous drinks, sodas, beer, and milk are not good substitutes for water because of their dehydrating effects.
- If urine is more colored than diluted lemonade, or the last urination cannot be remembered, there is probably insufficient water intake. Collect urine samples in field expedient containers and spot check the color as a guide to ensuring proper hydration. Very dark urine warns of dehydration. Soldiers/marines should observe their own urine, and use the buddy system to watch for signs of dehydration in others.
- Diseases, especially diarrheal diseases, will complicate and often prevent maintenance of proper hydration.

Salt, in correct proportions, is vital to the human body; however, the more a man sweats, the more salt he loses. The issue ration has enough salt for a soldier/marine drinking up to 4 quarts of water per day. Unacclimatized troops need additional salt during their first few days of exposure and all soldiers/marines need additional salt when sweating heavily. If the water demand to balance sweat loss rises, extra salt must be taken under medical direction. Salt, in excess of body requirements, may cause increased thirst and a feeling of sickness, and can be dangerous. Water must be tested before adding salt as some sources are already saline, especially those close to the sea.

What can individuals or small groups do when they are totally cut off from normal water supply? If you are totally cut off from the normal water supply, the first question you must consider is whether you should try to walk to safety or stay put and hope for rescue. Walking requires 1 gallon of water for every 20 miles covered at night, and 2 gallons for every 20 miles covered during the day. Without any water and walking only at night, you may be able to cover 20 to 25 miles before you collapse. If your chance of being rescued is not increased by walking 20 miles, you may be better off staying put and surviving one to three days longer. If you do not know where you are going, do not try to walk with a limited supply of water.

If you decide to walk to safety, follow the following guidelines in addition to the general conservation practices listed in the next section:

- Take as much water as you have and can carry, and carry little or no food.
- Drink as much as you can comfortably hold before you set out.
- Walk only at night.

Whether you decide to walk or not, you should follow the principles listed below to conserve water in emergency situations:

- Avoid the sun. Stay in shade as much as possible. If you are walking, rest in shade during the day. This may require some ingenuity. You may want to use standard or improvised tents, lie under vehicles, or dig holes in the ground.
- Cease activity. Do not perform any work that you do not have to for survival.
- Remain clothed. It will reduce the water lost to evaporation.
- Shield yourself from excessive winds. Winds, though they feel good also increase the evaporation rate.
- Drink any potable water you have as you feel the urge. Saving it will not reduce your body's need for it or the rate at which you use it.
- Do not drink contaminated water from such sources as car radiators or urine. It will actually require more water to remove the waste material. Instead, in emergencies, use such water to soak your clothing as this reduces sweating.
- Do not eat unless you have plenty of water.

Do not count on finding water if you are stranded in the desert. Still, in certain cases, some water can be found. It does rain sometimes in the desert (although it may be 20 years between showers) and some water will remain under the surface. Signs of possible water are green plants or dry lake beds. Sometimes water can be obtained in these places by digging down until the soil becomes moist and then waiting for water to seep into the hole. Desert trails lead from one water point to another, but they may be further apart than you can travel without water. Almost all soils contain some moisture.

Cold. The desert can be dangerously cold. The dry air, wind, and clear sky can combine to produce bone-chilling discomfort and even injury. The ability of the body to maintain body temperature within a narrow range is as important in the cold as in the heat. Loss of body heat to the environment can lead to cold injury; a general lowering of the body temperature can result in hypothermia, and local freezing of body tissues can lead to frostbite. Hypothermia is the major threat from the cold in the desert, but frostbite also occurs.

Troops must have enough clothing and shelter to keep warm. Remember, wood is difficult to find; any that is available is probably already in use. Troops may be tempted to leave clothing and equipment behind that seems unnecessary (and burdensome) during the heat of the day. Cold-wet injuries (immersion foot or trench foot) may be a problem for dismounted troops operating in the coastal marshes of the Persian Gulf during the winter. Some guidelines to follow when operating in the cold are:

- Anticipate an increased risk of cold-wet injuries if a proposed operation includes lowland or marshes. Prolonged exposure of the feet in cold water causes immersion foot injury, which is completely disabling.
- Check the weather—know what conditions you will be confronting. The daytime temperature is no guide to the nighttime temperature; 90-degree-Fahrenheit days can turn into 30-degree-Fahrenheit nights.
- The effects of the wind on the perception of cold is well known. Windchill charts contained in FM 21-10 allow estimation of the combined cooling power of air temperature and wind speed compared to the effects of an equally cooling still-air temperature.

Clothing. Uniforms should be worn to protect against sunlight and wind. Wear the uniform loosely. Use hats, goggles, and sunscreen. Standard lightweight clothing is suitable for desert operations but should be camouflaged in desert colors, not green. Wear nonstarched long-sleeved shirts, and full-length trousers tucked into combat boots. Wear a scarf or triangular bandanna loosely around the neck (as a sweat rag) to protect the face and neck during sandstorms against the sand and the sun. In extremely hot and dry conditions a wet sweat rag worn loosely around the neck will assist in body cooling.

Combat boots wear out quickly in desert terrain, especially if the terrain is rocky. The leather drys out and cracks unless a nongreasy mixture such as saddle soap is applied. Covering the ventilation holes on jungle boots with glue or epoxies prevents excessive sand from entering the boots. Although difficult to do, keep clothing relatively clean by washing in any surplus water that is available. When water is not available, air and sun clothing to help kill bacteria.

Change socks when they become wet. Prolonged wear of wet socks can lead to foot injury. Although dry desert air promotes evaporation of water from exposed clothing and may actually promote cooling, sweat tends to accumulate in boots.

Soldier/marines may tend to stay in thin clothing until too late in the desert day and become susceptible to chills—so respiratory infections may be common. Personnel should gradually add layers of clothing at night (such as sweaters), and gradually remove them in the morning. Where the danger of cold weather injury exists in the desert, commanders must guard against attempts by inexperienced troops to discard cold weather clothing during the heat of the day.

Compared to the desert battle dress uniform (DBDU) the relative impermeability of the battle dress overgarment (BDO) reduces evaporative cooling capacity. Wearing underwear and the complete DBDU, with sleeves rolled down and under the chemical protective garment, provides additional protection against chemical poisoning. However, this also increases the likelihood of heat stress casualties.

Hygiene and Sanitation. Personal hygiene is absolutely critical to sustaining physical fitness. Take every opportunity to wash. Poor personal hygiene and lack of attention to siting of latrines cause more casualties than actual combat. Field Marshal Rommel lost over 28,400 soldiers of his Afrika Corps to disease in 1942. During the desert campaigns of 1942, for every one combat injury, there were three hospitalized for disease.

Proper standards of personal hygiene must be maintained not only as a deterrent to disease but as a reinforcement to discipline and morale. If water is available, shave and bathe daily. Cleaning the areas of the body that sweat heavily is especially important; change underwear and socks frequently, and use foot powder often. Units deployed in remote desert areas must have a means of cutting hair therefore, barber kits should be maintained and inventoried prior to any deployment. If sufficient water is not available, troops should clean themselves with sponge baths, solution-impregnated pads, a damp rag, or even a dry, clean cloth. Ensure that waste water is disposed of in an approved area to prevent insect infestation. If sufficient water is not available for washing, a field expedient alternative is powder baths, that is, using talcum or baby powder to dry bathe.

Check troops for any sign of injury, no matter how slight, as dirt or insects can cause infection in minor cuts and scratches. Small quantities of disinfectant in washing water reduces the chance of infection. Minor sickness can have serious effects in the desert. Prickly heat for example, upsets the sweating mechanism and diarrhea increases water loss, making the soldier/marine more prone to heat illnesses. The buddy system helps to ensure that prompt attention is given to these problems before they incapacitate individuals.

Intestinal diseases can easily increase in the desert. Proper mess sanitation is essential. Site latrines well away and downwind of troop areas and lagers. Trench-type latrines should be used where the soil is suitable but must be dug deeply, as shallow latrines become exposed in areas of shifting sand. Funnels dug into a sump work well as urinals. Layer the bottom of slit trenches with lime and cover the top prior to being filled in. Ensure lime is available after each use of the latrine. Flies are a perpetual source of irritation and carry infections. Only good sanitation can keep the fly problem to a minimum. Avoid all local tribe camps since they are frequently a source of disease and vermin.

Desert Sickness. Diseases common to the desert include plague, typhus, malaria, dengue fever, dysentery, cholera, and typhoid. Diseases which adversely impact hydration, such as those which include nausea, vomiting, and diarrhea among their symptoms, can act to dramatically increase the risk of heat (and cold) illness or injury. Infectious diseases can result in a fever; this may make it difficult to diagnose heat illness. Occurrences of heat illness in troops suffering from other diseases complicate recovery from both ailments.

Many native desert animals and plants are hazardous. In addition to injuries as a result of bites, these natural inhabitants of the desert can be a source of infectious diseases.

Many desert plants and shrubs have a toxic resin that can cause blisters, or spines that can cause infection. Consider milky sap, all red beans, and smoke from burning oleander shrubs, poisonous. Poisonous snakes, scorpions, and spiders are common in all deserts. Coastal waters of the Persian Gulf contain hazardous marine animals including sea snakes, poisonous jellyfish, and sea urchins.

Skin diseases can result from polluted water so untreated water should not be used for washing clothes; although it can be used for vehicle cooling systems or vehicle decontamination.

The excessive sweating common in hot climates brings on prickly heat and some forms of fungus infections of the skin. The higher the humidity, the greater the possibility of their occurrence. Although many deserts are not humid, there are exceptions, and these ailments are common to humid conditions.

The following are additional health-related considerations when operating in a desert environment:

- The most common and significant diseases in deserts include diarrheal and insectborne febrile (i.e., fever causing) illnesses—both types of these diseases are preventable.
- Most diarrheal diseases result from ingestion of water or food contaminated with feces. Flies, mosquitoes, and other insects carry fever—causing illnesses such as malaria, sand fly fever, dengue (fever with severe pain in the joints), typhus, and tick fevers.
- There are no safe natural water sources in the desert. Standing water is usually infectious or too brackish to be safe for consumption. Units and troops must always know where and how to get safe drinking water.
- Avoid brackish water (i.e., salty). It, like sea water, increases thirst; it also dehydrates the soldier/ marine faster than were no water consumed. Brackish water is common even in public water supplies, Iodine tablets only kill germs, they do not reduce brackishness.
- Water supplies with insufficient chlorine residuals, native food and drink, and ice from all sources are common sources of infective organisms.

Preventive Measures. Both diarrheal and insectborne diseases can be prevented through a strategy which breaks the chain of transmission from infected sources to susceptible soldiers/marines by effectively applying preventive measures. Additional preventive measures are described below:

- Careful storage, handling and purification/preparation of water and food are the keys to prevention of diarrheal disease. Procure all food, water, ice, and beverages from US military approved sources and inspect them routinely.
- Well-cooked foods that are "steaming hot" when eaten are generally safe, as are peeled fruits and vegetables.
- Local dairy products and raw leafy vegetables are generally unsafe.
- Consider the food in native markets hazardous. Avoid local food unless approved by medical personnel officials.
- Assume raw ice and native water to be contaminated—raw ice cannot be properly disinfected. Ice has been a major source of illness in all prior conflicts; therefore, use ice only from approved sources.
- If any uncertainty exists concerning the quality of drinking water, troops should disinfect their supplies using approved field-expedient methods (e.g., hypochlorite for lister bags, iodine tablets for canteens, boiling).
- Untreated water used for washing or bathing risks infection.
- Hand washing facilities should be established at both latrines and mess facilities. Particular attention should be given to the cleanliness of hands and fingernails. Dirty hands are the primary means of transmitting disease.
- Dispose of human waste and garbage. Additional considerations regarding human waste and garbage are—

o   Sanitary disposal is important in preventing the spread of disease from insects, animals, and infected individuals, to healthy soldiers/marines.
o   Construction and maintenance of sanitary latrines are essential.
o   Burning is the best solution for waste.
o   Trench latrines can be used if the ground is suitable, but they must be dug deeply enough so that they are not exposed to shifting sand, and they must have protection against flies and other insects that can use them as breeding places.
o   Food and garbage attract animals—do not sleep where you eat and keep refuse areas away from living areas.
o   Survey the unit area for potential animal hazards.
o   Shakeout boots, clothing, and bedding before using them.

CHAPTER 3

Cold Weather Survival

One of the most difficult survival situations is a cold weather scenario. Remember, cold weather is an adversary that can be as dangerous as an enemy soldier. Every time you venture into the cold, you are pitting yourself against the elements. With a little knowledge of the environment, proper plans, and appropriate equipment, you can overcome the elements. As you remove one or more of these factors, survival becomes increasingly difficult. Remember, winter weather is highly variable. Prepare yourself to adapt to blizzard conditions even during sunny and clear weather.

Cold is a far greater threat to survival than it appears. It decreases your ability to think and weakens your will to do anything except to get warm. Cold is an insidious enemy; as it numbs the mind and body, it subdues the will to survive. Cold makes it very easy to forget your ultimate goal—to survive.

COLD REGIONS AND LOCATIONS

Cold regions include arctic and subarctic areas and areas immediately adjoining them. You can classify about 48 percent of the northern hemisphere's total landmass as a cold region due to the influence and extent of air temperatures. Ocean currents affect cold weather and cause large areas normally included in the temperate zone to fall within the cold regions during winter periods. Elevation also has a marked effect on defining cold regions.

Within the cold weather regions, you may face two types of cold weather environments—wet or dry. Knowing in which environment your area of operations falls will affect planning and execution of a cold weather operation.

Wet Cold Weather Environments. Wet cold weather conditions exist when the average temperature in a 24-hour period is –10 degrees C or above. Characteristics of this condition are freezing during the colder night hours and thawing during the day. Even though the temperatures are warmer during this condition, the terrain is usually very sloppy due to slush and mud. You must concentrate on protecting yourself from the wet ground and from freezing rain or wet snow.

Dry Cold Weather Environments. Dry cold weather conditions exist when the average temperature in a 24-hour period remains below –10 degrees C. Even though the temperatures in this condition are much lower than normal, you do not have to contend with the freezing and thawing. In these conditions, you need more layers of inner clothing to protect you from temperatures as low as –60 degrees C. Extremely hazardous conditions exist when wind and low temperature combine.

WINDCHILL

Windchill increases the hazards in cold regions. Windchill is the effect of moving air on exposed flesh. For instance, with a 27.8-kph (15-knot) wind and a temperature of –10 degrees C, the equivalent windchill temperature is –23 degrees C. Table 3-1 gives the windchill factors for various temperatures and wind speeds.

Remember, even when there is no wind, you will create the equivalent wind by skiing, running, being towed on skis behind a vehicle, working around aircraft that produce wind blasts.

BASIC PRINCIPLES OF COLD WEATHER SURVIVAL

It is more difficult for you to satisfy your basic water, food, and shelter needs in a cold environment than in a warm environment. Even if you have the basic requirements, you must also have adequate protective clothing and the will to survive. The will to survive is as important as the basic needs. There have been incidents when trained and well-equipped individuals have not survived cold weather situations because they lacked the will to live. Conversely, this will has sustained individuals less well-trained and equipped.

There are many different items of cold weather equipment and clothing issued by the U.S. Army today. Specialized units may have access to newer, lightweight underwear, outerwear and boots, and other special equipment. Remember, however, the older gear will keep you warm as long as you apply a few cold weather principles. If the newer types of clothing are available, use them. If not, then your clothing should be entirely wool, with the possible exception of a windbreaker.

You must not only have enough clothing to protect you from the cold, you must also know how to maximize the warmth you get from it. For example, always keep your head covered. You can lose 40 to 45 percent of body heat from an unprotected head and even more from the unprotected neck, wrist, and ankles. These areas of the body are good radiators of heat and have very little insulating fat. The brain is very susceptible to cold and can stand the least amount of cooling. Because there is much blood circulation in the head, most of which is on the surface, you can lose heat quickly if you do not cover your head.

There are four basic principles to follow to keep warm. An easy way to remember these basic principles is to use the word COLD—

    C—Keep clothing *clean*.
    O—Avoid *overheating*.
    L—Wear clothes loose and in *layers*.
    D—Keep clothing *dry*.

Table 3-1: Windchill table.

COOLING POWER OF WIND EXPRESSED AS "EQUIVALENT CHILL TEMPERATURE"																							
WIND SPEED		TEMPERATURE (DEGREES C)																					
CALM	CALM	4	2	-1	-4	-7	-9	-12	-15	-18	-21	-23	-26	-29	-32	-34	-37	-40	-43	-46	-48	-51	
KNOTS	KPH	EQUIVALENT CHILL TEMPERATURE																					
4	8	2	-1	-4	-7	-9	-12	-15	-18	-21	-23	-26	-29	-32	-34	-37	-40	-43	-46	-48	-54	-57
9	16	-1	-7	-9	-12	-15	-18	-23	-26	-29	-32	-37	-40	-43	-46	-51	-54	-57	-59	-62	-68	-71
13	24	-4	-9	-12	-18	-21	-23	-29	-32	-34	-40	-43	-46	-51	-54	-57	-62	-65	-68	-73	-76	-79
17	32	-7	-12	-15	-18	-23	-26	-32	-34	-37	-43	-46	-51	-54	-59	-62	-65	-71	-73	-79	-82	-84
22	40	-9	-12	-18	-21	-26	-29	-34	-37	-43	-46	-51	-54	-59	-62	-68	-71	-76	-79	-84	-87	-93
26	48	-12	-15	-18	-23	-29	-32	-34	-40	-46	-48	-54	-57	-62	-65	-71	-73	-79	-82	-87	-90	-96
30	56	-12	-15	-21	-23	-29	-34	-37	-40	-46	-51	-54	-59	-62	-68	-73	-76	-82	-84	-90	-93	-98
35	64	-12	-18	-21	-26	-29	-34	-37	-43	-48	-51	-57	-59	-65	-71	-73	-79	-82	-87	-90	-96	-101
(Higher winds have little additional effects)		LITTLE DANGER			INCREASING DANGER (Flesh may freeze within 1 minute)					GREAT DANGER (Flesh may freeze within 30 seconds)													
		DANGER OF FREEZING EXPOSED FLESH FOR PROPERLY CLOTHED PERSONS																					

C—Keep clothing clean. This principle is always important for sanitation and comfort. In winter, it is also important from the standpoint of warmth. Clothes matted with dirt and grease lose much of their insulation value. Heat can escape more easily from the body through the clothing's crushed or filled up air pockets.

O—Avoid overheating. When you get too hot, you sweat and your clothing absorbs the moisture. This affects your warmth in two ways: dampness decreases the insulation quality of clothing, and as sweat evaporates, your body cools. Adjust your clothing so that you do not sweat. Do this by partially opening your parka or jacket, by removing an inner layer of clothing, by removing heavy outer mittens, or by throwing back your parka hood or changing to lighter headgear. The head and hands act as efficient heat dissipaters when overheated.

L—Wear your clothing loose and in layers. Wearing tight clothing and footgear restricts blood circulation and invites cold injury. It also decreases the volume of air trapped between the layers, reducing its insulating value. Several layers of lightweight clothing are better than one equally thick layer of clothing, because the layers have dead-air space between them. The dead-air space provides extra insulation. Also, layers of clothing allow you to take off or add clothing layers to prevent excessive sweating or to increase warmth.

D—Keep clothing dry. In cold temperatures, your inner layers of clothing can become wet from sweat and your outer layer, if not water repellent, can become wet from snow and frost melted by body heat. Wear water repellent outer clothing, if available. It will shed most of the water collected from melting snow and frost. Before entering a heated shelter, brush off the snow and frost. Despite the precautions you take, there will be times when you cannot keep from getting wet. At such times, drying your clothing may become a major problem. On the march, hang your damp mittens and socks on your rucksack. Sometimes in freezing temperatures, the wind and sun will dry this clothing. You can also place damp socks or mittens, unfolded, near your body so that your body heat can dry them. In a campsite, hang damp clothing inside the shelter near the top, using drying lines or improvised racks. You may even be able to dry each item by holding it before an open fire. Dry leather items slowly. If no other means are available for drying your boots, put them between your sleeping bag shell and liner. Your body heat will help to dry the leather.

A heavy, down-lined sleeping bag is a valuable piece of survival gear in cold weather. Ensure the down remains dry. If wet, it loses a lot of its insulation value. If you do not have a sleeping bag, you can make one out of parachute cloth or similar material and natural dry material, such as leaves, pine needles, or moss. Place the dry material between two layers of the material.

Other important survival items are a knife; waterproof matches in a waterproof container, preferably one with a flint attached; a durable compass; map; watch; waterproof ground cloth and cover; flashlight; binoculars; dark glasses; fatty emergency foods; food gathering gear; and signaling items.

Remember, a cold weather environment can be very harsh. Give a good deal of thought to selecting the right equipment for survival in the cold. If unsure of an item you have never used, test it in an "overnight backyard" environment before venturing further. Once you have selected items that are essential for your survival, do not lose them after you enter a cold weather environment.

HYGIENE

Although washing yourself may be impractical and uncomfortable in a cold environment, you must do so. Washing helps prevent skin rashes that can develop into more serious problems.

In some situations, you may be able to take a snow bath. Take a handful of snow and wash your body where sweat and moisture accumulate, such as under the arms and between the legs, and then wipe yourself dry. If possible, wash your feet daily and put on clean, dry socks. Change your underwear at least twice a week. If you are unable to wash your underwear, take it off, shake it, and let it air out for an hour or two.

If you are using a previously used shelter, check your body and clothing for lice each night. If your clothing has become infested, use insecticide powder if you have any. Otherwise, hang your clothes in the cold, then beat and brush them. This will help get rid of the lice, but not the eggs.

If you shave, try to do so before going to bed. This will give your skin a chance to recover before exposing it to the elements.

MEDICAL ASPECTS

When you are healthy, your inner core temperature (torso temperature) remains almost constant at 37 degrees C (98.6 degrees F). Since your limbs and head have less protective body tissue than your torso, their temperatures vary and may not reach core temperature.

Your body has a control system that lets it react to temperature extremes to maintain a temperature balance. There are three main factors that affect this temperature balance—heat production, heat loss, and evaporation. The difference between the body's core temperature and the environment's temperature governs the heat production rate. Your body can get rid of heat better than it can produce it. Sweating helps to control the heat balance. Maximum sweating will get rid of heat about as fast as maximum exertion produces it.

Shivering causes the body to produce heat. It also causes fatigue that, in turn, leads to a drop in body temperature. Air movement around your body affects heat loss. It has been calculated that a naked man exposed to still air at or about 0 degrees C can maintain a heat balance if he shivers as hard as he can. However, he can't shiver forever.

It has also been calculated that a man at rest wearing the maximum arctic clothing in a cold environment can keep his internal heat balance during temperatures well below freezing. To withstand really cold conditions for any length of time, however, he will have to become active or shiver.

COLD INJURIES

The best way to deal with injuries and sicknesses is to take measures to prevent them from happening in the first place. Treat any injury or sickness that occurs as soon as possible to prevent it from worsening.

The knowledge of signs and symptoms and the use of the buddy system are critical in maintaining health. Following are cold injuries that can occur.

Hypothermia. Hypothermia is the lowering of the body temperature at a rate faster than the body can produce heat. Causes of hypothermia may be general exposure or the sudden wetting of the body by falling into a lake or spraying with fuel or other liquids.

The initial symptom is shivering. This shivering may progress to the point that it is uncontrollable and interferes with an individual's ability to care for himself. This begins when the body's core (rectal) temperature falls to about 35.5 degrees C (96 degrees F). When the core temperature reaches 35 to 32 degrees C (95 to 90 degrees F), sluggish thinking, irrational reasoning, and a false feeling of warmth may occur. Core temperatures of 32 to 30 degrees C (90 to 86 degrees F) and below result in muscle rigidity, unconsciousness, and barely detectable signs of life. If the victim's core temperature falls below 25 degrees C (77 degrees F), death is almost certain.

To treat hypothermia, rewarm the entire body. If there are means available, rewarm the person by first immersing the trunk area only in warm water of 37.7 to 43.3 degrees C (100 to 110 degrees F).

⚠ CAUTION

Rewarming the total body in a warm water bath should be done only in a hospital environment because of the increased risk of cardiac arrest and rewarming shock.

One of the quickest ways to get heat to the inner core is to give warm water enemas. Such an action, however, may not be possible in a survival situation. Another method is to wrap the victim in a warmed sleeping bag with another person who is already warm; both should be naked.

> ⚠ **CAUTION**
> The individual placed in the sleeping bag with victim could also become a hypothermia victim if left in the bag too long.

If the person is conscious, give him hot, sweetened fluids. One of the best sources of calories is honey or dextrose; if unavailable, use sugar, cocoa, or a similar soluble sweetener.

> ⚠ **CAUTION**
> Do not force an unconscious person to drink.

There are two dangers in treating hypothermia—rewarming too rapidly and "after drop." Rewarming too rapidly can cause the victim to have circulatory problems, resulting in heart failure. After drop is the sharp body core temperature drop that occurs when taking the victim from the warm water. Its probable muse is the return of previously stagnant limb blood to the core (inner torso) area as recirculation occurs. Concentrating on warming the core area and stimulating peripheral circulation will lessen the effects of after drop. Immersing the torso in a warm bath, if possible, is the best treatment.

Frostbite. This injury is the result of frozen tissues. Light frostbite involves only the skin that takes on a dull whitish pallor. Deep frostbite extends to a depth below the skin. The tissues become solid and immovable. Your feet, hands, and exposed facial areas are particularly vulnerable to frostbite.

The best frostbite prevention, when you are with others, is to use the buddy system. Check your buddy's face often and make sure that he checks yours. If you are alone, periodically cover your nose and lower part of your face with your mittened hand.

The following pointers will aid you in keeping warm and preventing frostbite when it is extremely cold or when you have less than adequate clothing:

- *Face.* Maintain circulation by twitching and wrinkling the skin on your face and making faces. Warm with your hands.
- *Ears.* Wiggle and move your ears. Warm with your hands.
- *Hands.* Move your hands inside your gloves. Warm by placing your hands close to your body.
- *Feet.* Move your feet and wiggle your toes inside your boots.

A loss of feeling in your hands and feet is a sign of frostbite. If you have lost feeling for only a short time, the frostbite is probably light. Otherwise, assume the frostbite is deep. To rewarm a light frostbite, use your hands or mittens to warm your face and ears. Place your hands under your armpits. Place your feet next to your buddy's stomach. A deep frostbite injury, if thawed and refrozen, will cause more damage than a nonmedically trained person can handle. Table 3-2 lists some do's and don'ts regarding frostbite.

Trench Foot and Immersion Foot. These conditions result from many hours or days of exposure to wet or damp conditions at a temperature just above freezing. The symptoms are a sensation of pins and needles, tingling, numbness, and then pain. The skin will initially appear wet, soggy, white, and shriveled. As it progresses and damage appears, the skin will take on a red and then a bluish or black discoloration. The feet become cold, swollen, and have a waxy appearance. Walking becomes difficult and the feet feel heavy and numb. The nerves and muscles sustain the main damage, but gangrene can occur. In extreme cases, the flesh dies and it may become necessary to have the foot or leg amputated. The best prevention is to keep your feet dry. Carry extra socks with you in a waterproof packet. You can dry wet socks against your torso (back or chest). Wash your feet and put on dry socks daily.

Dehydration. When bundled up in many layers of clothing during cold weather, you may be unaware that you are losing body moisture. Your heavy clothing absorbs the moisture that evaporates in the air. You

Table 3-2: Frostbite do's and dont's.

Do	Don't
• Periodically check for frostbite.	• Rub injury with snow.
• Rewarm light frostbite.	• Drink alcoholic beverages.
• Keep injured areas from refreezing.	• Smoke.
	• Try to thaw out a deep frostbite injury if you are away from definitive medical care.

must drink water to replace this loss of fluid. Your need for water is as great in a cold environment as it is in a warm environment (see Chapter 2). One way to tell if you are becoming dehydrated is to check the color of your urine on snow. If your urine makes the snow dark yellow, you are becoming dehydrated and need to replace body fluids. If it makes the snow light yellow to no color, your body fluids have a more normal balance.

Cold Diuresis. Exposure to cold increases urine output. It also decreases body fluids that you must replace.

Sunburn. Exposed skin can become sunburned even when the air temperature is below freezing. The sun's rays reflect at all angles from snow, ice, and water, hitting sensitive areas of skin—lips, nostrils, and eyelids. Exposure to the sun results in sunburn more quickly at high altitudes than at low altitudes. Apply sunburn cream or lip salve to your face when in the sun.

Snow Blindness. The reflection of the sun's ultraviolet rays off a snow-covered area causes this condition. The symptoms of snow blindness are a sensation of grit in the eyes, pain in and over the eyes that increases with eyeball movement, red and teary eyes, and a headache that intensifies with continued exposure to light. Prolonged exposure to these rays can result in permanent eye damage. To treat snow blindness, bandage your eyes until the symptoms disappear.

You can prevent snow blindness by wearing sunglasses. If you don't have sunglasses, improvise. Cut slits in a piece of cardboard, thin wood, tree bark, or other available material (Figure 15-3). Putting soot under your eyes will help reduce shine and glare.

Constipation. It is very important to relieve yourself when needed. Do not delay because of the cold condition. Delaying relieving yourself because of the cold, eating dehydrated foods, drinking too little liquid, and irregular eating habits can cause you to become constipated. Although not disabling, constipation can cause some discomfort. Increase your fluid intake to at least 2 liters above your normal 2 to 3 liters daily intake and, if available, eat fruit and other foods that will loosen the stool.

Insect Bites. Insect bites can become infected through constant scratching. Flies can carry various disease-producing germs. To prevent insect bites, use insect repellent, netting, and wear proper clothing.

SHELTERS

Your environment and the equipment you carry with you will determine the type of shelter you can build. You can build shelters in wooded areas, open country, and barren areas. Wooded areas usually provide the best location, while barren areas have only snow as building-material. Wooded areas provide timber for shelter construction, wood for fire, concealment from observation, and protection from the wind.

> ✍ **NOTE**
>
> In extreme cold, do not use metal, such as an aircraft fuselage, for shelter. The metal will conduct away from the shelter what little heat you can generate.

Shelters made from ice or snow usually require tools such as ice axes or saws. You must also expend much time and energy to build such a shelter. Be sure to ventilate an enclosed shelter, especially if you intend to build a fire in it. Always block a shelter's entrance, if possible, to keep the heat in and the wind out. Use a rucksack or snow block. Construct a shelter no larger than needed. This will reduce the amount of space to heat. A fatal error in cold weather shelter construction is making the shelter so large that it steals body heat rather than saving it. Keep shelter space small.

Never sleep directly on the ground. Lay down some pine boughs, grass, or other insulating material to keep the ground from absorbing your body heat.

Never fall asleep without turning out your stove or lamp. Carbon monoxide poisoning can result from a fire burning in an unventilated shelter. Carbon monoxide is a great danger. It is colorless and odorless. Anytime you have an open flame, it may generate carbon monoxide. Always check your ventilation. Even in a ventilated shelter, incomplete combustion can cause carbon monoxide poisoning. Usually, there are no symptoms. Unconsciousness and death can occur without warning. Sometimes, however, pressure at the temples, burning of the eyes, headache, pounding pulse, drowsiness, or nausea may occur. The one characteristic, visible sign of carbon monoxide poisoning is a cherry red coloring in the tissues of the lips, mouth, and inside of the eyelids. Get into fresh air at once if you have any of these symptoms.

There are several types of field-expedient shelters you can quickly build or employ. Many use snow for insulation. See Part III, Chapter 3, "Special Operations and Situations," for more information on field-expedient shelters.

FIRE

Fire is especially important in cold weather. It not only provides a means to prepare food, but also to get warm and to melt snow or ice for water. It also provides you with a significant psychological boost by making you feel a little more secure in your situation.

Use the techniques described in Part V, Chapter 1 to build and light your fire. If you are in enemy territory, remember that the smoke, smell, and light from your fire may reveal your location. Light reflects

Figure 3-1: Improvised sunglasses.

from surrounding trees or rocks, making even indirect light a source of danger. Smoke tends to go straight up in cold, calm weather, making it a beacon during the day, but helping to conceal the smell at night. In warmer weather, especially in a wooded area, smoke tends to hug the ground, making it less visible in the day, but making its odor spread.

If you are in enemy territory, cut low tree boughs rather than the entire tree for firewood. Fallen trees are easily seen from the air.

All wood will burn, but some types of wood create more smoke than others. For instance, coniferous trees that contain resin and tar create more and darker smoke than deciduous trees.

There are few materials to use for fuel in the high mountainous regions of the arctic. You may find some grasses and moss, but very little. The lower the elevation, the more fuel available. You may find some scrub willow and small, stunted spruce trees above the tree line. On sea ice, fuels are seemingly nonexistent. Driftwood or fats may be the only fuels available to a survivor on the barren coastlines in the arctic and subarctic regions.

Abundant fuels within the tree line are—

- Spruce trees are common in the interior regions. As a conifer, spruce makes a lot of smoke when burned in the spring and summer months. However, it burns almost smoke-free in late fall and winter.
- The tamarack tree is also a conifer. It is the only tree of the pine family that loses its needles in the fall. Without its needles, it looks like a dead spruce, but it has many knobby buds and cones on its bare branches. When burning, tamarack wood makes a lot of smoke and is excellent for signaling purposes.
- Birch trees are deciduous and the wood burns hot and fast, as if soaked with oil or kerosene. Most birches grow near streams and lakes, but occasionally you will find a few on higher ground and away from water.
- Willow and alder grow in arctic regions, normally in marsh areas or near lakes and streams. These woods burn hot and fast without much smoke.

Dried moss, grass, and scrub willow are other materials you can use for fuel. These are usually plentiful near streams in tundras (open, treeless plains). By bundling or twisting grasses or other scrub vegetation to form a large, solid mass, you will have a slower burning, more productive fuel.

If fuel or oil is available from a wrecked vehicle or downed aircraft, use it for fuel. Leave the fuel in the tank for storage, drawing on the supply only as you need it. Oil congeals in extremely cold temperatures, therefore, drain it from the vehicle or aircraft while still warm if there is no danger of explosion or fire. If you have no container, let the oil drain onto the snow or ice. Scoop up the fuel as you need it.

⚠ CAUTION

Do not expose flesh to petroleum, oil, and lubricants in extremely cold temperatures. The liquid state of these products is deceptive in that it can cause frostbite.

Some plastic products, such as MRE spoons, helmet visors, visor housings, and foam rubber will ignite quickly from a burning match. They will also burn long enough to help start a fire. For example, a plastic spoon will burn for about 10 minutes.

In cold weather regions, there are some hazards in using fires, whether to keep warm or to cook. For example—

- Fires have been known to burn underground, resurfacing nearby. Therefore, do not build a fire too close to a shelter.
- In snow shelters, excessive heat will melt the insulating layer of snow that may also be your camouflage.

- A fire inside a shelter lacking adequate ventilation can result in carbon monoxide poisoning.
- A person trying to get warm or to dry clothes may become careless and burn or scorch his clothing and equipment.
- Melting overhead snow may get you wet, bury you and your equipment, and possibly extinguish your fire.

In general, a small fire and some type of stove is the best combination for cooking purposes. A hobo stove (Figure 3-2) is particularly suitable to the arctic. It is easy to make out of a tin can, and it conserves fuel. A bed of hot coals provides the best cooking heat. Coals from a crisscross fire will settle uniformly. Make this type of fire by crisscrossing the firewood. A simple crane propped on a forked stick will hold a cooking container over a fire.

For heating purposes, a single candle provides enough heat to warm an enclosed shelter. A small fire about the size of a man's hand is ideal for use in enemy territory. It requires very little fuel, yet it generates considerable warmth and is hot enough to warm liquids.

WATER

There are many sources of water in the arctic and subarctic. Your location and the season of the year will determine where and how you obtain water.

Water sources in arctic and subarctic regions are more sanitary than in other regions due to the climatic and environmental conditions. However, always purify the water before drinking it. During the summer months, the best natural sources of water are freshwater lakes, streams, ponds, rivers, and springs. Water from ponds or lakes may be slightly stagnant, but still usable. Running water in streams, rivers, and bubbling springs is usually fresh and suitable for drinking.

The brownish surface water found in a tundra during the summer is a good source of water. However, you may have to filter the water before purifying it.

You can melt freshwater ice and snow for water. Completely melt both before putting them in your mouth. Trying to melt ice or snow in your mouth takes away body heat and may cause internal cold injuries. If on or near pack ice in the sea, you can use old sea ice to melt for water. In time, sea ice loses its salinity. You can identify this ice by its rounded corners and bluish color.

You can use body heat to melt snow. Place the snow in a water bag and place the bag between your layers of clothing. This is a slow process, but you can use it on the move or when you have no fire.

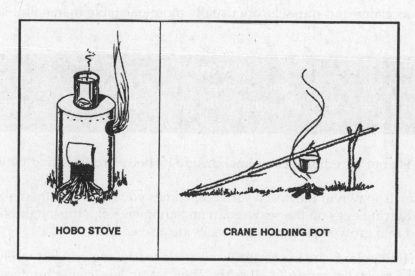

HOBO STOVE   CRANE HOLDING POT

Figure 3-2: Cooking fire/stove

> ✍ **NOTE**
>
> Do not waste fuel to melt ice or snow when drinkable water is available from other sources.

When ice is available, melt it, rather than snow. One cup of ice yields more water than one cup of snow. Ice also takes less time to melt. You can melt ice or snow in a water bag, MRE ration bag, tin can, or improvised container by placing the container near a fire. Begin with a small amount of ice or snow in the container and, as it turns to water, add more ice or snow.

Another way to melt ice or snow is by putting it in a bag made from porous material and suspending the bag near the fire. Place a container under the bag to catch the water.

During cold weather, avoid drinking a lot of liquid before going to bed.

Crawling out of a warm sleeping bag at night to relieve yourself means less rest and more exposure to the cold.

Once you have water, keep it next to you to prevent refreezing. Also, do not fill your canteen completely. Allowing the water to slosh around will help keep it from freezing.

FOOD

There are several sources of food in the arctic and subarctic regions. The type of food—fish, animal, fowl, or plant—and the ease in obtaining it depend on the time of the year and your location.

Fish. During the summer months, you can easily get fish and other water life from coastal waters, streams, rivers, and lakes. Use the techniques described in Part IV, Chapter 2 to catch fish.

The North Atlantic and North Pacific coastal waters are rich in seafood. You can easily find crawfish, snails, clams, oysters, and king crab. In areas where there is a great difference between the high and low tide water levels, you can easily find shellfish at low tide. Dig in the sand on the tidal flats. Look in tidal pools and on offshore reefs. In areas where there is a small difference between the high- and low-tide water levels, storm waves often wash shellfish onto the beaches.

The eggs of the spiny sea urchin that lives in the waters around the Aleutian Islands and southern Alaska are excellent food. Look for the sea urchins in tidal pools. Break the shell by placing it between two stones. The eggs are bright yellow in color.

Most northern fish and fish eggs are edible. Exceptions are the meat of the arctic shark and the eggs of the sculpins.

The bivalves, such as clams and mussels, are usually more palatable than spiral-shelled seafood, such as snails.

> **WARNING**
>
> The black mussel, a common mollusk of the far north, may be poisonous in any season. Toxins sometimes found in the mussel's tissue are as dangerous as strychnine.

The sea cucumber is another edible sea animal. Inside its body are five long white muscles that taste much like clam meat.

In early summer, smelt spawn in the beach surf. Sometimes you can scoop them up with your hands.

You can often find herring eggs on the seaweed in midsummer. Kelp, the long ribbonlike seaweed, and other smaller seaweed that grow among offshore rocks are also edible.

Sea Ice Animals. You find polar bears in practically all arctic coastal regions, but rarely inland. Avoid them if possible. They are the most dangerous of all bears. They are tireless, clever hunters with good sight and an extraordinary sense of smell. If you must kill one for food, approach it cautiously. Aim for the brain; a bullet elsewhere will rarely kill one. Always cook polar bear meat before eating it.

> ⚠ **CAUTION**
> Do not eat polar bear liver as it contains a toxic concentration of vitamin A.

Earless seal meat is some of the best meat available. You need considerable skill, however, to get close enough to an earless seal to kill it. In spring, seals often bask on the ice beside their breathing holes. They raise their heads about every 30 seconds, however, to look for their enemy, the polar bear.

To approach a seal, do as the Eskimos do—stay downwind from it, cautiously moving closer while it sleeps. If it moves, stop and imitate its movements by lying flat on the ice, raising your head up and down, and wriggling your body slightly. Approach the seal with your body sideways to it and your arms close to your body so that you look as much like another seal as possible. The ice at the edge of the breathing hole is usually smooth and at an incline, so the least movement of the seal may cause it to slide into the water. Therefore, try to get within 22 to 45 meters of the seal and kill it instantly (aim for the brain). Try to reach the seal before it slips into the water. In winter, a dead seal will usually float, but it is difficult to retrieve from the water.

Keep the seal blubber and skin from coming into contact with any scratch or broken skin you may have. You could get "spekk-finger," that is, a reaction that causes the hands to become badly swollen.

Keep in mind that where there are seals, there are usually polar bears, and polar bears have stalked and killed seal hunters.

You can find porcupines in southern subarctic regions where there are trees. Porcupines feed on bark; if you find tree limbs stripped bare, you are likely to find porcupines in the area.

Ptarmigans, owls, Canadian jays, grouse, and ravens are the only birds that remain in the arctic during the winter. They are scarce north of the tree line. Ptarmigans and owls are as good for food as any game bird. Ravens are too thin to be worth the effort it takes to catch them. Ptarmigans, which change color to blend with their surroundings, are hard to spot. Rock ptarmigans travel in pairs and you can easily approach them. Willow ptarmigans live among willow clumps in bottomlands. They gather in large flocks and you can easily snare them. During the summer months all arctic birds have a 2- to 3-week molting period during which they cannot fly and are easy to catch. Use one of the techniques described in Part IV, Chapter 2 to catch them.

Skin and butcher game (Part IV, Chapter 2) while it is still warm. If you do not have time to skin the game, at least remove its entrails, musk glands, and genitals before storing. If time allows, cut the meat into usable pieces and freeze each separately so that you can use the pieces as needed. Leave the fat on all animals except seals. During the winter, game freezes quickly if left in the open. During the summer, you can store it in underground ice holes.

Plants. Although tundras support a variety of plants during the warm months, all are small, however, when compared to plants in warmer climates. For instance, the arctic willow and birch are shrubs rather than trees. The following is a list of some plant foods found in arctic and subarctic regions (see Part IV, Chapter 6 for descriptions).

ARCTIC FOOD PLANTS

- Arctic raspberry and blueberry
- Arctic willow
- Bearberry
- Cranberry
- Crowberry
- Dandelion
- Eskimo potato

- Fireweed
- Iceland moss
- Marsh marigold
- Reindeer moss
- Rock tripe
- Spatterdock

There are some plants growing in arctic and subarctic regions that are poisonous if eaten (see Part IV, Chapter 7). Use the plants that you know are edible. When in doubt, follow the Universal Edibility Test in Part IV, Chapter 6.

TRAVEL

As a survivor or an evader in an arctic or subarctic region, you will face many obstacles. Your location and the time of the year will determine the types of obstacles and the inherent dangers. You should—

- Avoid traveling during a blizzard.
- Take care when crossing thin ice. Distribute your weight by lying flat and crawling.
- Cross streams when the water level is lowest. Normal freezing and thawing action may cause a stream level to vary as much as 2 to 2.5 meters per day. This variance may occur any time during the day, depending on the distance from a glacier, the temperature, and the terrain. Consider this variation in water level when selecting a campsite near a stream.
- Consider the clear arctic air. It makes estimating distance difficult. You more frequently underestimate than overestimate distances.
- Do not travel in "whiteout" conditions. The lack of contrasting colors makes it impossible to judge the nature of the terrain.
- Always cross a snow bridge at right angles to the obstacle it crosses. Find the strongest part of the bridge by poking ahead of you with a pole or ice axe. Distribute your weight by crawling or by wearing snowshoes or skis.
- Make camp early so that you have plenty of time to build a shelter.
- Consider frozen or unfrozen rivers as avenues of travel. However, some rivers that appear frozen may have soft, open areas that make travel very difficult or may not allow walking, skiing, or sledding.
- Use snowshoes if you are traveling over snow-covered terrain. Snow 30 or more centimeters deep makes traveling difficult. If you do not have snowshoes, make a pair using willow, strips of cloth, leather, or other suitable material.

It is almost impossible to travel in deep snow without snowshoes or skis. Traveling by foot leaves a well-marked trail for any pursuers to follow. If you must travel in deep snow, avoid snow-covered streams. The snow, which acts as an insulator, may have prevented ice from forming over the water. In hilly terrain, avoid areas where avalanches appear possible. Travel in the early morning in areas where there is danger of avalanches. On ridges, snow gathers on the lee side in overhanging piles called cornices. These often extend far out from the ridge and may break loose if stepped on.

WEATHER SIGNS

There are several good indicators of climatic changes.

Wind. You can determine wind direction by dropping a few leaves or grass or by watching the treetops. Once you determine the wind direction, you can predict the type of weather that is imminent. Rapidly shifting winds indicate an unsettled atmosphere and a likely change in the weather.

Clouds. Clouds come in a variety of shapes and patterns. A general knowledge of clouds and the atmospheric conditions they indicate can help you predict the weather.

Smoke. Smoke rising in a thin vertical column indicates fair weather. Low rising or "flattened out" smoke indicates stormy weather.

Birds and Insects. Birds and insects fly lower to the ground than normal in heavy, moisture-laden air. Such flight indicates that rain is likely. Most insect activity increases before a storm, but bee activity increases before fair weather.

Low-Pressure Front. Slow-moving or imperceptible winds and heavy, humid air often indicate a low-pressure front. Such a front promises bad weather that will probably linger for several days. You can "smell" and "hear" this front. The sluggish, humid air makes wilderness odors more pronounced than during high-pressure conditions. In addition, sounds are sharper and carry farther in low-pressure than high-pressure conditions.

CHAPTER 4

Survival in Mountain Terrain

SECTION 1: MOUNTAIN TERRAIN AND WEATHER

Operations in the mountains require soldiers to be physically fit and leaders to be experienced in operations in this terrain. Problems arise in moving men and transporting loads up and down steep and varied terrain in order to accomplish the mission. Chances for success in this environment are greater when a leader has experience operating under the same conditions as his men. Acclimatization, conditioning, and training are important factors in successful military mountaineering.

1-1. Definition. Mountains are land forms that rise more than 500 meters above the surrounding plain and are characterized by steep slopes. Slopes commonly range from 4 to 45 degrees. Cliffs and precipices may be vertical or overhanging. Mountains may consist of an isolated peak, single ridges, glaciers, snowfields, compartments, or complex ranges extending for long distances and obstructing movement. Mountains usually favor the defense; however, attacks can succeed by using detailed planning, rehearsals, surprise, and well-led troops.

1-2. Composition. All mountains are made up of rocks and all rocks of minerals (compounds that cannot be broken down except by chemical action). Of the approximately 2,000 known minerals, seven rock-forming minerals make up most of the earth's crust: quartz and feldspar make up granite and sandstone; olivene and pyroxene give basalt its dark color; and amphibole and biotite (mica) are the black crystalline specks in granitic rocks. Except for calcite, found in limestone, they all contain silicon and are often referred to as silicates.

1-3. Rock and Slope Types. Different types of rock and different slopes present different hazards. The following paragraphs discuss the characteristics and hazards of the different rocks and slopes.

a. Granite. Granite produces fewer rockfalls, but jagged edges make pulling rope and raising equipment more difficult. Granite is abrasive and increases the danger of ropes or accessory cords being cut. Climbers must beware of large loose boulders. After a rain, granite dries quickly. Most climbing holds are found in cracks. Face climbing can be found, however, it cannot be protected.

b. Chalk and Limestone. Chalk and limestone are slippery when wet. Limestone is usually solid; however, conglomerate type stones may be loose. Limestone has pockets, face climbing, and cracks.

c. Slate and Gneiss. Slate and gneiss can be firm and or brittle in the same area (red coloring indicates brittle areas). Rockfall danger is high, and small rocks may break off when pulled or when pitons are emplaced.

d. Sandstone. Sandstone is usually soft causing handholds and footholds to break away under pressure. Chocks placed in sandstone may or may not hold. Sandstone should be allowed to dry for a couple of days after a rain before climbing on it as wet sandstone is extremely soft. Most climbs follow a crack. Face climbing is possible, but any outward pull will break off handholds and foot holds, and it is usually difficult to protect.

e. Grassy Slopes. Penetrating roots and increased frost cracking cause a continuous loosening process. Grassy slopes are slippery after rain, new snow, and dew. After long, dry spells clumps of the slope tend to break away. Weight should be distributed evenly; for example, use flat hand push holds instead of finger pull holds.

f. **Firm Spring Snow (Firn Snow).** Stopping a slide on small, leftover snow patches in late spring can be difficult. Routes should be planned to avoid these dangers. Self-arrest should be practiced before encountering this situation. Beginning climbers should be secured with rope when climbing on this type surface. Climbers can glissade down firn snow if necessary. Firn snow is easier to ascend than walking up scree or talus.

g. **Talus.** Talus is rocks that are larger than a dinner plate, but smaller than boulders. They can be used as stepping-stones to ascend or descend a slope. However, if a talus rock slips away it can produce more injury than scree because of its size.

h. **Scree.** Scree is small rocks that are from pebble size to dinner plate size. Running down scree is an effective method of descending in a hurry. One can run at full stride without worry the whole scree field is moving with you. Climbers must beware of larger rocks that may be solidly planted under the scree. Ascending scree is a tedious task. The scree does not provide a solid platform and will only slide under foot. If possible, avoid scree when ascending.

1-4. Cross-Country Movement. Soldiers must know the terrain to determine the feasible routes for cross-country movement when no roads or trails are available.

a. A pre-operations intelligence effort should include topographic and photographic map coverage as well as detailed weather data for the area of operations. When planning mountain operations, additional information may be needed about size, location, and characteristics of landforms; drainage; types of rock and soil; and the density and distribution of vegetation. Control must be decentralized to lower levels because of varied terrain, erratic weather, and communication problems inherent to mountainous regions.

b. Movement is often restricted due to terrain and weather. The erratic weather requires that soldiers be prepared for wide variations in temperature, types, and amounts of precipitation.

   (1) Movement above the timberline reduces the amount of protective cover available at lower elevations. The logistical problem is important; therefore, each man must be self-sufficient to cope with normal weather changes using materials from his rucksack.

   (2) Movement during a storm is difficult due to poor visibility and bad footing on steep terrain. Although the temperature is often higher during a storm than during clear weather, the dampness of rain and snow and the penetration of wind cause soldiers to chill quickly. Although climbers should get off the high ground and seek shelter and warmth, during severe mountain storms, capable commanders may use reduced visibility to achieve tactical surprise.

c. When the tactical situation requires continued movement during a storm, the following precautions should be observed:
   - Maintain visual contact.
   - Keep warm. Maintain energy and body heat by eating and drinking often; carry food that can be eaten quickly and while on the move.
   - Keep dry. Wear wet-weather clothing when appropriate, but do not overdress, which can cause excessive perspiration and dampen clothing. As soon as the objective is reached and shelter secured, put on dry clothing.
   - Do not rush. Hasty movement during storms leads to breaks in contact and accidents.
   - If lost, stay warm, dry, and calm.
   - Do not use ravines as routes of approach during a storm as they often fill with water and are prone to flash floods.
   - Avoid high pinnacles and ridgelines during electrical storms.
   - Avoid areas of potential avalanche or rock-fall danger.

1-5. Cover and Concealment. When moving in the mountains, outcroppings, boulders, heavy vegetation, and intermediate terrain can provide cover and concealment. Digging fighting positions and temporary fortifications is difficult because soil is often thin or stony. The selection of dug-in positions requires detailed planning. Some rock types, such as volcanic tuff, are easily excavated. In other areas, boulders and other loose rocks can be used for building hasty fortifications. In alpine environments, snow and ice blocks may be cut and stacked to supplement dug-in positions. As in all operations, positions and routes must be camouflaged to blend in with the surrounding terrain to prevent aerial detection.

1-6. Observation. Observation in mountains varies because of weather and ground cover. The dominating height of mountainous terrain permits excellent long-range observation. However, rapidly changing weather with frequent periods of high winds, rain, snow, sleet, hail, and fog can limit visibility. The rugged nature of the terrain often produces dead space at midranges.

   a. Low cloud cover at higher elevations may neutralize the effectiveness of Observation Points (OPs) established on peaks or mountaintops. High wind speeds and sound often mask the noises of troop movement. Several OPs may need to be established laterally, in depth, and at varying altitudes to provide visual coverage of the battle area.

   b. Conversely, the nature of the terrain can be used to provide concealment from observation. This concealment can be obtained in the dead space. Mountainous regions are subject to intense shadowing effects when the sun is low in relatively clear skies. The contrast from lighted to shaded areas causes visual acuity in the shaded regions to be considerably reduced. These shadowed areas can provide increased concealment when combined with other camouflage and should be considered in maneuver plans.

MOUNTAIN WEATHER

Most people subconsciously "forecast" the weather. If they look outside and see dark clouds they may decide to take rain gear. If an unexpected wind strikes, people glance to the sky for other bad signs. A conscious effort to follow weather changes will ultimately lead to a more accurate forecast. An analysis of mountain weather and how it is affected by mountain terrain shows that such weather is prone to patterns and is usually severe, but patterns are less obvious in mountainous terrain than in other areas. Conditions greatly change with altitude, latitude, and exposure to atmospheric winds and air masses. Mountain weather can be extremely erratic. It varies from stormy winds to calm, and from extreme cold to warmth within a short time or with a minor shift in locality. The severity and variance of the weather causes it to have a major impact on military operations.

1-7. Considerations for Planning. Mountain weather can be either a dangerous obstacle to operations or a valuable aid, depending on how well it is understood and to what extent advantage is taken of its peculiar characteristics.

   a. Weather often determines the success or failure of a mission since it is highly changeable. Military operations plans must be flexible, especially in planning airmobile and airborne operations. The weather must be anticipated to allow enough time for planning so that the leaders of subordinate units can use their initiative in turning an important weather factor in their favor. The clouds that often cover the tops of mountains and the fogs that cover valleys are an excellent means of concealing movements that normally are made during darkness or in smoke. Limited visibility can be used as a combat multiplier.

   b. The safety or danger of almost all high mountain regions, especially in winter, depends upon a change of a few degrees of temperature above or below the freezing point. Ease and speed of travel depend mainly on the weather. Terrain that can be crossed swiftly and safely one day may become impassable or highly dangerous the next due to snowfall, rainfall, or a rise in temperature. The reverse can happen just as quickly. The prevalence of avalanches depends on terrain, snow conditions, and weather factors.

c. Some mountains, such as those found in desert regions, are dry and barren with temperatures ranging from extreme heat in the summer to extreme cold in the winter. In tropical regions, lush jungles with heavy seasonal rains and little temperature variation often cover mountains. High rocky crags with glaciated peaks can be found in mountain ranges at most latitudes along the western portion of the Americas and Asia.

d. Severe weather may decrease morale and increase basic survival problems. These problems can be minimized when men have been trained to accept the weather by being self-sufficient. Mountain soldiers properly equipped and trained can use the weather to their advantage in combat operations.

1-8. Mountain Air. High mountain air is dry and may be drier in the winter. Cold air has a reduced capacity to hold water vapor. Because of this increased dryness, equipment does not rust as quickly and organic material decomposes slowly. The dry air also requires soldiers to increase consumption of water. The reduced water vapor in the air causes an increase in evaporation of moisture from the skin and in loss of water through transpiration in the respiratory system. Due to the cold, most soldiers do not naturally consume the quantity of fluids they would at higher temperatures and must be encouraged to consciously increase their fluid intake.

a. Pressure is low in mountainous areas due to the altitude. The barometer usually drops 2.5 centimeters for every 300 meters gained in elevation (3 percent).

b. The air at higher altitudes is thinner as atmospheric pressure drops with the increasing altitude. The altitude has a natural filtering effect on the sun's rays. Rays are absorbed or reflected in part by the molecular content of the atmosphere. This effect is greater at lower altitudes. At higher altitudes, the thinner, drier air has a reduced molecular content and, consequently, a reduced filtering effect on the sun's rays. The intensity of both visible and ultraviolet rays is greater with increased altitude. These conditions increase the chance of sunburn, especially when combined with a snow cover that reflects the rays upward.

1-9. Weather Characteristics. The earth is surrounded by an atmosphere that is divided into several layers. The world's weather systems are in the lower of these layers known as the "troposphere." This layer reaches as high as 40,000 feet. Weather is a result of the atmosphere, oceans, land masses, unequal heating and cooling from the sun, and the earth's rotation. The weather found in any one place depends on many things such as the air temperature, humidity (moisture content), air pressure (barometric pressure), how it is being moved, and if it is being lifted or not.

a. Air pressure is the "weight" of the atmosphere at any given place. The higher the pressure, the better the weather will be. With lower air pressure, the weather will more than likely be worse. In order to understand this, imagine that the air in the atmosphere acts like a liquid. Areas with a high level of this "liquid" exert more pressure on an area and are called high-pressure areas. Areas with a lower level are called low-pressure areas. The average air pressure at sea level is 29.92 inches of mercury (hg) or 1,013 millibars (mb). The higher in altitude, the lower the pressure.

(1) *High Pressure.* The characteristics of a high-pressure area are as follows:
  - The airflow is clockwise and out.
  - Otherwise known as an "anticyclone."
  - Associated with clear skies.
  - Generally the winds will be mild.
  - Depicted as a blue "H" on weather maps.

(2) *Low Pressure.* The characteristics of a low-pressure area are as follows:
  - The airflow is counterclockwise and in.
  - Otherwise known as a "cyclone."
  - Associated with bad weather.
  - Depicted as a red "L" on weather maps.

b. Air from a high-pressure area is basically trying to flow out and equalize its pressure with the surrounding air. Low pressure, on the other hand, is building up vertically by pulling air in from outside itself, which causes atmospheric instability resulting in bad weather.

c. On a weather map, these differences in pressure are depicted as isobars. Isobars resemble contour lines and are measured in either millibars or inches of mercury. The areas of high pressure are called "ridges" and lows are called "troughs."

1-10. Wind. In high mountains, the ridges and passes are seldom calm; however, strong winds in protected valleys are rare. Normally, wind speed increases with altitude since the earth's frictional drag is strongest near the ground. This effect is intensified by mountainous terrain. Winds are accelerated when they converge through mountain passes and canyons. Because of these funneling effects, the wind may blast with great force on an exposed mountainside or summit. Usually, the local wind direction is controlled by topography.

a. The force exerted by wind quadruples each time the wind speed doubles; that is, wind blowing at 40 knots pushes four times harder than a wind blowing at 20 knots. With increasing wind strength, gusts become more important and may be 50 percent higher than the average wind speed. When wind strength increases to a hurricane force of 64 knots or more, soldiers should lay on the ground during gusts and continue moving during lulls. If a hurricane- force wind blows where there is sand or snow, dense clouds fill the air. The rocky debris or chunks of snow crust are hurled near the surface. During the winter season, or at high altitudes, commanders must be constantly aware of the wind-chill factor and associated cold-weather injuries.

b. Winds are formed due to the uneven heating of the air by the sun and rotation of the earth. Much of the world's weather depends on a system of winds that blow in a set direction.

c. Above hot surfaces, air expands and moves to colder areas where it cools and becomes denser, and sinks to the earth's surface. The results are a circulation of air from the poles along the surface of the earth to the equator, where it rises and moves to the poles again.

d. Heating and cooling together with the rotation of the earth causes surface winds. In the Northern Hemisphere, there are three prevailing winds:

　(1) *Polar Easterlies.* These are winds from the polar region moving from the east. This is air that has cooled and settled at the poles.

　(2) *Prevailing Westerlies.* These winds originate from approximately 30 degrees north latitude from the west. This is an area where prematurely cooled air, due to the earth's rotation, has settled to the surface.

　(3) *Northeast Tradewinds.* These are winds that originate from approximately 30° north from the northeast.

e. The jet stream is a long meandering current of high-speed winds often exceeding 250 miles per hour near the transition zone between the troposphere and the stratosphere known as the tropopause. These winds blow from a generally westerly direction dipping down and picking up air masses from the tropical regions and going north and bringing down air masses from the polar regions.

f. The patterns of wind mentioned above move air. This air comes in parcels called "air masses." These air masses can vary from the size of a small town to as large as a country. These air masses are named from where they originate:

- Maritime over water.
- Continental over land.
- Polar north of 60° north latitude.
- Tropical south of 60° north latitude.

Combining these parcels of air provides the names and description of the four types of air masses:

- Continental Polar cold, dry air mass.
- Maritime Polar cold, wet air mass.

- Maritime Tropical warm, wet air mass.
- Continental Tropical warm, dry air mass.

g. Two types of winds are peculiar to mountain environments, but do not necessarily affect the weather.

    (1) *Anabatic Wind (Valley Winds).* These winds blow up mountain valleys to replace warm rising air and are usually light winds.

    (2) *Katabatic Wind (Mountain Wind).* These winds blow down mountain valley slopes caused by the cooling of air and are occasionally strong winds.

1-11. Humidity. Humidity is the amount of moisture in the air. All air holds water vapor even if it cannot be seen. Air can hold only so much water vapor; however, the warmer the air, the more moisture it can hold. When air can hold all that it can the air is "saturated" or has 100 percent relative humidity.

    a. If air is cooled beyond its saturation point, the air will release its moisture in one form or another (clouds, fog, dew, rain, snow, and so on). The temperature at which this happens is called the "condensation point." The condensation point varies depending on the amount of water vapor contained in the air and the temperature of the air. If the air contains a great deal of water, condensation can occur at a temperature of 68 degrees Fahrenheit, but if the air is dry and does not hold much moisture, condensation may not form until the temperature drops to 32 degrees Fahrenheit or even below freezing.

    b. The adiabatic lapse rate is the rate at which air cools as it rises or warms as it descends. This rate varies depending on the moisture content of the air. Saturated (moist) air will warm and cool approximately 3.2 degrees Fahrenheit per 1,000 feet of elevation gained or lost. Dry air will warm and cool approximately 5.5 degrees Fahrenheit per 1,000 feet of elevation gained or lost.

1-12. Cloud Formation. Clouds are indicators of weather conditions. By reading cloud shapes and patterns, observers can forecast weather with little need for additional equipment such as a barometer, wind meter, and thermometer. Anytime air is lifted or cooled beyond its saturation point (100 percent relative humidity), clouds are formed. The four ways air gets lifted and cooled beyond its saturation point are as follows.

    a. *Convective Lifting.* This effect happens due to the sun's heat radiating off the Earth's surface causing air currents (thermals) to rise straight up and lift air to a point of saturation.

    b. *Frontal Lifting.* A front is formed when two air masses of different moisture content and temperature collide. Since air masses will not mix, warmer air is forced aloft over the colder air mass. From there it is cooled and then reaches its saturation point. Frontal lifting creates the majority of precipitation.

    c. *Cyclonic Lifting.* An area of low pressure pulls air into its center from all over in a counterclockwise direction. Once this air reaches the center of the low pressure, it has nowhere to go but up. Air continues to lift until it reaches the saturation point.

    d. *Orographic Lifting.* This happens when an air mass is pushed up and over a mass of higher ground such as a mountain. Air is cooled due to the adiabatic lapse rate until the air's saturation point is reached.

1-13. Types of Clouds. Clouds are one of the signposts to what is happening with the weather. Clouds can be described in many ways. They can be classified by height or appearance, or even by the amount of area covered vertically or horizontally. Clouds are classified into five categories: low-, mid-, and high-level clouds; vertically-developed clouds; and less common clouds.

a. *Low-Level Clouds.* Low-level clouds (0 to 6,500 feet) are either cumulus or stratus (Figures 1-1 and 1-2). Low-level clouds are mostly composed of water droplets since their bases lie below 6,500 feet. When temperatures are cold enough, these clouds may also contain ice particles and snow.

(1) The two types of precipitating low-level clouds are nimbostratus and stratocumulus (Figures 1-3 and 1-4).

Figure 1-1: Cumulus clouds.

Figure 1-2: Stratus clouds.

Figure 1-3: Nimbostratus clouds.

Figure 1-4: Stratocumulus clouds.

(a) Nimbostratus clouds are dark, low-level clouds accompanied by light to moderately falling precipitation. The sun or moon is not visible through nimbostratus clouds, which distinguishes them from mid-level altostratus clouds. Because of the fog and falling precipitation commonly found beneath and around nimbostratus clouds, the cloud base is typically extremely diffuse and difficult to accurately determine.

(b) Stratocumulus clouds generally appear as a low, lumpy layer of clouds that is sometimes accompanied by weak precipitation. Stratocumulus vary in color from dark gray to light gray and may appear as rounded masses with breaks of clear sky in between. Because the individual elements of stratocumulus are larger than those of altocumulus, deciphering between the two cloud types is easier. With your arm extended toward the sky, altocumulus elements are about the size of a thumbnail while stratocumulus are about the size of a fist.

(2) Low-level clouds may be identified by their height above nearby surrounding relief of known elevation. Most precipitation originates from low-level clouds because rain or snow usually evaporate before reaching the ground from higher clouds. Low-level clouds usually indicate impending precipitation, especially if the cloud is more than 3,000 feet thick. (Clouds that appear dark at their bases are more than 3,000 feet thick.)

b. *Mid-Level Clouds.* Mid-level clouds (between 6,500 to 20,000 feet) have a prefix of alto. Middle clouds appear less distinct than low clouds because of their height. Alto clouds with sharp edges are warmer because they are composed mainly of water droplets. Cold clouds, composed mainly of ice crystals and usually colder than –30 degrees F, have distinct edges that grade gradually into the surrounding sky. Middle clouds usually indicate fair weather, especially if they are rising over time. Lowering middle clouds indicate potential storms, though usually hours away. There are two types of mid-level clouds, altocumulus and altostratus clouds (Figures 1-5 and 1-6).

(1) Altocumulus clouds can appear as parallel bands or rounded masses. Typically a portion of an altocumulus cloud is shaded, a characteristic which makes them distinguishable from high-level

Figure 1-5: Altocumulus.

Figure 1-6: Altostratus.

cirrocumulus. Altocumulus clouds usually form in advance of a cold front. The presence of altocumulus clouds on a warm humid summer morning is commonly followed by thunderstorms later in the day. Altocumulus clouds that are scattered rather than even, in a blue sky, are called "fair weather" cumulus and suggest arrival of high pressure and clear skies.

(2) Altostratus clouds are often confused with cirrostratus. The one distinguishing feature is that a halo is not observed around the sun or moon. With altostratus, the sun or moon is only vaguely visible and appears as if it were shining through frosted glass.

c. *High-Level Clouds.* High-level clouds (more than 20,000 feet above ground level) are usually frozen clouds, indicating air temperatures at that elevation below -30 degrees Fahrenheit, with a fibrous structure and blurred outlines. The sky is often covered with a thin veil of cirrus that partly obscures the sun or, at night, produces a ring of light around the moon. The arrival of cirrus indicates moisture aloft and the approach of a traveling storm system. Precipitation is often 24 to 36 hours away. As the storm approaches, the cirrus thickens and lowers, becoming altostratus and eventually stratus. Temperatures are warm, humidity rises, and winds become southerly or southeasterly. The two types of high-level clouds are cirrus and cirrostratus (Figure 1-7 and Figure 1-8).

(1) Cirrus clouds are the most common of the high-level clouds. Typically found at altitudes greater than 20,000 feet, cirrus are composed of ice crystals that form when super-cooled water droplets freeze. Cirrus clouds generally occur in fair weather and point in the direction of air movement at their elevation. Cirrus can be observed in a variety of shapes and sizes. They can be nearly straight, shaped like a comma, or seemingly all tangled together. Extensive cirrus clouds are associated with an approaching warm front.

(2) Cirrostratus clouds are sheet-like, high-level clouds composed of ice crystals. They are relatively transparent and can cover the entire sky and be up to several thousand feet thick. The sun or moon can be seen through cirrostratus. Sometimes the only indication of cirrostratus clouds is a halo around the sun or moon. Cirrostratus clouds tend to thicken as a warm front approaches, signifying an increased production of ice crystals. As a result, the halo gradually disappears and the sun or moon becomes less visible.

d. *Vertical-Development Clouds.* Clouds with vertical development can grow to heights in excess of 39,000 feet, releasing incredible amounts of energy. The two types of clouds with vertical development are fair weather cumulus and cumulonimbus.

(1) Fair weather cumulus clouds have the appearance of floating cotton balls and have a lifetime of 5 to 40 minutes. Known for their flat bases and distinct outlines, fair weather cumulus exhibit only slight vertical growth, with the cloud tops designating the limit of the rising air. Given

Figure 1-7: Cirrus.

Figure 1-8: Cirrostratus.

suitable conditions, however, these clouds can later develop into towering cumulonimbus clouds associated with powerful thunderstorms. Fair weather cumulus clouds are fueled by buoyant bubbles of air known as thermals that rise up from the earth's surface. As the air rises, the water vapor cools and condenses forming water droplets. Young fair weather cumulus clouds have sharply defined edges and bases while the edges of older clouds appear more ragged, an artifact of erosion. Evaporation along the cloud edges cools the surrounding air, making it heavier and producing sinking motion outside the cloud. This downward motion inhibits further convection and growth of additional thermals from down below, which is why fair weather cumulus typically have expanses of clear sky between them. Without a continued supply of rising air, the cloud begins to erode and eventually disappears.

(2) Cumulonimbus clouds are much larger and more vertically developed than fair weather cumulus (Figure 1-9). They can exist as individual towers or form a line of towers called a squall line. Fueled by vigorous convective updrafts, the tops of cumulonimbus clouds can reach 39,000 feet or higher. Lower levels of cumulonimbus clouds consist mostly of water droplets while at higher elevations, where the temperatures are well below freezing, ice crystals dominate the composition. Under favorable conditions, harmless fair weather cumulus clouds can quickly develop into large cumulonimbus associated with powerful thunderstorms known as super-cells. Super-cells are large thunderstorms with deep rotating updrafts and can have a lifetime of several hours. Super-cells produce frequent lightning, large hail, damaging winds, and tornadoes. These storms tend to develop during the afternoon and early evening when the effects of heating from the sun are the strongest.

e. *Other Cloud Types.* These clouds are a collection of miscellaneous types that do not fit into the previous four groups. They are orographic clouds, lenticulars, and contrails.

(1) Orographic clouds develop in response to the forced lifting of air by the earth's topography. Air passing over a mountain oscillates up and down as it moves downstream. Initially, stable air encounters a mountain, is lifted upward, and cools. If the air cools to its saturation temperature during this process, the water vapor condenses and becomes visible as a cloud. Upon reaching the mountain top, the air is heavier than the environment and will sink down the other side, warming as it descends. Once the air returns to its original height, it has the same buoyancy as the surrounding air. However, the air does not stop immediately because it still has momentum

Figure 1-9: Cumulonimbus.

carrying it downward. With continued descent, the air becomes warmer then the surrounding air and accelerates back upwards towards its original height. Another name for this type of cloud is the lenticular cloud.

(2) Lenticular clouds are cloud caps that often form above pinnacles and peaks, and usually indicate higher winds aloft (Figure 1-10). Cloud caps with a lens shape, similar to a "flying saucer," indicate extremely high winds (over 40 knots). Lenticulars should always be watched for changes. If they grow and descend, bad weather can be expected.

(3) Contrails are clouds that are made by water vapor being inserted into the upper atmosphere by the exhaust of jet engines (Figure 1-11). Contrails evaporate rapidly in fair weather. If it takes longer than two hours for contrails to evaporate, then there is impending bad weather (usually about 24 hours prior to a front).

Figure 1-10: Lenticular.

Figure 1-11: Contrails.

f. *Cloud Interpretation.* Serious errors can occur in interpreting the extent of cloud cover, especially when cloud cover must be reported to another location. Cloud cover always appears greater on or near the horizon, especially if the sky is covered with cumulus clouds, since the observer is looking more at the sides of the clouds rather than between them. Cloud cover estimates should be restricted to sky areas more than 40 degrees above the horizon that is, to the local sky. Assess the sky by dividing the 360 degrees of sky around you into eighths. Record the coverage in eighths and the types of clouds observed.

1-14. Fronts. Fronts occur when two air masses of different moisture and temperature contents meet. One of the indicators that a front is approaching is the progression of the clouds. The four types of fronts are warm, cold, occluded, and stationary.

a. *Warm Front.* A warm front occurs when warm air moves into and over a slower or stationary cold air mass. Because warm air is less dense, it will rise up and over the cooler air. The cloud types seen when a warm front approaches are cirrus, cirrostratus, nimbostratus (producing rain), and fog. Occasionally, cumulonimbus clouds will be seen during the summer months.

b. *Cold Front.* A cold front occurs when a cold air mass overtakes a slower or stationary warm air mass. Cold air, being more dense than warm air, will force the warm air up. Clouds observed will be cirrus, cumulus, and then cumulonimbus producing a short period of showers.

c. *Occluded Front.* Cold fronts generally move faster than warm fronts. The cold fronts eventually overtake warm fronts and the warm air becomes progressively lifted from the surface. The zone of division between cold air ahead and cold air behind is called a "cold occlusion." If the air behind the front is warmer than the air ahead, it is a warm occlusion. Most land areas experience more occlusions than other types of fronts. The cloud progression observed will be cirrus, cirrostratus, altostratus, and nimbostratus. Precipitation can be from light to heavy.

d. *Stationary Front.* A stationary front is a zone with no significant air movement. When a warm or cold front stops moving, it becomes a stationary front. Once this boundary begins forward motion, it once again becomes a warm or cold front. When crossing from one side of a stationary front to another, there is typically a noticeable temperature change and shift in wind direction. The weather is usually clear to partly cloudy along the stationary front.

1-15. Temperature. Normally, a temperature drop of 3 to 5 degrees Fahrenheit for every 1,000 feet gain in altitude is encountered in motionless air. For air moving up a mountain with condensation occurring (clouds, fog, and precipitation), the temperature of the air drops 3.2 degrees Fahrenheit with every 1,000 feet of elevation gain. For air moving up a mountain with no clouds forming, the temperature of the air drops 5.5 degrees Fahrenheit for every 1,000 feet of elevation gain.

a. An expedient to this often occurs on cold, clear, calm mornings. During a troop movement or climb started in a valley, higher temperatures may often be encountered as altitude is gained. This reversal of the normal cooling with elevation is called temperature inversion. Temperature inversions are caused when mountain air is cooled by ice, snow, and heat loss through thermal radiation. This cooler, denser air settles into the valleys and low areas. The inversion continues until the sun warms the surface of the earth or a moderate wind causes a mixing of the warm and cold layers. Temperature inversions are common in the mountainous regions of the arctic, subarctic, and mid-latitudes.

b. At high altitudes, solar heating is responsible for the greatest temperature contrasts. More sunshine and solar heat are received above the clouds than below. The important effect of altitude is that the sun's rays pass through less of the atmosphere and more direct heat is received than at lower levels, where solar radiation is absorbed and reflected by dust and water vapor. Differences of 40 to 50 degrees Fahrenheit may occur between surface temperatures in the shade and surface temperatures in the sun. This is particularly true for dark metallic objects. The difference in temperature felt on the skin between the sun and shade is normally 7 degrees Fahrenheit. Special care must be taken to avoid sunburn and snow blindness. Besides permitting rapid heating, the clear air at high altitudes also favors rapid cooling at night. Consequently, the temperature rises fast after sunrise and drops quickly after sunset. Much of the chilled air drains downward, due to convection currents, so that the differences between day and night temperatures are greater in valleys than on slopes.

c. Local weather patterns force air currents up and over mountaintops. Air is cooled on the windward side of the mountain as it gains altitude, but more slowly (3.2 degrees Fahrenheit per 1,000 feet) if clouds are forming due to heat release when water vapor becomes liquid. On the leeward side of the mountain, this heat gained from the condensation on the windward side is added to the normal heating that occurs as the air descends and air pressure increases. Therefore, air and winds on the leeward slope are considerably warmer than on the windward slope, which is referred to as Chinook winds. The heating and cooling of the air affects planning considerations primarily with regard to the clothing and equipment needed for an operation.

1-16. Weather Forecasting. The use of a portable aneroid barometer, thermometer, wind meter, and hygrometer help in making local weather forecasts. Reports from other localities and from any weather service, including USAF, USN, or the National Weather Bureau, are also helpful. Weather reports should be used in conjunction with the locally observed current weather situation to forecast future weather patterns.

a. Weather at various elevations may be quite different because cloud height, temperature, and barometric pressure will all be different. There may be overcast and rain in a lower area, with mountains rising above the low overcast into warmer clear weather.

b. To be effective, a forecast must reach the small-unit leaders who are expected to utilize weather conditions for assigned missions. Several different methods can be used to create a forecast. The method a forecaster chooses depends upon the forecaster's experience, the amount of data available, the level of difficulty that the forecast situation presents, and the degree of accuracy needed to make the forecast. The five ways to forecast weather are:

(1) *Persistence Method.* "Today equals tomorrow" is the simplest way of producing a forecast. This method assumes that the conditions at the time of the forecast will not change; for example, if today was hot and dry, the persistence method predicts that tomorrow will be the same.

(2) *Trends Method.* "Nowcasting" involves determining the speed and direction of fronts, high- and low-pressure centers, and clouds and precipitation. For example, if a cold front moves 300 miles during a 24-hour period, we can predict that it will travel 300 miles in another 24-hours.

(3) *Climatology Method.* This method averages weather statistics accumulated over many years. This only works well when the pattern is similar to the following years.

(4) *Analog Method.* This method examines a day's forecast and recalls a day in the past when the weather looked similar (an analogy). This method is difficult to use because finding a perfect analogy is difficult.

(5) *Numerical Weather Prediction.* This method uses computers to analyze all weather conditions and is the most accurate of the five methods.

SECTION 2: MOUNTAIN HAZARDS

Hazards can be termed natural (caused by natural occurrence), man-made (caused by an individual, such as lack of preparation, carelessness, improper diet, equipment misuse), or as a combination (human trigger). There are two kinds of hazards while in the mountains: subjective and objective. Combinations of objective and subjective hazards are referred to as cumulative hazards.

2-1. Subjective Hazards. Subjective hazards are created by humans; for example, choice of route, companions, overexertion, dehydration, climbing above one's ability, and poor judgment.

a. *Falling.* Falling can be caused by carelessness, over-fatigue, heavy equipment, bad weather, overestimating ability, a hold breaking away, or other reasons.

b. *Bivouac Site.* Bivouac sites must be protected from rockfall, wind, lightning, avalanche run-out zones, and flooding (especially in gullies). If the possibility of falling exists, rope in; the tent and all equipment may have to be tied down.

c. *Equipment.* Ropes are not total security; they can be cut on a sharp edge or break due to poor maintenance, age, or excessive use. You should always pack emergency and bivouac equipment even if the weather situation, tour, or a short climb is seemingly low of dangers.

2-2. Objective Hazards. Objective hazards are caused by the mountain and weather and cannot be influenced by man; for example, storms, rockfalls, icefalls, lightning, and so on.

a. *Altitude.* At high altitudes (especially over 6,500 feet), endurance and concentration is reduced. Cut down on smoking and alcohol. Sleep well, acclimatize slowly, stay hydrated, and be aware of signs and symptoms of high-altitude illnesses. Storms can form quickly and lightning can be severe.

b. *Visibility.* Fog, rain, darkness, and or blowing snow can lead to disorientation. Take note of your exact position and plan your route to safety before visibility decreases. Cold combined with fog can cause a thin sheet of ice to form on rocks (verglas). Whiteout conditions can be extremely dangerous. If you must move under these conditions, it is best to rope up. Have the point man move to the end of the rope.

The second man will use the first man as an aiming point with the compass. Use a route sketch and march table. If the tactical situation does not require it, plan route so as not to get caught by darkness.

c. *Gullies.* Rock, snow, and debris are channeled down gullies. If ice is in the gully, climbing at night may be better because the warming of the sun will loosen stones and cause rockfalls.

d. *Rockfall.* Blocks and scree at the base of a climb can indicate recurring rockfall. Light colored spots on the wall may indicate impact chips of falling rock. Spring melt or warming by the sun of the rock/ice/snow causes rockfall.

e. *Avalanches.* Avalanches are caused by the weight of the snow overloading the slope. (Refer to paragraph 2-4 for more detailed information on avalanches.)

f. *Hanging Glaciers and Seracs.* Avoid, if at all possible, hanging glaciers and seracs. They will fall without warning regardless of the time of day or time of year. One cubic meter of glacier ice weighs 910 kilograms (about 2,000 pounds). If you must cross these danger areas, do so quickly and keep an interval between each person.

g. *Crevasses.* Crevasses are formed when a glacier flows over a slope and makes a bend, or when a glacier separates from the rock walls that enclose it. A slope of only two to three degrees is enough to form a crevasse. As this slope increases from 25 to 30 degrees, hazardous icefalls can be formed. Likewise, as a glacier makes a bend, it is likely that crevasses will form at the outside of the bend. Therefore, the safest route on a glacier would be to the inside of bends, and away from steep slopes and icefalls. Extreme care must be taken when moving off of or onto the glacier because of the moat that is most likely to be present.

2-3. Weather Hazards. Weather conditions in the mountains may vary from one location to another as little as 10 kilometers apart. Approaching storms may be hard to spot if masked by local peaks. A clear, sunny day in July could turn into a snowstorm in less than an hour. Always pack some sort of emergency gear.

a. Winds are stronger and more variable in the mountains; as wind doubles in speed, the force quadruples.

b. Precipitation occurs more on the windward side than the leeward side of ranges. This causes more frequent and denser fog on the windward slope.

c. Above approximately 8,000 feet, snow can be expected any time of year in the temperate climates.

d. Air is dryer at higher altitudes, so equipment does not rust as quickly, but dehydration is of greater concern.

e. Lightning is frequent, violent, and normally attracted to high points and prominent features in mountain storms. Signs indicative of thunderstorms are tingling of the skin, hair standing on end, humming of metal objects, crackling, and a bluish light (St. Elmo's fire) on especially prominent metal objects (summit crosses and radio towers).

(1) Avoid peaks, ridges, rock walls, isolated trees, fixed wire installations, cracks that guide water, cracks filled with earth, shallow depressions, shallow overhangs, and rock needles. Seek shelter around dry, clean rock without cracks; in scree fields; or in deep indentations (depressions, caves). Keep at least half a body's length away from a cave wall and opening.

(2) Assume a one-point-of-contact body position. Squat on your haunches or sit on a rucksack or rope. Pull your knees to your chest and keep both feet together. If half way up the rock face, secure yourself with more than one point—lightning can burn through rope. If already rappelling, touch the wall with both feet together and hurry to the next anchor.

f. During and after rain, expect slippery rock and terrain in general and adjust movement accordingly. Expect flash floods in gullies or chimneys. A climber can be washed away or even drowned if caught in a gully during a rainstorm. Be especially alert for falling objects that the rain has loosened.

g. Dangers from impending high winds include frostbite (from increased wind-chill factor), windburn, being blown about (especially while rappelling), and debris being blown about. Wear protective clothing and plan the route to be finished before bad weather arrives.

h. For each 100-meter rise in altitude, the temperature drops approximately one degree Fahrenheit. This can cause hypothermia and frostbite even in summer, especially when combined with wind, rain, and snow. Always wear or pack appropriate clothing.

i. If it is snowing, gullies may contain avalanches or snow sloughs, which may bury the trail. Snowshoes or skis may be needed in autumn or even late spring. Unexpected snowstorms may occur in the summer with accumulations of 12 to 18 inches; however, the snow quickly melts.

j. Higher altitudes provide less filtering effects, which leads to greater ultraviolet (UV) radiation intensity. Cool winds at higher altitudes may mislead one into underestimating the sun's intensity, which can lead to sunburns and other heat injuries. Use sunscreen and wear hat and sunglasses, even if overcast. Drink plenty of fluids.

2-4. Avalanche Hazards. Avalanches occur when the weight of accumulated snow on a slope exceeds the cohesive forces that hold the snow in place. (Table 2-1 shows an avalanche hazard evaluation checklist.)

Table 2-1: Avalanche hazard evaluation checklist.

AVALANCHE HAZARD EVALUATION CHECKLIST

Critical Data		Hazard Rating		
PARAMETERS:	KEY INFORMATION	G	Y	R
TERRAIN: *Is the terrain capable of producing an avalanche?*				
-Slope angle (steep enough to slide? prime time?)		☐	☐	☐
-Slope aspect (leeward, shadowed, or extremely sunny?)		☐	☐	☐
-Slope configuration (anchoring? shape?)		☐	☐	☐
	Overall Terrain Rating:	☐	☐	☐
SNOWPACK: *Could the snow fail?*				
-Slab Configuration (slab? depth and distribution?)		☐	☐	☐
-Bonding Ability (weak layer? tender spots?)		☐	☐	☐
-Sensitivity (how much force to fail? shear tests? clues?)		☐	☐	☐
	Overall Snowpack Rating:	☐	☐	☐
Weather: *Is the weather contributing to instability?*				
-Precipitation (type, amount, intensity? added weight?)		☐	☐	☐
-Wind (snow transport? amount and rate of deposition?)		☐	☐	☐
-Temperature (storm trends? effects on snowpack?)		☐	☐	☐
	Overall Weather Rating:	☐	☐	☐
Human: *What are your alternatives and their possible consequences?*				
-Attitude (toward life? risk? goals? assumptions?)		☐	☐	☐
-Technical Skill Level (traveling? evaluating aval. hazard?)		☐	☐	☐
-Strength/Equipment (strength? prepared for the worst?)		☐	☐	☐
	Overall Human Rating:	☐	☐	☐

Decision/Action:
Overall Hazard Rating/GO or NO Go? GO ☐ or NOGO ☐

*HAZARD LEVEL SYMBOLS:
  R = Red light (stop/dangerous)
  G = Green light (go/OK)
  Y = Yellow light (caution/potentially dangerous).

©Alaska Mountain Safety Center, Inc.

a. *Slope Stability.* Slope stability is the key factor in determining the avalanche danger.

    (1) *Slope Angle.* Slopes as gentle as 15 degrees have avalanched. Most avalanches occur on slopes between 30 and 45 degrees. Slopes above 60 degrees often do not build up significant quantities of snow because they are too steep.

    (2) *Slope Profile.* Dangerous slab avalanches are more likely to occur on convex slopes, but may occur on concave slopes.

    (3) *Slope Aspect.* Snow on north facing slopes is more likely to slide in midwinter. South facing slopes are most dangerous in the spring and on sunny, warm days. Slopes on the windward side are generally more stable than leeward slopes.

    (4) *Ground Cover.* Rough terrain is more stable than smooth terrain. On grassy slopes or scree, the snowpack has little to anchor to.

b. *Triggers.* Various factors trigger avalanches.

    (1) *Temperature.* When the temperature is extremely low, settlement and adhesion occur slowly. Avalanches that occur during extreme cold weather usually occur during or immediately following a storm. At a temperature just below freezing, the snowpack stabilizes quickly. At temperatures above freezing, especially if temperatures rise quickly, the potential for avalanche is high. Storms with a rise in temperature can deposit dry snow early, which bonds poorly with the heavier snow deposited later. Most avalanches occur during the warmer midday.

    (2) *Precipitation.* About 90 percent of avalanches occur during or within twenty-four hours after a snowstorm. The rate at which snow falls is important. High rates of snowfall (2.5 centimeters per hour or greater), especially when accompanied by wind, are usually responsible for major periods of avalanche activity. Rain falling on snow will increase its weight and weakens the snowpack.

    (3) *Wind.* Sustained winds of 15 miles per hour and over transport snow and form wind slabs on the lee side of slopes.

    (4) *Weight.* Most victims trigger the avalanches that kill them.

    (5) *Vibration.* Passing helicopters, heavy equipment, explosions, and earth tremors have triggered avalanches.

c. *Snow Pits.* Snow pits can be used to determine slope stability.

    (1) Dig the snow pit on the suspect slope or a slope with the same sun and wind conditions. Snow deposits may vary greatly within a few meters due to wind and sun variations. (On at least one occasion, a snow pit dug across the fall line triggered the suspect slope). Dig a 2-meter by 2-meter pit across the fall line, through all the snow, to the ground. Once the pit is complete, smooth the face with a shovel.

    (2) Conduct a shovel shear test.

        (a) A shovel shear test puts pressure on a representative sample of the snowpack. The core of this test is to isolate a column of the snowpack from three sides. The column should be of similar size to the blade of the shovel. Dig out the sides of the column without pressing against the column with the shovel (this affects the strength). To isolate the rear of the column, use a rope or string to saw from side to side to the base of the column.

        (b) If the column remained standing while cutting the rear, place the shovel face down on the top of the column. Tap with varying degrees of strength on the shovel to see what force it takes to create movement on the bed of the column. The surface that eventually slides will be the layer to look at closer. This test provides a better understanding of the snowpack strength. For greater results you will need to do this test in many areas and formulate a scale for the varying methods of tapping the shovel.

    (3) Conduct a Rutschblock test. To conduct the test, isolate a column slightly longer than the length of your snowshoes or skis (same method as for the shovel shear test). One person moves on their skis or snowshoes above the block without disturbing the block. Once above, the person carefully places one showshoe or ski onto the block with no body weight for the first stage of

the test. The next stage is adding weight to the first leg. Next, place the other foot on the block. If the block is still holding up, squat once, then twice, and so on. The remaining stage is to jump up and land on the block.

d. *Types of Snow Avalanches.* There are two types of snow avalanches: loose snow (point) and slab.

   (1) Loose snow avalanches start at one point on the snow cover and grow in the shape of an inverted "V." Although they happen most frequently during the winter snow season, they can occur at any time of the year in the mountains. They often fall as many small sluffs during or shortly after a storm. This process removes snow from steep upper slopes and either stabilize slower slopes or loads them with additional snow.

   (2) Wet loose snow avalanches occur in spring and summer in all mountain ranges. Large avalanches of this type, lubricated and weighed down by melt water or rain can travel long distances and have tremendous destructive power. Coastal ranges that have high temperatures and frequent rain are the most common areas for this type of avalanche.

   (3) Slab avalanches occur when cohesive snow begins to slide on a weak layer. The fracture line where the moving snow breaks away from the snowpack makes this type of avalanche easy to identify. Slab release is rapid. Although any avalanche can kill you, slab avalanches are generally considered more dangerous than loose snow avalanches.

   (a) Most slab avalanches occur during or shortly after a storm when slopes are loaded with new snow at a critical rate. The old rule of never travel in avalanche terrain for a few days after a storm still holds true.

   (b) As slabs become harder, their behavior becomes more unpredictable; they may allow several people to ski across before releasing. Many experts believe they are susceptible to rapid temperature changes. Packed snow expands and contracts with temperature changes. For normal density, settled snow, a drop in temperature of 10 degrees Celsius (18 degrees Fahrenheit) would cause a snow slope 300 meters wide to contract 2 centimeters. Early ski mountaineers in the Alps noticed that avalanches sometimes occurred when shadows struck a previously sun-warmed slope.

d. *Protective Measures.* Avoiding known or suspected avalanche areas is the easiest method of protection. Other measures include:

   (1) *Personal Safety.* Remove your hands from ski pole wrist straps. Detach ski runaway cords. Prepare to discard equipment. Put your hood on. Close up your clothing to prepare for hypothermia. Deploy avalanche cord. Make avalanche probes and shovels accessible. Keep your pack on at all times do not discard. Your pack can act as a flotation device, as well as protect your spine.

   (2) *Group Safety.* Send one person across the suspect slope at a time with the rest of the group watching. All members of the group should move in the same track from safe zone to safe zone.

e. *Route Selection.* Selecting the correct route will help avoid avalanche prone areas, which is always the best choice. Always allow a wide margin of safety when making your decision.

   (1) The safest routes are on ridge tops, slightly on the windward side; the next safest route is out in the valley, far from the bottom of slopes.

   (2) Avoid cornices from above or below. Should you encounter a dangerous slope, either climb to the top of the slope or descend to the bottom well out of the way of the run-out zone. If you must traverse, pick a line where you can traverse downhill as quickly as possible. When you must ascend a dangerous slope, climb to the side of the avalanche path, and not directly up the center.

   (3) Take advantage of dense timber, ridges, or rocky outcrops as islands of safety. Use them for lunch and rest stops. Spend as little time as possible on open slopes.

   (4) Since most avalanches occur within twenty-four hours of a storm and or at midday, avoid moving during these periods. Moving at night is tactically sound and may be safer.

f. *Stability Analysis.* Look for nature's billboards on slopes similar to the one you are on.

   (1) *Evidence of Avalanching.* Look for recent avalanches and for signs of wind-loading and wind-slabs.

    (2) *Fracture Lines.* Avoid any slopes showing cracks.

    (3) *Sounds.* Beware of hollow sounds such as a "whumping" noise. They may suggest a radical settling of the snowpack.

  g. *Survival.* People trigger avalanches that bury people. If these people recognized the hazard and chose a different route, they would avoid the avalanche. The following steps should be followed if caught in an avalanche.

    (1) Discard equipment. Equipment can injure or burden you; discarded equipment will indicate your position to rescuers.

    (2) Swim or roll to stay on tope of the snow. FIGHT FOR YOUR LIFE. Work toward the edge of the avalanche. If you feel your feet touch the ground, give a hard push and try to "pop out" onto the surface.

    (3) If your head goes under the snow, shut your mouth, hold your breath, and position your hands and arms to form an air pocket in front of your face. Many avalanche victims suffocate by having their mouths and noses plugged with snow.

    (4) When you sense the slowing of the avalanche, you must try your hardest to reach the surface. Several victims have been found quickly because a hand or foot was sticking above the surface.

    (5) When the snow comes to rest it sets up like cement and even if you are only partially buried, it may be impossible to dig yourself out. Don't shout unless you hear rescuers immediately above you; in snow, no one can hear you scream. Don't struggle to free yourself you will only waste energy and oxygen.

    (6) Try to relax. If you feel yourself about to pass out, do not fight it. The respiration of an unconscious person is shallower, their pulse rate declines, and the body temperature is lowered, all of which reduce the amount of oxygen needed.

2-5. Acute Mountain Sickness. In addition to dangers caused by dehydration, sunburn, hypothermia, and other cold-weather problems, a high altitude can have physiological effects in itself. Acute mountain sickness is a temporary illness that may affect both the beginner and experienced climber. Soldiers are subject to this sickness in altitudes as low as 5,000 feet. Incidence and severity increases with altitude, and when quickly transported to high altitudes. Disability and ineffectiveness can occur in 50 to 80 percent of the troops who are rapidly brought to altitudes above 10,000 feet. At lower altitudes, or where ascent to altitudes is gradual, most personnel can complete assignments with moderate effectiveness and little discomfort.

  a. Personnel arriving at moderate elevations (5,000 to 8,000 feet) usually feel well for the first few hours; a feeling of exhilaration or well-being is not unusual. There may be an initial awareness of breathlessness upon exertion and a need for frequent pauses to rest. Irregular breathing can occur, mainly during sleep; these changes may cause apprehension. Severe symptoms may begin 4 to 12 hours after arrival at higher altitudes with symptoms of nausea, sluggishness, fatigue, headache, dizziness, insomnia, depression, uncaring attitude, rapid and labored breathing, weakness, and loss of appetite.

  b. A headache is the most noticeable symptom and may be severe. Even when a headache is not present, some loss of appetite and a decrease in tolerance for food occurs. Nausea, even without food intake, occurs and leads to less food intake. Vomiting may occur and contribute to dehydration. Despite fatigue, personnel are unable to sleep. The symptoms usually develop and increase to a peak by the second day. They gradually subside over the next several days so that the total course of AMS may extend from five to seven days. In some instances, the headache may become incapacitating and the soldier should be evacuated to a lower elevation.

  c. Treatment for AMS includes the following:

    • Oral pain medications such as ibuprofen or aspirin.

    • Rest.

- Frequent consumption of liquids and light foods in small amounts.
- Movement to lower altitudes (at least 1,000 feet) to alleviate symptoms, which provides for a more gradual acclimatization.
- Realization of physical limitations and slow progression.
- Practice of deep-breathing exercises.
- Use of acetazolamide in the first 24 hours for mild to moderate cases.

   d. AMS is nonfatal, although if left untreated or further ascent is attempted, development of high-altitude pulmonary edema (HAPE) and or high-altitude cerebral edema (HACE) can be seen. A severe persistence of symptoms may identify soldiers who acclimatize poorly and, thus, are more prone to other types of mountain sickness.

2-6. Chronic Mountain Sickness. Although not commonly seen in mountaineers, chronic mountain sickness (CMS) (or Monge's disease) can been seen in people who live at sufficiently high altitudes (usually at or above 10,000 feet) over a period of several years. CMS is a right-sided heart failure characterized by chronic pulmonary edema that is caused by years of strain on the right ventricle.

2-7. Understanding High-Altitude Illnesses. As altitude increases, the overall atmospheric pressure decreases. Decreased pressure is the underlying source of altitude illnesses. Whether at sea level or 20,000 feet the surrounding atmosphere has the same percentage of oxygen. As pressure decreases the body has a much more difficult time passing oxygen from the lungs to the red blood cells and thus to the tissues of the body. This lower pressure means lower oxygen levels in the blood and increased carbon dioxide levels. Increased carbon dioxide levels in the blood cause a systemic vasodilatation, or expansion of blood vessels. This increased vascular size stretches the vessel walls causing leakage of the fluid portions of the blood into the interstitial spaces, which leads to cerebral edema or HACE. Unless treated, HACE will continue to progress due to the decreased atmospheric pressure of oxygen. Further ascent will hasten the progression of HACE and could possibly cause death.

   While the body has an overall systemic vasodilatation, the lungs initially experience pulmonary vasoconstriction. This constricting of the vessels in the lungs causes increased workload on the right ventricle, the chamber of the heart that receives de-oxygenated blood from the right atrium and pushes it to the lungs to be re-oxygenated. As the right ventricle works harder to force blood to the lungs, its overall output is decreased thus decreasing the overall pulmonary perfusion. Decreased pulmonary perfusion causes decreased cellular respiration, the transfer of oxygen from the alveoli to the red blood cells. The body is now experiencing increased carbon dioxide levels due to the decreased oxygen levels, which now causes pulmonary vasodilatation. Just as in HACE, this expanding of the vascular structure causes leakage into interstitial space resulting in pulmonary edema or HAPE. As the edema or fluid in the lungs increases, the capability to pass oxygen to the red blood cells decreases thus creating a vicious cycle, which can quickly become fatal if left untreated.

2-8. High-Altitude Pulmonary Edema. HAPE is a swelling and filling of the lungs with fluid, caused by rapid ascent. It occurs at high altitudes and limits the oxygen supply to the body.

   a. HAPE occurs under conditions of low oxygen pressure, is encountered at high elevations (over 8,000 feet), and can occur in healthy soldiers. HAPE may be considered a form of, or manifestation of AMS since it occurs during the period of susceptibility to this disorder.

   b. HAPE can cause death. Incidence and severity increase with altitude. Except for acclimatization to altitude, no known factors indicate resistance or immunity. Few cases have been reported after 10 days at high altitudes. When remaining at the same altitude, the incidence of HAPE is less frequent than that of AMS. No common indicator dictates how a soldier will react from one exposure to another. Contributing factors are:

- A history of HAPE.
- A rapid or abrupt transition to high altitudes.

- Strenuous physical exertion.
- Exposure to cold.
- Anxiety.

c. Symptoms of AMS can mask early pulmonary difficulties. Symptoms of HAPE include:

- Progressive dry coughing with frothy white or pink sputum (this is usually a later sign) and then coughing up of blood.
- Cyanosis a blue color to the face, hands, and feet.
- An increased ill feeling, labored breathing, dizziness, fainting, repeated clearing of the throat, and development of a cough.
- Respiratory difficulty, which may be sudden, accompanied by choking and rapid deterioration.
- Progressive shortness of breath, rapid heartbeat (pulse 120 to 160), and coughing (out of contrast to others who arrived at the same time to that altitude).
- Crackling, cellophane-like noises (rales) in the lungs caused by fluid buildup (a stethoscope is usually needed to hear them).
- Unconsciousness, if left untreated. Bubbles form in the nose and mouth, and death results.

d. HAPE is prevented by good nutrition, hydration, and gradual ascent to altitude (no more than 1,000 to 2,000 feet per day to an area of sleep). A rest day, with no gain in altitude or heavy physical exertion, is planned for every 3,000 feet of altitude gained. If a soldier develops symptoms despite precautions, immediate descent is mandatory where he receives prompt treatment, rest, warmth, and oxygen. He is quickly evacuated to lower altitudes as a litter patient. A descent of 300 meters may help; manual descent is not delayed to await air evacuation. If untreated, HAPE may become irreversible and cause death. Cases that are recognized early and treated promptly may expect to recover with no aftereffects. Soldiers who have had previous attacks of HAPE are prone to second attacks.

e. Treatment of HAPE includes:

- Immediate descent (2,000 to 3,000 feet minimum) if possible; if not, then treatment in a mono-place hyperbaric chamber.
- Rest (litter evacuation).
- Supplemental oxygen if available.
- Morphine for the systemic vasodilatation and reduction of preload. This should be carefully considered due to the respiratory depressive properties of the drug.
- Furosemide (Lasix), which is a diuretic, given orally can also be effective.
- The use of mannitol should not be considered due to the fact that it crystallizes at low temperatures. Since almost all high-altitude environments are cold, using mannitol could be fatal.
- Nifidipine (Procardia), which inhibits calcium ion flux across cardiac and smooth muscle cells, decreasing contractility and oxygen demand. It may also dilate coronary arteries and arterioles.
- Diphenhydramine (Benadryl), which can help alleviate the histamine response that increases mucosal secretions.

2-9. High-Altitude Cerebral Edema. HACE is the accumulation of fluid in the brain, which results in swelling and a depression of brain function that may result in death. It is caused by a rapid ascent to altitude without progressive acclimatization. Prevention of HACE is the same as for HAPE. HAPE and HACE may occur in experienced, well-acclimated mountaineers without warning or obvious predisposing conditions. They can be fatal; when the first symptoms occur, immediate descent is mandatory.

a. Contributing factors include rapid ascent to heights over 8,000 feet and aggravation by overexertion.

b. Symptoms of HACE include mild personality changes, paralysis, stupor, convulsions, coma, inability to concentrate, headaches, vomiting, decrease in urination, and lack of coordination. The main symptom of HACE is a severe headache. A headache combined with any other physical or psychological disturbances should be assumed to be manifestations of HACE. Headaches may be accompanied by a loss of coordination, confusion, hallucinations, and unconsciousness. These may be combined with symptoms of HAPE. The victim is often mistakenly left alone since others may think he is only irritable or temperamental; no one should ever be ignored. The symptoms may rapidly progress to death. Prompt descent to a lower altitude is vital.

c. Preventive measures include good eating habits, maintaining hydration, and using a gradual ascent to altitude. Rest, warmth, and oxygen at lower elevations enhance recovery. Left untreated, HACE can cause death.

d. Treatment for HACE includes:
- Dexamethasone injection immediately followed by oral dexamethasone.
- Supplemental oxygen.
- Rapid descent and medical attention.
- Use of a hyberbaric chamber if descent is delayed.

2-10. Hydration In Hape and Hace. HAPE and HACE cause increased proteins in the plasma, or the fluid portion of the blood, which in turn increases blood viscosity. Increased viscosity increases vascular pressure. Vascular leakage caused by stretching of the vessel walls is made worse because of this increased vascular pressure. From this, edema, both cerebral and pulmonary, occurs. Hydration simply decreases viscosity.

SECTION 3: MOUNTAINEERING EQUIPMENT

EQUIPMENT DESCRIPTION AND MAINTENANCE

With mountainous terrain encompassing a large portion of the world's land mass, the proper use of mountaineering equipment will enhance a unit's combat capability and provide a combat multiplier. The equipment described in this chapter is produced by many different manufacturers; however, each item is produced and tested to extremely high standards to ensure safety when being used correctly. The weak link in the safety chain is the user. Great care in performing preventative maintenance checks and services and proper training in the use of the equipment is paramount to ensuring safe operations. The manufacturers of each and every piece of equipment provide recommendations on how to use and care for its product. It is imperative to follow these instructions explicitly.

3-1. Footwear. In temperate climates a combination of footwear is most appropriate to accomplish all tasks.

a. The hot weather boot provides an excellent all-round platform for movement and climbing techniques and should be the boot of choice when the weather permits. The intermediate cold weather boot provides an acceptable platform for operations when the weather is less than ideal. These two types of boots issued together will provide the unit with the footwear necessary to accomplish the majority of basic mountain missions.

b. Mountain operations are encumbered by extreme cold, and the extreme cold weather boot (with vapor barrier) provides an adequate platform for many basic mountain missions. However, plastic mountaineering boots should be incorporated into training as soon as possible. These boots provide a more versatile platform for any condition that would be encountered in the mountains, while keeping the foot dryer and warmer.

c. Level 2 and level 3 mountaineers will need mission-specific footwear that is not currently available in the military supply system. The two types of footwear they will need are climbing shoes and plastic mountaineering boots.

(1) Climbing shoes are made specifically for climbing vertical or near vertical rock faces. These shoes are made with a soft leather upper, a lace-up configuration, and a smooth "sticky rubber" sole (Figure 3-1). The smooth "sticky rubber" sole is the key to the climbing shoe, providing greater friction on the surface of the rock, allowing the climber access to more difficult terrain.

Figure 3-1: Climbing shoes and plastic mountaineering boots.

(2) The plastic mountaineering boot is a double boot system (Figure 3-1). The inner boot provides support, as well as insulation against the cold. The inner boot may or may not come with a breathable membrane. The outer boot is a molded plastic (usually with a lace-up configuration) with a lug sole. The welt of the boot is molded in such a way that crampons, ski bindings, and snowshoes are easily attached and detached.

Note: Maintenance of all types of footwear must closely follow the manufacturers' recommendations.

3-2. Clothing. Clothing is perhaps the most underestimated and misunderstood equipment in the military inventory. The clothing system refers to every piece of clothing placed against the skin, the insulation layers, and the outer most garments, which protect the soldier from the elements. When clothing is worn properly, the soldier is better able to accomplish his tasks. When worn improperly, he is, at best, uncomfortable and, at worst, develops hypothermia or frostbite.

a. *Socks.* Socks are one of the most under-appreciated part of the entire clothing system. Socks are extremely valuable in many respects, if worn correctly. As a system, socks provide cushioning for the foot, remove excess moisture, and provide insulation from cold temperatures. Improper wear and excess moisture are the biggest causes of hot spots and blisters. Regardless of climatic conditions, socks should always be worn in layers.

(1) The first layer should be a hydrophobic material that moves moisture from the foot surface to the outer sock.

(2) The outer sock should also be made of hydrophobic materials, but should be complimented with materials that provide cushioning and abrasion resistance.

(3) A third layer can be added depending upon the climatic conditions.
   (a) In severe wet conditions, a waterproof type sock can be added to reduce the amount of water that would saturate the foot. This layer would be worn over the first two layers if conditions were extremely wet.
   (b) In extremely cold conditions a vapor barrier sock can be worn either over both of the original pairs of socks or between the hydrophobic layer and the insulating layer. If the user is wearing VB boots, the vapor barrier sock is not recommended.

b. *Underwear.* Underwear should also be made of materials that move moisture from the body. Many civilian companies manufacture this type of underwear. The primary material in this product is polyester, which moves moisture from the body to the outer layers keeping the user drier and more comfortable in all climatic conditions. In colder environments, several pairs of long underwear of different thickness should be made available. A lightweight set coupled with a heavyweight set will provide a multitude of layering combinations.

c. *Insulating Layers.* Insulating layers are those layers that are worn over the underwear and under the outer layers of clothing. Insulating layers provide additional warmth when the weather turns bad. For the most part, today's insulating layers will provide for easy moisture movement as well as trap air to increase the insulating factor. The insulating layers that are presently available are referred to as pile or fleece. The ECWCS (Figure 3-2) also incorporates the field jacket and field pants liner as additional insulating layers. However, these two components do not move moisture as effectively as the pile or fleece.

d. *Outer Layers.* The ECWCS provides a jacket and pants made of a durable waterproof fabric. Both are constructed with a nylon shell with a laminated breathable membrane attached. This membrane allows the garment to release moisture to the environment while the nylon shell provides a degree of water resistance during rain and snow. The nylon also acts as a barrier to wind, which helps the garment retain the warm air trapped by the insulating layers. Leaders at all levels must understand the importance of wearing the ECWCS correctly.

Note: Cotton layers must not be included in any layer during operations in a cold environment.

e. *Gaiters.* Gaiters are used to protect the lower leg from snow and ice, as well as mud, twigs, and stones. The use of waterproof fabrics or other breathable materials laminated to the nylon makes the

Figure 3-2: Extreme cold weather clothing system.

gaiter an integral component of the cold weather clothing system. Gaiters are not presently fielded in the standard ECWCS and, in most cases, will need to be locally purchased. Gaiters are available in three styles (Figure 3-3).

(1) The most common style of gaiter is the open-toed variety, which is a nylon shell that may or may not have a breathable material laminated to it. The open front allows the boot to slip easily into it and is closed with a combination of zipper, hook-pile tape, and snaps. It will have an adjustable neoprene strap that goes under the boot to keep it snug to the boot. The length should reach to just below the knee and will be kept snug with a drawstring and cord lock.

(2) The second type of gaiter is referred to as a full or randed gaiter. This gaiter completely covers the boot down to the welt. It can be laminated with a breathable material and can also be insulated if necessary. This gaiter is used with plastic mountaineering boots and should be glued in place and not removed.

(3) The third type of gaiter is specific to high-altitude mountaineering or extremely cold temperatures and is referred to as an overboot. It is worn completely over the boot and must be worn with crampons because it has no traction sole.

f. *Hand Wear.* During operations in mountainous terrain the use of hand wear is extremely important. Even during the best climatic conditions, temperatures in the mountains will dip below the freezing point. While mittens are always warmer than gloves, the finger dexterity needed to do most tasks makes gloves the primary cold weather hand wear (Figure 3-4).

(1) The principals that apply to clothing also apply to gloves and mittens. They should provide moisture transfer from the skin to the outer layers. The insulating layer must insulate the hand from the cold and move moisture to the outer layer. The outer layer must be weather resistant and breathable. Both gloves and mittens should be required for all soldiers during mountain operations, as well as replacement liners for both. This will provide enough flexibility to accomplish all tasks and keep the users' hands warm and dry.

(2) Just as the clothing system is worn in layers, gloves and mittens work best using the same principle. Retention cords that loop over the wrist work extremely well when the wearer needs to remove the outer layer to accomplish a task that requires fine finger dexterity. Leaving the glove or mitten dangling from the wrist ensures the wearer knows where it is at all times.

g. *Headwear.* A large majority of heat loss (25 percent) occurs through the head and neck area. The most effective way to counter heat loss is to wear a hat. The best hat available to the individual soldier through the military supply system is the black watch cap. Natural fibers, predominately wool, are acceptable but can be bulky and difficult to fit under a helmet. As with clothes and hand

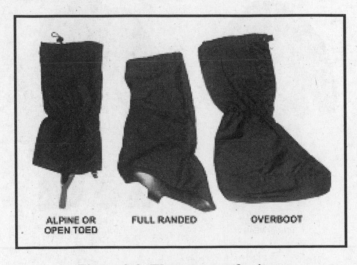

Figure 3-3: Three types of gaiters.

Figure 3-4: Hand wear.

wear, man-made fibers are preferred. For colder climates a neck gaiter can be added. The neck gaiter is a tube of man-made material that fits around the neck and can reach up over the ears and nose (Figure 3-5). For extreme cold, a balaclava can be added. This covers the head, neck, and face leaving only a slot for the eyes (Figure 3-5). Worn together the combination is warm and provides for moisture movement, keeping the wearer drier and warmer.

h. *Helmets.* The Kevlar ballistic helmet can be used for most basic mountaineering tasks. It must be fitted with parachute retention straps and the foam impact pad (Figure 3-6). The level 2 and 3 mountaineer will need a lighter weight helmet for specific climbing scenarios. Several civilian manufacturers produce an effective helmet. Whichever helmet is selected, it should be designed specifically for mountaineering and adjustable so the user can add a hat under it when needed.

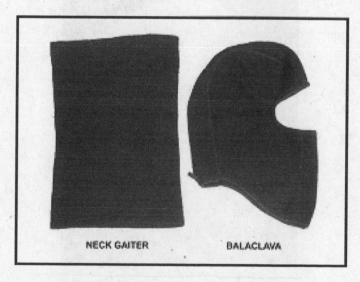

Figure 3-5: Neck gaiter and balaclava.

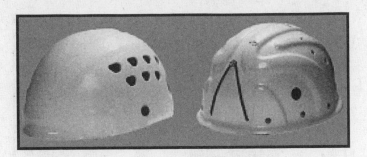

Figure 3-6: Helmets.

i. *Eyewear.* The military supply system does not currently provide adequate eyewear for mountaineering. Eyewear is divided into two categories: glacier glasses and goggles (Figure 3-7). Glacier glasses are sunglasses that cover the entire eye socket. Many operations in the mountains occur above the tree line or on ice and snow surfaces where the harmful UV rays of the sun can bombard the eyes from every angle increasing the likelihood of snow blindness. Goggles for mountain operations should be antifogging. Double or triple lenses work best. UV rays penetrate clouds so the goggles should be UV protected. Both glacier glasses and goggles are required equipment in the mountains. The lack of either one can lead to severe eye injury or blindness.

j. *Maintenance of Clothing.* Clothing and equipment manufacturers provide specific instructions for proper care. Following these instructions is necessary to ensure the equipment works as intended.

3-3. Climbing Software. Climbing software refers to rope, cord, webbing, and harnesses. All mountaineering specific equipment, to include hardware (see paragraph 3-4), should only be used if it has the UIAA certificate of safety. UIAA is the organization that oversees the testing of mountaineering equipment. It is based in Paris, France, and comprises several commissions. The safety commission has established standards for mountaineering and climbing equipment that have become well recognized throughout the world. Their work continues as new equipment develops and is brought into common use. Community Europe (CE) recognizes UIAA testing standards and, as the broader-based testing facility for the combined European economy, meets or exceeds the UIAA standards for all climbing and mountaineering equipment

Figure 3-7: Glacier glasses and goggles.

produced in Europe. European norm (EN) and CE have been combined to make combined European norm (CEN). While the United States has no specific standards, American manufacturers have their equipment tested by UIAA to ensure safe operating tolerances.

a. *Ropes and Cord.* Ropes and cords are the most important pieces of mountaineering equipment and proper selection deserves careful thought. These items are your lifeline in the mountains, so selecting the right type and size is of the utmost importance. All ropes and cord used in mountaineering and climbing today are constructed with the same basic configuration. The construction technique is referred to as Kernmantle, which is, essentially, a core of nylon fibers protected by a woven sheath, similar to parachute or 550 cord (Figure 3-8).

(1) Ropes come in two types: static and dynamic. This refers to their ability to stretch under tension. A static rope has very little stretch, perhaps as little as one to two percent, and is best used in rope installations. A dynamic rope is most useful for climbing and general mountaineering. Its ability to stretch up to 1/3 of its overall length makes it the right choice any time the user might take a fall. Dynamic and static ropes come in various diameters and lengths. For most military applications, a standard 10.5- or 11-millimeter by 50-meter dynamic rope and 11-millimeter by 45-meter static rope will be sufficient.

(2) When choosing dynamic rope, factors affecting rope selection include intended use, impact force, abrasion resistance, and elongation. Regardless of the rope chosen, it should be UIAA certified.

(3) Cord or small diameter rope is indispensable to the mountaineer. Its many uses make it a valuable piece of equipment. All cord is static and constructed in the same manner as larger rope. If used for Prusik knots, the cord's diameter should be 5 to 7 millimeters when used on an 11-mm rope.

b. *Webbing and Slings.* Loops of tubular webbing or cord, called slings or runners, are the simplest pieces of equipment and some of the most useful. The uses for these simple pieces are endless, and they are a critical link between the climber, the rope, carabiners, and anchors. Runners are predominately made from either 9/16-inch or 1-inch tubular webbing and are either tied or sewn by a manufacturer (Figure 3-9). Runners can also be made from a high-performance fiber known as spectra, which is stronger, more durable, and less susceptible to ultraviolet deterioration. Runners should be retired regularly following the same considerations used to retire a rope. For most military applications, a combination of different lengths of runners is adequate.

(1) Tied runners have certain advantages over sewn runners: they are inexpensive to make, can be untied and threaded around natural anchors, and can be untied and retied to other pieces of webbing to create extra long runners.

(2) Sewn runners have their own advantages: they tend to be stronger, are usually lighter, and have less bulk than the tied version. They also eliminate a major concern with the homemade knotted runner—the possibility of the knot untying. Sewn runners come in four standard lengths: 2 inches, 4 inches, 12 inches, and 24 inches. They also come in three standard widths: 9/16 inch, 11/16 inch, and 1 inch.

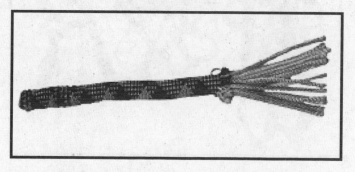

Figure 3-8: Kernmantle construction.

Figure 3-9: Tied or sewn runners.

c. *Harnesses*. Years ago climbers secured themselves to the rope by wrapping the rope around their bodies and tying a bowline-on-a-coil. While this technique is still a viable way of attaching to a rope, the practice is no longer encouraged because of the increased possibility of injury from a fall. The bowline-on-a-coil is best left for low-angle climbing or an emergency situation where harness material is unavailable. Climbers today can select from a wide range of manufactured harnesses. Fitted properly, the harness should ride high on the hips and have snug leg loops to better distribute the force of a fall to the entire pelvis. This type of harness, referred to as a seat harness, provides a comfortable seat for rappelling (Figure 3-10).

(1) Any harness selected should have one very important feature: a double-passed buckle. This is a safety standard that requires the waist belt to be passed over and back through the main buckle a second time. At least 2 inches of the strap should remain after double-passing the buckle.

(2) Another desirable feature on a harness is adjustable leg loops, which allows a snug fit regardless of the number of layers of clothing worn. Adjustable leg loops allow the soldier to make a latrine call without removing the harness or untying the rope.

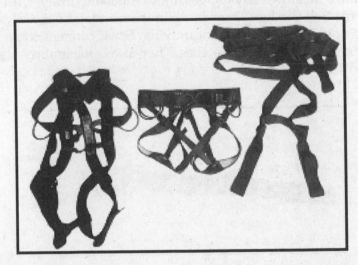

Figure 3-10: Seat harness, field-expedient harness, and full body harness.

(3) Equipment loops are desirable for carrying pieces of climbing equipment. For safety purposes always follow the manufacturer's directions for tying-in.

(4) A field-expedient version of the seat harness can be constructed by using 22 feet of either 1-inch or 2-inch (preferred) tubular webbing (Figure 3-10). Two double-overhand knots form the leg loops, leaving 4 to 5 feet of webbing coming from one of the leg loops. The leg loops should just fit over the clothing. Wrap the remaining webbing around the waist ensuring the first wrap is routed through the 6- to 10-inch long strap between the double-overhand knots. Finish the waist wrap with a water knot tied as tightly as possible. With the remaining webbing, tie a square knot without safeties over the water knot ensuring a minimum of 4 inches remains from each strand of webbing.

(5) The full body harness incorporates a chest harness with a seat harness (Figure 3-10). This type of harness has a higher tie-in point and greatly reduces the chance of flipping backward during a fall. This is the only type of harness that is approved by the UIAA. While these harnesses are safer, they do present several disadvantages: they are more expensive, are more restrictive, and increase the difficulty of adding or removing clothing. Most mountaineers prefer to incorporate a separate chest harness with their seat harness when warranted.

(6) A separate chest harness can be purchased from a manufacturer, or a field-expedient version can be made from either two runners or a long piece of webbing. Either chest harness is then attached to the seat harness with a carabiner and a length of webbing or cord.

3-4. Climbing Hardware. Climbing hardware refers to all the parts and pieces that allow the trained mountain soldier to accomplish many tasks in the mountains. The importance of this gear to the mountaineer is no less than that of the rifle to the infantryman.

a. *Carabiners.* One of the most versatile pieces of equipment available to the mountaineer is the carabiner. This simple piece of gear is the critical connection between the climber, his rope, and the protection attaching him to the mountain. Carabiners must be strong enough to hold hard falls, yet light enough for the climber to easily carry a quantity of them. Today's high tech metal alloys allow carabiners to meet both of these requirements. Steel is still widely used, but is not preferred for general mountaineering, given other options. Basic carabiner construction affords the user several different shapes. The oval, the D-shaped, and the pear-shaped carabiner are just some of the types currently available. Most models can be made with or without a locking mechanism for the gate opening (Figure 3-11). If the carabiner does have a locking mechanism, it is usually referred to as a locking carabiner. When using a carabiner, great care should be taken to avoid loading the carabiner on its minor axis and to avoid three-way loading (Figure 3-12).

Note: Great care should be used to ensure all carabiner gates are closed and locked during use.

(1) The major difference between the oval and the D-shaped carabiner is strength. Because of the design of the D-shaped carabiner, the load is angled onto the spine of the carabiner thus keeping it off the gate. The down side is that racking any gear or protection on the D-shaped carabiner is difficult because the angle of the carabiner forces all the gear together making it impossible to separate quickly.

(2) The pear-shaped carabiner, specifically the locking version, is excellent for clipping a descender or belay device to the harness. They work well with the munter hitch belaying knot.

(3) Regardless of the type chosen, all carabiners should be UIAA tested. This testing is extensive and tests the carabiner in three ways along its major axis, along its minor axis, and with the gate open.

b. *Pitons.* A piton is a metal pin that is hammered into a crack in the rock. They are described by their thickness, design, and length (Figure 3-13). Pitons provide a secure anchor for a rope attached by a carabiner. The many different kinds of pitons include: vertical, horizontal, wafer, and angle. They are made of malleable steel, hardened steel, or other alloys. The strength of the piton is determined by its placement rather than its rated tensile strength. The two most common types of pitons are:

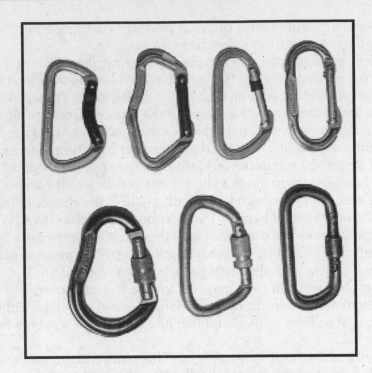

Figure 3-11: Nonlocking and locking carabiners.

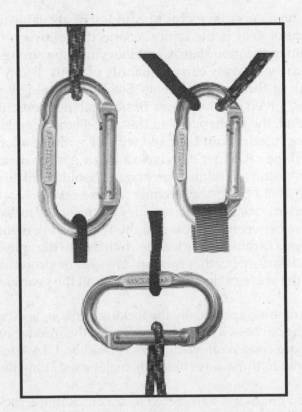

Figure 3-12: Major and minor axes and three-way loading.

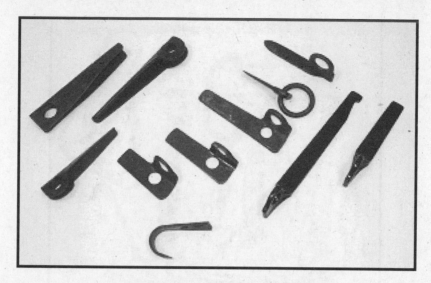

Figure 3-13: Various pitons.

blades, which hold when wedged into tight-fitting cracks, and angles, which hold blade compression when wedged into a crack.

(1) Vertical Pitons. On vertical pitons, the blade and eye are aligned. These pitons are used in flush, vertical cracks.

(2) Horizontal Pitons. On horizontal pitons, the eye of the piton is at right angles to the blade. These pitons are used in flush, horizontal cracks and in offset or open-book type vertical or horizontal cracks. They are recommended for use in vertical cracks instead of vertical pitons because the torque on the eye tends to wedge the piton into place. This provides more holding power than the vertical piton under the same circumstances.

(3) Wafer Pitons. These pitons are used in shallow, flush cracks. They have little holdingpower and their weakest points are in the rings provided for the carabiner.

(4) Knife Blade Pitons. These are used in direct-aid climbing. They are small and fit into thin, shallow cracks. They have a tapered blade that is optimum for both strength and holding power.

(5) Realized Ultimate Reality Pitons. Realized ultimate reality pitons (RURPs) are hatchet-shaped pitons about 1-inch square. They are designed to bite into thin, shallow cracks.

(6) Angle Pitons. These are used in wide cracks that are flush or offset. Maximum strength is attained only when the legs of the piton are in contact with the opposite sides of the crack.

(7) Bong Pitons. These are angle pitons that are more than 3.8 centimeters wide. Bongs are commonly made of steel or aluminum alloy and usually contain holes to reduce weight and accommodate carabiners. They have a high holding power and require less hammering than other pitons.

(8) Skyhook (Cliffhangers). These are small hooks that cling to tiny rock protrusions, ledges, or flakes. Skyhooks require constant tension and are used in a downward pull direction. The curved end will not straighten under body weight. The base is designed to prevent rotation and aid stability.

c. *Piton Hammers.* A piton hammer has a flat metal head; a handle made of wood, metal, or fiberglass; and a blunt pick on the opposite side of the hammer (Figure 3-14). A safety lanyard of nylon cord, webbing, or leather is used to attach it to the climber The lanyard should be long enough to allow for full range of motion. Most hammers are approximately 25.5 centimeters long and weigh 12 to 25 ounces. The primary use for a piton hammer is to drive pitons, to be used as anchors, into the rock. The piton hammer can also be used to assist in removing pitons, and in cleaning cracks and rock surfaces to prepare for inserting the piton. The type selected should suit individual preference and the intended use.

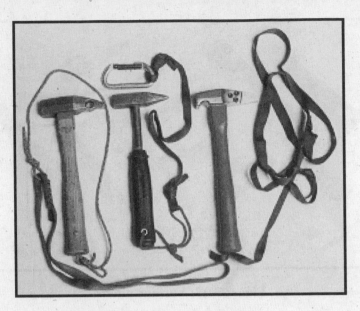

Figure 3-14: Piton hammer.

d. *Chocks.* "Chocks" is a generic term used to describe the various types of artificial protection other than bolts or pitons. Chocks are essentially a tapered metal wedge constructed in various sizes to fit different sized openings in the rock (Figure 3-15). The design of a chock will determine whether it fits into one of two categories: wedges or cams. A wedge holds by wedging into a constricting crack in the rock. A cam holds by slightly rotating in a crack, creating a camming action that lodges the chock in the crack or pocket. Some chocks are manufactured to perform either in the wedging mode or the camming mode. One of the chocks that falls into the category of both a wedge and cam is the hexagonal-shaped or "hex" chock. This type of chock is versatile and comes with either a cable loop or is tied with cord or webbing. All chocks come in different sizes to fit varying widths of cracks. Most chocks come with a wired loop that is stronger than cord and allows for easier placement. Bigger chocks can be threaded with cord or webbing if the user ties the chock himself. Care should be taken to place tubing in the chock before threading the cord. The cord used with

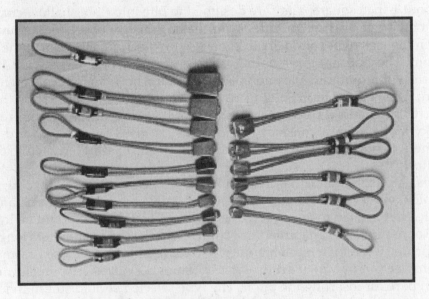

Figure 3-15: Chocks.

chocks is designed to be stiffer and stronger than regular cord and is typically made of Kevlar. The advantage of using a chock rather than a piton is that a climber can carry many different sizes and use them repeatedly.

e. *Three-Point Camming Device.* The three-point camming device's unique design allows it to be used both as a camming piece and a wedging piece (Figure 3-16). Because of this design it is extremely versatile and, when used in the camming mode, will fit a wide range of cracks. The three-point camming device comes in several different sizes with the smaller sizes working in pockets that no other piece of gear would fit in.

f. *Spring-Loaded Camming Devices.* Spring-loaded camming devices (SLCDs) (Figure 3-17) provide convenient, reliable placement in cracks where standard chocks are not practical (parallel or flaring cracks or cracks under roofs). SLCDs have three or four cams rotating around a single or double axis with a rigid or semi-rigid point of attachment. These are placed quickly and easily, saving time and effort. SLCDs are available in many sizes to accommodate different size cracks. Each fits a wide range of crack widths due to the rotating cam heads. The shafts may be rigid metal or semi-rigid cable loops. The flexible cable reduces the risk of stem breakage over an edge in horizontal placements.

g. *Chock Picks.* Chock picks are primarily used to extract chocks from rock when the they become severely wedged (Figure 3-18). They are also handy to clean cracks with. Made from thin metal, they can be purchased or homemade. When using a chock pick to extract a chock be sure no force is applied directly to the cable juncture. One end of the chock pick should have a hook to use on jammed SLCDs.

h. *Bolts.* Bolts are screw-like shafts made from metal that are drilled into rock to provide protection (Figure 3-19). The two types are contraction bolts and expansion bolts. Contraction bolts are squeezed together when driven into a rock. Expansion bolts press around a surrounding sleeve to form a snug fit into a rock. Bolts require drilling a hole into a rock, which is time-consuming, exhausting, and extremely noisy. Once emplaced, bolts are the most secure protection for a multidirectional pull. Bolts should be used only when chocks and pitons cannot be emplaced. A bolt is hammered only when it is the nail or self-driving type.

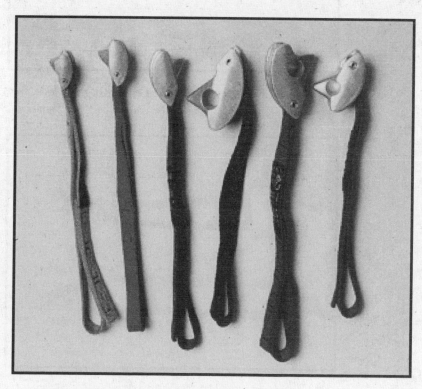

Figure 3-16: Three-point camming device.

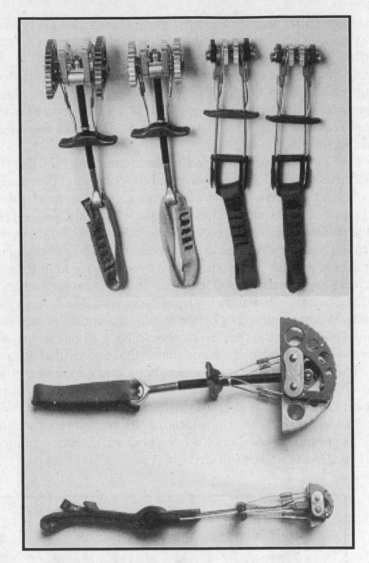

Figure 3-17: Spring-loaded camming devices.

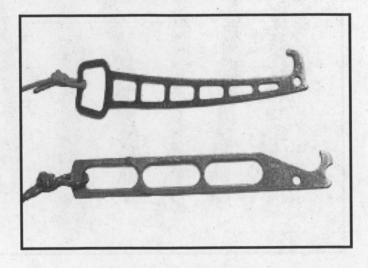

Figure 3-18: Chock picks.

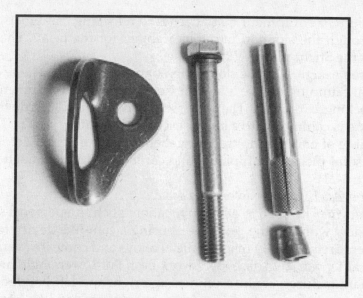

Figure 3-19: Bolts and hangers.

(1) A hanger (for carabiner attachment) and nut are placed on the bolt. The bolt is then inserted and driven into the hole. Because of this requirement, a hand drill must be carried in addition to a piton hammer. Hand drills (also called star drills) are available in different sizes, brands, and weights. A hand drill should have a lanyard to prevent loss.

(2) Self-driving bolts are quicker and easier to emplace. These require a hammer, bolt driver, and drilling anchor, which is driven into the rock. A bolt and carrier are then secured to the emplaced drilling anchor. All metal surfaces should be smooth and free of rust, corrosion, dirt, and moisture. Burrs, chips, and rough spots should be filed smooth and wire-brushed or rubbed clean with steel wool. Items that are cracked or warped indicate excessive wear and should be discarded.

i. *Belay Devices.* Belay devices range from the least equipment intensive (the body belay) to high-tech metal alloy pieces of equipment. Regardless of the belay device chosen, the basic principal remains the same: friction around or through the belay device controls the ropes' movement. Belay devices are divided into three categories: the slot, the tuber, and the mechanical camming device (Figure 3-20).

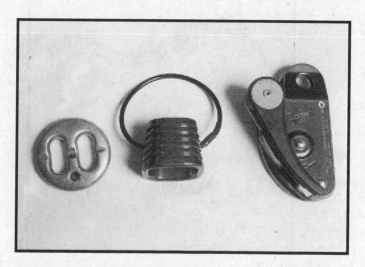

Figure 3-20: Slot, tuber, mechanical camming device.

(1) The slot is a piece of equipment that attaches to a locking carabiner in the harness; a bight of rope slides through the slot and into the carabiner for the belay. The most common slot type belay device is the Sticht plate.

(2) The tuber is used exactly like the slot but its shape is more like a cone or tube.

(3) The mechanical camming device is a manufactured piece of equipment that attaches to the harness with a locking carabiner. The rope is routed through this device so that when force is applied the rope is cammed into a highly frictioned position.

j. *Descenders.* One piece of equipment used for generations as a descender is the carabiner. A figure-eight is another useful piece of equipment and can be used in conjunction with the carabiner for descending (Figure 3-21).

Note: *All belay devices can also be used as descending devices.*

k. *Ascenders.* Ascenders may be used in other applications such as a personal safety or hauling line cam. All modern ascenders work on the principle of using a cam-like device to allow movement in one direction. Ascenders are primarily made of metal alloys and come in a variety of sizes (Figure 3-22). For difficult vertical terrain, two ascenders work best. For lower angle movement, one ascender is sufficient. Most manufacturers make ascenders as a right and left-handed pair.

l. *Pulleys.* Pulleys are used to change direction in rope systems and to create mechanical advantage in hauling systems. A pulley should be small, lightweight, and strong. They should accommodate the largest diameter of rope being used. Pulleys are made with several bearings, different-sized sheaves (wheel), and metal alloy sideplates (Figure 3-23). Plastic pulleys should always be avoided. The sideplate should rotate on the pulley axle to allow the pulley to be attached at any point along the rope. For best results, the sheave diameter must be at least four times larger than the rope's diameter to maintain high rope strength.

3-5. Snow and Ice Climbing Hardware. Snow and ice climbing hardware is the equipment that is particular to operations in some mountainous terrain. Specific training on this type of equipment is essential for safe use. Terrain that would otherwise be inaccessible—snowfields, glaciers, frozen waterfalls—can now be considered avenues of approach using the snow and ice climbing gear listed in this paragraph.

a. *Ice Ax.* The ice ax is one of the most important tools for the mountaineer operating on snow or ice. The climber must become proficient in its use and handling. The versatility of the ax lends itself to

Figure 3-21: Figure-eights.

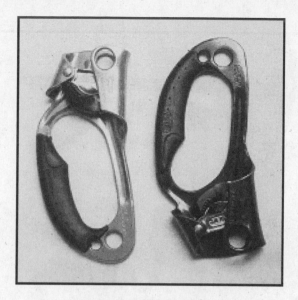

Figure 3-22: Ascenders.

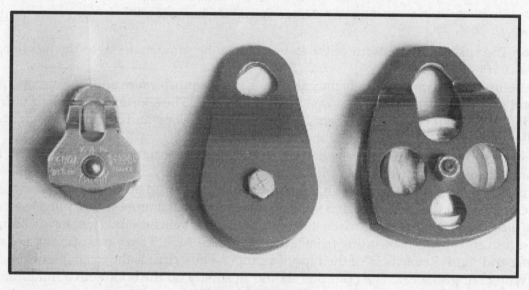

Figure 3-23: Pulley.

balance, step cutting, probing, self-arrest, belays, anchors, direct-aid climbing, and ascending and descending snow and ice covered routes.

(1) Several specific parts comprise an ice ax: the shaft, head (pick and adze), and spike (Figure 3-24).

    (a) The shaft (handle) of the ax comes in varying lengths (the primary length of the standard mountaineering ax is 70 centimeters). It can be made of fiberglass, hollow aluminum, or wood; the first two are stronger, therefore safer for mountaineering.

    (b) The head of the ax, which combines the pick and the adze, can have different configurations. The pick should be curved slightly and have teeth at least one-fourth of its length. The adze, used for chopping, is perpendicular to the shaft. It can be flat or curved along its length and straight or rounded from side to side. The head can be of one-piece construction or have replaceable picks and adzes. The head should have a hole directly above the shaft to allow for a leash to be attached.

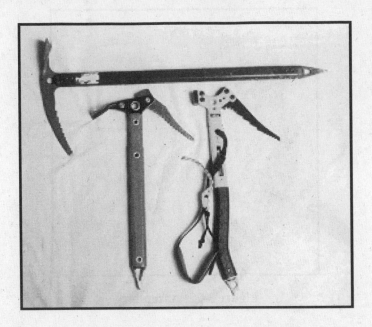

Figure 3-24: Ice ax and ice hammers.

    (c) The spike at the bottom of the ax is made of the same material as the head and comes in a variety of shapes.

    (2) As climbing becomes more technical, a shorter ax is much more appropriate, and adding a second tool is a must when the terrain becomes vertical. The shorter ax has all the attributes of the longer ax, but it is anywhere from 40 to 55 centimeters long and can have a straight or bent shaft depending on the preference of the user.

  b. *Ice Hammer.* The ice hammer is as short or shorter than the technical ax (Figure 3-24). It is used for pounding protection into the ice or pitons into the rock. The only difference between the ice ax and the ice hammer is the ice hammer has a hammerhead instead of an adze. Most of the shorter ice tools have a hole in the shaft to which a leash is secured, which provides a more secure purchase in the ice.

  c. *Crampons.* Crampons are used when the footing becomes treacherous. They have multiple spikes on the bottom and spikes protruding from the front (Figure 3-25). Two types of crampons are available: flexible and rigid. Regardless of the type of crampon chosen, fit is the most important factor associated with crampon wear. The crampon should fit snugly on the boot with a minimum of 1 inch of front point protruding. Straps should fit snugly around the foot and any long, loose ends should be trimmed. Both flexible and rigid crampons come in pairs, and any tools needed for adjustment will be provided by the manufacturer.

    (1) The hinged or flexible crampon is best used when no technical ice climbing will be done. It is designed to be used with soft, flexible boots, but can be attached to plastic mountaineering boots. The flexible crampon gets its name from the flexible hinge on the crampon itself. All flexible crampons are adjustable for length while some allow for width adjustment. Most flexible crampons will attach to the boot by means of a strap system. The flexible crampon can be worn with a variety of boot types.

    (2) The rigid crampon, as its name implies, is rigid and does not flex. This type of crampon is designed for technical ice climbing, but can be used on less vertical terrain. The rigid crampon can only be worn with plastic mountaineering boots. Rigid crampons will have a toe and heel bail attachment with a strap that wraps around the ankle.

  d. *Ice Screws.* Ice screws provide artificial protection for climbers and equipment for operations in icy terrain. They are screwed into ice formations. Ice screws are made of chrome-molybdenum steel

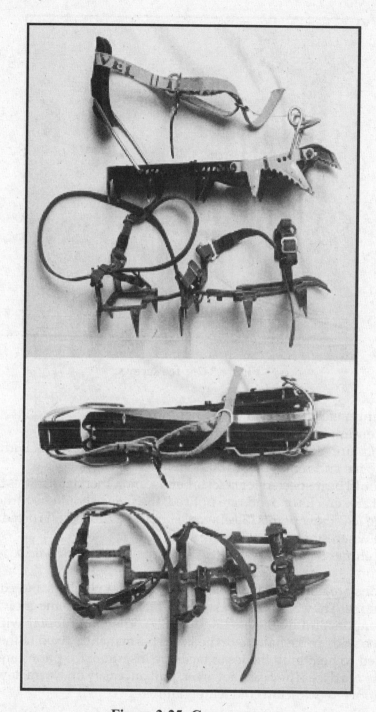

Figure 3-25: Crampons.

and vary in lengths from 11 centimeters to 40 centimeters (Figure 3-26). The eye is permanently affixed to the top of the ice screw. The tip consists of milled or hand-ground teeth, which create sharp points to grab the ice when being emplaced. The ice screw has right-hand threads to penetrate the ice when turned clockwise.

(1) When selecting ice screws, choose a screw with a large thread count and large hollow opening. The close threads will allow for ease in turning and better strength. The large hollow opening will allow snow and ice to slide through when turning.

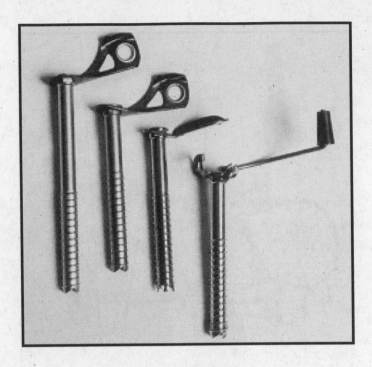

Figure 3-26: Ice screws.

- Type I is 17 centimeters in length with a hollow inner tube.
- Type II is 22 centimeters in length with a hollow inner tube.
- Other variations are hollow alloy screws that have a tapered shank with external threads, which are driven into ice and removed by rotation.

(2) Ice screws should be inspected for cracks, bends, and other deformities that may impair strength or function. If any cracks or bends are noticed, the screw should be turned in. A file may be used to sharpen the ice screw points. Steel wool should be rubbed on rusted surfaces and a thin coat of oil applied when storing steel ice screws.

Note: Ice screws should always be kept clean and dry. The threads and teeth should be protected and kept sharp for ease of application.

e. *Ice Pitons.* Ice pitons are used to establish anchor points for climbers and equipment when conducting operations on ice. They are made of steel or steel alloys (chrome-molybdenum), and are available in various lengths and diameters (Figure 3-27). They are tubular with a hollow core and are hammered into ice with an ice hammer. The eye is permanently fixed to the top of the ice piton. The tip may be beveled to help grab the ice to facilitate insertion. Ice pitons are extremely strong when placed properly in hard ice. They can, however, pull out easily on warm days and require a considerable amount of effort to extract in cold temperatures.

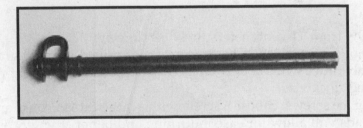

Figure 3-27: Ice piton.

f. *Wired Snow Anchors.* The wired snow anchor (or fluke) provides security for climbers and equipment in operations involving steep ascents by burying the snow anchor into deep snow (Figure 3-28). The fluted anchor portion of the snow anchor is made of aluminum. The wired portion is made of either galvanized steel or stainless steel. Fluke anchors are available in various sizes; their holding ability generally increases with size. They are available with bent faces, flanged sides, and fixed cables. Common types are:

- Type I is 22 by 14 centimeters. Minimum breaking strength of the swaged wire loop is 600 kilograms.
- Type II is 25 by 20 centimeters. Minimum breaking strength of the swaged wire loop is 1,000 kilograms.

The wired snow anchor should be inspected for cracks, broken wire strands, and slippage of the wire through the swage. If any cracks, broken wire strands, or slippage is noticed, the snow anchor should be turned in.

g. *Snow Picket.* The snow picket is used in constructing anchors in snow and ice (Figure 3-28). The snow picket is made of a strong aluminum alloy 3 millimeters thick by 4 centimeters wide, and 45 to 90 centimeters long. They can be angled or T-section stakes. The picket should be inspected for bends, chips, cracks, mushrooming ends, and other deformities. The ends should be filed smooth. If bent or cracked, the picket should be turned in for replacement.

3-6. Sustainability Equipment. This paragraph describes all additional equipment not directly involved with climbing. This equipment is used for safety (avalanche equipment, wands), bivouacs, movement, and carrying gear. While not all of it will need to be carried on all missions, having the equipment available and knowing how to use it correctly will enhance the unit's capability in mountainous terrain.

a. *Snow Saw.* The snow saw is used to cut into ice and snow. It can be used in step cutting, in shelter construction, for removing frozen obstacles, and for cutting snow stability test pits. The special tooth design of the snow saw easily cuts into frozen snow and ice. The blade is a rigid aluminum alloy

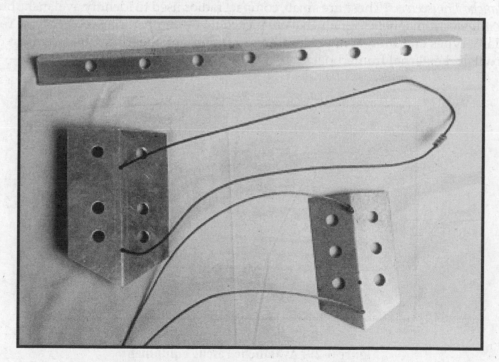

Figure 3-28: Snow anchors, flukes, and pickets.

of high strength about 3 millimeters thick and 38 centimeters long with a pointed end to facilitate entry on the forward stroke. The handle is either wooden or plastic and is riveted to the blade for a length of about 50 centimeters. The blade should be inspected for rust, cracks, warping, burrs, and missing or dull teeth. A file can repair most defects, and steel wool can be rubbed on rusted areas. The handle should be inspected for cracks, bends, and stability. On folding models, the hinge and nuts should be secure. If the saw is beyond repair, it should not be used.

b. *Snow Shovel*. The snow shovel is used to cut and remove ice and snow. It can be used for avalanche rescue, shelter construction, step cutting, and removing obstacles. The snow shovel is made of a special, lightweight aluminum alloy. The handle should be telescopic, folding, or removable to be compact when not in use. The shovel should have a flat or rounded bottom and be of strong construction. The shovel should be inspected for cracks, bends, rust, and burrs. A file and steel wool can remove rust and put an edge on the blade of the shovel. The handle should be inspected for cracks, bends, and stability. If the shovel is beyond repair, it should be turned in.

c. *Wands*. Wands are used to identify routes, crevasses, snow-bridges, caches, and turns on snow and glaciers. Spacing of wands depends on the number of turns, number of hazards identified, weather conditions (and visibility), and number of teams in the climbing party. Carrying too many wands is better than not having enough if they become lost. Wands are 1 to 1.25 meters long and made of lightweight bamboo or plastic shafts pointed on one end with a plastic or nylon flag (bright enough in color to see at a distance) attached to the other end. The shafts should be inspected for cracks, bends, and deformities. The flag should be inspected for tears, frays, security to the shaft, fading, and discoloration. If any defects are discovered, the wands should be replaced.

d. *Avalanche rescue equipment*. Avalanche rescue equipment (Figure 3-29) includes the following:

(1) *Avalanche Probe*. Although ski poles may be used as an emergency probe when searching for a victim in an avalanche, commercially manufactured probes are better for a thorough search. They are 9-millimeter thick shafts made of an aluminum alloy, which can be joined to probe up to 360 centimeters. The shafts must be strong enough to probe through avalanche debris. Some manufacturers of ski poles design poles that are telescopic and mate with other poles to create an avalanche probe.

(2) *Avalanche Transceivers*. These are small, compact radios used to identify avalanche burial sites. They transmit electromagnetic signals that are picked up by another transceiver on the receive mode.

e. *Packs*. Many types and brands of packs are used for mountaineering. The two most common types are internal and external framed packs.

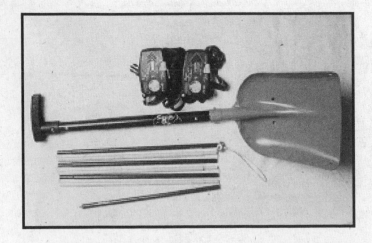

Figure 3-29: Avalanche rescue equipment.

(1) Internal framed packs have a rigid frame within the pack that help it maintain its shape and hug the back. This assists the climber in keeping their balance as they climb or ski. The weight in an internal framed pack is carried low on the body assisting with balance. The body-hugging nature of this type pack also makes it uncomfortable in warm weather.

(2) External framed packs suspend the load away from the back with a ladder-like frame. The frame helps transfer the weight to the hips and shoulders easier, but can be cumbersome when balance is needed for climbing and skiing.

(3) Packs come in many sizes and should be sized appropriately for the individual according to manufacturer's specifications. Packs often come with many unneeded features. A good rule of thumb is: The simpler the pack, the better it will be.

f. *Stoves.* When selecting a stove one must define its purpose: will the stove be used for heating, cooking or both? Stoves or heaters for large elements can be large and cumbersome. Stoves for smaller elements might just be used for cooking and making water, and are simple and lightweight. Stoves are a necessity in mountaineering for cooking and making water from snow and ice. When choosing a stove, factors that should be considered are weight, altitude and temperature where it will be used, fuel availability, and its reliability.

(1) There are many choices in stove design and in fuel types. White gas, kerosene, and butane are the common fuels used. All stoves require a means of pressurization to force the fuel to the burner. Stoves that burn white gas or kerosene have a hand pump to generate the pressurization and butane stoves have pressurized cartridges. All stoves need to vaporize the liquid fuel before it is burned. This can be accomplished by burning a small amount of fuel in the burner cup assembly, which will vaporize the fuel in the fuel line.

(2) Stoves should be tested and maintained prior to a mountaineering mission. They should be easy to clean and repair during an operation. The reliability of the stove has a huge impact on the success of the mission and the morale of personnel.

g. *Tents.* When selecting a tent, the mission must be defined to determine the number of people the tent will accommodate. The climate the tents will be used in is also of concern. A tent used for warmer temperatures will greatly differ from tents used in a colder, harsher environment. Manufacturers of tents offer many designs of different sizes, weights, and materials.

(1) Mountaineering tents are made out of a breathable or weatherproof material. A single-wall tent allows for moisture inside the tent to escape through the tent's material. A double-wall tent has a second layer of material (referred to as a fly) that covers the tent. The fly protects against rain and snow and the space between the fly and tent helps moisture to escape from inside. Before using a new tent, the seams should be treated with seam sealer to prevent moisture from entering through the stitching.

(2) The frame of a tent is usually made of an aluminum or carbon fiber pole. The poles are connected with an elastic cord that allows them to extend, connect, and become long and rigid. When the tent poles are secured into the tent body, they create the shape of the tent.

(3) Tents are rated by a "relative strength factor," the speed of wind a tent can withstand before the frame deforms. Temperature and expected weather for the mission should be determined before choosing the tent.

h. *Skis.* Mountaineering skis are wide and short. They have a binding that pivots at the toe and allows for the heel to be free for uphill travel or locked for downhill. Synthetic skins with fibers on the bottom can be attached to the bottom of the ski and allow the ski to travel forward and prevent slipping backward. The skins aid in traveling uphill and slow down the rate of descents. Wax can be applied to the ski to aid in ascents instead of skins. Skis can decrease the time needed to reach an objective depending on the ability of the user. Skis can make crossing crevasses easier because of the load distribution, and they can become a makeshift stretcher for casualties. Ski techniques can be complicated and require thorough training for adequate proficiency.

i. *Snowshoes.* Snowshoes are the traditional aid to snow travel that attach to most footwear and have been updated into small, lightweight designs that are more efficient than older models. Snowshoes offer a large displacement area on top of soft snow preventing tiresome post-holing. Some snowshoes come equipped with a crampon like binding that helps in ascending steep snow and ice. Snowshoes are slower than skis, but are better suited for mixed terrain, especially if personnel are not experienced with the art of skiing. When carrying heavy packs, snowshoes can be easier to use than skis.

j. *Ski poles.* Ski poles were traditionally designed to assist in balance during skiing. They have become an important tool in mountaineering for aid in balance while hiking, snowshoeing, and carrying heavy packs. They can take some of the weight off of the lower body when carrying a heavy pack. Some ski poles are collapsible for ease of packing when not needed (Figure 3-30). The basket at the bottom prevents the pole from plunging deep into the snow and, on some models, can be detached so the pole becomes an avalanche or crevasse probe. Some ski poles come with a self-arrest grip, but should not be the only means of protection on technical terrain.

k. *Sleds.* Sleds vary greatly in size, from the squad-size Ahkio, a component of the 10-man arctic tent system, to the one-person skow. Regardless of the size, sleds are an invaluable asset during mountainous operations when snow and ice is the primary surface on which to travel. Whichever sled is chosen, it must be attachable to the person or people that will be pulling it. Most sleds are constructed using fiberglass bottoms with or without exterior runners. Runners will aid the sled's ability to maintain a true track in the snow. The sled should also come with a cover of some sort—whether nylon or canvas, a cover is essential for keeping the components in the sled dry. Great care should be taken when packing the sled, especially when hauling fuel. Heavier items should be carried towards the rear of the sled and lighter items towards the front.

l. *Headlamps.* A headlamp is a small item that is not appreciated until it is needed. It is common to need a light source and the use of both hands during limited light conditions in mountaineering operations. A flashlight can provide light, but can be cumbersome when both hands are needed. Most headlamps attach to helmets by means of elastic bands.

(1) When choosing a headlamp, ensure it is waterproof and the battery apparatus is small. All components should be reliable in extreme weather conditions. When the light is being packed, care should be taken that the switch doesn't accidentally activate and use precious battery life.

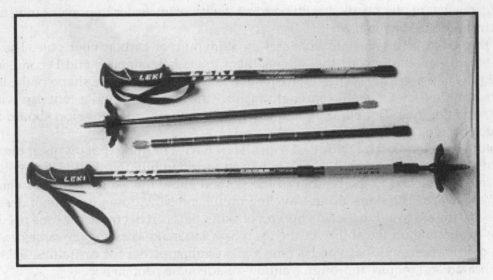

Figure 3-30: Collapsible ski poles.

(2) The battery source should compliment the resupply available. Most lights will accept alkaline, nickel-cadmium, or lithium batteries. Alkaline battery life diminishes quickly in cold temperatures, nickel-cadmium batteries last longer in cold but require a recharging unit, and lithium batteries have twice the voltage so modifications are required.

EQUIPMENT PACKING

Equipment brought on a mission is carried in the pack, worn on the body, or hauled in a sled (in winter). Obviously, the rucksack and sled (or Ahkio) can hold much more than a climber can carry. They would be used for major bivouac gear, food, water, first aid kits, climbing equipment, foul weather shells, stoves, fuel, ropes, and extra ammunition and demolition materials, if needed.

3-7. Choice of Equipment. Mission requirements and unit SOP will influence the choice of gear carried but the following lists provide a sample of what should be considered during mission planning.

   a. *Personal Gear.* Personal gear includes emergency survival kit containing signaling material, fire starting material, food procurement material, and water procurement material. Pocket items should include a knife, whistle, pressure bandage, notebook with pen or pencil, sunglasses, sunblock and lip protection, map, compass and or altimeter.

   b. *Standard Gear.* Standard gear that can be individually worn or carried includes cushion sole socks; combat boots or mountain boots, if available; BDU and cap; LCE with canteens, magazine pouches, and first aid kit; individual weapon; a large rucksack containing waterproof coat and trousers, polypropylene top, sweater, or fleece top; helmet; poncho; and sleeping bag.

> ⚠ **CAUTION**
>
> Cotton clothing, due to its poor insulating and moisture-wicking characteristics, is virtually useless in most mountain climates, the exception being hot, desert, or jungle mountain environments. Cotton clothing should be replaced with synthetic fabric clothing.

   c. *Mountaineering Equipment and Specialized Gear.* This gear includes:
   • Sling rope or climbing harness.
   • Utility cord(s).
   • Nonlocking carabiners.
   • Locking carabiner(s).
   • Rappelling gloves.
   • Rappel/belay device.
   • Ice ax.
   • Crampons.
   • Climbing rope, one per climbing team.
   • Climbing rack, one per climbing team.
   d. *Day Pack.* When the soldier plans to be away from the bivouac site for the day on a patrol or mountaineering mission, he carries a light day pack. This pack should contain the following items:
   • Extra insulating layer: polypropylene, pile top, or sweater.
   • Protective layer: waterproof jacket and pants, rain suit, or poncho.
   • First aid kit.
   • Flashlight or headlamp.
   • Canteen.
   • Cold weather hat or scarf.
   • Rations for the time period away from the base camp.
   • Survival kit.

- Sling rope or climbing harness.
- Carabiners.
- Gloves.
- Climbing rope, one per climbing team.
- Climbing rack, one per climbing team.

e. *Squad or Team Safety Pack.* When a squad-sized element leaves the bivouac site, squad safety gear should be carried in addition to individual day packs. This can either be loaded into one rucksack or cross-loaded among the squad members. In the event of an injury, casualty evacuation, or unplanned bivouac, these items may make the difference between success and failure of the mission.

- Sleeping bag.
- Sleeping mat.
- Squad stove.
- Fuel bottle.

f. *The Ten Essentials.* Regardless of what equipment is carried, the individual military mountaineer should always carry the "ten essentials" when moving through the mountains.

(1) *Map.*

(2) *Compass, Altimeter, and or GPS.*

(3) *Sunglasses and Sunscreen.*

   (a) In alpine or snow-covered sub-alpine terrain, sunglasses are a vital piece of equipment for preventing snow blindness. They should filter 95 to 100 percent of ultraviolet light. Side shields, which minimize the light entering from the side, should permit ventilation to help prevent lens fogging. At least one extra pair of sunglasses should be carried by each independent climbing team.

   (b) Sunscreens should have an SPF factor of 15 or higher. For lip protection, a total UV blocking lip balm that resists sweating, washing, and licking is best. This lip protection should be carried in the chest pocket or around the neck to allow frequent reapplication.

(4) *Extra Food.* One day's worth extra of food should be carried in case of delay caused by bad weather, injury, or navigational error.

(5) *Extra Clothing.* The clothing used during the active part of a climb, and considered to be the basic climbing outfit, includes socks, boots, underwear, pants, blouse, sweater or fleece jacket, hat, gloves or mittens, and foul weather gear (waterproof, breathable outerwear or waterproof rain suit).

   (a) Extra clothing includes additional layers needed to make it through the long, inactive hours of an unplanned bivouac. Keep in mind the season when selecting this gear.

   - Extra underwear to switch out with sweat-soaked underwear.
   - Extra hats or balaclavas.
   - Extra pair of heavy socks.
   - Extra pair of insulated mittens or gloves.
   - In winter or severe mountain conditions, extra insulation for the upper body and the legs.

   (b) To back up foul weather gear, bring a poncho or extra-large plastic trash bag. A reflective emergency space blanket can be used for hypothermia first aid and emergency shelter. Insulated foam pads prevent heat loss while sitting or lying on snow. Finally, a bivouac sack can help by protecting insulating layers from the weather, cutting the wind, and trapping essential body heat inside the sack.

(6) *Headlamp and or Flashlight.* Headlamps provide the climber a hands-free capability, which is important while climbing, working around the camp, and employing weapons systems. Miniature flashlights can be used, but commercially available headlamps are best. Red lens covers can be fabricated for tactical conditions. Spare batteries and spare bulbs should also be carried.

(7) *First-aid Kit.* Decentralized operations, the mountain environment steep, slick terrain and loose rock combined with heavy packs, sharp tools, and fatigue requires each climber to carry his own first-aid kit. Common mountaineering injuries that can be expected are punctures and abrasions with severe bleeding, a broken bone, serious sprain, and blisters. Therefore, the kit should contain at least enough material to stabilize these conditions. Pressure dressings, gauze pads, elastic compression wrap, small adhesive bandages, butterfly bandages, moleskin, adhesive tape, scissors, cleanser, latex gloves and splint material (if above tree line) should all be part of the kit.

(8) *Fire Starter.* Fire starting material is key to igniting wet wood for emergency campfires. Candles, heat tabs, and canned heat all work. These can also be used for quick warming of water or soup in a canteen cup. In alpine zones above tree line with no available firewood, a stove works as an emergency heat source.

(9) *Matches and Lighter.* Lighters are handy for starting fires, but they should be backed up by matches stored in a waterproof container with a strip of sandpaper.

(10) *Knife.* A multipurpose pocket tool should be secured with cord to the belt, harness, or pack.

g. Other Essential Gear. Other essential gear may be carried depending on mission and environmental considerations.

(1) *Water and Water Containers.* These include wide-mouth water bottles for water collection; camel-back type water holders for hands-free hydration; and a small length of plastic tubing for water procurement at snow-melt seeps and rainwater puddles on bare rock.

(2) *Ice Ax.* The ice ax is essential for travel on snowfields and glaciers as well as snow-covered terrain in spring and early summer. It helps for movement on steep scree and on brush and heather covered slopes, as well as for stream crossings.

(3) *Repair Kit.* A repair kit should include:
- Stove tools and spare parts.
- Duct tape.
- Patches.
- Safety pins.
- Heavy-duty thread.
- Awl and or needles.
- Cord and or wire.
- Small pliers (if not carrying a multipurpose tool).
- Other repair items as needed.

(4) *Insect Repellent.*

(5) *Signaling Devices.*

(6) *Snow Shovel.*

3-8. Tips on Packing. When loading the internal frame pack the following points should be considered.

a. In most cases, speed and endurance are enhanced if the load is carried more by the hips (using the waist belt) and less by the shoulders and back. This is preferred for movement over trails or less difficult terrain. By packing the lighter, more compressible items (sleeping bag, clothing) in the bottom of the rucksack and the heavier gear (stove, food, water, rope, climbing hardware, extra ammunition) on top, nearer the shoulder blades, the load is held high and close to the back, thus placing the most weight on the hips.

b. In rougher terrain it pays to modify the pack plan. Heavy articles of gear are placed lower in the pack and close to the back, placing more weight on the shoulders and back. This lowers the climber's center of gravity and helps him to better keep his balance.

c. Equipment that may be needed during movement should be arranged for quick access using either external pockets or placing immediately underneath the top flap of the pack. As much as possible, this placement should be standardized across the team so that necessary items can be quickly reached without unnecessary unpacking of the pack in emergencies.

d. The pack and its contents should be soundly waterproofed. Clothing and sleeping bag are separately sealed and then placed in the larger wet weather bag that lines the rucksack. Zip-lock plastic bags can be used for small items, which are then organized into color-coded stuff sacks. A few extra-large plastic garbage bags should be carried for a variety of uses: spare waterproofing, emergency bivouac shelter, and water procurement, among others.

e. The ice ax, if not carried in hand, should be stowed on the outside of the pack with the spike up and the adze facing forward or to the outside, and be securely fastened. Mountaineering packs have ice ax loops and buckle fastening systems for this. If not, the ice ax is placed behind one of the side pockets, as stated above, and then tied in place.

f. Crampons should be secured to the outside rear of the pack with the points covered.

SECTION 4: ROPE MANAGEMENT AND KNOTS

The rope is a vital piece of equipment to the mountaineer. When climbing, rappelling, or building various installations, the mountaineer must know how to properly use and maintain this piece of equipment. If the rope is not managed or maintained properly, serious injury may occur. This chapter discusses common rope terminology, management techniques, care and maintenance procedures, and knots.

PREPARATION, CARE AND MAINTENANCE, INSPECTION, TERMINOLOGY

The service life of a rope depends on the frequency of use, applications (rappelling, climbing, rope installations), speed of descent, surface abrasion, terrain, climate, and quality of maintenance. Any rope may fail under extreme conditions (shock load, sharp edges, misuse).

4-1. Preparation. The mountaineer must select the proper rope for the task to be accomplished according to type, diameter, length, and tensile strength. It is important to prepare all ropes before departing on a mission. Avoid rope preparation in the field.

a. *Packaging.* New rope comes from the manufacturer in different configurations boxed on a spool in various lengths, or coiled and bound in some manner. Precut ropes are usually packaged in a protective cover such as plastic or burlap . Do not remove the protective cover until the rope is ready for use.

b. *Securing the Ends of the Rope:* If still on a spool, the rope must be cut to the desired length. All ropes will fray at the ends unless they are bound or seared. Both static and dynamic rope ends are secured in the same manner. The ends must be heated to the melting point so as to attach the inner core strands to the outer sheath. By fusing the two together, the sheath cannot slide backward or forward. Ensure that this is only done to the ends of the rope. If the rope is exposed to extreme temperatures, the sheath could be weakened, along with the inner core, reducing overall tensile strength. The ends may also be dipped in enamel or lacquer for further protection.

4-2. Care and Maintenance. The rope is a climber's lifeline. It must be cared for and used properly. These general guidelines should be used when handling ropes.

a. Do not step on or drag ropes on the ground unnecessarily. Small particles of dirt will be ground between the inner strands and will slowly cut them.

b. While in use, do not allow the rope to come into contact with sharp edges. Nylon rope is easily cut, particularly when under tension. If the rope must be used over a sharp edge, pad the edge for protection.

c. Always keep the rope as dry as possible. Should the rope become wet, hang it in large loops off the ground and allow it to dry. Never dry a rope with high heat or in direct sunlight.

d. Never leave a rope knotted or tightly stretched for longer than necessary. Over time it will reduce the strength and life of the rope.

e. Never allow one rope to continuously rub over or against another. Allowing rope-on-rope contact with nylon rope is extremely dangerous because the heat produced by the friction will cause the nylon to melt.

f. Inspect the rope before each use for frayed or cut spots, mildew or rot, or defects in construction (new rope).

g. The ends of the rope should be whipped or melted to prevent unraveling.

h. Do not splice ropes for use in mountaineering.

i. Do not mark ropes with paints or allow them to come in contact with oils or petroleum products. Some of these will weaken or deteriorate nylon.

j. Never use a mountaineering rope for any purpose except mountaineering.

k. Each rope should have a corresponding rope log, which is also a safety record. It should annotate use, terrain, weather, application, number of falls, dates, and so on, and should be annotated each time the rope is used (Figure 4-1).

l. Never subject the rope to high heat or flame. This will significantly weaken it.

m. All ropes should be washed periodically to remove dirt and grit, and rinsed thoroughly. Commercial rope washers are made from short pieces of modified pipe that connect to any faucet. Pinholes within the pipe force water to circulate around and scrub the rope as you slowly feed it through the washer. Another method is to machine wash, on a gentle cycle, in cold water with a nylon safe soap, never bleach or harsh cleansers. Ensure that only front loading washing machines are used to wash ropes.

n. Ultraviolet radiation (sunlight) tends to deteriorate nylon over long periods of time. This becomes important if rope installations are left in place over a number of months.

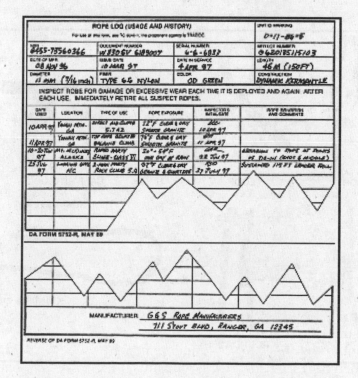

Figure 4-1: Example of completed DA Form 5752-R.

o. When not in use, ropes should be loosely coiled and hung on wooden pegs rather than nails or other metal objects. Storage areas should be relatively cool with low humidity levels to prevent mildew or rotting. Rope may also be loosely stacked and placed in a rope bag and stored on a shelf. Avoid storage in direct sunlight, as the ultraviolet radiation will deteriorate the nylon over long periods.

4-3. Inspection. Ropes should be inspected before and after each use, especially when working around loose rock or sharp edges.

a. Although the core of the kernmantle rope cannot be seen, it is possible to damage the core without damaging the sheath. Check a kernmantle rope by carefully inspecting the sheath before and after use while the rope is being coiled. When coiling, be aware of how the rope feels as it runs through the hands. Immediately note and tie off any lumps or depressions felt.

b. Damage to the core of a kernmantle rope usually consists of filaments or yarn breakage that results in a slight retraction. If enough strands rupture, a localized reduction in the diameter of the rope results in a depression that can be felt or even seen.

c. Check any other suspected areas further by putting them under tension (the weight of one person standing on a Prusik tensioning system is about maximum). This procedure will emphasize the lump or depression by separating the broken strands and enlarging the dip. If a noticeable difference in diameter is obvious, retire the rope immediately.

d. Many dynamic kernmantle ropes are quite soft. They may retain an indention occasionally after an impact or under normal use without any trauma to the core. When damage is suspected, patiently inspect the sheath for abnormalities. Damage to the sheath does not always mean damage to the core. Inspect carefully.

4-4. Terminology. When using ropes, understanding basic terminology is important. The terms explained in this section are the most commonly used in military mountaineering. (Figure 4-2 illustrates some of these terms.)

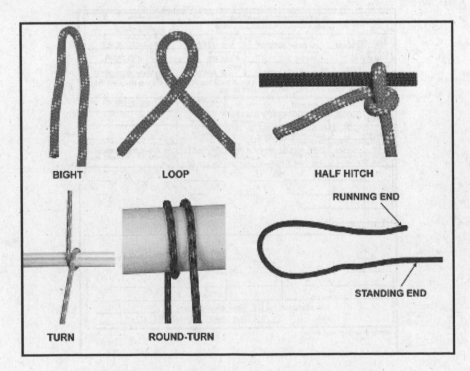

Figure 4-2: Examples of roping terminology.

a. *Bight.* A bight of rope is a simple bend of rope in which the rope does not cross itself.

b. *Loop.* A loop is a bend of a rope in which the rope does cross itself.

c. *Half Hitch.* A half hitch is a loop that runs around an object in such a manner as to lock or secure itself.

d. *Turn.* A turn wraps around an object, providing 360-degree contact.

e. *Round Turn.* A round turn wraps around an object one and one-half times. A round turn is used to distribute the load over a small diameter anchor (3 inches or less). It may also be used around larger diameter anchors to reduce the tension on the knot, or provide added friction.

f. *Running End.* A running end is the loose or working end of the rope.

g. *Standing Part.* The standing part is the static, stationary, or nonworking end of the rope.

h. *Lay.* The lay is the direction of twist used in construction of the rope.

i. *Pigtail.* The pigtail (tail) is the portion of the running end of the rope between the safety knot and the end of the rope.

j. *Dress.* Dress is the proper arrangement of all the knot parts, removing unnecessary kinks, twists, and slack so that all rope parts of the knot make contact.

COILING, CARRYING, THROWING

The ease and speed of rope deployment and recovery greatly depends upon technique and practice.

4-5. Coiling and Carrying The Rope. Use the butterfly or mountain coil to coil and carry the rope. Each is easy to accomplish and results in a minimum amount of kinks, twists, and knots later during deployment.

a. *Mountain Coil.* To start a mountain coil, grasp the rope approximately 1 meter from the end with one hand. Run the other hand along the rope until both arms are outstretched. Grasping the rope firmly, bring the hands together forming a loop, which is laid in the hand closest to the end of the rope. This is repeated, forming uniform loops that run in a clockwise direction, until the rope is completely coiled. The rope may be given a 1/4 twist as each loop is formed to overcome any tendency for the rope to twist or form figure-eights.

(1) In finishing the mountain coil, form a bight approximately 30 centimeters long with the starting end of the rope and lay it along the top of the coil. Uncoil the last loop and, using this length of the rope, begin making wraps around the coil and the bight, wrapping toward the closed end of the bight and making the first wrap bind across itself so as to lock it into place. Make six to eight wraps to adequately secure the coil, and then route the end of the rope through the closed end of the bight. Pull the running end of the bight tight, securing the coil.

(2) The mountain coil may be carried either in the pack (by forming a figure eight), doubling it and placing it under the flap, or by placing it over the shoulder and under the opposite arm, slung across the chest. (Figure 4-3 shows how to coil a mountain coil.)

b. *Butterfly Coil.* The butterfly coil is the quickest and easiest technique for coiling (Figure 4-4).

(1) Coiling. To start the double butterfly, grasp both ends of the rope and begin back feeding. Find the center of the rope forming a bight. With the bight in the left hand, grasp both ropes and slide the right hand out until there is approximately one arms length of rope. Place the doubled rope over the head, draping it around the neck and on top of the shoulders. Ensure that it hangs no lower than the waist. With the rest of the doubled rope in front of you, make doubled bights placing them over the head in the same manner as the first bight. Coil alternating from side to side (left to right, right to left) while maintaining equal-length bights. Continue coiling until approximately two arm-lengths of rope remain. Remove the coils from the neck and shoulders carefully, and hold the center in one hand. Wrap the two ends around the coils a minimum of three doubled wraps, ensuring that the first wrap locks back on itself.

(2) Tie-off and Carrying. Take a doubled bight from the loose ends of rope and pass it through the apex of the coils. Pull the loose ends through the doubled bight and dress it down. Place an

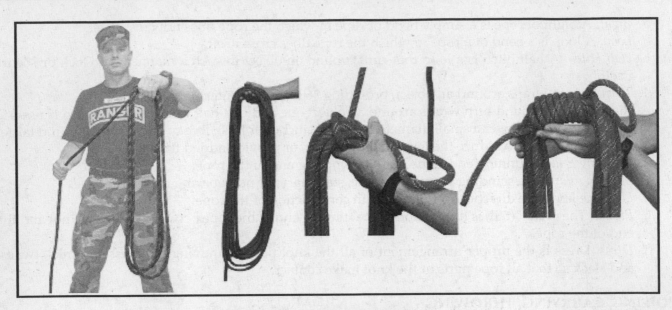

Figure 4-3: Mountain coil.

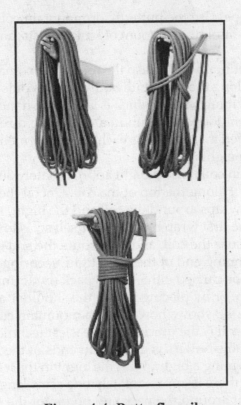

Figure 4-4: Butterfly coil.

overhand knot in the loose ends, dressing it down to the apex of the bight securing coils. Ensure that the loose ends do not exceed the length of the coils. In this configuration the coiled rope is secure enough for hand carrying or carrying in a rucksack, or for storage. (Figure 4-5 shows a butterfly coil tie-off.)

c. *Coiling Smaller Diameter Rope.* Ropes of smaller diameters may be coiled using the butterfly or mountain coil depending on the length of the rope. Pieces 25 feet and shorter (also known as cordage, sling

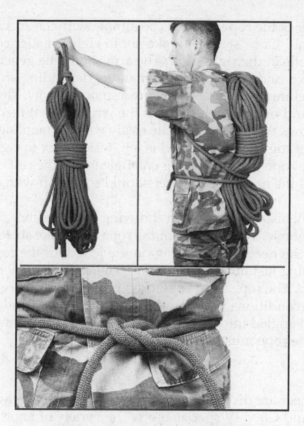

Figure 4-5: Butterfly coil tic-off.

rope, utility cord) may be coiled so that they can be hung from the harness. Bring the two ends of the rope together, ensuring no kinks are in the rope. Place the ends of the rope in the left hand with the two ends facing the body. Coil the doubled rope in a clockwise direction forming 6- to 8-inch coils (coils may be larger depending on the length of rope) until an approximate 12-inch bight is left. Wrap that bight around the coil, ensuring that the first wrap locks on itself. Make three or more wraps. Feed the bight up through the bights formed at the top of the coil. Dress it down tightly. Now the piece of rope may be hung from a carabiner on the harness.

d. *Uncoiling, Back-feeding, and Stacking.* When the rope is needed for use, it must be uncoiled and stacked on the ground properly to avoid kinks and snarls.

(1) Untie the tie-off and lay the coil on the ground. Back-feed the rope to minimize kinks and snarls. (This is also useful when the rope is to be moved a short distance and coiling is not desired.) Take one end of the rope in the left hand and run the right hand along the rope until both arms are outstretched. Next, lay the end of the rope in the left hand on the ground. With the left hand, re-grasp the rope next to the right hand and continue laying the rope on the ground.

(2) The rope should be laid or stacked in a neat pile on the ground to prevent it from becoming tangled and knotted when throwing the rope, feeding it to a lead climber, and so on. This technique can also be started using the right hand.

4-6. Throwing The Rope. Before throwing the rope, it must be properly managed to prevent it from tangling during deployment. The rope should first be anchored to prevent complete loss of the rope over the edge when it is thrown. Several techniques can be used when throwing a rope. Personal preference and situational and environmental conditions should be taken into consideration when determining which technique is best.

a. Back feed and neatly stack the rope into coils beginning with the anchored end of the rope working toward the running end. Once stacked, make six to eight smaller coils in the left hand. Pick up the rest of the larger coils in the right hand. The arm should be generally straight when throwing. The rope may be thrown underhanded or overhanded depending on obstacles around the edge of the site. Make a few preliminary swings to ensure a smooth throw. Throw the large coils in the right hand first. Throw up and out. A slight twist of the wrist, so that the palm of the hand faces up as the rope is thrown, allows the coils to separate easily without tangling. A smooth follow through is essential. When a slight tug on the left hand is felt, toss the six to eight smaller coils out. This will prevent the ends of the rope from becoming entangled with the rest of the coils as they deploy. As soon as the rope leaves the hand, the thrower should sound off with a warning of "ROPE" to alert anyone below the site.

b. Another technique may also be used when throwing rope. Anchor, back feed, and stack the rope properly as described above. Take the end of the rope and make six to eight helmet-size coils in the right hand (more may be needed depending on the length of the rope). Assume a "quarterback" simulated stance. Aiming just above the horizon, vigorously throw the rope overhanded, up and out toward the horizon. The rope must be stacked properly to ensure smooth deployment.

c. When windy weather conditions prevail, adjustments must be made. In a strong cross wind, the rope should be thrown angled into the wind so that it will land on the desired target. The stronger the wind, the harder the rope must be thrown to compensate.

KNOTS

All knots used by a mountaineer are divided into four classes: Class I joining knots, Class II anchor knots, Class III middle rope knots, and Class IV special knots. The variety of knots, bends, bights, and hitches is almost endless. These classes of knots are intended only as a general guide since some of the knots discussed may be appropriate in more than one class. The skill of knot tying can perish if not used and practiced. With experience and practice, knot tying becomes instinctive and helps the mountaineer in many situations.

4-7. Square Knot. The square knot is used to tie the ends of two ropes of equal diameter (Figure 4-6). It is a joining knot.

a. *Tying the Knot.*
STEP 1. Holding one working end in each hand, place the working end in the right hand over the one in the left hand.

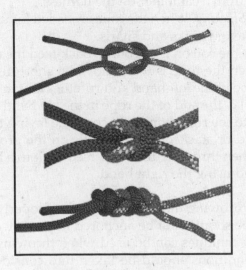

Figure 4-6: Square knot.

STEP 2. Pull it under and back over the top of the rope in the left hand.

STEP 3. Place the working end in the left hand over the one in the right hand and repeat STEP 2.

STEP 4. Dress the knot down and secure it with an overhand knot on each side of the square knot.

b. *Check points.*

(1) There are two interlocking bights.

(2) The running end and standing part are on the same side of the bight formed by the other rope.

(3) The running ends are parallel to and on the same side of the standing ends with 4-inch minimum pig tails after the overhand safeties are tied.

4-8. Fisherman's Knot. The fisherman's knot is used to tie two ropes of the same or approximately the same diameter (Figure 4-7). It is a joining knot.

a. Tying the Knot.

STEP 1. Tie an overhand knot in one end of the rope.

STEP 2. Pass the working end of the other rope through the first overhand knot. Tie an overhand knot around the standing part of the first rope with the working end of the second rope.

STEP 3. Tightly dress down each overhand knot and tightly draw the knots together.

b. Checkpoints.

(1) The two separate overhand knots are tied tightly around the long, standing part of the opposing rope.

(2) The two overhand knots are drawn snug.

(3) Ends of rope exit knot opposite each other with 4-inch pigtails.

4-9. Double Fisherman's Knot. The double fisherman's knot (also called double English or grapevine) is used to tie two ropes of the same or approximately the same diameter (Figure 4-8). It is a joining knot.

a. *Tying the Knot.*

STEP 1. With the working end of one rope, tie two wraps around the standing part of another rope.

STEP 2. Insert the working end (STEP 1) back through the two wraps and draw it tight.

STEP 3. With the working end of the other rope, which contains the standing part (STEPS 1 and 2), tie two wraps around the standing part of the other rope (the working end in STEP 1). Insert the working end back through the two wraps and draw tight.

STEP 4. Pull on the opposing ends to bring the two knots together.

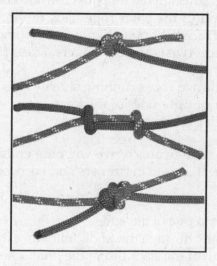

Figure 4-7: Fisherman's knot.

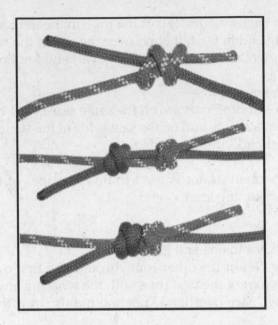

Figure 4-8: Double fisherman's knot.

b. *Checkpoints.*
  (1) Two double overhand knots securing each other as the standing parts of the rope are pulled apart.
  (2) Four rope parts on one side of the knot form two "x" patterns, four rope parts on the other side of the knot are parallel.
  (3) Ends of rope exit knot opposite each other with 4-inch pigtails.

4-10. Figure-Eight Bend. The figure-eight bend is used to join the ends of two ropes of equal or unequal diameter within 5-mm difference (Figure 4-9).

a. *Tying the Knot.*
STEP 1. Grasp the top of a 2-foot bight.
STEP 2. With the other hand, grasp the running end (short end) and make a 360-degree turn around the standing end.
STEP 3. Place the running end through the loop just formed creating an in-line figure eight.
STEP 4. Route the running end of the other rope back through the figure eight starting from the original rope's running end. Trace the original knot to the standing end.
STEP 5. Remove all unnecessary twists and crossovers. Dress the knot down.
b. *Checkpoints.*
  (1) There is a figure eight with two ropes running side by side.
  (2) The running ends are on opposite sides of the knot.
  (3) There is a minimum 4-inch pigtail.

4-11. Water Knot. The water knot is used to attach two webbing ends (Figure 4-10). It is also called a ring bend, overhand retrace, or tape knot. It is used in runners and harnesses and is a joining knot.

a. *Tying the Knot.*
STEP 1. Tie an overhand knot in one of the ends.
STEP 2. Feed the other end back through the knot, following the path of the first rope in reverse.
STEP 3. Draw tight and pull all of the slack out of the knot. The remaining tails must extend at least 4 inches beyond the knot in both directions.

Figure 4-9: Figure-eight bend.

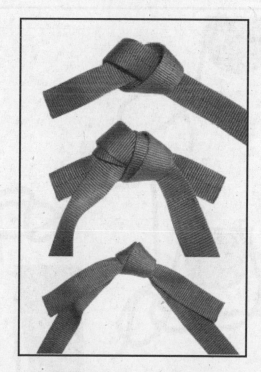

Figure 4-10: Water knot.

b. *Checkpoints*.
    (1) There are two overhand knots, one retracing the other.
    (2) There is no slack in the knot, and the working ends come out of the knot in opposite directions.
    (3) There is a minimum 4-inch pigtail.

4-12. Bowline. The bowline is used to tie the end of a rope around an anchor. It may also be used to tie a single fixed loop in the end of a rope (Figure 4-11). It is an anchor knot.

    a. *Tying the Knot.*
        STEP 1. Bring the working end of the rope around the anchor, from right to left (as the climber faces the anchor).
        STEP 2. Form an overhand loop in the standing part of the rope (on the climber's right) toward the anchor.
        STEP 3. Reach through the loop and pull up a bight.
        STEP 4. Place the working end of the rope (on the climber's left) through the bight, and bring it back onto itself. Now dress the knot down.
        STEP 5. Form an overhand knot with the tail from the bight.
    b. *Checkpoints.*
        (1) The bight is locked into place by a loop.
        (2) The short portion of the bight is on the inside and on the loop around the anchor (or inside the fixed loop).
        (3) There is a minimum 4-inch pigtail after tying the overhand safety.

4-13. Round Turn and Two Half Hitches. This knot is used to tie the end of a rope to an anchor, and it must have constant tension (Figure 4-12). It is an anchor knot.

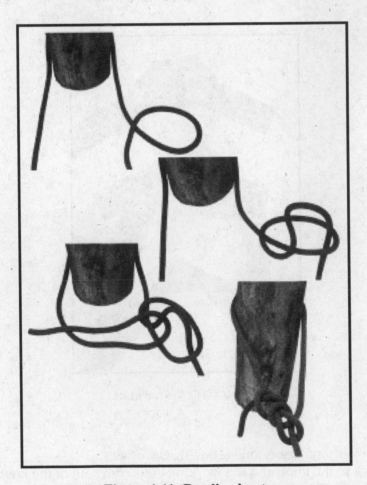

Figure 4-11: Bowline knot.

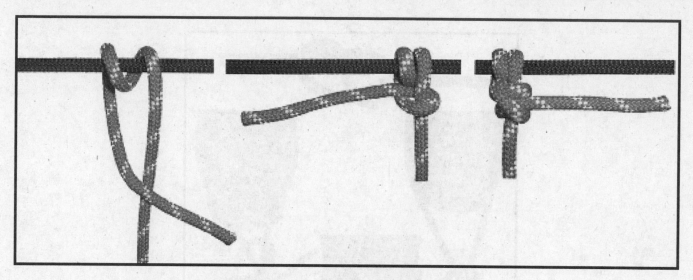

Figure 4-12: Round turn and two half hitches.

a. *Tying the Knot.*
   STEP 1. Route the rope around the anchor from right to left and wrap down (must have two wraps in the rear of the anchor, and one in the front). Run the loop around the object to provide 360-degree contact, distributing the load over the anchor.
   STEP 2. Bring the working end of the rope left to right and over the standing part, forming a half hitch (first half hitch).
   STEP 3. Repeat STEP 2 (last half hitch has a 4 inch pigtail).
   STEP 4. Dress the knot down.
b. *Checkpoints.*
   (1) A complete round turn should exist around the anchor with no crosses.
   (2) Two half hitches should be held in place by a diagonal locking bar with no less than a 4-inch pigtail remaining.

4-14. Figure-Eight Retrace (Rerouted Figure-Eight). The figure-eight retrace knot produces the same result as a figure-eight loop. However, by tying the knot in a retrace, it can be used to fasten the rope to trees or to places where the loop cannot be used (Figure 4-13). It is also called a rerouted figure-eight and is an anchor knot.

a. *Tying the Knot.*
   STEP 1. Use a length of rope long enough to go around the anchor, leaving enough rope to work with.
   STEP 2. Tie a figure-eight knot in the standing part of the rope, leaving enough rope to go around the anchor. To tie a figure-eight knot form a loop in the rope, wrap the working end around the standing part, and route the working end through the loop. The finished knot is dressed loosely.
   STEP 3. Take the working end around the anchor point.
   STEP 4. With the working end, insert the rope back through the loop of the knot in reverse.
   STEP 5. Keep the original figure eight as the outside rope and retrace the knot around the wrap and back to the long-standing part.
   STEP 6. Remove all unnecessary twists and crossovers; dress the knot down.
b. *Checkpoints*
   (1) A figure eight with a doubled rope running side by side, forming a fixed loop around a fixed object or harness.
   (2) There is a minimum 4-inch pigtail.

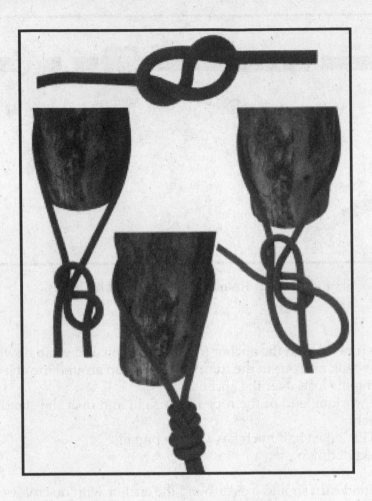

Figure 4-13: Figure-eight retrace.

4-15. Clove Hitch. The clove hitch is an anchor knot that can be used in the middle of the rope as well as at the end (Figure 4-14). The knot must have constant tension on it once tied to prevent slipping. It can be used as either an anchor or middle of the rope knot, depending on how it is tied.

   a. *Tying the Knot.*
     (1) *Middle of the Rope.*
       STEP 1. Hold rope in both hands, palms down with hands together. Slide the left hand to the left from 20 to 25 centimeters.
       STEP 2. Form a loop away from and back toward the right.
       STEP 3. Slide the right hand from 20 to 25 centimeters to the right. Form a loop inward and back to the left hand.
       STEP 4. Place the left loop on top of the right loop. Place both loops over the anchor and pull both ends of the rope in opposite directions. The knot is tied.
     (2) *End of the Rope.*
  Note: *For instructional purposes, assume that the anchor is horizontal.*
       STEP 1. Place 76 centimeters of rope over the top of the anchor. Hold the standing end in the left hand. With the right hand, reach under the horizontal anchor, grasp the working end, and bring it inward.
       STEP 2. Place the working end of the rope over the standing end (to form a loop). Hold the loop in the left hand. Place the working end over the anchor from 20 to 25 centimeters to the left of the loop.

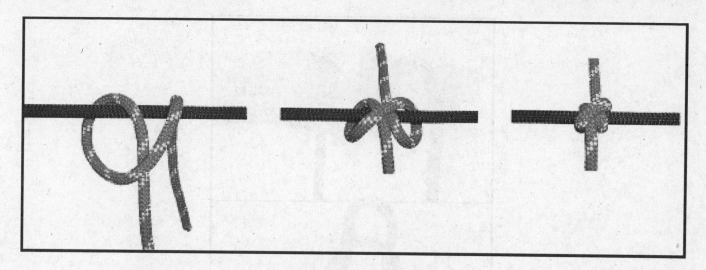

Figure 4-14: Clove hitch.

STEP 3. With the right hand, reach down to the left hand side of the loop under the anchor. Grasp the working end of the rope. Bring the working end up and outward.
STEP 4. Dress down the knot.
b. *Checkpoints.*
   (1) The knot has two round turns around the anchor with a diagonal locking bar.
   (2) The locking bar is facing 90 degrees from the direction of pull.
   (3) The ends exit 180 degrees from each other.
   (4) The knot has more than a 4-inch pigtail remaining.

4-16. Wireman's Knot. The wireman's knot forms a single, fixed loop in the middle of the rope (Figure 4-15). It is a middle rope knot.

a. *Tying the Knot.*
   STEP 1. When tying this knot, face the anchor that the tie-off system will be tied to. Take up the slack from the anchor, and wrap two turns around the left hand (palm up) from left to right.
   STEP 2. A loop of 30 centimeters is taken up in the second round turn to create the fixed loop of the knot.
   STEP 3. Name the wraps from the palm to the fingertips: heel, palm, and fingertip.
   STEP 4. Secure the palm wrap with the right thumb and forefinger, and place it over the heel wrap.
   STEP 5. Secure the heel wrap and place it over the fingertip wrap.
   STEP 6. Secure the fingertip wrap and place it over the palm wrap.
   STEP 7. Secure the palm wrap and pull up to form a fixed loop.
   STEP 8. Dress the knot down by pulling on the fixed loop and the two working ends.
   STEP 9. Pull the working ends apart to finish the knot.
b. *Checkpoints.*
   (1) The completed knot should have four separate bights locking down on themselves with the fixed loop exiting from the top of the knot and laying toward the near side anchor point.
   (2) Both ends should exit opposite each other without any bends.

4-17. Directional Figure-Eight. The directional figure-eight knot forms a single, fixed loop in the middle of the rope that lays back along the standing part of the rope (Figure 4-16). It is a middle rope knot.

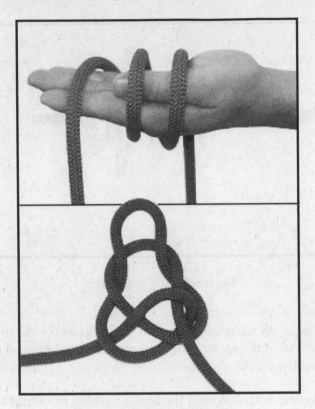

Figure 4-15: Wireman's knot.

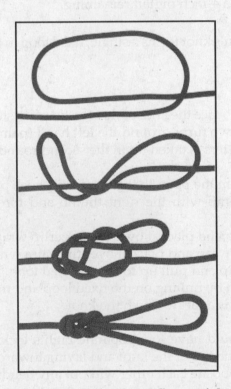

Figure 4-16: Directional figure-eight.

a. *Tying the Knot.*
STEP 1. Face the far side anchor so that when the knot is tied, it lays inward.
STEP 2. Lay the rope from the far side anchor over the left palm. Make one wrap around the palm.
STEP 3. With the wrap thus formed, tie a figure-eight knot around the standing part that leads to the far side anchor.
STEP 4. When dressing the knot down, the tail and the bight must be together.
b. *Checkpoints.*
(1) The loop should be large enough to accept a carabiner but no larger than a helmet-size loop.
(2) The tail and bight must be together.
(3) The figure eight is tied tightly.
(4) The bight in the knot faces back toward the near side.

4-18. Bowline-on-a-Bight (Two-Loop Bowline)

The bowline-on-a-bight is used to form two fixed loops in the middle of a rope (Figure 4-17). It is a middle rope knot.

a. *Tying the Knot.*
STEP 1. Form a bight in the rope about twice as long as the finished loops will be.
STEP 2. Tie an overhand knot on a bight.
STEP 3. Hold the overhand knot in the left hand so that the bight is running down and outward.
STEP 4. Grasp the bight with the right hand; fold it back over the overhand knot so that the overhand knot goes through the bight.

Figure 4-17: Bowline-on-a-bight.

STEP 5. From the end (apex) of the bight, follow the bight back to where it forms the cross in the overhand knot. Grasp the two ropes that run down and outward and pull up, forming two loops.

STEP 6. Pull the two ropes out of the overhand knot and dress the knot down.

STEP 7. A final dress is required: grasp the ends of the two fixed loops and pull, spreading them apart to ensure the loops do not slip.

b. *Checkpoints.*
(1) There are two fixed loops that will not slip.
(2) There are no twists in the knot.
(3) A double loop is held in place by a bight.

4-19. Two-Loop Figure-Eight. The two-loop figure-eight is used to form two fixed loops in the middle of a rope (Figure 4-18.) It is a middle rope knot.

a. *Tying the Knot.*
STEP 1. Using a doubled rope, form an 18-inch bight in the left hand with the running end facing to the left.

STEP 2. Grasp the bight with the right hand and make a 360-degree turn around the standing end in a counterclockwise direction.

STEP 3. With the working end, form another bight and place that bight through the loop just formed in the left hand.

STEP 4. Hold the bight with the left hand, and place the original bight (moving toward the left hand) over the knot.

STEP 5. Dress the knot down.

b. *Checkpoints.*
(1) There is a double figure-eight knot with two loops that share a common locking bar.
(2) The two loops must be adjustable by means of a common locking bar.
(3) The common locking bar is on the bottom of the double figure-eight knot.

4-20. Figure-Eight Loop (Figure-Eight-on-a-Bight). The figure-eight loop, also called the figure-eight-on-a-bight, is used to form a fixed loop in a rope (Figure 4-19). It is a middle of the rope knot.

a. *Tying the Knot.*
STEP 1. Form a bight in the rope about as large as the diameter of the desired loop.

STEP 2. With the bight as the working end, form a loop in rope (standing part).

STEP 3. Wrap the working end around the standing part 360 degrees and feed the working end through the loop. Dress the knot tightly.

Figure 4-18: Two-loop figure-eight.

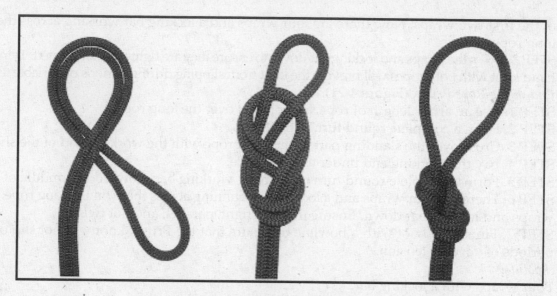

Figure 4-19: Figure-eight loop.

b. *Checkpoints.*
   (1) The loop is the desired size.
   (2) The ropes in the loop are parallel and do not cross over each other.
   (3) The knot is tightly dressed.

4-21. Prusik Knot. The Prusik knot is used to put a moveable rope on a fixed rope such as a Prusik ascent or a tightening system. This knot can be tied as a middle or end of the rope Prusik. It is a specialty knot.

a. *Tying the Knot.*
   (1) *Middle-of-the-Rope Prusik.* The middle-of-the-rope Prusik knot can be tied with a short rope to a long rope as follows (Figure 4-20):
      STEP 1. Double the short rope, forming a bight, with the working ends even. Lay it over the long rope so that the closed end of the bight is 12 inches below the long rope and the remaining part of the rope (working ends) is the closest to the climber; spread the working end apart.
      STEP 2. Reach down through the 12-inch bight. Pull up both of the working ends and lay them over the long rope. Repeat this process making sure that the working ends pass in the middle

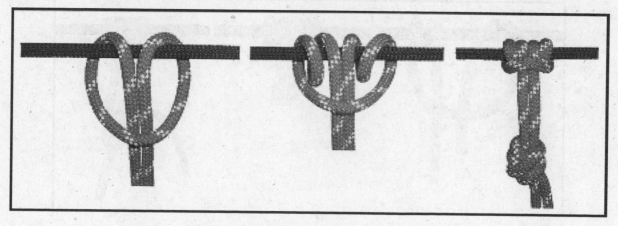

Figure 4-20: Middle-of-the-rope Prusik.

of the first two wraps. Now there are four wraps and a locking bar working across them on the long rope.

STEP 3. Dress the wraps and locking bar down to ensure they are tight and not twisted. Tying an over-hand knot with both ropes will prevent the knot from slipping during periods of variable tension.

(2) *End-of-the-Rope Prusik* (Figure 4-21).

STEP 1. Use an arm's length of rope, and place it over the long rope.

STEP 2. Form a complete round turn in the rope.

STEP 3. Cross over the standing part of the short rope with the working end of the short rope.

STEP 4. Lay the working end under the long rope.

STEP 5. Form a complete round turn in the rope, working back toward the middle of the knot.

STEP 6. There are four wraps and a locking bar running across them on the long rope. Dress the wraps and locking bar down. Ensure they are tight, parallel, and not twisted.

STEP 7. Finish the knot with a bowline to ensure that the Prusik knot will not slip out during periods of varying tension.

b. *Checkpoints.*

(1) Four wraps with a locking bar.

(2) The locking bar faces the climber.

(3) The knot is tight and dressed down with no ropes twisted or crossed.

(4) Other than a finger Prusik, the knot should contain an overhand or bowline to prevent slipping.

4-22. Bachman Knot. The Bachman knot provides a means of using a makeshift mechanized ascender (Figure 4-22). It is a specialty knot.

a. *Tying the Knot.*

STEP 1. Find the middle of a utility rope and insert it into a carabiner.

STEP 2. Place the carabiner and utility rope next to a long climbing rope.

STEP 3. With the two ropes parallel from the carabiner, make two or more wraps around the climbing rope and through the inside portion of the carabiner.

Note: *The rope can be tied into an etrier (stirrup) and used as a Prusik-friction principle ascender.*

b. *Checkpoints.*

(1) The bight of the climbing rope is at the top of the carabiner.

(2) The two ropes run parallel without twisting or crossing.

(3) Two or more wraps are made around the long climbing rope and through the inside portion of the carabiner.

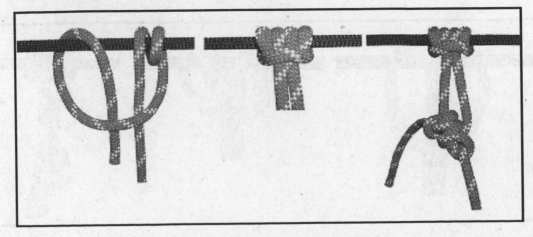

Figure 4-21: End-of-the-rope Prusik knot.

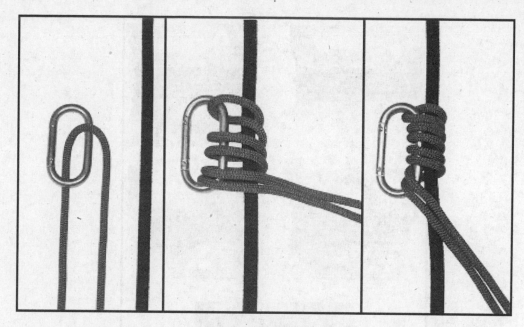

Figure 4-22: Bachman knot.

4-23. Bowline-on-a-Coil. The bowline-on-a-coil is an expedient tie-in used by climbers when a climbing harness is not available (Figure 4-23). It is a specialty knot.

   a. *Tying the Knot.*
      STEP 1. With the running end, place 3 feet of rope over your right shoulder. The running end is to the back of the body.
      STEP 2. Starting at the bottom of your rib cage, wrap the standing part of the rope around your body and down in a clockwise direction four to eight times.
      STEP 3. With the standing portion of the rope in your left hand, make a clockwise loop toward the body. The standing portion is on the bottom.
      STEP 4. Ensuring the loop does not come uncrossed, bring it up and under the coils between the rope and your body.
      STEP 5. Using the standing part, bring a bight up through the loop. Grasp the running end of the rope with the right hand. Pass it through the bight from right to left and back on itself.
      STEP 6. Holding the bight loosely, dress the knot down by pulling on the standing end.
      STEP 7. Safety the bowline with an overhand around the top, single coil. Then, tie an overhand around all coils, leaving a minimum 4-inch pigtail.
   b. *Checkpoints.*
      (1) A minimum of four wraps, not crossed, with a bight held in place by a loop.
      (2) The loop must be underneath all wraps.
      (3) A minimum 4-inch pigtail after the second overhand safety is tied.
      (4) Must be centered on the mid-line of the body.

4-24. Three-Loop Bowline. The three-loop bowline is used to form three fixed loops in the middle of a rope (Figure 4-24). It is used in a self-equalizing anchor system. It is a specialty knot.

   a. *Tying the Knot.*
      STEP 1. Form an approximate 24-inch bight.
      STEP 2. With the right thumb facing toward the body, form a doubled loop in the standing part by turning the wrist clockwise. Lay the loops to the right.

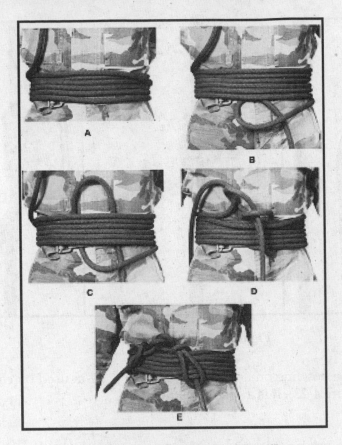

Figure 4-23: Bowline-on-a-coil.

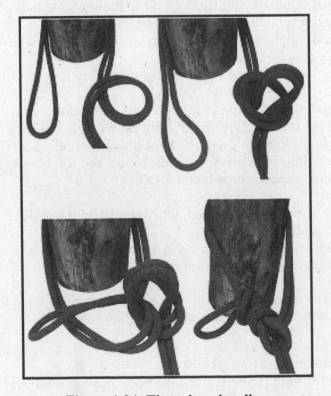

Figure 4-24: Three-loop bowline.

STEP 3. With the right hand, reach down through the loops and pull up a doubled bight from the standing part of the rope.

STEP 4. Place the running end (bight) of the rope (on the left) through the doubled bight from left to right and bring it back on itself. Hold the running end loosely and dress the knot down by pulling on the standing parts.

STEP 5. Safety it off with a doubled overhand knot.

b. *Checkpoints.*
  (1) There are two bights held in place by two loops.
  (2) The bights form locking bars around the standing parts.
  (3) The running end (bight) must be on the inside of the fixed loops.
  (4) There is a minimum 4-inch pigtail after the double overhand safety knot is tied.

4-25. Figure-Eight Slip Knot. The figure-eight slip knot forms an adjustable bight in a rope (Figure 4-25). It is a specialty knot.

a. *Tying the Knot.*
  STEP 1. Form a 12-inch bight in the end of the rope.
  STEP 2. Hold the center of the bight in the right hand. Hold the two parallel ropes from the bight in the left hand about 12 inches up the rope.
  STEP 3. With the center of the bight in the right hand, twist two complete turns clockwise.
  STEP 4. Reach through the bight and grasp the long, standing end of the rope. Pull another bight (from the long standing end) back through the original bight.
  STEP 5. Pull down on the short working end of the rope and dress the knot down.
  STEP 6. If the knot is to be used in a transport tightening system, take the working end of the rope and form a half hitch around the loop of the figure eight knot.

b. *Checkpoints.*
  (1) The knot is in the shape of a figure eight.
  (2) Both ropes of the bight pass through the same loop of the figure eight.
  (3) The sliding portion of the rope is the long working end of the rope.

4-26. Transport Knot (Overhand Slip Knot/Mule Knot). The transport knot is used to secure the transport tightening system (Figure 4-26). It is simply an overhand slip knot.

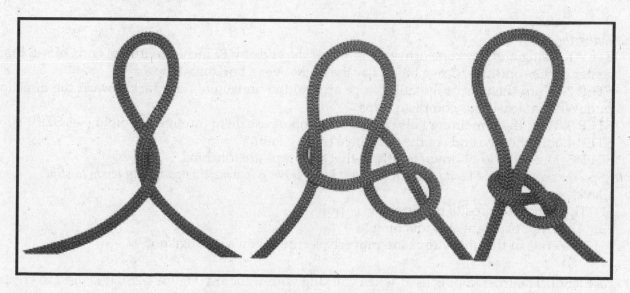

Figure 4-25: Figure-eight slip knot.

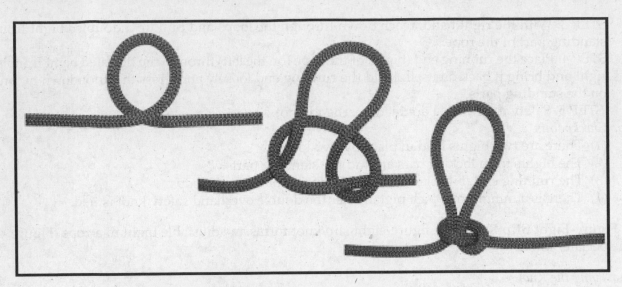

Figure 4-26: Transport knot.

a. *Tying the Knot.*
STEP 1. Pass the running end of the rope around the anchor point passing it back under the standing portion (leading to the far side anchor) forming a loop.
STEP 2. Form a bight with the running end of the rope. Pass over the standing portion and down through the loop and dress it down toward the anchor point.
STEP 3. Secure the knot by tying a half hitch around the standing portion with the bight.
b. *Check Points.*
(1) There is a single overhand slip knot.
(2) The knot is secured using a half hitch on a bight.
(3) The bight is a minimum of 12 inches long.

4-27. Kleimhiest Knot. The Kleimhiest knot provides a moveable, easily adjustable, high-tension knot capable of holding extremely heavy loads while being pulled tight (Figure 4-27). It is a special-purpose knot.

a. *Tying the Knot.*
STEP 1. Using a utility rope or webbing, offset the ends by 12 inches. With the ends offset, find the center of the rope and form a bight. Lay the bight over a horizontal rope.
STEP 2. Wrap the tails of the utility rope around the horizontal rope back toward the direction of pull. Wrap at least four complete turns.
STEP 3. With the remaining tails of the utility rope, pass them through the bight (see STEP 1).
STEP 4. Join the two ends of the tail with a joining knot.
STEP 5. Dress the knot down tightly so that all wraps are touching.
Note: *Spectra should not be used for the Kleimhiest knot. It has a low melting point and tends to slip.*
b. *Checkpoints.*
(1) The bight is opposite the direction of pull.
(2) All wraps are tight and touching.
(3) The ends of the utility rope are properly secured with a joining knot.

4-28. Frost Knot. The frost knot is used when working with webbing (Figure 4-28). It is used to create the top loop of an etrier. It is a special-purpose knot.

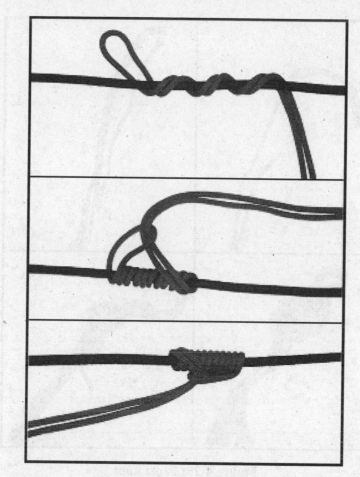

Figure 4-27: Kleimhiest knot.

a. *Tying the Knot.*
STEP 1. Lap one end (a bight) of webbing over the other about 10 to 12 inches.
STEP 2. Tie an overhand knot with the newly formed triple-strand webbing; dress tightly.
b. *Checkpoints.*
(1) The tails of the webbing run in opposite directions.
(2) Three strands of webbing are formed into a tight overhand knot.
(3) There is a bight and tail exiting the top of the overhand knot.

4-29. Girth Hitch. The girth hitch is used to attach a runner to an anchor or piece of equipment (Figure 4-29). It is a special-purpose knot.

a. *Tying the Knot.*
STEP 1: Form a bight.
STEP 2: Bring the runner back through the bight.
STEP 3: Cinch the knot tightly.
b. *Checkpoint.*
(1) Two wraps exist with a locking bar running across the wraps.
(2) The knot is dressed tightly.

4-30. Munter Hitch. The munter hitch, when used in conjunction with a pear-shaped locking carabiner, is used to form a mechanical belay (Figure 4-30).

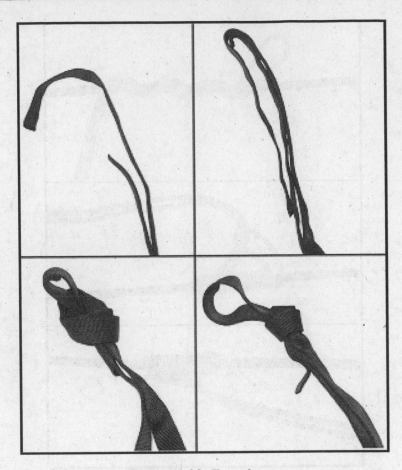

Figure 4-28: Frost knot.

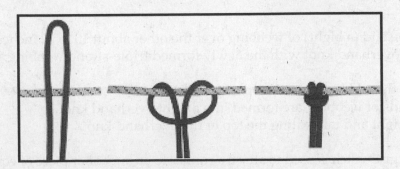

Figure 4-29: Girth hitch.

a. *Tying the Knot.*
STEP 1. Hold the rope in both hands, palms down about 12 inches apart.
STEP 2. With the right hand, form a loop away from the body toward the left hand. Hold the loop with the left hand.
STEP 3. With the right hand, place the rope that comes from the bottom of the loop over the top of the loop.
STEP 4. Place the bight that has just been formed around the rope into the pear shaped carabiner. Lock the locking mechanism.

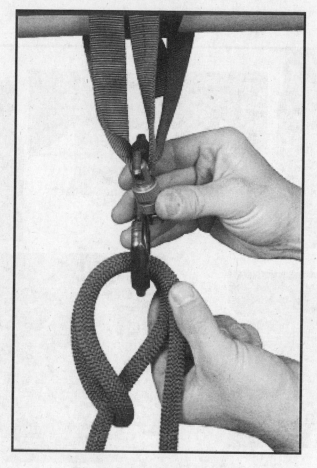

Figure 4-30: Munter hitch.

b. *Check Points.*
   (1) A bight passes through the carabiner, with the closed end around the standing or running part of the rope.
   (2) The carabiner is locked.

4-31. Rappel Seat. The rappel seat is an improvised seat rappel harness made of rope (Figure 4-31). It usually requires a sling rope 14 feet or longer.

   a. *Tying the Knot.*
   STEP 1. Find the middle of the sling rope and make a bight.
   STEP 2. Decide which hand will be used as the brake hand and place the bight on the opposite hip.
   STEP 3. Reach around behind and grab a single strand of rope. Bring it around the waist to the front and tie two overhands on the other strand of rope, thus creating a loop around the waist.
   STEP 4. Pass the two ends between the legs, ensuring they do not cross.
   STEP 5. Pass the two ends up under the loop around the waist, bisecting the pocket flaps on the trousers. Pull up on the ropes, tightening the seat.
   STEP 6. From rear to front, pass the two ends through the leg loops creating a half hitch on both hips.
   STEP 7. Bring the longer of the two ends across the front to the nonbrake hand hip and secure the two ends with a square knot safetied with overhand knots. Tuck any excess rope in the pocket below the square knot.

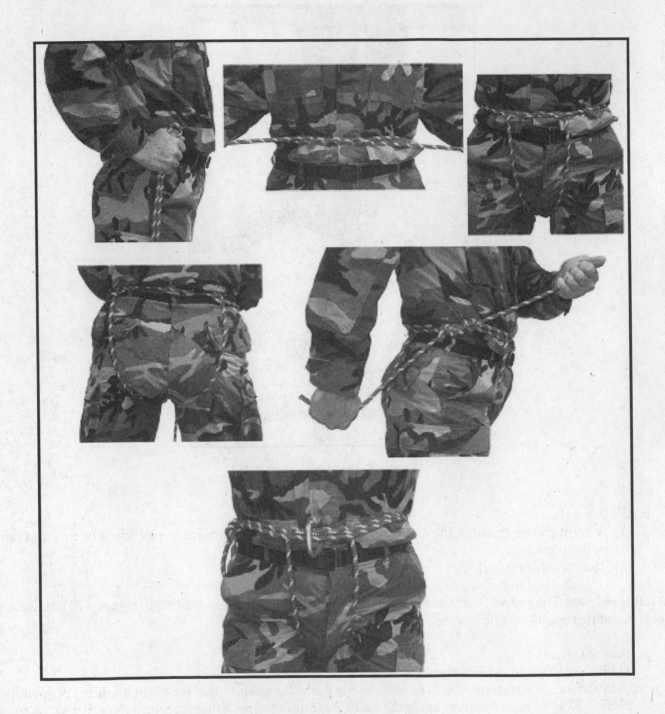

Figure 4-31: Rappel seat.

b. *Check Points*.
  (1) There are two overhand knots in the front.
  (2) The ropes are not crossed between the legs.
  (3) A half hitch is formed on each hip.
  (4) Seat is secured with a square knot with overhand safeties on the non-brake hand side.
  (5) There is a minimum 4-inch pigtail after the overhand safeties are tied.

4-32. Guarde Knot. The guarde knot (ratchet knot, alpine clutch) is a special purpose knot primarily used for hauling systems or rescue (Figure 4-32). The knot works in only one direction and cannot be reversed while under load.

a. *Tying the Knot.*
STEP 1. Place a bight of rope into the two anchored carabiners (works best with two like carabiners, preferably ovals).
STEP 2. Take a loop of rope from the non-load side and place it down into the opposite cararabiner so that the rope comes out between the two carabiners.

b. *Check Points.*
(1) When properly dressed, rope can only be pulled in one direction.
(2) The knot will not fail when placed under load.

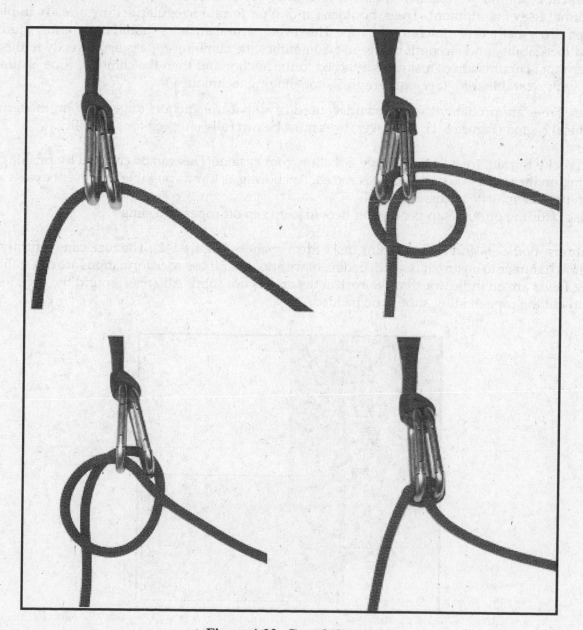

Figure 4-32: Guarde knot.

SECTION 5: ANCHORS

This chapter discusses different types of anchors and their application in rope systems and climbing. Proper selection and placement of anchors is a critical skill that requires a great deal of practice. Failure of any system is most likely to occur at the anchor point. If the anchor is not strong enough to support the intended load, it will fail. Failure is usually the result of poor terrain features selected for the anchor point, or the equipment used in rigging the anchor was placed improperly or in insufficient amounts. When selecting or constructing anchors, always try to make sure the anchor is "bombproof." A bombproof anchor is stronger than any possible load that could be placed on it. An anchor that has more strength than the climbing rope is considered bombproof.

NATURAL ANCHORS

Natural anchors should be considered for use first. They are usually strong and often simple to construct with minimal use of equipment. Trees, boulders, and other terrain irregularities are already in place and simply require a method of attaching the rope. However, natural anchors should be carefully studied and evaluated for stability and strength before use. Sometimes the climbing rope is tied directly to the anchor, but under most circumstances a sling is attached to the anchor and then the climbing rope is attached to the sling with a carabiner(s). (See paragraph 5-7 for slinging techniques.)

5-1. Trees. Trees are probably the most widely used of all natural anchors depending on the terrain and geographical region (Figure 5-1). However, trees must be carefully checked for suitability.

    a. In rocky terrain, trees usually have a shallow root system. This can be checked by pushing or tugging on the tree to see how well it is rooted. Anchoring as low as possible to prevent excess leverage on the tree may be necessary.

    b. Use padding on soft, sap producing trees to keep sap off ropes and slings.

5-2. Boulders. Boulders and rock nubbins make ideal anchors (Figure 5-2). The rock can be firmly tapped with a piton hammer to ensure it is solid. Sedimentary and other loose rock formations are not stable. Talus and scree fields are an indicator that the rock in the area is not solid. All areas around the rock formation that could cut the rope or sling should be padded.

Figure 5-1: Trees used as anchors.

Figure 5-2: Boulders used as anchors.

5-3. Chockstones. A chockstone is a rock that is wedged in a crack because the crack narrows downward (Figure 5-3). Chockstones should be checked for strength, security, and crumbling and should always be tested before use. All chockstones must be solid and strong enough to support the load. They must have maximum surface contact and be well tapered with the surrounding rock to remain in position.

    a. Chockstones are often directional: they are secure when pulled in one direction but may pop out if pulled in another direction.

    b. A creative climber can often make his own chockstone by wedging a rock into position, tying a rope to it, and clipping on a carabiner.

    c. Slings should not be wedged between the chockstone and the rock wall since a fall could cut the webbing runner.

5-4. Rock Projections. Rock projections (sometimes called nubbins) often provide suitable protection (Figure 5-4). These include blocks, flakes, horns, and spikes. If rock projections are used, their firmness

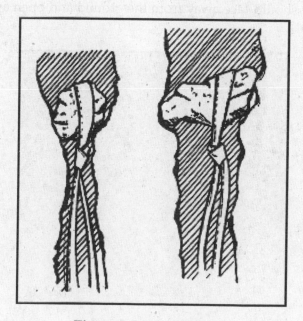

Figure 5-3: Chockstones.

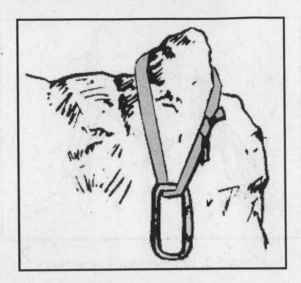

Figure 5-4: Rock projections.

is important. They should be checked for cracks or weathering that may impair their firmness. If any of these signs exist, the projection should be avoided.

5-5. Tunnels and Arches. Tunnels and arches are holes formed in solid rock and provide one of the more secure anchor points because they can be pulled in any direction. A sling is threaded through the opening hole and secured with a joining knot or girth hitch. The load-bearing hole must be strong and free of sharp edges (pad if necessary).

5-6. Bushes and Shrubs. If no other suitable anchor is available, the roots of bushes can be used by routing a rope around the bases of several bushes (Figure 5-5). As with trees, the anchoring rope is placed as low as possible to reduce leverage on the anchor. All vegetation should be healthy and well rooted to the ground.

5-7. Slinging Techniques. Three methods are used to attach a sling to a natural anchor: drape, wrap, and girth. Whichever method is used, the knot is set off to the side where it will not interfere with normal carabiner movement. The carabiner gate should face away from the ground and open away from the anchor for easy

Figure 5-5: Bushes and shrubs.

insertion of the rope. When a locking carabiner cannot be used, two carabiners are used with gates opposed. Correctly opposed gates should open on opposite sides and form an "X" when opened (Figure 5-6).

a. *Drape.* Drape the sling over the anchor (Figure 5-7). Untying the sling and routing it around the anchor and then retying is still considered a drape.
b. *Wrap.* Wrap the sling around the anchor and connect the two ends together with a carabiner(s) or knot (Figure 5-8).
c. *Girth.* Tie the sling around the anchor with a girth hitch (Figure 5-9). Although a girth hitch reduces the strength of the sling, it allows the sling to remain in position and not slide on the anchor.

ANCHORING WITH THE ROPE

The climbing or installation rope can be tied directly to the anchor using several different techniques. This requires less equipment, but also sacrifices some rope length to tie the anchor. The rope can be tied to the

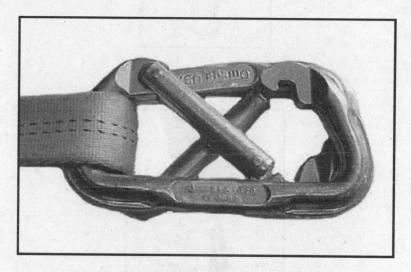

Figure 5-6: Correctly opposed carabiners.

Figure 5-7: Drape.

Figure 5-8: Wrap.

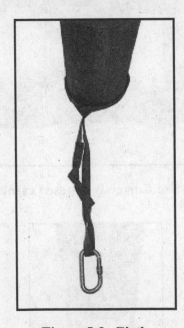

Figure 5-9: Girth.

anchor using an appropriate anchor knot such as a bowline or a rerouted figure eight. Round turns can be used to help keep the rope in position on the anchor. A tensionless anchor can be used in high-load installations where tension on the attachment point and knot is undesirable.

5-8. Rope Anchor. When tying the climbing or installation rope around an anchor, the knot should be placed approximately the same distance away from the anchor as the diameter of the anchor (Figure 5-10). The knot shouldn't be placed up against the anchor because this can stress and distort the knot under tension.

5-9. Tensionless Anchor. The tensionless anchor is used to anchor the rope on high-load installations such as bridging and traversing (Figure 5-11). The wraps of the rope around the anchor absorb the tension of the installation and keep the tension off the knot and carabiner. The anchor is usually tied with a

Figure 5-10: Rope tied to anchor with anchor knot.

Figure 5-11: Tensionless anchor.

minimum of four wraps, more if necessary, to absorb the tension. A smooth anchor may require several wraps, whereas a rough barked tree might only require a few. The rope is wrapped from top to bottom. A fixed loop is placed into the end of the rope and attached loosely back onto the rope with a carabiner.

ARTIFICIAL ANCHORS

Using artificial anchors becomes necessary when natural anchors are unavailable. The art of choosing and placing good anchors requires a great deal of practice and experience. Artificial anchors are available in many different types such as pitons, chocks, hexcentrics, and SLCDs. Anchor strength varies greatly; the type used depends on the terrain, equipment, and the load to be placed on it.

5-10. Deadman. A "deadman" anchor is any solid object buried in the ground and used as an anchor.

a. An object that has a large surface area and some length to it works best. (A hefty timber, such as a railroad tie, would be ideal.) Large boulders can be used, as well as a bundle of smaller tree limbs or poles. As with natural anchors, ensure timbers and tree limbs are not dead or rotting and that boulders are solid. Equipment, such as skis, ice axes, snowshoes, and ruck sacks, can also be used if necessary.

b. In extremely hard, rocky terrain (where digging a trench would be impractical, if not impossible) a variation of the deadman anchor can be constructed by building above the ground. The sling is attached to the anchor, which is set into the ground as deeply as possible. Boulders are then stacked on top of it until the anchor is strong enough for the load. Though normally not as strong as when buried, this method can work well for light-load installations as in anchoring a hand line for a stream crossing.

Note: Artificial anchors, such as pitons and bolts, are not widely accepted for use in all areas because of the scars they leave on the rock and the environment. Often they are left in place and become unnatural, unsightly fixtures in the natural environment. For training planning, local laws and courtesies should be taken into consideration for each area of operation.

5-11. Pitons. Pitons have been in use for over 100 years. Although still available, pitons are not used as often as other types of artificial anchors due primarily to their impact on the environment. Most climbers prefer to use chocks, SLCDs and other artificial anchors rather than pitons because they do not scar the rock and are easier to remove. Eye protection should always be worn when driving a piton into rock.

Note: The proper use and placement of pitons, as with any artificial anchor, should be studied, practiced, and tested while both feet are firmly on the ground and there is no danger of a fall.

a. *Advantages.* Some advantages in using pitons are:
- Depending on type and placement, pitons can support multiple directions of pull.
- Pitons are less complex than other types of artificial anchors.
- Pitons work well in thin cracks where other types of artificial anchors do not.

b. *Disadvantages.* Some disadvantages in using pitons are:
- During military operations, the distinct sound created when hammering pitons is a tactical disadvantage.
- Due to the expansion force of emplacing a piton, the rock could spread apart or break causing an unsafe condition.
- Pitons are more difficult to remove than other types of artificial anchors.
- Pitons leave noticeable scars on the rock.
- Pitons are easily dropped if not tied off when being used.

c. *Piton Placement.* The proper positioning or placement of pitons is critical. (Figure 5-12 shows examples of piton placement.) Usually a properly sized piton for a rock crack will fit one half to two thirds into the crack before being driven with the piton hammer. This helps ensure the depth of the crack is adequate for the size piton selected. As pitons are driven into the rock the pitch or sound that is made will change with each hammer blow, becoming higher pitched as the piton is driven in.

(1) Test the rock for soundness by tapping with the hammer. Driving pitons in soft or rotten rock is not recommended. When this type of rock must be used, clear the loose rock, dirt, and debris from the crack before driving the piton completely in.

(2) While it is being driven, attach the piton to a sling with a carabiner (an old carabiner should be used, if available) so that if the piton is knocked out of the crack, it will not be lost. The greater the resistance overcome while driving the piton, the firmer the anchor will be. The holding

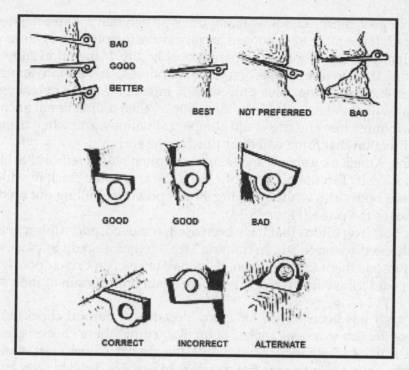

Figure 5-12: Examples of piton placements.

power depends on the climber placing the piton in a sound crack, and on the type of rock. The piton should not spread the rock, thereby loosening the emplacement.

Note: Pitons that have rings as attachment points might not display much change in sound as they are driven in as long as the ring moves freely.

(3) Military mountaineers should practice emplacing pitons using either hand. Sometimes a piton cannot be driven completely into a crack, because the piton is too long. Therefore, it should be tied off using a hero-loop (an endless piece of webbing) (Figure 5-13). Attach this loop to the piton using a girth hitch at the point where the piton enters the rock so that the girth hitch is snug against the rock. Clip a carabiner into the loop.

d. Testing. To test pitons pull up about 1 meter of slack in the climbing rope or use a sling. Insert this rope into a carabiner attached to the piton, then grasp the rope at least 1/2 meter from the carabiner. Jerk vigorously upward, downward, to each side, and then outward while observing the piton

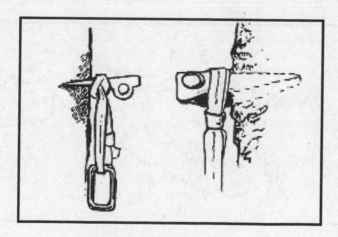

Figure 5-13: Hero-loop.

for movement. Repeat these actions as many times as necessary. Tap the piton to determine if the pitch has changed. If the pitch has changed greatly, drive the piton in as far as possible. If the sound regains its original pitch, the emplacement is probably safe. If the piton shows any sign of moving or if, upon driving it, there is any question of its soundness, drive it into another place. Try to be in a secure position before testing. This procedure is intended for use in testing an omni-directional anchor (one that withstands a pull in any direction). When a directional anchor (pull in one direction) is used, as in most free and direct-aid climbing situations, and when using chocks, concentrate the test in the direction that force will be applied to the anchor.

e. Removing Pitons. Attach a carabiner and sling to the piton before removal to eliminate the chance of dropping and losing it. Tap the piton firmly along the axis of the crack in which it is located. Alternate tapping from both sides while applying steady pressure. Pulling out on the attached carabiner eventually removes the piton (Figure 5-14).

f. Reusing Pitons. Soft iron pitons that have been used, removed, and straightened may be reused, but they must be checked for strength. In training areas, pitons already in place should not be trusted since weather loosens them in time. Also, they may have been driven poorly the first time. Before use, test them as described above and drive them again until certain of their soundness.

5-12. Chocks. Chock craft has been in use for many decades. A natural chockstone, having fallen and wedged in a crack, provides an excellent anchor point. Sometimes these chockstones are in unstable positions, but can be made into excellent anchors with little adjustment. Chock craft is an art that requires time and technique to master—simple in theory, but complex in practice. Imagination and resourcefulness are key principles to chock craft. The skilled climber must understand the application of mechanical advantage, vectors, and other forces that affect the belay chain in a fall.

a. *Advantages.* The advantages of using chocks are:
- Tactically quiet installation and recovery.
- Usually easy to retrieve and, unless severely damaged, are reusable.
- Light to carry.
- Easy to insert and remove.
- Minimal rock scarring as opposed to pitons.
- Sometimes can be placed where pitons cannot (expanding rock flakes where pitons would further weaken the rock).

b. *Disadvantages.* The disadvantages of using chocks are:
- May not fit in thin cracks, which may accept pitons.
- Often provide only one direction of pull.
- Practice and experience necessary to become proficient in proper placement.

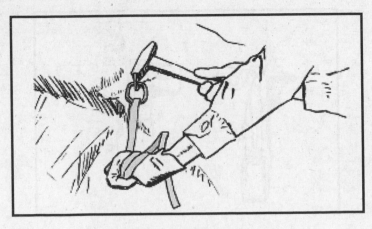

Figure 5-14: Piton removal.

c. *Placement.* The principles of placing chocks are to find a crack with a constriction at some point, place a chock of appropriate size above and behind the constriction, and set the chock by jerking down on the chock loop (Figure 5-15). Maximum surface contact with a tight fit is critical. Chocks are usually good for a single direction of pull.

(1) Avoid cracks that have crumbly (soft) or deteriorating rock, if possible. Some cracks may have loose rock, grass, and dirt, which should be removed before placing the chock. Look for a constriction point in the crack, then select a chock to fit it.

(2) When selecting a chock, choose one that has as much surface area as possible in contact with the rock. A chock resting on one small crystal or point of rock is likely to be unsafe. A chock that sticks partly out of the crack is avoided. Avoid poor protection. Ensure that the chock has a wire or runner long enough; extra ropes, cord, or webbing may be needed to extend the length of the runner.

(3) End weighting of the placement helps to keep the protection in position. A carabiner often provides enough weight.

(4) Parallel-sided cracks without constrictions are a problem. Chocks designed to be used in this situation rely on camming principles to remain emplaced. Weighting the emplacement with extra hardware is often necessary to keep the chocks from dropping out.

(a) Emplace the wedge-shaped chock above and behind the constriction; seat it with a sharp downward tug.

(b) Place a camming chock with its narrow side into the crack, then rotate it to the attitude it will assume under load; seat it with a sharp downward tug.

d. *Testing.* After seating a chock, test it to ensure it remains in place. A chock that falls out when the climber moves past it is unsafe and offers no protection. To test it, firmly pull the chock in every anticipated direction of pull. Some chock placements fail in one or more directions; therefore, use pairs of chocks in opposition.

5-13. Spring-Loaded Camming Device. The SLCD offers quick and easy placement of artificial protection. It is well suited in awkward positions and difficult placements, since it can be emplaced with one hand. It can usually be placed quickly and retrieved easily (Figure 5-16).

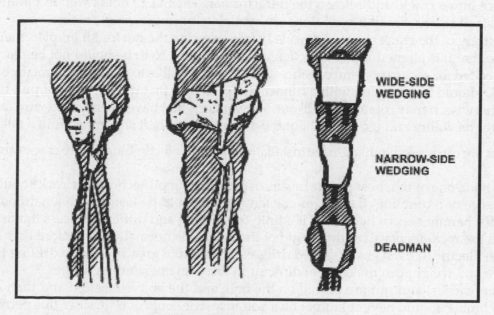

WIDE-SIDE
WEDGING

NARROW-SIDE
WEDGING

DEADMAN

Figure 5-15: Chock placements.

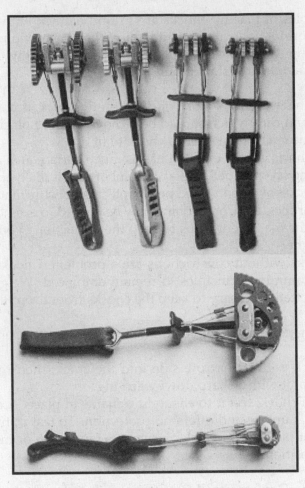

Figure 5-16: SLCD placements.

a. To emplace an SLCD hold the device in either hand like a syringe, pull the retractor bar back, place the device into a crack, and release the retractor bar. The SLCD holds well in parallel-sided hand- and fist-sized cracks. Smaller variations are available for finger-sized cracks.

b. Careful study of the crack should be made before selecting the device for emplacement. It should be placed so that it is aligned in the direction of force applied to it. It should not be placed any deeper than is needed for secure placement, since it may be impossible to reach the extractor bar for removal. An SLCD should be extended with a runner and placed so that the direction of pull is parallel to the shaft; otherwise, it may rotate and pull out. The versions that have a semi-rigid wire cable shaft allow for greater flexibility and usage, without the danger of the shaft snapping off in a fall.

5-14. Bolts. Bolts are often used in fixed-rope installations and in aid climbing where cracks are not available.

a. Bolts provide one of the most secure means of establishing protection. The rock should be inspected for evidence of crumbling, flaking, or cracking, and should be tested with a hammer. Emplacing a bolt with a hammer and a hand drill is a time-consuming and difficult process that requires drilling a hole in the rock deeper than the length of the bolt. This normally takes more than 20 minutes for one hole. Electric or even gas-powered drills can be used to greatly shorten drilling time. However, their size and weight can make them difficult to carry on the climbing route.

b. A hanger (carrier) and nut are placed on the bolt, and the bolt is inserted and then driven into the hole. A climber should never hammer on a bolt to test or "improve" it, since this permanently weakens it. Bolts should be used with carriers, carabiners, and runners.

c. When using bolts, the climber uses a piton hammer and hand drill with a masonry bit for drilling holes. Some versions are available in which the sleeve is hammered and turned into the rock (self-drilling), which bores the hole. Split bolts and expanding sleeves are common bolts used to secure hangers and carriers (Figure 5-17). Surgical tubing is useful in blowing dust out of the holes. Nail type bolts are emplaced by driving the nail with a hammer to expand the sleeve against the wall of the drilled hole. Safety glasses should always be worn when emplacing bolts.

5-15. Equalizing Anchors. Equalizing anchors are made up of more than one anchor point joined together so that the intended load is shared equally. This not only provides greater anchor strength, but also adds redundancy or backup because of the multiple points.

a. *Self-equalizing Anchor.* A self-equalizing anchor will maintain an equal load on each individual point as the direction of pull changes (Figure 5-18). This is sometimes used in rappelling when the route must change left or right in the middle of the rappel. A self-equalizing anchor should only be used when necessary because if any one of the individual points fail, the anchor will extend and shock-load the remaining points or even cause complete anchor failure.

b. *Pre-equalized Anchor.* A pre-equalized anchor distributes the load equally to each individual point (Figure 5-19). It is aimed in the direction of the load. A pre-equalized anchor prevents extension and shock-loading of the anchor if an individual point fails. An anchor is pre-equalized by tying an overhand or figure-eight knot in the webbing or sling.

Note: *When using webbing or slings, the angles of the webbing or slings directly affect the load placed on an anchor. An angle greater than 90 degrees can result in anchor failure (Figure 5-20).*

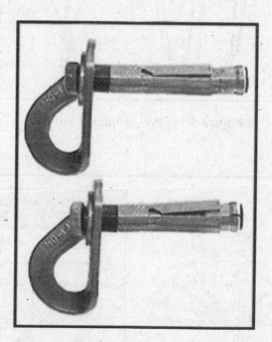

Figure 5-17: Bolt with expanding sleeve.

Figure 5-18: Self-equalizing anchors.

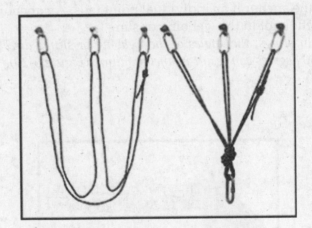

Figure 5-19: Pre-equalized anchor.

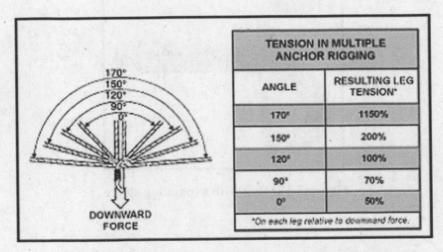

Figure 5-20: Effects of angles on an anchor.

SECTION 6: CLIMBING

A steep rock face is a terrain feature that can be avoided most of the time through prior planning and good route selection. Rock climbing can be time consuming, especially for a larger unit with a heavy combat load. It can leave the climbing party totally exposed to weather, terrain hazards, and the enemy for the length of the climb.

Sometimes steep rock cannot be avoided. Climbing relatively short sections of steep rock (one or two pitches) may prove quicker and safer than using alternate routes. A steep rock route would normally be considered an unlikely avenue of approach and, therefore, might be weakly defended or not defended at all.

All personnel in a unit preparing for deployment to mountainous terrain should be trained in the basics of climbing. Forward observers, reconnaissance personnel, and security teams are a few examples of small units who may require rock climbing skills to gain their vantage points in mountainous terrain. Select personnel demonstrating the highest degree of skill and experience should be trained in roped climbing techniques. These personnel will have the job of picking and "fixing" the route for the rest of the unit.

Rock climbing has evolved into a specialized "sport" with a wide range of varying techniques and styles. This chapter focuses on the basics most applicable to military operations.

CLIMBING FUNDAMENTALS

A variety of refined techniques are used to climb different types of rock formations. The foundation for all of these styles is the art of climbing. Climbing technique stresses climbing with the weight centered over the feet, using the hands primarily for balance. It can be thought of as a combination of the balanced movement required to walk a tightrope and the technique used to ascend a ladder. No mountaineering equipment is required; however, the climbing technique is also used in roped climbing.

6-1. Route Selection. The experienced climber has learned to climb with the "eyes." Even before getting on the rock, the climber studies all possible routes, or "lines," to the top looking for cracks, ledges, nubbins, and other irregularities in the rock that will be used for footholds and handholds, taking note of any larger ledges or benches for resting places. When picking the line, he mentally climbs the route, rehearsing the step-by-step sequence of movements that will be required to do the climb, ensuring himself that the route has an adequate number of holds and the difficulty of the climb will be well within the limit of his ability.

6-2. Terrain Selection for Training. Route selection for military climbing involves picking the easiest and quickest possible line for all personnel to follow. However, climbing skill and experience can only be developed by increasing the length and difficulty of routes as training progresses. In the training environment, beginning lessons in climbing should be performed CLOSE to the ground on lower-angled rock with plenty of holds for the hands and feet. Personnel not climbing can act as "otters" for those climbing. In later lessons, a "top-rope" belay can be used for safety, allowing individuals to increase the length and difficulty of the climb under the protection of the climbing rope.

6-3. Preparation. In preparation for climbing, the boot soles should be dry and clean. A small stick can be used to clean out dirt and small rocks that might be caught between the lugs of the boot sole. If the soles are wet or damp, dry them off by stomping and rubbing the soles on clean, dry rock. All jewelry should be removed from the fingers. Watches and bracelets can interfere with hand placements and may become damaged if worn while climbing. Helmets should be worn to protect the head from injury if an object, such as a rock or climbing gear, falls from climbers above. Most climbing helmets are not designed to provide protection from impact to the head if the wearer falls, but will provide a minimal amount of protection if a climber comes in contact with the rock during climbing.

> ⚠ **CAUTION**
>
> Rings can become caught on rock facial features and or lodged into cracks, which could cause injuries during a slip or fall.

6-4. Spotting. Spotting is a technique used to add a level of safety to climbing without a rope. A second man stands below and just outside of the climbers fall path and helps (spots) the climber to land safely if he should fall. Spotting is only applicable if the climber is not going above the spotters head on the rock. Beyond that height a roped climbing should be conducted. If an individual climbs beyond the effective range of the spotter(s), he has climbed TOO HIGH for his own safety. The duties of the spotter are to help prevent the falling climber from impacting the head and or spine, help the climber land feet first, and reduce the impact of a fall.

> ⚠ **CAUTION**
>
> The spotter should not catch the climber against the rock because additional injuries could result. If the spotter pushes the falling climber into the rock, deep abrasions of the skin or knee may occur. Ankle joints could be twisted by the fall if the climber's foot remained high on the rock. The spotter might be required to fully support the weight of the climber causing injury to the spotter.

6-5. Climbing Technique. Climbing involves linking together a series of movements based on foot and hand placement, weight shift, and movement. When this series of movements is combined correctly, a smooth climbing technique results. This technique reduces excess force on the limbs, helping to minimize fatigue. The basic principle is based on the five body parts described here.

a. *Five Body Parts.* The five body parts used for climbing are the right hand, left hand, right foot, left foot, and body (trunk). The basic principle to achieve smooth climbing is to move only one of the five body parts at a time. The trunk is not moved in conjunction with a foot or in conjunction with a hand, a hand is not moved in conjunction with a foot, and so on. Following this simple technique forces both legs to do all the lifting simultaneously.

b. *Stance or Body Position.* Body position is probably the single most important element to good technique. A relaxed, comfortable stance is essential. (Figure 6-1 shows a correct climbing stance, and Figure 6-2 shows an incorrect stance.) The body should be in a near vertical or erect stance with the weight centered over the feet. Leaning in towards the rock will cause the feet to push outward, away from the rock, resulting in a loss of friction between the boot sole and rock surface. The legs are straight and the heels are kept low to reduce fatigue. Bent legs and tense muscles tire quickly. If strained for too long, tense muscles may vibrate uncontrollably. This vibration, known as "Elvising" or "sewing-machine leg" can be cured by straightening the leg, lowering the heel, or moving on to a more restful position. The hands are used to maintain balance. Keeping the hands between waist and shoulder level will reduce arm fatigue.

(1) Whenever possible, three points of contact are maintained with the rock. Proper positioning of the hips and shoulders is critical. When using two footholds and one handhold, the hips and shoulders should be centered over both feet. In most cases, as the climbing progresses, the body is resting on one foot with two handholds for balance. The hips and shoulders must be centered over the support foot to maintain balance, allowing the "free" foot to maneuver.

(2) The angle or steepness of the rock also determines how far away from the rock the hips and shoulders should be. On low-angle slopes, the hips are moved out away from the rock to keep the body in balance with the weight over the feet. The shoulders can be moved closer to the rock to reach handholds. On steep rock, the hips are pushed closer to the rock. The shoulders

Figure 6-1: Correct climbing stance-balanced over both feet.

Figure 6-2: Incorrect stance-stretched out.

are moved away from the rock by arching the back. The body is still in balance over the feet and the eyes can see where the hands need to go. Sometimes, when footholds are small, the hips are moved back to increase friction between the foot and the rock. This is normally done on quick, intermediate holds. It should be avoided in the rest position as it places more weight on the arms and hands. When weight must be placed on handholds, the arms should be kept straight to reduce fatigue. Again, flexed muscles tire quickly.

c. *Climbing Sequence.* The steps defined below provide a complete sequence of events to move the entire body on the rock. These are the basic steps to follow for a smooth climbing technique. Performing these steps in this exact order will not always be necessary because the nature of the route will dictate the availability of hand and foot placements. The basic steps are weight, shift, and movement (movement being either the foot, hand, or body). (A typical climbing sequence is shown in Figure 6-3).

STEP ONE: Shift the weight from both feet to one foot. This will allow lifting of one foot with no effect on the stance.

STEP TWO: Lift the unweighted foot and place it in a new location, within one to two feet of the starting position, with no effect on body position or balance (higher placement will result in a potentially higher lift for the legs to make, creating more stress, and is called a high step). The trunk does not move during foot movement.

STEP THREE: Shift the weight onto both feet. (Repeat steps 1 through 3 for remaining foot.)

STEP FOUR: Lift the body into a new stance with both legs.

STEP FIVE: Move one hand to a new position between waist and head height. During this movement, the trunk should be completely balanced in position and the removed hand should have no effect on stability.

STEP SIX: Move the remaining hand as in Step 5.

Now the entire body is in a new position and ready to start the process again. Following these steps will prevent lifting with the hands and arms, which are used to maintain stance and balance. If both legs are bent, leg extension can be performed as soon as one foot has been moved. Hand movements can be delayed until numerous foot movements have been made, which not only creates shorter lifts with the legs, but may allow a better choice for the next hand movements because the reach will have increased.

(1) Many climbers will move more than one body part at a time, usually resulting in lifting the body with one leg or one leg and both arms. This type of lifting is inefficient, requiring one leg to perform the work of two or using the arms to lift the body. Proper climbing technique is lifting the body with the legs, not the arms, because the legs are much stronger.

(2) When the angle of the rock increases, these movements become more critical. Holding or pulling the body into the rock with the arms and hands may be necessary as the angle increases (this is still not lifting with the arms). Many climbing routes have angles greater than ninety degrees (overhanging) and the arms are used to support partial body weight. The same technique applies even at those angles.

(3) The climber should avoid moving on the knees and elbows. Other than being uncomfortable, even painful, to rest on, these bony portions of the limbs offer little friction and "feel" on the rock.

6-6. Safety Precautions. The following safety precautions should be observed when rock climbing.

a. While ascending a seldom or never traveled route, you may encounter precariously perched rocks. If the rock will endanger your second, it may be possible to remove it from the route and trundle it, tossing it down. This is extremely dangerous to climbers below and should not be attempted unless you are absolutely sure no men are below. If you are not sure that the flight path is clear, do not do it. Never dislodge loose rocks carelessly. Should a rock become loose accidentally, immediately

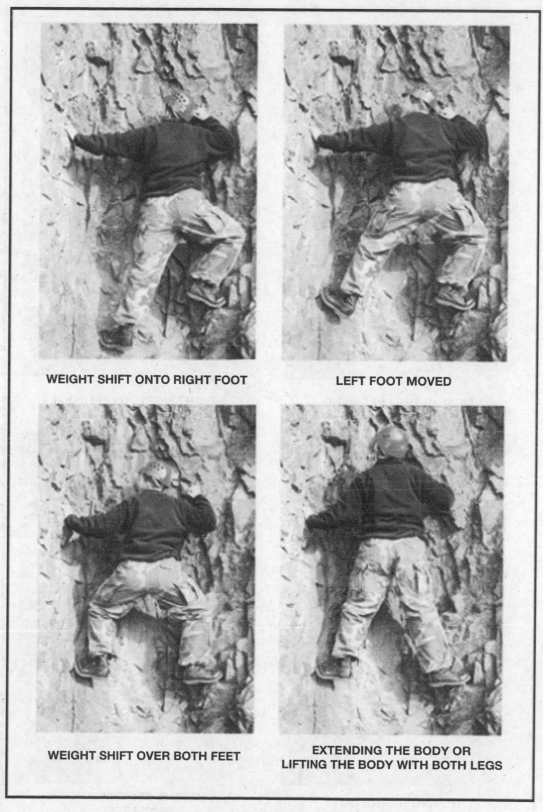

WEIGHT SHIFT ONTO RIGHT FOOT

LEFT FOOT MOVED

WEIGHT SHIFT OVER BOTH FEET

EXTENDING THE BODY OR
LIFTING THE BODY WITH BOTH LEGS

Figure 6-3: Typical climbing sequence.

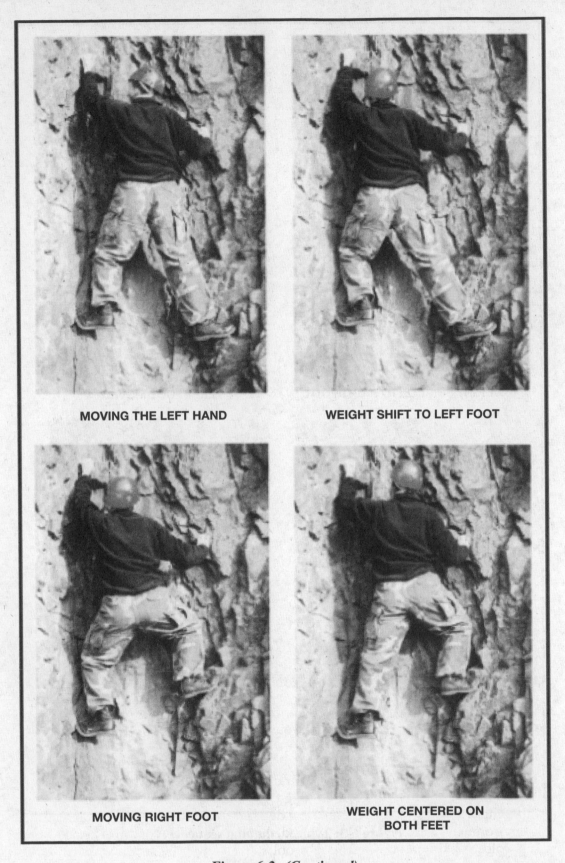

MOVING THE LEFT HAND

WEIGHT SHIFT TO LEFT FOOT

MOVING RIGHT FOOT

WEIGHT CENTERED ON
BOTH FEET

Figure 6-3: *(Continued)*

RIGHT HAND MOVED

Figure 6-3: *(Continued)*

shout the warning "ROCK" to alert climbers below. Upon hearing the warning, personnel should seek immediate cover behind any rock bulges or overhangs available, or flatten themselves against the rock to minimize exposure.

b. Should a climber fall, he should do his utmost to maintain control and not panic. If on a low-angle climb, he may be able to arrest his own fall by staying in contact with the rock, grasping for any possible hold available. He should shout the warning "FALLING" to alert personnel below.

> ⚠ **CAUTION**
>
> Grasping at the rock in a fall can result in serious injuries to the upper body. If conducting a roped climb, let the rope provide protection.

c. When climbing close to the ground and without a rope, a spotter can be used for safety. The duties of the spotter are to ensure the falling climber does not impact the head or spine, and to reduce the impact of a fall.

d. Avoid climbing directly above or below other climbers (with the exception of spotters). When personnel must climb at the same time, following the same line, a fixed rope should be installed.

e. Avoid climbing with gloves on because of the decreased "feel" for the rock. The use of gloves in the training environment is especially discouraged, while their use in the mountains is often mandatory when it is cold. A thin polypropylene or wool glove is best for rock climbing, although heavier cotton or leather work gloves are often used for belaying.

f. Be extremely careful when climbing on wet or moss-covered rock; friction on holds is greatly reduced.

g. Avoid grasping small vegetation for handholds; the root systems can be shallow and will usually not support much weight.

6-7. Margin of Safety. Besides observing the standard safety precautions, the climber can avoid catastrophe by climbing with a wide margin of safety. The margin of safety is a protective buffer the climber places between himself and potential climbing hazards. Both subjective (personnel-related) and objective (environmental) hazards must be considered when applying the margin of safety. The leader must apply the margin of safety taking into account the strengths and weaknesses of the entire team or unit.

a. When climbing, the climber increases his margin of safety by selecting routes that are well within the limit of his ability. When leading a group of climbers, he selects a route well within the ability of the weakest member.

b. When the rock is wet, or when climbing in other adverse weather conditions, the climber's ability is reduced and routes are selected accordingly. When the climbing becomes difficult or exposed, the climber knows to use the protection of the climbing rope and belays. A lead climber increases his margin of safety by placing protection along the route to limit the length of a potential fall.

USE OF HOLDS

The climber should check each hold before use. This may simply be a quick, visual inspection if he knows the rock to be solid. When in doubt, he should grab and tug on the hold to test it for soundness BEFORE depending on it. Sometimes, a hold that appears weak can actually be solid as long as minimal force is applied to it, or the force is applied in a direction that strengthens it. A loose nubbin might not be strong enough to support the climber's weight, but it may serve as an adequate handhold. Be especially careful when climbing on weathered, sedimentary-type rock.

6-8. Climbing with the Feet. "Climb with the feet and use the hands for balance" is extremely important to remember. In the early learning stages of climbing, most individuals will rely heavily on the arms, forgetting to use the feet properly. It is true that solid handholds and a firm grip are needed in some combination techniques; however, even the most strenuous techniques require good footwork and a quick return to a balanced position over one or both feet. Failure to climb any route, easy or difficult, is usually the result of poor footwork.

a. The beginning climber will have a natural tendency to look up for handholds. Try to keep the hands low and train your eyes to look down for footholds. Even the smallest irregularity in the rock can support the climber once the foot is positioned properly and weight is committed to it.

b. The foot remains on the rock as a result of friction. Maximum friction is obtained from a correct stance over a properly positioned foot. The following describes a few ways the foot can be positioned on the rock to maximize friction.

(1) *Maximum Sole Contact.* The principle of using full sole contact, as in mountain walking, also applies in climbing. Maximum friction is obtained by placing as much of the boot sole on the rock as possible. Also, the leg muscles can relax the most when the entire foot is placed on the rock. (Figure 6-4 shows examples of maximum and minimum sole contact.)

(a) Smooth, low-angled rock (slab) and rock containing large "bucket" holds and ledges are typical formations where the entire boot sole should be used.

(b) On some large holds, like bucket holds that extend deep into the rock, the entire foot cannot be used. The climber may not be able to achieve a balanced position if the foot is stuck too far underneath a bulge in the rock. In this case, placing only part of the foot on the hold may allow the climber to achieve a balanced stance. The key is to use as much of the

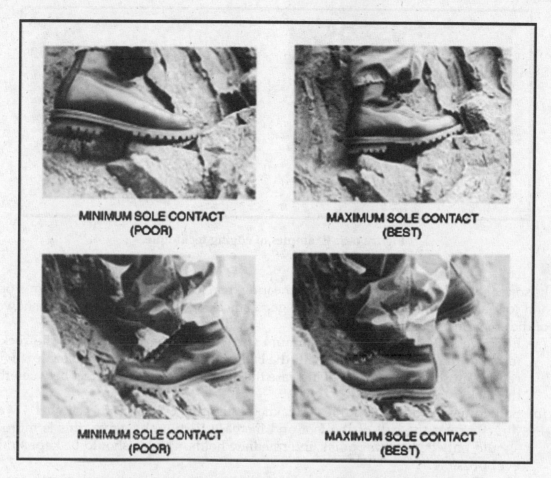

Figure 6-4: Examples of maximum and minimum sole contact.

boot sole as possible. Remember to keep the heels low to reduce strain on the lower leg muscles.

(2) *Edging*. The edging technique is used where horizontal crack systems and other irregularities in the rock form small, well-defined ledges. The edge of the boot sole is placed on the ledge for the foothold. Usually, the inside edge of the boot or the edge area around the toes is used. Whenever possible, turn the foot sideways and use the entire inside edge of the boot. Again, more sole contact equals more friction and the legs can rest more when the heel is on the rock. (Figure 6-5 shows examples of the edging technique).

(a) On smaller holds, edging with the front of the boot, or toe, may be used. Use of the toe is most tiring because the heel is off the rock and the toes support the climber's weight. Remember to keep the heel low to reduce fatigue. Curling and stiffening the toes in the boot increases support on the hold. A stronger position is usually obtained on small ledges by turning the foot at about a 45-degree angle, using the strength of the big toe and the ball of the foot.

(b) Effective edging on small ledges requires stiff-soled footwear. The stiffer the sole, the better the edging capability. Typical mountain boots worn by the US military have a relatively flexible lugged sole and, therefore, edging ability on smaller holds will be somewhat limited.

(3) *Smearing*. When footholds are too small to use a good edging technique, the ball of the foot can be "smeared" over the hold. The smearing technique requires the boot to adhere to the rock by deformation of the sole and by friction. Rock climbing shoes are specifically designed to maximize friction for smearing; some athletic shoes also work well. The Army mountain boot, with

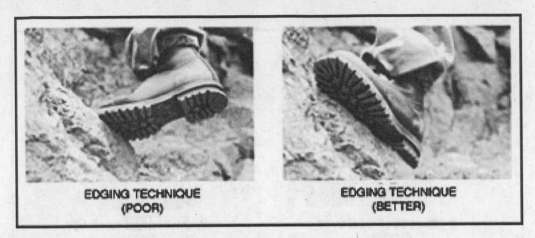

Figure 6-5: Examples of edging technique.

its softer sole, usually works better for smearing than for edging. Rounded, down-sloping ledges and low-angled slab rock often require good smearing technique. (Figure 6-6 shows examples of the smearing technique.)

(a) Effective smearing requires maximum friction between the foot and the rock. Cover as much of the hold as possible with the ball of the foot. Keeping the heel low will not only reduce muscle strain, but will increase the amount of surface contact between the foot and the rock.

(b) Sometimes flexing the ankles and knees slightly will place the climber's weight more directly over the ball of the foot and increase friction; however, this is more tiring and should only be used for quick, intermediate holds. The leg should be kept straight whenever possible.

(4) *Jamming.* The jamming technique works on the same principal as chock placement. The foot is set into a crack in such a way that it "jams" into place, resisting a downward pull. The jamming technique is a specialized skill used to climb vertical or near vertical cracks when no other holds are available on the rock face. The technique is not limited to just wedging the feet; fingers, hands, arms, even the entire leg or body are all used in the jamming technique, depending on the size of the crack. Jam holds are described in this text to broaden the range of climbing skills. Jamming holds can be used in a crack while other hand/foot holds are used on the face of the

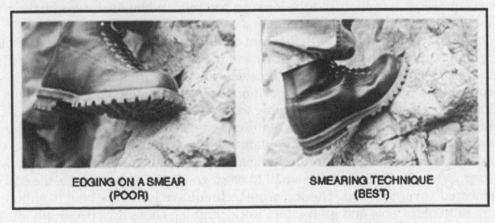

Figure 6-6: Examples of the smearing technique.

rock. Many cracks will have facial features, such as edges, pockets, and soon, inside and within reach. Always look or feel for easier to use features. (Figure 6-7 shows examples of jamming.)

(a) The foot can be jammed in a crack in different ways. It can be inserted above a constriction and set into the narrow portion, or it can be placed in the crack and turned, like a camming device, until it locks in place tight enough to support the climber's weight. Aside from these two basic ideas, the possibilities are endless. The toes, ball of the foot, or the entire foot can be used. Try to use as much of the foot as possible for maximum surface contact. Some positions are more tiring, and even more painful on the foot, than others. Practice jamming the foot in various ways to see what offers the most secure, restful position.

(b) Some foot jams may be difficult to remove once weight has been committed to them, especially if a stiffer sole boot is used. The foot is less likely to get stuck when it is twisted or "cammed" into position. When removing the boot from a crack, reverse the way it was placed to prevent further constriction.

6-9. Using the Hands. The hands can be placed on the rock in many ways. Exactly how and where to position the hands and arms depends on what holds are available, and what configuration will best support the current stance as well as the movement to the next stance. Selecting handholds between waist and shoulder level helps in different ways. Circulation in the arms and hands is best when the arms are kept low. Secondly, the climber has less tendency to "hang" on his arms when the handholds are at shoulder level and below. Both of these contribute to a relaxed stance and reduce fatigue in the hands and arms.

a. As the individual climbs, he continually repositions his hands and arms to keep the body in balance, with the weight centered over the feet. On lower-angled rock, he may simply need to place the hands up against the rock and extend the arm to maintain balance; just like using an ice ax as a third point of contact in mountain walking. Sometimes, he will be able to push directly down on a large hold with the palm of the hand. More often though, he will need to "grip" the rock in some fashion and then push or pull against the hold to maintain balance.

b. As stated earlier, the beginner will undoubtedly place too much weight on the hands and arms. If we think of ourselves climbing a ladder, our body weight is on our legs. Our hands grip, and our arms pull on each rung only enough to maintain our balance and footing on the ladder. Ideally, this is the amount of grip and pull that should be used in climbing. Of course, as the

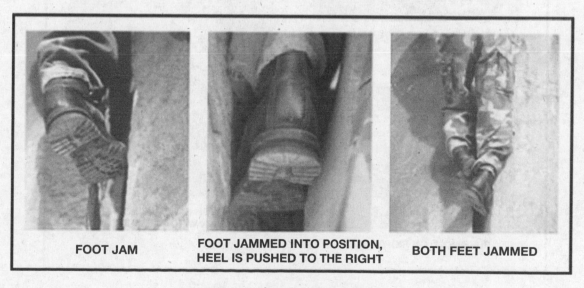

FOOT JAM   FOOT JAMMED INTO POSITION,   BOTH FEET JAMMED
                  HEEL IS PUSHED TO THE RIGHT

Figure 6-7: Examples of jamming.

size and availability of holds decreases, and the steepness of the rock approaches the vertical, the grip must be stronger and more weight might be placed on the arms and handholds for brief moments. The key is to move quickly from the smaller, intermediate holds to the larger holds where the weight can be placed back on the feet allowing the hands and arms to relax. The following describes some of the basic handholds and how the hand can be positioned to maximize grip on smaller holds.

(1) *Push Holds.* Push holds rely on the friction created when the hand is pushed against the rock. Most often a climber will use a push hold by applying "downward pressure" on a ledge or nubbin. This is fine, and works well; however, the climber should not limit his use of push holds to the application of down pressure. Pushing sideways, and on occasion, even upward on less obvious holds can prove quite secure. Push holds often work best when used in combination with other holds. Pushing in opposite directions and "push-pull" combinations are excellent techniques. (Figure 6-8 shows examples of push holds).

   (a) An effective push hold does not necessarily require the use of the entire hand. On smaller holds, the side of the palm, the fingers, or the thumb may be all that is needed to support the stance. Some holds may not feel secure when the hand is initially placed on them. The hold may improve or weaken during the movement. The key is to try and select a hold that will improve as the climber moves past it.

   (b) Most push holds do not require much grip; however, friction might be increased by taking advantage of any rough surfaces or irregularities in the rock. Sometimes the strength of the

PUSH HOLD - SIDE PRESSURE

PUSH HOLD ON RIGHT HAND

PUSH HOLD - USING DOWNWARD PRESSURE

Figure 6-8: Examples of push holds.

hold can be increased by squeezing, or "pinching," the rock between the thumb and fingers (see paragraph on pinch holds).

(2) *Pull Holds.* Pull holds, also called "cling holds," which are grasped and pulled upon, are probably the most widely used holds in climbing. Grip plays more of a role in a pull hold, and, therefore, it normally feels more secure to the climber than a push hold. Because of this increased feeling of security, pull holds are often overworked. These are the holds the climber has a tendency to hang from. Most pull holds do not require great strength, just good technique. Avoid the "death grip" syndrome by climbing with the feet. (Figure 6-9 shows examples of pull holds.)

(a) Like push holds, pressure on a pull hold can be applied straight down, sideways, or upward. Again, these are the holds the climber tends to stretch and reach for, creating an unbalanced stance. Remember to try and keep the hands between waist and shoulder level, making use of intermediate holds instead of reaching for those above the head.

(b) Pulling sideways on vertical cracks can be very secure. There is less tendency to hang from "side-clings" and the hands naturally remain lower. The thumb can often push against one

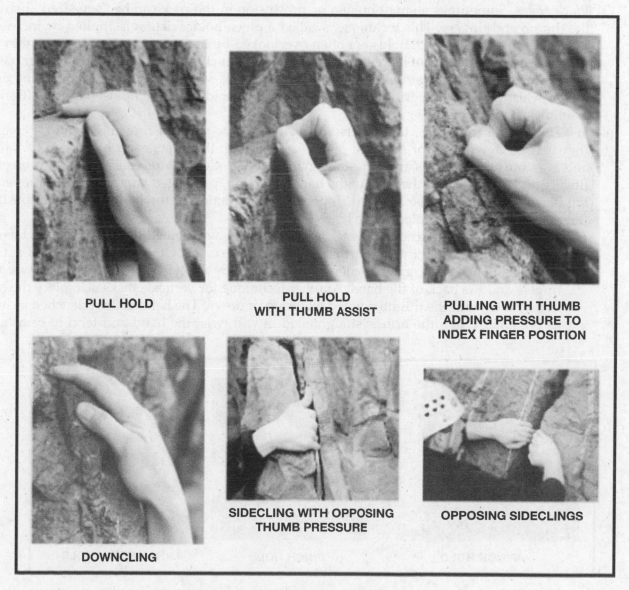

PULL HOLD

PULL HOLD
WITH THUMB ASSIST

PULLING WITH THUMB
ADDING PRESSURE TO
INDEX FINGER POSITION

DOWNCLING

SIDECLING WITH OPPOSING
THUMB PRESSURE

OPPOSING SIDECLINGS

Figure 6-9: Examples of pull holds.

side of the crack, in opposition to the pull by the fingers, creating a stronger hold. Both hands can also be placed in the same crack, with the hands pulling in opposite directions. The number of possible combinations is limited only by the imagination and experience of the climber.

(c) Friction and strength of a pull hold can be increased by the way the hand grips the rock. Normally, the grip is stronger when the fingers are closed together; however, sometimes more friction is obtained by spreading the fingers apart and placing them between irregularities on the rock surface. On small holds, grip can often be improved by bending the fingers upward, forcing the palm of the hand to push against the rock. This helps to hold the finger tips in place and reduces muscle strain in the hand. Keeping the forearm up against the rock also allows the arm and hand muscles to relax more.

(d) Another technique that helps to strengthen a cling hold for a downward pull is to press the thumb against the side of the index finger, or place it on top of the index finger and press down. This hand configuration, known as a "ring grip," works well on smaller holds.

(3) *Pinch Holds.* Sometimes a small nubbin or protrusion in the rock can be "squeezed" between the thumb and fingers. This technique is called a pinch hold. Friction is applied by increasing the grip on the rock. Pinch holds are often overlooked by the novice climber because they feel insecure at first and cannot be relied upon to support much body weight. If the climber has his weight over his feet properly, the pinch hold will work well in providing balance. The pinch hold can also be used as a gripping technique for push holds and pull holds. (Figure 6-10 shows examples of pinch holds.)

(4) *Jam Holds.* Like foot jams, the fingers and hands can be wedged or cammed into a crack so they resist a downward or outward pull. Jamming with the fingers and hands can be painful and may cause minor cuts and abrasions to tender skin. Cotton tape can be used to protect the fingertips, knuckles, and the back of the hand; however, prolonged jamming technique requiring hand taping should be avoided. Tape also adds friction to the hand in jammed position. (Figure 6-11 shows examples of jam holds.)

(a) The hand can be placed in a crack a number of ways. Sometimes an open hand can be inserted and wedged into a narrower portion of the crack. Other times a clenched fist will provide the necessary grip. Friction can be created by applying cross pressure between the fingers and the back of the hand. Another technique for vertical cracks is to place the hand in the crack with the thumb pointed either up or down. The hand is then clenched as much as possible. When the arm is straightened, it will twist the hand and tend to cam it into

PINCH HOLD   PINCH HOLD   LARGE PINCH HOLD

Figure 6-10: Examples of pinch holds.

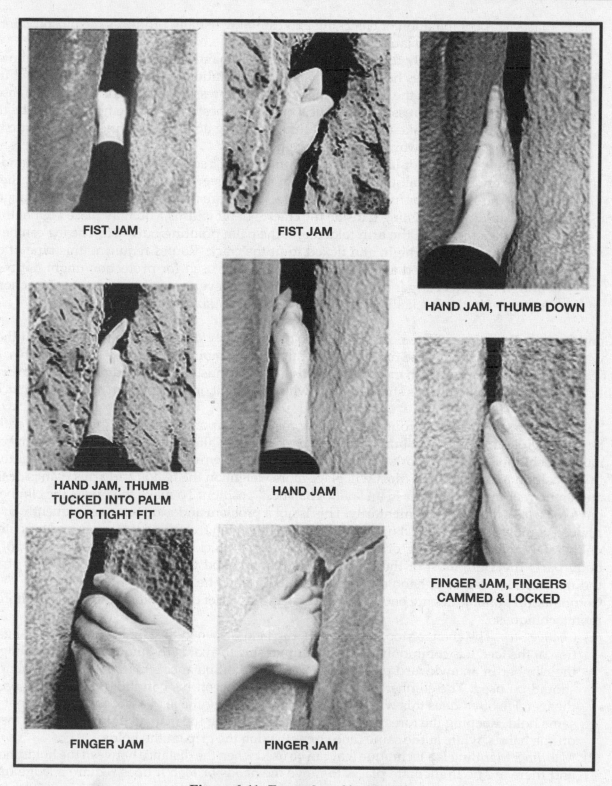

Figure 6-11: Examples of jam holds.

place. This combination of clenching and camming usually produces the most friction, and the most secure hand jam in vertical cracks.

(b) In smaller cracks, only the fingers will fit. Use as many fingers as the crack will allow. The fingers can sometimes be stacked in some configuration to increase friction. The thumb is usually kept outside the crack in finger jams and pressed against the rock to increase friction or create cross pressure. In vertical cracks it is best to insert the fingers with the thumb pointing down to make use of the natural camming action of the fingers that occurs when the arm is twisted towards a normal position.

(c) Jamming technique for large cracks, or "off widths," requiring the use of arm, leg, and body jams, is another technique. To jam or cam an arm, leg, or body into an off width, the principle is the same as for fingers, hands, or feet—you are making the jammed appendage "fatter" by folding or twisting it inside the crack. For off widths, you may place your entire arm inside the crack with the arm folded and the palm pointing outward. The leg can be used, from the calf to the thigh, and flexed to fit the crack. Routes requiring this type of climbing should be avoided as the equipment normally used for protection might not be large enough to protect larger cracks and openings. However, sometimes a narrower section may be deeper in the crack allowing the use of "normal" size protection.

6-10. Combination Techniques. The positions and holds previously discussed are the basics and the ones most common to climbing. From these fundamentals, numerous combination techniques are possible. As the climber gains experience, he will learn more ways to position the hands, feet, and body in relation to the holds available; however, he should always strive to climb with his weight on his feet from a balanced stance.

a. Sometimes, even on an easy route, the climber may come upon a section of the rock that defies the basic principles of climbing. Short of turning back, the only alternative is to figure out some combination technique that will work. Many of these type problems require the hands and feet to work in opposition to one another. Most will place more weight on the hands and arms than is desirable, and some will put the climber in an "out of balance" position. To make the move, the climber may have to "break the rules" momentarily. This is not a problem and is done quite frequently by experienced climbers. The key to using these type of combination techniques is to plan and execute them deliberately, without lunging or groping for holds, yet quickly, before the hands, arms, or other body parts tire. Still, most of these maneuvers require good technique more than great strength, though a certain degree of hand and arm strength certainly helps.

b. Combination possibilities are endless. The following is a brief description of some of the more common techniques.

(1) *Change Step.* The change step, or hop step, can be used when the climber needs to change position of the feet. It is commonly used when traversing to avoid crossing the feet, which might put the climber in an awkward position. To prevent an off balance situation, two solid handholds should be used. The climber simply places his weight on his handholds while here positions the feet. He often does this with a quick "hop," replacing the lead foot with the trail foot on the same hold. Keeping the forearms against the rock during the maneuver takes some of the strain off the hands, while at the same time strengthening the grip on the holds.

(2) *Mantling.* Mantling is a technique that can be used when the distance between the holds increases and there are no immediate places to move the hands or feet. It does require a ledge (mantle) or projection in the rock that the climber can press straight down upon. (Figure 6-12 shows the mantling sequence.)

(a) When the ledge is above head height, mantling begins with pull holds, usually "hooking" both hands over the ledge. The climber pulls himself up until his head is above the hands, where the pull holds become push holds. He elevates himself until the arms are straight and he can lock the elbows to relax the muscles. Rotating the hands inward during the

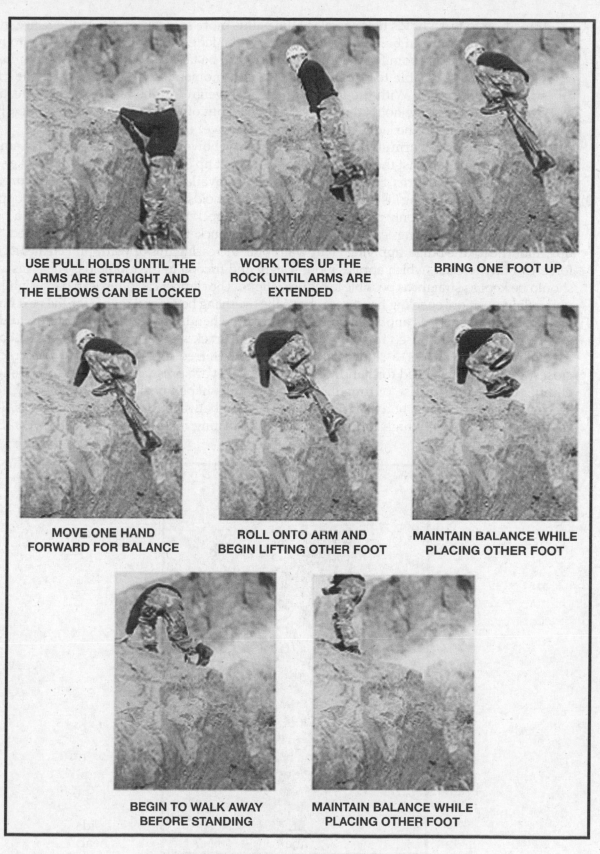

Figure 6-12: Mantling sequence.

transition to push holds helps to place the palms more securely on the ledge. Once the arms are locked, a foot can be raised and placed on the ledge. The climber may have to remove one hand to make room for the foot. Mantling can be fairly strenuous; however, most individuals should be able to support their weight, momentarily, on one arm if they keep it straight and locked. With the foot on the ledge, weight can be taken off the arms and the climber can grasp the holds that were previously out of reach. Once balanced over the foot, he can stand up on the ledge and plan his next move.

(b) Pure mantling uses arm strength to raise the body; however, the climber can often smear the balls of the feet against the rock and "walk" the feet up during the maneuver to take some of the weight off the arms. Sometimes edges will be available for short steps in the process.

(3) *Undercling.* An "undercling" is a classic example of handholds and footholds working in opposition (Figure 6-13). It is commonly used in places where the rock projects outward, forming a bulge or small overhang. Underclings can be used in the tops of buckets, also. The hands are placed "palms-up" underneath the bulge, applying an upward pull. Increasing this upward pull creates a counter-force, or body tension, which applies more weight and friction to the footholds. The arms and legs should be kept as straight as possible to reduce fatigue. The climber can often lean back slightly in the undercling position, enabling him to see above the overhang better and search for the next hold.

(4) *Lieback.* The "lieback" is another good example of the hands working in opposition to the feet. The technique is often used in a vertical or diagonal crack separating two rock faces that come together at, more or less, a right angle (commonly referred to as a dihedral). The crack edge closest to the body is used for handholds while the feet are pressed against the other edge. The climber bends at the waist, putting the body into an L-shaped position. Leaning away from the crack on two pull holds, body tension creates friction between the feet and the hands. The feet must be kept relatively high to maintain weight, creating maximum friction between the sole

Figure 6-13: Undercling.

and the rock surface. Either full sole contact or the smearing technique can be used, whichever seems to produce the most friction.

(a) The climber ascends a dihedral by alternately shuffling the hands and feet upward. The lieback technique can be extremely tiring, especially when the dihedral is near vertical. If the hands and arms tire out before completing the sequence, the climber will likely fall. The arms should be kept straight throughout the entire maneuver so the climber's weight is pulling against bones and ligaments, rather than muscle. The legs should be straightened whenever possible.

(b) Placing protection in a lieback is especially tiring. Look for edges or pockets for the feet in the crack or on the face for a better position to place protection from, or for a rest position. Often, a lieback can be avoided with closer examination of the available face features. The lieback can be used alternately with the jamming technique, or vice versa, for variation or to get past a section of a crack with difficult or nonexistent jam possibilities. The lieback can sometimes be used as a face maneuver (Figure 6-14).

(5) *Stemming.* When the feet work in opposition from a relatively wide stance, the maneuver is known as stemming. The stemming technique can sometimes be used on faces, as well as in a dihedral in the absence of solid handholds for the lieback (Figure 6-15).

(a) The classic example of stemming is when used in combination with two opposing push holds in wide, parallel cracks, known as chimneys. Chimneys are cracks in which the walls are at least 1 foot apart and just big enough to squeeze the body into. Friction is created by pushing outward with the hands and feet on each side of the crack. The climber ascends the chimney by alternately moving the hands and feet up the crack (Figure 6-16). Applying pressure with the back and bottom is usually necessary in wider chimneys. Usually, full sole contact of the shoes will provide the most friction, although smearing may work best in some instances. Chimneys that do not allow a full stemming position can be negotiated

Figure 6-14: Lieback on a face.

Figure 6-15: Stemming on a face.

using the arms, legs, or body as an integral contact point. This technique will often feel more secure since there is more body to rock contact.

(b) The climber can sometimes rest by placing both feet on the same side of the crack, forcing the body against the opposing wall. The feet must be kept relatively high up under the body so the force is directed sideways against the walls of the crack. The arms should be straightened with the elbows locked whenever possible to reduce muscle strain. The climber must ensure that the crack does not widen beyond the climbable width before committing to the maneuver. Remember to look for face features inside chimneys for more security in the climb.

(c) Routes requiring this type of climbing should be avoided as the equipment normally used for protection might not be large enough to protect chimneys. However, face features, or a much narrower crack irotection.

(6) *Slab Technique.* A slab is a relatively smooth, low-angled rock formation that requires a slightly modified climbing technique (Figure 6-17). Since slab rock normally contains few, if any holds, the technique requires maximum friction and perfect balance over the feet.

(a) On lower-angled slab, the climber can often stand erect and climb using full sole contact and other mountain walking techniques. On steeper slab, the climber will need to apply good smearing technique. Often, maximum friction cannot be attained on steeper slab from an erect stance. The climber will have to flex the ankles and knees so his weight is placed more directly over the balls of the feet. He may then have to bend at the waist to place the hands on the rock, while keeping the hips over his feet.

(b) The climber must pay attention to any changes in slope angle and adjust his body accordingly. Even the slightest change in the position of the hips over the feet can mean the

**BEGIN WITH ONE FOOT
AND THE BACK**

**BRING OTHER FOOT
INTO POSITION**

FEET OPPOSING BACK

REST POSITION

Figure 6-16: Chimney sequence.

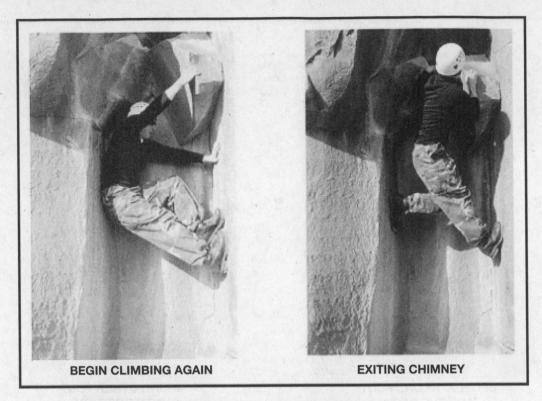

BEGIN CLIMBING AGAIN   **EXITING CHIMNEY**

Figure 6-16: *(Continued)*

Figure 6-17: Slab technique.

difference between a good grip or a quick slip. The climber should also take advantage of any rough surfaces, or other irregularities in the rock he can place his hands or feet on, to increase friction.

(7) *Down Climbing.* Descending steep rock is normally performed using a roped method; however, the climber may at some point be required to down climb a route. Even if climbing ropes and related equipment are on hand, down climbing easier terrain is often quicker than taking the time to rig a rappel point. Also, a climber might find himself confronted with difficulties part way up a route that exceed his climbing ability, or the abilities of others to follow. Whatever the case may be, down climbing is a skill well worth practicing.

> ⚠ **CAUTION**
>
> 1. Down climbing can inadvertently lead into an unforeseen dangerous position on a descent. When in doubt, use a roped descent.
> 2. Down climbing is accomplished at a difficulty level well below the ability of the climber. When in doubt, use a roped descent.

(a) On easier terrain, the climber can face outward, away from the rock, enabling him to see the route better and descend quickly. As the steepness and difficulty increase, he can often turn sideways, still having a good view of the descent route, but being better able to use the hands and feet on the holds available. On the steepest terrain, the climber will have to face the rock and down climb using good climbing techniques.

(b) Down climbing is usually more difficult than ascending a given route. Some holds will be less visible when down climbing, and slips are more likely to occur. The climber must often lean well away from the rock to look for holds and plan his movements. More weight is placed on the arms and handholds at times to accomplish this, as well as to help lower the climber to the next foothold. Hands should be moved to holds as low as waist level to give the climber more range of movement with each step. If the handholds are too high, he may have trouble reaching the next foothold. The climber must be careful not to overextend himself, forcing a release of his handholds before reaching the next foothold.

> ⚠ **CAUTION**
>
> Do not drop from good handholds to a standing position. A bad landing could lead to injured ankles or a fall beyond the planned landing area.

(c) Descending slab formations can be especially tricky. The generally lower angle of slab rock may give the climber a false sense of security, and a tendency to move too quickly. Down climbing must be slow and deliberate, as in ascending, to maintain perfect balance and weight distribution over the feet. On lower-angle slab the climber may be able to stand more or less erect, facing outward or sideways, and descend using good flat foot technique. The climber should avoid the tendency to move faster, which can lead to uncontrollable speed.

(d) On steeper slab, the climber will normally face the rock and down climb, using the same smearing technique as for ascending. An alternate method for descending slab is to face away from the rock in a "crab" position (Figure 6-18). Weight is still concentrated over the feet, but may be shifted partly onto the hands to increase overall friction. The climber is able to maintain full sole contact with the rock and see the entire descent route. Allowing the buttocks to "drag behind" on the rock will decrease the actual weight on the footholds, reducing

Figure 6-18: Descending slab in the crab position.

friction, and leading to the likelihood of a slip. Facing the rock, and down-climbing with good smearing technique, is usually best on steeper slab.

ROPED CLIMBING

When the angle, length, and difficulty of the proposed climbing route surpasses the ability of the climbers' safety margin (possibly on class 4 and usually on class 5 terrain), ropes must be used to proceed. Roped climbing is only safe if accomplished correctly. Reading this manual does not constitute skill with ropes—much training and practice is necessary. Many aspects of roped climbing take time to understand and learn. Ropes are normally not used in training until the basic principles of climbing are covered.

Note: A rope is completely useless for climbing unless the climber knows how to use it safely.

6-11. Typing-in to the Climbing Rope. Over the years, climbers have developed many different knots and procedures for tying-in to the climbing rope. Some of the older methods of tying directly into the rope require minimal equipment and are relatively easy to inspect; however, they offer little support to the climber, may induce further injuries, and may even lead to strangulation in a severe fall. A severe fall, where the climber might fall 20 feet or more and be left dangling on the end of the rope, is highly unlikely in most instances, especially for most personnel involved in military climbing. Tying directly into the rope is perfectly safe for many roped party climbs used in training on lower-angled rock. All climbers should know how to properly tie the rope around the waist in case a climbing harness is unavailable.

6-12. Presewn Harnesses. Although improvised harnesses are made from readily available materials and take little space in the pack or pocket, presewn harnesses provide other aspects that should be considered. No assembly is required, which reduces preparation time for roped movement. All presewn harnesses provide a range of adjustability. These harnesses have a fixed buckle that, when used correctly, will not fail before the nylon materials connected to it. However, specialized equipment, such as a presewn harness, reduce the flexibility of gear. Presewn harnesses are bulky, also.

a. *Seat Harness.* Many presewn seat harnesses are available with many different qualities separating them, including cost.

  (1) The most notable difference will be the amount and placement of padding. The more padding the higher the price and the more comfort. Gear loops sewn into the waist belt on the sides and in the back are a common feature and are usually strong enough to hold quite a few carabiners and or protection. The gear loops will vary in number from one model/manufacturer to another.

  (2) Although most presewn seat harnesses have a permanently attached belay loop connecting the waist belt and the leg loops, the climbing rope should be run around the waist belt and leg loop connector. The presewn belay loop adds another link to the chain of possible failure points and only gives one point of security whereas running the rope through the waist belt and leg loop connector provides two points of contact.

  (3) If more than two men will be on the rope, connect the middle position(s) to the rope with a carabiner routed the same as stated in the previous paragraph.

  (4) Many manufactured seat harnesses will have a presewn loop of webbing on the rear. Although this loop is much stronger than the gear loops, it is not for a belay anchor. It is a quick attachment point to haul an additional rope.

b. *Chest Harness.* The chest harness will provide an additional connecting point for the rope, usually in the form of a carabiner loop to attach a carabiner and rope to. This type of additional connection will provide a comfortable hanging position on the rope, but otherwise provides no additional protection from injury during a fall (if the seat harness is fitted correctly).

  (1) A chest harness will help the climber remain upright on the rope during rappelling or ascending a fixed rope, especially while wearing a heavy pack. (If rappelling or ascending long or multiple pitches, let the pack hang on a drop cord below the feet and attached to the harness tie-in point.)

  (2) The presewn chest harnesses available commercially will invariably offer more comfort or performance features, such as padding, gear loops, or ease of adjustment, than an improvised chest harness.

c. *Full-Body Harness.* Full-body harnesses incorporate a chest and seat harness into one assembly. This is the safest harness to use as it relocates the tie-in point higher, at the chest, reducing the chance of an inverted position when hanging on the rope. This is especially helpful when moving on ropes with heavy packs. A full-body harness only affects the body position when hanging on the rope and will not prevent head injury in a fall.

⚠ **CAUTION**

This type of harness does not prevent the climber from falling headfirst. Body position during a fall is affected only by the forces that generated the fall, and this type of harness promotes an upright position only when hanging on the rope from the attachment point.

6-13. Improvised Harnesses. Without the use of a manufactured harness, many methods are still available for attaching oneself to a rope. Harnesses can be improvised using rope or webbing and knots.

a. *Swami Belt.* The swami belt is a simple, belt-only harness created by wrapping rope or webbing around the waistline and securing the ends. One-inch webbing will provide more comfort. Although an effective swami belt can be assembled with a minimum of one wrap, at least two wraps are recommended for comfort, usually with approximately ten feet of material. The ends are secured with an appropriate knot.

b. *Bowline-on-a-Coil.* Traditionally, the standard method for attaching oneself to the climbing rope was with a bowline-on-a-coil around the waist. The extra wraps distribute the force of a fall over a larger area of the torso than a single bowline would, and help prevent the rope from riding up over the rib cage and under the armpits. The knot must be tied snugly around the narrow part of the waist, just above the bony portions of the hips (pelvis). Avoid crossing the wraps by keeping them spread over the waist area. "Sucking in the gut" a bit when making the wraps will ensure a snug fit.

(1) The bowline-on-a-coil can be used to tie-in to the end of the rope (Figure 6-19). The end man should have a minimum of four wraps around the waist before completing the knot.

(2) The bowline-on-a-coil is a safe and effective method for attaching to the rope when the terrain is low-angled, WITHOUT THE POSSIBILITY OF A SEVERE FALL. When the terrain becomes steeper, a fall will generate more force on the climber and this will be felt through the coils of this type of attachment. A hard fall will cause the coils to ride up against the ribs. In a severe fall, any tie-in around the waist only could place a "shock load" on the climber's lower back. Even in a relatively short fall, if the climber ends up suspended in mid-air and unable to regain footing on the rock, the rope around the waist can easily cut off circulation and breathing in a relatively short time.

(3) The climbing harness distributes the force of a fall over the entire pelvic region, like a parachute harness. Every climber should know how to tie some sort of improvised climbing harness from sling material. A safe, and comfortable, seat/chest combination harness can be tied from one-inch tubular nylon.

c. *Improvised Seat Harness.* A seat harness can be tied from a length of webbing approximately 25 feet long (Figure 6-20).

(1) Locate the center of the rope. Off to one side, tie two fixed loops approximately 6 inches apart (overhand loops). Adjust the size of the loops so they fit snugly around the thigh. The loops are tied into the sling "off center" so the remaining ends are different lengths. The short end should be approximately 4 feet long (4 to 5 feet for larger individuals).

(2) Slip the leg loops over the feet and up to the crotch, with the knots to the front. Make one complete wrap around the waist with the short end, wrapping to the outside, and hold it in place on the hip. Keep the webbing flat and free of twists when wrapping.

(3) Make two to three wraps around the waist with the long end in the opposite direction (wrapping to the outside), binding down on the short end to hold it in place. Grasping both ends,

Figure 6-19: Tying-in with a bowline-on-a-coil.

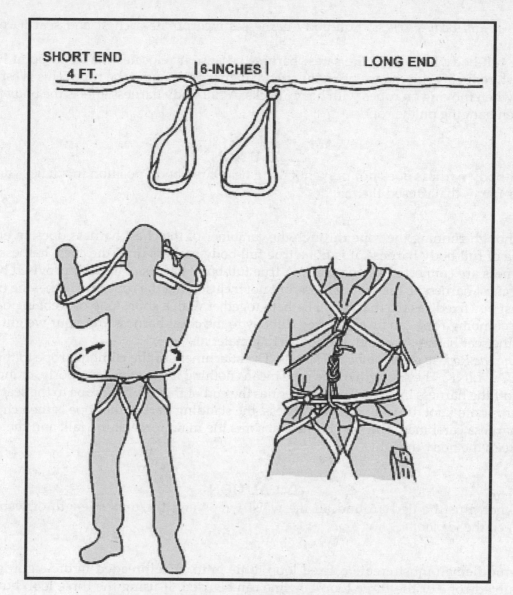

Figure 6-20: Improvised seat and chest harness.

adjust the waist wraps to a snug fit. Connect the ends with the appropriate knot between the front and one side so you will be able to see what you are doing.

d. *Improvised Chest Harness.* The chest harness can be tied from rope or webbing, but remember that with webbing, wider is better and will be more comfortable when you load this harness. Remember as you tie this harness that the remaining ends will need to be secured so choose the best length. Approximately 6 to 10 feet usually works.

(1) Tie the ends of the webbing together with the appropriate knot, making a sling 3 to 4 feet long.

(2) Put a single twist into the sling, forming two loops.

(3) Place an arm through each loop formed by the twist, just as you would put on a jacket, and drape the sling over the shoulders. The twist, or cross, in the sling should be in the middle of the back.

(4) Join the two loops at the chest with a carabiner. The water knot should be set off to either side for easy inspection (if a pack is to be worn, the knot will be uncomfortable if it gets between the body and the pack). The chest harness should fit just loose enough to allow necessary clothing and not to restrict breathing or circulation. Adjust the size of the sling if necessary.

e. *Improvised Full-Body Harness.* Full-body harnesses incorporate a chest and seat harness into one assembly.

   (1) The full-body harness is the safest harness because it relocates the tie-in point higher, at the chest, reducing the chance of an inverted hanging position on the rope. This is especially helpful when moving on ropes with heavy packs. A full-body harness affects the body position only when hanging on the rope.

> ⚠ CAUTION
>
> A full-body harness does not prevent falling head first; body position in a fall is caused by the forces that caused the fall.

   (2) Although running the rope through the carabiner of the chest harness does, in effect, create a type of full-body harness, it is not a true full-body harness until the chest harness and the seat harness are connected as one piece. A true full-body harness can be improvised by connecting the chest harness to the seat harness, but not by just tying the rope into both—the two harnesses must be "fixed" as one harness. Fix them together with a short loop of webbing or rope so that the climbing rope can be connected directly to the chest harness and your weight is supported by the seat harness through the connecting material.

f. *Attaching the Rope to the Improvised Harness.* The attachment of the climbing rope to the harness is a CRITICAL LINK. The strength of the rope means nothing if it is attached poorly, or incorrectly, and comes off the harness in a fall. The climber ties the end of the climbing rope to the seat harness with an appropriate knot. If using a chest harness, the standing part of the rope is then clipped into the chest harness carabiner. The seat harness absorbs the main force of the fall, and the chest harness helps keep the body upright.

> ⚠ CAUTION
>
> The knot must be tied around all the waist wraps and the 6-inch length of webbing between the leg loops.

   (1) A middleman must create a fixed loop to tie in to. A rethreaded figure-eight loop tied on a doubled rope or the three loop bowline can be used. If using the three loop bowline, ensure the end, or third loop formed in the knot, is secured around the tie-in loops with an overhand knot. The standing part of the rope going to the lead climber is clipped into the chest harness carabiner.

Note: The climbing rope is not clipped into the chest harness when belaying.

   (2) The choice of whether to tie-in with a bowline-on-a-coil or into a climbing harness depends entirely on the climber's judgment, and possibly the equipment available. A good rule of thumb is: "Wear a climbing harness when the potential for severe falls exists and for all travel over snow-covered glaciers because of the crevasse fall hazard."

   (3) Under certain conditions many climbers prefer to attach the rope to the seat harness with a locking carabiner, rather than tying the rope to it. This is a common practice for moderate snow and ice climbing, and especially for glacier travel where wet and frozen knots become difficult to untie.

> ⚠ CAUTION
>
> Because the carabiner gate may be broken or opened by protruding rocks during a fall, tie the rope directly to the harness for maximum safety.

BELAY TECHNIQUES

Tying-in to the climbing rope and moving as a member of a rope team increases the climber's margin of safety on difficult, exposed terrain. In some instances, such as when traveling over snow-covered glaciers, rope team members can often move at the same time, relying on the security of a tight rope and "team arrest" techniques to halt a fall by any one member. On steep terrain, however, simultaneous movement only helps to ensure that if one climber falls, he will jerk the other rope team members off the slope. For the climbing rope to be of any value on steep rock climbs, the rope team must incorporate "belays" into the movement.

Belaying is a method of managing the rope in such a way that, if one person falls, the fall can be halted or "arrested" by another rope team member (belayer). One person climbs at a time, while being belayed from above or below by another. The belayer manipulates the rope so that friction, or a "brake," can be applied to halt a fall. Belay techniques are also used to control the descent of personnel and equipment on fixed rope installations, and for additional safety on rappels and stream crossings.

Belaying is a skill that requires practice to develop proficiency. Setting up a belay may at first appear confusing to the beginner, but with practice, the procedure should become "second nature." If confronted with a peculiar problem during the setup of a belay, try to use common sense and apply the basic principles stressed throughout this text.

Remember the following key points:

- Select the best possible terrain features for the position and use terrain to your advantage.
- Use a well braced, sitting position whenever possible.
- Aim and anchor the belay for all possible load directions.
- Follow the "minimum" rule for belay anchors-2 for a downward pull, 1 for an upward pull.
- Ensure anchor attachments are aligned, independent, and snug.
- Stack the rope properly.
- Choose a belay technique appropriate for the climbing.
- Use a guide carabiner for rope control in all body belays.
- Ensure anchor attachments, guide carabiner (if applicable), and rope running to the climber are all on the guidehand side.
- The brake hand remains on the rope when belaying.

> ⚠ CAUTION
>
> Never remove the brake hand from the rope while belaying. If the brake hand is removed, there is no belay.

- Ensure you are satisfied with your position before giving the command "BELAY ON."
- The belay remains in place until the climber gives the command "OFF BELAY."

> ⚠ CAUTION
>
> The belay remains in place from the time the belayer commands *"BELAY ON"* until the climber commands *"OFF BELAY."*

6-14. Procedure for Managing the Rope. A number of different belay techniques are used in modern climbing, ranging from the basic "body belays" to the various "mechanical belays," which incorporate some type of friction device.

a. Whether the rope is wrapped around the body, or run through a friction device, the rope management procedure is basically the same. The belayer must be able to perform three basic functions: manipulate the rope to give the climber slack during movement, take up rope to remove excess slack, and apply the brake to halt a fall.

b. The belayer must be able to perform all three functions while maintaining "total control" of the rope at all times. Total control means the brake hand is NEVER removed from the rope. When giving slack, the rope simply slides through the grasp of the brake hand, at times being fed to the climber with the other "feeling" or guide hand. Taking up rope, however, requires a certain technique to ensure the brake hand remains on the rope at all times. The following procedure describes how to take up excess rope and apply the brake in a basic body belay.

   (1) Grasping the rope with both hands, place it behind the back and around the hips. The hand on the section of rope between the belayer and the climber would be the guide hand. The other hand is the brake hand.

   (2) Take in rope with the brake hand until the arm is fully extended. The guide hand can also help to pull in the rope (Figure 6-21, step 1).

   (3) Holding the rope in the brake hand, slide the guide hand out, extending the arm so the guide hand is farther away from the body than the brake hand (Figure 6-21, step 2).

   (4) Grasp both parts of the rope, to the front of the brake hand, with the guide hand (Figure 6-21, step 3).

   (5) Slide the brake hand back towards the body (Figure 6-21, step 4).

   (6) Repeat step 5 of Figure 6-21. The brake can be applied at any moment during the procedure. It is applied by wrapping the rope around the front of the hips while increasing grip with the brake hand (Figure 6-21, step 6).

6-15. Choosing a Belay Technique. The climber may choose from a variety of belay techniques. A method that works well in one situation may not be the best choice in another. The choice between body belays and mechanical belays depends largely on equipment available, what the climber feels most comfortable with, and the amount of load, or fall force, the belay may have to absorb. The following describes a few of the more widely used techniques, and the ones most applicable to military mountaineering.

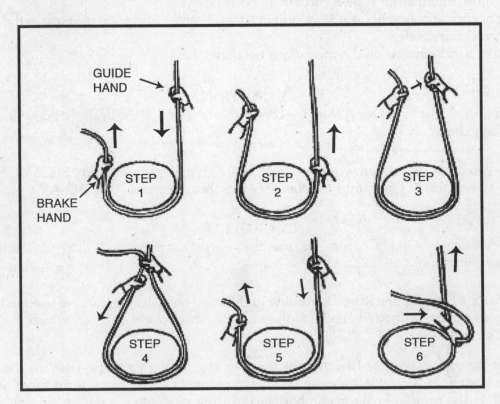

Figure 6-21: Managing the rope.

a. *Body Belay.* The basic body belay is the most widely used technique on moderate terrain. It uses friction between the rope and the clothed body as the rope is pressured across the clothing. It is the simplest belay, requiring no special equipment, and should be the first technique learned by all climbers. A body belay gives the belayer the greatest "feel" for the climber, letting him know when to give slack or take up rope. Rope management in a body belay is quick and easy, especially for beginners, and is effective in snow and ice climbing when ropes often become wet, stiff, and frozen. The body belay, in its various forms, will hold low to moderate impact falls well. It has been known to arrest some severe falls, although probably not without inflicting great pain on the belayer.

> ⚠ **CAUTION**
> The belayer must ensure he is wearing adequate clothing to protect his body from rope burns when using a body belay. Heavy duty cotton or leather work gloves can also be worn to protect the hands.

(1) *Sitting Body Belay.* The sitting body belay is the preferred position and is usually the most secure (Figure 6-22). The belayer sits facing the direction where the force of a fall will likely come from, using terrain to his advantage, and attempts to brace both feet against the rock to support his position. It is best to sit in a slight depression, placing the buttocks lower than the feet, and straightening the legs for maximum support. When perfectly aligned, the rope running to the climber will pass between the belayer's feet, and both legs will equally absorb the force of a fall. Sometimes, the belayer may not be able to sit facing the direction he would like, or both feet cannot be braced well. The leg on the "guide hand" side should then point towards the load, bracing the foot on the rock when possible. The belayer can also "straddle" a large tree or rock nubbin for support, as long as the object is solid enough to sustain the possible load.

Figure 6-22: Sitting body belay.

(2) *Standing Body Belay.* The standing body belay is used on smaller ledges where there is no room for the belayer to sit (Figure 6-23). What appears at first to be a fairly unstable position can actually be quite secure when belay anchors are placed at or above shoulder height to support the stance when the force will be downward.

  (a) For a body belay to work effectively, the belayer must ensure that the rope runs around the hips properly, and remains there under load when applying the brake. The rope should run around the narrow portion of the pelvic girdle, just below the bony high points of the hips. If the rope runs too high, the force of a fall could injure the belayer's midsection and lower rib cage. If the rope runs too low, the load may pull the rope below the buttocks, dumping the belayer out of position. It is also possible for a strong upward or downward pull to strip the rope away from the belayer, rendering the belay useless.

  (b) To prevent any of these possibilities from happening, the belay rope is clipped into a carabiner attached to the guide hand side of the seat harness (or bowline-on-a-coil). This "guide carabiner" helps keep the rope in place around the hips and prevents loss of control in upward or downward loads (Figure 6-24).

b. *Mechanical Belay.* A mechanical belay must be used whenever there is potential for the lead climber to take a severe fall. The holding power of a belay device is vastly superior to any body belay under high loads. However, rope management in a mechanical belay is more difficult to master and requires more practice. For the most part, the basic body belay should be totally adequate on a typical military route, as routes used during military operations should be the easiest to negotiate.

(1) *Munter Hitch.* The Munter hitch is an excellent mechanical belay technique and requires only a rope and a carabiner (Figure 6-25). The Munter is actually a two-way friction hitch. The Munter hitch will flip back and forth through the carabiner as the belayer switches from giving slack to taking up rope. The carabiner must be large enough, and of the proper design, to allow this function. The locking pear-shaped carabiner, or pearabiner, is designed for the Munter hitch.

Figure 6-23: Standing body belay.

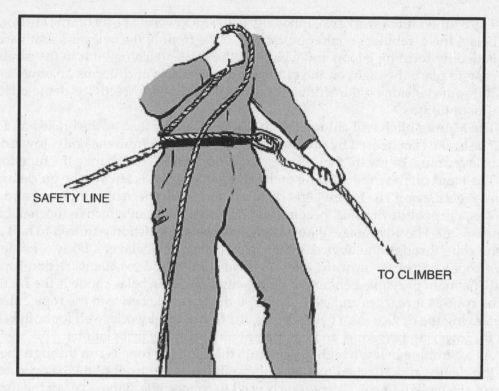

Figure 6-24: Guide carabiner for rope control in a body belay.

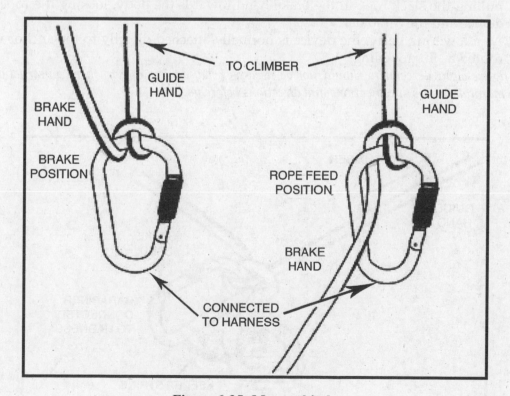

Figure 6-25: Munter hitch.

(a) The Munter hitch works exceptionally well as a lowering belay off the anchor. As a climbing belay, the carabiner should be attached to the front of the belayer's seat harness. The hitch is tied by forming a loop and a bight in the rope, attaching both to the carabiner. It's fairly easy to place the bight on the carabiner backwards, which forms an obvious, useless hitch. Put some tension on the Munter to ensure it is formed correctly, as depicted in the following illustrations.

(b) The Munter hitch will automatically "lock-up" under load as the brake hand grips the rope. The brake is increased by pulling the slack rope away from the body, towards the load. The belayer must be aware that flipping the hitch DOES NOT change the function of the hands. The hand on the rope running to the climber, or load, is always the guide hand.

(2) *Figure-Eight Device.* The figure-eight device is a versatile piece of equipment and, though developed as a rappel device, has become widely accepted as an effective mechanical belay device (Figure 6-26). The advantage of any mechanical belay is friction required to halt a fall is applied on the rope through the device, rather than around the belayer's body. The device itself provides rope control for upward and downward pulls and excellent friction for halting severe falls. The main principle behind the figure-eight device in belay mode is the friction developing on the rope as it reaches and exceeds the 90-degree angle between the rope entering the device and leaving the device. As a belay device, the figure-eight works well for both belayed climbing and for lowering personnel and equipment on fixed-rope installations.

(a) As a climbing belay, a bight placed into the climbing rope is run through the "small eye" of the device and attached to a locking carabiner at the front of the belayer's seat harness. A short, small diameter safety rope is used to connect the "large eye" of the figure eight to the locking carabiner for control of the device. The guide hand is placed on the rope running to the climber. Rope management is performed as in a body belay. The brake is applied by pulling the slack rope in the brake hand towards the body, locking the rope between the device and the carabiner.

(b) As a lowering belay, the device is normally attached directly to the anchor with the rope routed as in rappelling.

Note: Some figure-eight descenders should not be used as belay devices due to their construction and design. Always refer to manufacturer's specifications and directions before use.

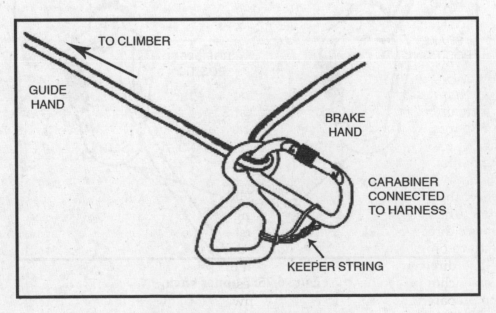

Figure 6-26: Figure-eight device.

(3) *Mechanical Camming Device*. The mechanical camming device has an internal camming action that begins locking the rope in place as friction is increased. Unlike the other devices, the mechanical camming device can stop a falling climber without any input from the belayer. A few other devices perform similarly to this, but have no moving parts. Some limitations to these type devices are minimum and maximum rope diameters.

(4) *Other Mechanical Belay Devices*. There are many other commercially available mechanical belay devices. Most of these work with the same rope movement direction and the same braking principle. The air traffic controller (ATC), slotted plate, and other tube devices are made in many different shapes. These all work on the same principle as the figure-eight device friction increases on the rope as it reaches and exceeds the 90-degree angle between the rope entering the device and leaving the device.

6-16. Establishing a Belay. A belay can be established using either a direct or indirect connection. Each type has advantages and disadvantages. The choice will depend on the intended use of the belay.

a. *Direct Belay*. The direct belay removes any possible forces from the belayer and places this force completely on the anchor. Used often for rescue installations or to bring a second climber up to a new belay position in conjunction with the Munter hitch, the belay can be placed above the belayer's stance, creating a comfortable position and ease of applying the brake. Also, if the second falls or weights the rope, the belayer is not locked into a position. Direct belays provide no shock-absorbing properties from the belayer's attachment to the system as does the indirect belay; therefore, the belayer is apt to pay closer attention to the belaying process.

b. *Indirect Belay*. An indirect belay, the most commonly used, uses a belay device attached to the belayer's harness. This type of belay provides dynamic shock or weight absorption by the belayer if the climber falls or weights the rope, which reduces the direct force on the anchor and prevents a severe shock load to the anchor.

6-17. Setting Up a Belay. In rock climbing, climbers must sometimes make do with marginal protection placements along a route, but belay positions must be made as "bombproof" as possible. Additionally, the belayer must set up the belay in relation to where the fall force will come from and pay strict attention to proper rope management for the belay to be effective. All belay positions are established with the anchor connection to the front of the harness. If the belay is correctly established, the belayer will feel little or no force if the climber falls or has to rest on the rope. Regardless of the actual belay technique used, five basic steps are required to set up a sound belay.

a. *Select Position and Stance*. Once the climbing line is picked, the belayer selects his position. It's best if the position is off to the side of the actual line, putting the belayer out of the direct path of a potential fall or any rocks kicked loose by the climber. The position should allow the belayer to maintain a comfortable, relaxed stance, as he could be in the position for a fairly long time. Large ledges that allow a well braced, sitting stance are preferred. Look for belay positions close to bombproof natural anchors. The position must at least allow for solid artificial placements.

b. *Aim the Belay*. With the belay position selected, the belay must now be "aimed." The belayer determines where the rope leading to the climber will run and the direction the force of a fall will likely come from. When a lead climber begins placing protection, the fall force on the belayer will be in some upward direction, and in line with the first protection placement. If this placement fails under load, the force on the belay could be straight down again. The belayer must aim his belay for all possible load directions, adjusting his position or stance when necessary. The belay can be aimed through an anchor placement to immediately establish an upward pull; however, the belayer must always be prepared for the more severe downward fall force in the event intermediate protection placements fail.

c. *Anchor the Belay.* For a climbing belay to be considered bombproof, the belayer must be attached to a solid anchor capable of withstanding the highest possible fall force. A solid natural anchor would be ideal, but more often the belayer will have to place pitons or chocks. A single artificial placement should never be considered adequate for anchoring a belay (except at ground level). Multiple anchor points capable of supporting both upward and downward pulls should be placed. The rule of thumb is to place two anchors for a downward pull and one anchor for an upward pull as a MINIMUM. The following key points also apply to anchoring belays.

(1) Each anchor must be placed in line with the direction of pull it is intended to support.

(2) Each anchor attachment must be rigged "independently" so a failure of one will not shock load remaining placements or cause the belayer to be pulled out of position.

(3) The attachment between the anchor and the belayer must be snug to support the stance. Both belayer's stance and belay anchors should absorb the force of a fall.

(4) It is best for the anchors to be placed relatively close to the belayer with short attachments. If the climber has to be tied-off in an emergency, say after a severe fall, the belayer can attach a Prusik sling to the climbing rope, reach back, and connect the sling to one of the anchors. The load can be placed on the Prusik and the belayer can come out of the system to render help.

(5) The belayer can use either a portion of the climbing rope or slings of the appropriate length to connect himself to the anchors. It's best to use the climbing rope whenever possible, saving the slings for the climb. The rope is attached using either figure eight loops or clove hitches. Clove hitches have the advantage of being easily adjusted. If the belayer has to change his stance at some point, he can reach back with the guide hand and adjust the length of the attachment through the clove hitch as needed.

(6) The anchor attachments should also help prevent the force of a fall from "rotating" the belayer out of position. To accomplish this, the climbing rope must pass around the "guide-hand side" of the body to the anchors. Sling attachments are connected to the belayer's seat harness (or bowline-on-a-coil) on the guide-hand side.

(7) Arrangement of rope and sling attachments may vary according to the number and location of placements. Follow the guidelines set forth and remember the key points for belay anchors; "in line", "independent", and "snug". Figure 6-27 shows an example of a common arrangement, attaching the rope to the two "downward" anchors and a sling to the "upward" anchor. Note how the rope is connected from one of the anchors back to the belayer. This is not mandatory, but often helps "line-up" the second attachment.

d. *Stack the Rope.* Once the belayer is anchored into position, he must stack the rope to ensure it is free of twists and tangles that might hinder rope management in the belay. The rope should be stacked on the ground, or on the ledge, where it will not get caught in cracks or nubbins as it is fed out to the climber.

(1) On small ledges, the rope can be stacked on top of the anchor attachments if there is no other place to lay it, but make sure to stack it carefully so it won't tangle with the anchored portion of the rope or other slings. The belayer must also ensure that the rope will not get tangled around his legs or other body parts as it "feeds" out.

(2) The rope should never be allowed to hang down over the ledge. If it gets caught in the rock below the position, the belayer may have to tie-off the climber and come out of the belay to free the rope; a time-consuming and unnecessary task. The final point to remember is the rope must be stacked "from the belayer's end" so the rope running to the climber comes off the "top" of the stacked pile.

e. *Attach the Belay.* The final step of the procedure is to attach the belay. With the rope properly stacked, the belayer takes the rope coming off the top of the pile, removes any slack between himself and the climber, and applies the actual belay technique. If using a body belay, ensure the rope is clipped into the guide carabiner.

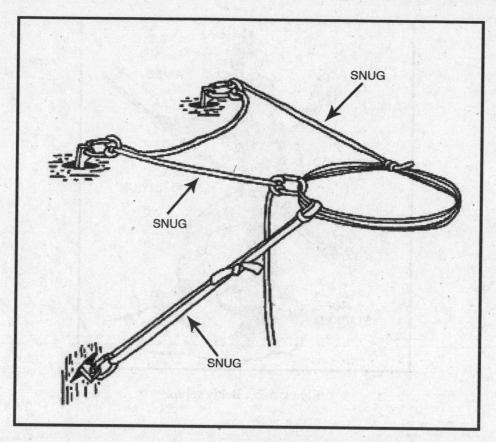

Figure 6-27: Anchoring a belay.

(1) The belayer should make one quick, final inspection of his belay. If the belay is set up correctly, the anchor attachments, guide carabiner if applicable, and the rope running to the climber will all be on the "guide hand" side, which is normally closest to the rock (Figure 6-28). If the climber takes a fall, the force, if any, should not have any negative effect on the belayer's involvement in the system. The brake hand is out away from the slope where it won't be jammed between the body and the rock. The guide hand can be placed on the rock to help support the stance when applying the brake.

(2) When the belayer is satisfied with his position, he gives the signal, "BELAY ON!". When belaying the "second", the same procedure is used to set up the belay. Unless the belay is aimed for an upward pull, the fall force is of course downward and the belayer is usually facing away from the rock, the exception being a hanging belay on a vertical face. If the rope runs straight down to the climber and the anchors are directly behind the position, the belayer may choose to brake with the hand he feels most comfortable with. Anchor attachments, guide carabiner, and rope running to the climber through the guide hand must still be aligned on the same side to prevent the belayer from being rotated out of position, unless the belayer is using an improvised harness and the anchor attachment is at the rear.

6-18. Top-Rope Belay. A "top-rope" is a belay setup used in training to protect a climber while climbing on longer, exposed routes. A solid, bombproof anchor is required at the top of the pitch. The belayer is positioned either on the ground with the rope running through the top anchor and back to the climber, or at the top at the anchor. The belayer takes in rope as the climber proceeds up the rock. If this is accomplished with the belayer at the bottom, the instructor can watch the belayer while he coaches the climber through the movements.

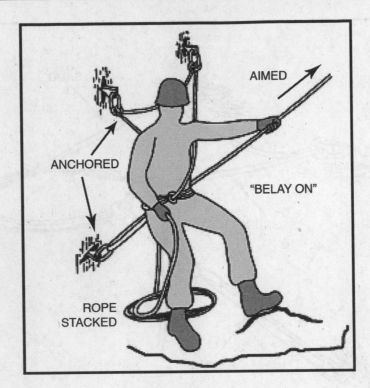

Figure 6-28: Belay setup.

⚠ CAUTION

Do not use a body belay for top-rope climbing. The rope will burn the belayer if the climber has to be lowered.

CLIMBING COMMANDS

Communication is often difficult during a climb. As the distance between climber and belayer increases, it becomes harder to distinguish one word from another and the shortest sentence may be heard as nothing more than jumbled syllables. A series of standard voice commands were developed over the years to signal the essential rope management functions in a belayed climb. Each command is concise and sounds a bit different from another to reduce the risk of a misunderstanding between climber and belayer. They must be pronounced clearly and loudly so they can be heard and understood in the worst conditions.

6-19. Verbal Commands. Table 6-1 lists standard rope commands and their meanings in sequence as they would normally be used on a typical climb. (Note how the critical "BELAY" commands are reversed so they sound different and will not be confused.)

6-20. Rope Tug Commands. Sometimes the loudest scream cannot be heard when the climber and belayer are far apart. This is especially true in windy conditions, or when the climber is around a corner, above an overhang, or at the back of a ledge. It may be necessary to use a series of "tugs" on the rope in place of the standard voice commands. To avoid any possible confusion with interpretation of multiple rope tug commands, use only one.

    a. While a lead climb is in progress, the most important command is "BELAY ON." This command is given only by the climber when the climber is anchored and is prepared for the second to begin

Table 6-1: Rope commands.

BELAYER	CLIMBER	MEANING/ACTION TAKEN
"BELAY ON"		The belay is on; you may climb when ready; the rope will be managed as needed.
	"CLIMBING" (as a courtesy)	I am ready to climb.
"CLIMB" (as a courtesy)		Proceed, and again, the rope will be managed as necessary.
"ROCK"	"ROCK"	**PROTECT YOURSELF FROM FALLING OBJECTS.** Signal will be echoed by all climbers in the area. If multipitch climbing, ensure climbers below hear.
	"TAKE ROPE"	Take in excess rope between us without pulling me off the route. Belayer takes in rope.
	"SLACK"	Release all braking/tension on the rope so I can have slack without pulling the rope. Belayer removes brake/tension.
	"TENSION"	Take all the slack, apply brake, and hold me. My weight will be on the rope. Belayer removes slack and applies brake.
	"FALLING"	I am falling. Belayer applies brake to arrest the fall.
"TWEN-TY-FIVE"		You have approximately 25 feet of rope left. Start looking for the next belay position. Climber selects a belay position.
"FIF-TEEN"		You have approximately 15 feet of rope left. Start looking for the next belay position. Climber selects a belay position within the next few feet.
"FIVE"	Set up the belay.	You have 5 feet of rope left. Set up the belay position. You have no more rope. Climber sets up the belay.
Removes the belay, remains anchored. Prepares to climb.	"OFF BELAY"	I have finished climbing and I am anchored. You may remove the belay. Belayer removes the belay and, remaining anchored, prepares to climb.

climbing. With the issue of this command, the second knows the climber is anchored and the second prepares to climb.

b. For a rope tug command, the leader issues three distinct tugs on the rope AFTER anchoring and putting the second on belay. This is the signal for "BELAY ON" and signals the second to climb when ready. The new belayer keeps slack out of the rope.

ROPED CLIMBING METHODS

In military mountaineering, the primary mission of a roped climbing team is to "fix" a route with some type of rope installation to assist movement of less trained personnel in the unit. This duty falls upon the most experienced climbers in the unit, usually working in two- or three-man groups or teams called assault climbing teams. Even if the climbing is for another purpose, roped climbing should be performed whenever the terrain becomes difficult and exposed.

6-21. Top-Roped Climbing. Top-roped climbing is used for training purposes only. This method of climbing is not used for movement due to the necessity of pre-placing anchors at the top of a climb. If you can easily access the top of a climb, you can easily avoid the climb itself.

a. For training, top-roped climbing is valuable because it allows climbers to attempt climbs above their skill level and or to hone present skills without the risk of a fall. Top-roped climbing may be used to increase the stamina of a climber training to climb longer routes as well as for a climber practicing protection placements.

b. The belayer is positioned either at the base of a climb with the rope running through the top anchor and back to the climber or at the top at the anchor. The belayer takes in rope as the climber moves up the rock, giving the climber the same protection as a belay from above. If this is accomplished with the belayer at the bottom, the instructor is able to keep an eye on the belayer while he coaches the climber through the movements.

6-22. Lead Climbing. A lead climb consists of a belayer, a leader or climber, rope(s), and webbing or hardware used to establish anchors or protect the climb. As he climbs the route, the leader emplaces "intermediate" anchors, and the climbing rope is connected to these anchors with a carabiner. These "intermediate" anchors protect the climber against a fall—thus the term "protecting the climb."

Note: Intermediate anchors are commonly referred to as "protection," "pro," "pieces," "pieces of pro," "pro placements," and so on. For standardization within this publication, these specific anchors will be referred to as "protection;" anchors established for other purposes, such as rappel points, belays, or other rope installations, will be referred to as "anchors."

> ⚠ CAUTION
> During all lead climbing, each climber in the team is either anchored or being belayed.

a. Lead climbing with two climbers is the preferred combination for movement on technically difficult terrain. Two climbers are at least twice as fast as three climbers, and are efficient for installing a "fixed rope," probably the most widely used rope installation in the mountains. A group of three climbers are typically used on moderate snow, ice, and snow-covered glaciers where the rope team can often move at the same time, stopping occasionally to set up belays on particularly difficult sections. A group or team of three climbers is sometimes used in rock climbing because of an odd number of personnel, a shortage of ropes (such as six climbers and only two ropes), or to protect and assist an individual who has little or no experience in climbing and belaying.

Whichever technique is chosen, a standard roped climbing procedure is used for maximum speed and safety.

b. When the difficulty of the climbing is within the "leading ability" of both climbers, valuable time can be saved by "swinging leads." This is normally the most efficient method for climbing multi-pitch routes. The second finishes cleaning the first pitch and continues climbing, taking on the role of lead climber. Unless he requires equipment from the other rack or desires a break, he can climb past the belay and immediately begin leading. The belayer simply adjusts his position, re-aiming the belay once the new leader begins placing protection. Swinging leads, or "leapfrogging," should be planned before starting the climb so the leader knows to anchor the upper belay for both upward and downward pulls during the setup.

c. The procedures for conducting a lead climb with a group of two are relatively simple. The most experienced individual is the "lead" climber or leader, and is responsible for selecting the route. The leader must ensure the route is well within his ability and the ability of the second. The lead climber carries most of the climbing equipment in order to place protection along the route and set up the next belay. The leader must also ensure that the second has the necessary equipment, such as a piton hammer, nut tool, etc., to remove any protection that the leader may place.

(1) The leader is responsible for emplacing protection frequently enough and in such a manner that, in the event that either the leader or the second should fall, the fall will be neither long enough nor hard enough to result in injury. The leader must also ensure that the rope is routed in a way that will allow it to run freely through the protection placements, thus minimizing friction, or "rope drag".

(2) The other member of the climbing team, the belayer (sometimes referred to as the "second"), is responsible for belaying the leader, removing the belay anchor, and retrieving the protection placed by the leader between belay positions (also called "cleaning the pitch").

(3) Before the climb starts, the second will normally set up the first belay while the leader is arranging his rack. When the belay is ready, the belayer signals, "BELAY ON", affirming that the belay is "on" and the rope will be managed as necessary. When the leader is ready, he double checks the belay. The leader can then signal, "CLIMBING", only as a courtesy, to let the belayer know he is ready to move. The belayer can reply with "CLIMB", again, only as a courtesy, reaffirming that the belay is "on" and the rope will be managed as necessary. The leader then begins climbing.

(4) While belaying, the second must pay close attention to the climber's every move, ensuring that the rope runs free and does not inhibit the climber's movements. If he cannot see the climber, he must "feel" the climber through the rope. Unless told otherwise by the climber, the belayer can slowly give slack on the rope as the climber proceeds on the route. The belayer should keep just enough slack in the rope so the climber does not have to pull it through the belay. If the climber wants a tighter rope, it can be called for. If the belayer notices too much slack developing in the rope, the excess rope should be taken in quickly. It is the belayer's responsibility to manage the rope, whether by sight or feel, until the climber tells him otherwise.

(5) As the leader protects the climb, slack will sometimes be needed to place the rope through the carabiner (clipping), in a piece of protection above the tie-in point on the leader's harness. In this situation, the leader gives the command "SLACK" and the belayer gives slack, (if more slack is needed the command will be repeated). The leader is able to pull a bight of rope above the tie-in point and clip it into the carabiner in the protection above. When the leader has completed the connection, or the clip, the command "TAKE ROPE" is given by the leader and the belayer takes in the remaining slack.

(6) The leader continues on the route until either a designated belay location is reached or he is at the end of or near the end of the rope. At this position, the leader sets an anchor, connects to the anchor and signals "OFF BELAY". The belayer prepares to climb by removing all but at least one of his anchors and secures the remaining equipment. The belayer remains attached to at least one anchor until the command "BELAY ON" is given.

d. When the leader selects a particular route, he must also determine how much, and what types, of equipment might be required to safely negotiate the route. The selected equipment must be carried by the leader. The leader must carry enough equipment to safely protect the route, additional anchors for the next belay, and any other items to be carried individually such as rucksacks or individual weapons.

(1) The leader will assemble, or "rack," the necessary equipment onto his harness or onto slings around the head and shoulder. A typical leader "rack" consists of:

- Six to eight small wired stoppers on a carabiner.
- Four to six medium to large wired stoppers on a carabiner.
- Assorted hexentrics, each on a separate carabiner.
- SLCDs of required size, each on a separate carabiner.
- Five to ten standard length runners, with two carabiners on each.
- Two to three double length runners, with two carabiners on each.
- Extra carabiners.
- Nut tool.

Note: The route chosen will dictate, to some degree, the necessary equipment. Members of a climbing team may need to consolidate gear to climb a particular route.

(2) The belayer and the leader both should carry many duplicate items while climbing.

- Short Prusik sling.
- Long Prusik sling.
- Cordellette.
- 10 feet of 1-inch webbing.
- 20 feet of 1-inch webbing.
- Belay device (a combination belay/rappel device is multifunctional).
- Rappel device (a combination belay/rappel device is multifunctional).
- Large locking carabiner (pear shape carabiners are multifunctional).
- Extra carabiners.
- Nut tool (if stoppers are carried).

Note: If using an over the shoulder gear sling, place the items in order from smallest to the front and largest to the rear.

e. Leading a difficult pitch is the most hazardous task in roped climbing. The lead climber may be exposed to potentially long, hard falls and must exercise keen judgment in route selection, placement of protection, and routing of the climbing rope through the protection. The leader should try to keep the climbing line as direct as possible to the next belay to allow the rope to run smoothly through the protection with minimal friction. Protection should be placed whenever the leader feels he needs it, and BEFORE moving past a difficult section.

⚠ CAUTION

The climber must remember he will fall twice the distance from his last piece of protection before the rope can even begin to stop him.

---

(1) *Placing Protection.* Generally, protection is placed from one stable position to the next. The anchor should be placed as high as possible to reduce the potential fall distance between placements. If the climbing is difficult, protection should be placed more frequently. If the climbing becomes easier, protection can be placed farther apart, saving hardware for difficult sections. On some routes an extended diagonal or horizontal movement, known as a traverse, is required. As the leader begins this type of move, he must consider the second's safety as well as his own. The potential fall of the second will result in a pendulum swing if protection is not adequate to prevent this. The danger comes from any objects in the swinging path of the second.

Leader should place protection prior to, during, and upon completion of any traverse. This will minimize the potential swing, or pendulum, for both the leader and second if either should fall.

(2) *Correct Clipping Technique.* Once an anchor is placed, the climber "clips" the rope into the carabiner (Figure 6-29). As a carabiner hangs from the protection, the rope can be routed through the carabiner in two possible ways. One way will allow the rope to run smoothly as the climber moves past the placement; the other way will often create a dangerous situation in which the rope could become "unclipped" from the carabiner if the leader were to fall on this piece of protection. In addition, a series of incorrectly clipped carabiners may contribute to rope drag. When placing protection, the leader must ensure the carabiner on the protection does not hang with the carabiner gate facing the rock; when placing protection in a crack ensure the carabiner gate is not facing into the crack.

- Grasp the rope with either hand with the thumb pointing down the rope towards the belayer.
- Pull enough rope to reach the carabiner with a bight.
- Note the direction the carabiner is hanging from the protection.
- Place the bight into the carabiner so that, when released, the rope does not cause the carabiner to twist.

(a) If the route changes direction, clipping the carabiner will require a little more thought. Once leaving that piece of protection, the rope may force the carabiner to twist if not correctly clipped. If the clip is made correctly, a rotation of the clipped carabiner to ensure that the gate is not resting against the rock may be all that is necessary.

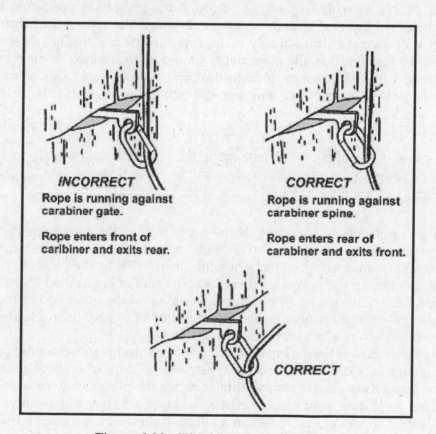

Figure 6-29: Clipping on to protection.

⚠ **CAUTION**

Ensure the carabiner gate is not resting against a protrusion or crack edge in the rock surface; the rock may cause the gate to open.

(b) Once the rope is clipped into the carabiner, the climber should check to see that it is routed correctly by pulling on the rope in the direction it will travel when the climber moves past that position.

(c) Another potential hazard peculiar to leading should be eliminated before the climber continues. The carabiner is attached to the anchor or runner with the gate facing away from the rock and opening down for easy insertion of the rope. However, in a leader fall, it is possible for the rope to run back over the carabiner as the climber falls below the placement. If the carabiner is left with the gate facing the direction of the route there is a chance that the rope will open the gate and unclip itself entirely from the placement. To prevent this possibility, the climber should ensure that after the clip has been made, the gate is facing away from the direction of the route. There are two ways to accomplish this: determine which direction the gate will face before the protection or runner is placed or once clipped, rotate the carabiner upwards 180 degrees. This problem is more apt to occur if bent gate carabiners are used. Straight gate ovals or "Ds" are less likely to have this problem and are stronger and are highly recommended. Bent gate carabiners are easier to clip the rope into and are used mostly on routes with bolts preplaced for protection. Bent gate carabiners are not recommended for many climbing situations.

(3) *Reducing Rope Drag; Using Runners.* No matter how direct the route, the climber will often encounter problems with "rope drag" through the protection positions. The friction created by rope drag will increase to some degree every time the rope passes through a carabiner, or anchor. It will increase dramatically if the rope begins to "zigzag" as it travels through the carabiners. To prevent this, the placements should be positioned so the rope creates a smooth, almost straight line as it passes through the carabiners (Figure 6-30). Minimal rope drag is an inconvenience; severe rope drag may actually pull the climber off balance, inducing a fall.

⚠ **CAUTION**

Rope drag can cause confusion when belaying the second or follower up to a new belay position. Rope drag can be mistaken for the climber, causing the belayer to not take in the necessary slack in the rope and possibly resulting in a serious fall.

(a) If it is not possible to place all the protection so the carabiners form a straight line as the rope moves through, you should "extend" the protection (Figure 6-31). Do this by attaching an appropriate length sling, or runner, to the protection to extend the rope connection in the necessary direction. The runner is attached to the protection's carabiner while the rope is clipped into a carabiner at the other end of the runner. Extending placements with runners will allow the climber to vary the route slightly while the rope continues to run in a relatively straight line.

(b) Not only is rope drag a hindrance, it can cause undue movement of protection as the rope tightens between any "out of line" placements. Rope drag through chock placements can be dangerous. As the climber moves above the placements, an outward or upward pull from rope drag may cause correctly set chocks to pop out, even when used "actively". Most all chocks placed for leader protection should be extended with a runner, even if the line is direct, to eliminate the possibility of movement.

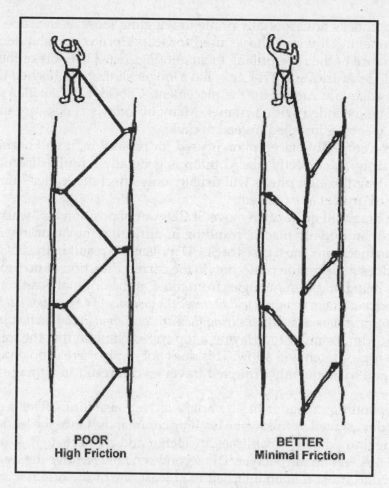

Figure 6-30: Use of slings on protection.

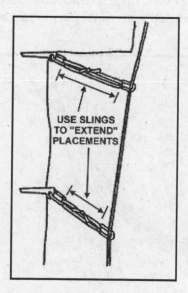

Figure 6-31: Use of slings to extend placement positions.

(c) Wired chocks are especially prone to wiggling loose as the rope pulls on the stiff cable attachment. All wired chocks used for leader protection should be extended to reduce the chance of the rope pulling them out (Figure 6-32). Some of the larger chocks, such as roped Hexentrics and Tri-Cams, have longer slings pre-attached that will normally serve as an adequate runner for the placement. Chocks with smaller sling attachments must often be extended with a runner. Many of today's chocks are manufactured with pre-sewn webbing installed instead of cable.

(d) When a correctly placed piton is used for protection, it will normally not be affected by rope drag. A correctly placed piton is generally a multi-directional anchor, therefore, rope drag through pitons will usually only affect the leader's movements but will continue to protect as expected.

(e) Rope drag will quite often move SLCDs out of position, or "walk" them deeper into the crack than initially placed, resulting in difficult removal or inability to remove them at all. Furthermore, most cases of SLCD movement result in the SLCD moving to a position that does not provide protection in the correct direction or no protection at all due to the lobes being at different angles from those at the original position.

Note: Any placement extended with a runner will increase the distance of a potential fall by the actual length of the sling. Try to use the shortest runners possible, ensuring they are long enough to function properly.

f. Belaying the follower is similar to belaying a top-roped climb in that the follower is not able to fall any farther than rope stretch will allow. This does not imply there is no danger in following. Sharp rocks, rock fall, and inadequately protected traverses can result in damage to equipment or injury to the second.

g. Following, or seconding, a leader has a variety of responsibilities. The second has to issue commands to the leader, as well as follow the leader's commands. Once the lead climber reaches a good belay position, he immediately establishes an anchor and connects to it. When this is completed he can signal "OFF BELAY" to the belayer. The second can now remove the leader's belay and prepare to climb. The second must remain attached to at least one of the original anchors while the leader is preparing the next belay position. The removed materials and hardware can be organized and secured on the second's rack in preparation to climb.

(1) When the leader has established the new belay position and is ready to belay the follower, the "new" belayer signals "BELAY ON." The second, now the climber, removes any remaining anchor hardware/materials and completes any final preparations. The belayer maintains tension on the

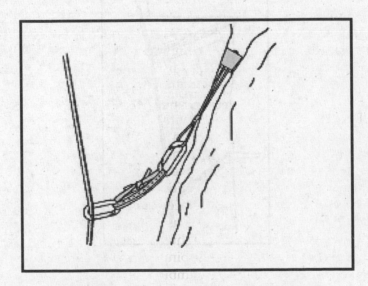

Figure 6-32: Use of sling on a wired stopper.

rope, unless otherwise directed, while the final preparations are taking place, since removal of these remaining anchors can introduce slack into the rope. When the second is ready, he can, as a courtesy, signal "CLIMBING," and the leader can, again as a courtesy, reply with "CLIMB."

(2) Upon signaling "BELAY ON," the belayer must remove and keep all slack from the rope. (This is especially important as in many situations the belayer cannot see the follower. A long pitch induces weight and sometimes "drag" on the rope and the belayer above will have difficulty distinguishing these from a rope with no slack.)

h. When removing protection, the man cleaning the pitch should rack it properly to facilitate the exchange and or arrangement of equipment at the end of the pitch. When removing the protection, or "cleaning the pitch," SLCDs or chocks may be left attached to the rope to prevent loss if they are accidentally dropped during removal. If necessary, the hardware can remain on the rope until the second reaches a more secure stance. If removing a piton, the rope should be unclipped from the piton to avoid the possibility of damaging the rope with a hammer strike.

(1) The second may need to place full body weight on the rope to facilitate use of both hands for protection removal by giving the command "TENSION." The second must also ensure that he does not climb faster than the rope is being taken in by the belayer. If too much slack develops, he should signal "TAKE ROPE" and wait until the excess is removed before continuing the climb. Once the second completes the pitch, he should immediately connect to the anchor. Once secured, he can signal "OFF BELAY." The leader removes the belay, while remaining attached to an anchor. The equipment is exchanged or organized in preparation for the next pitch or climb.

(2) When the difficulty of the climbing is within the "leading ability" of both climbers, valuable time can be saved by "swinging leads." This is normally the most efficient method for climbing multi-pitch routes. The second finishes cleaning the first pitch and continues climbing, taking on the role of lead climber. Unless he requires equipment from the belayer or desires a break, he can climb past the belay and immediately begin leading. The belayer simply adjusts his position, re-aiming the belay once the new leader begins placing protection. Swinging leads, or "leap frogging," should be planned before starting the climb so the leader knows to anchor the upper belay for both upward and downward pulls during the setup.

6-23. Aid Climbing. When a route is too difficult to free climb and is unavoidable, if the correct equipment is available you might aid climb the route. Aid climbing consists of placing protection and putting full body weight on the piece. This allows you to hang solely on the protection you place, giving you the ability to ascend more difficult routes than you can free climb. Clean aid consists of using SLCDs and chocks, and is the simplest form of aid climbing.

a. *Equipment.* Aid climbing can be accomplished with various types of protection. Regardless of the type of protection used, the method of aid climbing is the same. In addition to the equipment for free climbing, other specialized equipment will be needed.

(1) *Pitons.* Pitons are used the same as for free climbing. Most piton placements will require the use of both hands. Piton usage will usually leave a scar in the rock just by virtue of the hardness of the piton and the force required to set it with a hammer. Swinging a hammer to place pitons will lead to climber fatigue sooner than clean aid. Since pitons are multidirectional, the strength of a well-placed piton is more secure than most clean aid protection. Consider other forms of protection when noise could be hazardous to tactics.

(2) *Bolts.* Bolts are used when no other protection will work. They are a more permanent form of protection and more time is needed to place them. Placing bolts creates more noise whether drilled by hand or by motorized drill. Bolts used in climbing are a multi-part expanding system pounded into predrilled holes and then tightened to the desired torque with a wrench or other tool. Bolts are used in many ways in climbing today. The most common use is with a hanger attached and placed for anchors in face climbing. However, bolts can be used for aid climbing, with or without the hanger.

(a) Placing bolts for aid climbing takes much more time than using pitons or clean aid. Bolting for aid climbing consists of consecutive bolts about 2 feet apart. Drilling a deep enough hole takes approximately thirty minutes with a hand drill and up to two minutes with a powered hammer drill. A lot of time and work is expended in a short distance no matter how the hole is drilled. (The weight of a powered hammer drill becomes an issue in itself.) Noise will also be a factor in both applications. A constant pounding with a hammer on the hand drill or the motorized pounding of the powered drill may alert the enemy to the position. The typical climbing bolt/hanger combination normally is left in the hole where it was placed.

(b) Other items that can be used instead of the bolt/hanger combination are the removable and reusable "spring-loaded removable bolts" such as rivets (hex head threaded bolts sized to fit tightly into the hole and pounded in with a hammer), split-shaft rivets, and some piton sizes that can be pounded into the holes. When using rivets or bolts without a hanger, place a loop of cable over the head and onto the shaft of the rivet or bolt and attach a carabiner to the other end of the loop (a stopper with the chock slid back will suffice). Rivet hangers are available that slide onto the rivet or bolt after it is placed and are easily removed for reuse. Easy removal means a slight loss of security while in use.

(3) *SLCDs.* SLCDs are used the same as for free climbing, although in aid climbing, full bodyweight is applied to the SLCD as soon as it is placed.

(4) Chocks. Chocks are used the same as for free climbing, although in aid climbing, weight is applied to the chock as soon as it is placed.

(5) Daisy Chains. Daisy chains are tied or presewn loops of webbing with small tied or presewn loops approximately every two inches. The small loops are just large enough for two or three carabiners. Two daisy chains should be girth-hitched to the tie-in point in the harness.

(6) Etriers (or Aiders). Etriers (aiders) are tied or presewn webbing loops with four to six tied or presewn internal loops, or steps, approximately every 12 inches. The internal loops are large enough to easily place one booted foot into. At least two etriers (aiders) should be connected by carabiner to the free ends of the daisy chains.

(7) Fifi Hook. A fifi hook is a small, smooth-surfaced hook strong enough for body weight. The fifi hook should be girth-hitched to the tie-in point in the harness and is used in the small loops of the daisy chain. A carabiner can be used in place of the fifi hook, although the fifi hook is simpler and adequate.

(8) Ascenders. Ascenders are mechanical devices that will move easily in one direction on the rope, but will lock in place if pushed or pulled the other direction. (Prusiks can be used but are more difficult than ascenders.)

b. *Technique.* The belay will be the same as in normal lead climbing and the rope will be routed through the protection the same way also. The big difference is the movement up the rock. With the daisy chains, aiders, and fifi hook attached to the rope tie-in point of the harness as stated above, and secured temporarily to a gear loop or gear sling, the climb continues as follows:

(1) The leader places the first piece of protection as high as can safely be reached and attaches the appropriate sling/carabiner.

(2) Attach one daisy chain/aider group to the newly placed protection.

(3) Clip the rope into the protection, (the same as for normal lead climbing).

(4) Insure the protection is sound by weighting it gradually; place both feet, one at a time, into the steps in the aider, secure your balance by grasping the top of the aider with your hands.

(5) When both feet are in the aider, move up the steps until your waist is no higher than the top of the aider.

(6) Place the fifi hook (or substituted carabiner) into the loop of the daisy chain closest to the daisy chain/aider carabiner, this effectively shortens the daisy chain; maintain tension on the daisy chain as the hook can fall out of the daisy chain loop if it is unweighted.

Note: Moving the waist higher than the top of the aider is possible, but this creates a potential for a fall to occur even though you are on the aider and "hooked" close to the protection with the daisy chain. As the daisy chain tie-in point on the harness moves above the top of the aider, you are no longer supported from above by the daisy chain, you are now standing above your support. From this height, the fifi hook can easily fall out of the daisy chain loop if it is unweighted. If this happens, you could fall the full length of the daisy chain resulting in a static fall on the last piece of protection placed.

    (7) Release one hand from the aider and place the next piece of protection, again, as high as you can comfortably reach; if using pitons or bolts you may need both hands free- "lean" backwards slowly, and rest your upper body on the daisy chain that you have "shortened" with the fifi hook.

    (8) Clip the rope into the protection.

    (9) Attach the other daisy chain/aider group to the next piece of protection.

    (10) Repeat entire process until climb is finished.

  c. *Seconding.* When the pitch is completed, the belayer will need to ascend the route. To ascend the route, use ascenders instead of Prusiks; ascenders are much faster and safer than Prusiks. Attach each ascender to a daisy chain/aider group with carabiners. To adjust the maximum reach/height of the ascenders on the rope, adjust the effective length of the daisy chains with a carabiner the same as with the fifi hook; the typical height will be enough to hold the attached ascender in the hand at nose level. When adjusted to the correct height, the arms need not support much body weight. If the ascender is too high, you will have difficulty reaching and maintaining a grip on the handle.

    (1) Unlike lead climbing, there will be a continuous load on the rope during the cleaning of the route; this would normally increase the difficulty of removing protection. To make this easier, as you approach the protection on the ascenders, move the ascenders, one at a time, above the piece. When your weight is on the rope above the piece, you can easily unclip and remove the protection.

> ### ⚠ CAUTION
> If both ascenders should fail while ascending the pitch, a serious fall could result. To prevent this possibility, tie-in short on the rope every 10-20 feet by tying a figure eight loop and clipping it into the harness with a separate locking carabiner as soon as the ascent is started. After ascending another 20 feet, repeat this procedure. Do not unclip the previous figure eight until the new knot is attached to another locking carabiner. Clear each knot as you unclip it.

Notes: 1. Ensure the loops formed by the short tie-ins do not catch on anything below as you ascend.
      2. If the nature of the rock will cause the "hanging loop" of rope, formed by tying in at the end of the rope, to get caught as you move upward, do not tie into the end of the rope.

    (2) Seconding an aid pitch can be done in a similar fashion as seconding free-climbed pitches. The second can be belayed from above as the second "climbs" the protection. However, the rope is unclipped from the protection before the aider/daisy chain is attached.

  d. *Seconding Through a Traverse.* While leading an aid traverse, the climber is hanging on the protection placed in front of the current position. If the second were to clean the section by hanging on the rope while cleaning, the protection will be pulled in more than one direction, possibly resulting in the protection failing. To make this safer and easier, the second should hang on the protection just as the leader did. As the second moves to the beginning of the traverse, one ascender/daisy chain/aider group is removed from the rope and clipped to the protection with a carabiner, (keep the ascenders attached to the daisy chain/aider group for convenience when the traverse ends). The second will negotiate the traverse by leapfrogging the daisy chain/aider groups on the next protection just as the leader did. Cleaning is accomplished by removing the protection; a sit is passed when all weight is removed from it. This is in effect a self-belay. The second maintains a shorter safety tie-in on the rope than for vertical movement to reduce the possibility of a lengthy pendulum if the protection should pull before intended.

e. *Clean Aid Climbing.* Clean aid climbing consists of using protection placed without a hammer or drill involvement: chocks, SLCDs, hooks, and other protection placed easily by hand. This type of aid climbing will normally leave no trace of the climb when completed. When climbing the aiders on clean aid protection, ensure the protection does not "move" from its original position.

(1) Hooks are any device that rests on the rock surface without a camming or gripping action. Hooks are just what the name implies, a curved piece of hard steel with a hole in one end for webbing attachment. The hook blade shape will vary from one model to another, some have curved or notched "blades" to better fit a certain crystal shape on a face placement. These types of devices, due to their passive application, are only secure while weighted by the climber.

(2) Some featureless sections of rock can be negotiated with hook use, although bolts can be used. Hook usage is faster and quieter but the margin of safety is not there unless hooks are alternated with more active forms of protection. If the last twenty foot section of a route is negotiated with hooks, a forty foot fall could result.

6-24. Three-Man Climbing Team. Often times a movement on steep terrain will require a team of more than two climbers, which involves more difficulties. A four-man team (or more) more than doubles the difficulty found in three men climbing together. A four-man team should be broken down into two groups of two unless prevented by a severe lack of gear.

a. Given one rope, a three-man team is at a disadvantage on a steep, belayed climb. It takes at least twice as long to climb an average length pitch because of the third climber and the extra belaying required. The distance between belay positions will be halved if only one rope is used because one climber must tie in at the middle of the rope. Two ropes are recommended for a team of three climbers.

Note: *Time and complications will increase when a three-man team uses only one rope. For example: a 100-foot climb with a 150-foot rope would normally require two belays for two climbers; a 100-foot climb with a 150-foot rope would require six belays for three climbers.*

b. At times a three-man climb may be unavoidable and personnel should be familiar with the procedure. Although a team of three may choose from many different methods, only two are described below. If the climb is only one pitch, the methods will vary.

> ⚠ **CAUTION**
>
> When climbing with a team of three, protected traverses will require additional time. The equipment used to protect the traverse must be left in place to protect both the second and third climbers.

(1) The first method can be used when the belay positions are not large enough for three men. If using one rope, two climbers tie in at each end and the other at the mid point. When using two ropes, the second will tie in at one end of both ropes, and the other two climbers will each tie in to the other ends. The most experienced individual is the leader, or number 1 climber. The second, or number 2 climber, is the stronger of the remaining two and will be the belayer for both number 1 and number 3. Number 3 will be the last to climb. Although the number 3 climber does no belaying in this method, each climber should be skilled in the belay techniques required. The sequence for this method (in one pitch increments) is as follows (repeated until the climb is complete):

(a) Number 1 ascends belayed by number 2. Number 2 belays the leader up the first pitch while number 3 is simply anchored to the rock for security (unless starting off at ground level) and manages the rope between himself and number 2. When the leader completes the pitch, he sets up the next belay and belays number 2 up.

(b) Number 2 ascends belayed by number 1, and cleans the route (except for traverses). Number 2 returns the hardware to the leader and belays him up the next pitch. When the leader completes this pitch, he again sets up a new belay. When number 2 receives "OFF BELAY" from the leader, he changes ropes and puts number 3 on belay. He should not have to change anchor attachments because the position was already aimed for a downward as well as an upward pull when he belayed the leader.

(c) Number 3 ascends belayed by number 2. When number 3 receives "BELAY ON," he removes his anchor and climbs to number 2's position. When the pitch is completed he secures himself to one of number 2's belay anchors. When number 1's belay is ready, he brings up number 2 while number 3 remains anchored for security. Number 2 again cleans the pitch and the procedure is continued until the climb is completed.

(d) In this method, number 3 performs no belay function. He climbs when told to do so by number 2. When number 3 is not climbing, he remains anchored to the rock for security. The standard rope commands are used; however, the number 2 climber may include the trailing climber's name or number in the commands to avoid confusion as to who should be climbing.

(e) Normally, only one climber would be climbing at a time; however, the number 3 climber could ascend a fixed rope to number 2's belay position using proper ascending technique, with no effect on the other two members of the team. This would save time for a team of three, since number 2 would not have to belay number 3 and could be either belaying number 1 to the next belay or climbing to number 1. If number 3 is to ascend a fixed rope to the next belay position, the rope will be loaded with number 3's weight, and positioned directly off the anchors established for the belay. The rope should be located so it does not contact any sharp edges. The rope to the ascending number 3 could be secured to a separate anchor, but this would require additional time and gear.

(2) The second method uses either two ropes or a doubled rope, and number 2 and number 3 climb simultaneously. This requires either a special belay device that accepts two ropes, such as the tuber type, or with two Munter hitches. The ropes must travel through the belay device(s) without affecting each other.

(a) As the leader climbs the pitch, he will trail a second rope or will be tied in with a figure eight in the middle of a doubled rope. The leader reaches the next belay position and establishes the anchor and then places both remaining climbers on belay. One remaining climber will start the ascent toward the leader and the other will start when a gap of at least 10 feet is created between the two climbers. The belayer will have to remain alert for differences in rope movement and the climbers will have to climb at the same speed. One of the "second" climbers also cleans the pitch.

(b) Having at least two experienced climbers in this team will also save time. The belayer will have additional requirements to meet as opposed to having just one second. The possible force on the anchor will be twice that of one second. The second that is not cleaning the pitch can climb off route, but staying on route will usually prevent a possible swing if stance is not maintained.

CHAPTER 5

Sea Survival

Perhaps the most difficult survival situation to be in is sea survival. Short- or long-term survival depends upon rations and equipment available and your ingenuity. You must be resourceful to survive.

Water covers about 75 percent of the earth's surface, with about 70 percent being oceans and seas. You can assume that you will sometime cross vast expanses of water. There is always the chance that the plane or ship you are on will become crippled by such hazards as storms, collision, fire, or war.

THE OPEN SEA

As a survivor on the open sea, you will face waves and wind. You may also face extreme heat or cold. To keep these environmental hazards from becoming serious problems, take precautionary measures as soon as possible. Use the available resources to protect yourself from the elements and from heat or extreme cold and humidity.

Protecting yourself from the elements meets only one of your basic needs. You must also be able to obtain water and food. Satisfying these three basic needs will help prevent serious physical and psychological problems. However, you must know how to treat health problems that may result from your situation.

PRECAUTIONARY MEASURES

Your survival at sea depends upon—

- Your knowledge of and ability to use the available survival equipment.
- Your special skills and ability to apply them to cope with the hazards you face.
- Your will to live.

When you board a ship or aircraft, find out what survival equipment is on board, where it is stowed, and what it contains. For instance, how many life preservers and lifeboats or rafts are on board? Where are they located? What type of survival equipment do they have? How much food, water, and medicine do they contain? How many people are they designed to support?

If you are responsible for other personnel on board, make sure you know where they are and they know where you are.

DOWN AT SEA

If you are in an aircraft that goes down at sea, take the following actions once you clear the aircraft. Whether you are in the water or in a raft—

- Get clear and upwind of the aircraft as soon as possible, but stay in the vicinity until the aircraft sinks.
- Get clear of fuel-covered water in case the fuel ignites.
- Try to find other survivors.

A search for survivors usually takes place around the entire area of and near the crash site. Missing personnel may be unconscious and floating low in the water. Figure 5-1 illustrates rescue procedures.

The best technique for rescuing personnel from the water is to throw them a life preserver attached to a line. Another is to send a swimmer (rescuer) from the raft with a line attached to a flotation device that will support the rescuer's weight. This device will help conserve a rescuer's energy while recovering the survivor. The least acceptable technique is to send an attached swimmer without flotation devices to retrieve a survivor. In all cases, the rescuer wears a life preserver. A rescuer should not underestimate the strength of a panic-stricken person in the water. A careful approach can prevent injury to the rescuer.

When the rescuer approaches a survivor in trouble from behind, there is little danger the survivor will kick, scratch, or grab him. The rescuer swims to a point directly behind the survivor and grasps the life preserver's backstrap. The rescuer uses the sidestroke to drag the survivor to the raft.

Figure 5-1: Rescue from Water.

If you are in the water, make your way to a raft. If no rafts are available, try to find a large piece of floating debris to cling to. Relax; a person who knows how to relax in ocean water is in very little danger of drowning. The body's natural buoyancy will keep at least the top of the head above water, but some movement is needed to keep the face above water.

Floating on your back takes the least energy. Lie on your back in the water, spread your arms and legs, and arch your back. By controlling your breathing in and out, your face will always be out of the water and you may even sleep in this position for short periods. Your head will be partially submerged, but your face will be above water. If you cannot float on your back or if the sea is too rough, float facedown in the water as shown in Figure 5-2.

The following are the best swimming strokes during a survival situation:

- *Dog paddle.* This stroke is excellent when clothed or wearing a life jacket. Although slow in speed, it requires very little energy.
- *Breaststroke.* Use this stroke to swim underwater, through oil or debris, or in rough seas. It is probably the best stroke for long-range swimming: it allows you to conserve your energy and maintain a reasonable speed.
- *Sidestroke.* It is a good relief stroke because you use only one arm to maintain momentum and buoyancy.
- *Backstroke.* This stroke is also an excellent relief stroke. It relieves the muscles that you use for other strokes. Use it if an underwater explosion is likely.

If you are in an area where surface oil is burning—

- Discard your shoes and buoyant life preserver.
  Note: If you have an uninflated life preserver, keep it.
- Cover your nose, mouth, and eyes and quickly go underwater.
- Swim underwater as far as possible before surfacing to breathe.

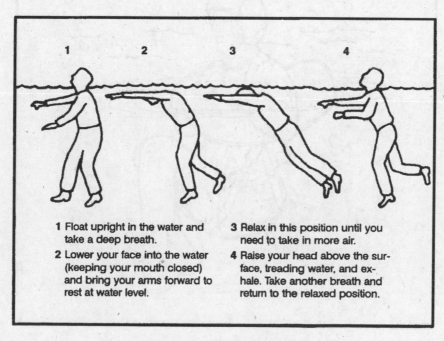

Figure 5-2: Floating position.

- Before surfacing to breathe and while still underwater, use your hands to push burning fluid away from the area where you wish to surface. Once an area is clear of burning liquid, you can surface and take a few breaths. Try to face downwind before inhaling.
- Submerge feet first and continue as above until clear of the flames.

If you are in oil-covered water that is free of fire, hold your head high to keep the oil out of your eyes. Attach your life preserver to your wrist and then use it as a raft.

If you have a life preserver, you can stay afloat for an indefinite period. In this case, use the "HELP" body position: Heat Escaping Lessening Posture (HELP). Remain still and assume the fetal position to help you retain body heat. You lose about 50 percent of your body heat through your head. Therefore, keep your head out of the water. Other areas of high heat loss are the neck, the sides, and the groin. Figure 5-3 illustrates the HELP position.

If you are in a raft—

- Check the physical condition of all on board. Give first aid if necessary. Take seasickness pills if available. The best way to take these pills is to place them under the tongue and let them dissolve. There are also suppositories or injections against seasickness. Vomiting, whether from seasickness or other causes, increases the danger of dehydration.
- Try to salvage all floating equipment—rations; canteens, thermos jugs, and other containers; clothing; seat cushions; parachutes; and anything else that will be useful to you. Secure the salvaged items in or to your raft. Make sure the items have no sharp edges that can puncture the raft.
- If there are other rafts, lash the rafts together so they are about 7.5 meters apart. Be ready to draw them closer together if you see or hear an aircraft. It is easier for an aircrew to spot rafts that are close together rather than scattered.

Figure 5-3: HELP position.

- Remember, rescue at sea is a cooperative effort. Use all available visual or electronic signaling devices to signal and make contact with rescuers. For example, raise a flag or reflecting material on an oar as high as possible to attract attention.
- Locate the emergency radio and get it into operation. Operating instructions are on it. Use the emergency transceiver only when friendly aircraft are likely to be in the area.
- Have other signaling devices ready for instant use. If you are in enemy territory, avoid using a signaling device that will alert the enemy. However, if your situation is desperate, you may have to signal the enemy for rescue if you are to survive.
- Check the raft for inflation, leaks, and points of possible chafing. Make sure the main buoyancy chambers are firm (well rounded) but not overly tight (Figure 5-4). Check inflation regularly. Air expands with heat; therefore, on hot days, release some air and add air when the weather cools.
- Decontaminate the raft of all fuel. Petroleum will weaken its surfaces and break down its glued joints.
- Throw out the sea anchor, or improvise a drag from the raft's case, bailing bucket, or a roll of clothing. A sea anchor helps you stay close to your ditching site, making it easier for searchers to find you if you have relayed your location. Without a sea anchor, your raft may drift over 160 kilometers in a day, making it much harder to find you. You can adjust the sea anchor to act as a drag to slow down the rate of travel with the current, or as a means to travel with the current. You make this adjustment by opening or closing the sea anchor's apex. When open, the sea anchor (Figure 5-5) acts as a drag that keeps you in the general area. When closed, it forms a pocket for the current to strike and propels the raft in the current's direction.

Additionally, adjust the sea anchor so that when the raft is on the wave's crest, the sea anchor is in the wave's trough (Figure 5-6).

- Wrap the sea anchor rope with cloth to prevent its chafing the raft. The anchor also helps to keep the raft headed into the wind and waves.
- In stormy water, rig the spray and windshield at once. In a 20-man raft, keep the canopy erected at all times. Keep your raft as dry as possible. Keep it properly balanced. All personnel should stay seated, the heaviest one in the center.

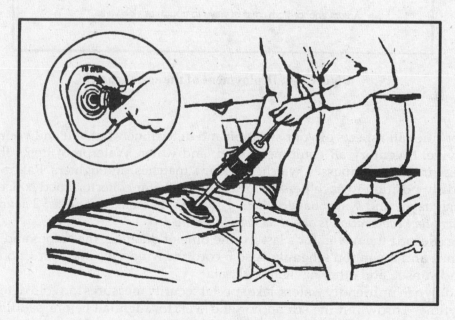

Figure 5-4: Inflating raft.

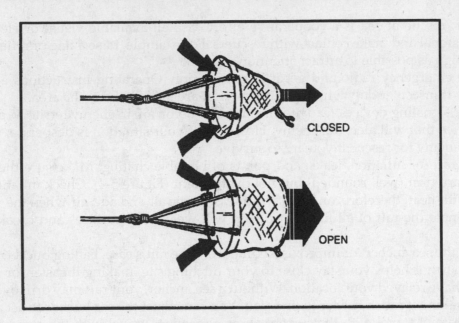

Figure 5-5: Sea anchor.

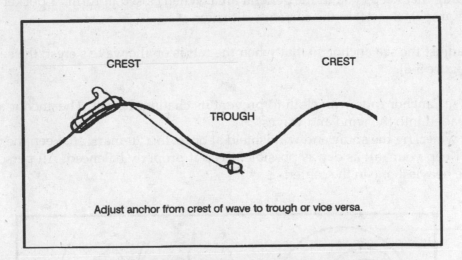

Adjust anchor from crest of wave to trough or vice versa.

Figure 5-6: Deployment of the sea anchor.

- Calmly consider all aspects of your situation and determine what you and your companions must do to survive. Inventory all equipment, food, and water. Waterproof items that salt water may affect. These include compasses, watches, sextant, matches, and lighters. Ration food and water.
- Assign a duty position to each person: for example, water collector, food collector, lookout, radio operator, signaler, and water bailers. *Note: Lookout duty should not exceed 2 hours. Keep in mind and remind others that cooperation is one of the keys to survival.*
- Keep a log. Record the navigator's last fix, the time of ditching, the names and physical condition of personnel, and the ration schedule. Also record the winds, weather, direction of swells, times of sunrise and sunset, and other navigational data.
- If you are down in unfriendly waters, take special security measures to avoid detection. Do not travel in the daytime. Throw out the sea anchor and wait for nightfall before paddling or hoisting sail. Keep low in the raft; stay covered with the blue side of the camouflage cloth up. Be sure a passing

ship or aircraft is friendly or neutral before trying to attract its attention. If the enemy detects you and you are close to capture, destroy the log book, radio, navigation equipment, maps, signaling equipment, and firearms. Jump overboard and submerge if the enemy starts strafing.

- Decide whether to stay in position or to travel. Ask yourself, "How much information was signaled before the accident? Is your position known to rescuers? Do you know it yourself? Is the weather favorable for a search? Are other ships or aircraft likely to pass your present position? How many days supply of food and water do you have?"

COLD WEATHER CONSIDERATIONS

If you are in a cold climate—

- Put on an antiexposure suit. If unavailable, put on any extra clothing available. Keep clothes loose and comfortable.
- Take care not to snag the raft with shoes or sharp objects. Keep the repair kit where you can readily reach it.
- Rig a windbreak, spray shield, and canopy.
- Try to keep the floor of the raft dry. Cover it with canvas or cloth for insulation.
- Huddle with others to keep warm, moving enough to keep the blood circulating. Spread an extra tarpaulin, sail, or parachute over the group.
- Give extra rations, if available, to men suffering from exposure to cold.

The greatest problem you face when submerged in cold water is death due to hypothermia. When you are immersed in cold water, hypothermia occurs rapidly due to the decreased insulating quality of wet clothing and the result of water displacing the layer of still air that normally surrounds the body. The rate of heat exchange in water is about 25 times greater than it is in air of the same temperature. Table 5-1 lists life expectancy times for immersion in water.

Your best protection against the effects of cold water is to get into the life raft, stay dry, and insulate your body from the cold surface of the bottom of the raft. If these actions are not possible, wearing an antiexposure suit will extend your life expectancy considerably. Remember, keep your head and neck out of the water and well insulated from the cold water's effects when the temperature is below 19 degrees C. Wearing life preservers increases the predicted survival time as body position in the water increases the chance of survival.

HOT WEATHER CONSIDERATIONS

If you are in a hot climate—

- Rig a sunshade or canopy. Leave enough space for ventilation.
- Cover your skin, where possible, to protect it from sunburn. Use sunburn cream, if available, on all exposed skin. Your eyelids, the back of your ears, and the skin under your chin sunburn easily.

RAFT PROCEDURES

Most of the rafts in the U.S. Army and Air Force inventories can satisfy the needs for personal protection, mode of travel, and evasion and camouflage.

Note: Before boarding any raft, remove and tether (attach) your life preserver to yourself or the raft. Ensure there are no other metallic or sharp objects on your clothing or equipment that could damage the raft. After boarding the raft, don your life preserver again.

One-Man Raft. The one-man raft has a main cell inflation. If the CO_2 bottle should malfunction or if the raft develops a leak, you can inflate it by mouth.

Table 5-1: Life expectancy times for immersion in water.

Water Temperature	Time
21.0–15.5 degrees C (70–60 degrees F)	12 hours
15.5–10.0 degrees C (60–50 degrees F)	6 hours
10.0–4.5 degrees C (50–40 degrees F)	1 hour
4.5 degrees C (40 degrees F) and below	less than 1 hour
Note: Wearing an antiexposure suit may increase these times up to a maximum of 24 hours.	

The spray shield acts as a shelter from the cold, wind, and water. In some cases, this shield serves as insulation. The raft's insulated bottom limits the conduction of cold thereby protecting you from hypothermia (Figure 5-7).

You can travel more effectively by inflating or deflating the raft to take advantage of the wind or current. You can use the spray shield as a sail while the ballast buckets serve to increase drag in the water. You may use the sea anchor to control the raft's speed and direction.

There are rafts developed for use in tactical areas that are black. These rafts blend with the sea's background. You can further modify these rafts for evasion by partially deflating them to obtain a lower profile.

A lanyard connects the one-man raft to a parachutist (survivor) landing in the water. You (the survivor) inflate it upon landing. You do not swim to the raft, but pull it to you via the lanyard. The raft may hit the water upside down, but you can right it by approaching the side to which the bottle is attached and flipping the raft over. The spray shield must be in the raft to expose the boarding handles. Follow the steps outlined in the note under raft procedures above when boarding the raft (Figure 5-8).

If you have an arm injury, the best way to board is by turning your back to the small end of the raft, pushing the raft under your buttocks, and lying back. Another way to board the raft is to push down on its small end until one knee is inside and lie forward (Figure 5-9).

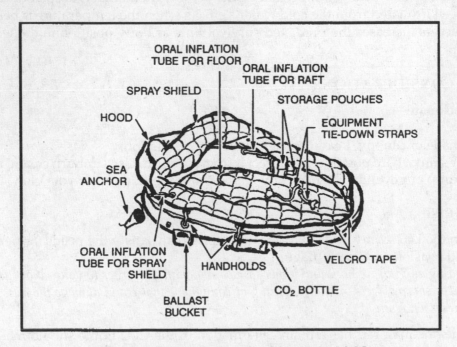

Figure 5-7: One-man raft with spray shield.

Figure 5-8: Boarding the one-man raft.

In rough seas, it may be easier for you to grasp the small end of the raft and, in a prone position, to kick and pull yourself into the raft. When you are lying face down in the raft, deploy and adjust the sea anchor. To sit upright, you may have to disconnect one side of the seat kit and roll to that side. Then you adjust the spray shield. There are two variations of the one-man raft; the improved model incorporates an inflatable spray shield and floor that provide additional insulation. The spray shield helps keep you dry and warm in cold oceans and protects you from the sun in the hot climates (Figure 5-10).

Seven-Man Raft. Some multiplace aircraft carry the seven-man raft. It is a component of the survival drop kit (Figure 5-11). This raft may inflate upside down and require you to right the raft before boarding. Always work from the bottle side to prevent injury if the raft turns over. Facing into the wind, the wind provides additional help in righting the raft. Use the handles on the inside bottom of the raft for boarding (Figure 5-12).

Use the boarding ramp if someone holds down the raft's opposite side. If you don't have help, again work from the bottle side with the wind at your back to help hold down the raft. Follow the steps outlined in the note under raft procedures above. Then grasp an oarlock and boarding handle, kick your legs to get your body prone on the water, and then kick and pull yourself into the raft. If you are weak or injured, you may partially deflate the raft to make boarding easier (Figure 5-13).

Use the hand pump to keep the buoyancy chambers and cross seat firm. Never overinflate the raft.

Twenty- or Twenty-Five-Man Rafts. You may find 20- or 25-man rafts in multiplace aircraft (Figures 5-14 and 5-15). You will find them in accessible areas of the fuselage or in raft compartments. Some may be automatically deployed from the cockpit, while others may need manual deployment. No matter how the raft lands in the water, it is ready for boarding. A lanyard connects the accessory kit to the raft and you retrieve the kit by hand. You must manually inflate the center chamber with the hand pump. Board the 20- or 25-man raft from the aircraft, if possible. If not, board in the following manner:

Figure 5-9: Boarding the one-man raft (other methods).

Figure 5-10: One-man raft with spray shield inflated.

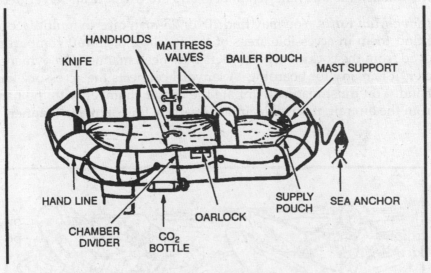

Figure 5-11: Seven-man raft.

Figure 5-12: Method of righting craft.

Figure 5-13: Method of barding seven-man raft.

- Approach the lower boarding ramp.
- Remove your life preserver and tether it to yourself so that it trails behind you.
- Grasp the boarding handles and kick your legs to get your body into a prone position on the water's surface; then kick and pull until you are inside the raft.

An incompletely inflated raft will make boarding easier. Approach the intersection of the raft and ramp, grasp the upper boarding handle, and swing one leg onto the center of the ramp, as in mounting a horse (Figure 5-16).

Immediately tighten the equalizer clamp upon entering the raft to prevent deflating the entire raft in case of a puncture (Figure 5-17).

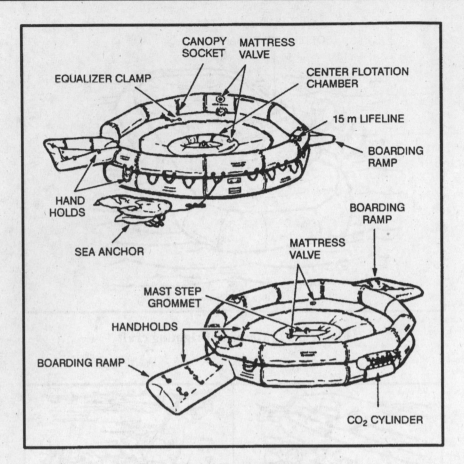

Figure 5-14: Twenty-man raft.

Use the pump to keep these rafts' chambers and center ring firm. They should be well rounded but not overly tight.

SAILING RAFTS

Rafts do not have keels, therefore, you can't sail them into the wind. However, anyone can sail a raft downwind. You can successfully sail multiplace (except 20- to 25-man) rafts 10 degrees off from the direction of the wind. Do not try to sail the raft unless land is near. If you decide to sail and the wind is blowing toward a desired destination, fully inflate the raft, sit high, take in the sea anchor, rig a sail, and use an oar as a rudder.

In a multiplace (except 20- to 25-man) raft, erect a square sail in the bow using the oars and their extensions as the mast and crossbar (Figure 5-18). You may use a waterproof tarpaulin or parachute material for the sail.

If the raft has no regular mast socket and step, erect the mast by tying it securely to the front cross seat using braces. Pad the bottom of the mast to prevent it from chafing or punching a hole through the floor, whether or not there is a socket. The heel of a shoe, with the toe wedged under the seat, makes a good improvised mast step. Do not secure the corners of the lower edge of the sail. Hold the lines attached to the corners with your hands so that a gust of wind will not rip the sail, break the mast, or capsize the raft.

Take every precaution to prevent the raft from turning over. In rough weather, keep the sea anchor away from the bow. Have the passengers sit low in the raft, with their weight distributed to hold the upwind side down. To prevent falling out, they should also avoid sitting on the sides of the raft or standing up. Avoid sudden movements without warning the other passengers. When the sea anchor is not in use, tie it to the raft and stow it in such a manner that it will hold immediately if the raft capsizes.

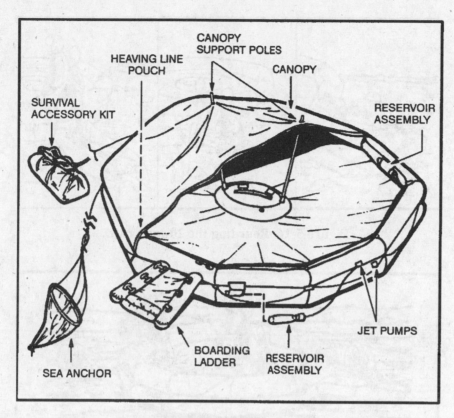

Figure 5-15: Twenty-five-man raft.

WATER

Water is your most important need. With it alone, you can live for ten days or longer, depending on your will to live. When drinking water, moisten your lips, tongue, and throat before swallowing.

Short Water Rations. When you have a limited water supply and you can't replace it by chemical or mechanical means, use the water efficiently. Protect freshwater supplies from seawater contamination. Keep your body well shaded, both from overhead sun and from reflection off the sea surface. Allow ventilation of air; dampen your clothes during the hottest part of the day. Do not exert yourself. Relax and sleep when possible. Fix your daily water ration after considering the amount of water you have, the output of solar stills and desalting kit, and the number and physical condition of your party.

If you don't have water, don't eat. If your water ration is two liters or more per day, eat any part of your ration or any additional food that you may catch, such as birds, fish, shrimp. Anxiety and the life raft's motion may cause nausea. If you eat when nauseated, you may lose your food immediately. If nauseated, rest and relax as much as you can, and take only water.

To reduce your loss of water through perspiration, soak your clothes in the sea and wring them out before putting them on again. Don't overdo this during hot days when no canopy or sun shield is available. This is a trade-off between cooling and saltwater boils and rashes that will result.

Be careful not to get the bottom of the raft wet.

Watch the clouds and be ready for any chance of showers. Keep the tarpaulin handy for catching water. If it is encrusted with dried salt, wash it in seawater. Normally, a small amount of seawater mixed with rain will hardly be noticeable and will not cause any physical reaction. In rough seas you cannot get uncontaminated fresh water.

At night, secure the tarpaulin like a sunshade, and turn up its edges to collect dew. It is also possible to collect dew along the sides of the raft using a sponge or cloth. When it rains, drink as much as you can hold.

Figure 5-16: Boarding the 20-man raft.

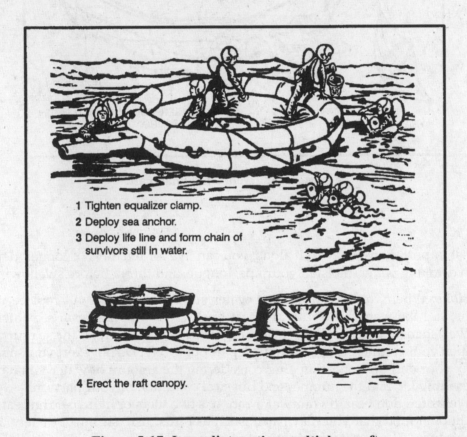

1 Tighten equalizer clamp.
2 Deploy sea anchor.
3 Deploy life line and form chain of survivors still in water.

4 Erect the raft canopy.

Figure 5-17: Immediate action-multiplace raft.

Solar Still. When solar stills are available, read the instructions and set them up immediately. Use as many stills as possible, depending on the number of men in the raft and the amount of sunlight available. Secure solar stills to the raft with care. This type of solar still only works on flat, calm seas.

Desalting Kits. When desalting kits are available in addition to solar stills, use them only for immediate water needs or during long overcast periods when you cannot use solar stills. In any event, keep desalting kits and emergency water stores for periods when you cannot use solar stills or catch rainwater.

Water from Fish. Drink the aqueous fluid found along the spine and in the eyes of large fish. Carefully cut the fish in half to get the fluid along the spine and suck the eye. If you are so short of water that you need

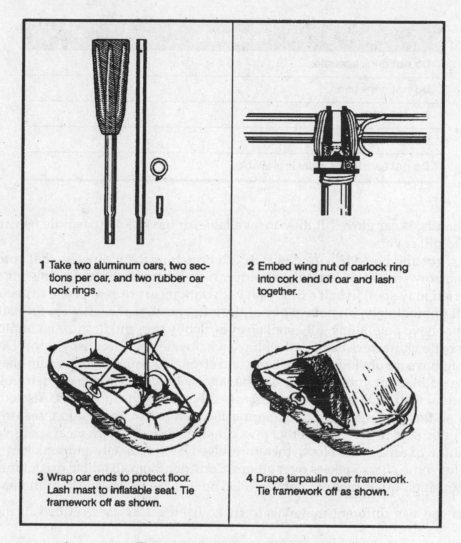

1 Take two aluminum oars, two sections per oar, and two rubber oar lock rings.

2 Embed wing nut of oarlock ring into cork end of oar and lash together.

3 Wrap oar ends to protect floor. Lash mast to inflatable seat. Tie framework off as shown.

4 Drape tarpaulin over framework. Tie framework off as shown.

Figure 5-18: Sail construction.

to do this, then do not drink any of the other body fluids. These other fluids are rich in protein and fat and will use up more of your reserve water in digestion than they supply.

Sea Ice. In arctic waters, use old sea ice for water. This ice is bluish, has rounded corners, and splinters easily. It is nearly free of salt. New ice is gray, milky, hard, and salty. Water from icebergs is fresh, but icebergs are dangerous to approach. Use them as a source of water only in emergencies.

Sleep and rest are the best ways of enduring periods of reduced water and food intake. However, make sure that you have enough shade when napping during the day. If the sea is rough, tie yourself to the raft, close any cover, and ride out the storm as best you can. Relax is the keyword—at least try to relax.

FOOD PROCUREMENT

In the open sea, fish will be the main food source. There are some poisonous and dangerous ocean fish, but, in general, when out of sight of land, fish are safe to eat. Nearer the shore there are fish that are both dangerous and poisonous to eat. There are some fish, such as the red snapper and barracuda, that are normally edible but poisonous when taken from the waters of atolls and reefs. Flying fish will even jump into your raft!

Fish. When fishing, do not handle the fishing line with bare hands and never wrap it around your hands or tie it to a life raft. The salt that adheres to it can make it a sharp cutting edge, an edge dangerous both to

REMEMBER!
Do not drink seawater.
Do not drink urine.
Do not drink alcohol.
Do not smoke.
Do not eat, unless water is available.

the raft and your hands. Wear gloves, if they are available, or use a cloth to handle fish and to avoid injury from sharp fins and gill covers.

In warm regions, gut and bleed fish immediately after catching them. Cut fish that you do not eat immediately into thin, narrow strips and hang them to dry. A well-dried fish stays edible for several days. Fish not cleaned and dried may spoil in half a day. Fish with dark meat are very prone to decomposition. If you do not eat them all immediately, do not eat any of the leftovers. Use the leftovers for bait.

Never eat fish that have pale, shiny gills, sunken eyes, flabby skin and flesh, or an unpleasant odor. Good fish show the opposite characteristics. Sea fish have a saltwater or clean fishy odor. Do not confuse eels with sea snakes that have an obviously scaly body and strongly compressed, paddle-shaped tail. Both eels and sea snakes are edible, but you must handle the latter with care because of their poisonous bites. The heart, blood, intestinal wall, and liver of most fish are edible. Cook the intestines. Also edible are the partly digested smaller fish that you may find in the stomachs of large fish. In addition, sea turtles are edible.

Shark meat is a good source of food whether raw, dried, or cooked. Shark meat spoils very rapidly due to the high concentration of urea in the blood, therefore, bleed it immediately and soak it in several changes of water. People prefer some shark species over others. Consider them all edible except the Greenland shark whose flesh contains high quantities of vitamin A. Do not eat the livers, due to high vitamin A content.

Fishing Aids. You can use different materials to make fishing aids as described in the following paragraphs:

- *Fishing line.* Use pieces of tarpaulin or canvas. Unravel the threads and tie them together in short lengths in groups of three or more threads. Shoelaces and parachute suspension line also work well.
- *Fish hooks.* No survivor at sea should be without fishing equipment but if you are, improvise hooks as shown in Part IV, Chapter 2.
- *Fish lures.* You can fashion lures by attaching a double hook to any shiny piece of metal.
- *Grapple.* Use grapples to hook seaweed. You may shake crabs, shrimp, or small fish out of the seaweed. These you may eat or use for bait. You may eat seaweed itself, but only when you have plenty of drinking water. Improvise grapples from wood. Use a heavy piece of wood as the main shaft, and lash three smaller pieces to the shaft as grapples.
- *Bait.* You can use small fish as bait for larger ones. Scoop the small fish up with a net. If you don't have a net, make one from cloth of some type. Hold the net under the water and scoop upward. Use all the guts from birds and fish for bait. When using bait, try to keep it moving in the water to give it the appearance of being alive.

Helpful Fishing Hints. Your fishing should be successful if you remember the following important hints:

- Be extremely careful with fish that have teeth and spines.
- Cut a large fish loose rather than risk capsizing the raft. Try to catch small rather than large fish.

- Do not puncture your raft with hooks or other sharp instruments.
- Do not fish when large sharks are in the area.
- Watch for schools of fish; try to move close to these schools.
- Fish at night using a light. The light attracts fish.
- In the daytime, shade attracts some fish. You may find them under your raft.
- Improvise a spear by tying a knife to an oar blade. This spear can help you catch larger fish, but you must get them into the raft quickly or they will slip off the blade. Also, tie the knife very securely or you may lose it.
- Always take care of your fishing equipment. Dry your fishing lines, clean and sharpen the hooks, and do not allow the hooks to stick into the fishing lines.

Birds. As stated in Part IV, Chapter 2, all birds are edible. Eat any birds you can catch. Sometimes birds may land on your raft, but usually they are cautious. You may be able to attract some birds by towing a bright piece of metal behind the raft. This will bring the bird within shooting range, provided you have a firearm.

If a bird lands within your reach, you may be able to catch it. If the birds do not land close enough or land on the other end of the raft, you may be able to catch them with a bird noose. Bait the center of the noose and wait for the bird to land. When the bird's feet are in the center of the noose, pull it tight.

Use all parts of the bird. Use the feathers for insulation, the entrails and feet for bait, and so on. Use your imagination.

Medical Problems Associated With Sea Survival. At sea, you may become seasick, get saltwater sores, or face some of the same medical problems that occur on land, such as dehydration or sunburn. These problems can become critical if left untreated.

Seasickness. Seasickness is the nausea and vomiting caused by the motion of the raft. It can result in—

- Extreme fluid loss and exhaustion.
- Loss of the will to survive.
- Others becoming seasick.
- Attraction of sharks to the raft.
- Unclean conditions.

To treat seasickness—

- Wash both the patient and the raft to remove the sight and odor of vomit.
- Keep the patient from eating food until his nausea is gone.
- Have the patient lie down and rest.
- Give the patient seasickness pills if available. If the patient is unable to take the pills orally, insert them rectally for absorption by the body.

Note: Some survivors have said that erecting a canopy or using the horizon as a focal point helped overcome seasickness. Others have said that swimming alongside the raft for short periods helped, but extreme care must be taken if swimming.

Saltwater Sores. These sores result from a break in skin exposed to saltwater for an extended period. The sores may form scabs and pus. Do not open or drain. Flush the sores with fresh water, if available, and allow to dry. Apply an antiseptic, if available.

Immersion Rot, Frostbite, and Hypothermia. These problems are similar to those encountered in cold weather environments. Symptoms and treatment are the same as covered in Chapter 3, "Cold Weather Survival."

Blindness/Headache. If flame, smoke, or other contaminants get in the eyes, flush them immediately with salt water, then with fresh water, if available. Apply ointment, if available. Bandage both eyes 18 to 24 hours, or longer if damage is severe. If the glare from the sky and water causes your eyes to become bloodshot and inflamed, bandage them lightly. Try to prevent this problem by wearing sunglasses. Improvise sunglasses if necessary.

Constipation. This condition is a common problem on a raft. Do not take a laxative, as this will cause further dehydration. Exercise as much as possible and drink an adequate amount of water, if available.

Difficult Urination. This problem is not unusual and is due mainly to dehydration. It is best not to treat it, as it could cause further dehydration.

Sunburn. Sunburn is a serious problem in sea survival. Try to prevent sunburn by staying in shade and keeping your head and skin covered. Use cream or Chap Stick from your first aid kit. Remember, reflection from the water also causes sunburn.

SHARKS

Whether you are in the water or in a boat or raft, you may see many types of sea life around you. Some may be more dangerous than others. Generally, sharks are the greatest danger to you. Other animals such as whales, porpoises, and stingrays may look dangerous, but really pose little threat in the open sea.

Of the many hundreds of shark species, only about 20 species are known to attack man. The most dangerous are the great white shark, the hammerhead, the make, and the tiger shark. Other sharks known to attack man include the gray, blue, lemon, sand, nurse, bull, and oceanic white tip sharks. Consider any shark longer than 1 meter dangerous.

There are sharks in all oceans and seas of the world. While many live and feed in the depths of the sea, others hunt near the surface. The sharks living near the surface are the ones you will most likely see. Their dorsal fins frequently project above the water. Sharks in the tropical and subtropical seas are far more aggressive than those in temperate waters.

All sharks are basically eating machines. Their normal diet is live animals of any type, and they will strike at injured or helpless animals. Sight, smell, or sound may guide them to their prey. Sharks have an acute sense of smell and the smell of blood in the water excites them. They are also very sensitive to any abnormal vibrations in the water. The struggles of a wounded animal or swimmer, underwater explosions, or even a fish struggling on a fish line will attract a shark.

Sharks can bite from almost any position; they do not have to turn on their side to bite. The jaws of some of the larger sharks are so far forward that they can bite floating objects easily without twisting to the side.

Sharks may hunt alone, but most reports of attacks cite more than one shark present. The smaller sharks tend to travel in schools and attack in mass. Whenever one of the sharks finds a victim, the other sharks will quickly join it. Sharks will eat a wounded shark as quickly as their prey.

Sharks feed at all hours of the day and night. Most reported shark contacts and attacks were during daylight, and many of these have been in the late afternoon. Some of the measures that you can take to protect yourself against sharks when you are in the water are—

- *Stay with other swimmers.* A group can maintain a 360-degree watch. A group can either frighten or fight off sharks better than one man.
- *Always watch for sharks.* Keep all your clothing on, to include your shoes. Historically, sharks have attacked the unclothed men in groups first, mainly in the feet. Clothing also protects against abrasions should the shark brush against you.
- *Avoid urinating.* If you must, only do so in small amounts. Let it dissipate between discharges. If you must defecate, do so in small amounts and throw it as far away from you as possible. Do the same if you must vomit.

If a shark attack is imminent while you are in the water, splash and yell just enough to keep the shark at bay. Sometimes yelling underwater or slapping the water repeatedly will scare the shark away. Conserve your strength for fighting in case the shark attacks.

If attacked, kick and strike the shark. Hit the shark on the gills or eyes if possible. If you hit the shark on the nose, you may injure your hand if it glances off and hits its teeth.

When you are in a raft and see sharks—

- Do not fish. If you have hooked a fish, let it go. Do not clean fish in the water.
- Do not throw garbage overboard.
- Do not let your arms, legs, or equipment hang in the water.
- Keep quiet and do not move around.
- Dispose of all dead as soon as possible. If there are many sharks in the area, conduct the disposal at night.

When you are in a raft and a shark attack is imminent, hit the shark with anything you have, except your hands. You will do more damage to your hands than the shark. If you strike with an oar, be careful not to lose or break it.

DETECTING LAND

You should watch carefully for any signs of land. There are many indicators that land is near.

A fixed cumulus cloud in a clear sky or in a sky where all other clouds are moving often hovers over or slightly downwind from an island.

In the tropics, the reflection of sunlight from shallow lagoons or shelves of coral reefs often causes a greenish tint in the sky.

In the arctic, light-colored reflections on clouds often indicate ice fields or snow-covered land. These reflections are quite different from the dark gray ones caused by open water.

Deep water is dark green or dark blue. Lighter color indicates shallow water, which may mean land is near.

At night, or in fog, mist, or rain, you may detect land by odors and sounds. The musty odor of mangrove swamps and mud flats carry a long way. You hear the roar of surf long before you see the surf. The continued cries of seabirds coming from one direction indicate their roosting place on nearby land.

There usually are more birds near land than over the open sea. The direction from which flocks fly at dawn and to which they fly at dusk may indicate the direction of land. During the day, birds are searching for food and the direction of flight has no significance.

Mirages occur at any latitude, but they are more likely in the tropics, especially during the middle of the day. Be careful not to mistake a mirage for nearby land. A mirage disappears or its appearance and elevation change when viewed from slightly different heights.

You may be able to detect land by the pattern of the waves (refracted) as they approach land (Figure 5-19). By traveling with the waves and parallel to the slightly turbulent area marked "X" on the illustration, you should reach land.

RAFTING OR BEACHING TECHNIQUES

Once you have found land, you must get ashore safely. To raft ashore, you can usually use the one-man raft without danger. However, going ashore in a strong surf is dangerous. Take your time. Select your landing point carefully.

Try not to land when the sun is low and straight in front of you. Try to land on the lee side of an island or on a point of land jutting out into the water. Keep your eyes open for gaps in the surf line, and head for them. Avoid coral reefs and rocky cliffs. There are no coral reefs near the mouths of freshwater streams.

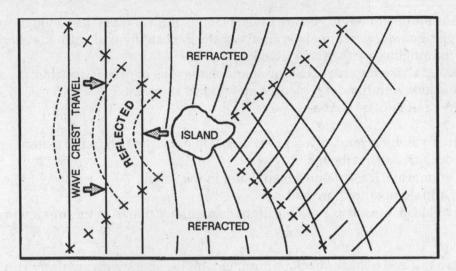

Figure 5-19: Wave patterns about an island.

Avoid rip currents or strong tidal currents that may carry you far out to sea. Either signal ashore for help or sail around and look for a sloping beach where the surf is gentle.

If you have to go through the surf to reach shore, take down the mast. Keep your clothes and shoes on to avoid severe cuts. Adjust and inflate your life vest. Trail the sea anchor over the stern using as much line as you have. Use the oars or paddles and constantly adjust the sea anchor to keep a strain on the anchor line. These actions will keep the raft pointed toward shore and prevent the sea from throwing the stern around and capsizing you. Use the oars or paddles to help ride in on the seaward side of a large wave.

The surf may be irregular and velocity may vary, so modify your procedure as conditions demand. A good method of getting through the surf is to have half the men sit on one side of the raft, half on the other, facing away from each other. When a heavy sea bears down, half should row (pull) toward the sea until the crest passes; then the other half should row (pull) toward the shore until the next heavy sea comes along.

Against a strong wind and heavy surf, the raft must have all possible speed to pass rapidly through the oncoming crest to avoid being turned broadside or thrown end over end. If possible, avoid meeting a large wave at the moment it breaks.

If in a medium surf with no wind or offshore wind, keep the raft from passing over a wave so rapidly that it drops suddenly after topping the crest. If the raft turns over in the surf, try to grab hold of it and ride it in.

As the raft nears the beach, ride in on the crest of a large wave. Paddle or row hard and ride in to the beach as far as you can. Do not jump out of the raft until it has grounded, then quickly get out and beach it.

If you have a choice, do not land at night. If you have reason to believe that people live on the shore, lay away from the beach, signal, and wait for the inhabitants to come out and bring you in.

If you encounter sea ice, land only on large, stable floes. Avoid icebergs that may capsize and small floes or those obviously disintegrating. Use oars and hands to keep the raft from rubbing on the edge of the ice. Take the raft out of the water and store it well back from the floe's edge. You may be able to use it for shelter. Keep the raft inflated and ready for use. Any floe may break up without warning.

SWIMMING ASHORE

If rafting ashore is not possible and you have to swim, wear your shoes and at least one thickness of clothing. Use the sidestroke or breaststroke to conserve strength.

If the surf is moderate, ride in on the back of a small wave by swimming forward with it. Dive to a shallow depth to end the ride just before the wave breaks.

In high surf, swim toward shore in the trough between waves. When the seaward wave approaches, face it and submerge. After it passes, work toward shore in the next trough. If caught in the undertow of a large wave, push off the bottom or swim to the surface and proceed toward shore as above.

If you must land on a rocky shore, look for a place where the waves rush up onto the rocks. Avoid places where the waves explode with a high, white spray. Swim slowly when making your approach. You will need your strength to hold on to the rocks. You should be fully clothed and wear shoes to reduce injury.

After selecting your landing point, advance behind a large wave into the breakers. Face toward shore and take a sitting position with your feet in front, 60 to 90 centimeters (2 or 3 feet) lower than your head. This position will let your feet absorb the shock when you land or strike submerged boulders or reefs. If you do not reach shore behind the wave you picked, swim with your hands only. As the next wave approaches, take a sitting position with your feet forward. Repeat the procedure until you land.

Water is quieter in the lee of a heavy growth of seaweed. Take advantage of such growth. Do not swim through the seaweed; crawl over the top by grasping the vegetation with overhand movements.

Cross a rocky or coral reef as you would land on a rocky shore. Keep your feet close together and your knees slightly bent in a relaxed sitting posture to cushion the blows against the coral.

PICKUP OR RESCUE

On sighting rescue craft approaching for pickup (boat, ship, conventional aircraft, or helicopter), quickly clear any lines (fishing lines, desalting kit lines) or other gear that could cause entanglement during rescue. Secure all loose items in the raft. Take down canopies and sails to ensure a safer pickup. After securing all items, put on your helmet, if available. Fully inflate your life preserver. Remain in the raft, unless otherwise instructed, and remove all equipment except the preservers. If possible, you will receive help from rescue personnel lowered into the water. Remember, follow all instructions given by the rescue personnel.

- If the helicopter recovery is unassisted, do the following before pickup:
- Secure all the loose equipment in the raft, accessory bag, or in pockets.
- Deploy the sea anchor, stability bags, and accessory bag.
- Partially deflate the raft and fill it with water.
- Unsnap the survival kit container from the parachute harness.
- Grasp the raft handhold and roll out of the raft.
- Allow the recovery device or the cable to ground out on the water's surface.
- Maintain the handhold until the recovery device is in your other hand.
- Mount the recovery device, avoiding entanglement with the raft.
- Signal the hoist operator for pickup.

SEASHORES

Search planes or ships do not always spot a drifting raft or swimmer. You may have to land along the coast before being rescued. Surviving along the seashore is different from open sea survival. Food and water are more abundant and shelter is obviously easier to locate and construct.

If you are in friendly territory and decide to travel, it is better to move along the coast than to go inland. Do not leave the coast except to avoid obstacles (swamps and cliffs) or unless you find a trail that you know leads to human habitation.

In time of war, remember that the enemy patrols most coastlines. These patrols may cause problems for you if you land on a hostile shore. You will have extremely limited travel options in this situation. Avoid all contact with other humans, and make every effort to cover all tracks you leave on the shore.

SPECIAL HEALTH HAZARDS

Coral, poisonous and aggressive fish, crocodiles, sea urchins, sea biscuits, sponges, anemones, and tides and undertow pose special health hazards.

Coral. Coral, dead or alive, can inflict painful cuts. There are hundreds of water hazards that can cause deep puncture wounds, severe bleeding, and the danger of infection. Clean all coral cuts thoroughly. Do not use iodine to disinfect any coral cuts. Some coral polyps feed on iodine and may grow inside your flesh if you use iodine.

Poisonous Fish. Many reef fish have toxic flesh. For some species, the flesh is always poisonous, for other species, only at certain times of the year. The poisons are present in all parts of the fish, but especially in the liver, intestines, and eggs.

Fish toxins are water soluble—no amount of cooking will neutralize them. They are tasteless, therefore the standard edibility tests are useless. Birds are least susceptible to the poisons. Therefore, do not think that because a bird can eat a fish, it is a safe species for you to eat.

The toxins will produce a numbness of the lips, tongue, toes, and tips of the fingers, severe itching, and a clear reversal of temperature sensations. Cold items appear hot and hot items cold. There will probably also be nausea, vomiting, loss of speech, dizziness, and a paralysis that eventually brings death.

In addition to fish with poisonous flesh, there are those that are dangerous to touch. Many stingrays have a poisonous barb in their tail. There are also species that can deliver an electric shock. Some reef fish, such as stonefish and toadfish, have venomous spines that can cause very painful although seldom fatal injuries. The venom from these spines causes a burning sensation or even an agonizing pain that is out of proportion to the apparent severity of the wound. Jellyfish, while not usually fatal, can inflict a very painful sting if it touches you with its tentacles. See Part IV, Chapter 5 for details on particularly dangerous fish of the sea and seashore.

Aggressive Fish. You should also avoid some ferocious fish. The bold and inquisitive barracuda has attacked men wearing shiny objects. It may charge lights or shiny objects at night. The sea bass, which can grow to 1.7 meters, is another fish to avoid. The moray eel, which has many sharp teeth and grows to 1.5 meters, can also be aggressive if disturbed.

Sea Snakes. Sea snakes are venomous and sometimes found in mid ocean. They are unlikely to bite unless provoked. Avoid them.

Crocodiles. Crocodiles inhabit tropical saltwater bays and mangrove-bordered estuaries and range up to 65 kilometers into the open sea. Few remain near inhabited areas. You commonly find crocodiles in the remote areas of the East Indies and Southeast Asia. Consider specimens over 1 meter long dangerous, especially females guarding their nests. Crocodile meat is an excellent source of food when available.

Sea Urchins, Sea Biscuits, Sponges, and Anemones. These animals can cause extreme, though seldom fatal, pain. Usually found in tropical shallow water near coral formations, sea urchins resemble small, round porcupines. If stepped on, they slip fine needles of lime or silica into the skin, where they break off and fester. If possible, remove the spines and treat the injury for infection. The other animals mentioned inflict injury similarly.

Tides and Undertow. These are another hazard to contend with. If caught in a large wave's undertow, push off the bottom or swim to the surface and proceed shoreward in a trough between waves. Do not fight against the pull of the undertow. Swim with it or perpendicular to it until it loses strength, then swim for shore.

Food. Obtaining food along a seashore should not present a problem. There are many types of seaweed and other plants and animal life you can easily find and eat. See Part IV, Chapters 2 and 6 for more information.

Mollusks. Mussels, limpets, clams, sea snails, octopuses, squids, and sea slugs are all edible. Shellfish will usually supply most of the protein eaten by coastal survivors. Avoid the blue-ringed octopus and cone shells (described in Part IV, Chapter 5). Also beware of "red tides" that make mollusks poisonous. Apply the edibility test on each species before eating.

Worms. Coastal worms are generally edible, but it is better to use them for fish bait. Avoid bristle worms that look like fuzzy caterpillars. Also avoid tubeworms that have sharp-edged tubes. Arrow worms, alias amphioxus, are not true worms. You find them in the sand and are excellent either fresh or dried.

Crabs, Lobsters, and Barnacles. These animals are seldom dangerous to man and are an excellent food source. The pincers of larger crabs or lobsters can crush a man's finger. Many species have spines on their shells, making it preferable to wear gloves when catching them. Barnacles can cause scrapes or cuts and are difficult to detach from their anchor, but the larger species are an excellent food source.

Sea Urchins. These are common and can cause painful injuries when stepped on or touched. They are also a good source of food. Handle them with gloves, and remove all spines.

Sea Cucumbers. This animal is an important food source in the Indo-Pacific regions. Use them whole after evisceration or remove the five muscular strips that run the length of its body. Eat them smoked, pickled, or cooked.

CHAPTER 6

Water Crossings

In a survival situation, you may have to cross a water obstacle. It may be in the form of a river, a stream, a lake, a bog, quicksand, quagmire, or muskeg. Even in the desert, flash floods occur, making streams an obstacle. Whatever it is, you need to know how to cross it safely.

RIVERS AND STREAMS

You can apply almost every description to rivers and streams. They may be shallow or deep, slow or fast moving, narrow or wide. Before you try to cross a river or stream, develop a good plan.

Your first step is to look for a high place from which you can get a good view of the river or stream. From this place, you can look for a place to cross. If there is no high place, climb a tree. Good crossing locations include—

- A level stretch where it breaks into several channels. Two or three narrow channels are usually easier to cross than a wide river.
- A shallow bank or sandbar. If possible, select a point upstream from the bank or sandbar so that the current will carry you to it if you lose your footing.
- A course across the river that leads downstream so that you will cross the current at about a 45-degree angle.

The following areas possess potential hazards; avoid them, if possible:

- *Obstacles on the opposite side of the river that might hinder your travel.* Try to select the spot from which travel will be the safest and easiest.
- *A ledge of rocks that crosses the river.* This often indicates dangerous rapids or canyons.
- *A deep or rapid waterfall or a deep channel.* Never try to ford a stream directly above or even close to such hazards.
- *Rocky places.* You may sustain serious injuries from slipping or falling on rocks. Usually, submerged rocks are very slick, making balance extremely difficult. An occasional rock that breaks the current, however, may help you.
- *An estuary of a river.* An estuary is normally wide, has strong currents, and is subject to tides. These tides can influence some rivers many kilometers from their mouths. Go back upstream to an easier crossing site.
- *Eddies.* An eddy can produce a powerful backward pull downstream of the obstruction causing the eddy and pull you under the surface.

The depth of a fordable river or stream is no deterrent if you can keep your footing. In fact, deep water sometimes runs more slowly and is therefore safer than fast-moving shallow water. You can always dry your clothes later, or if necessary, you can make a raft to carry your clothing and equipment across the river.

You must not try to swim or wade across a stream or river when the water is at very low temperatures. This swim could be fatal. Try to make a raft of some type. Wade across if you can get only your feet wet. Dry them vigorously as soon as you reach the other bank.

RAPIDS

If necessary, you can safely cross a deep, swift river or rapids. To swim across a deep, swift river, swim with the current, never fight it. Try to keep your body horizontal to the water. This will reduce the danger of being pulled under.

In fast, shallow rapids, lie on your back, feet pointing downstream, finning your hands alongside your hips. This action will increase buoyancy and help you steer away from obstacles. Keep your feet up to avoid getting them bruised or caught by rocks.

In deep rapids, lie on your stomach, head downstream, angling toward the shore whenever you can. Watch for obstacles and be careful of backwater eddies and converging currents, as they often contain dangerous swirls. Converging currents occur where new watercourses enter the river or where water has been diverted around large obstacles such as small islands.

To ford a swift, treacherous stream, apply the following steps:

- Remove your pants and shirt to lessen the water's pull on you. Keep your footgear on to protect your feet and ankles from rocks. It will also provide you with firmer footing.
- Tie your pants and other articles to the top of your rucksack or in a bundle, if you have no pack. This way, if you have to release your equipment, all your articles will be together. It is easier to find one large pack than to find several small items.
- Carry your pack well up on your shoulders and be sure you can easily remove it, if necessary. Not being able to get a pack off quickly enough can drag even the strongest swimmers under.
- Find a strong pole about 7.5 centimeters in diameter and 2.1 to 2.4 meters long to help you ford the stream. Grasp the pole and plant it firmly on your upstream side to break the current. Plant your feet firmly with each step, and move the pole forward a little downstream from its previous position, but still upstream from you. With your next step, place your foot below the pole. Keep the pole well slanted so that the force of the current keeps the pole against your shoulder (Figure 6-1).
- Cross the stream so that you will cross the downstream current at a 45-degree angle.

Using this method, you can safely cross currents usually too strong for one person to stand against. Do not concern yourself about your pack's weight, as the weight will help rather than hinder you in fording the stream.

If there are other people with you, cross the stream together. Ensure that everyone has prepared their pack and clothing as outlined above. Position the heaviest person on the downstream end of the pole and the lightest on the upstream end. In using this method, the upstream person breaks the current, and those below can move with relative ease in the eddy formed by the upstream person. If the upstream person gets temporarily swept off his feet, the others can hold steady while he regains his footing (Figure 6-2).

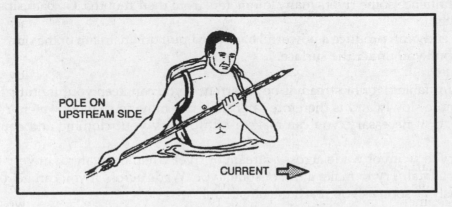

POLE ON
UPSTREAM SIDE

CURRENT

Figure 6-1: One man crossing swift stream.

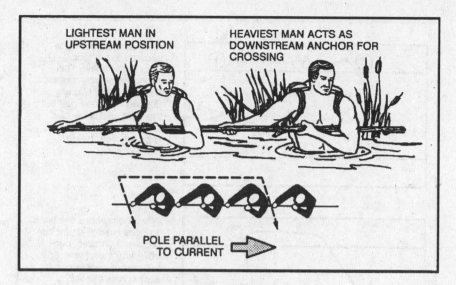

Figure 6-2: Several men crossing swift stream.

If you have three or more people and a rope available, you can use the technique shown in Figure 6-3 to cross the stream. The length of the rope must be three times the width of the stream.

RAFTS

If you have two ponchos, you can construct a brush raft or an Australian poncho raft. With either of these rafts, you can safely float your equipment across a slow-moving stream or river.

Brush Raft. The brush raft, if properly constructed, will support about 115 kilograms. To construct it, use ponchos, fresh green brush, two small saplings, and rope or vine as follows (Figure 6-4):

- Push the hood of each poncho to the inner side and tightly tie off the necks using the drawstrings.
- Attach the ropes or vines at the corner and side grommets of each poncho. Make sure they are long enough to cross to and tie with the others attached at the opposite corner or side.
- Spread one poncho on the ground with the inner side up. Pile fresh, green brush (no thick branches) on the poncho until the brush stack is about 45 centimeters high. Pull the drawstring up through the center of the brush stack.
- Make an X-frame from two small saplings and place it on top of the brush stack. Tie the X-frame securely in place with the poncho drawstring.
- Pile another 45 centimeters of brush on top of the X-frame, then compress the brush slightly.
- Pull the poncho sides up around the brush and, using the ropes or vines attached to the corner or side grommets, tie them diagonally from corner to corner and from side to side.
- Spread the second poncho, inner side up, next to the brush bundle.
- Roll the brush bundle onto the second poncho so that the tied side is down. Tie the second poncho around the brush bundle in the same manner as you tied the first poncho around the brush.
- Place it in the water with the tied side of the second poncho facing up.

Australian Poncho Raft. If you do not have time to gather brush for a brush raft, you can make an Australian poncho raft. This raft, although more waterproof than the poncho brush raft, will only float about 35 kilograms of equipment. To construct this raft, use two ponchos, two rucksacks, two 1.2-meter poles or branches, and ropes, vines, bootlaces, or comparable material as follows (Figure 6-5):

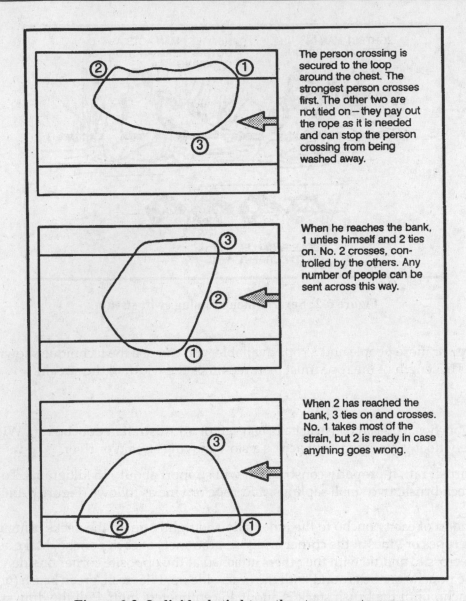

The person crossing is secured to the loop around the chest. The strongest person crosses first. The other two are not tied on—they pay out the rope as it is needed and can stop the person crossing from being washed away.

When he reaches the bank, 1 unties himself and 2 ties on. No. 2 crosses, controlled by the others. Any number of people can be sent across this way.

When 2 has reached the bank, 3 ties on and crosses. No. 1 takes most of the strain, but 2 is ready in case anything goes wrong.

Figure 6-3: Individuals tied together to cross stream.

- Push the hood of each poncho to the inner side and tightly tie off the necks using the drawstrings.
- Spread one poncho on the ground with the inner side up. Place and center the two 1.2-meter poles on the poncho about 45 centimeters apart.
- Place your rucksacks or packs or other equipment between the poles. Also place other items that you want to keep dry between the poles. Snap the poncho sides together.
- Use your buddy's help to complete the raft. Hold the snapped portion of the poncho in the air and roll it tightly down to the equipment. Make sure you roll the full width of the poncho.
- Twist the ends of the roll to form pigtails in opposite directions. Fold the pigtails over the bundle and tie them securely in place using ropes, bootlaces, or vines.
- Spread the second poncho on the ground, inner side up. If you need more buoyancy, place some fresh green brush on this poncho.
- Place the equipment bundle, tied side down, on the center of the second poncho. Wrap the second poncho around the equipment bundle following the same procedure you used for wrapping the equipment in the first poncho.

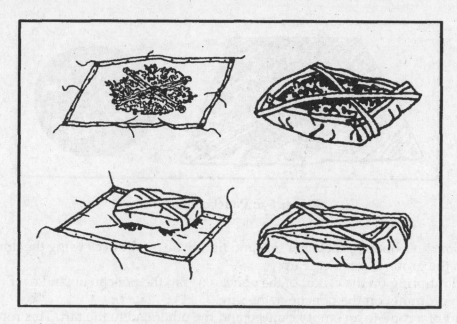

Figure 6-4: Brush raft.

- Tie ropes, bootlaces, vines, or other binding material around the raft about 30 centimeters from the end of each pigtail. Place and secure weapons on top of the raft.
- Tie one end of a rope to an empty canteen and the other end to the raft. This will help you to tow the raft.

Poncho Donut Raft. Another type of raft is the poncho donut raft. It takes more time to construct than the brush raft or Australian poncho raft, but it is effective. To construct it, use one poncho, small saplings, willow or vines, and rope, bootlaces, or other binding material (Figure 6-6) as follows:
- Make a framework circle by placing several stakes in the ground that roughly outline an inner and outer circle.
- Using young saplings, willow, or vines, construct a donut ring within the circles of stakes.
- Wrap several pieces of cordage around the donut ring about 30 to 60 centimeters apart and tie them securely.

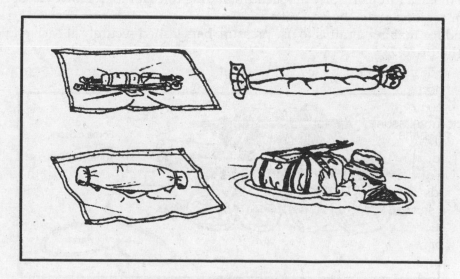

Figure 6-5: Australian poncho raft.

Figure 6-6: Poncho donut raft.

- Push the poncho's hood to the inner side and tightly tie off the neck using the drawstring. Place the poncho on the ground, inner side up.
- Place the donut ring on the center of the poncho. Wrap the poncho up and over the donut ring and tie off each grommet on the poncho to the ring.
- Tie one end of a rope to an empty canteen and the other end to the raft. This rope will help you to tow the raft.

When launching any of the above rafts, take care not to puncture or tear it by dragging it on the ground. Before you start to cross the river or stream, let the raft lay on the water a few minutes to ensure that it floats.

If the river is too deep to ford, push the raft in front of you while you are swimming. The design of the above rafts does not allow them to carry a person's full body weight. Use them as a float to get you and your equipment safely across the river or stream.

Be sure to check the water temperature before trying to cross a river or water obstacle. If the water is extremely cold and you are unable to find a shallow fording place in the river, do not try to ford it. Devise other means for crossing. For instance, you might improvise a bridge by felling a tree over the river. Or you might build a raft large enough to carry you and your equipment. For this, however, you will need an axe, a knife, a rope or vines, and time.

Log Raft. You can make a raft using any dry, dead, standing trees for logs. However, spruce trees found in polar and subpolar regions make the best rafts.

A simple method for making a raft is to use pressure bars lashed securely at each end of the raft to hold the logs together (Figure 6-7).

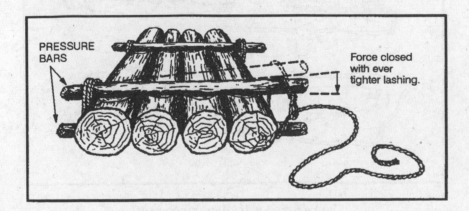

Figure 6-7: Use of pressure bars.

FLOTATION DEVICES

If the water is warm enough for swimming and you do not have the time or materials to construct one of the poncho-type rafts, you can use various flotation devices to negotiate the water obstacle. Some items you can use for flotation devices are—

- *Trousers.* Knot each trouser leg at the bottom and close the fly. With both hands, grasp the waistband at the sides and swing the trousers in the air to trap air in each leg. Quickly press the sides of the waistband together and hold it underwater so that the air will not escape. You now have water wings to keep you afloat as you cross the body of water. *Note: Wet the trousers before inflating to trap the air better. You may have to reinflate the trousers several times when crossing a large body of water.*
- *Empty containers.* Lash together empty gas cans, water jugs, ammo cans, boxes, or other items that will trap or hold air. Use them as water wings. Use this type of flotation device only in a slow-moving river or stream.
- *Plastic bags and ponchos.* Fill two or more plastic bags with air and secure them together at the opening. Use your poncho and roll green vegetation tightly inside it so that you have a roll at least 20 centimeters in diameter. Tie the ends of the roll securely. You can wear it around your waist or across one shoulder and under the opposite arm.
- *Logs.* Use a stranded drift log if one is available, or find a log near the water to use as a float. Be sure to test the log before starting to cross. Some tree logs, palm for example, will sink even when the wood is dead. Another method is to tie two logs about 60 centimeters apart. Sit between the logs with your back against one and your legs over the other (Figure 6-8).
- *Cattails.* Gather stalks of cattails and tie them in a bundle 25 centimeters or more in diameter. The many air cells in each stalk cause a stalk to float until it rots. Test the cattail bundle to be sure it will support your weight before trying to cross a body of water.

There are many other flotation devices that you can devise by using some imagination. Just make sure to test the device before trying to use it.

OTHER WATER OBSTACLES

Other water obstacles that you may face are bogs, quagmire, muskeg, or quicksand. Do not try to walk across these. Trying to lift your feet while standing upright will make you sink deeper. Try to bypass these obstacles. If you are unable to bypass them, you may be able to bridge them using logs, branches, or foliage.

A way to cross a bog is to lie face down, with your arms and legs spread. Use a flotation device or form pockets of air in your clothing. Swim or pull your way across moving slowly and trying to keep your body horizontal.

Figure 6-8: Log flotation.

In swamps, the areas that have vegetation are usually firm enough to support your weight. However, vegetation will usually not be present in open mud or water areas. If you are an average swimmer, however, you should have no problem swimming, crawling, or pulling your way through miles of bog or swamp.

Quicksand is a mixture of sand and water that forms a shifting mass. It yields easily to pressure and sucks down and engulfs objects resting on its surface. It varies in depth and is usually localized. Quicksand commonly occurs on flat shores, in silt-choked rivers with shifting watercourses, and near the mouths of large rivers. If you are uncertain whether a sandy area is quicksand, toss a small stone on it. The stone will sink in quicksand. Although quicksand has more suction than mud or muck, you can cross it just as you would cross a bog. Lie face down, spread your arms and legs, and move slowly across.

VEGETATION OBSTACLES

Some water areas you must cross may have underwater and floating plants that will make swimming difficult. However, you can swim through relatively dense vegetation if you remain calm and do not thrash about. Stay as near the surface as possible and use the breaststroke with shallow leg and arm motion. Remove the plants around you as you would clothing. When you get tired, float or swim on your back until you have rested enough to continue with the breaststroke.

The mangrove swamp is another type of obstacle that occurs along tropical coastlines. Mangrove trees or shrubs throw out many prop roots that form dense masses. To get through a mangrove swamp, wait for low tide. If you are on the inland side, look for a narrow grove of trees and work your way seaward through these. You can also try to find the bed of a waterway or creek through the trees and follow it to the sea. If you are on the seaward side, work inland along streams or channels. Be on the lookout for crocodiles that you find along channels and in shallow water. If there are any near you, leave the water and scramble over the mangrove roots. While crossing a mangrove swamp, it is possible to gather food from tidal pools or tree roots.

To cross a large swamp area, construct some type of raft.

CHAPTER 7

Survival in Nuclear, Biological, and Chemical Environments

Nuclear, chemical, and biological weapons have become potential realities on any modern battlefield. Recent experience in Afghanistan, Cambodia, and other areas of conflict has proved the use of chemical and biological weapons (such as mycotoxins). The war fighting doctrine of the NATO and Warsaw Pact nations addresses the use of both nuclear and chemical weapons. The potential use of these weapons intensifies the problems of survival because of the serious dangers posed by either radioactive fallout or contamination produced by persistent biological or chemical agents.

You must use special precautions if you expect to survive in these man-made hazards. If you are subjected to any of the effects of nuclear, chemical, or biological warfare, the survival procedures recommended in this chapter may save your life. This chapter presents some background information on each type of hazard so that you may better understand the true nature of the hazard. Awareness of the hazards, knowledge of this chapter, and application of common sense should keep you alive.

THE NUCLEAR ENVIRONMENT

Prepare yourself to survive in a nuclear environment. Know how to react to a nuclear hazard.

Effects of Nuclear Weapons. The effects of nuclear weapons are classified as either initial or residual. Initial effects occur in the immediate area of the explosion and are hazardous in the first minute after the explosion. Residual effects can last for days or years and cause death. The principal initial effects are blast and radiation.

Blast. Defined as the brief and rapid movement of air away from the explosion's center and the pressure accompanying this movement. Strong winds accompany the blast. Blast hurls debris and personnel, collapses lungs, ruptures eardrums, collapses structures and positions, and causes immediate death or injury with its crushing effect.

Thermal Radiation. The heat and light radiation a nuclear explosion's fireball emits. Light radiation consists of both visible light and ultraviolet and infrared light. Thermal radiation produces extensive fires, skin burns, and flash blindness.

Nuclear Radiation. Nuclear radiation breaks down into two categories—initial radiation and residual radiation.

Initial nuclear radiation consists of intense gamma rays and neutrons produced during the first minute after the explosion. This radiation causes extensive damage to cells throughout the body. Radiation damage may cause headaches, nausea, vomiting, diarrhea, and even death, depending on the radiation dose received. The major problem in protecting yourself against the initial radiation's effects is that you may have received a lethal or incapacitating dose before taking any protective action. Personnel exposed to lethal amounts of initial radiation may well have been killed or fatally injured by blast or thermal radiation.

Residual radiation consists of all radiation produced after one minute from the explosion. It has more effect on you than initial radiation. A discussion of residual radiation follows.

Types of Nuclear Bursts. There are three types of nuclear bursts—airburst, surface burst, and subsurface burst. The type of burst directly affects your chances of survival. A subsurface burst occurs completely underground or underwater. Its effects remain beneath the surface or in the immediate area where the surface collapses into a crater over the burst's location. Subsurface bursts cause you little or no radioactive hazard unless you enter the immediate area of the crater. No further discussion of this type of burst will take place.

An airburst occurs in the air above its intended target. The airburst provides the maximum radiation effect on the target and is, therefore, most dangerous to you in terms of immediate nuclear effects.

A surface burst occurs on the ground or water surface. Large amounts of fallout result, with serious long-term effects for you. This type of burst is your greatest nuclear hazard.

Nuclear Injuries. Most injuries in the nuclear environment result from the initial nuclear effects of the detonation. These injuries are classed as blast, thermal, or radiation injuries. Further radiation injuries may occur if you do not take proper precautions against fallout. Individuals in the area near a nuclear explosion will probably suffer a combination of all three types of injuries.

Blast Injuries. Blast injuries produced by nuclear weapons are similar to those caused by conventional high-explosive weapons. Blast overpressure can produce collapsed lungs and ruptured internal organs. Projectile wounds occur as the explosion's force hurls debris at you. Large pieces of debris striking you will cause fractured limbs or massive internal injuries. Blast overpressure may throw you long distances, and you will suffer severe injury upon impact with the ground or other objects. Substantial cover and distance from the explosion are the best protection against blast injury. Cover blast injury wounds as soon as possible to prevent the entry of radioactive dust particles.

Thermal Injuries. The heat and light the nuclear fireball emits causes thermal injuries. First-, second-, or third-degree burns may result. Flash blindness also occurs. This blindness may be permanent or temporary depending on the degree of exposure of the eyes. Substantial cover and distance from the explosion can prevent thermal injuries. Clothing will provide significant protection against thermal injuries. Cover as much exposed skin as possible before a nuclear explosion. First aid for thermal injuries is the same as first aid for burns. Cover open burns (second- or third-degree) to prevent the entry of radioactive particles. Wash all burns before covering.

Radiation Injuries. Neutrons, gamma radiation, alpha radiation, and beta radiation cause radiation injuries. Neutrons are high-speed, extremely penetrating particles that actually smash cells within your body. Gamma radiation is similar to X rays and is also a highly penetrating radiation. During the initial fireball stage of a nuclear detonation, initial gamma radiation and neutrons are the most serious threat. Beta and alpha radiation are radioactive particles normally associated with radioactive dust from fallout.

They are short-range particles and you can easily protect yourself against them if you take precautions. See Bodily Reactions to Radiation, below, for the symptoms of radiation injuries.

Residual Radiation. Residual radiation is all radiation emitted after 1 minute from the instant of the nuclear explosion. Residual radiation consists of induced radiation and fallout.

Induced Radiation. It describes a relatively small, intensely radioactive area directly underneath the nuclear weapon's fireball. The irradiated earth in this area will remain highly radioactive for an extremely long time. You should not travel into an area of induced radiation.

Fallout. Fallout consists of radioactive soil and water particles, as well as weapon fragments. During a surface detonation, or if an airburst's nuclear fireball touches the ground, large amounts of soil and water are vaporized along with the bomb's fragments, and forced upward to altitudes of 25,000 meters or more. When these vaporized contents cool, they can form more than 200 different radioactive products. The vaporized bomb contents condense into tiny radioactive particles that the wind carries and they fall back to earth as radioactive dust. Fallout particles emit alpha, beta, and gamma radiation. Alpha and beta radiation are relatively easy to counteract, and residual gamma radiation is much less intense than the gamma radiation emitted during the first minute after the explosion. Fallout is your most significant radiation hazard, provided you have not received a lethal radiation dose from the initial radiation.

Bodily Reactions to Radiation. The effects of radiation on the human body can be broadly classed as either chronic or acute. Chronic effects are those that occur some years after exposure to radiation. Examples are cancer and genetic defects. Chronic effects are of minor concern in so far as they affect your immediate survival in a radioactive environment. On the other hand, acute effects are of primary importance to your survival. Some acute effects occur within hours after exposure to radiation. These effects result from the radiation's direct physical damage to tissue. Radiation sickness and beta burns are examples of acute effects. Radiation sickness symptoms include nausea, diarrhea, vomiting, fatigue, weakness, and loss of hair. Penetrating beta rays cause radiation burns; the wounds are similar to fire burns.

Recovery Capability. The extent of body damage depends mainly on the part of the body exposed to radiation and how long it was exposed, as well as its ability to recover. The brain and kidneys have little recovery capability. Other parts (skin and bone marrow) have a great ability to recover from damage.

Usually, a dose of 600 centigrams (cgys) to the entire body will result in almost certain death. If only your hands received this same dose, your overall health would not suffer much, although your hands would suffer severe damage.

External and Internal Hazards. An external or an internal hazard can cause body damage. Highly penetrating gamma radiation or the less penetrating beta radiation that causes burns can cause external damage. The entry of alpha or beta radiation-emitting particles into the body can cause internal damage. The external hazard produces overall irradiation and beta burns. The internal hazard results in irradiation of critical organs such as the gastrointestinal tract, thyroid gland, and bone. A very small amount of radioactive material can cause extreme damage to these and other internal organs. The internal hazard can enter the body either through consumption of contaminated water or food or by absorption through cuts or abrasions. Material that enters the body through breathing presents only a minor hazard. You can greatly reduce the internal radiation hazard by using good personal hygiene and carefully decontaminating your food and water.

Symptoms. The symptoms of radiation injuries include nausea, diarrhea, and vomiting. The severity of these symptoms is due to the extreme sensitivity of the gastrointestinal tract to radiation. The severity of the symptoms and the speed of onset after exposure are good indicators of the degree of radiation damage. The gastrointestinal damage can come from either the external or the internal radiation hazard.

Countermeasures Against Penetrating External Radiation. Knowledge of the radiation hazards discussed earlier is extremely important in surviving in a fallout area. It is also critical to know how to protect yourself from the most dangerous form of residual radiation—penetrating external radiation.

The means you can use to protect yourself from penetrating external radiation are time, distance, and shielding. You can reduce the level of radiation and help increase your chance of survival by controlling the duration of exposure. You can also get as far away from the radiation source as possible. Finally you can place some radiation-absorbing or shielding material between you and the radiation.

Time. Time is important to you, as the survivor, in two ways. First, radiation dosages are cumulative. The longer you are exposed to a radioactive source, the greater the dose you will receive. Obviously, spend as little time in a radioactive area as possible. Second, radioactivity decreases or decays over time. This concept is known as radioactive half-life. Thus, a radioactive element decays or loses half of its radioactivity within a certain time. The rule of thumb for radioactivity decay is that it decreases in intensity by a factor of ten for every sevenfold increase in time following the peak radiation level. For example, if a nuclear fallout area had a maximum radiation rate of 200 cgys per hour when fallout is complete, this rate would fall to 20 cgys per hour after 7 hours; it would fall still further to 2 cgys per hour after 49 hours. Even an untrained observer can see that the greatest hazard from fallout occurs immediately after detonation, and that the hazard decreases quickly over a relatively short time. As a survivor, try to avoid fallout areas until the radioactivity decays to safe levels. If you can avoid fallout areas long enough for most of the radioactivity to decay, you enhance your chance of survival.

Distance. Distance provides very effective protection against penetrating gamma radiation because radiation intensity decreases by the square of the distance from the source. For example, if exposed to 1,000 cgys of radiation standing 30 centimeters from the source, at 60 centimeters, you would only receive 250 cgys. Thus, when you double the distance, radiation decreases to $(0.5)^2$ or 0.25 the amount. While this formula is valid for concentrated sources of radiation in small areas, it becomes more complicated for large areas of radiation such as fallout areas.

Shielding. Shielding is the most important method of protection from penetrating radiation. Of the three countermeasures against penetrating radiation, shielding provides the greatest protection and is the easiest to use under survival conditions. Therefore, it is the most desirable method. If shielding is not possible, use the other two methods to the maximum extent practical.

Shielding actually works by absorbing or weakening the penetrating radiation, thereby reducing the amount of radiation reaching your body. The denser the material, the better the shielding effect. Lead, iron, concrete, and water are good examples of shielding materials.

Special Medical Aspects. The presence of fallout material in your area requires slight changes in first aid procedures. You must cover all wounds to prevent contamination and the entry of radioactive particles. You must first wash burns of beta radiation, then treat them as ordinary burns. Take extra measures to prevent infection. Your body will be extremely sensitive to infections due to changes in your blood chemistry. Pay close attention to the prevention of colds or respiratory infections. Rigorously practice personal hygiene to prevent infections. Cover your eyes with improvised goggles to prevent the entry of particles.

Shelter. As stated earlier, the shielding material's effectiveness depends on its thickness and density. An ample thickness of shielding material will reduce the level of radiation to negligible amounts.

The primary reason for finding and building a shelter is to get protection against the high-intensity radiation levels of early gamma fallout as fast as possible. Five minutes to locate the shelter is a good guide. Speed in finding shelter is absolutely essential. Without shelter, the dosage received in the first few hours will exceed that received during the rest of a week in a contaminated area. The dosage received in this first week will exceed the dosage accumulated during the rest of a lifetime spent in the same contaminated area.

Shielding Materials. The thickness required to weaken gamma radiation from fallout is far less than that needed to shield against initial gamma radiation. Fallout radiation has less energy than a nuclear detonation's initial radiation. For fallout radiation, a relatively small amount of shielding material can provide adequate protection. Figure 7-1 gives an idea of the thickness of various materials needed to reduce residual gamma radiation transmission by 50 percent. The principle of half-value layer thickness is useful in understanding the absorption of gamma radiation by various materials. According to this principle, if 5 centimeters of brick reduce the gamma radiation level by one-half, adding another 5 centimeters of brick (another half-value layer) will reduce the intensity by another half, namely, to one-fourth the original amount. Fifteen centimeters will reduce gamma radiation fallout levels to one-eighth its original amount, 20 centimeters to one-sixteenth, and so on. Thus, a shelter protected by 1 meter of dirt would reduce a radiation intensity of 1,000 cgys per hour on the outside to about 0.5 cgy per hour inside the shelter.

Natural Shelters. Terrain that provides natural shielding and easy shelter construction is the ideal location for an emergency shelter. Good examples are ditches, ravines, rocky outcropping, hills, and river banks. In level areas without natural protection, dig a fighting position or slit trench.

Trenches. When digging a trench, work from inside the trench as soon as it is large enough to cover part of your body thereby not exposing all your body to radiation. In open country, try to dig the trench from a prone position, stacking the dirt carefully and evenly around the trench. On level ground, pile the dirt around your body for additional shielding.

Depending upon soil conditions, shelter construction time will vary from a few minutes to a few hours. If you dig as quickly as possible, you will reduce the dosage you receive.

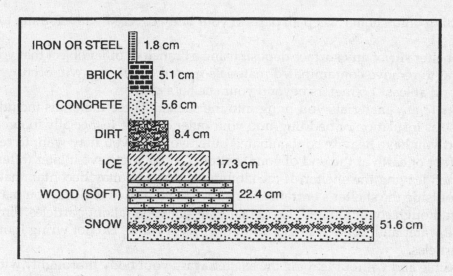

Figure 7-1: Thickness of materials to reduce gamma radiation

Other Shelters. While an underground shelter covered by 1 meter or more of earth provides the best protection against fallout radiation, the following unoccupied structures (in order listed) offer the next best protection:

- Caves and tunnels covered by more than 1 meter of earth.
- Storm or storage cellars.
- Culverts.
- Basements or cellars of abandoned buildings.
- Abandoned buildings made of stone or mud.

Roofs. It is not mandatory that you build a roof on your shelter. Build one only if the materials are readily available with only a brief exposure to outside contamination. If building a roof would require extended exposure to penetrating radiation, it would be wiser to leave the shelter roofless. A roof's sole function is to reduce radiation from the fallout source to your body. Unless you use a thick roof, a roof provides very little shielding.

You can construct a simple roof from a poncho anchored down with dirt, rocks, or other refuse from your shelter. You can remove large particles of dirt and debris from the top of the poncho by beating it off from the inside at frequent intervals. This cover will not offer shielding from the radioactive particles deposited on the surface, but it will increase the distance from the fallout source and keep the shelter area from further contamination.

Shelter Site Selection and Preparation. To reduce your exposure time and thereby reduce the dosage received, remember the following factors when selecting and setting up a shelter:

- Where possible, seek a crude, existing shelter that you can improve. If none is available, dig a trench.
- Dig the shelter deep enough to get good protection, then enlarge it as required for comfort.
- Cover the top of the fighting position or trench with any readily available material and a thick layer of earth, if you can do so without leaving the shelter. While a roof and camouflage are both desirable, it is probably safer to do without them than to expose yourself to radiation outside your fighting position.

- While building your shelter, keep all parts of your body covered with clothing to protect it against beta burns.
- Clean the shelter site of any surface deposit using a branch or other object that you can discard. Do this cleaning to remove contaminated materials from the area you will occupy. The cleaned area should extend at least 1.5 meters beyond your shelter's area.
- Decontaminate any materials you bring into the shelter. These materials include grass or foliage that you use as insulation or bedding, and your outer clothing (especially footgear). If the weather permits and you have heavily contaminated outer clothing, you may want to remove it and bury it under a foot of earth at the end of your shelter. You may retrieve it later (after the radioactivity decays) when leaving the shelter. If the clothing is dry, you may decontaminate it by beating or shaking it outside the shelter's entrance to remove the radioactive dust. You may use any body of water, even though contaminated, to rid materials of excess fallout particles. Simply dip the material into the water and shake it to get rid of the excess water. Do not wring it out, this action will trap the particles.
- If at all possible and without leaving the shelter, wash your body thoroughly with soap and water, even if the water on hand may be contaminated. This washing will remove most of the harmful radioactive particles that are likely to cause beta burns or other damage. If water is not available, wipe your face and any other exposed skin surface to remove contaminated dust and dirt. You may wipe your face with a clean piece of cloth or a handful of uncontaminated dirt. You get this uncontaminated dirt by scraping off the top few inches of soil and using the "clean" dirt.
- Upon completing the shelter, lie down, keep warm, and sleep and rest as much as possible while in the shelter.
- When not resting, keep busy by planning future actions, studying your maps, or making the shelter more comfortable and effective.
- Don't panic if you experience nausea and symptoms of radiation sickness. Your main danger from radiation sickness is infection. There is no first aid for this sickness. Resting, drinking fluids, taking any medicine that prevents vomiting, maintaining your food intake, and preventing additional exposure will help avoid infection and aid recovery. Even small doses of radiation can cause these symptoms which may disappear in a short time.

Exposure Timetable. The following timetable provides you with the information needed to avoid receiving serious dosage and still let you cope with survival problems:

- Complete isolation from 4 to 6 days following delivery of the last weapon.
- A very brief exposure to procure water on the third day is permissible, but exposure should not exceed 30 minutes.
- One exposure of not more than 30 minutes on the seventh day.
- One exposure of not more than 1 hour on the eighth day.
- Exposure of 2 to 4 hours from the ninth day through the twelfth day.
- Normal operation, followed by rest in a protected shelter, from the thirteenth day on.
- In all instances, make your exposures as brief as possible. Consider only mandatory requirements as valid reasons for exposure. Decontaminate at every stop.

The times given above are conservative. If forced to move after the first or second day, you may do so, Make sure that the exposure is no longer than absolutely necessary.

Water Procurement. In a fallout-contaminated area, available water sources may be contaminated. If you wait at least 48 hours before drinking any water to allow for radioactive decay to take place and select the safest possible water source, you will greatly reduce the danger of ingesting harmful amounts of radioactivity.

Although many factors (wind direction, rainfall, sediment) will influence your choice in selecting water sources, consider the following guidelines.

Safest Water Sources. Water from springs, wells, or other underground sources that undergo natural filtration will be your safest source. Any water found in the pipes or containers of abandoned houses or stores will also be free from radioactive particles. This water will be safe to drink, although you will have to take precautions against bacteria in the water.

Snow taken from 15 or more centimeters below the surface during the fallout is also a safe source of water.

Streams and Rivers. Water from streams and rivers will be relatively free from fallout within several days after the last nuclear explosion because of dilution. If at all possible, filter such water before drinking to get rid of radioactive particles. The best filtration method is to dig sediment holes or seepage basins along the side of a water source. The water will seep laterally into the hole through the intervening soil that acts as a filtering agent and removes the contaminated fallout particles that settled on the original body of water. This method can remove up to 99 percent of the radioactivity in water. You must cover the hole in some way in order to prevent further contamination. See Part IV, Chapter 1, Illustration 1-7 for examples of water filters.

Standing Water. Water from lakes, pools, ponds, and other standing sources is likely to be heavily contaminated, though most of the heavier, long-lived radioactive isotopes will settle to the bottom. Use the settling technique to purify this water. First, fill a bucket or other deep container three-fourths full with contaminated water. Then take dirt from a depth of 10 or more centimeters below the ground surface and stir it into the water. Use about 2.5 centimeters of dirt for every 10 centimeters of water. Stir the water until you see most dirt particles suspended in the water. Let the mixture settle for at least 6 hours. The settling dirt particles will carry most of the suspended fallout particles to the bottom and cover them. You can then dip out the clear water. Purify this water using a filtration device.

Additional Precautions. As an additional precaution against disease, treat all water with water purification tablets from your survival kit or boil it.

Food Procurement. Although it is a serious problem to obtain edible food in a radiation-contaminated area, it is not impossible to solve. You need to follow a few special procedures in selecting and preparing rations and local foods for use. Since secure packaging protects your combat rations, they will be perfectly safe for use. Supplement your rations with any food you can find on trips outside your shelter. Most processed foods you may find in abandoned buildings are safe for use after decontaminating them. These include canned and packaged foods after removing the containers or wrappers or washing them free of fallout particles. These processed foods also include food stored in any closed container and food stored in protected areas (such as cellars), if you wash them before eating. Wash all food containers or wrappers before handling them to prevent further contamination.

If little or no processed food is available in your area, you may have to supplement your diet with local food sources. Local food sources are animals and plants.

Animals as a Food Source. Assume that all animals, regardless of their habitat or living conditions, were exposed to radiation. The effects of radiation on animals are similar to those on humans. Thus, most of the wild animals living in a fallout area are likely to become sick or die from radiation during the first month after the nuclear explosion. Even though animals may not be free from harmful radioactive materials, you can and must use them in survival conditions as a food source if other foods are not available. With careful preparation and by following several important principles, animals can be safe food sources.

First, do not eat an animal that appears to be sick. It may have developed a bacterial infection as a result of radiation poisoning. Contaminated meat, even if thoroughly cooked, could cause severe illness or death if eaten.

Carefully skin all animals to prevent any radioactive particles on the skin or fur from entering the body. Do not eat meat close to the bones and joints as an animal's skeleton contains over 90 percent of the radioactivity. The remaining animal muscle tissue, however, will be safe to eat. Before cooking it, cut the meat away from the bone, leaving at least a 3-millimeter thickness of meat on the bone. Discard all internal organs (heart, liver, and kidneys) since they tend to concentrate beta and gamma radioactivity.

Cook all meat until it is very well done. To be sure the meat is well done, cut it into less than 13-millimeter-thick pieces before cooking. Such cuts will also reduce cooking time and save fuel.

The extent of contamination in fish and aquatic animals will be much greater than that of land animals. This is also true for water plants, especially in coastal areas. Use aquatic food sources only in conditions of extreme emergency.

All eggs, even if laid during the period of fallout, will be safe to eat. Completely avoid milk from any animals in a fallout area because animals absorb large amounts of radioactivity from the plants they eat.

Plants as a Food Source. Plant contamination occurs by the accumulation of fallout on their outer surfaces or by absorption of radioactive elements through their roots. Your first choice of plant food should be vegetables such as potatoes, turnips, carrots, and other plants whose edible portion grows underground. These are the safest to eat once you scrub them and remove their skins.

Second in order of preference are those plants with edible parts that you can decontaminate by washing and peeling their outer surfaces. Examples are bananas, apples, tomatoes, prickly pears, and other such fruits and vegetables.

Any smooth-skinned vegetable, fruit, or plant that you cannot easily peel or effectively decontaminate by washing will be your third choice of emergency food.

The effectiveness of decontamination by scrubbing is inversely proportional to the roughness of the fruit's surface. Smooth-surfaced fruits have lost 90 percent of their contamination after washing, while washing rough-surfaced plants removes only about 50 percent of the contamination.

You eat rough-surfaced plants (such as lettuce) only as a last resort because you cannot effectively decontaminate them by peeling or washing.

Other difficult foods to decontaminate by washing with water include dried fruits (figs, prunes, peaches, apricots, pears) and soya beans.

In general, you can use any plant food that is ready for harvest if you can effectively decontaminate it. Growing plants, however, can absorb some radioactive materials through their leaves as well as from the soil, especially if rains have occurred during or after the fallout period. Avoid using these plants for food except in an emergency.

BIOLOGICAL ENVIRONMENTS

The use of biological agents is real. Prepare yourself for survival by being proficient in the tasks identified in your Soldier's Manuals of Common Tasks (SMCTs). Know what to do to protect yourself against these agents.

Biological Agents and Effects. Biological agents are microorganisms that can cause disease among personnel, animals, or plants. They can also cause the deterioration of material. These agents fall into two broad categories—pathogens (usually called germs) and toxins. Pathogens are living microorganisms that cause lethal or incapacitating diseases. Bacteria, rickettsiae, fungi, and viruses are included in the pathogens. Toxins are poisons that plants, animals, or microorganisms produce naturally. Possible biological warfare toxins include a variety of neurotoxic (affecting the central nervous system) and cytotoxic (causing cell death) compounds.

Germs. Germs are living organisms. Some nations have used them in the past as weapons. Only a few germs can start an infection, especially if inhaled into the lungs. Because germs are so small and weigh so little, the wind can spread them over great distances; they can also enter unfiltered or nonairtight places. Buildings and bunkers can trap them thus causing a higher concentration. Germs do not affect the body immediately. They must multiply inside the body and overcome the body's defenses—a process called the incubation period. Incubation periods vary from several hours to several months, depending on the germ.

Most germs must live within another living organism (host), such as your body, to survive and grow. Weather conditions such as wind, rain, cold, and sunlight rapidly kill germs.

Some germs can form protective shells, or spores, to allow survival outside the host. Spore-producing agents are a long-term hazard you must neutralize by decontaminating infected areas or personnel. Fortunately, most live agents are not spore-producing. These agents must find a host within roughly a day of their delivery or they die. Germs have three basic routes of entry into your body: through the respiratory tract, through a break in the skin, and through the digestive tract. Symptoms of infection vary according to the disease.

Toxins. Toxins are substances that plants, animals, or germs produce naturally. These toxins are what actually harm man, not bacteria. Botulin, which produces botulism, is an example. Modern science has allowed large-scale production of these toxins without the use of the germ that produces the toxin. Toxins may produce effects similar to those of chemical agents. Toxic victims may not, however, respond to first aid measures used against chemical agents. Toxins enter the body in the same manner as germs. However, some toxins, unlike germs, can penetrate unbroken skin. Symptoms appear almost immediately, since there is no incubation period. Many toxins are extremely lethal, even in very small doses. Symptoms may include any of the following:

- Dizziness.
- Mental confusion.
- Blurred or double vision.
- Numbness or tingling of skin.
- Paralysis.
- Convulsions.
- Rashes or blisters.
- Coughing.
- Fever.
- Aching muscles.
- Tiredness.
- Nausea, vomiting, and/or diarrhea.
- Bleeding from body openings.
- Blood in urine, stool, or saliva.
- Shock.
- Death.

Detection of Biological Agents. Biological agents are, by nature, difficult to detect. You cannot detect them by any of the five physical senses. Often, the first sign of a biological agent will be symptoms of the victims exposed to the agent. Your best chance of detecting biological agents before they can affect you is to recognize their means of delivery. The three main means of delivery are—

- *Bursting-type munitions.* These may be bombs or projectiles whose burst causes very little damage. The burst will produce a small cloud of liquid or powder in the immediate impact area. This cloud will disperse eventually; the rate of dispersion depends on terrain and weather conditions.
- *Spray tanks or generators.* Aircraft or vehicle spray tanks or ground-level aerosol generators produce an aerosol cloud of biological agents.
- *Vectors.* Insects such as mosquitoes, fleas, lice, and ticks deliver pathogens. Large infestations of these insects may indicate the use of biological agents.

Another sign of a possible biological attack is the presence of unusual substances on the ground or on vegetation, or sick-looking plants, crops, or animals.

Influence of Weather and Terrain. Your knowledge of how weather and terrain affect the agents can help you avoid contamination by biological agents. Major weather factors that affect biological agents are sunlight, wind, and precipitation. Aerosol sprays will tend to concentrate in low areas of terrain, similar to early morning mist.

Sunlight contains visible and ultraviolet solar radiation that rapidly kills most germs used as biological agents. However, natural or man-made cover may protect some agents from sunlight. Other man-made mutant strains of germs may be resistant to sunlight.

High wind speeds increase the dispersion of biological agents, dilute their concentration, and dehydrate them. The further downwind the agent travels, the less effective it becomes due to dilution and death of the pathogens. However, the downwind hazard area of the biological agent is significant and you cannot ignore it.

Precipitation in the form of moderate to heavy rain tends to wash biological agents out of the air, reducing downwind hazard areas. However, the agents may still be very effective where they were deposited on the ground.

Protection Against Biological Agents. While you must maintain a healthy respect for biological agents, there is no reason for you to panic. You can reduce your susceptibility to biological agents by maintaining current immunizations, avoiding contaminated areas, and controlling rodents and pests. You must also use proper first aid measures in the treatment of wounds and only safe or properly decontaminated sources of food and water. You must ensure that you get enough sleep to prevent a run-down condition. You must always use proper field sanitation procedures.

Assuming you do not have a protective mask, always try to keep your face covered with some type of cloth to protect yourself against biological agent aerosols. Dust may contain biological agents; wear some type of mask when dust is in the air.

Your uniform and gloves will protect you against bites from vectors (mosquitoes and ticks) that carry diseases. Completely button your clothing and tuck your trousers tightly into your boots. Wear a chemical protective overgarment, if available, as it provides better protection than normal clothing. Covering your skin will also reduce the chance of the agent entering your body through cuts or scratches. Always practice high standards of personal hygiene and sanitation to help prevent the spread of vectors.

Bathe with soap and water whenever possible. Use germicidal soap, if available. Wash your hair and body thoroughly, and clean under your fingernails. Clean teeth, gums, tongue, and the roof of your mouth frequently. Wash your clothing in hot, soapy water if you can. If you cannot wash your clothing, lay it out in an area of bright sunlight and allow the light to kill the microorganisms. After a toxin attack, decontaminate yourself as if for a chemical attack using the M258A2 kit (if available) or by washing with soap and water.

Shelter. You can build expedient shelters under biological contamination conditions using the same techniques described in Part III. However, you must make slight changes to reduce the chance of biological contamination. Do not build your shelter in depressions in the ground. Aerosol sprays tend to concentrate in these depressions. Avoid building your shelter in areas of vegetation, as vegetation provides shade and some degree of protection to biological agents. Avoid using vegetation in constructing your shelter. Place your shelter's entrance at a 90-degree angle to the prevailing winds. Such placement will limit the entry of airborne agents and prevent air stagnation in your shelter. Always keep your shelter clean.

Water Procurement. Water procurement under biological conditions is difficult but not impossible. Whenever possible, try to use water that has been in a sealed container. You can assume that the water inside the sealed container is not contaminated. Wash the water container thoroughly with soap and water or boil it for at least 10 minutes before breaking the seal.

If water in sealed containers is not available, your next choice, only under emergency conditions, is water from springs. Again, boil the water for at least 10 minutes before drinking. Keep the water covered while boiling to prevent contamination by airborne pathogens. Your last choice, only in an extreme emergency, is to use standing water. Vectors and germs can survive easily in stagnant water. Boil this water as long as practical to kill all organisms. Filter this water through a cloth to remove the dead vectors. Use water purification tablets in all cases.

Food Procurement. Food procurement, like water procurement, is not impossible, but you must take special precautions. Your combat rations are sealed, and you can assume they are not contaminated. You can also assume that sealed containers or packages of processed food are safe. To ensure safety, decontaminate all food containers by washing with soap and water or by boiling the container in water for 10 minutes.

You consider supplementing your rations with local plants or animals only in extreme emergencies. No matter what you do to prepare the food, there is no guarantee that cooking will kill all the biological agents. Use local food only in life or death situations. Remember, you can survive for a long time without food, especially if the food you eat may kill you!

If you must use local food, select only healthy-looking plants and animals. Do not select known carriers of vectors such as rats or other vermin. Select and prepare plants as you would in radioactive areas. Prepare animals as you do plants. Always use gloves and protective clothing when handling animals or plants. Cook all plant and animal food by boiling only. Boil all food for at least 10 minutes to kill all pathogens. Do not try to fry, bake, or roast local food. There is no guarantee that all infected portions have reached the required temperature to kill all pathogens. Do not eat raw food.

CHEMICAL ENVIRONMENTS

Chemical agent warfare is real. It can create extreme problems in a survival situation, but you can overcome the problems with the proper equipment, knowledge, and training. As a survivor, your first line of defense against chemical agents is your proficiency in individual nuclear, biological, and chemical (NBC) training, to include donning and wearing the protective mask and overgarment, personal decontamination, recognition of chemical agent symptoms, and individual first aid for chemical agent contamination. The SMCTs cover these subjects. If you are not proficient in these skills, you will have little chance of surviving a chemical environment.

Detection of Chemical Agents. The best method for detecting chemical agents is the use of a chemical agent detector. If you have one, use it. However, in a survival situation, you will most likely have to rely solely on the use of all of your physical senses. You must be alert and able to detect any clues indicating the use of chemical warfare. General indicators of the presence of chemical agents are tears, difficult breathing, choking, itching, coughing, and dizziness. With agents that are very hard to detect, you must watch for symptoms in fellow survivors. Your surroundings will provide valuable clues to the presence of chemical agents; for example, dead animals, sick people, or people and animals displaying abnormal behavior.

Your sense of smell may alert you to some chemical agents, but most will be odorless. The odor of newly cut grass or hay may indicate the presence of choking agents. A smell of almonds may indicate blood agents.

Sight will help you detect chemical agents. Most chemical agents in the solid or liquid state have some color. In the vapor state, you can see some chemical agents as a mist or thin fog immediately after the bomb or shell bursts. By observing for symptoms in others and by observing delivery means, you may be able to have some warning of chemical agents. Mustard gas in the liquid state will appear as oily patches on leaves or on buildings.

The sound of enemy munitions will give some clue to the presence of chemical weapons. Muffled shell or bomb detonations are a good indicator.

Irritation in the nose or eyes or on the skin is an urgent warning to protect your body from chemical agents. Additionally, a strange taste in food, water, or cigarettes may serve as a warning that they have been contaminated.

Protection Against Chemical Agents. As a survivor, always use the following general steps, in the order listed, to protect yourself from a chemical attack:

- Use protective equipment.
- Give quick and correct self-aid when contaminated.
- Avoid areas where chemical agents exist.
- Decontaminate your equipment and body as soon as possible.

Your protective mask and overgarment are the key to your survival. Without these, you stand very little chance of survival. You must take care of these items and protect them from damage. You must practice and know correct self-aid procedures before exposure to chemical agents. The detection of chemical agents and the avoidance of contaminated areas is extremely important to your survival. Use whatever detection kits may be available to help in detection. Since you are in a survival situation, avoid contaminated areas at all costs. You can expect no help should you become contaminated. If you do become contaminated, decontaminate yourself as soon as possible using proper procedures.

Shelter. If you find yourself in a contaminated area; try to move out of the area as fast as possible. Travel crosswind or upwind to reduce the time spent in the downwind hazard area. If you cannot leave the area immediately and have to build a shelter, use normal shelter construction techniques, with a few changes. Build the shelter in a clearing, away from all vegetation. Remove all topsoil in the area of the shelter to decontaminate the area. Keep the shelter's entrance closed and oriented at a 90-degree angle to the prevailing wind. Do not build a fire using contaminated wood—the smoke will be toxic. Use extreme caution when entering your shelter so that you will not bring contamination inside.

Water Procurement. As with biological and nuclear environments, getting water in a chemical environment is difficult. Obviously, water in sealed containers is your best and safest source. You must protect this water as much as possible. Be sure to decontaminate the containers before opening.

If you cannot get water in sealed containers, try to get it from a closed source such as underground water pipes. You may use rainwater or snow if there is no evidence of contamination. Use water from slow-moving streams, if necessary, but always check first for signs of contamination, and always filter the water as described under nuclear conditions. Signs of water source contamination are foreign odors such as garlic, mustard, geranium, or bitter almonds; oily spots on the surface of the water or nearby; and the presence of dead fish or animals. If these signs are present, do not use the water. Always boil or purify the water to prevent bacteriological infection.

Food Procurement. It is extremely difficult to eat while in a contaminated area. You will have to break the seal on your protective mask to eat. If you eat, find an area in which you can safely unmask. The safest source of food is your sealed combat rations. Food in sealed cans or bottles will also be safe. Decontaminate all sealed food containers before opening, otherwise you will contaminate the food.

If you must supplement your combat rations with local plants or animals, do not use plants from contaminated areas or animals that appear to be sick. When handling plants or animals, always use protective gloves and clothing.